Springer Handbook
of Crystal Growth

Govindhan Dhanaraj, Kullaiah Byrappa,
Vishwanath Prasad, Michael Dudley (Eds.)

Springer Handbook of Crystal Growth
Organization of the Handbook

Part A Fundamentals of Crystal Growth and Defect Formation
1 Crystal Growth Techniques and Characterization: An Overview
2 Nucleation at Surfaces
3 Morphology of Crystals Grown from Solutions
4 Generation and Propagation of Defects During Crystal Growth
5 Single Crystals Grown Under Unconstrained Conditions
6 Defect Formation During Crystal Growth from the Melt

Part B Crystal Growth from Melt Techniques
7 Indium Phosphide: Crystal Growth and Defect Control by Applying Steady Magnetic Fields
8 Czochralski Silicon Single Crystals for Semiconductor and Solar Cell Applications
9 Czochralski Growth of Oxide Photorefractive Crystals
10 Bulk Crystal Growth of Ternary III – V Semiconductors
11 Growth and Characterization of Antimony-Based Narrow-Bandgap III – V Semiconductor Crystals for Infrared Detector Applications
12 Crystal Growth of Oxides by Optical Floating Zone Technique
13 Laser-Heated Pedestal Growth of Oxide Fibers
14 Synthesis of Refractory Materials by Skull Melting Technique
15 Crystal Growth of Laser Host Fluorides and Oxides
16 Shaped Crystal Growth

Part C Solution Growth of Crystals
17 Bulk Single Crystals Grown from Solution on Earth and in Microgravity
18 Hydrothermal Growth of Polyscale Crystals
19 Hydrothermal and Ammonothermal Growth of ZnO and GaN
20 Stoichiometry and Domain Structure of KTP-Type Nonlinear Optical Crystals
21 High-Temperature Solution Growth: Application to Laser and Nonlinear Optical Crystals
22 Growth and Characterization of KDP and Its Analogs

Part D Crystal Growth from Vapor
23 Growth and Characterization of Silicon Carbide Crystals
24 AlN Bulk Crystal Growth by Physical Vapor Transport
25 Growth of Single-Crystal Organic Semiconductors
26 Growth of III –Nitrides with Halide Vapor Phase Epitaxy(HVPE)
27 Growth of Semiconductor Single Crystals from Vapor Phase

Part E Epitaxial Growth and Thin Films

28 Epitaxial Growth of Silicon Carbide by Chemical Vapor Deposition
29 Liquid-Phase Electroepitaxy of Semiconductors
30 Epitaxial Lateral Overgrowth of Semiconductors
31 Liquid-Phase Epitaxy of Advanced Materials
32 Molecular-Beam Epitaxial Growth of HgCdTe
33 Metalorganic Vapor-Phase Epitaxy of Diluted Nitrides and Arsenide Quantum Dots
34 Formation of SiGe Heterostructures and Their Properties
35 Plasma Energetics in Pulsed Laser and Pulsed Electron Deposition

Part F Modeling in Crystal Growth and Defects

36 Convection and Control in Melt Growth of Bulk Crystals
37 Vapor Growth of III Nitrides
38 Continuum-Scale Quantitative Defect Dynamics in Growing Czochralski Silicon Crystals
39 Models for Stress and Dislocation Generation in Melt Based Compound Crystal Growth
40 Mass and Heat Transport in BS and EFG Systems

Part G Defects Characterization and Techniques

41 Crystalline Layer Structures with X-Ray Diffractometry
42 X-Ray Topography Techniques for Defect Characterization of Crystals
43 Defect-Selective Etching of Semiconductors
44 Transmission Electron Microscopy Characterization of Crystals
45 Electron Paramagnetic Resonance Characterization of Point Defects
46 Defect Characterization in Semiconductors with Positron Annihilation Spectroscopy

Part H Special Topics in Crystal Growth

47 Protein Crystal Growth Methods
48 Crystallization from Gels
49 Crystal Growth and Ion Exchange in Titanium Silicates
50 Single-Crystal Scintillation Materials
51 Silicon Solar Cells: Materials, Devices, and Manufacturing
52 Wafer Manufacturing and Slicing Using Wiresaw

Subject Index

使 用 说 明

1.《晶体生长手册》原版为一册,分为A~H部分。考虑到使用方便以及内容一致,影印版分为6册:第1册—Part A,第2册—Part B,第3册—Part C,第4册—Part D、E,第5册—Part F、G,第6册—Part H。

2.各册在页脚重新编排页码,该页码对应中文目录。保留了原书页眉及页码,其页码对应原书目录及主题索引。

3.各册均给出完整6册书的章目录。

4.作者及其联系方式、缩略语表各册均完整呈现。

5.主题索引安排在第6册。

6.文前介绍基本采用中英文对照形式,方便读者快速浏览。

材料科学与工程图书工作室

联系电话　0451-86412421
　　　　　0451-86414559

邮　　箱　yh_bj@yahoo.com.cn
　　　　　xuyaying81823@gmail.com
　　　　　zhxh6414559@yahoo.com.cn

Springer 手册精选系列

晶体生长手册

蒸发及外延法晶体生长技术

【第4册】

Springer
Handbook of

Crystal

Growth

［美］Govindhan Dhanaraj 等主编

（影印版）

黑版贸审字08-2012-047号

Reprint from English language edition:
Springer Handbook of Crystal Growth
by Govindhan Dhanaraj, Kullaiah Byrappa, Vishwanath Prasad
and Michael Dudley
Copyright © 2010 Springer Berlin Heidelberg
Springer Berlin Heidelberg is a part of Springer Science+Business Media
All Rights Reserved

This reprint has been authorized by Springer Science & Business Media for distribution in China Mainland only and not for export there from.

图书在版编目（CIP）数据

晶体生长手册. 4, 蒸发及外延法晶体生长技术 =Handbook of Crystal Growth. 4, Crystal Growth from Vapor and Epitaxial Techniques：英文 / (美) 德哈纳拉 (Dhanaraj,D.) 等主编.—影印本.—哈尔滨：哈尔滨工业大学出版社, 2013.1
（Springer手册精选系列）
ISBN 978-7-5603-3869-9

Ⅰ.①晶… Ⅱ.①德… Ⅲ.①晶体生长－蒸发－手册－英文②晶体生长－外延生长－手册－英文 Ⅳ.①O78-62

中国版本图书馆CIP数据核字(2012)第292323号

材料科学与工程
图书工作室

责任编辑 杨 桦 许雅莹 张秀华
出版发行 哈尔滨工业大学出版社
社　　址 哈尔滨市南岗区复华四道街10号 邮编 150006
传　　真 0451-86414749
网　　址 http://hitpress.hit.edu.cn
印　　刷 哈尔滨市石桥印务有限公司
开　　本 787mm×960mm 1/16 印张 28.5
版　　次 2013年1月第1版 2013年1月第1次印刷
书　　号 ISBN 978-7-5603-3869-9
定　　价 88.00元

（如因印刷质量问题影响阅读，我社负责调换）

序 言

多年以来，有很多探索研究已经成功地描述了晶体生长的生长工艺和科学，有许多文章、专著、会议文集和手册对这一领域的前沿成果做了综合评述。这些出版物反映了人们对体材料晶体和薄膜晶体的兴趣日益增长，这是由于它们的电子、光学、机械、微结构以及不同的科学和技术应用引起的。实际上，大部分半导体和光器件的现代成果，如果没有基本的、二元的、三元的及其他不同特性和大尺寸的化合物晶体的发展则是不可能的。这些文章致力于生长机制的基本理解、缺陷形成、生长工艺和生长系统的设计，因此数量是庞大的。

本手册针对目前备受关注的体材料晶体和薄膜晶体的生长技术水平进行阐述。我们的目的是使读者了解经常使用的生长工艺、材料生产和缺陷产生的基本知识。为完成这一任务，我们精选了50多位顶尖科学家、学者和工程师，他们的合作者来自于22个不同国家。这些作者根据他们的专业所长，编写了关于晶体生长和缺陷形成共计52章内容：从熔体、溶液到气相体材料生长；外延生长；生长工艺和缺陷的模型；缺陷特性的技术以及一些现代的特别课题。

本手册分为七部分。Part A介绍基础理论：生长和表征技术综述，表面成核工艺，溶液生长晶体的形态，生长过程中成核的层错，缺陷形成的形态。

Part B介绍体材料晶体的熔体生长，一种生长大尺寸晶体的关键方法。这一部分阐述了直拉单晶工艺、泡生法、布里兹曼法、浮区熔融等工艺，以及这些方法的最新进展，例如应用磁场的晶体生长、生长轴的取向、增加底基和形状控制。本部分涉及材料从硅和Ⅲ–Ⅴ族化合物到氧化物和氟化物的广泛内容。

第三部分，本书的Part C关注了溶液生长法。在前两章里讨论了水热生长法的不同方面，随后的三章介绍了非线性和激光晶体、KTP和KDP。通过在地球上和微重力环境下生长的比较给出了重力对溶液生长法的影响的知识。

Part D的主题是气相生长。这一部分提供了碳化硅、氮化镓、氮化铝和有机半导体的气相生长的内容。随后的Part E是关于外延生长和薄膜的，主要包括从液相的化学气相淀积到脉冲激光和脉冲电子淀积。

Part F介绍了生长工艺和缺陷形成的模型。这些章节验证了工艺参数和产生晶体质量问题包括缺陷形成的直接相互作用关系。随后的Part G展示了结晶材料特性和分析的发展。Part F和G说明了预测工具和分析技术在帮助高质量的大尺寸晶体生长工艺的设计和控制方面是非常好用的。

最后的Part H致力于精选这一领域的部分现代课题，例如蛋白质晶体生长、凝胶结晶、原位结构、单晶闪烁材料的生长、光电材料和线切割大晶体薄膜。

我们希望这本施普林格手册对那些学习晶体生长的研究生，那些从事或即将从事这一领域研究的来自学术界和工业领域的研究人员、科学家和工程师以及那些制备晶体的人是有帮助的。

我们对施普林格的Dr. Claus Acheron，Dr. Werner Skolaut和le-tex的Ms Anne Strobach的特别努力表示真诚的感谢，没有他们本书将无法呈现。

我们感谢我们的作者编写了详尽的章节内容和在本书出版期间对我们的耐心。一位编者（GD）感谢他的家庭成员和Dr. Kedar Gupta(ARC Energy 的CEO)，感谢他们在本书编写期间的大力支持和鼓励。还对Peter Rudolf, David Bliss, Ishwara Bhat和Partha Dutta在A、B、E部分的编写中所给予的帮助表示感谢。

Nashua, New Hampshire, April 2010	G. Dhanaraj
Mysore, India	K. Byrappa
Denton, Texas	V. Prasad
Stony Brook, New York	M. Dudley

Preface

Over the years, many successful attempts have been made to describe the art and science of crystal growth, and many review articles, monographs, symposium volumes, and handbooks have been published to present comprehensive reviews of the advances made in this field. These publications are testament to the growing interest in both bulk and thin-film crystals because of their electronic, optical, mechanical, microstructural, and other properties, and their diverse scientific and technological applications. Indeed, most modern advances in semiconductor and optical devices would not have been possible without the development of many elemental, binary, ternary, and other compound crystals of varying properties and large sizes. The literature devoted to basic understanding of growth mechanisms, defect formation, and growth processes as well as the design of growth systems is therefore vast.

The objective of this Springer Handbook is to present the state of the art of selected topical areas of both bulk and thin-film crystal growth. Our goal is to make readers understand the basics of the commonly employed growth processes, materials produced, and defects generated. To accomplish this, we have selected more than 50 leading scientists, researchers, and engineers, and their many collaborators from 22 different countries, to write chapters on the topics of their expertise. These authors have written 52 chapters on the fundamentals of crystal growth and defect formation; bulk growth from the melt, solution, and vapor; epitaxial growth; modeling of growth processes and defects; and techniques of defect characterization, as well as some contemporary special topics.

This Springer Handbook is divided into seven parts. Part A presents the fundamentals: an overview of the growth and characterization techniques, followed by the state of the art of nucleation at surfaces, morphology of crystals grown from solutions, nucleation of dislocation during growth, and defect formation and morphology.

Part B is devoted to bulk growth from the melt, a method critical to producing large-size crystals. The chapters in this part describe the well-known processes such as Czochralski, Kyropoulos, Bridgman, and floating zone, and focus specifically on recent advances in improving these methodologies such as application of magnetic fields, orientation of the growth axis, introduction of a pedestal, and shaped growth. They also cover a wide range of materials from silicon and III–V compounds to oxides and fluorides.

The third part, Part C of the book, focuses on solution growth. The various aspects of hydrothermal growth are discussed in two chapters, while three other chapters present an overview of the nonlinear and laser crystals, KTP and KDP. The knowledge on the effect of gravity on solution growth is presented through a comparison of growth on Earth versus in a microgravity environment.

The topic of Part D is vapor growth. In addition to presenting an overview of vapor growth, this part also provides details on vapor growth of silicon carbide, gallium nitride, aluminum nitride, and organic semiconductors. This is followed by chapters on epitaxial growth and thin films in Part E. The topics range from chemical vapor deposition to liquid-phase epitaxy to pulsed laser and pulsed electron deposition.

Modeling of both growth processes and defect formation is presented in Part F. These chapters demonstrate the direct correlation between the process parameters and quality of the crystal produced, including the formation of defects. The subsequent Part G presents the techniques that have been developed for crystalline material characterization and analysis. The chapters in Parts F and G demonstrate how well predictive tools and analytical techniques have helped the design and control of growth processes for better-quality crystals of large sizes.

The final Part H is devoted to some selected contemporary topics in this field, such as protein crystal growth, crystallization from gels, in situ structural studies, growth of single-crystal scintillation materials, photovoltaic materials, and wire-saw slicing of large crystals to produce wafers.

We hope this Springer Handbook will be useful to graduate students studying crystal growth and to researchers, scientists, and engineers from academia and industry who are conducting or intend to conduct research in this field as well as those who grow crystals.

We would like to express our sincere thanks to Dr. Claus Acheron and Dr. Werner Skolaut of Springer and Ms Anne Strohbach of le-tex for their extraordinary efforts without which this handbook would not have taken its final shape.

We thank our authors for writing comprehensive chapters and having patience with us during the publication of this Handbook. One of the editors (GD) would like to thank his family members and Dr. Kedar Gupta (CEO of ARC Energy) for their generous support and encouragement during the entire course of editing this handbook. Acknowledgements are also due to Peter Rudolf, David Bliss, Ishwara Bhat, and Partha Dutta for their help in editing Parts A, B, E, and H, respectively.

Nashua, New Hampshire, April 2010	G. Dhanaraj
Mysore, India	K. Byrappa
Denton, Texas	V. Prasad
Stony Brook, New York	M. Dudley

About the Editors

Govindhan Dhanaraj is the Manager of Crystal Growth Technologies at Advanced Renewable Energy Company (ARC Energy) at Nashua, New Hampshire (USA) focusing on the growth of large size sapphire crystals for LED lighting applications, characterization and related crystal growth furnace development. He received his PhD from the Indian Institute of Science, Bangalore and his Master of Science from Anna University (India). Immediately after his doctoral degree, Dr. Dhanaraj joined a National Laboratory, presently known as Rajaramanna Center for Advanced Technology in India, where he established an advanced Crystal Growth Laboratory for the growth of optical and laser crystals. Prior to joining ARC Energy, Dr. Dhanaraj served as a Research Professor at the Department of Materials Science and Engineering, Stony Brook University, NY, and also held a position of Research Assistant Professor at Hampton University, VA. During his 25 years of focused expertise in crystal growth research, he has developed optical, laser and semiconductor bulk crystals and SiC epitaxial films using solution, flux, Czochralski, Bridgeman, gel and vapor methods, and characterized them using x-ray topography, synchrotron topography, chemical etching and optical and atomic force microscopic techniques. He co-organized a symposium on Industrial Crystal Growth under the 17th American Conference on Crystal Growth and Epitaxy in conjunction with the 14th US Biennial Workshop on Organometallic Vapor Phase Epitaxy held at Lake Geneva, WI in 2009. Dr. Dhanaraj has delivered invited lectures and also served as session chairman in many crystal growth and materials science meetings. He has published over 100 papers and his research articles have attracted over 250 rich citations.

Kullaiah Byrappa received his Doctor's degree in Crystal Growth from the Moscow State University, Moscow in 1981. He is Professor of Materials Science, Head of the Crystal Growth Laboratory, and Director of the Internal Quality Assurance Cell of the University of Mysore, India. His current research is in crystal engineering of polyscale materials through novel solution processing routes, particularly covering hydrothermal, solvothermal and supercritical methods. Professor Byrappa has co-authored the Handbook of Hydrothermal Technology, and edited 4 books as well as two special editions of Journal of Materials Science, and published 180 research papers including 26 invited reviews and book chapters on various aspects of novel routes of solution processing. Professor Byrappa has delivered over 60 keynote and invited lectures at International Conferences, and several hundreds of colloquia and seminars at various institutions around the world. He has also served as chair and co-chair for numerous international conferences. He is a Fellow of the World Academy of Ceramics. Professor Byrappa is serving in several international committees and commissions related to crystallography, crystal growth, and materials science. He is the Founder Secretary of the International Solvothermal and Hydrothermal Association. Professor Byrappa is a recipient of several awards such as the Sir C.V. Raman Award, Materials Research Society of India Medal, and the Golden Jubilee Award of the University of Mysore.

About the Editors

Vishwanath "Vish" Prasad is the Vice President for Research and Economic Development and Professor of Mechanical and Energy Engineering at the University of North Texas (UNT), one of the largest university in the state of Texas. He received his PhD from the University of Delaware (USA), his Masters of Technology from the Indian Institute of Technology, Kanpur, and his bachelor's from Patna University in India all in Mechanical Engineering. Prior to joining UNT in 2007, Dr. Prasad served as the Dean at Florida International University (FIU) in Miami, where he also held the position of Distinguished Professor of Engineering. Previously, he has served as a Leading Professor of Mechanical Engineering at Stony Brook University, New York, as an Associate Professor and Assistant Professor at Columbia University. He has received many special recognitions for his contributions to engineering education. Dr. Prasad's research interests include thermo-fluid sciences, energy systems, electronic materials, and computational materials processing. He has published over 200 articles, edited/co-edited several books and organized numerous conferences, symposia, and workshops. He serves as the lead editor of the Annual Review of Heat Transfer. In the past, he has served as an Associate Editor of the ASME Journal of Heat. Dr. Prasad is an elected Fellow of the American Society of Mechanical Engineers (ASME), and has served as a member of the USRA Microgravity Research Council. Dr. Prasad's research has focused on bulk growth of silicon, III-V compounds, and silicon carbide; growth of large diameter Si tube; design of crystal growth systems; and sputtering and chemical vapor deposition of thin films. He is also credited to initiate research on wire saw cutting of large crystals to produce wafers with much reduced material loss. Dr. Prasad's research has been well funded by US National Science Foundation (NSF), US Department of Defense, US Department of Energy, and industry.

Michael Dudley received his Doctoral Degree in Engineering from Warwick University, UK, in 1982. He is Professor and Chair of the Materials Science and Engineering Department at Stony Brook University, New York, USA. He is director of the Stony Brook Synchrotron Topography Facility at the National Synchrotron Light Source at Brookhaven National Laboratory, Upton New York. His current research focuses on crystal growth and characterization of defect structures in single crystals with a view to determining their origins. The primary technique used is synchrotron topography which enables analysis of defects and generalized strain fields in single crystals in general, with particular emphasis on semiconductor, optoelectronic, and optical crystals. Establishing the relationship between crystal growth conditions and resulting defect distributions is a particular thrust area of interest to Dudley, as is the correlation between electronic/optoelectronic device performance and defect distribution. Other techniques routinely used in such analysis include transmission electron microscopy, high resolution triple-axis x-ray diffraction, atomic force microscopy, scanning electron microscopy, Nomarski optical microscopy, conventional optical microscopy, IR microscopy and fluorescent laser scanning confocal microscopy. Dudley's group has played a prominent role in the development of SiC and AlN growth, characterizing crystals grown by many of the academic and commercial entities involved enabling optimization of crystal quality. He has co-authored some 315 refereed articles and 12 book chapters, and has edited 5 books. He is currently a member of the Editorial Board of Journal of Applied Physics and Applied Physics Letters and has served as Chair or Co-Chair for numerous international conferences.

List of Authors

Francesco Abbona
Università degli Studi di Torino
Dipartimento di Scienze Mineralogiche
e Petrologiche
via Valperga Caluso 35
10125 Torino, Italy
e-mail: *francesco.abbona@unito.it*

Mohan D. Aggarwal
Alabama A&M University
Department of Physics
Normal, AL 35762, USA
e-mail: *mohan.aggarwal@aamu.edu*

Marcello R.B. Andreeta
University of São Paulo
Crystal Growth and Ceramic Materials Laboratory,
Institute of Physics of São Carlos
Av. Trabalhador Sãocarlense, 400
São Carlos, SP 13560-970, Brazil
e-mail: *marcello@if.sc.usp.br*

Dino Aquilano
Università degli Studi di Torino
Facoltà di Scienze Matematiche, Fisiche e Naturali
via P. Giuria, 15
Torino, 10126, Italy
e-mail: *dino.aquilano@unito.it*

Roberto Arreguín-Espinosa
Universidad Nacional Autónoma de México
Instituto de Química
Circuito Exterior, C.U. s/n
Mexico City, 04510, Mexico
e-mail: *arrespin@unam.mx*

Jie Bai
Intel Corporation
RA3-402, 5200 NE Elam Young Parkway
Hillsboro, OR 97124-6497, USA
e-mail: *jie.bai@intel.com*

Stefan Balint
West University of Timisoara
Department of Computer Science
Blvd. V. Parvan 4
Timisoara, 300223, Romania
e-mail: *balint@math.uvt.ro*

Ashok K. Batra
Alabama A&M University
Department of Physics
4900 Meridian Street
Normal, AL 35762, USA
e-mail: *ashok.batra@aamu.edu*

Handady L. Bhat
Indian Institute of Science
Department of Physics
CV Raman Avenue
Bangalore, 560012, India
e-mail: *hlbhat@physics.iisc.ernet.in*

Ishwara B. Bhat
Rensselaer Polytechnic Institute
Electrical Computer
and Systems Engineering Department
110 8th Street, JEC 6031
Troy, NY 12180, USA
e-mail: *bhati@rpi.edu*

David F. Bliss
US Air Force Research Laboratory
Sensors Directorate Optoelectronic Technology
Branch
80 Scott Drive
Hanscom AFB, MA 01731, USA
e-mail: *david.bliss@hanscom.af.mil*

Mikhail A. Borik
Russian Academy of Sciences
Laser Materials and Technology Research Center,
A.M. Prokhorov General Physics Institute
Vavilov 38
Moscow, 119991, Russia
e-mail: *borik@lst.gpi.ru*

Liliana Braescu
West University of Timisoara
Department of Computer Science
Blvd. V. Parvan 4
Timisoara, 300223, Romania
e-mail: *lilianabraescu@balint1.math.uvt.ro*

Kullaiah Byrappa
University of Mysore
Department of Geology
Manasagangotri
Mysore, 570 006, India
e-mail: *kbyrappa@gmail.com*

Dang Cai
CVD Equipment Corporation
1860 Smithtown Ave.
Ronkonkoma, NY 11779, USA
e-mail: *dcai@cvdequipment.com*

Michael J. Callahan
GreenTech Solutions
92 Old Pine Drive
Hanson, MA 02341, USA
e-mail: *mjcal37@yahoo.com*

Joan J. Carvajal
Universitat Rovira i Virgili (URV)
Department of Physics and Crystallography
of Materials and Nanomaterials (FiCMA-FiCNA)
Campus Sescelades, C/ Marcel·lí Domingo, s/n
Tarragona 43007, Spain
e-mail: *joanjosep.carvajal@urv.cat*

Aaron J. Celestian
Western Kentucky University
Department of Geography and Geology
1906 College Heights Blvd.
Bowling Green, KY 42101, USA
e-mail: *aaron.celestian@wku.edu*

Qi-Sheng Chen
Chinese Academy of Sciences
Institute of Mechanics
15 Bei Si Huan Xi Road
Beijing, 100190, China
e-mail: *qschen@imech.ac.cn*

Chunhui Chung
Stony Brook University
Department of Mechanical Engineering
Stony Brook, NY 11794-2300, USA
e-mail: *chuchung@ic.sunysb.edu*

Ted Ciszek
Geolite/Siliconsultant
31843 Miwok Trl.
Evergreen, CO 80437, USA
e-mail: *ted_ciszek@siliconsultant.com*

Abraham Clearfield
Texas A&M University
Distinguished Professor of Chemistry
College Station, TX 77843-3255, USA
e-mail: *clearfield@chem.tamu.edu*

Hanna A. Dabkowska
Brockhouse Institute for Materials Research
Department of Physics and Astronomy
1280 Main Str W.
Hamilton, Ontario L8S 4M1, Canada
e-mail: *dabkoh@mcmaster.ca*

Antoni B. Dabkowski
McMaster University, BIMR
Brockhouse Institute for Materials Research,
Department of Physics and Astronomy
1280 Main Str W.
Hamilton, Ontario L8S 4M1, Canada
e-mail: *dabko@mcmaster.ca*

Rafael Dalmau
HexaTech Inc.
991 Aviation Pkwy Ste 800
Morrisville, NC 27560, USA
e-mail: *rdalmau@hexatechinc.com*

Govindhan Dhanaraj
ARC Energy
18 Celina Avenue, Unit 77
Nashua, NH 03063, USA
e-mail: *dhanaraj@arc-energy.com*

Ramasamy Dhanasekaran
Anna University Chennai
Crystal Growth Centre
Chennai, 600 025, India
e-mail: *rdhanasekaran@annauniv.edu;
rdcgc@yahoo.com*

Ernesto Diéguez
Universidad Autónoma de Madrid
Department Física de Materiales
Madrid 28049, Spain
e-mail: *ernesto.dieguez@uam.es*

Vijay K. Dixit
Raja Ramanna Center for Advance Technology
Semiconductor Laser Section,
Solid State Laser Division
Rajendra Nagar, RRCAT.
Indore, 452013, India
e-mail: *dixit@rrcat.gov.in*

Sadik Dost
University of Victoria
Crystal Growth Laboratory
Victoria, BC V8W 3P6, Canada
e-mail: *sdost@me.uvic.ca*

Michael Dudley
Stony Brook University
Department of Materials Science and Engineering
Stony Brook, NY 11794-2275, USA
e-mail: *mdudley@notes.cc.sunysb.edu*

Partha S. Dutta
Rensselaer Polytechnic Institute
Department of Electrical, Computer
and Systems Engineering
110 Eighth Street
Troy, NY 12180, USA
e-mail: *duttap@rpi.edu*

Francesc Díaz
Universitat Rovira i Virgili (URV)
Department of Physics and Crystallography
of Materials and Nanomaterials (FiCMA-FiCNA)
Campus Sescelades, C/ Marcel·lí Domingo, s/n
Tarragona 43007, Spain
e-mail: *f.diaz@urv.cat*

Paul F. Fewster
PANalytical Research Centre,
The Sussex Innovation Centre
Research Department
Falmer
Brighton, BN1 9SB, UK
e-mail: *paul.fewster@panalytical.com*

Donald O. Frazier
NASA Marshall Space Flight Center
Engineering Technology Management Office
Huntsville, AL 35812, USA
e-mail: *donald.o.frazier@nasa.gov*

James W. Garland
EPIR Technologies, Inc.
509 Territorial Drive, Ste. B
Bolingbrook, IL 60440, USA
e-mail: *jgarland@epir.com*

Thomas F. George
University of Missouri-St. Louis
Center for Nanoscience,
Department of Chemistry and Biochemistry,
Department of Physics and Astronomy
One University Boulevard
St. Louis, MO 63121, USA
e-mail: *tfgeorge@umsl.edu*

Andrea E. Gutiérrez-Quezada
Universidad Nacional Autónoma de México
Instituto de Química
Circuito Exterior, C.U. s/n
Mexico City, 04510, Mexico
e-mail: *30111390@escolar.unam.mx*

Carl Hemmingsson
Linköping University
Department of Physics, Chemistry
and Biology (IFM)
581 83 Linköping, Sweden
e-mail: *cah@ifm.liu.se*

Antonio Carlos Hernandes
University of São Paulo
Crystal Growth and Ceramic Materials Laboratory,
Institute of Physics of São Carlos
Av. Trabalhador Sãocarlense
São Carlos, SP 13560-970, Brazil
e-mail: *hernandes@if.sc.usp.br*

Koichi Kakimoto
Kyushu University
Research Institute for Applied Mechanics
6-1 Kasuga-kouen, Kasuga
816-8580 Fukuoka, Japan
e-mail: *kakimoto@riam.kyushu-u.ac.jp*

Imin Kao
State University of New York at Stony Brook
Department of Mechanical Engineering
Stony Brook, NY 11794-2300, USA
e-mail: imin.kao@stonybrook.edu

John J. Kelly
Utrecht University,
Debye Institute for Nanomaterials Science
Department of Chemistry
Princetonplein 5
3584 CC, Utrecht, The Netherlands
e-mail: j.j.kelly@uu.nl

Jeonggoo Kim
Neocera, LLC
10000 Virginia Manor Road #300
Beltsville, MD, USA
e-mail: kim@neocera.com

Helmut Klapper
Institut für Kristallographie
RWTH Aachen University
Aachen, Germany
e-mail: klapper@xtal.rwth-aachen.de;
helmut-klapper@web.de

Christine F. Klemenz Rivenbark
Krystal Engineering LLC
General Manager and Technical Director
1429 Chaffee Drive
Titusville, FL 32780, USA
e-mail: ckr@krystalengineering.com

Christian Kloc
Nanyang Technological University
School of Materials Science and Engineering
50 Nanyang Avenue
639798 Singapore
e-mail: ckloc@ntu.edu.sg

Solomon H. Kolagani
Neocera LLC
10000 Virginia Manor Road
Beltsville, MD 20705, USA
e-mail: harsh@neocera.com

Akinori Koukitu
Tokyo University of Agriculture and Technology (TUAT)
Department of Applied Chemistry
2-24-16 Naka-cho, Koganei
184-8588 Tokyo, Japan
e-mail: koukitu@cc.tuat.ac.jp

Milind S. Kulkarni
MEMC Electronic Materials
Polysilicon and Quantitative Silicon Research
501 Pearl Drive
St. Peters, MO 63376, USA
e-mail: mkulkarni@memc.com

Yoshinao Kumagai
Tokyo University of Agriculture and Technology
Department of Applied Chemistry
2-24-16 Naka-cho, Koganei
184-8588 Tokyo, Japan
e-mail: 4470kuma@cc.tuat.ac.jp

Valentin V. Laguta
Institute of Physics of the ASCR
Department of Optical Materials
Cukrovarnicka 10
Prague, 162 53, Czech Republic
e-mail: laguta@fzu.cz

Ravindra B. Lal
Alabama Agricultural and Mechanical University
Physics Department
4900 Meridian Street
Normal, AL 35763, USA
e-mail: rblal@comcast.net

Chung-Wen Lan
National Taiwan University
Department of Chemical Engineering
No. 1, Sec. 4, Roosevelt Rd.
Taipei, 106, Taiwan
e-mail: cwlan@ntu.edu.tw

Hongjun Li
Chinese Academy of Sciences
R & D Center of Synthetic Crystals,
Shanghai Institute of Ceramics
215 Chengbei Rd., Jiading District
Shanghai, 201800, China
e-mail: lh_li@mail.sic.ac.cn

Elena E. Lomonova
Russian Academy of Sciences
Laser Materials and Technology Research Center,
A.M. Prokhorov General Physics Institute
Vavilov 38
Moscow, 119991, Russia
e-mail: *lomonova@lst.gpi.ru*

Ivan V. Markov
Bulgarian Academy of Sciences
Institute of Physical Chemistry
Sofia, 1113, Bulgaria
e-mail: *imarkov@ipc.bas.bg*

Bo Monemar
Linköping University
Department of Physics, Chemistry and Biology
58183 Linköping, Sweden
e-mail: *bom@ifm.liu.se*

Abel Moreno
Universidad Nacional Autónoma de México
Instituto de Química
Circuito Exterior, C.U. s/n
Mexico City, 04510, Mexico
e-mail: *carcamo@unam.mx*

Roosevelt Moreno Rodriguez
State University of New York at Stony Brook
Department of Mechanical Engineering
Stony Brook, NY 11794-2300, USA
e-mail: *roosevelt@dove.eng.sunysb.edu*

S. Narayana Kalkura
Anna University Chennai
Crystal Growth Centre
Sardar Patel Road
Chennai, 600025, India
e-mail: *kalkura@annauniv.edu*

Mohan Narayanan
Reliance Industries Limited
1, Rich Branch court
Gaithersburg, MD 20878, USA
e-mail: *mohan.narayanan@ril.com*

Subramanian Natarajan
Madurai Kamaraj University
School of Physics
Palkalai Nagar
Madurai, India
e-mail: *s_natarajan50@yahoo.com*

Martin Nikl
Academy of Sciences of the Czech Republic (ASCR)
Department of Optical Crystals, Institute of Physics
Cukrovarnicka 10
Prague, 162 53, Czech Republic
e-mail: *nikl@fzu.cz*

Vyacheslav V. Osiko
Russian Academy of Sciences
Laser Materials and Technology Research Center,
A.M. Prokhorov General Physics Institute
Vavilov 38
Moscow, 119991, Russia
e-mail: *osiko@lst.gpi.ru*

John B. Parise
Stony Brook University
Chemistry Department
and Department of Geosciences
ESS Building
Stony Brook, NY 11794-2100, USA
e-mail: *john.parise@stonybrook.edu*

Srinivas Pendurti
ASE Technologies Inc.
11499, Chester Road
Cincinnati, OH 45246, USA
e-mail: *spendurti@asetech.com*

Benjamin G. Penn
NASA/George C. Marshall Space Flight Center
ISHM and Sensors Branch
Huntsville, AL 35812, USA
e-mail: *benjamin.g.penndr@nasa.gov*

Jens Pflaum
Julius-Maximilians Universität Würzburg
Institute of Experimental Physics VI
Am Hubland
97078 Würzburg, Germany
e-mail: *jpflaum@physik.uni-wuerzburg.de*

Jose Luis Plaza
Universidad Autónoma de Madrid
Facultad de Ciencias,
Departamento de Física de Materiales
Madrid 28049, Spain
e-mail: *joseluis.plaza@uam.es*

Udo W. Pohl
Technische Universität Berlin
Institut für Festkörperphysik EW5-1
Hardenbergstr. 36
10623 Berlin, Germany
e-mail: *pohl@physik.tu-berlin.de*

Vishwanath (Vish) Prasad
University of North Texas
1155 Union Circle
Denton, TX 76203-5017, USA
e-mail: *vish.prasad@unt.edu*

Maria Cinta Pujol
Universitat Rovira i Virgili
Department of Physics and Crystallography
of Materials and Nanomaterials (FiCMA-FiCNA)
Campus Sescelades, C/ Marcel·lí Domingo
Tarragona 43007, Spain
e-mail: *mariacinta.pujol@urv.cat*

Balaji Raghothamachar
Stony Brook University
Department of Materials Science and Engineering
310 Engineering Building
Stony Brook, NY 11794-2275, USA
e-mail: *braghoth@notes.cc.sunysb.edu*

Michael Roth
The Hebrew University of Jerusalem
Department of Applied Physics
Bergman Bld., Rm 206, Givat Ram Campus
Jerusalem 91904, Israel
e-mail: *mroth@vms.huji.ac.il*

Peter Rudolph
Leibniz Institute for Crystal Growth
Technology Development
Max-Born-Str. 2
Berlin, 12489, Germany
e-mail: *rudolph@ikz-berlin.de*

Akira Sakai
Osaka University
Department of Systems Innovation
1-3 Machikaneyama-cho, Toyonaka-shi
560-8531 Osaka, Japan
e-mail: *sakai@ee.es.osaka-u.ac.jp*

Yasuhiro Shiraki
Tokyo City University
Advanced Research Laboratories,
Musashi Institute of Technology
8-15-1 Todoroki, Setagaya-ku
158-0082 Tokyo, Japan
e-mail: *yshiraki@tcu.ac.jp*

Theo Siegrist
Florida State University
Department of Chemical
and Biomedical Engineering
2525 Pottsdamer Street
Tallahassee, FL 32310, USA
e-mail: *siegrist@eng.fsu.edu*

Zlatko Sitar
North Carolina State University
Materials Science and Engineering
1001 Capability Dr.
Raleigh, NC 27695, USA
e-mail: *sitar@ncsu.edu*

Sivalingam Sivananthan
University of Illinois at Chicago
Department of Physics
845 W. Taylor St. M/C 273
Chicago, IL 60607-7059, USA
e-mail: *siva@uic.edu; siva@epir.com*

Mikhail D. Strikovski
Neocera LLC
10000 Virginia Manor Road, suite 300
Beltsville, MD 20705, USA
e-mail: *strikovski@neocera.com*

Xun Sun
Shandong University
Institute of Crystal Materials
Shanda Road
Jinan, 250100, China
e-mail: *sunxun@icm.sdu.edu.cn*

Ichiro Sunagawa
University Tohoku University (Emeritus)
Kashiwa-cho 3-54-2, Tachikawa
Tokyo, 190-0004, Japan
e-mail: *i.sunagawa@nifty.com*

Xu-Tang Tao
Shandong University
State Key Laboratory of Crystal Materials
Shanda Nanlu 27, 250100
Jinan, China
e-mail: *txt@sdu.edu.cn*

Vitali A. Tatartchenko
Saint – Gobain, 23 Rue Louis Pouey
92800 Puteaux, France
e-mail: *vitali.tatartchenko@orange.fr*

Filip Tuomisto
Helsinki University of Technology
Department of Applied Physics
Otakaari 1 M
Espoo TKK 02015, Finland
e-mail: *filip.tuomisto@tkk.fi*

Anna Vedda
University of Milano-Bicocca
Department of Materials Science
Via Cozzi 53
20125 Milano, Italy
e-mail: *anna.vedda@unimib.it*

Lu-Min Wang
University of Michigan
Department of Nuclear Engineering
and Radiological Sciences
2355 Bonisteel Blvd.
Ann Arbor, MI 48109-2104, USA
e-mail: *lmwang@umich.edu*

Sheng-Lai Wang
Shandong University
Institute of Crystal Materials,
State Key Laboratory of Crystal Materials
Shanda Road No. 27
Jinan, Shandong, 250100, China
e-mail: *slwang@icm.sdu.edu.cn*

Shixin Wang
Micron Technology Inc.
TEM Laboratory
8000 S. Federal Way
Boise, ID 83707, USA
e-mail: *shixinwang@micron.com*

Jan L. Weyher
Polish Academy of Sciences Warsaw
Institute of High Pressure Physics
ul. Sokolowska 29/37
01/142 Warsaw, Poland
e-mail: *weyher@unipress.waw.pl*

Jun Xu
Chinese Academy of Sciences
Shanghai Institute of Ceramics
Shanghai, 201800, China
e-mail: *xujun@mail.shcnc.ac.cn*

Hui Zhang
Tsinghua University
Department of Engineering Physics
Beijing, 100084, China
e-mail: *zhhui@tsinghua.edu.cn*

Lili Zheng
Tsinghua University
School of Aerospace
Beijing, 100084, China
e-mail: *zhenglili@tsinghua.edu.cn*

Mary E. Zvanut
University of Alabama at Birmingham
Department of Physics
1530 3rd Ave S
Birmingham, AL 35294-1170, USA
e-mail: *mezvanut@uab.edu*

Zbigniew R. Zytkiewicz
Polish Academy of Sciences
Institute of Physics
Al. Lotnikow 32/46
02668 Warszawa, Poland
e-mail: *zytkie@ifpan.edu.pl*

Acknowledgements

D.24 AlN Bulk Crystal Growth by Physical Vapor Transport
by Rafael Dalmau, Zlatko Sitar

This work was supported by the DoD Multidisciplinary University Research Initiative (MURI) administered by the Office of Naval Research (ONR) under grant N00014-01-1-0716, monitored by Dr. C.E. Wood.

D.26 Growth of III-Nitrides with Halide Vapor Phase Epitaxy (HVPE)
by Carl Hemmingsson, Bo Monemar, Yoshinao Kumagai, Akinori Koukitu

The authors gratefully acknowledge contributions from P.P. Paskov, T. Paskova, and V. Darakchieva, concerning new data cited and some illustrations. Our collaboration with the Epigress/Aixtron company (F. Wischmeyer, M. Heuken) in development of HVPE growth procedures and related equipment has been most helpful. We also have benefiting from a collaboration with A. Usui at Furukawa KK.

D.27 Growth of Semiconductor Single Crystals from Vapor Phase
by Ramasamy Dhanasekaran

The author is highly grateful to his research co-workers Dr. O. Senthil Kumar, Dr. S. Soundeswaran, Dr. M. J. Tafreshi, Dr. E. Varadarajan, Dr. P. Prabukanthan, and K. Senthilkumar for useful discussions and support in understanding the concepts presented in this chapter. Thanks are also due to Dr. B. Vengatesan and Dr. K. Balakrishnan, who initiated the vapor growth activities at Crystal Growth Centre, Anna University. The help rendered by R. Arunkumar, G. Bhagyaraj, T. Shabi, and M. Senthil Kumar in preparing this manuscript is duly acknowledged.

E.28 Epitaxial Growth of Silicon Carbide by Chemical Vapor Deposition
by Ishwara B. Bhat

The author would like to thank Canhua Li and Rongjun Wang for carrying out a major portion of the work described herein as part of their Ph.D. theses. Financial supports from DARPA contract #DAAD19-02-1-026 and the ERC program of the NSF are acknowledged.

E.30 Epitaxial Lateral Overgrowth of Semiconductors
by Zbigniew R. Zytkiewicz

The author thanks Dr. D. Dobosz for her assistance in the LPE growth of the ELO structures, Dr. E. Papis and K. Babska for the photolithography and processing of the substrates, and Dr. J. Domagala for XRD analysis of the samples. Contributions of Prof. T. Tuomi, Prof. P. McNally, Dr. R. Rantamaki, and Dr. D. Danilewsky to synchrotron x-ray topography experiments and Dr. A. Rocher to TEM studies of the GaAs structures are highly appreciated. The author is also very grateful to Prof. S. Dost for his valuable comments and feedback. This work was carried out with partial financial support from the Polish Committee for Scientific Research under grant 3T08A 021 26. Partial support from the Natural Sciences and Engineering Research Council of Canada (NSERC) is also gratefully acknowledged.

目 录

缩略语

Part D 晶体的气相生长

23 SiC晶体的生长与表征 3
23.1 SiC——背景与历史 3
23.2 气相生长 5
23.3 高温溶液生长 7
23.4 籽晶升华的产业化体材料生长 8
23.5 结构缺陷及其构造 11
23.6 结 语 22
参考文献 23

24 物理气相传输法生长体材料AlN晶体 27
24.1 物理气相传输法晶体生长 28
24.2 高温材料兼容 31
24.3 AlN体材料晶体的自籽晶生长 33
24.4 AlN体材料晶体的籽晶生长 35
24.5 高质量晶体表征 38
24.6 结论与展望 45
参考文献 45

25 单晶有机半导体的生长 51
25.1 基 础 51
25.2 成核与晶体生长理论 53
25.3 对半导体单晶有机材料的兴趣 54
25.4 提纯预生长 56
25.5 晶体生长 60
25.6 有机半导体单晶的质量 68
25.7 有机单晶场效应晶体管 69
25.8 结 论 70
参考文献 71

26 卤化物气相外延生长Ⅲ族氮化物 ... 75
26.1 生长化学和热力学 ... 75
26.2 HVPE生长设备 ... 78
26.3 体材料GaN的生长衬底和模版 ... 81
26.4 衬底除去技术 ... 85
26.5 HVPE中GaN的掺杂方法 ... 88
26.6 缺陷密度、位错和残留杂质 ... 89
26.7 HVPE生长的体材料GaN的一些重要性能 ... 93
26.8 通过HVPE生长AlN：一些初步的结论 ... 94
26.9 通过HVPE生长InN：一些初步的结论 ... 96
参考文献 ... 97

27 半导体单晶的气相生长 ... 103
27.1 气相生长分类 ... 105
27.2 化学气相传输——传输动力学 ... 107
27.3 热力学讨论 ... 111
27.4 CVT法Ⅱ-Ⅵ化合物半导体的生长 ... 118
27.5 纳米材料的气相生长 ... 122
27.6 I-Ⅲ-Ⅵ$_2$化合物生长 ... 123
27.7 VPE法生长氮化镓 ... 131
27.8 结 论 ... 135
参考文献 ... 136

Part E 外延生长和薄膜

28 化学气相沉积的碳化硅外延生长 ... 145
28.1 碳化硅极化类型 ... 147
28.2 碳化硅的缺陷 ... 148
28.3 碳化硅外延生长 ... 150
28.4 图形衬底上的外延生长 ... 158
28.5 结 论 ... 167
参考文献 ... 167

29 半导体的液相电外延 ... 173
29.1 背 景 ... 173
29.2 早期理论和模型的研究 ... 177

29.3 二维连续模型 183
　　29.4 静态磁场下的LPEE生长法 184
　　29.5 三维仿真 187
　　29.6 LPEE的高生长率：电磁场下迁移率 198
　　参考文献 202

30 半导体的外延横向增生 205
　　30.1 概　述 206
　　30.2 液相外延横向增生的机制 208
　　30.3 ELO层中的位错 217
　　30.4 ELO层张力 222
　　30.5 半导体结构横向增生的最新进展 232
　　30.6 结　语 240
　　参考文献 241

31 新材料的液相外延 247
　　31.1 LPE的发展历史 248
　　31.2 LPE 的基础和溶液生长 248
　　31.3 液相外延的要求 250
　　31.4 新材料研究：外延淀积法的选择 250
　　31.5 高温超导体的LPE法 252
　　31.6 锗酸钙镓的LEP 261
　　31.7 氮化物的液相外延 265
　　31.8 结　论 269
　　参考文献 270

32 分子束外延的HgCdTe生长 275
　　32.1 综　述 276
　　32.2 MBE生长理论 279
　　32.3 衬底材料 282
　　32.4 生长硬件的设计 294
　　32.5 监测和控制生长的原位表征工具 296
　　32.6 成核和生长过程 307
　　32.7 掺杂和掺杂激活 310
　　32.8 MBE法生长的HgCdTe外延层的特性 313

32.9 HgTe/CdTe超晶格 … 318
32.10 先进红外探测器的结构 … 321
32.11 红外焦平面阵列（FPAs） … 324
32.12 结　论 … 325
参考文献 … 327

33 稀释氮化物的金属有机物气相外延和砷化物量子点 … 339
33.1 MOVPE原则 … 339
33.2 稀释氮化物InGaAsN量子阱 … 343
33.3 InAs/GaAs量子点 … 348
33.4 结　语 … 354
参考文献 … 354

34 锗硅异质结的形成及其特性 … 359
34.1 背　景 … 359
34.2 Si/Ge异质结的能带结构 … 360
34.3 生长技术 … 362
34.4 表面隔离 … 363
34.5 临界厚度 … 367
34.6 应力松弛机理 … 369
34.7 松弛SiGe层的形成 … 371
34.8 量子阱的形成、超晶格、量子线 … 379
34.9 点形成 … 383
34.10 结语与展望 … 390
参考文献 … 390

35 脉冲激光的等离子能量和脉冲电子淀积 … 399
35.1 薄膜淀积的能量聚集 … 399
35.2 PLD和PED技术 … 400
35.3 PLD和PED中的原子能转换 … 401
35.4 薄膜生长的等离子体熔体的优化 … 410
35.5 结　论 … 414
参考文献 … 415

Contents

List of Abbreviations

Part D Crystal Growth from Vapor

23 Growth and Characterization of Silicon Carbide Crystals
Govindhan Dhanaraj, Balaji Raghothamachar, Michael Dudley 797
- 23.1 Silicon Carbide – Background and History 797
- 23.2 Vapor Growth 799
- 23.3 High-Temperature Solution Growth 801
- 23.4 Industrial Bulk Growth by Seed Sublimation 802
- 23.5 Structural Defects and Their Configurations 805
- 23.6 Concluding Remarks 816
- **References** 817

24 AlN Bulk Crystal Growth by Physical Vapor Transport
Rafael Dalmau, Zlatko Sitar 821
- 24.1 PVT Crystal Growth 822
- 24.2 High-Temperature Materials Compatibility 825
- 24.3 Self-Seeded Growth of AlN Bulk Crystals 827
- 24.4 Seeded Growth of AlN Bulk Crystals 829
- 24.5 Characterization of High-Quality Bulk Crystals 832
- 24.6 Conclusions and Outlook 839
- **References** 839

25 Growth of Single-Crystal Organic Semiconductors
Christian Kloc, Theo Siegrist, Jens Pflaum 845
- 25.1 Basics 845
- 25.2 Theory of Nucleation and Crystal Growth 847
- 25.3 Organic Materials of Interest for Semiconducting Single Crystals 848
- 25.4 Pregrowth Purification 850
- 25.5 Crystal Growth 854
- 25.6 Quality of Organic Semiconducting Single Crystals 862
- 25.7 Organic Single-Crystalline Field-Effect Transistors 863
- 25.8 Conclusions 864
- **References** 865

26 Growth of III-Nitrides with Halide Vapor Phase Epitaxy (HVPE)
Carl Hemmingsson, Bo Monemar, Yoshinao Kumagai, Akinori Koukitu 869
- 26.1 Growth Chemistry and Thermodynamics 869
- 26.2 HVPE Growth Equipment 872

26.3	Substrates and Templates for Bulk GaN Growth	875
26.4	Substrate Removal Techniques	879
26.5	Doping Techniques for GaN in HVPE	882
26.6	Defect Densities, Dislocations, and Residual Impurities	883
26.7	Some Important Properties of HVPE-Grown Bulk GaN Material	887
26.8	Growth of AlN by HVPE: Some Preliminary Results	888
26.9	Growth of InN by HVPE: Some Preliminary Results	890
	References	891

27 Growth of Semiconductor Single Crystals from Vapor Phase
Ramasamy Dhanasekaran 897

27.1	Classifications of Vapor Growth	899
27.2	Chemical Vapor Transport – Transport Kinetics	901
27.3	Thermodynamic Considerations	905
27.4	Growth of II–VI Compound Semiconductors by CVT	912
27.5	Growth of Nanomaterial from Vapor Phase	916
27.6	Growth of I–III–VI$_2$ Compounds	917
27.7	Growth of GaN by VPE	925
27.8	Conclusion	929
	References	930

Part E Epitaxial Growth and Thin Films

28 Epitaxial Growth of Silicon Carbide by Chemical Vapor Deposition
Ishwara B. Bhat 939

28.1	Polytypes of Silicon Carbide	941
28.2	Defects in SiC	942
28.3	Epitaxial Growth of Silicon Carbide	944
28.4	Epitaxial Growth on Patterned Substrates	952
28.5	Conclusions	961
	References	961

29 Liquid-Phase Electroepitaxy of Semiconductors
Sadik Dost 967

29.1	Background	967
29.2	Early Theoretical and Modeling Studies	971
29.3	Two-Dimensional Continuum Models	977
29.4	LPEE Growth Under a Stationary Magnetic Field	978
29.5	Three-Dimensional Simulations	981
29.6	High Growth Rates in LPEE: Electromagnetic Mobility	992
	References	996

30 Epitaxial Lateral Overgrowth of Semiconductors
Zbigniew R. Zytkiewicz 999
- 30.1 Overview 1000
- 30.2 Mechanism of Epitaxial Lateral Overgrowth from the Liquid Phase ... 1002
- 30.3 Dislocations in ELO Layers 1011
- 30.4 Strain in ELO Layers 1016
- 30.5 Recent Progress in Lateral Overgrowth of Semiconductor Structures . 1026
- 30.6 Concluding Remarks 1034
- **References** 1035

31 Liquid-Phase Epitaxy of Advanced Materials
Christine F. Klemenz Rivenbark 1041
- 31.1 Historical Development of LPE 1042
- 31.2 Fundamentals of LPE and Solution Growth 1042
- 31.3 Requirements for Liquid-Phase Epitaxy 1044
- 31.4 Developing New Materials: On the Choice of the Epitaxial Deposition Method 1044
- 31.5 LPE of High-Temperature Superconductors 1046
- 31.6 LPE of Calcium Gallium Germanates 1055
- 31.7 Liquid-Phase Epitaxy of Nitrides 1059
- 31.8 Conclusions 1063
- **References** 1064

32 Molecular-Beam Epitaxial Growth of HgCdTe
James W. Garland, Sivalingam Sivananthan 1069
- 32.1 Overview 1070
- 32.2 Theory of MBE Growth 1073
- 32.3 Substrate Materials 1076
- 32.4 Design of the Growth Hardware 1088
- 32.5 In situ Characterization Tools for Monitoring and Controlling the Growth 1090
- 32.6 Nucleation and Growth Procedure 1101
- 32.7 Dopants and Dopant Activation 1104
- 32.8 Properties of HgCdTe Epilayers Grown by MBE 1107
- 32.9 HgTe/CdTe Superlattices 1112
- 32.10 Architectures of Advanced IR Detectors 1115
- 32.11 IR Focal-Plane Arrays (FPAs) 1118
- 32.12 Conclusions 1119
- **References** 1121

33 Metalorganic Vapor-Phase Epitaxy of Diluted Nitrides and Arsenide Quantum Dots
Udo W. Pohl 1133
- 33.1 Principle of MOVPE 1133
- 33.2 Diluted Nitride InGaAsN Quantum Wells 1137

33.3 InAs/GaAs Quantum Dots 1142
33.4 Concluding Remarks 1148
References 1148

34 Formation of SiGe Heterostructures and Their Properties
Yasuhiro Shiraki, Akira Sakai 1153
34.1 Background 1153
34.2 Band Structures of Si/Ge Heterostructures 1154
34.3 Growth Technologies 1156
34.4 Surface Segregation 1157
34.5 Critical Thickness 1161
34.6 Mechanism of Strain Relaxation 1163
34.7 Formation of Relaxed SiGe Layers 1165
34.8 Formation of Quantum Wells, Superlattices, and Quantum Wires 1173
34.9 Dot Formation 1177
34.10 Concluding Remarks and Future Prospects 1184
References 1184

35 Plasma Energetics in Pulsed Laser and Pulsed Electron Deposition
Mikhail D. Strikovski, Jeonggoo Kim, Solomon H. Kolagani 1193
35.1 Energetic Condensation in Thin Film Deposition 1193
35.2 PLD and PED Techniques 1194
35.3 Transformations of Atomic Energy in PLD and PED 1195
35.4 Optimization of Plasma Flux for Film Growth 1204
35.5 Conclusions 1208
References 1209

List of Abbreviations

μ-PD	micro-pulling-down
1S-ELO	one-step ELO structure
2-D	two-dimensional
2-DNG	two-dimensional nucleation growth
2S-ELO	double layer ELO
3-D	three-dimensional
4T	quaterthiophene
6T	sexithienyl
8MR	eight-membered ring
8T	hexathiophene

A

a-Si	amorphous silicon
A/D	analogue-to-digital
AA	additional absorption
AANP	2-adamantylamino-5-nitropyridine
AAS	atomic absorption spectroscopy
AB	Abrahams and Burocchi
ABES	absorption-edge spectroscopy
AC	alternate current
ACC	annular capillary channel
ACRT	accelerated crucible rotation technique
ADC	analog-to-digital converter
ADC	automatic diameter control
ADF	annular dark field
ADP	ammonium dihydrogen phosphate
AES	Auger electron spectroscopy
AFM	atomic force microscopy
ALE	arbitrary Lagrangian Eulerian
ALE	atomic layer epitaxy
ALUM	aluminum potassium sulfate
ANN	artificial neural network
AO	acoustooptic
AP	atmospheric pressure
APB	antiphase boundaries
APCF	advanced protein crystallization facility
APD	avalanche photodiode
APPLN	aperiodic poled LN
APS	Advanced Photon Source
AR	antireflection
AR	aspect ratio
ART	aspect ratio trapping
ATGSP	alanine doped triglycine sulfo-phosphate
AVT	angular vibration technique

B

BA	Born approximation
BAC	band anticrossing
BBO	BaB_2O_4
BCF	Burton–Cabrera–Frank
BCT	$Ba_{0.77}Ca_{0.23}TiO_3$
BCTi	$Ba_{1-x}Ca_xTiO_3$
BE	bound exciton
BF	bright field
BFDH	Bravais–Friedel–Donnay–Harker
BGO	$Bi_{12}GeO_{20}$
BIBO	BiB_3O_6
BLIP	background-limited performance
BMO	$Bi_{12}MO_{20}$
BN	boron nitride
BOE	buffered oxide etch
BPD	basal-plane dislocation
BPS	Burton–Prim–Slichter
BPT	bipolar transistor
BS	Bridgman–Stockbarger
BSCCO	Bi–Sr–Ca–Cu–O
BSF	bounding stacking fault
BSO	$Bi_{20}SiO_{20}$
BTO	$Bi_{12}TiO_{20}$
BU	building unit
BaREF	barium rare-earth fluoride
BiSCCO	$Bi_2Sr_2CaCu_2O_n$

C

C–V	capacitance–voltage
CALPHAD	calculation of phase diagram
CBED	convergent-beam electron diffraction
CC	cold crucible
CCC	central capillary channel
CCD	charge-coupled device
CCVT	contactless chemical vapor transport
CD	convection diffusion
CE	counterelectrode
CFD	computational fluid dynamics
CFD	cumulative failure distribution
CFMO	Ca_2FeMoO_6
CFS	continuous filtration system
CGG	calcium gallium germanate
CIS	copper indium diselenide
CL	cathode-ray luminescence
CL	cathodoluminescence
CMM	coordinate measuring machine
CMO	$CaMoO_4$
CMOS	complementary metal–oxide–semiconductor
CMP	chemical–mechanical polishing
CMP	chemomechanical polishing

COD	calcium oxalate dihydrate	DS	directional solidification
COM	calcium oxalate-monohydrate	DSC	differential scanning calorimetry
COP	crystal-originated particle	DSE	defect-selective etching
CP	critical point	DSL	diluted Sirtl with light
CPU	central processing unit	DTA	differential thermal analysis
CRSS	critical-resolved shear stress	DTGS	deuterated triglycine sulfate
CSMO	$Ca_{1-x}Sr_xMoO_3$	DVD	digital versatile disk
CST	capillary shaping technique	DWBA	distorted-wave Born approximation
CST	crystalline silico titanate	DWELL	dot-in-a-well
CT	computer tomography		
CTA	$CsTiOAsO_4$		
CTE	coefficient of thermal expansion		

E

CTF	contrast transfer function		
CTR	crystal truncation rod	EADM	extended atomic distance mismatch
CV	Cabrera–Vermilyea	EALFZ	electrical-assisted laser floating zone
CVD	chemical vapor deposition	EB	electron beam
CVT	chemical vapor transport	EBIC	electron-beam-induced current
CW	continuous wave	ECE	end chain energy
CZ	Czochralski	ECR	electron cyclotron resonance
CZT	Czochralski technique	EDAX	energy-dispersive x-ray analysis
		EDMR	electrically detected magnetic resonance
		EDS	energy-dispersive x-ray spectroscopy
		EDT	ethylene dithiotetrathiafulvalene

D

		EDTA	ethylene diamine tetraacetic acid
D/A	digital to analog	EELS	electron energy-loss spectroscopy
DBR	distributed Bragg reflector	EFG	edge-defined film-fed growth
DC	direct current	EFTEM	energy-filtered transmission electron microscopy
DCAM	diffusion-controlled crystallization apparatus for microgravity	ELNES	energy-loss near-edge structure
DCCZ	double crucible CZ	ELO	epitaxial lateral overgrowth
DCPD	dicalcium-phosphate dihydrate	EM	electromagnetic
DCT	dichlorotetracene	EMA	effective medium theory
DD	dislocation dynamics	EMC	electromagnetic casting
DESY	Deutsches Elektronen Synchrotron	EMCZ	electromagnetic Czochralski
DF	dark field	EMF	electromotive force
DFT	density function theory	ENDOR	electron nuclear double resonance
DFW	defect free width	EO	electrooptic
DGS	diglycine sulfate	EP	EaglePicher
DI	deionized	EPD	etch pit density
DIA	diamond growth	EPMA	electron microprobe analysis
DIC	differential interference contrast	EPR	electron paramagnetic resonance
DICM	differential interference contrast microscopy	erfc	error function
		ES	equilibrium shape
DKDP	deuterated potassium dihydrogen phosphate	ESP	edge-supported pulling
		ESR	electron spin resonance
DLATGS	deuterated L-alanine-doped triglycine sulfate	EVA	ethyl vinyl acetate

F

DLTS	deep-level transient spectroscopy		
DMS	discharge mass spectroscopy		
DNA	deoxyribonucleic acid		
DOE	Department of Energy	F	flat
DOS	density of states	FAM	free abrasive machining
DPH-BDS	2,6-diphenylbenzo[1,2-*b*:4,5-*b*']diselenophene	FAP	$Ca_5(PO_4)_3F$
		FCA	free carrier absorption
DPPH	2,2-diphenyl-1-picrylhydrazyl	fcc	face-centered cubic
DRS	dynamic reflectance spectroscopy	FEC	full encapsulation Czochralski

FEM	finite element method
FES	fluid experiment system
FET	field-effect transistor
FFT	fast Fourier transform
FIB	focused ion beam
FOM	figure of merit
FPA	focal-plane array
FPE	Fokker–Planck equation
FSLI	femtosecond laser irradiation
FT	flux technique
FTIR	Fourier-transform infrared
FWHM	full width at half-maximum
FZ	floating zone
FZT	floating zone technique

G

GAME	gel acupuncture method
GDMS	glow-discharge mass spectrometry
GE	General Electric
GGG	gadolinium gallium garnet
GNB	geometrically necessary boundary
GPIB	general purpose interface bus
GPMD	geometric partial misfit dislocation
GRI	growth interruption
GRIIRA	green-radiation-induced infrared absorption
GS	growth sector
GSAS	general structure analysis software
GSGG	$Gd_3Sc_2Ga_3O_{12}$
GSMBE	gas-source molecular-beam epitaxy
GSO	Gd_2SiO_5
GU	growth unit

H

HA	hydroxyapatite
HAADF	high-angle annular dark field
HAADF-STEM	high-angle annular dark field in scanning transmission electron microscope
HAP	hydroxyapatite
HB	horizontal Bridgman
HBM	Hottinger Baldwin Messtechnik GmbH
HBT	heterostructure bipolar transistor
HBT	horizontal Bridgman technique
HDPCG	high-density protein crystal growth
HE	high energy
HEM	heat-exchanger method
HEMT	high-electron-mobility transistor
HF	hydrofluoric acid
HGF	horizontal gradient freezing
HH	heavy-hole
HH-PCAM	handheld protein crystallization apparatus for microgravity
HIV	human immunodeficiency virus
HIV-AIDS	human immunodeficiency virus–acquired immunodeficiency syndrome
HK	high potassium content
HLA	half-loop array
HLW	high-level waste
HMDS	hexamethyldisilane
HMT	hexamethylene tetramine
HNP	high nitrogen pressure
HOE	holographic optical element
HOLZ	higher-order Laue zone
HOMO	highest occupied molecular orbital
HOPG	highly oriented pyrolytic graphite
HOT	high operating temperature
HP	Hartman–Perdok
HPAT	high-pressure ammonothermal technique
HPHT	high-pressure high-temperature
HRTEM	high-resolution transmission electron microscopy
HRXRD	high-resolution x-ray diffraction
HSXPD	hemispherically scanned x-ray photoelectron diffraction
HT	hydrothermal
HTS	high-temperature solution
HTSC	high-temperature superconductor
HVPE	halide vapor-phase epitaxy
HVPE	hydride vapor-phase epitaxy
HWC	hot-wall Czochralski
HZM	horizontal ZM

I

IBAD	ion-beam-assisted deposition
IBE	ion beam etching
IC	integrated circuit
IC	ion chamber
ICF	inertial confinement fusion
ID	inner diameter
ID	inversion domain
IDB	incidental dislocation boundary
IDB	inversion domain boundary
IF	identification flat
IG	inert gas
IK	intermediate potassium content
ILHPG	indirect laser-heated pedestal growth
IML-1	International Microgravity Laboratory
IMPATT	impact ionization avalanche transit-time
IP	image plate
IPA	isopropyl alcohol
IR	infrared
IRFPA	infrared focal plane array
IS	interfacial structure
ISS	ion-scattering spectroscopy
ITO	indium-tin oxide
ITTFA	iterative target transform factor analysis
IVPE	iodine vapor-phase epitaxy

J

JDS	joint density of states
JFET	junction FET

K

K	kinked
KAP	potassium hydrogen phthalate
KDP	potassium dihydrogen phosphate
KGW	$KY(WO_4)_2$
KGdP	$KGd(PO_3)_4$
KLYF	$KLiYF_5$
KM	Kubota–Mullin
KMC	kinetic Monte Carlo
KN	$KNbO_3$
KNP	$KNd(PO_3)_4$
KPZ	Kardar–Parisi–Zhang
KREW	$KRE(WO_4)_2$
KTA	potassium titanyl arsenate
KTN	potassium niobium tantalate
KTP	potassium titanyl phosphate
KTa	$KTaO_3$
KTaN	$KTa_{1-x}Nb_xO_3$
KYF	KYF_4
KYW	$KY(WO_4)_2$

L

LACBED	large-angle convergent-beam diffraction
LAFB	L-arginine tetrafluoroborate
LAGB	low-angle grain boundary
LAO	$LiAlO_2$
LAP	L-arginine phosphate
LBIC	light-beam induced current
LBIV	light-beam induced voltage
LBO	LiB_3O_5
LBO	$LiBO_3$
LBS	laser-beam scanning
LBSM	laser-beam scanning microscope
LBT	laser-beam tomography
LCD	liquid-crystal display
LD	laser diode
LDT	laser-induced damage threshold
LEC	liquid encapsulation Czochralski
LED	light-emitting diode
LEEBI	low-energy electron-beam irradiation
LEM	laser emission microanalysis
LEO	lateral epitaxial overgrowth
LES	large-eddy simulation
LG	$LiGaO_2$
LGN	$La_3Ga_{5.5}Nb_{0.5}O_{14}$
LGO	$LaGaO_3$
LGS	$La_3Ga_5SiO_{14}$
LGT	$La_3Ga_{5.5}Ta_{0.5}O_{14}$
LH	light hole
LHFB	L-histidine tetrafluoroborate
LHPG	laser-heated pedestal growth
LID	laser-induced damage
LK	low potassium content
LLNL	Lawrence Livermore National Laboratory
LLO	laser lift-off
LLW	low-level waste
LN	$LiNbO_3$
LP	low pressure
LPD	liquid-phase diffusion
LPE	liquid-phase epitaxy
LPEE	liquid-phase electroepitaxy
LPS	$Lu_2Si_2O_7$
LSO	Lu_2SiO_5
LST	laser scattering tomography
LST	local shaping technique
LT	low-temperature
LTa	$LiTaO_3$
LUMO	lowest unoccupied molecular orbital
LVM	local vibrational mode
LWIR	long-wavelength IR
LY	light yield
LiCAF	$LiCaAlF_6$
LiSAF	lithium strontium aluminum fluoride

M

M–S	melt–solid
MAP	magnesium ammonium phosphate
MASTRAPP	multizone adaptive scheme for transport and phase change processes
MBE	molecular-beam epitaxy
MBI	multiple-beam interferometry
MC	multicrystalline
MCD	magnetic circular dichroism
MCT	HgCdTe
MCZ	magnetic Czochralski
MD	misfit dislocation
MD	molecular dynamics
ME	melt epitaxy
ME	microelectronics
MEMS	microelectromechanical system
MESFET	metal-semiconductor field effect transistor
MHP	magnesium hydrogen phosphate-trihydrate
MI	morphological importance
MIT	Massachusetts Institute of Technology
ML	monolayer
MLEC	magnetic liquid-encapsulated Czochralski

MLEK	magnetically stabilized liquid-encapsulated Kyropoulos		NTRS	National Technology Roadmap for Semiconductors
MMIC	monolithic microwave integrated circuit		NdBCO	$NdBa_2Cu_3O_{7-x}$

O

MNA	2-methyl-4-nitroaniline
MNSM	modified nonstationary model
MOCVD	metalorganic chemical vapor deposition
MOCVD	molecular chemical vapor deposition
MODFET	modulation-doped field-effect transistor
MOMBE	metalorganic MBE
MOS	metal–oxide–semiconductor
MOSFET	metal–oxide–semiconductor field-effect transistor
MOVPE	metalorganic vapor-phase epitaxy
mp	melting point
MPMS	mold-pushing melt-supplying
MQSSM	modified quasi-steady-state model
MQW	multiple quantum well
MR	melt replenishment
MRAM	magnetoresistive random-access memory
MRM	melt replenishment model
MSUM	monosodium urate monohydrate
MTDATA	metallurgical thermochemistry database
MTS	methyltrichlorosilane
MUX	multiplexor
MWIR	mid-wavelength infrared
MWRM	melt without replenishment model
MXRF	micro-area x-ray fluorescence

OCP	octacalcium phosphate
ODE	ordinary differential equation
ODLN	opposite domain LN
ODMR	optically detected magnetic resonance
OEIC	optoelectronic integrated circuit
OF	orientation flat
OFZ	optical floating zone
OLED	organic light-emitting diode
OMVPE	organometallic vapor-phase epitaxy
OPO	optical parametric oscillation
OSF	oxidation-induced stacking fault

P

PAMBE	photo-assisted MBE
PB	proportional band
PBC	periodic bond chain
pBN	pyrolytic boron nitride
PC	photoconductivity
PCAM	protein crystallization apparatus for microgravity
PCF	primary crystallization field
PCF	protein crystal growth facility
PCM	phase-contrast microscopy
PD	Peltier interface demarcation
PD	photodiode
PDE	partial differential equation
PDP	programmed data processor
PDS	periodic domain structure
PE	pendeo-epitaxy
PEBS	pulsed electron beam source
PEC	polyimide environmental cell
PECVD	plasma-enhanced chemical vapor deposition
PED	pulsed electron deposition
PEO	polyethylene oxide
PET	positron emission tomography
PID	proportional–integral–differential
PIN	positive intrinsic negative diode
PL	photoluminescence
PLD	pulsed laser deposition
PMNT	$Pb(Mg,Nb)_{1-x}Ti_xO_3$
PPKTP	periodically poled KTP
PPLN	periodic poled LN
PPLN	periodic poling lithium niobate
ppy	polypyrrole
PR	photorefractive
PSD	position-sensitive detector
PSF	prismatic stacking fault

N

N	nucleus
N	nutrient
NASA	National Aeronautics and Space Administration
NBE	near-band-edge
NBE	near-bandgap emission
NCPM	noncritically phase matched
NCS	neighboring confinement structure
NGO	$NdGaO_3$
NIF	National Ignition Facility
NIR	near-infrared
NIST	National Institute of Standards and Technology
NLO	nonlinear optic
NMR	nuclear magnetic resonance
NP	no-phonon
NPL	National Physical Laboratory
NREL	National Renewable Energy Laboratory
NS	Navier–Stokes
NSF	National Science Foundation
nSLN	nearly stoichiometric lithium niobate
NSLS	National Synchrotron Light Source
NSM	nonstationary model

PSI	phase-shifting interferometry	RTV	room temperature vulcanizing
PSM	phase-shifting microscopy	R&D	research and development
PSP	pancreatic stone protein		
PSSM	pseudo-steady-state model		
PSZ	partly stabilized zirconium dioxide		
PT	pressure–temperature		
PV	photovoltaic		
PVA	polyvinyl alcohol		
PVD	physical vapor deposition		
PVE	photovoltaic efficiency		
PVT	physical vapor transport		
PWO	$PbWO_4$		
PZNT	$Pb(Zn, Nb)_{1-x}Ti_xO_3$		
PZT	lead zirconium titanate		

S

S	stepped
SAD	selected area diffraction
SAM	scanning Auger microprobe
SAW	surface acoustical wave
SBN	strontium barium niobate
SC	slow cooling
SCBG	slow-cooling bottom growth
SCC	source-current-controlled
SCF	single-crystal fiber
SCF	supercritical fluid technology
SCN	succinonitrile
SCW	supercritical water
SD	screw dislocation
SE	spectroscopic ellipsometry
SECeRTS	small environmental cell for real-time studies
SEG	selective epitaxial growth
SEM	scanning electron microscope
SEM	scanning electron microscopy
SEMATECH	Semiconductor Manufacturing Technology
SF	stacking fault
SFM	scanning force microscopy
SGOI	SiGe-on-insulator
SH	second harmonic
SHG	second-harmonic generation
SHM	submerged heater method
SI	semi-insulating
SIA	Semiconductor Industry Association
SIMS	secondary-ion mass spectrometry
SIOM	Shanghai Institute of Optics and Fine Mechanics
SL	superlattice
SL-3	Spacelab-3
SLI	solid–liquid interface
SLN	stoichiometric LN
SM	skull melting
SMB	stacking mismatch boundary
SMG	surfactant-mediated growth
SMT	surface-mount technology
SNR	signal-to-noise ratio
SNT	sodium nonatitanate
SOI	silicon-on-insulator
SP	sputtering
sPC	scanning photocurrent
SPC	Scientific Production Company
SPC	statistical process control
SR	spreading resistance
SRH	Shockley–Read–Hall
SRL	strain-reducing layer
SRS	stimulated Raman scattering

Q

QD	quantum dot
QDT	quantum dielectric theory
QE	quantum efficiency
QPM	quasi-phase-matched
QPMSHG	quasi-phase-matched second-harmonic generation
QSSM	quasi-steady-state model
QW	quantum well
QWIP	quantum-well infrared photodetector

R

RAE	rotating analyzer ellipsometer
RBM	rotatory Bridgman method
RC	reverse current
RCE	rotating compensator ellipsometer
RE	rare earth
RE	reference electrode
REDG	recombination enhanced dislocation glide
RELF	rare-earth lithium fluoride
RF	radiofrequency
RGS	ribbon growth on substrate
RHEED	reflection high-energy electron diffraction
RI	refractive index
RIE	reactive ion etching
RMS	root-mean-square
RNA	ribonucleic acid
ROIC	readout integrated circuit
RP	reduced pressure
RPI	Rensselaer Polytechnic Institute
RSM	reciprocal space map
RSS	resolved shear stress
RT	room temperature
RTA	$RbTiOAsO_4$
RTA	rapid thermal annealing
RTCVD	rapid-thermal chemical vapor deposition
RTP	$RbTiOPO_4$
RTPL	room-temperature photoluminescence
RTR	ribbon-to-ribbon

SRXRD	spatially resolved XRD	TTV	total thickness variation
SS	solution-stirring	TV	television
SSL	solid-state laser	TVM	three-vessel solution circulating method
SSM	sublimation sandwich method	TVTP	time-varying temperature profile
ST	synchrotron topography	TWF	transmitted wavefront
STC	standard testing condition	TZM	titanium zirconium molybdenum
STE	self-trapped exciton	TZP	tetragonal phase
STEM	scanning transmission electron microscopy		

U

STM	scanning tunneling microscopy
STOS	sodium titanium oxide silicate
STP	stationary temperature profile
STS	space transportation system
SWBXT	synchrotron white beam x-ray topography
SWIR	short-wavelength IR
SXRT	synchrotron x-ray topography

UC	universal compliant
UDLM	uniform-diffusion-layer model
UHPHT	ultrahigh-pressure high-temperature
UHV	ultrahigh-vacuum
ULSI	ultralarge-scale integrated circuit
UV	ultraviolet
UV-vis	ultraviolet–visible
UVB	ultraviolet B

T

TCE	trichloroethylene
TCNQ	tetracyanoquinodimethane
TCO	thin-film conducting oxide
TCP	tricalcium phosphate
TD	Tokyo Denpa
TD	threading dislocation
TDD	threading dislocation density
TDH	temperature-dependent Hall
TDMA	tridiagonal matrix algorithm
TED	threading edge dislocation
TEM	transmission electron microscopy
TFT-LCD	thin-film transistor liquid-crystal display
TGS	triglycine sulfate
TGT	temperature gradient technique
TGW	Thomson–Gibbs–Wulff
TGZM	temperature gradient zone melting
THM	traveling heater method
TMCZ	transverse magnetic-field-applied Czochralski
TMOS	tetramethoxysilane
TO	transverse optic
TPB	three-phase boundary
TPRE	twin-plane reentrant-edge effect
TPS	technique of pulling from shaper
TQM	total quality management
TRAPATT	trapped plasma avalanche-triggered transit
TRM	temperature-reduction method
TS	titanium silicate
TSC	thermally stimulated conductivity
TSD	threading screw dislocation
TSET	two shaping elements technique
TSFZ	traveling solvent floating zone
TSL	thermally stimulated luminescence
TSSG	top-seeded solution growth
TSSM	Tatarchenko steady-state model
TSZ	traveling solvent zone

V

VAS	void-assisted separation
VB	valence band
VB	vertical Bridgman
VBT	vertical Bridgman technique
VCA	virtual-crystal approximation
VCSEL	vertical-cavity surface-emitting laser
VCZ	vapor pressure controlled Czochralski
VDA	vapor diffusion apparatus
VGF	vertical gradient freeze
VLS	vapor–liquid–solid
VLSI	very large-scale integrated circuit
VLWIR	very long-wavelength infrared
VMCZ	vertical magnetic-field-applied Czochralski
VP	vapor phase
VPE	vapor-phase epitaxy
VST	variable shaping technique
VT	Verneuil technique
VTGT	vertical temperature gradient technique
VUV	vacuum ultraviolet

W

WBDF	weak-beam dark-field
WE	working electrode

X

XP	x-ray photoemission
XPS	x-ray photoelectron spectroscopy
XPS	x-ray photoemission spectroscopy
XRD	x-ray diffraction
XRPD	x-ray powder diffraction
XRT	x-ray topography

Y

YAB	$YAl_3(BO_3)_4$
YAG	yttrium aluminum garnet
YAP	yttrium aluminum perovskite
YBCO	$YBa_2Cu_3O_{7-x}$
YIG	yttrium iron garnet
YL	yellow luminescence
YLF	$LiYF_4$
YOF	yttrium oxyfluoride
YPS	$(Y_2)Si_2O_7$
YSO	Y_2SiO_5

Z

ZA	$Al_2O_3\text{-}ZrO_2(Y_2O_3)$
ZLP	zero-loss peak
ZM	zone-melting
ZNT	ZN-Technologies
ZOLZ	zero-order Laue zone

Part D Crystal Growth from Vapor

23 Growth and Characterization of Silicon Carbide Crystals
Govindhan Dhanaraj, Nashua, USA
Balaji Raghothamachar, Stony Brook, USA
Michael Dudley, Stony Brook, USA

24 AlN Bulk Crystal Growth by Physical Vapor Transport
Rafael Dalmau, Morrisville, USA
Zlatko Sitar, Raleigh, USA

25 Growth of Single-Crystal Organic Semiconductors
Christian Kloc, Singapore
Theo Siegrist, Tallahassee, USA
Jens Pflaum, Würzburg, Germany

26 Growth of III-Nitrides with Halide Vapor Phase Epitaxy (HVPE)
Carl Hemmingsson, Linköping, Sweden
Bo Monemar, Linköping, Sweden
Yoshinao Kumagai, Tokyo, Japan
Akinori Koukitu, Tokyo, Japan

27 Growth of Semiconductor Single Crystals from Vapor Phase
Ramasamy Dhanasekaran, Chennai, India

796

23. Growth and Characterization of Silicon Carbide Crystals

Govindhan Dhanaraj, Balaji Raghothamachar, Michael Dudley

Silicon carbide is a semiconductor that is highly suitable for various high-temperature and high-power electronic technologies due to its large energy bandgap, thermal conductivity, and breakdown voltage, among other outstanding properties. Large-area high-quality single-crystal wafers are the chief requirement to realize the potential of silicon carbide for these applications. Over the past 20 years, considerable advances have been made in silicon carbide single-crystal growth technology through understanding of growth mechanisms and defect nucleation. Wafer sizes have been greatly improved from wafer diameters of a few millimeters to 100 mm, with overall dislocation densities steadily reducing over the years. Device-killing micropipe defects have almost been eliminated, and the reduction in defect densities has facilitated enhanced understanding of various defect configurations in bulk and homoepitaxial layers. Silicon carbide electronics is expected to continue to grow and steadily replace silicon, particularly for applications under extreme conditions, as higher-quality, lower-priced large wafers become readily available.

23.1 Silicon Carbide – Background and History 797
    23.1.1 Applications of SiC 798
    23.1.2 Historical Development of SiC Crystal Growth 798
23.2 Vapor Growth 799
    23.2.1 Acheson Method 799
    23.2.2 Lely Method 799
    23.2.3 Modified Lely Method 800
    23.2.4 Sublimation Sandwich Method 800
    23.2.5 Chemical Vapor Deposition 800
23.3 High-Temperature Solution Growth 801
    23.3.1 Bulk Growth 801
    23.3.2 Liquid-Phase Epitaxy................... 802
23.4 Industrial Bulk Growth by Seed Sublimation 802
    23.4.1 Growth System............................. 803
    23.4.2 Seeding and Growth Process.......... 804
23.5 Structural Defects and Their Configurations...................... 805
    23.5.1 Micropipes and Closed-Core Screw Dislocations 806
    23.5.2 Basal Plane Dislocations in 4H-SiC.. 809
    23.5.3 Threading Edge Dislocations (TEDs) in 4H-SiC...................................... 814
23.6 Concluding Remarks 816
References 817

23.1 Silicon Carbide – Background and History

Silicon carbide (SiC), one of the oldest known semiconductor materials, has received special attention in recent years because of its suitability for electronic and optoelectronic devices operating under high-temperature, high-power, high-frequency, and/or strong radiation conditions, where conventional semiconductor materials such as silicon, GaAs, and InP are considered to have reached their limits. SiC exists as a family of crystals with more than 200 polytypes and a bandgap range of 2.4–3.3 eV. As a wide-bandgap material, SiC possesses many superior properties, e.g., a larger operating temperature range, a high critical breakdown field (E_{cr}), high resistance to radiation, and the ability to construct visible-range light-emitting devices [23.1]. It also distinguishes itself by a combination of high thermal conductivity (higher than that of copper), hardness second only to diamond, high thermal stability, and chemical inertness.

23.1.1 Applications of SiC

High-Temperature Applications
Current and future applications of electronic components have placed much more critical environmental requirements on semiconductors [23.2]; for example, high-temperature electronic components and systems can play an important role in many areas, e.g., aircraft, spacecraft, automotive, defense equipment, power systems, etc. For reliable functioning of electronic devices under extreme conditions they need to withstand high temperatures. SiC appears to be a desirable candidate because of its high working temperature as well as Debye temperature. As reported by *Chelnokov* and *Syrkin* [23.2], 6H-SiC is superior to Si, GaAs, GaN, and AlN for high-temperature application. SiC can also find applications in sensors for high-temperature, high-pressure, and highly corrosive environments (e.g., combustion systems, gas turbines, and in the oil industry) [23.3]; for example, pressure sensors based on SiC thin layers deposited on an insulator structure have been successfully used to measure the pressure in a combustion engine up to 200 bar at temperatures up to 300 °C [23.4].

High-Power Devices
Power semiconductor devices are important for regulation and distribution of electricity. Since the efficient use of electricity depends on the performance of power rectifiers and switches, further improvements in efficiency, size, and weight of these devices are desirable. SiC has a high breakdown strength, and therefore it is possible to dope it at higher concentration while still having thinner layers for a given blocking voltage compared with corresponding Si devices [23.4]. Indeed, power losses can decrease dramatically with the use of SiC-based devices. Another desirable property of SiC for power application is its high thermal conductivity, which can facilitate quick dissipation of heat generated in the component. SiC power metal–oxide–semiconductor field-effect transistors (MOSFETs), diode rectifiers, and thyristors are expected to function over wider voltage and temperature ranges with superior switching characteristics.

High-Frequency Devices
Cellular phones, digital television (TV), telecommunication systems, and radars have made microwave technology an essential part of everyday life. Although some high-power microwave semiconductor components have existed for a long time, e.g., Gunn, impact ionization avalanche transit time (IMPATT), and trapped plasma avalanche transit time (TRAPATT) diodes, these devices can only operate in parametric amplifiers, which are much more difficult to manufacture and tune. SiC-based microwave transistors are predicted to produce more efficient microwave systems and further expand their existing applications [23.4]. Silicon carbide static induction transistors (SITs) and metal semiconductor field effect transistors (MESFETs) have already been developed for these applications.

Optoelectronic Applications
The special physical and optical properties of SiC have been further exploited to fabricate bright blue and green light-emitting diodes (LEDs) [23.5]. In terms of manufacturing, there are several advantages to using SiC as a substrate material, such as easier handling and cheaper processing. SiC is also being used as a substrate for the growth of GaN, an important material for LEDs. Compared with GaN growth on sapphire substrates, it is possible to obtain structurally more perfect epitaxial GaN layers on SiC due to the smaller lattice mismatch and closer match of thermal expansion coefficients.

The primary requirement for SiC-based devices is the production of high-quality thin films, which in turn requires high-quality substrates.

23.1.2 Historical Development of SiC Crystal Growth

SiC has been known in the materials world since 1824. It was recognized as a silicide of carbon in 1895 and could be synthesized successfully by the Acheson process [23.6] using sand and coke. SiC-based LEDs were made as early as 1907 using small SiC crystals obtained from the cavities formed in the Acheson system. In 1955, *Lely* demonstrated the growth of SiC on a porous SiC cylinder by vapor condensation [23.7]. This method was further refined by *Hamilton* [23.8] and *Novikov* and *Ionov* [23.9], and is commonly referred to as the Lely method. Based on this method SiC platelets were prepared in the laboratory for several different applications. *Halden* [23.10] grew single crystals of SiC using Si melt solutions, but this method was not continued because of the difficulty in obtaining larger crystals. *Kendal* [23.11] later proposed a method of cracking of gaseous compounds containing C and Si at high temperature to form SiC crystallites, which is probably the basis for today's SiC chemical vapor deposition (CVD) technology.

A real breakthrough occurred in 1978 when *Tairov* and *Tsvetkov* [23.12] demonstrated seeded growth of

SiC using the sublimation method. Since Tairov and Tsvetkov used Lely's concept of vapor condensation, their method is commonly known as the modified Lely method. Further research on bulk growth is only a refinement and improvement of this technology. Commercially SiC wafers were first made available by Cree Research, Inc., in 1991 [23.13]. The availability of SiC wafers in recent years has spurred extensive research on epitaxial growth. *Matsunami* group's [23.14] research in establishing step-controlled epitaxy is a notable development in optimizing SiC epitaxial growth morphology. Today, 100 mm SiC wafers are routinely available commercially, and overall defect densities show gradual improvement.

23.2 Vapor Growth

Unlike most semiconductor crystals, melt growth methods cannot be adopted for growth of silicon carbide since it is not possible to melt SiC under easily achievable process conditions. The calculated values show that stoichiometric SiC would melt only at above 10 000 atm and 3200 °C [23.15]. Because of these reasons, single crystals of silicon carbide are grown using techniques based on vapor growth, high-temperature solution growth, and their variants. Since SiC readily sublimes, physical vapor growth can be easily adapted and has become the primary method for growing large-size SiC boules. On the other hand, SiC can also dissolve in certain melts, e.g., silicon, which makes melt solution growth a possible technique. This method is predominantly used for growing single-crystal films.

23.2.1 Acheson Method

Commercial production of SiC was established as early as in 1892 [23.6] using the Acheson method. This process is primarily used for synthesis of low-purity polycrystalline material. The Acheson method also yields spontaneously nucleated SiC platelets of incomplete habit. In this process, a predetermined mixture of silica, carbon, sawdust, and common salt [23.16] (e.g., 50% silica, 40% coke, 7% sawdust, 3% salt) is heated by resistive heating of the core of graphite and coke placed at the center of the furnace. The furnace is heated to 2700 °C and maintained at that temperature for a certain amount of time, and then the temperature is gradually decreased. During the thermal cycle, different regions of the reactants are subjected to different temperatures. In between the outermost and innermost regions, the temperature reaches above 1800 °C and the mixture transforms to amorphous SiC. In the core region, SiC is formed first but as the temperature increases it decomposes into graphite and silicon. The decomposed graphite remains at the core and the silicon vapor reacts with the carbon in the adjacent cooler regions to form SiC. Crystalline SiC is therefore formed outside the graphite layers. The common salt reacts with metallic impurities and escapes in the form of chloride vapors, improving the overall purity of the charge. The reaction yields predominantly 6H-SiC polycrystalline materials. Platelet crystals up to 2–3 cm are formed in some hollow cavities. This method does not yield reproducible quality and dimensions of single crystals and hence is not suitable for commercial production, although one can obtain SiC platelets suitable for use as seeds in physical vapor growth.

23.2.2 Lely Method

In the Lely method, developed in 1955 [23.7], SiC lumps are filled between two concentric graphite tubes [23.8]. After proper packing, the inner tube is carefully withdrawn, leaving a porous SiC layer inside the outer graphite tube called the crucible. The crucible with the charge is closed with a graphite or SiC lid and is loaded vertically into a furnace. The furnace is then heated to $\approx 2500\,°C$ in an argon environment at atmospheric pressure. The SiC powder near the crucible wall sublimes and decomposes because of a higher temperature in this region. Since the temperature at the inner surface of the charge is slightly lower, SiC crystals start nucleating at the inner surface of the porous SiC cylinder. These thin platelets subsequently grow larger in areas if the heating is prolonged at this temperature. Since crystals are nucleated on the lumps of SiC (at the inner surface) and it is difficult to impose higher supersaturation, there is no control over the nucleation process, leading to platelets of incomplete hexagonal habit. The original Lely method was later improved by *Hamilton* [23.8] and others [23.9], where SiC charge is packed in between the two annular graphite cylinders. The outer cylinder (crucible) is thick whereas the inner cylinder is thin and porous and acts as a diaphragm. The sublimed SiC vapor passes through the small holes in the diaphragm, and crystals are nucleated at the inner surface. Thick layers of SiC are also deposited on the lids at both ends. This modification offers slightly better control over the number of nucleation sites and yield, and crystals up to $20 \times 20\,\mathrm{mm}^2$ have been grown using this method. Good-quality, larger crystals are obtained when the temperature variation in the cavity is small and the Ar pressure is maintained at about 1 atm [23.9]. Similar to the Acheson process, crystals of 6H polytype are predominantly produced by this method. The amount of crystals of other polytypes, such as 15R and 4H, depends on the growth temperature and dopant. Even though Lely platelets show good structural perfection, they have nonuniform physical and electrical characteristics. Also, since the yield is low ($\approx 3\%$), this method is not suitable for industrial production. The Lely method is, however, ideal for producing platelets of high structural perfection that can be used as seed crystals in bulk growth using other methods.

23.2.3 Modified Lely Method

In 1978, *Tairov* and *Tsvetkov* [23.12] developed the seeded sublimation growth technique, commonly known as the modified Lely method. They succeeded in suppressing the widespread spontaneous nucleation occurring on the (inner) graphite cylinder wall and achieved controlled growth on the seed (Fig. 23.1). This method also led to the control of polytypes to some extent. In the modified Lely method, growth occurs in argon environment at 10^{-4}–760 Torr in the temperature range of 1800–2600 °C and the vapor transport is facilitated by a temperature differential, $\Delta T = T_2 \approx T_1$, between the seed and the source material. The seed temperature T_1 is maintained slightly lower than the source temperature. The kinetics of the transport of Si- and C-containing species are primarily controlled by the diffusion process.

There are two different designs of the seeded sublimation growth system based on the locations of the charge and seed. In earlier work [23.17–19], the source SiC was placed in the upper half of the graphite crucible in a circular hollow cylindrical configuration between the crucible and a thin-walled porous cylinder (Fig. 23.1a). The seed platelet was held on a pedestal in the lower half of the crucible. Using this configuration, *Ziegler* et al. [23.17] grew 20 mm-diameter 24 mm-long crystals while *Barrett* et al. [23.18] succeeded in growing 6H-SiC of 33 mm diameter and 18 mm length. In the second configuration [23.15, 20–24] (Fig. 23.1b), which is commonly used today, the source material is held at the bottom of the crucible and the seed plate on the top. No graphite diaphragm is used in this configuration. This arrangement has a high yield (90%) [23.25] and has therefore become the industry standard for production of SiC boules.

23.2.4 Sublimation Sandwich Method

The sublimation sandwich method (SSM) is another variant of physical vapor transport (PVT) growth where the growth cell is partially open and the environment containing Si vapor may be used to control the gas-phase stoichiometry [23.26–30]. The source material consists of a SiC single-crystal or polycrystalline plate with small source-to-crystal distance (0.02–3 mm). There are several parameters, such as the source-to-substrate distance, small temperature gradient, and presence of Ta for gettering of excess carbon, that can be used to control the growth process. A high growth rate is achieved mainly due to the small source-to-seed distance and a large heat flux onto a small amount of source material with a low to slightly moderate temperature differential between the substrate and the source (0.5–10 °C). While growth of large boules is quite difficult, this method is quite promising for better-quality epitaxial films with uniform polytype structures.

23.2.5 Chemical Vapor Deposition

Chemical vapor deposition (CVD) is a popular method for growing thin crystalline layers directly from the gas phase [23.14, 31–33]. In this process a mixture of gases (source gases for Si and C, and carrier gas) is injected into the growth chamber with substrate temperatures above 1300 °C. Silane is the common Si source, and a hydrocarbon is used for C. Propane is quite popular, but methane is of interest because of its availability with very high purity, although it has lower carbon

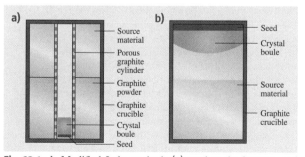

Fig. 23.1a,b Modified Lely method: (**a**) seed at the bottom, and (**b**) seed at the top

cracking efficiency. The carrier gas is high-purity H_2, which also acts as a co-reactant. Conventional Si and C source molecules, called multiple-source precursors, have been used successfully, and reproducible CVD epitaxial films have been produced. However, the single-source CVD SiC precursor shows several advantages over the multiple-source precursors, including a lower growth temperature (less than 1100 °C).

The first successful large-area, heteroepitaxially grown 3C-SiC was obtained by *Nishinov* et al. [23.34] on a high-quality commercial Si wafer. With the availability of 6H-SiC wafers grown by the modified Lely method, homoepitaxial growth of 6H-SiC and heteroepitaxial growth of 3C-SiC have been achieved with good success [23.14]. SiC epitaxial growth is mainly performed on 4H and 6H substrates. Growth on the Si face is preferred because of the superior quality and well-understood doping behavior. In general, the quality of films obtained on the (0001) face is poor, but this problem can be overcome by using a wafer misoriented by 3–8° from the basal plane to control the morphology of the deposited epilayer [23.14], a process referred to as step-controlled epitaxy. Doping of the film is obtained in situ during the growth of each epilayer by flowing either a p-type or n-type source gas. The growth rates of CVD processes are low, a few tens of microns per hour, generally making it unsuitable for boule production. The rate can be increased by increasing the deposition temperature, but this makes control of the process much more difficult and also results in many other problems such as homogeneous nucleation in the gas phase. The high-temperature CVD (HTCVD) process [23.35, 36] is an improved version that can yield thicker films and higher growth rates. In HTCVD the substrate is placed at the top of a vertically held graphite susceptor, similar to the crucible used in the modified Lely method, with holes at the bottom and top. The gaseous reactants are passed from the bottom of the susceptor upwards through the hole at the bottom. To maintain the growth for a long time and obtain the maximum deposition on the substrate, the temperature of the susceptor wall is kept high and the substrate temperature is kept slightly lower. In principle, SiC growth by this method can be continued for longer periods, and bulk crystals can be obtained. The hot-wall CVD reported by *Kordina* et al. [23.37] is probably a good method to produce uniform epitaxial films on large-diameter wafers. In this approach, the CVD reactor is made of a single graphite block with a protective SiC layer. It has an elliptical outer cross-section with a rectangular tapered hole which runs through the entire length. The substrate wafers are placed appropriately on both sides of the rectangular slit. The mixture of gaseous precursors with carrier gas is passed through the reactor from the end containing the larger hole. The tapered hole compensates for severe depletion of Si and C content in the reactant. The flow rate of the gas can be sufficiently large. This design provides good temperature homogeneity, and epitaxial films of excellent thickness uniformity can be obtained. The surface morphology of the film can be controlled by choosing off-*c*-oriented substrate. Recent technological developments have allowed the growth of uniform epitaxial films on wafers as large as 100 mm in diameter [23.38–41].

23.3 High-Temperature Solution Growth

23.3.1 Bulk Growth

Carbon is soluble in a Si melt, which enables growth of SiC from high-temperature solution. The solubility ranges from 0.01% to 19% in the temperature interval 1412–2830 °C [23.20], although at high temperatures the evaporation of silicon makes the growth unstable. The solubility of carbon can be increased by adding certain transition metals to the Si melt [23.42]. In principle, this can enable growth of SiC from saturated solution by seeded solution growth. Unfortunately there is no crucible material that can remain stable at the required temperatures and with these melts, and also the evaporation of the Si melt poses a serious problem at higher temperatures. It is also speculated that the incorporation of the added metals into the growing crystal is too high to be acceptable for semiconductor applications [23.20]. These difficulties restrict the application of this method to bulk growth of SiC. *Halden* [23.10], however, has grown SiC platelets from Si melt at 1665 °C on a graphite tip in Czochralski configuration. *Epelbaum* et al. [23.43] have obtained SiC boules of 20–25 mm diameter and 20 mm length at pull rate of 5–15 mm/h in the temperature range of 1900–2400 °C at Ar pressure of 100–120 bar. Even though solution-grown crystals free of micropipes have been produced, they contain a number of flat silicon inclusions and show rather high rocking-curve width.

Because of the high temperature and high pressure involved in this process, the method is not considered economic for large-scale production.

23.3.2 Liquid-Phase Epitaxy

Even though the high-temperature solution method for SiC poses enormous difficulties and is not popular for bulk growth, it has been successfully adopted for growth of thin films by liquid-phase epitaxy (LPE). Indeed, LPE has been used for production of several SiC-based optoelectronic devices [23.17]. In LPE, semiconductor-grade silicon is used as a solvent in a graphite crucible. Carbon from the graphite crucible dissolves in the Si melt and is transported to the surface of the SiC substrate, which is placed at the bottom of the crucible at a relatively lower temperature. Bright LEDs have been fabricated using this process. However this method suffers from several setbacks such as cracking of the film due to differential thermal contraction while solidifying and the cumbersome process of extraction of the substrate containing the epitaxial film by etching off the solidified Si. These problems can be overcome using a dipping technique [23.44]. In this improved process a SiC substrate attached to a graphite holder is dipped into the molten Si heated by an induction furnace and is kept in the lower-temperature region of the crucible. The substrate is withdrawn from the melt after obtaining growth of the epitaxial film of a desired thickness. The growth is performed in an Ar environment at 1650–1800 °C, and the typical growth rate is 2–7 μm/h. Doping is obtained by adding Al to the Si melt for p-type layer, and Si_3N_4 powder to Si for n-type. It is also possible to obtain n-type film by passing N_2 gas along with Ar. Epitaxial films obtained on (0001) 6H-SiC substrate have shown degradation in LED performance. This problem can be resolved by using substrates misoriented with respect to the c-axis by a few degrees (3–10°), which leads to better polytype control and improved surface morphologies [23.14]. The method is known as step-controlled epitaxy and has led to improved quality of the film and thereby reliable device performance. A better epitaxial surface morphology is obtained by this process, which has been explained based on the nucleation concept [23.14].

It is possible to lower the growth temperature range by selecting an alternative melt which has a higher solubility than Si. *Tairov* et al. [23.45] have used Sn and Ga melts in the temperature range of 1100–1400 °C and have produced LPE layers by using a sliding boat technique. *Dmitriev* et al. [23.46] have grown p–n junctions in the temperature range of 1100–1200 °C. In addition they have demonstrated container-free epitaxial growth of 6H- and 3C-SiC films in which the melt is held by an electromagnetic field. This method has the advantage of mixing induced by the electromagnetic forces [23.47]. Recently, *Syvajarvi* et al. [23.48] used a special sandwich configuration in LPE and succeeded in obtaining growth rates as high as 300 μm/h. One of the main attractions of LPE of SiC is the potential for filling of micropipes. The substrate, after filling of micropipes, does not reveal the presence of micropipes by either optical microscopy or etching [23.35]. However, detailed analysis is needed to understand the defect configuration around the filled core of the micropipes.

23.4 Industrial Bulk Growth by Seed Sublimation

Seeded sublimation growth, commonly known as the modified Lely method, is the only method that has been implemented by industry. The method has spurred intense worldwide research activity in recent years and has become a standard method for growing SiC crystals [23.12, 17–24, 49–51]. The instrumentation and technology involved in bulk growth of SiC are complex, and hence the availability of large-size crystals is still limited. This is primarily due to the fact that the operating temperatures are extreme, and monitoring and control are difficult [23.20]. Even today, only a few companies are successful in producing SiC boules of reasonable quality and size. The main constraint is the difficulty in determining the optimum growth conditions for the modified Lely method, such as the right combination of pressure, temperature, temperature gradient, charge size, geometric configuration, etc. It is not feasible to determine the exact thermal conditions in the growth zone experimentally due to high operating temperatures and opacity of the graphite crucible. In spite of these limitations, great success has been achieved in the industrial production of SiC crystals in terms of crystal perfection and size [23.52–55]. Numerical modeling and simulation have been of great help in this endeavor [23.56].

23.4.1 Growth System

As described earlier, there are two main configurations for seeded sublimation growth (Fig. 23.1). The second configuration [23.20, 50, 54, 55], where the source material is held at the bottom of the crucible and the seed plate is fixed onto the crucible lid, is the system commonly used today (Fig. 23.1b). This arrangement yields a higher growth rate compared with the other approach, because of the smaller source-to-seed distance. Also, since the growing surface is facing downwards, there is no danger of incorporation of charge particles into the growing crystal as in the first case (Fig. 23.1a) where particulates can fall from the top. The main disadvantage of this configuration is that, in a system for growing larger-diameter boules, maintaining temperature uniformity in the source material becomes difficult. The first configuration is slightly less susceptible to temperature and pressure fluctuations. The operating temperature range of seeded sublimation growth is 1800–2600 °C [23.12], with the actual temperature for growth depending on many different process conditions. Induction furnaces operating at lower frequencies (4–300 kHz) [23.17, 58] are commonly used for the modified Lely method. The optimum operating frequency of the induction furnace is 10 kHz, which corresponds to a reasonably high skin depth for the graphite crucible. Recently, efficient solid-state induction furnaces have become readily available, and these low-frequency generators preferentially couple with the susceptor and crucible, with minimum induction on the graphite insulator. Additionally, in an induction furnace, it is possible to vary the temperature gradient at the initial stage as well as during the run. Another advantage of the induction furnace over the resistive furnace is the minimal thermal insulation required as the heat is generated directly on the susceptor and crucible. The dimensions and number of turns of the induction coil are selected based on geometric considerations, the temperature and temperature gradients desired, and heat losses.

The main chamber of the SiC growth system (Fig. 23.2) is a vertically mounted double-walled water-cooled assembly that consists of two concentric quartz tubes sealed with vacuum-tight end flanges using double O-ring seals on ground surfaces of the quartz tubes. Cooling water is circulated between the concentric quartz tubes from bottom to top, although Yakimova et al. [23.49] have shown that it is also possible to use an air-cooled quartz enclosure. The hot zone consists of a high-density graphite crucible

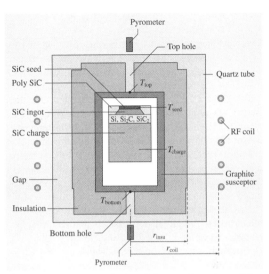

Fig. 23.2 Schematic of the SiC growth system (after [23.57])

and susceptor surrounded by rigid graphite insulation. Because of the higher resistivity of the rigid insulation compared with that of the graphite susceptor and crucible, heat is generated primarily on the susceptor by eddy currents induced by the low-frequency magnetic field. The graphite components of the growth chamber, particularly the crucible and susceptor, are treated at high temperatures in fluorine atmosphere to remove metallic impurities. The design of the hot zone is modified based on the requirements of the axial and radial gradient. This is accomplished with the help of computer modeling [23.56] and prediction of the temperature profile as a function of the growth front.

The normally practiced measurement procedure is to monitor temperatures of the top (T_1) and bottom (T_2) surfaces of the graphite crucible using two color pyrometers (Fig. 23.2). The temperatures are controlled by varying the output power of the induction furnace. Often, the induction coil is mounted on a motorized linear vertical translation stage, and the position of the coil is changed during the growth to vary the temperature gradient and the seed temperature. The vertical growth chamber is connected to a high-vacuum system to obtain initial degassing (at 10^{-7} Torr) as well as to maintain the required vacuum (e.g., 10–100 Torr) conditions during growth. Maintaining vacuum at a predetermined value, as closely as possible, is essential to control the growth rate.

23.4.2 Seeding and Growth Process

After repeated degassing and baking of the growth zone, the chamber is filled with Ar gas. The Ar partial pressure is maintained at about 600 Torr–1 atm while heating to the maximum required growth temperature (2200–2400 °C). The coil position is adjusted such that a desirable temperature gradient of 10–20 °C/cm is obtained. The seed temperature T_1 and the temperature differential ΔT can be varied by changing the coil position; however, T_1 and T_2 cannot be controlled independently. The Ar pressure is brought down to a lower value between 1 and 40 Torr at a predetermined pumping speed to initiate the growth smoothly. The axial temperature gradient influences the growth rate, whereas the radial temperature gradient changes the diameter of the crystal [23.56].

The main stages of the growth are:

1. Dissociative sublimation of SiC source
2. Mass transfer of gaseous species
3. Crystallization onto the seed

At a high temperature, the SiC source material decomposes into several Si- and C-containing species such as Si, C, SiC_2, and Si_2C. Since the crucible is made of graphite, vapor species will react with the graphite wall to form Si_2C and SiC_2, with the graphite crucible acting like a catalyst. Details of the reaction kinetics are described by *Chen* et al. [23.56]. The temperature difference between the seed and source ΔT works as a driving force and facilitates transport of vapor species, mainly Si, Si_2C, and SiC_2. The presence of the temperature gradient leads to supersaturation of vapor, and controlled growth occurs at the seed. Initially, a high-quality Lely plate is used as the seed crystal, and the diameter of the growing crystal is increased by properly adjusting the thermal conditions. To grow larger boules of approximately uniform diameter, wafers from previously grown boules are used as seed discs.

The seed crystal is attached to the graphite top using sugar melt [23.59], which decomposes into carbon and gets bonded to the graphite lid. Optimizing this bonding process is quite important, since the differential thermal expansion between the seed and the graphite lid can cause bending of the seed plate, leading to formation of domain-like structure, low-angle boundaries, and polygonization [23.60]. Micropipes can form at such low-angle boundaries. Any nonuniformity in seed attachment, such as a void between the seed and the lid, can cause variation in the temperature distribution, and the heat dissipation through the seed may be altered. This can result in uneven surfaces and depressions in the growth front corresponding to the void. Evaporation of the back surface of the seed crystal can create thermally decomposed voids which can propagate further into the bulk [23.59]. These voids can then become sources for the generation of micropipes. Protecting the back surface of the seed with a suitable coating eliminates these voids. Seed platelet attachment to the graphite lid is one of the important technical aspects of industrial growth.

If the growth process is not optimized, polycrystalline deposition due to uncontrolled nucleation may occur. In addition to optimizing the Ar pressure and temperature gradient to achieve controlled nucleation, removal of a thin layer of the seed surface by thermal etching, obtained by imposing a reverse temperature gradient [23.61], has been found to be helpful. Etching is also possible by oscillatory motion of the induction coil. The in situ thermal etching helps in cleaning the surface of the seed crystal before starting growth. In some cases, a small amount of excess silicon is added to the charge in order to maintain the Si vapor concentration and stabilize the growth of certain polytypes. The growth of boule is initiated at a very slow rate and is increased progressively by decreasing the pressure.

Depending on the design of the crucible and supersaturation ratio, simultaneous growth of polycrystalline SiC, predominantly 3C, occurs particularly on the graphite lid surrounding the seed crystal. If the growth conditions are not optimal, the polycrystalline SiC can get incorporated into the boule near the periphery, leading to cracking due to high stresses. If the growth rate of the boule is higher than the growth rate of polycrystalline SiC, smooth growth of the boule dominating over the polycrystalline mass is favored. Design of the crucible for increasing the diameter of the boule is normally accomplished through modeling [23.56].

Bahng et al. [23.62] have proposed a method of rapid enlargement of the boule using a cone-shaped platform, where enlargement depends on the taper angle of the cone. It has been reported that, in this technique, the broadening of the boule is not affected by the growth of polycrystalline SiC. After obtaining the required diameter of boule, seed discs of larger diameter are prepared from these boules for further growth in a specifically designed hot zone suitable for promoting predominantly axial growth. As growth of the boule progresses, the temperature of the growing surface changes, which can be compensated by moving the induction coil. It is evident that the process parameters must be optimized for a particular crucible design, system geometry, and boule dimension.

Among the numerous SiC polytypes, only 6H, 4H, 15R, and 3C have been studied for different applications. Polytypes 6H and 4H have been studied extensively in bulk crystal as well as epitaxial form, whereas 3C has been investigated predominantly in epitaxial form. Recently, work on bulk growth of 15R has been initiated for MOSFET applications [23.63]. Crystals of 4H polytype are grown in a narrow temperature range of 2350–2375 °C at 5 mbar using (0001)C face of 4H seed plates [23.35]. A lower growth rate (0.1 mm/h) is used in the beginning and then increased to 0.5 mm/h after growing a 1 mm-long boule. Above 2375 °C the 4H polytype transforms into 6H; below 2350 °C, crystal quality becomes a limiting factor. Among the SiC polytypes, 6H is the most extensively studied, and the reported growth temperature ranges vary widely, although this may be due to differences in growth cell configuration and temperature measurement convention. *Snyder* et al. [23.53] have reported the growth of 100 mm 6H boules at 2100–2200 °C and 5–30 Torr Ar pressure with 10–30 °C/cm temperature gradient. The clearly established result is that (0001)Si face should be used for growth of 6H, whereas (0001)C face is needed for growth of 4H. It seems that, for bulk growth of 15R, seed platelets of the same polytype are required. *Schulze* et al. [23.64] have demonstrated growth of 15R crystals on (0001)Si seed face at 2150–2180 °C with 5 °C/cm gradient. However, *Nishiguchi* et al. [23.58] have shown that 15R polytypes can grow stably on both C and Si of (0001) face at seed temperature not exceeding 2000 °C with growth rate controlled between 0.1 mm/h to 0.5 mm/h. In addition to temperature there are several other parameters that can be used to control polytype formation.

Growth of SiC boule depends on many parameters, such as growth temperature, temperature gradient, Ar pressure, crystal temperature, source-to-crystal distance, and the porosity of the source material [23.56]. Preparative conditions of the source material alter the vapor species concentrations and vary the growth conditions. Deviation from stoichiometry can lead to a lower growth rate. The growth rate increases as the seed crystal temperature increases. It also increases with the temperature differential ($T_2 \propto T_1$) and temperature gradient but decreases with the source-to-seed distance. The growth rate varies almost inversely with the Ar pressure, and the trend is consistent with $1/P$ dependence on the molecular diffusion coefficient [23.50]. There exists a saturation of growth rate at very low pressures, and one would tend to select this growth regime, but then control of the vapor composition becomes more difficult.

The growth rates have been measured by inducing growth bands by simultaneously introducing N_2 gas along with the Ar flow at different intervals and subsequent post mortem studies. In general, (0001) plate is used as a seed, and growth proceeds along the c-direction. Even though the crystal grows smoothly on (0001) plate, this is also the favorable orientation for nucleation of micropipes. There have been several attempts [23.65, 66] to grow crystals on non-(0001) orientation. Even though the micropipe density was reduced in the bulk, the generation of other types of defects such as stacking faults on the basal plane, which hinder electron transport in device applications, increased. Presently, seeding is restricted to (0001) orientation for industrial production of SiC boules.

Monitoring and controlling growth of SiC is very difficult because of the use of opaque graphite materials in the hot zone. Recently, radiography has been employed to study the growth interface during the growth process [23.67]. This imaging technique has also revealed the graphitization of the SiC source material, which could reduce the growth rate as well as affect the structural perfection of the growing boule. Attempts have also been made to study defect generation during the growth process using in situ x-ray topography [23.68].

23.5 Structural Defects and Their Configurations

Assessment of crystalline imperfections and growth inhomogeneities in grown crystals is necessary to understand how they are formed and for the development of engineering methods to eliminate them or minimize their effect in order to obtain high-quality crystals required for electronic applications. SiC crystals grown using different techniques can contain crystalline imperfections such as growth dislocations of screw character with closed or hollow cores (micropipes), deformation-induced basal plane dislocations, parasitic polytype inclusions, planar defects (stacking faults, microscopic twins, and small-angle boundaries), hexagonal voids, etc. that affect device performance. A review of SiC defect characterization efforts reveals that x-ray topography, and in particularly synchrotron white-beam x-ray topography (SWBXT) [23.69–71], is superior to other techniques such as chemical etching, atomic force microscopy (AFM), scanning electron microscopy (SEM), transmission electron microscopy (TEM), and optical microscopy-based methods, although these other techniques can be used in a complementary manner. Defects imaged by x-ray topography are primarily discussed in this section.

23.5.1 Micropipes and Closed-Core Screw Dislocations

Origin of the Hollow Core and Frank's Theory

Among the various defects that exist in SiC crystals, screw dislocations lying along the [0001] axis are the most significant and are generally considered to be one of the major factors limiting the extent of the application of SiC. These screw dislocations (SDs) have been shown to have Burgers vectors equal to nc (where c is the lattice parameter along the [0001] direction in the hexagonal coordinate space and n is an integer), with hollow cores becoming evident with $n \geq 2$ for 6H-SiC and $n \geq 3$ for 4H-SiC [23.69]. These latter screw dislocations are generally referred to as micropipes (MPs), and their hollow cores can be understood from Frank's theory [23.73], which predicts that a screw dislocation whose Burgers vector exceeds a critical value in crystals with large shear modulus should have a hollow core with equilibrium diameter D related to the magnitude of the Burgers vector b by

$$D = \frac{\mu b^2}{4\pi^2 \gamma}, \quad (23.1)$$

where μ is the shear modulus and γ is the specific surface energy. Experimentally, the diameter D can be measured directly using SEM or AFM, while the Burgers vector magnitude b can be obtained by determining the step height of the growth spiral on the as-grown surface using optical interferometry or AFM, or directly using x-ray topography. Detailed experimental results indicate a directly proportional relationship between D and b^2 for micropipes in both 6H- and 4H-SiC [23.69–71].

Growth Spirals and Screw Dislocations

Growth spirals observed on habit faces of as-grown SiC crystals [23.74, 75] are a clear manifestation of screw dislocations emerging on the growth surface. The emergence of screw dislocations on a habit face produces a ledge of height equal to the Burgers vector. The crystal grows by attachment of molecules to the edge of this ledge. The ledge is self-perpetuating and continues to be present on the surface as long as the dislocation line intersects the surface. The ledge winds itself into a circular or polygonal spiral with a dislocation line at the center and, as the growth proceeds, the spiral apparently revolves. The step height of these spirals is equal to an integer times c. Depending on the sign of the Burgers vector the spiral can revolve in a clockwise or anticlockwise direction [23.69]. These spirals have been studied using phase-contrast microscopy [23.76], scanning electron microscopy (SEM) [23.76], and recently by atomic force microscopy (AFM) [23.77]. When two screw dislocations of the same sign are present very close to each other, their spirals spin without intersecting each other, which are called cooperative spirals. Two dislocations of opposite sign can form a closed loop.

Back-Reflection Observation of Screw Dislocations

Screw dislocations in (0001)SiC wafers, of both closed and hollow core (micropipes), can be effectively characterized using back-reflection geometry in SWBXT [23.78]. Figure 23.3, a typical back-reflection topograph taken from a (0001)SiC wafer (grown by Cree Research, Inc.), clearly reveals the screw dislocations, both hollow and closed core, as white circular spots surrounded by black rings. The distribution of micropipes and screw dislocations as well as their detailed structures can be obtained from such images.

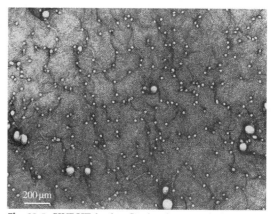

Fig. 23.3 SWBXT back-reflection images of closed-core (*smaller white spots*) and hollow-core (*large white spots*) screw dislocation in a (0001) 6H-SiC wafer. The *faint lines* connecting these screw dislocation images are basal plane dislocation images [23.72]

The circular white spots in Fig. 23.3 are not images of micropipes and closed-core screw dislocations but are actually related to diffraction effects associated with the long-range strain fields of the screw dislocations. Using a ray-tracing simulation based on the orientation contrast mechanism, a model for qualitative and quantitative interpretation of topographic observations in SiC has been developed [23.79–82]. It has been successfully used in back-reflection XRT to clarify the screw character of MPs and also to reveal the dislocation sense of threading screw dislocations (TSDs)/MPs, the Burgers vectors of threading edge dislocations (TEDs), the core structure of Shockley partial dislocations, and the sign of Frank partial dislocations [23.83–85]. Based on this ray-tracing principle, images of micropipes and screw dislocations can be rigorously simulated. Figure 23.4a shows a magnified image of an $8c$ micropipe in 6H-SiC, while Fig. 23.4b shows the simulated image of a screw dislocation with Burgers vector of $8c$ ($b \approx 12.1$ nm). It is apparent that the simulation is in excellent agreement with the recorded micropipe image. This proves that micropipes in SiC are indeed pure screw dislocations. The magnitude of the Burgers vector can be estimated from the diameter of the screw dislocation image, while the twist direction unambiguously indicates the dislocation sense, i.e., the direction of the Burgers vector [23.81]. Back-reflection section topographs of micropipes recorded with a 20 μm-wide slit-limited synchrotron beam can also reveal the sense of the screw dislocation (Fig. 23.5a–d).

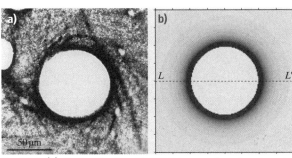

Fig. 23.4 (a) Back-reflection SWBXT image of an $8c$ micropipe taken from a (0001) surface of a 6H-SiC crystal, with sample-to-film distance of 20 cm. (b) Simulation of a screw dislocation with Burgers vector $|b| = 8c$ [23.80]

Grazing-Incidence Imaging of Screw Dislocations

In the grazing-reflection geometry, the incident beam makes an extremely small angle (less than 1°) with respect to the (0001) plane, but there is no limitation for the exit angle of the diffracted beam. This is an especially useful geometry for characterizing micropipes and screw dislocations as well as threading edge dislocations in SiC epitaxial films, since one can control the penetration depth of x-rays at will by adjusting the incidence angle. Figure 23.6a shows a $11\bar{2}8$ topograph taken with the recording x-ray film parallel to the (0001) surface, in which the oval-shaped white spots are images of micropipes. The simulated images based on the ray-tracing principle are shown in Fig. 23.6b, c [23.84].

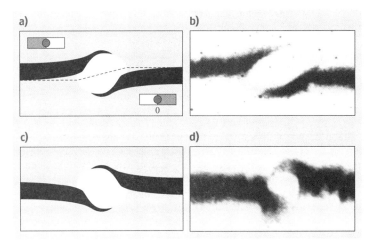

Fig. 23.5a–d Section topographs showing the senses of screw dislocations associated to micropipes. (a) and (b) are simulated and recorded images of a right-handed screw dislocation, respectively; (c) and (d) are simulated and recorded images of a left-handed screw dislocation [23.80]

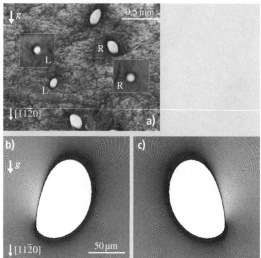

Fig. 23.6 (**a**) A (11$\bar{2}$8) grazing-incidence SWBXT image of 4H-SiC. MPs appear as roughly oval shaped with various orientations and dimensions. The MPs can be divided into two groups according to the orientation of the white elliptical contrast. Examples of each kind are shown (marked by "L" and "R"). Their images in back-reflection geometry appear as complete circular shapes (*insets*, on a different scale), indicating that they are isolated MPs (not MP pairs or groups). Simulated (11$\bar{2}$8) grazing-incidence x-ray topographic images of left-handed (**b**) and right-handed (**c**) 8*c* MPs at a specimen-to-film distance of 35 cm. Both images appear as roughly white ellipses canted clockwise (**a**) or counterclockwise (**b**) from vertical configuration [23.84]

The observed (Fig. 23.6a) and simulated (Fig. 23.6b,c) images correlate very well. Simulation shows that left-handed MPs appear as nearly elliptical features canted clockwise, while right-handed MPs are canted counter-clockwise.

Model for the Origin of Screw Dislocations and Micropipes

Systematic observations of screw dislocation or micropipe formation processes using a variety of techniques [23.77, 86–89] suggest that a possible mechanism for nucleation of micropipes in SiC involves the incorporation of inclusions, which could be, for example, graphite particles, silicon droplets, or even voids, into the crystal lattice. Nucleation of screw dislocations at inclusions has been observed in several systems [23.90–92]. The formation of screw dislocation pairs from inclusions has been briefly proposed by *Chernov* [23.93] for crystals grown from solution. This approach has been extended to explain the observed nucleation of screw dislocations in SiC [23.94]. This model assumes that two macrosteps of different heights, approaching each other on the growth surface, trap a layer of foreign material (solvent, void or impurity) on the growth surface. As a result of the higher rate of feeding of the protruding edge than the re-entrant edge, an overhanging ledge can subsequently be produced as the crystal attempts to overgrow the inclusion and incorporate it into its lattice. This overhanging ledge is vulnerable to deformation and vibrations and, when the macrostep meets the approaching macrostep, horizontal atomic planes which were at the same original height may no longer meet along the line where the two steps meet. If the layer of foreign material constituted a void (or transport gases), downward depression of the overhanging ledge may be expected, whereas if it constituted an impurity, deformation of the opposite-sense dislocations might be expected. In order to accommodate this misalignment, screw dislocations of opposite sign are created with Burgers vector magnitudes equal to the magnitude of the misalignment. Since the degree of misalignment depends on the relative size of the two approaching steps and the lateral and vertical extent of the inclusion, the production of dislocations with a range of Burgers vectors becomes possible. In fact, micropipe-related screw dislocations in SiC can have Burgers vectors as large as several tens of times the basic lattice constant along the *c*-axis [23.95]. In addition, in cases of large inclusions or groups of inclusions, the deformation of the protruding ledge may be spread over the length of the line along which they meet, resulting in the creation of distributed groups of opposite-sign screw dislocations. These groups may not necessarily be distributed symmetrically, but in all cases, the sum of all the Burgers vectors of the dislocations created must equal zero.

In SiC the sources of the growth steps involved in the above-described model can be the vicinal nature of the growth surface (which tends to be slightly dome shaped), two-dimensional (2-D) nucleation, as well as spiral steps associated with intersections of screw dislocations with the surface. Differences in step height are certainly conceivable when a vicinal step meets a spiral step and can even occur when two spiral steps associated with screw dislocations of different Burgers vector magnitude meet or if step bunching occurs for a group of dislocations. Moreover, the merging of 2-D grown islands also plays an important role in the formation of growth steps as well as voids. These various sources in conjunction with the formation of inclusions thus provide opportunities for micropipes as well as closed-core screw dislocations to be created during growth.

23.5.2 Basal Plane Dislocations in 4H-SiC

The primary slip plane for hexagonal 4H- and 6H-SiC is the basal plane. It is therefore not surprising that deformation-induced dislocations on the basal plane are observed in both structures [23.97, 98]. An example is shown in Fig. 23.7, which shows a transmission topograph recorded from a 4H-SiC basal plane wafer. Also shown in this figure is a grazing-incidence topograph recorded from the same region of the 4H-SiC crystal. Detailed Burgers vector analysis of these dislocations can be easily performed. Observation of the morphologies of the basal plane dislocation (BPD) loops clearly indicates that they are deformation induced and appear to have been nucleated both at the crystal edges and at the sites of micropipes/screw dislocations.

The Burgers vectors of the BPDs in SiC are $1/3\langle11\bar{2}0\rangle$, and two extra $(11\bar{2}0)$ half-planes are associated with an edge-oriented BPD, as the magnitude of its Burgers vector is twice the d-spacing of $(11\bar{2}0)$ plane. The BPD is energetically favorable to be dissociated

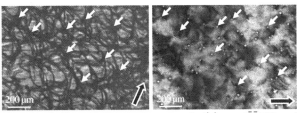

Fig. 23.7a,b Transmission x-ray topograph ((**a**), $g = [\bar{1}\bar{1}20]$) and grazing incidence ((**b**), $g = [11\bar{2}8]$) of 20 μm epilayer on 8° off-cut 4H-SiC substrate. Circular basal plane dislocations (BPDs) anchored by SDs are seen. Some anchor points are marked by *arrows*

into two Shockley partial dislocations with a stacking fault (SF) area in between, and the equilibrium separation d of the two Shockley partial dislocations is given by $d = Gb^2/4\pi\gamma$, where G is the shear modulus, b is the magnitude of the Burgers vector of the Shockley partial dislocation, and γ is the energy of the SF. This equilibrium separation is $\approx 330\,\text{Å}$ for 4H-SiC, assuming the SF energy to be $14.7\,\text{mJ/m}^2$ [23.99].

The dislocation character of a BPD is determined by the angle between its line direction and the Burgers vector. Figure 23.8 illustrates the various cases when

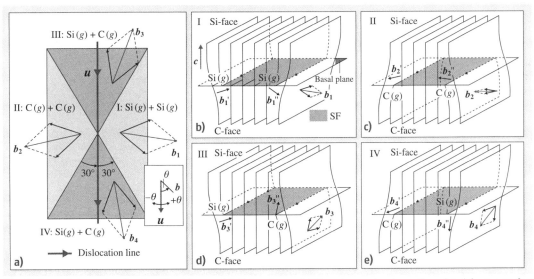

Fig. 23.8a–e Schematics showing Shockley partial dislocations of different core structures dissociated from a perfect BPD. (**a**) Four regions defined according to the direction of Burgers vector with respect to the line direction of the dislocation, assuming Si-face is facing up: (**b**) $30° < \theta < 150°$, the BPD is dissociated into two Si-core partials; (**c**) $210° < \theta < 330°$, the BPD is dissociated into two C-core partials; (**d**) $-30° < \theta < +30°$, one Si-core and one C-core; (**e**) $150° < \theta < 210°$, one Si-core and one C-core. θ is defined in the *inset* of (**a**) (after [23.96])

a perfect BPD is dissociated into Shockley partial dislocations [23.96]. The angle θ is the counterclockwise angle from the line direction to the Burgers vector direction (clockwise angles are negative); see the inset of Fig. 23.8a. In region I, in which $30° < \theta < 150°$, the extra half-plane(s) associated with the BPD extend toward the Si-face determined by the right-hand rule, $\boldsymbol{u} \times \boldsymbol{b}$. If $30° < \theta < 150°$, when the BPD with Burgers vector $\boldsymbol{b}_1$ is dissociated into two partial dislocations of Burgers vectors $\boldsymbol{b}'_1$ and $\boldsymbol{b}''_1$ (Fig. 23.8b), the angles between $\boldsymbol{b}'_1/\boldsymbol{b}''_1$ and their line directions (assuming the same line direction as the BPD) will be in the range $(0°, 180°)$. The extra half-planes associated with the two partials can be subsequently determined again by right-hand rule: $\boldsymbol{u} \times \boldsymbol{b}'_1$ and $\boldsymbol{u} \times \boldsymbol{b}''_1$; both are extending toward the Si-face. Since the Shockley partials in SiC are glide set dislocations, both partials are Si-core. Under this circumstance, the SF will expand toward both directions, as both partials are mobile. Similar mechanism can be applied for regions II, III, and IV. In region II (Fig. 23.8c), both dissociated partials are C-core and neither of them advances, while for regions III (Fig. 23.8d) and IV (Fig. 23.8e), a Si-core and a C-core partial are formed and the SF expands as the Si-core partial advances. Notice that, if the Burgers vector of the BPD is at $30°/150°/210°/330°$ to its line direction, it is dissociated into a screw-oriented and a C-core or Si-core partial.

Susceptibility of Basal Plane Dislocations in 4H Silicon Carbide

Dissociation of basal plane dislocations (BPDs) into mobile silicon-core (Si-core) partial dislocations and subsequent advancement of these partial dislocations under forward bias pose a large challenge for the lifetime of SiC-based bipolar devices [23.100] since the expansion of Shockley stacking faults (SFs) associated with the advancement of Si-core partials causes the forward voltage to drop. Such expansion of basal SFs is activated by the electron–hole recombination-enhanced dislocation glide (REDG) process [23.101–103]. Through detailed x-ray topography analysis of a dislocation configuration formed after stacking fault expansion under forward bias, the susceptibility of basal plane dislocations to REDG has been determined. Figure 23.9 shows x-ray topographs (Fig. 23.9a–e) and the schematic configuration after advancement of the mobile partial. During the advancement of the partial, it interacted with a few threading screw dislocations (TSDs), and the final configuration obtained is shown in Fig. 23.9f. The Burgers vector $\boldsymbol{b}$ of the original BPD is determined to be $1/3(\bar{1}\bar{1}20)$, as indicated in Fig. 23.9f. Thus, the original BPD is screw oriented ($\theta = 0°$) and it is dissociated into a C-core and a Si-core partial. This corresponds to the case in region IV

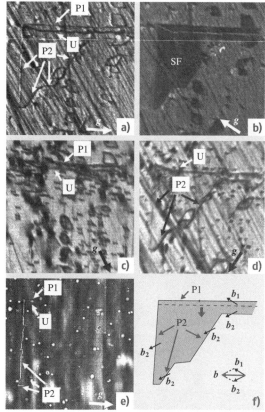

Fig. 23.9a–f REDG-activated SF after forward bias from a screw-oriented BPD. (**a**) $(11\bar{2}0)$ transmission topograph showing the partials (P1 and P2) bounding the SF. The SF area is out of contrast since $\boldsymbol{g} \cdot \boldsymbol{R}$ is equal to an integer; (**b**) $(\bar{1}010)$ transmission topograph showing the SF; (**c**) P2 is out of contrast in the $(2\bar{1}\bar{1}0)$ transmission topograph, indicating its Burgers vector of $1/3(0\bar{1}10)$; (**d**) P1 is out of contrast in the $(\bar{1}2\bar{1}0)$ transmission topograph, indicating its Burgers vector of $1/3(\bar{1}010)$; (**e**) (0008) back-reflection topograph. The sign of P1 and P2 can be determined; (**f**) schematics showing the SF configuration. The SF is obtained via expansion of Si-core partial toward the bottom edge of the view (*dashed line*) and interaction with TSDs. The Burgers vector of each partial segment is labeled; the Burgers vector $\boldsymbol{b}$ of the original BPD can be obtained and it is screw-oriented (after [23.96])

of Fig. 23.8, in which θ is between $-30°$ and $+30°$. Thus, the susceptibility of the basal plane dislocations to REDG process is determined by the counterclockwise angle θ from the line direction to its Burgers vector. Basal plane dislocations with $30° < \theta < 150°$ are most detrimental, as both partials will advance under forward bias. If $-30° < \theta < 30°$ or $150° < \theta < 210°$, only one partial advances. Both partials are immobile if $210° < \theta < 330°$ [23.96].

The Nucleation Mechanism in 4H-SiC Homoepitaxial Layers

BPDs in 4H-SiC homoepitaxial layers result largely from replication, during growth, of BPDs which intersect the surface of the offcut SiC substrates, a process which can be mitigated by the conversion of the BPDs into threading edge dislocations (TEDs) [23.105], which are not susceptible to REDG. While various schemes have been developed to increase the conversion rate to nearly 100% [23.106–109], BPDs which intersect the surface in screw orientation are observed to persist [23.110], and furthermore, they are observed to nucleate half-loop arrays. By recording the behavior of a half-loop array (HLA) from a Si-face epilayer using ultraviolet photoluminescence (UVPL) imaging, a model has been developed to explain the formation of HLAs. Figure 23.10a shows a screw-type BPD with Burgers vector $1/3[11\bar{2}0]$ intersecting the surface of the substrate, which is expected to be replicated during epitaxy, in contrast to those with significant edge components, which are likely to be converted into TEDs. As soon as the epilayer exceeds a critical thickness, as per

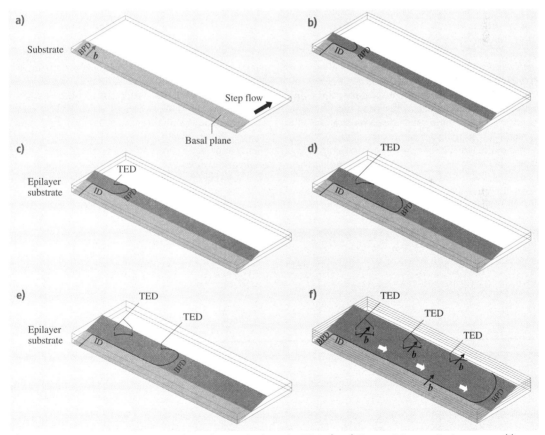

Fig. 23.10a–f Schematic showing the formation mechanism of a HLA. (**a–e**) Sequential stages in the process; (**f**) summary of process. The *lighter-shaded planes* in (**a–f**) indicate the basal plane on which the BPD lies in the substrate, while the *darker one* lies in the epilayer. See text for details of the mechanism (after [23.104])

the predictions of *Matthews* and *Blakeslee* [23.111], the threading segment of the screw-oriented BPD will be forced to glide sideways, leaving a trailing interfacial segment in its wake at or near the substrate–epilayer interface. During this glide process, the mobile threading segment adopts more edge character near the growth surface (Fig. 23.10b), rendering it susceptible to conversion to a TED during continued growth. Slip in SiC is confined to the basal plane, so that the sessile TED segment pins the surface intersection of the mobile BPD segment. During further growth, the TED segment is replicated while the mobile basal segment of dislocation pivots about the pinning point, as shown in Fig. 23.10c. At this juncture, part of the mobile BPD segment can escape through the epilayer surface (creating a surface step of magnitude equal to the Burgers vector), as shown in Fig. 23.10d, leaving two further BPD surface intersections which, since they are not in screw orientation, are susceptible to conversion to TEDs. Upon conversion, one of these TEDs is connected via a short BPD segment to the TED segment created in Fig. 23.10b, thus creating a half-loop comprising two TEDs and a connecting BPD. The other TED again acts as a pinning point for the still mobile segment of threading BPD, as shown in Fig. 23.10e, as the process repeats during continued growth as the TED segments further replicate and the threading BPD segment continues to glide. The net result of this process is an array of half-loops with short, large-edge-component BPD segments, all deposited on the exact same basal plane. The direction of the array is nearly perpendicular to the offcut direction, as summarized in Fig. 23.10f.

The value of this angle depends on the competition between the growth rate and the rate of sideways glide of the threading BPD segment.

Dislocation Behavior in SiC Single Crystals

Recently, the availability of 76 and 100 mm-diameter 4H-SiC wafers with extremely low BPD densities ($3-4 \times 10^2$ cm^{-2}) has provided a unique opportunity to discern details of BPD behavior which were previously mostly unresolvable [23.113]. Figure 23.11a–d shows typical x-ray topographs from a section of such a 4H-SiC wafer (4° offcut towards [11$\bar{3}$0]). BPDs ($g = \bar{1}\bar{1}20$) belonging to the three $1/3\langle 11\bar{2}0\rangle(0001)$ slip systems can be observed. For the dislocation inside the dashed frame pinned at points a–d, at point a, there is a TSD close to the surface intersection of the BPD which appears to be responsible for the pinning at that point, while no TSD is present at point d. Conversion of the BPD into a sessile TED at point d during growth creates an effective single-ended pinning point for the BPD. As the crystal grows, the BPD continues to glide, forming a spiral configuration around the TED pinning point, as shown schematically in Fig. 23.11e, thus operating as a single-ended Frank–Read source [23.112].

Figure 23.12a–d shows images of a $1/3[11\bar{2}0]$ basal plane dislocation exhibiting several loophole configurations that is pinned at five points a–e (Fig. 23.12a). The pinning at point a appears to have occurred close to the wafer surface at a TSD, while no TSDs are observed at any of the other pinning points. In this case, conversion of BPD segments into TEDs occurs during growth but also includes reconversion of the same TED into a BPD

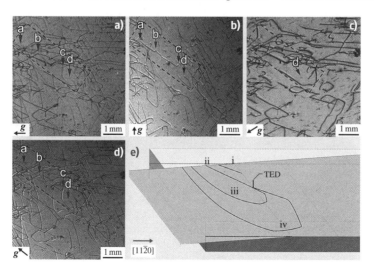

Fig. 23.11a–e SWBXT images showing BPD of interest inside dashed frame. Offcut direction is horizontal towards the right: (a) $g = \bar{1}\bar{1}20$; (b) $g = \bar{1}100$; (c) $g = 0\bar{1}10$; (d) $g = \bar{1}010$. (e) Schematic showing the originally screw-oriented BPD at position i being converted into a TED at its surface intersection at position ii and beginning to act as a single-ended Frank–Read source at positions iii and iv (after [23.112])

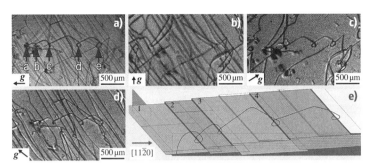

Fig. 23.12a–e SWBXT images showing dislocation loophole configurations. Offcut direction is horizontal towards the right: (a) $g = \bar{1}\bar{1}20$; (b) $g = \bar{1}100$; (c) $g = 01\bar{1}0$; (d) $g = \bar{1}010$. (e) Schematic showing the deflection of the BPD gliding on basal plane 1 into a TED, which then acts as a pinning point as the BPD continues to glide, followed by deflection of the TED onto basal plane 2 through overgrowth by a macrostep. The process repeats through basal planes 3 and 4 (after [23.112])

in a process that repeats throughout the growth process. This is shown schematically in Fig. 23.12e. The initial BPD segment may have had screw orientation, but continued glide may cause it to move away from screw orientation at its growth surface intersection, rendering it susceptible to conversion into a TED. During further growth, this short TED segment acts as a single-ended pinning point for the BPD which continues to glide under thermal stress. This TED can be redirected back into the basal plane as a screw-oriented BPD through overgrowth by a macrostep traveling from left to right. Once back in the basal plane, the screw-oriented BPD, being glissile, begins to glide in spiral configuration about its single-ended TED pinning point. Again, once the BPD moves away from screw orientation, it becomes more susceptible to conversion into a TED and the whole process repeats, as shown schematically in Fig. 23.12e, leading to the type of configuration observed in Fig. 23.12a,c,d [23.112].

In Fig. 23.13a, the transmission topograph from near the edge of a 75 mm wafer reveals several long, mostly straight dislocation images running approximately in the radial direction, e.g., at AB and CD. These dislocations are growth induced and may have been redirected from an originally threading orientation onto the basal plane, for example, due to overgrowth by a macrostep. These dislocations are associated with several overlapping stacking faults (Fig. 23.13b), the contrast from which arises from the phase shift experienced by the x-ray wave fields as they cross the fault plane [23.115]. Detailed analysis of the fault contrast on different reflections [23.114] indicates that there are three different types of faults present here. The first type is a pure Shockley fault, the second is a fault comprising the sum of a Frank fault ($c/4$) and a Shockley fault, and the third is a pure Frank fault ($c/4$). Figure 23.14a shows the surface intersection of the TED core, where the two extra ($1\bar{2}10$) half-planes correspond to the $1/3[1\bar{2}10]$ Burgers vector. Note that, since the SiC structure comprises corner-sharing tetrahedra, overgrowth can only occur if the stacking position of the underside of the overgrowing step is able to maintain tetrahedral bonding with the top side of the terrace being overgrown, i.e., the stacking sequence rules, as, for example, described in [23.116], must be obeyed. As the macrostep advances over the surface outcrop, it is not able to admit the dislocation into itself so that the dislocation is necessarily deflected into the direction of step flow,

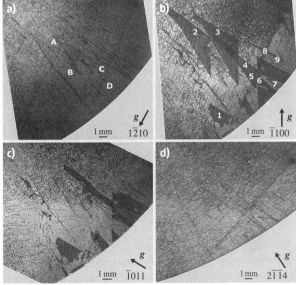

Fig. 23.13a–d SWBXT images recorded from a region near the edge of a 75 mm wafer cut with 4° offcut towards [$11\bar{2}0$]: (a) $1\bar{2}10$ reflection showing long straight dislocations, for example, at AB and CD; (b) $\bar{1}100$ reflection from the same area, showing stacking fault contrast. Faults of interest are numbered 1–9; (c) $\bar{1}011$ reflection from same area. Some of the fault images have disappeared on this image; (d) $2\bar{1}\bar{1}4$ reflection showing absence of all fault contrast [23.114]

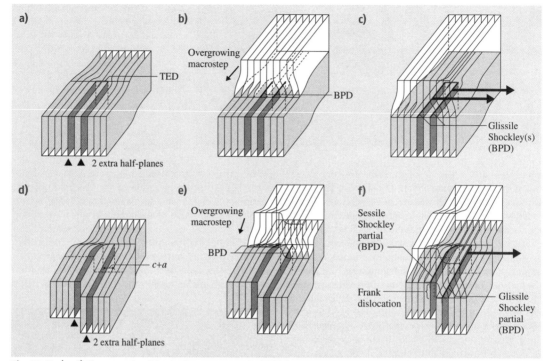

Fig. 23.14 (a–f) Overgrowth of the surface outcrop of a TED by a macrostep converting it to a BPD. Once on the basal plane, both Shockley partials are glissile and move in tandem; (b) overgrowth of a $c+a$ dislocation with a c-height step. After deflection, one Shockley is sessile and the other is glissile (after [23.114])

onto the basal plane. As this happens, the surface intersections of the two extra half-planes are frozen into the crystal and, in fact, define the line direction of the resulting BPD, as shown in Fig. 23.13b,c. If these surface intersections occur on an atomically flat terrace, the resulting BPD will comprise two extra half-planes on the exact same basal plane. Once on the basal plane, the BPD can become glissile (i.e., mobile) if sufficient basal plane shear stresses are available, and the glissile partials (comprising one extra half-plane each) will most likely separate to their equilibrium value of 20 nm and track each other as they move; i.e., the leading partial will fault the slip plane and the trailing partial will *unfault* the plane. If the surface intersections of the two extra half-planes occur on a region of surface such that they straddle the riser of a surface step which is parallel to them, it becomes possible for the overgrowth to result in two partials lying on slip planes separated by the height of the step, provided again that the overgrowth process does not breach the stacking rules [23.116].

Stacking faults are observed in the vicinity of what appear to be deflected threading dislocations. Detailed contrast analysis carried out on the faults is consistent with the Burgers vectors for the original threading dislocations of type $c+a$. The surface step (associated with the screw component) created at the surface intersection of the deflected dislocations creates a separation between the slip planes of the partials associated with the a component of the dislocations, causing one to be sessile and the other glissile. Glide of the glissile partial creates the stacking faults, which can be pure Shockley (if the original dislocation step is one unit cell high), Shockley plus $c/4$ (if the step is split into $c/4$ and $3c/4$ components), or $c/4$ if a second $c+a$ dislocation becomes involved [23.114].

23.5.3 Threading Edge Dislocations (TEDs) in 4H-SiC

In 4H-SiC, threading edge dislocations (TEDs) are dislocations with line directions roughly parallel to the c-axis and Burgers vectors in the c-plane. TEDs have been observed to be one of the major components of the LAGBs, and they play critical roles in the defect structures in SiC, e.g., they act as a barrier for gliding BPDs if the spacing between adjacent TEDs is less than a critical value [23.117]. On {0001} wafer, TEDs have a $1/3\langle11\bar{2}0\rangle$ Burgers vector. Thus six different directions of the TED Burgers vector, $[\bar{1}\bar{1}20]$, $[12\bar{1}0]$, $[2\bar{1}\bar{1}0]$, $[11\bar{2}0]$, $[\bar{1}2\bar{1}0]$, and $[\bar{2}1\bar{1}0]$, exist on the 4H-SiC epilayers, although all with the same Burgers vector magnitude. Figure 23.15a shows a $(11\bar{2}8)$ topograph recorded from a 4H-SiC wafer [23.118]. The large roughly white circles are images of TSDs, as marked in the figure. Other than the TSDs, smaller features are also seen, corresponding to the TEDs with various Burgers vectors. They appear as two dark arcs, either separated by a white spot or canted to one side or the other of the g vector. By carefully examining the images of the TEDs in more than 50 topographs recorded, six different configurations of TEDs were observed and their highly magnified images are shown in Fig. 23.15b–g. The topographic images of the expected six types of TEDs have been simulated by ray-tracing method and are shown in Fig. 23.16. The schematics of the six types of TEDs are illustrated at the top of the figure, according to the extra atomic half-

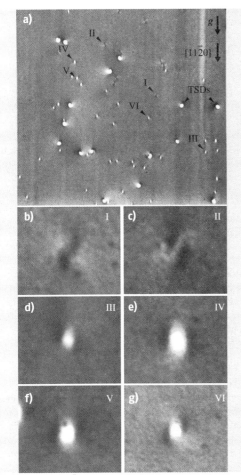

Fig. 23.15 (a) $(11\bar{2}8)$ topograph showing various images of TEDs. (b–g) Six different types of images of TEDs observed in the topographs, probably corresponding to the six types of TEDs

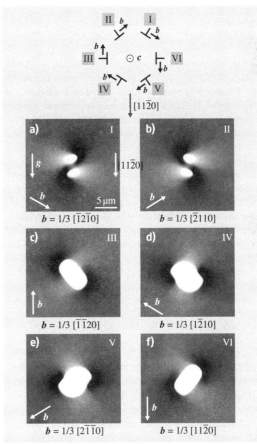

Fig. 23.16a–f Simulated $(11\bar{2}8)$ grazing-incidence XRT images of TEDs with six different Burgers vectors (a–f) *Top*: six types of TEDs are illustrated according to the position of the extra atomic half-planes associated with them

planes associated with them. They appear as two dark arcs canted to one side or the other of the *g*-vector, and these two dark arcs are either shifted vertically (Fig. 23.15a, b) or separated by an area of white contrast (Fig. 23.15c–f). Their Burgers vectors are at 60°, 120°, 180°, 240°, 300°, and 0° counterclockwise from the *g*-vector on the x-ray films for Fig. 23.15a–f, respectively. By comparing the simulated images in Fig. 23.15 and the observation in Fig. 23.16, one-to-one correspondence can be seen. Thus, the Burgers vectors of the TEDs can be revealed from a single-reflection $(11\bar{2}8)$ grazing-incidence x-ray topograph, based on their topographic contrast [23.118]. TED arrays in 4H-SiC prefer to be oriented along $\langle 1\bar{1}00 \rangle$, and their Burgers vectors are perpendicular to the arrays. This is consistent with the fundamental dislocation theory that edge-oriented dislocations tend to align perpendicular to their Burgers vector to minimize the total strain energy.

23.6 Concluding Remarks

An extraordinary combination of physical and electronic properties makes silicon carbide a unique material for devices in high-power, high-frequency, high-temperature, and intense-radiation applications. Recent developments in SiC bulk growth and epitaxial film technology have greatly advanced SiC-based device technology. The modified Lely method has now become a standard process for industrial production of SiC boules. Wafers of 75 and 100 mm diameter are commercially available from numerous vendors. Dislocation densities have been reduced by optimizing the crystal growth technology in conjunction with modeling and computer simulation.

Defects, particularly micropipes, have nearly been eliminated by improving the growth technique, optimizing the process parameters, and developing better understanding of defect generation and propagation. Undesirable polytype inclusions have been understood reasonably well, and it is now possible to grow a single polytype using the modified Lely method. With the availability of 6H-SiC wafers, research on epitaxial growth has increased tremendously. High-quality films are being produced for different device applications using CVD techniques; in particular, HTCVD and hot-wall CVD have yielded films of good uniformity. Even though LPE of SiC is not as successful as CVD, recent developments such as container-free LPE growth show promise for better-quality films. In addition, the quality of the epitaxial film and thereby the functioning of the device have been greatly improved by using step-controlled epitaxy. High dopant incorporation can be achieved using the site-competition technique.

Defects present in SiC crystals have been characterized using x-ray topography and microscopy-based techniques such as chemical etching, AFM, SEM, TEM, and reflection and transmission optical microscopy. Even though many of these techniques are used in a complementary manner to obtain detailed information on defects present in the crystal, x-ray topography, particularly SWBXT, is quite superior to other methods in revealing defects present in SiC crystals. Indeed, SWBXT has provided complete quantitative characterization of both closed-core and hollow-core (micropipes) screw dislocations as well as basal plane dislocations and threading edge dislocations. It has also given insights into the formation mechanisms of these defects and various interesting configurations that can be formed. Since SWBXT is capable of imaging defects in a full-size wafer with devices fabricated on it, this technique can be successfully used to study the influence of various defects on device performance.

The challenges in SiC growth still remain quality, size, and cost. To enable widespread use of SiC as a semiconductor, it is important that further progress be made in all of these areas. While micropipe defects have been nearly eliminated, other defects such as basal plane dislocations and screw dislocations need to be substantially reduced. Another important area is polytype control and infringement of polycrystalline growth directly under the seed. Since different polytypes of SiC have different properties and also polytype inclusions can become potential sources for defect generation, it is critical that a single polytype is maintained throughout the growth of the boule.

Control of the growth process requires sensing, measurement, and control strategies. However, the process does not allow much measurement, in fact, nothing inside the growth zone. Measurement of temperatures at only two locations, far away from the growth surface and the bulk of the SiC charge, gives very little information on the actual growth temperature and temperature gradient, two critical growth parameters. Also, it is difficult to determine experimentally the rate of sublimation, the chemical composition of the vapor, the growth interface shape, etc., which makes control of the process very difficult.

References

23.1 A.A. Lebedev, V.E. Chelnokov: Wide-gap semiconductors for high-power electronics, Semiconductors **33**, 999–1001 (1999)

23.2 V.E. Chelnokov, A.L. Syrkin: High temperature electroics using SiC: Actual situation and unsolved problems, Mater. Sci. Eng. B **46**, 248–253 (1997)

23.3 A. Lloyd, P. Tobias, A. Baranzahi, P. Martensson, I. Lundström: Current status of silicon carbide based high-temperature gas sensors, IEEE Trans. Electron. Dev. **46**, 561–566 (1999)

23.4 C.I. Harris, A.O. Konstantinov: Recent developments in SiC device research, Phys. Scr. **79**, 27–31 (1999)

23.5 V. Harle, N. Hiller, S. Kugler, B. Hahn, N. Stath: Industrial aspects of GaN/SiC blue light emitting diodes in Europe, Mater. Sci. Eng. B **61–62**, 310–313 (1999)

23.6 A.G. Acheson: Br. Pat. **17**, 911 (1892)

23.7 A.J. Lely: Darstellung von Einkristallen von Siliziumcarbid und Beherrschung von Art und Menge der eingebauten Verunreinigungen, Ber. Dtsch. Keram. Ges. **32**, 229–231 (1955)

23.8 D.R. Hamilton: The growth of silicon carbide by sublimation. In: *Silicon Carbide. A High Temperature Semiconductor*, ed. by J.R. Connor, J. Smilestens (Pergamon, Oxford 1960) pp. 45–51

23.9 V.P. Novikov, V.I. Ionov: Production of monocrystals of alpha silicon carbide, Growth Cryst. **6**, 9–21 (1968)

23.10 F.A. Halden: The growth of silicon carbide from solution. In: *Silicon Carbide. A High Temperature Semiconductor*, ed. by J.R. Connor, J. Smilestens (Pergamon, Oxford 1960) pp. 115–123

23.11 J.T. Kendal: The growth of silicon carbide from gaseous cracking. In: *Silicon Carbide. A High Temperature Semiconductor*, ed. by J.R. Connor, J. Smilestens (Pergamon, Oxford 1960) pp. 67–72

23.12 Y.M. Tairov, V.F. Tsvetkov: Investigation of growth processes of ingots of silicon carbide single crystals, J. Cryst. Growth **43**, 209–212 (1978)

23.13 Cree Research, Inc., 2810 Meridian Parkway, Durham, NC 27713, USA

23.14 H. Matsunami, T. Kimoto: Step controlled epitaxial growth of SiC: High quality homoepitaxy, Mater. Sci. Eng. R **20**, 125–166 (1997)

23.15 R.C. Glass, D. Henshall, V.F. Tsvetkov, C.H. Carter Jr.: SiC-seeded crystal growth, MRS Bulletin **22**, 30–35 (1997)

23.16 W.F. Knippenberg: Growth phenomena in silicon carbide: Preparative procedures, Philips Res. Rep. **18**, 170–179 (1966)

23.17 G. Ziegler, P. Lanig, D. Theis, C. Weyrich: Single crystal growth of SiC substrate material for blue light emitting diodes, IEEE Trans. Electron. Dev. **30**, 277–281 (1983)

23.18 D.L. Barrett, R.G. Seidensticker, W. Gaida, R.H. Hopkins: SiC boule growth by sublimation vapor transport, J. Cryst. Growth **109**, 17–23 (1991)

23.19 R.A. Stein, P. Lanig, S. Leibenzeder: Influence of surface energy on the growth of 6H and 4H-SiC polytypes sublimation, Mater. Sci. Eng. B **11**, 69–71 (1992)

23.20 R.C. Glass, D. Henshall, V.F. Tsvetkov, C.H. Carter Jr.: SiC seeded crystal growth, Phys. Status Solidi (b) **202**, 149–162 (1997)

23.21 D.L. Barrett, J.P. Mchugh, H.M. Hobgood, R.H. Hopkins, P.G. McMullin, R.C. Clarke: Growth of large SiC single crystals, J. Cryst. Growth **128**, 358–362 (1993)

23.22 A.R. Powell, S. Wang, G. Fechko, G.R. Brandes: Sublimation growth of 50 mm diameter SiC wafers, Mater. Sci. Forum **264–268**, 13–16 (1998)

23.23 I. Garcon, A. Rouault, M. Anikin, C. Jaussaud, R. Madar: Study of SiC single-crystal sublimation growth conditions, Mater. Sci. Eng. B **29**, 90–93 (1995)

23.24 Y.M. Tairov, Y.V.F. Tsvetkov: General principles of growing large-size single crystals of various silicon carbide polytypes, J. Cryst. Growth **52**, 146–150 (1981)

23.25 Y.M. Tairov: Growth of bulk SiC, Mater. Sci. Eng. B **29**, 83–89 (1995)

23.26 S.Y. Karpov, Y.N. Makarov, E.N. Mokhov, M.G. Ramm, M.S. Ramm, A.D. Roenkov, R.A. Talalaev, Y.A. Vodakov: Analysis of silicon carbide growth by sublimation sandwich method, J. Cryst. Growth **173**, 408–416 (1997)

23.27 S.Y. Karpov, Y.N. Makarov, M.S. Ramm: Simulation of sublimation growth of SiC single crystals, Phys. Status Solidi (b) **202**, 201–220 (1997)

23.28 Y.A. Vodakov, A.D. Roenkov, M.G. Ramm, E.N. Mokhov, Y.N. Makarov: Use of Ta-container for sublimation growth and doping of SiC bulk crystals and epitaxial layers, Phys. Status Solidi (b) **202**, 177–200 (1997)

23.29 E.N. Mokhov, M.G. Ramm, A.D. Roenkov, Y.A. Vodakov: Growth of silicon carbide bulk crystals by the sublimation sandwich method, Mater. Sci. Eng. B **46**, 317–323 (1997)

23.30 S.Y. Karpov, Y.N. Makarov, M.S. Ramm, R.A. Talalaev: Control of SiC growth and graphitization in sublimation sandwich system, Mater. Sci. Eng. B **46**, 340–344 (1997)

23.31 A. Henry, I.G. Ivanov, T. Egilsson, C. Hallin, A. Ellison, O. Kordina, U. Lindefelt, E. Janzen: High quality 4H-SiC grown on various substrate orientations, Diam. Relat. Mater. **6**, 1289–1292 (1997)

23.32 D.J. Larkin: An overview of SiC epitaxial growth, MRS Bulletin **22**, 36–41 (1997)

23.33 H. Matsunami: Progress in epitaxial growth of SiC, Physica B **185**, 65–74 (1993)

23.34 S. Nishinov, J.A. Powell, H.A. Will: Production of large-are single-crystal wafers of cubic SiC for semiconductors, Appl. Phys. Lett. **42**, 460–462 (1983)

23.35 R. Yakimova, R.E. Janzen: Current status and advances in the growth of SiC, Diam. Relat. Mater. **9**, 432–438 (2000)

23.36 O. Kordina, C. Hallin, A. Ellison, A.S. Bakin, I.G. Ivanov, A. Henry, R. Yakimova, M. Touminen, A. Vehanen, E. Janzen: High temperature chemical vapor deposition of SiC, Appl. Phys. Lett. **69**, 14561458 (1996)

23.37 O. Kordina, C. Hallin, A. Henry, J.P. Bergman, I. Ivanov, A. Ellison, N.T. Son, E. Janzen: Growth of SiC by "hot-wall" CVD and HTCVD, Phys. Status Solidi (b) **202**, 321–334 (1997)

23.38 C.H. Carter, V.F. Tsvetkov, R.C. Glass, D. Henshall, M. Brady, S.G. Muller, O. Kordina, K. Irvine, J.A. Edmond, H.S. Kong, R. Singh, S.T. Allen, J.A. Palmour: Progress in SiC: From material growth to commericial device development, Mater. Sci. Eng. B **61–62**, 1–8 (1999)

23.39 E. Janzen, O. Kordina: SiC Material for high-power applications, Mater. Sci. Eng. B **46**, 203–209 (1997)

23.40 M. Nakabayashi, T. Fujimoto, M. Katsuno, N. Ohtani, H. Tsuge, H. Yashiro, T. Aigo, T. Hoshino, H. Hirano, K. Tatsumi: Growth of crack-free 100 mm diameter 4H-SiC crystals with low micropipe densities, Mater. Sci. Forum **600–603**, 3–6 (2009)

23.41 R.T. Leonard, Y. Khlebnikov, A.R. Powell, C. Basceri, M.F. Brady, I. Khlebnikov, J.R. Jenny, D.P. Malta, M.J. Paisley, V.F. Tsvetkov, R. Zilli, E. Deyneka, H. Hobgood, V. Balakrishna, C.H. Carter Jr.: 100 mm 4HN-SiC wafers with zero micropipe density, Mater. Sci. Forum **600–603**, 7–10 (2009)

23.42 M. Syväjärvi, R. Yakimova, I.G. Ivanov, E. Janzen: Growth of 4H SiC from liquid phase, Mater. Sci. Eng. B **46**, 329–332 (1997)

23.43 B.M. Epelbaum, D. Hofmann, M. Muller, A. Winnacker: Top-seeded soloution growth of bulk SiC: Search for the fast growth regimes, Mater. Sci. Forum **338–342**, 107–110 (2000)

23.44 A. Suzuki, M. Ikeda, N.T. Nagoa, H. Matsunami, T.J. Tanaka: Liquid phase epitaxial growth of 6H-SiC by the dipping technique for preparation of blue-light-emitting diodes, J. Appl. Phys. **47**, 4546–4550 (1976)

23.45 Y.M. Tairov, F.I. Raihel, V.F. Tsvetkov: Silicon solubility in tin and gallium, Neorg. Mater. **3**, 1390–1391 (1982)

23.46 V.A. Dmitriev, L.B. Elfimov, N.D. Il'inskaya, S.V. Rendakova: Liquid phase epitaxy of silicon carbide at temperatures of 1100–1200 °C, Springer Proc. Phys. **56**, 307–311 (1996)

23.47 V.A. Dmitriev, A. Cherenov: Growth of SiC and SiC-AlN solid solution by container by container free liquid phase epitaxy, J. Cryst. Growth **128**, 343–348 (1993)

23.48 M. Syväjärvi, R. Yakimova, H.H. Radamsom, N.T. Son, Q. Wahab, I.G. Ivanov, E. Janzen: Liquid phase epitaxial growth of SiC, J. Cryst. Growth **197**, 147–154 (1999)

23.49 R. Yakimova, M. Syväjärvi, M. Tuominen, T. Iakimov, R. Råback, A. Vehanen, E. Janz: Seeded sublimation growth of 6H and 4H-SiC crystals, Mater. Sci. Eng. B **61–62**, 54–57 (1999)

23.50 G. Augustine, H.M. Hobgood, V. Balakrishna, G. Dunne, R.H. Hopkins: Physical vapor transport growth and properties of SiC monocrystals of 4H polytype, Phys. Status Solidi (b) **202**, 137–148 (1997)

23.51 S.I. Nishizawa, Y. Kitou, W. Bahng, N. Oyanagi, M.N. Khan, K. Arai: Shape of SiC bulk single crystal grown by sublimation, Mater. Sci. Forum **338–342**, 99–102 (2000)

23.52 D. Hobgood, M. Brady, W. Brixius, G. Fechko, R. Glass, D. Henshall, J.R. Jenny, R. Leonard, D. Malta, S.G. Muller, V. Tsvetkov, C.H. Carter Jr.: Status of large diameter SiC crystal growth for electronic and optical applications, Mater. Sci. Forum **338–342**, 3–8 (2000)

23.53 D.W. Snyder, V.D. Heydemann, W.J. Everson, D.L. Barret: Large diameter PVT growth of bulk 6H SiC crystals, Mater. Sci. Forum **338–342**, 9–12 (2000)

23.54 R.R. Siergiej, R.C. Clarke, S. Sriram, A.K. Aggarwal, R.J. Bojko, A.W. Morse, V. Balakrishana, M.F. MacMillan, A.A. Burk, C.D. Brandt: Advances in SiC materials and devices: An industrial point of view, Mater. Sci. Eng. B **61–62**, 9–17 (1999)

23.55 S.G. Muller, R.C. Glass, H.M. Hobgood, V.F. Tsvetkov, M. Brady, D. Henshall, J.R. Jenny, D. Malta, C.H. Carter Jr.: The status of SiC bulk growth from an industrial point of view, J. Cryst. Growth **211**, 325–332 (2000)

23.56 Q.-S. Chen, V. Prasad, H. Zhang, M. Dudley: Silicon carbide crystals – Part II: Process physics and modeling. In: *Crystal Growth Technology*, ed. by K. Byrappa, T. Ohachi (Springer, Berlin, Heidelberg 2001) pp. 233–269

23.57 G. Dhanaraj, X.R. Huang, M. Dudley, V. Prasad, R.-H. Ma: Silicon carbide crystals – Part II: Process physics and modeling. In: *Crystal Growth Technology*, ed. by K. Byrappa, T. Ohachi (Springer, Berlin, Heidelberg 2001) pp. 181–232

23.58 T. Nishiguchi, S. Okada, M. Sasaki, H. Harima, S. Nishino: Crystal growth of 15R-SiC boules by sublimation method, Mater. Sci. Forum **338–342**, 115–118 (2000)

23.59 E.K. Sanchez, T. Kuhr, D. Heydemann, W. Snyder, S. Rohrer, M. Skowronski: Formation of thermal decomposition cavities in physical vapor transport of silicon carbide, J. Electron. Mater. **29**, 347–351 (2000)

23.60 M. Tuominen, R. Yakimova, R.C. Glass, T. Tuomi, E. Janzen: Crystalline imperfections in 4H SiC grown with a seeded Lely method, J. Cryst. Growth **144**, 267–276 (1994)

23.61 M. Anikin, R. Madar: Temperature gradient controlled SiC crystal growth, Mater. Sci. Eng. B **46**, 278–286 (1997)

23.62 W. Bahng, Y. Kitou, S. Nishizawa, H. Yamaguchi, M. Nasir Khan, N. Oyanagi, S. Nishino, K. Arai: Rapid enlargement of SiC single crystal using a cone-shaped platform, J. Cryst. Growth **209**, 767–772 (2000)

23.63 R. Schorner, P. Friedrichs, D. Peters, D. Stephani: Significantly improved performance of MOSFETs on silicon carbide using the 15R-SiC, IEEE Electron. Device Lett. **20**, 241–244 (1999)

23.64 N. Schulze, D. Barrett, M. Weidner, G. Pensl: Controlled growth of bulk 15R SiC single crystals by the modified Lely method, Mater. Sci. Forum **338–342**, 111–114 (2000)

23.65 J. Takahashi, M. Kanaya, Y. Fujiwara: Sublimation growth of SiC single crystalline ingots on faces perpendicular to the (0001) basal-plane, J. Cryst. Growth **135**, 61–70 (1994)

23.66 M. Touminen, R. Yakimova, E. Prieur, A. Ellison, T. Toumi, A. Vehanen, E. Janzen: Growth-related structural defects in seeded sublimation-grown SiC, Diam. Relat. Mater. **6**, 1272–1275 (1997)

23.67 N. Oyanagi, S.I. Nishizawa, T. Kato, H. Yamaguchi, K. Arai: SiC Single crystal growth rate by in-situ observation using the transmission x-ray technique, Mater. Sci. Forum **338–342**, 75–78 (2000)

23.68 T. Kato, N. Oyangi, H. Yamaguchi, Y. Takano, S. Nishizawa, K. Arai: In-situ observation of SiC bulk single crystal growth by x-ray topography, Mater. Sci. Forum **338–342**, 457–460 (2000)

23.69 W. Si, M. Dudley, R. Glass, V. Tsvetkov, C.H. Carter Jr.: Experimental studies of hollow-core screw dislocations in 6H-SiC and 4H-SiC single crystals, Mater. Sci. Forum **264–268**, 429–432 (1998)

23.70 W. Si, M. Dudley, R. Glass, V. Tsvetkov, C.H. Carter Jr.: Hollow-core screw dislocations in 6H-SiC single crystals. A test of Frank's theory, J. Electron. Mater. **26**, 128–133 (1997)

23.71 M. Dudley, W. Si, S. Wang, C.H. Carter Jr., R. Glass, V.F. Tsvetkov: Quantitative analysis of screw dislocations in 6H-SiC single crystals, Nuovo Cim. D **19**, 153–164 (1997)

23.72 M. Dudley, S. Wang, W. Huang, C.H. Carter Jr., V.F. Tsvetkov, C. Fazi: White beam synchrotron topographic studies of defects in 6H-SiC single crystals, J. Phys. D **28**, A63–A68 (1995)

23.73 F.C. Frank: Capillary equillibria of dislocated crystals, Acta Cryst. **4**, 497–501 (1951)

23.74 A.R. Verma: Spiral growth on carborundum crystal faces, Nature **167**, 939 (1951)

23.75 S. Amelinck: Spiral growth on carborundum crystal faces, Nature **167**, 939–940 (1951)

23.76 A.R. Verma: *Crystal Growth and Dislocations* (Butterworths Scientific, London 1953)

23.77 J. Giocondi, G.S. Rohrer, M. Skowronski, V. Balakrishna, G. Augustine, H.M. Hobgood, R.H. Hopkins: An AFM study of super-dislocation/micropipe complexes on the 6H-SiC(0001) growth surfaces, J. Cryst. Growth **181**, 351–362 (1997)

23.78 M. Dudley, W. Huang, S. Wang, J.A. Powell, P. Neudeck, C. Fazi: White beam synchrotron topographic analysis of multipolytype SiC device configurations, J. Phys. D **28**, A56–A62 (1995)

23.79 X.R. Huang, M. Dudley, W.M. Vetter, W. Huang, C.H. Carter Jr.: Contrast mechanism in superscrew dislocation images on synchrotron back-reflection topographs, Mater. Res. Soc. Symp. Proc. **524**, 71–76 (1998)

23.80 X.R. Huang, M. Dudley, W.M. Vetter, W. Huang, S. Wang, C.H. Carter Jr.: Direct evidence of micropipe-related pure superscrew dislocations in SiC, Appl. Phys. Lett. **74**, 353–356 (1999)

23.81 X.R. Huang, M. Dudley, W.M. Vetter, W. Huang, W. Si, C.H. Carter Jr.: Superscrew dislocation contrast on synchrotron white-beam topographs: An accurate description of the direct dislocation image, J. Appl. Crystallogr. **32**, 516–524 (1999)

23.82 M. Dudley, X.R. Huang, W. Huang: Assessment of orientation and extinction contrast contributions to the direct dislocation image, J. Phys. D **32**, A139–A144 (1999)

23.83 Y. Chen, M. Dudley: Direct determination of dislocation sense of closed-core threading screw dislocations using synchrotron white beam x-ray topography in 4H silicon carbide, Appl. Phys. Lett. **91**, 141918 (2007)

23.84 Y. Chen, G. Dhanaraj, M. Dudley, E.K. Sanchez, M.F. MacMillan: Sense determination of micropipes via grazing-incidence synchrotron white beam x-ray topography in 4H silicon carbide, Appl. Phys. Lett. **91**, 071917 (2007)

23.85 I. Kamata, M. Nagano, H. Tsuchida, Y. Chen, M. Dudley: Investigation of character and spatial distribution of threading edge dislocations in 4H-SiC epilayers by high-resolution topography, J. Cryst. Growth **311**, 1416–1422 (2009)

23.86 N. Schulze, D.L. Barret, G. Pensl: Near-equilibrium growth of micropipe free 6H-SiC single crystals by physical vapor transport, Appl. Phys. Lett. **72**, 1632–1634 (1998)

23.87 V. Tsvetkov, R. Glass, D. Henshall, D. Asbury, C.H. Carter Jr.: SiC seeded boule growth, Mater. Sci. Forum **264–268**, 3–8 (1998)

23.88 V.D. Heydemann, E.K. Sanchez, G.S. Rohrer, M. Skowronski: Structural evolution of Lely seeds during the initial stages of SiC sublimation growth, Mater. Res. Soc. Symp. Proc. **483**, 295–300 (1998)

23.89 V. Balakrishna, R.H. Hopkins, G. Augustine, G.T. Donne, R.N. Thomas: Characterization of 4H-SiC monocrystals grown by physical vapor transport, Inst. Phys. Conf. Ser. **160**, 321–331 (1998)

23.90 S. Gits-Leon, F. Lefaucheux, M.C. Robert: Effect of stirring on crystalline quality of solution grown

23.91 H. Klapper, H. Kuppers: Directions of dislocation lines in crystals of ammonium hydrogen oxalate hemihydrate grown from solution, Acta Cryst. A **29**, 495–503 (1973)

crystals – case of potash aluminum, J. Cryst. Growth **44**, 345–355 (1978)

23.92 G. Neuroth: Der Einfluß von Einschlußbildung und mechanischer Verletzung auf das Wachstum und die Perfektion von Kristallen. Ph.D. Thesis (University of Bonn, Bonn 1996)

23.93 A.A. Chernov: Formation of crystals in solutions, Contemp. Phys. **30**, 251–276 (1989)

23.94 M. Dudley, X.R. Huang, W. Huang, A. Powell, S. Wang, P. Neudeck, M. Skowronski: The mechanism of micropipe nucleation at inclusions in silicon carbide, Appl. Phys. Lett. **75**, 784–786 (1999)

23.95 P. Krishna, S.-S. Jiang, A.R. Lang: An optical and x-ray topographic study of giant screw dislocations in silicon-carbide, J. Cryst. Growth **71**, 41–56 (1985)

23.96 Y. Chen, M. Dudley, K.X. Liu, R.E. Stahlbush: Observations of the influence of threading dislocations on the recombination enhanced partial dislocation glide in 4H-silicon carbide epitaxial layers, Appl. Phys. Lett. **90**, 171930 (2007)

23.97 S. Wang: Characterization of growth defects in silicon carbide single crystals by synchrotron x-ray topography. Ph.D. Thesis (State University of New York at Stony Brook, Stony Brook 1995)

23.98 W.M. Vetter: Characterization of dislocation structures in silicon carbide crystals. Ph.D. Thesis (State University of New York at Stony Brook, Stony Brook 1999)

23.99 M.H. Hong, A.V. Samant, P. Pirouz: Stacking fault energy of 6H-SiC and 4H-SiC single crystals, Philos. Mag. A **80**, 919–935 (2000)

23.100 H. Lendenmann, F. Dahlquist, N. Johansson, R. Soderholm, P.A. Nilsson, J.P. Bergman, P. Skytt: Long term operation of 4.5 kV PiN and 2.5 kV JBS diodes, Mater. Sci. Forum **353–356**, 727 (2001)

23.101 A. Galeckas, J. Linnros, P. Pirouz: Recombination-enhanced extension of stacking faults in 4H-SiC p-i-n diodes under forward bias, Appl. Phys. Lett. **81**, 883 (2002)

23.102 J.D. Weeks, J.C. Tully, L.C. Kimerling: Theory of recombination-enhanced defect reactions in semiconductors, Phys. Rev. B **12**, 3286–3292 (1975)

23.103 H. Sumi: Dynamic defect reactions induced by multiphonon nonradiative recombination of injected carriers at deep levels in semiconductors, Phys. Rev. B **29**, 4616–4630 (1984)

23.104 N. Zhang, Y. Chen, Y. Zhang, M. Dudley, R.E. Stahlbush: Nucleation mechanism of dislocation half-loop arrays in 4H-silicon carbide homoepitaxial layers, Appl. Phys. Lett. **94**, 122108 (2009)

23.105 S. Ha, P. Mieszkowski, M. Skowronski, L.B. Rowland: Dislocation conversion in 4H silicon carbide epitaxy, J. Cryst. Growth **244**, 257–266 (2002)

23.106 Z. Zhang, T.S. Sudarshan: Basal plane dislocation-free epitaxy of silicon carbide, Appl. Phys. Lett. **87**, 151913 (2005)

23.107 J.J. Sumakeris, J.P. Bergman, M.K. Das, C. Hallin, B.A. Hull, E. Janzen, H. Lendenmann, M.J. O'Loughlin, M.J. Paisley, S. Ha, M. Skowronski, J.W. Palmour, C.H. Carter Jr.: Techniques for minimizing the basal plane dislocation density in SiC epilayers to reduce V_f drift in SiC bipolar power devices, Mater. Sci. Forum **527–529**, 141 (2006)

23.108 E.R. Stahlbush, B.L. VanMil, R.L. Myers-Ward, K.-K. Lew, D.K. Gaskill, C.R. Eddy Jr.: Basal plane dislocation reduction in 4H-SiC epitaxy by growth interruptions, Appl. Phys. Lett. **94**, 041916 (2009)

23.109 R.E. Stahlbush, B.L. VanMil, K.X. Liu, K.K. Lew, R.L. Myers-Ward, D.K. Gaskill, C.R. Eddy Jr., X. Zhang, M. Skowronski: Evolution of basal plane dislocations during 4H-SiC epitaxial growth, Mater. Sci. Forum **600–603**, 317–320 (2009)

23.110 H. Tsuchida, I. Kamata, M. Nagano: Investigation of defect formation in 4H-SiC(0001) and (000$\bar{1}$) epitaxy, Mater. Sci. Forum **600–603**, 267–272 (2009)

23.111 J.W. Matthews, A.E. Blakeslee: Defects in epitaxial multilayers: I. Misfit dislocations, J. Cryst. Growth **27**, 118–125 (1974)

23.112 M. Dudley, N. Zhang, Y. Zhang, B. Raghothamachar, S. Byrappa, G. Choi, E.K. Sanchez, D. Hansen, R. Drachev, M.J. Loboda: Characterization of 100 mm diameter 4H-Silicon carbide crystals with extremely low basal plane dislocation density, Mater. Sci. Forum **645–648**, 291 (2010)

23.113 M. Selder, L. Kadinski, F. Durst, T.L. Straubinder, P.L. Wellmann, D. Hofmann: Numerical simulation of thermal stress formation during PVT-growth of SiC bulk crystals, Mater. Sci. Forum **353–356**, 65 (2001)

23.114 M. Dudley, S. Byrappa, H. Wang, F. Wu, Y. Zhang, B. Raghothamachar, G. Choi, E.K. Sanchez, D. Hansen, R. Drachev, M.J. Loboda: Analysis of dislocation behavior in low dislocation density, PVT-grown, four-inch silicon carbide single crystals, Mater. Res. Soc. Symp. Proc. **1246**, (2010) in press

23.115 A. Authier, Y. Epelboin: Variation of stacking fault contrast with the value of the phase shift in x-ray topography, Phys. Status Solidi (a) **41**, K9–K12 (1977)

23.116 P. Pirouz, J.W. Yang: Polytypic transformations in SiC: The role of TEM, Ultramicroscopy **51**, 189–214 (1993)

23.117 Y. Chen, H. Chen, N. Zhang, M. Dudley, R. Ma: Investigation and properties of grain boundaries in silicon carbide, Mater. Res. Soc. Symp. Proc. E **955**, 0955I0750 (2007)

23.118 I. Kamata, M. Nagano, H. Tsuchida, Y. Chen, M. Dudley: Investigation of character and spatial distribution of threading edge dislocations in 4H-SiC epilayers by high-resolution topography, J. Cryst. Growth **311**, 1416–1422 (2009)

24. AlN Bulk Crystal Growth by Physical Vapor Transport

Rafael Dalmau, Zlatko Sitar

Despite considerable research in thin-film growth of wide-bandgap group III nitride semiconductors, substrate technology remains a critical issue for the improvement of nitride devices. With applications ranging from high-power electronics to optoelectronics, an increasing number of nitride semiconductor devices are becoming commercially available. Currently, many of these devices are being grown heteroepitaxially on nonnative substrates, leading to a high defect density in the active layers, which limits device performance and lifetime. Aluminum nitride (AlN) is considered a highly desirable candidate as a native substrate material for III-nitride epitaxy, especially for AlGaN devices with high Al concentrations. AlN crystals have been grown by a variety of methods. High-temperature growth of AlN bulk crystals by physical vapor transport (PVT) has emerged as the most promising growth technique to date for production of large, high-quality single crystals. This chapter reviews recent growth and characterization results of AlN bulk crystals grown by PVT and discusses several issues that remain to be addressed for continued development of this technology.

24.1	PVT Crystal Growth	822
24.2	High-Temperature Materials Compatibility	825
24.3	Self-Seeded Growth of AlN Bulk Crystals	827
24.4	Seeded Growth of AlN Bulk Crystals	829
	24.4.1 Growth on SiC Seeds	829
	24.4.2 Growth on AlN Seeds	831
24.5	Characterization of High-Quality Bulk Crystals	832
	24.5.1 Structural Properties	832
	24.5.2 Fundamental Optical Properties of AlN	835
	24.5.3 Impurities	838
24.6	Conclusions and Outlook	839
References		839

Wide-bandgap nitride semiconductors, AlN, GaN, and InN, have been identified as promising materials for a broad range of applications in electronics and optoelectronics [24.1]. Currently, epitaxial heterostructures involving these semiconductors are being grown by various techniques on a number of substrates [24.2–6]. The two most commonly used substrates, sapphire and SiC, are not closely lattice-matched to the III-nitride overlayers, leading to a high defect density in overgrown active layers, which limits device performance and lifetime. Additional limitations of the currently available substrates include cracking of the device layers due to the large thermal mismatch, and poor thermal conductivity. Thus, the performance of III-nitride semiconductor devices would be greatly improved by the availability of native substrates. High-quality, single-crystalline AlN substrates with low dislocation densities are expected to decrease defect density in the overgrown device structures by several orders of magnitude and, thereby, greatly improve the performance and lifetime of III-nitride devices. AlN has a number of excellent properties that make it a highly desirable candidate as a substrate for III-nitride epitaxy. Its crystalline structure is the same as that of GaN, with a lattice mismatch in the c-plane of approximately 2.5%. Since AlN makes a continuous range of solid solutions with GaN, it plays an important role in GaN-based devices and is highly suited as a substrate for AlGaN devices with high Al concentrations or structures with graded layers. Its high thermal conductivity makes it desirable for high-temperature electronic and high-power microwave devices where heat dissipation is critical. The direct, large optical bandgap of 6.1 eV [24.7] makes it suitable for ultraviolet applications down to wavelengths as short as 200 nm. This chapter reviews recent growth and characterization results of AlN bulk crystals grown by physical vapor transport (PVT).

24.1 PVT Crystal Growth

The vast majority of commercially grown semiconductor bulk crystals are grown from the melt using one of several methods, such as the Czochralski, Bridgman, and vertical gradient freeze methods. However, bulk crystal growth by physical vapor transport is an alternative when melt growth is not possible, such as when the melting point is too high, the material decomposes before it melts or the melt reacts with the crucible. Because of the high melting temperature and large dissociation pressure at the melting point of the III-nitrides [24.8], bulk crystal growth from the melt is precluded unless very high pressure is applied. Although the first AlN was synthesized in 1862 by *Briegleb* and *Geuther* [24.9] by the reaction between molten aluminum and nitrogen, it took more than a century before any sizeable single crystals of AlN were grown [24.10]. Past efforts to grow AlN bulk crystals have explored sublimation of AlN, vaporization of Al, and solution routes, with sublimation yielding the most voluminous AlN crystals to date. Crystal growth by other methods, such as hydride vapor-phase epitaxy [24.11], ammonothermal growth [24.12], or solution growth [24.13, 14], has been reported, but only crystals of either inferior quality or size have been produced thus far. In recent years, several research groups [24.15–18] independently developed processes and models for growth of AlN crystals which all converge to the same basic growth principle and process parameter space. All these efforts clearly demonstrate that AlN bulk crystals of very high quality and of sizes appropriate for use as III-nitride substrates can be produced by PVT.

In a typical PVT process an AlN powder source is sublimed within a closed or semi-open crucible, and the vapors are subsequently transported in nitrogen (N_2) atmosphere through a temperature gradient to a region held at a lower temperature than the source, where they recrystallize. The region where recrystallization takes place can consist of the crucible walls, in which case we speak of self-seeded growth, or it may consist of an intentionally selected seed crystal, in which case we have seeded growth. AlN growth can be achieved at temperatures as low as 1800 °C; however, temperatures in excess of 2200 °C are required to achieve commercially viable growth rates. This high growth temperature, in combination with the highly reactive Al vapor, creates a challenge for the identification of appropriate crucible materials and has been a major obstacle in growth of high-purity, large-size AlN crystals.

Early kinetic theory formulation for the sublimation growth of AlN indicated that the useful growth temperature range was 2000–2400 °C, yielding growth rates ranging from 0.3 to 15 mm/h [24.19]. A two-dimensional model of mass transport in the gas phase was analyzed by *Liu* and *Edgar* [24.20], who determined that the activation energy for AlN growth was 681 kJ/mol, which is close to the heat of sublimation of AlN, 630 kJ/mol. To better describe growth at pressures below 100 Torr, a refined model [24.21] included the influence of surface kinetics (N_2 sticking coefficient), which is not a limiting factor at higher pressures. *Karpov* et al. [24.22] and *Segal* et al. [24.23] identified two mechanisms of vapor transport in AlN sublimation: at high pressure (760 Torr), vapor transport was controlled by diffusion in the gas phase, while at low pressure (10^{-4} Torr), it was dominated by drift of the reactive species, Al and N_2. Growth at low pressure required 350–400 °C lower temperature to achieve the same growth rate.

A one-dimensional model was developed for the high-temperature growth by *Noveski* et al. [24.24]. Gas-phase mass transfer of Al species was assumed to limit the overall growth rate. Thus an equation describing the temperature (T) and pressure (p) dependence of the growth rate (v_G) was derived by considering the transport of Al species through the N_2 gas,

$$v_G = k \frac{\exp\left(\Delta S - \frac{\Delta H}{T}\right)}{RT^{1.2} p^{1.5}} \frac{\Delta T}{\delta}, \qquad (24.1)$$

where the pre-exponential term k contains the diffusion coefficient of Al; ΔS and ΔH are the entropy and enthalpy of sublimation, respectively; R is the universal gas constant; and $\Delta T/\delta$ is the temperature gradient in the crucible. An apparent activation energy of

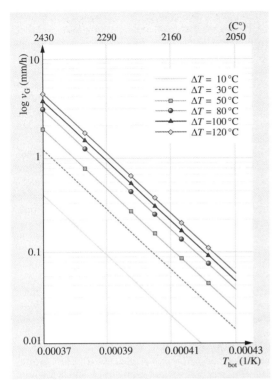

Fig. 24.1 Predicted growth rate as a function of temperature at the bottom of the crucible (T_{bot}) for different temperature differences (ΔT) along the crucible, 600 Torr pressure, and 10 mm source-to-seed distance (after [24.24], with permission)

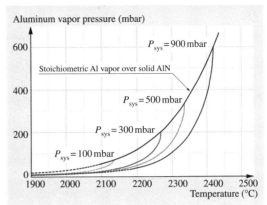

Fig. 24.2 Dependence of aluminum vapor pressure on temperature and system pressure (after [24.25], with permission of Trans Tech)

638.1 kJ/mol was predicted. The theoretical model was experimentally validated by selecting growth parameters for which the model predicted a growth rate of 1 mm/h (Fig. 24.1). Sustained growth rates on the order of 1 mm/h were demonstrated, clearly showing that the growth rate is Al transport limited at total N_2 pressures in the range of 400–800 Torr.

The dependence of growth rate on temperature and pressure was also studied by *Epelbaum* et al. [24.25], who found that PVT transport of AlN was possible starting at 1850 °C, but temperatures exceeding 2100 °C were necessary to obtain stable growth of well-faceted crystals. The vapor pressure of Al as a function of temperature was calculated for different total system pressures (Fig. 24.2) and used to determine the corresponding AlN growth rates, under the assumption that reaction of adsorbed Al and N_2 species to form AlN is the rate-limiting growth step. According to the calculations, growth temperatures in excess of 2100 °C are required to obtain growth rates in excess of 1 mm/h for typical PVT process conditions (e.g., total pressure, thermal gradient). Experimentally observed growth rates in the range of 0.3–3 mm/h were achieved during growth of polycrystalline AlN boules up to 51 mm in diameter and 15 mm in length [24.25].

Two-dimensional simulations [24.26] demonstrated that at a given temperature both the powder source and the seed sublime below a critical pressure when the sum of the Al and N_2 partial pressures at the seed and source are greater than the ambient pressure. Under growth conditions below this critical pressure, the simulations showed that the gas phase is transported out of the growth cell and the sublimation growth fails. The simulations were used to explain the experimentally observed effect of growth temperature (T_g), source-to-seed temperature difference (ΔT), and ambient N_2 pressure on the growth rate [24.27]: For high ΔT, the source sublimation and crystal growth rates increased exponentially with temperature; as ΔT decreased, the sublimation rate continued to exhibit an exponential dependence with temperature, but the growth rate became a decreasing function of temperature, dropping sharply at temperatures greater than 2130 °C; finally, as the ambient nitrogen pressure was decreased, the growth rate initially increased, but then sharply dropped at a critical pressure. This critical pressure was found to increase from ≈ 50 Torr at 2100 °C to ≈ 120 Torr at 2200 °C. The simulation results are presented in Fig. 24.3, where the N_2 molar fraction distributions and the velocity vector fields in the crucible are shown for two differ-

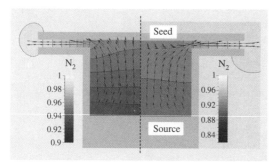

Fig. 24.3 Nitrogen molar fraction distributions and velocity vector fields in the crucible under different thermal conditions: (*left*) $\Delta T \approx 35$ K, $T_g \approx 2150\,°C$; (*right*) $\Delta T \approx 10$ K, $T_g \approx 2215\,°C$ (after [24.27], with permission of Trans Tech)

ent growth regimes. In the first regime, T_g is 2150 °C and ΔT is about 35 °C. The Al/N$_2$ gas mixture evaporated from the source is divided into two flows, one of which deposits on the growing crystal, while the other is transported out of the crucible. In the second regime, T_g is higher, 2215 °C, and ΔT is lower, about 10 °C. Now, the gas flows evaporated from the source and seed are both transported out of the crucible, and the growth fails. When AlN crystal growth experiments were performed using the optimized conditions determined from the simulations, a distinct dependence of the growth morphology on system pressure was observed. Crystals grown at close to atmospheric N$_2$ pressure had a hexagonal facet shape, with a nominally *c*-plane growth surface. At much lower pressures, the presence of many competing growth centers resulted in a rough, porous surface with a rounded shape. Under optimal growth conditions, growth proceeded from a single growth center and the surface exhibited distinct macroscopic steps, suggesting a layer growth mechanism.

The role of oxygen in the sublimation growth was also analyzed [24.28]. Closed-box thermodynamics calculations indicated that at elevated temperature Al$_2$O and AlO are the only major Al–O gaseous compounds, and that at temperatures less than $\approx 2350\,°C$ the only condensed phase, solid AlN, is thermodynamically stable. Model calculations were performed for growth conditions where the growth rate is determined by the transport of Al-containing species from the source to the seed. After solving for the fluxes and partial pressures of all reactive species as a function of the source temperature (T_s), ΔT, total pressure (p_{tot}), and oxygen atomic fraction in the vapor averaged over the gas volume (X_O), the AlN growth rate was determined by the total aluminum flux from five Al-containing gaseous species: Al, Al$_2$, AlN, Al$_2$O, and AlO. The AlN growth rate as a function of temperature computed for pure and oxygen-contaminated N$_2$ atmospheres is shown in Fig. 24.4 for various oxygen atomic fractions and total

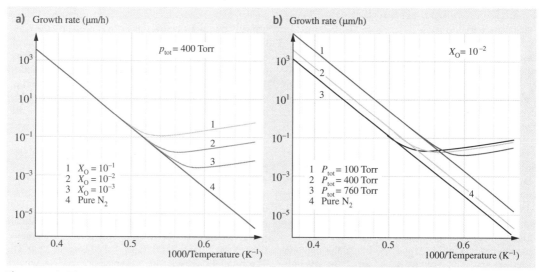

Fig. 24.4a,b The temperature variation of AlN growth rate computed for pure and oxygen-contaminated N$_2$ atmospheres, $\Delta T = 30$ K, and various X_O (**a**) or P_{tot} (**b**) (after [24.28], with permission of Wiley-VCH)

pressures. It is seen that the effect of oxygen on the growth rate is appreciable only for low growth temperatures and becomes negligible in the temperature range $1850\,°C < T < 2350\,°C$ typical for sublimation growth. In addition, the critical oxygen fraction in the vapor corresponding to Al_2O_3 inclusion generation on the AlN surface was determined [24.28]. It was found that, for typical growth temperatures, sufficient purification of AlN source powder is required to produce less than a 10^{-3} oxygen fraction in the vapor in order to avoid inclusion formation. High growth temperatures are also favorable for obtaining crystals free of inclusions.

The above analyses tend to assume, especially at high temperature, that the only gas-phase species of any significance are Al and N_2. First-principles gas-phase composition calculations have indicated that Al_nN ($n = 2, 3, 4$) species, though present in much smaller mole fractions than Al and N_2, are supersaturated with respect to the AlN crystal (Fig. 24.5) and may contribute to the growth [24.29]. Additional analysis of the model was used to show how these trace precursors contribute to mass transport and the growth rate [24.30]. The model predicts the existence of a small mass transport barrier whose height is dependent primarily on the amount of Al_3N in the vapor, and is sensitive to changes in the source temperature and total pressure. Results were used to predict the effective range of ΔT as a function of T_s and p_{tot}, yielding good agreement with published experimental data. However, little is known about the kinetics of these trace species. Their existence and the effect they have on AlN bulk crystal growth have yet to be determined experimentally.

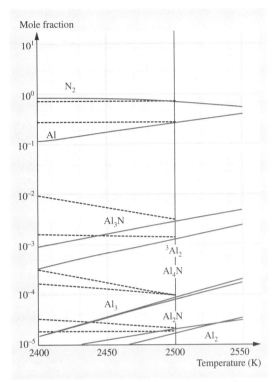

Fig. 24.5 Calculated equilibrium mole fractions of precursors for AlN sublimation growth at a nitrogen pressure of 400 Torr: boundary layer of an Al surface (*solid lines*); and boundary layer of the bulk gas (*dashed lines*) sublimed at $T = 2500$ K (*vertical line*) (after [24.29], with permission of APS)

24.2 High-Temperature Materials Compatibility

By necessity, commercially viable growth rates are achieved at very high process temperatures (typically $> 2200\,°C$), making furnace design and materials selection critical to the success of the overall process, both for achieving durability of growth hardware and keeping crystal impurity levels low. In particular, crucible materials must be refractory and compatible with elevated growth temperatures, inert to chemically aggressive Al vapor, a negligible source of contamination to the growth process, reusable for multiple growth runs, relatively inexpensive, and manufacturable in various shapes and dimensions [24.31].

In his early work, *Slack* [24.10] demonstrated successful growth using W crucibles; however, crucible lifetime was limited and ≈ 50 ppm of W incorporation was reported. Some efforts have employed graphite or coated (SiC, NbC, TaC) graphite crucibles. Several independent studies show that pure graphite crucibles should be avoided due to incompatibility with Al, high levels of carbon in crystals, and the detrimental influence of carbon on growth morphology. Coated graphite crucibles reduce these shortcomings for low-temperature growth; however, these coatings deteriorate quickly above $2000\,°C$, regardless of their thickness or deposition process, and thus do not offer a long-term stable growth environment.

There are several reports of crystal growth in boron nitride (BN) crucibles [24.16, 32, 33]. Sizeable transparent crystals with very low dislocation densities were grown; however, it seems that a BN growth environment produces highly anisotropic growth rates at high temperatures, where the growth rate in the *a*-direction is almost completely inhibited. As a result, coalescence and crystal-size expansion are difficult to achieve.

Compatibility of reactor materials at the high temperatures needed for crystal growth was addressed by *Epelbaum* et al. [24.35]. Crystals were grown from AlN powder with approximately 1% aluminum oxide impurity in a resistively heated reactor at temperatures of 1800–2200 °C using W or graphite heating elements. Different combinations of crucible materials and heating elements yielded results similar to those reported previously [24.36]. Crucibles made of graphite were readily attacked by the Al vapor, while graphite crucibles coated with SiC were unstable at temperatures above 1950 °C, leading to the formation of mixed AlN–SiC crystals varying in color from dark blue to light green. Problems were also associated with the combination of W crucible and heating element, namely degradation of W by aluminum vapor or by oxygen from impurities in the source. The most flexible reactor design was deemed to be a combination of W crucible and graphite heating element.

More recently, efforts with sintered tantalum nitride and tantalum carbide crucibles confirmed that these materials are more stable than any of the aforementioned crucible materials, with crucible lifetimes exceeding 500 h at growth temperatures exceeding 2200 °C [24.31, 37]. Well-faceted crystals with isotropic growth and very low dislocation densities were obtained. These materials have melting points around 3100 °C and 3900 °C, respectively, and thermodynamic calculations [24.34, 38] of the partial pressure of Ta over the solid carbide or nitride have indicated that these materials possess excellent high-temperature stability. Figure 24.6 shows the calculated partial pressures present over several solid crucible materials at 2300 °C in one atmosphere of N_2.

Several groups are developing proprietary processes for fabrication of suitable growth crucibles. Optimization of the sintering process parameters yielded better than 96% dense TaC shapes, which were successfully employed in PVT growth of 38 mm diameter AlN boules [24.31]. Elemental analysis demonstrated that Ta incorporation was below the detection limit (sub-ppm level). However, since TaC appears to possess a larger thermal expansion than AlN, TaC crucibles may exert a compressive stress upon AlN boules during cool-down, leading to the formation of stress-related defects. Thus, use of these crucibles requires that wall contact between the AlN boule and the crucible be avoided. In other work, chemically passivated TaC crucible surfaces were formed by carburization of Ta metal shapes [24.39, 40]. *Mokhov* et al. [24.39] carburized Ta crucibles with 1–2 mm thick walls at 2200–2500 °C in a carbon-containing atmosphere. Carburized crucibles were stable in an atmosphere of Al vapor and N_2 gas at 2300 °C for up to 500 h, after which surface cracks appeared, making them permeable to the vapor. *Hartmann*

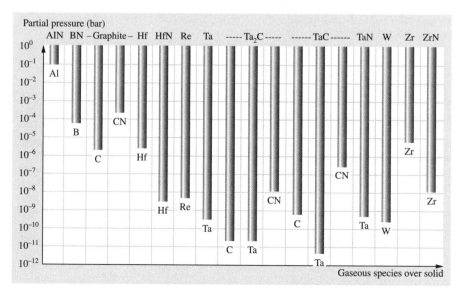

Fig. 24.6 Possible crucible contaminants: partial pressures present over solid crucible materials at 2300 °C in 1 atm nitrogen (after [24.34], with permission)

et al. [24.40] studied the reaction kinetics of Ta carburization, showing that the diffusion-controlled phase transformation follows the sequence Ta → Ta_2C → TaC_{1-x}. Incomplete carburization led to a mechanically stable layered structure of different phases, with Ta_2C at the center and TaC at the surface. When the carburization was allowed to proceed completely to TaC, grain coarsening and anisotropic lattice expansion of the polycrystalline material led to cracking along grain boundaries, rendering fully carburized crucibles unsuitable for PVT growth. Further evaluation of other inert, high-temperature materials may prove fruitful.

24.3 Self-Seeded Growth of AlN Bulk Crystals

Edgar et al. [24.17] reported on AlN crystal growth by sublimation in resistively heated furnaces with W or graphite heating elements. AlN needles and platelets freely nucleated from AlN source material in the cold zone of BN crucibles. Source temperatures were varied from 2000 to 2200 °C and reactor nitrogen pressures ranged from 300 to 800 Torr. Crystals grown in a furnace with W heating elements were either colorless or amber in color, while those grown in a furnace with a graphite heater were colorless. Observed morphologies included needles up to 4 mm in length and 0.5 mm in diameter, and thin plates. The plates were as large as 60 mm^2 and contained growth striations running the length of the crystals along the c-direction. These striations appear to be characteristic of crystals grown in BN environments. They are not seen in AlN grown in other types of crucibles [24.17, 33].

AlN boules up to 10 mm in diameter were produced at Crystal IS [24.41] in conical crucibles. No seed crystals were used in the growth process and, typically, several nuclei formed on the crucible walls during the early stages of growth. As the crystal grew, growth competition between different nuclei resulted in single-crystal regions of varying sizes and orientations. A driving rate for growth was set by translation of the crucible relative to the thermal gradient in the reactor. Under adequate growth conditions (e.g., thermal gradient, reactor pressure) the crystal growth rate was equal to the driving rate, which was varied between 0.65 and 0.9 mm/h. Atomic force microscopy (AFM) imaging of the as-grown crystals revealed 0.25 nm high monolayer steps with straight segments. Step flow resulted from screw dislocations intersecting the growing surface. Screw dislocation density was estimated at 5×10^4 cm^{-2}.

Wafering of these boules revealed several large grains and polycrystalline regions, or single-crystalline regions exhibiting severe cracking around the periphery [24.15, 42, 43]. Chemomechanical polishing (CMP) was used to obtain surfaces suitable for epitaxial growth. Final etching of vicinal surfaces in a mixture of phosphoric and sulfuric acids or potassium hydroxide solution revealed that the N-terminated face was etched much faster and was rougher than the Al-terminated face.

More recently, growth of single-crystal boules up to 15 mm in diameter and several centimeters in length was reported [24.44]. These were used to prepare 0.5 mm thick wafers which were polished by CMP. Wafers exhibited some color variation, which was attributed to absorption by nitrogen vacancies, but were free of cracks. Synchrotron white-beam x-ray topography (SWBXT) was used to characterize the wafers' defect content. Contrast arising from surface damage was observed near the edges, probably due to imperfect polishing, while a high density of small inclusions was observed near the center. Typical dislocation densities were in the 800–5000 cm^{-2} range; dislocations were distributed inhomogeneously, with higher concentrations near the wafer edge. Narrow x-ray rocking curve widths attested to the high quality of the material obtained.

Schlesser et al. [24.32] and *Schlesser* and *Sitar* [24.16] reported on growth of AlN by vaporization of metallic Al in a nitrogen atmosphere and by sublimation of an AlN source [24.32, 45]. Growth temperatures ranged from 1800 °C to 2300 °C at reactor pressures of 250–750 Torr. Temperature gradients of 10–100 K/cm between the source material and crystal growth region were employed. In Al vaporization experiments, the crystal shape and fastest growth direction was found to depend strongly on the growth temperature: at relatively low temperatures (1800–1900 °C) long needles were grown, temperatures around 1900–2000 °C yielded twinned platelets, while c-platelets were formed at temperatures above 2100 °C. These c-plates grew at a rate of 5 mm/h in the c-plane and 0.2 mm/h along the c-axis. Vaporization experiments were performed for 2 h each at a constant growth temperature. Longer growth times did not yield substantially larger crystals. The observed slowdown in growth rate with time was attributed to a decreasing Al flux from the Al source over time, which was due to the progressive formation of an Al-rich, polycrystalline AlN coating over the molten Al.

In order to overcome problems with the Al source instability in vaporization experiments, crystals were grown for longer periods of time by subliming AlN source material [24.32]. Sublimation yielded a stable Al flux over several days of growth. Experiments were carried out at higher temperatures of 2200–2300 °C in order to obtain vapor pressures of Al above AlN comparable with those above metallic Al in the vaporization experiments. Transparent AlN single crystals with dimensions as large as 13 mm were grown with growth rates exceeding 500 μm/h. These sublimation experiments were performed in BN crucibles and typically yielded fastest growth along the c-axis and crystals with surface striations along the c-direction, similar to those observed by *Edgar* et al. [24.33].

The natural growth habit of AlN bulk crystals was investigated by *Epelbaum* et al. [24.46]. Crystals were grown in the 2050–2250 °C temperature range using a low temperature gradient of 3–5 K/cm, intended to facilitate free nucleation of separate single crystals under conditions enhancing formation of natural crystal habit planes. A distinct dependence of morphology on growth temperature was observed: crystals grown at 2050 °C were nearly transparent six-sided prismatic needles, 0.1–0.3 mm in diameter and 5–15 mm in length; columnar crystals 3 mm in length and 1.5 mm in diameter characterized primarily by $\{10\bar{1}1\}$ and $\{10\bar{1}2\}$ rhombohedral facets were grown at 2150 °C; finally, thick platelets of dark amber or brownish color grew at 2250 °C. The largest of these latter crystals was approximately $14 \times 7 \times 2$ mm^3 (Fig. 24.7). Their morphology, orientation, and polarity was studied in greater detail. X-ray diffraction was used to index individual facets and their polarity was confirmed by etching in a molten KOH/NaOH eutectic at 250 °C for 3 min. The Al-polar (0001) and positive rhombohedral faces were characterized by mirror-like facets and transparent material of high crystalline quality, while the N-polar c-plane face contained micrometer-sized inclusions and was opaque. Since these freestanding crystals were grown under approximately isothermal conditions, they possessed a zonal structure corresponding to the simultaneous growth on multiple facets. Examination of polished cuts prepared from them revealed that the zones all belonged to the same single crystal, but exhibited different coloration, etching response, and optical properties (see the discussion in Sect. 24.5.2). Structural, optical, thermal, elemental, and electrical characterization results on single crystals and polycrystalline AlN boules were also reported [24.47–49].

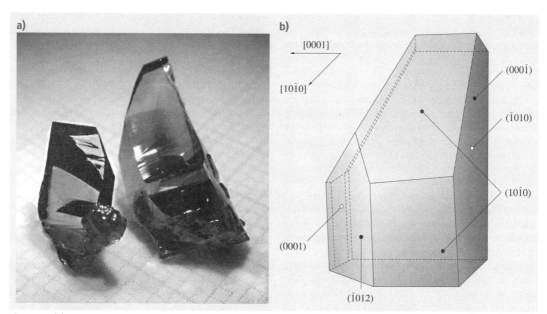

Fig. 24.7 (a) Selected crystals grown by spontaneous nucleation at 2250 °C showing typical growth habit (millimeter grid) and (b) characteristic growth faces of best formed AlN platelets (after [24.46], with permission of Elsevier)

24.4 Seeded Growth of AlN Bulk Crystals

Seeded growth of AlN on SiC has been studied as a way to exploit the availability of large, high-quality SiC substrates and to control the polarity and orientation of AlN crystals [24.50]. Due to the high process temperatures involved and the difference in thermal expansion between AlN and SiC, the stability of the SiC seeds and the cracking of the AlN layers are important issues that need to be addressed in establishing a reproducible seeded growth process on SiC. In contrast, seeded growth on recently available AlN native seeds eliminates many of the problems associated with heteroepitaxial growth, but is only beginning to be investigated. Techniques to avoid renucleation and maintain stable growth have been developed and led to considerable single-crystal size expansion [24.51, 52]. However, several questions, such as which orientation is most favorable for growth, still need to be addressed.

24.4.1 Growth on SiC Seeds

The growth of AlN crystals seeded on SiC substrates was first reported by *Balkas* et al. [24.36]. Single-crystal platelets were grown in a resistively heated graphite furnace by PVT. Growth temperature was varied from 1900 °C to 2250 °C. SiC-coated graphite crucibles were used in 10–15 h experiments. The source material was 99% dense sintered AlN, chosen to allow controllable source-to-seed separation, which was crucial for good crystal growth. Optimal separation was found to be between 1 and 5 mm. Single-crystal 6H-SiC(0001) substrates 10×10 mm^2 were used as seeds. Growth in a high temperature range (2100–2250 °C) and a low temperature range (1950–2050 °C) was investigated. Single crystals of ≈ 1 mm thickness that covered the entire SiC seed were grown at 2150 °C and 4 mm separation distance. The growth rate was estimated at 0.5 mm/h. Due to the degradation of the SiC substrates at higher temperatures, isolated nucleation sites were formed on the seeds at temperatures above 2150 °C, and 2×2 mm^2 hexagonal AlN crystals were grown. The crystals were colored from green to blue, indicating the incorporation of impurities. Secondary-ion mass spectrometry (SIMS) analysis confirmed the presence of Si and C in these crystals. Crystals grown in the low-temperature range were colorless and transparent, but growth rates were significantly lower, 30–50 µm/h. Cracking was always observed in as-grown crystals, due to the thermal expansion coefficient mismatch between SiC and AlN. X-ray diffraction (XRD) patterns confirmed the single-crystal nature of all crystals. Bright-field plan-view transmission-electron microscopy (TEM) and associated selected-area diffraction (SAD) indicated the high quality of the single crystals.

The growth of AlN crystals by sublimation on 6H-SiC seeds was more extensively investigated by *Edgar* et al. [24.17], *Shi* et al. [24.53–55], and *Liu* et al. [24.56–58]. Experiments were carried out in tungsten crucibles placed within the axial temperature gradient of a resistively heated furnace. The growth temperature was typically 1800 °C. SiC wafers (on-axis and 3.5° off-axis) with silicon and carbon terminations were used as substrates. Direct growth [24.53] on as-received Si-terminated SiC resulted in the formation of discontinuous hexagonal subgrains of 1 mm^2 average size. No growth was observed on C-terminated as-received SiC. In order to promote two-dimensional growth on Si-terminated substrates, a 2 µm thick AlN buffer layer was deposited by metalorganic chemical vapor deposition (MOCVD). Continuous growth was achieved by the use of the buffer layer, although cracks formed during cool-down due to stress resulting from the thermal expansion coefficient mismatch. AFM images indicated that AlN grew by the step-flow growth mode.

The initial stages of AlN growth on SiC were studied by *Liu* et al. [24.58]. Fifteen minute growth runs were performed on as-received, on-axis, Si-terminated 6H-SiC(0001) substrates under various temperature and pressure conditions. During the initial stages of growth, AlN nucleated as individual hexagonal hillocks and platelets in an island-like growth mode. Nuclei size and density increased at constant pressure with increased growth temperature in the range of 1800–1900 °C. At constant temperature, growth under reduced pressures yielded coalesced, irregularly shaped platelet crystals. Scanning Auger microprobe (SAM) measurements indicated varying relative compositions of Al, N, Si, and C on different crystal facets of the AlN nuclei. The surface morphology and stress in AlN crystals grown on SiC substrates were also characterized [24.56,57]. AFM images revealed scratches and steps on as-received 6H-SiC substrates, which served as nucleation sites for individual AlN grains grown in a three-dimensional mode. On SiC substrates with an AlN MOCVD epilayer, however, AlN deposited in a two-dimensional growth mode without island formation. Surface morphology varied across the sample, from flat surfaces to regions with large steps (120 nm) separated by large terraces (up to 5 μm). Root-mean-square (RMS) roughness for samples grown with an AlN epilayer was less than 5 nm, compared with 40 nm for crystals grown on as-received substrates. Stress-induced cracks were always observed in the AlN crystals. It was predicted [24.59] that AlN grown on 6H-SiC should be at least 2 mm thick in order to avoid cracking during cool-down from a growth temperature of 2000 °C. Raman spectroscopy revealed that crystals were under compressive stress at the surface and tensile stress (1 GPa) at the interface. Raman spectra indicated improved crystal quality with increasing AlN thickness.

The above method was modified in order to reduce cracking of AlN [24.54, 55]. After deposition of the MOCVD AlN epilayer, an $AlN_{0.8}SiC_{0.2}$ alloy layer was deposited by sublimation from a source mixture of AlN–SiC powders. Pure AlN was then sublimed on the alloy seed as above. The intermediate properties of the alloy layer helped reduce cracking in the overgrown AlN. In addition, the SiC powder source decreased the degradation of the SiC substrates during sublimation growth, allowing for longer growth times. Single-crystal AlN, $4 \times 6 \times 0.5$ mm^3, was obtained after 100 h of growth. Characterization by XRD and Raman spectroscopy confirmed the high quality of the grown material. Thus, three problems identified with growth of AlN on SiC seeds were addressed by this method: (1) the presence of Si and C in the vapor helped suppress the decomposition at high temperature of the SiC seed, (2) an AlN epilayer promoted two-dimensional growth, and (3) cracking of the AlN bulk layer was greatly reduced by an AlN–SiC alloy interlayer.

Dalmau et al. [24.50, 61] developed a two-step process for deposition of thick AlN layers on SiC, and for reduction of cracks in the grown layers. AlN layers up to 3 mm thick were grown on on-axis and off-axis, (0001)-oriented, Si-face SiC seeds by PVT from an AlN powder source. During the growth, the SiC seeds were gradually decomposed at high temperature, yielding freestanding AlN crystals up to 25 mm in diameter. In other samples, the AlN was delaminated from the SiC seeds. As-grown surfaces of layers grown on on-axis SiC seeds were characterized by sharp hexagonal hillocks, suggesting a dislocation-mediated island growth mode, while layers grown on off-axis seeds exhibited steps aligned perpendicular to the off-axis direction, characteristic of the step-flow growth. Crack-free AlN crystals were obtained from these layers and used to fabricate AlN wafers, as shown in Fig. 24.8. High-resolution rocking curves and reciprocal space maps of the (0002) reflection showed that the full-width at half-maximum intensity (FWHM) ranged from 282 to 1440 arcsec, indicating a tilt distribution in the grown layers caused by strain and/or the formation of low-angle grain boundaries as the AlN coalesced [24.60, 62]; nevertheless, these values were comparable to or better than those of the SiC seeds used in these experiments. SWBXT indicated that some strain-free crystals

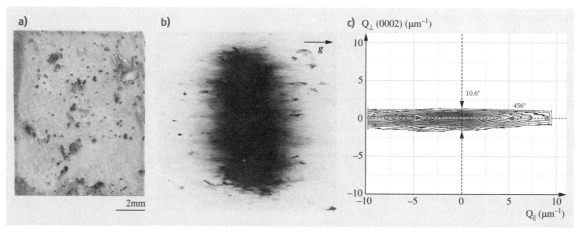

Fig. 24.8 (a) Bulk AlN crystal grown by gradually decomposing the SiC substrate; (b) transmission topograph ($g = [20\bar{2}1]$) showing inhomogeneous strain and high dislocation density; (c) reciprocal space map showing very low triple-axis ω–2Θ scan width (10.6 arcsec) (after [24.60], with permission of Trans Tech)

were obtained, but in all samples the density of dislocations was significantly higher ($> 10^6\,\text{cm}^{-2}$) than in self-seeded AlN crystals. Elemental characterization showed impurity concentrations comparable to those found in the AlN source powder (300 ppmw C and 200 ppmw Si), indicating negligible incorporation of C and Si during growth; the low triple axis ω–2θ scan widths typically observed (≈ 11 arcsec) were characteristic of high-purity crystals and consistent with these results.

In other efforts, *Sarney* et al. [24.63] grew bulk AlN on on-axis and 3.5° off-axis, *c*-oriented 6H-SiC seeds. Sublimation from an AlN powder source in N_2 atmosphere was performed in the temperature range 2150–2200 °C with 4 mm separation between the source and seed. The AlN grew well aligned with the substrate. As in previous work [24.36], cracks were observed in the AlN. *Epelbaum* et al. [24.64] studied AlN crystal growth on SiC substrates of different orientations. Layers of 200–500 μm thickness were deposited at seed temperatures around 2000 °C in 350 mbar N_2 pressure. Growth on Si-face, *c*-oriented substrates was characterized by many hexagonal hillocks on the surface. In contrast, 10° off-axis and on-axis, *a*-plane substrates resulted in more stable growth. The smoothest morphology, typical of step-flow growth, was obtained with on-axis, *a*-plane substrates, however, cracks were also observed in the AlN layers. *Epelbaum* et al. [24.65] and *Heimann* et al. [24.66] suggested a possible vapor–liquid–solid mechanism, mediated by the presence of a molten AlOC$_x$ layer on the surface of SiC, during seeded growth on SiC. Finally, successful seeded growth on C-face SiC was reported recently [24.67, 68].

24.4.2 Growth on AlN Seeds

Reports on AlN growth on native seeds are limited, as these seeds have only recently become available. Seeded growth of AlN on native seeds by PVT was reported for the first time by *Schlesser* et al. [24.45]. Transparent, single-crystal *c*-platelets prepared by vaporization of Al in N_2 were used as seeds. They were mounted into the top of a BN growth crucible filled with AlN source material. Growth was carried out at 2200 °C with a temperature gradient between the source and seed of approximately 3 K/mm. A small seed, 4 mm tall and 0.5 mm thick, grew over a total of 34 h into a 5 mm tall and 7 mm wide single crystal. Growth rates were highly anisotropic, with the fastest growth direction along the *c*-axis. Also, growth rates on the two *c*-faces of opposite polarity differed by a factor of 2–3, with the Al polarity showing slower and smoother growth. Crystal quality of the grown crystals was characterized by XRD. X-ray rocking curves around the (0002) reflection varied from 25 to 45 arcsec, indicating very high single-crystal quality of the material grown by seeded growth.

Noveski et al. [24.51, 69] demonstrated a process for continuous growth of AlN on previously deposited material, resulting in significant expansion of single-crystal grains. Growth was performed in a radiofrequency (RF)-heated reactor at temperatures between 2050 and 2150 °C and pressures of 450–500 Torr, yielding growth rates in the range of 0.1 to 0.3 mm/h. In this process, a starting layer of coalesced polycrystalline AlN was grown into boules up to 38 mm in length and 32 mm in diameter over the course of

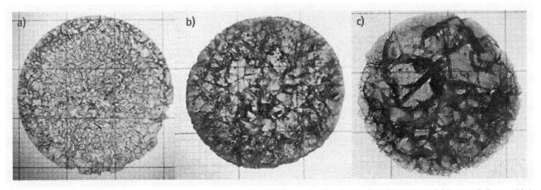

Fig. 24.9a–c Grain evolution observed in different cross-sections of a boule with a diameter of 32 mm: 0.5 mm thick slice, cut after the first 1 mm of growth (**a**); 1 mm thick slice cut at 22 mm boule length (**b**); 2 mm thick slice cut at 35 mm boule length, showing centimeter-size grains (**c**) (after [24.51])

several growth runs. Sublimation growth from a presintered AlN source was interrupted several times in order to replenish the source and keep the source-to-seed distance constant. Renucleation of AlN on the previous growth front after exposure of the boules to air was suppressed by using the inverted temperature gradient method. During the early stages of each run, the crystal growth region was maintained in an inverted temperature gradient, effectively desorbing surface contamination and part of the previously grown layer. In this manner, continuous expansion of previously formed single-crystalline grains was achieved (Fig. 24.9). Centimeter-sized, single-crystal grains were observed in polished cross-sections of boules, and epitaxial regrowth was demonstrated regardless of the orientation of individual grains.

In order to overcome problems associated with the formation of cracks in these polycrystalline boules, large single-crystal grains were harvested and used to prepare seeds for subsequent seeded growth [24.52, 70]. To ensure epitaxial regrowth, seeds were etched in a phosphoric and sulfuric acid mixture, followed by dilute hydrofluoric acid, prior to loading into the system, and were maintained in an inverted temperature gradient during the ramp to the growth temperature. Crystal growth was performed in an axial temperature gradient of 5–10 °C/cm, maintaining the source temperature at 2200–2250 °C and reactor pressure at 400–900 Torr. A $(41\bar{5}0)$-oriented seed was expanded from 10 to 18 mm after several consecutive growth runs, representing a 45° crystal expansion angle (Fig. 24.10).

In related work, *Herro* et al. [24.71, 72] investigated seeded growth along the ⟨0001⟩ polar directions. {0001}-Oriented AlN single crystals were successfully

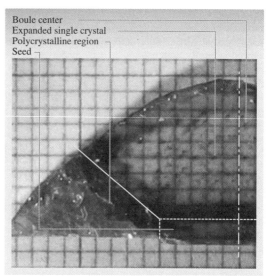

Fig. 24.10 Longitudinal cut of a crystal boule after several growth runs showing the crystal expansion angle (millimeter grid) (after [24.70])

grown along both polar directions, but more stable growth results were obtained on N-polar seeds. The growth surface of N-polar crystals was controlled by a single growth center, leading to a mirror-like growth facet. In contrast, the surfaces of crystals grown on Al-polar seeds showed numerous growth centers, leading to a deterioration of crystal quality, even though the same growth parameters were used for both types of seeds. These observations suggest that lower supersaturation is required to obtain stable growth in the Al-polar direction.

24.5 Characterization of High-Quality Bulk Crystals

As the size of AlN single crystals continues to increase, several characterization techniques are being used to assess the structural and optical properties of large bulk crystals, as well as to determine impurity incorporation, so that the concentration of extended defects and adverse growth contaminants may be minimized by adjusting the growth conditions. Availability of large, strain-free, high-quality bulk crystals has allowed investigators to study the fundamental properties of this material in more detail than has previously been possible with AlN ceramics or thin films. Recent experimental findings have been used to confirm theoretical predictions about the band structure of AlN, leading to revised values for the fundamental bandgap.

24.5.1 Structural Properties

The equilibrium crystal structure (α-phase) of the III-nitrides, AlN, GaN, and InN, is the wurtzite (2H) structure. The stacking sequence of the (0001) close-packed wurtzite planes is ABAB..., comprising bilayer sheets of nitrogen and III-metal atoms; this structure consists of two interpenetrating sublattices that contain the nitrogen and III-metal atoms in tetrahedral coordination. The space group of the wurtzite nitrides is $P6_3mc$, the same as that of the hexagonal (4H and 6H) polytypes of SiC.

X-ray diffraction topography is used to study a crystal's internal diffracting planes in order to discern local changes in the spacing and relative rotation of the planes [24.74]. X-ray topographs, two-dimensional projections of the distribution of diffracted intensity as a function of position in the sample, can be used to map defect structures in large, nearly perfect single crystals and to identify the crystallographic orientations of the diffracting planes. Images are produced by scattering a low-divergence area-filling beam from a set of Bragg planes onto a two-dimensional detector, typically high-resolution x-ray film. In addition, high-resolution x-ray diffraction (HRXRD) is also commonly employed to provide information about the orientation and perfection of single crystals. This technique can be used to generate a reciprocal space map (RSM) representing the two-dimensional intensity contour of the diffracted intensity about a given lattice reflection. These maps provide much more information than typical x-ray rocking curves, since the distribution of lattice tilts (i.e., orientations) and lattice dilations (i.e., d-spacings) can be read independently from the RSM. The lattice dilation distribution is correlated with the concentration of point defects (e.g., vacancies or impurities) in the sampled crystal volume.

SWBXT studies of self-seeded crystals grown by sublimation of AlN powder revealed crystals to be virtually dislocation free [24.37, 60, 73, 75]. Overall dislocation densities were estimated to be around 10^3 cm^{-2}. Defects such as inclusions, growth sector boundaries, and growth dislocations were detected. The presence of Pendellösung fringes in the topographs (Fig. 24.11) was indicative of the high crystalline perfection attained in several samples. Triple-axis x-ray rocking curve FWHM of $\omega-2\theta$ scans of several large crystals (≈ 10 mm) were as low as 7.2 arcsec, marginally larger than the theoretical limit of 6 arcsec, indicating a low density of point defects in these samples.

Characterization results reported for bulk single crystals grown at Crystal IS demonstrated that crystals were of high quality. X-ray topographs of an unevenly shaped polished wafer [24.76], approximately 7×9 mm^2 in size, indicated no significant strain in the wafer and showed an overall dislocation density of 800–1000 cm^{-2}. The density of detected inclusions (presumably oxygen related) was on the order of 10^5 cm^{-3}. FWHM of high-resolution rocking curves ranged from 9 to 12 arcsec, indicating very good crystalline quality. The edge of the wafer in contact with the crucible wall contained cracks and slip bands, probably due to the thermal expansion mismatch between

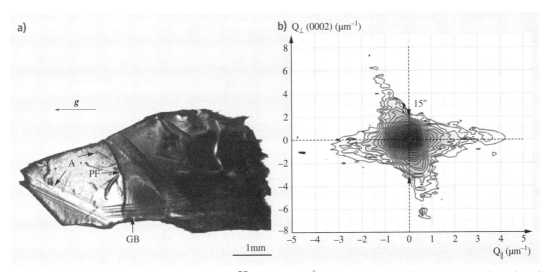

Fig. 24.11 (a) Transmission x-ray topograph ($g = [\bar{1}\bar{1}20]$, $\lambda = 0.54$ Å) from a dislocation-free, spontaneously nucleated AlN single crystal (A – surface artifacts, GB – growth bands, PF – pendellösung fringes); (b) (0002) reciprocal space map (triple-axis $\omega-2\Theta$ scan width: 15 arcsec) (after [24.73], with permission)

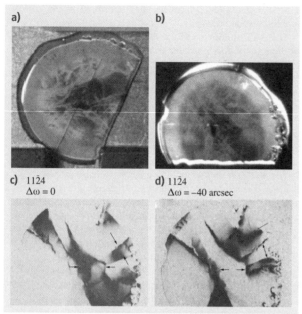

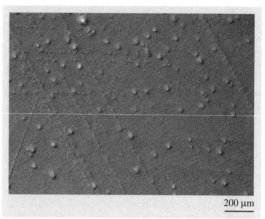

Fig. 24.12 (a) Optical image of an AlN substrate. The length of the flat is 22 mm. (b) The same substrate taken through crossed polarizers. (c,d) Two x-ray topography images of the same substrate taken at an angular distance of 40 arcsec. Small-angle grain boundaries are shown (*black arrows*). $\Delta\omega = 0$ refers to the maximum of the rocking curve. The picture at the right shows the intensity observed at 40 arcsec from the maximum of the rocking curve. Now the areas adjacent to the original black arrows show stronger diffraction intensity. The misalignment between these areas is small, on the order of 40 arcsec (after [24.77], with permission of Wiley-VCH)

Fig. 24.13 Nomarski image of prismatic slip bands as seen on on-axis oriented AlN after chemical etching (after [24.78], with permission)

the boule and crucible. Figure 24.12 shows a picture of a polished AlN substrate with a flat width of 22 mm, an image of the same substrate taken through crossed polarizers, and x-ray topography images taken at two different rocking angles [24.77]. X-ray topographs revealed the presence of small-angle grain boundaries throughout the wafer and individual dislocations at the wafer edge. No diffraction was observed from inclusions that were optically visible. The nature of these inclusions was not discussed.

Defects in polished AlN wafers were studied by XRD, optical microscopy, etch pit pattern delineation, and AFM [24.78]. A triple-axis rocking curve with a FWHM of 10 arcsec was reported. Figure 24.13 shows an optical image of an AlN surface where chemical etching was used to reveal dislocations (viewed as single dots) and slip bands (viewed as straight lines). These slip bands were oriented parallel to the $\langle 1\bar{1}00 \rangle$ directions, demonstrating the activation of prismatic glide in AlN single crystals. The density and shape of etch pits in chemically etched AlN were also studied. Etch pit densities for *c*-plane wafers varied from 1×10^3 to 3×10^4 cm^{-2}, while etch pit patterns were used to distinguish between screw and edge dislocations. In addition, it was shown how subgrain boundaries may propagate as cracks into another grain. Also, etch pits on $\{1\bar{1}00\}$ prismatic planes were reported for the first time; the etch pit density was higher than on basal planes, averaging from 6×10^4 to 4×10^5 cm^{-2}. AFM images of Al-polar, *c*-plane, epiready substrate surfaces prepared by CMP revealed atomic-level steps for near-on-axis substrates and for off-axis orientations up to 6° off-axis [24.79]. RMS roughness of $5 \times 5 \ \mu m^2$ scans was as low as 2.15 Å. The presence of shallow pits, whose density decreased with increasing off-axis orientation, was observed on these Al-polar *c*-plane samples. However, the origin of the pits is still not well understood, but may be related to the polishing process. These pits were not observed in AFM scans of substrates with nonpolar orientations.

AFM studies [24.72] of Al- and N-polar *c*-plane AlN surfaces, obtained by seeded growth on AlN seeds, showed a significant difference in the step-and-terrace structures observed on the two different polar surfaces (Fig. 24.14). Single-unit-cell-high steps were observed on both surfaces, but the terrace width on N-polar AlN, 200–250 nm, was much larger than that observed on Al-polar AlN, 50–70 nm. Since seeded growth

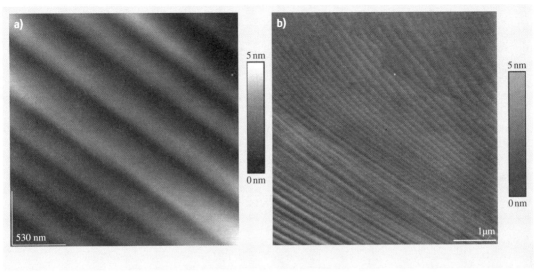

Fig. 24.14a,b AFM micrographs taken on as-grown AlN surfaces: (**a**) N-polar orientation; (**b**) Al-polar orientation. In both cases, single-unit-cell-high steps were observed (after [24.72], with permission)

performed on seeds of both polarities using the same growth conditions, thus ensuring equal supersaturations of Al vapor, these differences were attributed to the surface energy difference of the N- and Al-polar facets.

24.5.2 Fundamental Optical Properties of AlN

AlN has a direct bandgap at the center (Γ-point) of the Brillouin zone exceeding 6 eV. In general, there are still many details concerning the band structure and optical properties of AlN that require further investigation. For example, the band structure parameters near the Γ-point and the fundamental optical transitions were until very recently not well known. In the past, measurements of the bandgap have been performed by optical absorption [24.80–82] and ellipsometry [24.83]. Variations in the measured values were likely due to differences in crystal quality (e.g., impurity and defect concentrations). The room-temperature value commonly quoted in the literature was 6.2 eV. Band-edge luminescence has been investigated using cathodoluminescence (CL) [24.84] and photoluminescence (PL) [24.85].

Recently, though, measurements on high-quality bulk crystals have provided a more complete picture of the band structure. The conduction band has a single minimum (Γ_{7c}) at the Γ-point. The valence band, on the other hand, is split at the Γ-point by the crystal field and the spin–orbit interaction. According to calculations [24.86], the spin–orbit splitting ranges from 11 to 20 meV. The crystal field splitting at the top of the valence band in AlN was predicted [24.87, 88] to be negative, in contrast to the other III-nitrides, but calculated values have ranged widely. However, this information gives a qualitative picture of the valence-band ordering at the Γ-point, and of the associated intrinsic free-exciton transitions. In order of increasing transition energies, these are Γ_{7v} (upper, A-exciton), Γ_{9v} (B-exciton), and Γ_{7v} (lower, C-exciton). The square of the dipole transition matrix elements between the conduction band and the three Γ-point valence states calculated by *Li* et al. [24.89] indicated that the A-exciton transition is nearly forbidden for light polarized perpendicular ($\perp$) to the wurtzite c-axis, while the B- and C-exciton transitions are nearly forbidden for light polarized parallel ($\parallel$) to the c-axis. This picture has recently been confirmed by *Li* et al. [24.89] and optical reflectivity measurements [24.7, 90] (Figs. 24.15 and 24.16) which have provided experimental values for the exciton resonances, the crystal field splitting parameter, and the fundamental bandgap of AlN. The fundamental bandgap energy of unstrained AlN was determined by *Chen* et al. [24.7] to be 6.096 eV at 1.7 K, while the crystal field splitting parameter was −230 meV and

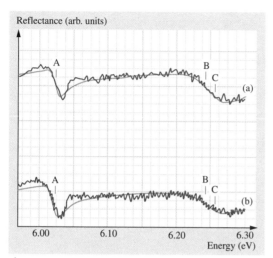

Fig. 24.15 Low-temperature partially polarized optical reflectance spectra from the *m*-face of a bulk AlN crystal with the *c*-axis parallel to the spectrometer slits (*graph* a) and the *c*-axis perpendicular to the slits (*graph* b). *Solid lines* are theoretical fits to the spectra. The bottom of the graph is a (different) finite signal level in each case, to show the features more clearly (after [24.7], with permission of AIP)

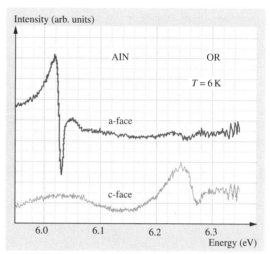

Fig. 24.16 Low-temperature optical reflectivity data at near-normal incidence for AlN samples oriented in two different crystallographic orientations (after [24.90], with permission of APS)

the exciton energies were 6.025, 6.243, and 6.257 eV for the A-, B-, and C-excitons, respectively. Note that the A transition is allowed for light polarized parallel to the *c*-axis. This may explain why earlier absorption measurements consistently resulted in larger values for the bandgap; these measurements were typically performed with light polarized perpendicular to the *c*-axis, and likely probed the B- or C-transitions. These results imply that (0001)-oriented devices grown with AlN or high-Al-content AlGaN alloys will be better edge emitters than surface emitters, and other orientations should be investigated for surface-emitting devices [24.7].

In other studies, the crystalline quality and orientation of a sample oriented with the *c*-axis in the plane of the crystal were evaluated by testing the selection rules for the $A_1(TO)$, $E_1(TO)$, and E_2(high) Raman modes [24.91, 92]. In the $x(zz)x$ geometry, the allowed $A_1(TO)$ and $E_1(TO)$ modes were observed and the forbidden E_2(high) mode was not, while in the $x(zy)x$ geometry, the $A_1(TO)$ mode was suppressed and the $E_1(TO)$ mode was enhanced, confirming the crystal's orientation and high crystalline quality. The superposition of different Raman modes and the presence of quasi-LO modes was observed in the spectra of a randomly oriented sample.

The dependence of phonon spectra on crystal orientation was also observed by *Bickermann* et al. [24.93]. Raman spectra of optical phonons in AlN were taken in backscattering geometry on different well-developed facets of a self-seeded bulk crystal (such as that shown in Fig. 24.7). The results indicated that facets belonging to the same crystal class showed very similar Raman spectra, while the appearance or absence of the $A_1(LO)$, $A_1(TO)$, $E_1(LO)$, and $E_1(TO)$ phonon bands in the spectra could be used to identify the basal *c*-plane facets unambiguously from the prismatic $\{10\bar{1}0\}$ facets (Fig. 24.17). *c*-Plane facets showed, in addition to the E_2 modes, only the $A_1(LO)$ and $E_1(TO)$ bands, while prismatic facets showed the $A_1(TO)$ and both $E_1(LO)$ and $E_1(TO)$ bands. When facets with crystallographic orientations between the basal and prismatic planes were studied, i.e., rhombohedral facets, features corresponding to quasi-LO and quasi-TO phonons with mixed A_1–E_1 symmetry appeared in the spectra.

The zonal dependence of the optical absorption and CL spectra of self-seeded crystals was also investigated [24.94]. Crystals exhibiting natural crystal habits of AlN with well-developed facets [24.46] were selected, cut in different orientations, and polished on both sides. The resulting samples included a number of different zones, which corresponded to volumes of the crystal grown on a different facet. Thus, differences in impurity incorporation and/or defect formation, as reflected in the optical spectra, were correlated to growth on different facets or polar orientations of AlN. In the near-ultraviolet (near-UV) and visible range, Al-polar zones had the lowest absorption, followed by zones grown in the r- and a-directions, and finally by N-polar zones. These differences corresponded to differences in crystal coloration, with N-polar zones exhibiting the deepest amber coloration, which is primarily caused by a broad absorption band at 2.8 eV whose origin is still unclear but has in the past been assigned to nitrogen vacancies. In the mid-UV range, the peak position and intensity of the broad absorption bands observed varied depending on the zone. Zones grown on the Al-polar c-face exhibited a strong band at 4.6 eV, which was nearly absent in all other samples. Finally, in the 5.0–5.8 eV range, an increase in absorption was observed for zones grown in the c-direction regardless of polarity, while zones grown in r- or a-directions exhibited a local minimum; absorption in this range was attributed to nitrogen vacancies. The CL spectra of bulk crystals and polished cuts also showed intensity variations that were dependent on the investigated area. As shown in Fig. 24.18, r- and a-plane facets exhibited intense luminescence peaking at 3.8 eV, while this feature was absent on the Al-polar c-plane facet, where broad, weak bands at 2.5–2.8, 3.3, and 4.3 eV were present.

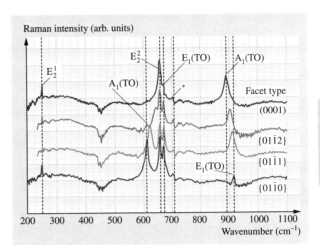

Fig. 24.17 Raman spectra taken in normal incidence on different facets of an AlN single crystal. A logarithmic intensity scale is used to show weak features. The *symbols* denote AlN phonon bands (after [24.93], with permission of AIP)

Optical transitions with energies in the 3–5 eV range are likely due to Al vacancies and their complexes with oxygen.

In other work, *Silveira* et al. [24.84, 95, 96] used CL to study self-seeded AlN crystals and homoepitaxial thin films of AlN grown on these crystals. Both

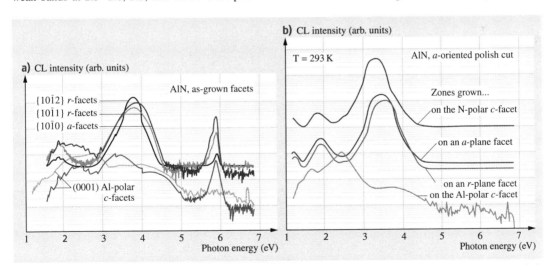

Fig. 24.18a,b Cathodoluminescence spectra of bulk AlN single crystals taken at room temperature. Logarithmic scale is used to show weak features; small jumps in the spectra (*left*) are measurement artifacts. (**a**) Spectra taken on different facet surfaces of an as-grown boule. (**b**) Spectra taken on different zones of an a-oriented polished cut (after [24.94], with permission of Wiley-VCH)

c-plane and a-plane wafers were cut from bulk crystals and polished by CMP. Strong emission was observed at the near-band-edge (NBE) region around 6 eV, and two additional bands were observed in the energy range between 2 and 5.5 eV. One of these bands (VB), at about 3.5 eV, was attributed to oxygen-related defects, while the second band (UVB), at about 4.4 eV, was tentatively assigned to oxygen-related complexes. The integrated intensity of the NBE emission showed a linear dependence with beam current up to 5 µA, and this band was therefore related to exciton recombination processes. Analysis of the NBE CL spectrum of an a-plane sample revealed five transitions in this energy range. Based on the temperature dependence of the NBE spectra, features at 6.026 and 6.041 eV (6 K) were attributed to the free-exciton A- and B-transitions [24.96]. Similar assignments, based on thermal quenching studies, were made for features observed at 6.023 and 6.036 eV in the NBE spectra of a nominally c-plane AlN homoepitaxial film [24.84]. However, recent reflectivity measurements [24.90, 91] demonstrate that these assignments are likely incorrect.

24.5.3 Impurities

Growth of high-purity AlN is a challenging task. As a result of the high affinity of Al for oxygen (the standard Gibbs free energy of formation of Al_2O_3 at 298.15 K is highly negative, -1.582×10^6 J/mol [24.97]), oxygen is a common contaminant in AlN, and influences, among other things, the lattice parameters, thermal conductivity, luminescence, and defect structure of AlN. Early material property measurements on oxygen-contaminated AlN have been revised as higher-purity single-crystal material has become available. In addition, since commercially available AlN powder contains approximately 1% oxygen impurities, obtaining a high-purity AlN source powder is of interest to the crystal growth community. Calculations by *Karpov* et al. [24.28] have shown that AlN source purification is favorable for obtaining bulk crystals free of Al_2O_3 inclusions. Carbon is also a common contaminant in AlN, while metallic and other impurities are typically found at trace levels. Knowledge of the influence these different contaminants have on the growth process and fundamental properties of AlN is still limited, but recent investigations have provided some valuable data.

Dalmau et al. [24.36] reported that sintering of a commercially purchased AlN source powder at a temperature of 2200 °C resulted in a significant reduction of nearly all impurities, with the largest reductions observed for oxygen and carbon impurities (Table 24.1). Several investigators have incorporated source presin-

Table 24.1 Published GDMS analysis results for AlN (all values in ppmw; T: sintering/growth temperature, pressure: system pressure, NA: not available)

Sample	T (°C)	Pressure (Torr)	Crucible material	O	C	Si	W	Reference
1. As-received powder	NA	NA	NA	≈ 1000	≈ 3000	200	< 50	[24.37]
2. Sintered powder	2200	400	TaN	≤ 300	≤ 200	200	< 20	[24.37]
3. Self-seeded single crystal	1950/2070*	500/400*	BN	≤ 500	≤ 300	5.5	< 1	[24.91]
4. Self-seeded single crystal	1885/2030*	600/600*	TaN	≤ 1200	≤ 160	130	< 0.05	[24.91]
5. Self-seeded polycrystal	2100	500	TaC	≤ 50	≤ 50	40	< 1	[24.91]
6. Self-seeded polycrystal	2100	500	TaC	≤ 400	≤ 30	40	< 1	[24.91]
7. Self-seeded polycrystal	2250	< 750	W	≈ 86	≈ 100	2.5	7.9	[24.18]
8. Seeded single crystal	2200	400–900	NA	< 100	< 100	80	< 10	[24.52]

* Samples 3 and 4 were grown using a two-stage growth process; temperatures were gradually ramped between the two stages

tering as part of the PVT growth process [24.46, 70, 72]. *Epelbaum* et al. [24.60] identified how low-temperature transport of Al as suboxides during the ramp to growth temperature can lead to poisoning of the AlN seed surface during seeded growth on native seeds. Even when the AlN source is relatively pure (< 300 ppmw O), evaporation of a thin oxide layer on the source may occur at temperatures as low as 1750 °C, leading to an accumulation of aluminum oxynitrides on the seed surface, which make the surface unsuitable for growth. However, the inverted temperature gradient method developed by *Noveski* et al. [24.51, 69] and used by others [24.52, 70–72] during seeded growth on native seeds represents a practical solution to the problem. This surface poisoning does not appear to be a problem during seeded growth on SiC seeds, possibly due to the different growth mechanism involved [24.64, 65].

Bulk AlN with low oxygen content was reported by *Bickermann* et al. [24.18, 98]. Polycrystals were grown in a vertical, cold-wall reactor equipped with W heating elements. Growth was performed in N_2 atmosphere, at pressures below 1000 mbar, using almost sealed crucibles. Source and crystal growth temperatures ranged from 2200 to 2350 °C, and from 2100 to 2250 °C, respectively. Dense, polycrystalline AlN boules up to 15 mm high and 51 mm in diameter were produced with growth rates between 0.2 and 2 mm/h. Boules were composed of c-textured crystalline grains, some as large as 5×5 mm^2. Although the AlN source material contained significant amounts of various impurities (6000 ppmw oxygen, 300 ppmw carbon, and 500 ppmw metals), as determined by glow discharge mass spectrometry (GDMS), impurity incorporation into grown material was significantly lower (86 ppmw oxygen, 100 ppmw carbon, ≈ 22 ppmw metals). GDMS is a mass-spectrometric technique for the analysis of trace elements in bulk solid samples. It offers several advantages over other trace analysis techniques, including detection limits down to the sub-ppb range, wide dynamic range ($\approx 10^{11}$ range between minor and major components), relative matrix insensitivity, and applicability to a wide variety of materials systems. Table 24.1 shows the results of GDMS analysis on different AlN samples reported in the literature.

A number of researchers have used compositional analysis of AlN crystals together with measurements of optical properties (Sect. 24.5.2) in order to identify impurities that correlate with specific optical transitions. *Slack* et al. [24.99] reported an oxygen-related absorption region between 3.5 and 5.2 eV, with peak positions varying from 4.3 to 4.8 eV depending on the amount of oxygen impurity. These features are likely due to transitions involving Al vacancy–oxygen complexes. The influence of different impurities on the absorption and luminescence spectra of AlN has also been studied by other investigators [24.31, 91, 92, 98]. However, more work is needed before features observed in the absorption and luminescence spectra of high-quality bulk crystals can be unambiguously assigned to specific defect transitions.

24.6 Conclusions and Outlook

AlN crystal growth is a challenging task that has been attempted in the past via a variety of growth methods. Although several issues remain, PVT growth of AlN at high temperatures shows the most promising results and is the only growth technique that can produce high-quality low-dislocation-density crystals. This method has yielded AlN crystals of very high quality and of sufficient size for fabrication of the first devices. The recent demonstration of seeded growth with subsequent crystal-size expansion is certainly a crucial milestone for future development of this technology that will lead to further expansion of single-crystal size. Lifetime and stability of growth crucibles and reactor parts remain a challenge that will need to be addressed in the quest for high-purity crystals and lower production cost. Further research is also needed to improve the understanding of the detrimental effects of various impurities in AlN, so that technologically relevant properties of AlN may be fully exploited by reducing the concentration of adverse growth contaminants.

References

24.1 R. Dalmau, Z. Sitar: Sublimation growth of AlN crystals. In: *Encyclopedia of Materials: Science and Technology*, ed. by K.H.J. Buschow, R.W. Cahn, M.C. Flemings, B. Ilschner, E.J. Kramer, S. Mahajan, P. Veyssière (Elsevier, Oxford 2005) pp. 1–9

24.2 O. Ambacher: Growth and applications of group III-nitrides, J. Phys. D **31**, 2653–26710 (1998)

24.3 R. Gaska, C. Chen, J. Yang, E. Kuokstis, A. Khan, G. Tamulaitis, I. Yilmaz, M.S. Shur, J.C. Rojo, L.J. Schowalter: Deep-ultraviolet emission of AlGaN/AlN quantum wells on bulk AlN, Appl. Phys. Lett. **81**, 4658–4660 (2002)

24.4 L. Liu, J.H. Edgar: Substrates for gallium nitride epitaxy, Mater. Sci. Eng. R **37**, 61–127 (2002)

24.5 B. Monemar: III-V nitrides – important future electronic materials, J. Mater. Sci. **10**, 227–254 (1999)

24.6 L.J. Schowalter, Y. Shusterman, R. Wang, I. Bhat, G. Arunmodhi, G.A. Slack: Epitaxial growth of AlN and $Al_{0.5}Ga_{0.5}N$ layers on aluminum nitride substrates, Appl. Phys. Lett. **76**, 985–987 (2000)

24.7 L. Chen, B.J. Skromme, R.F. Dalmau, R. Schlesser, Z. Sitar, C. Chen, W. Sun, J. Yang, M.A. Khan, M.L. Nakarmi, J.Y. Lin, H.-X. Jiang: Band-edge exciton states in AlN single crystals and epitaxial layers, Appl. Phys. Lett. **85**, 4334–4336 (2004)

24.8 I. Grzegory, J. Jun, M. Boćkowski, S. Krukowski, M. Wróblewski, B. Lucznik, S. Porowski: III-V-nitrides – Thermodynamics and crystal growth at high N_2 pressure, J. Phys. Chem. Solids **56**, 639–647 (1995)

24.9 F. Briegleb, A. Geuther: Ueber das Stickstoffmagnesium und die Affinitäten des Stickgases zu Metallen, Ann. Chem. **123**, 228–241 (1862)

24.10 G. A. Slack: Aluminum nitride crystal growth, Final Report, Contract No. F49620-78-C-0021 (1980) 1–31

24.11 A. Nikolaev, I. Nikitina, A. Zubrilov, M. Mynbaeva, Y. Melnik, V. Dmitriev: AlN wafers fabricated by hydride vapor phase epitaxy, MRS Internet J. Nitride Semicond. Res. **5S1**, W6.5.1–W6.5.5 (2000)

24.12 R. Dwiliński, R. Doradziński, J. Garczyński, L. Sierzputowski, M. Palczewska, A. Wysmolek, M. Kamińska: AMMONO method of BN, AlN and GaN synthesis and crystal growth, MRS Internet J. Nitride Semicond. Res. **3**, 25.1–25.4 (1998)

24.13 C.O. Dugger: Synthesis of AlN single crystals, Mater. Res. Bull. **9**, 331–336 (1974)

24.14 M. Bockowski: Growth and doping of GaN and AlN single crystals under high nitrogen pressure, Cryst. Res. Technol. **36**, 771–787 (2001)

24.15 J.C. Rojo, G.A. Slack, K. Morgan, B. Raghothamachar, M. Dudley, L.J. Schowalter: Report on the growth of bulk aluminum nitride and subsequent substrate preparation, J. Cryst. Growth **231**, 317–321 (2001)

24.16 R. Schlesser, Z. Sitar: Growth of bulk AlN crystals by vaporization of aluminum in a nitrogen atmosphere, J. Cryst. Growth **234**, 349–353 (2002)

24.17 J.H. Edgar, L. Liu, B. Liu, D. Zhuang, J. Chaudhuri, M. Kuball, S. Rajasingam: Bulk AlN crystal growth: self-seeding and seeding on 6H-SiC substrates, J. Cryst. Growth **246**, 187–193 (2002)

24.18 M. Bickermann, B.M. Epelbaum, A. Winnacker: PVT growth of bulk AlN crystals with low oxygen contamination, Phys. Status Solidi (c) **0**, 1993–1996 (2003)

24.19 P.M. Dryburgh: The estimation of maximum growth rate for aluminum nitride crystals grown by direct sublimation, J. Cryst. Growth **125**, 65–68 (1992)

24.20 L. Liu, J.H. Edgar: Transport effects in the sublimation growth of aluminum nitride, J. Cryst. Growth **220**, 243–253 (2000)

24.21 L. Liu, J.H. Edgar: A global growth rate model for aluminum nitride sublimation, J. Electrochem. Soc. **149**, G12–G15 (2002)

24.22 S.Y. Karpov, D.V. Zimina, Y.N. Makarov, E.N. Mokhov, A.D. Roenkov, M.G. Ramm, Y.A. Vodakov: Sublimation growth of AlN in vacuum and in a gas atmosphere, Phys. Status Solidi (a) **176**, 435–438 (1999)

24.23 A.S. Segal, S.Y. Karpov, Y.N. Makarov, E.N. Mokhov, A.D. Roenkov, M.G. Ramm, Y.A. Vodakov: On mechanisms of sublimation growth of AlN bulk crystals, J. Cryst. Growth **211**, 68–72 (2000)

24.24 V. Noveski, R. Schlesser, S. Mahajan, S. Beaudoin, Z. Sitar: Mass transfer in AlN crystal growth at high temperatures, J. Cryst. Growth **264**, 369–378 (2004)

24.25 B.M. Epelbaum, M. Bickermann, A. Winnacker: Sublimation growth of bulk AlN crystals: Process temperature and growth rate, Mater. Sci. Forum **457-460**, 1537–1540 (2004)

24.26 S.Y. Karpov, A.V. Kulik, M.S. Ramm, E.N. Mokhov, A.D. Roenkov, Y.A. Vodakov, Y.N. Makarov: AlN crystal growth by sublimation technique, Mater. Sci. Forum **353-356**, 779–782 (2001)

24.27 E. Mokhov, S. Smirnov, A. Segal, D. Bazarevskiy, Y. Makarov, M. Ramm, H. Helava: Experimental and theoretical analysis of sublimation growth of bulk AlN crystals, Mater. Sci. Forum **457-460**, 1545–1548 (2004)

24.28 S.Y. Karpov, A.V. Kulik, I.N. Przhevalskii, M.S. Ramm, Y.N. Makarov: Role of oxygen in AlN sublimation growth, Phys. Status Solidi (c) **0**, 1989–1992 (2003)

24.29 Y. Li, D.W. Brenner: First principles prediction of the gas-phase precursors for AlN sublimation growth, Phys. Rev. Lett. **92**, 75503.1–75503.4 (2004)

24.30 Y. Li, D.W. Brenner: Influence of trace precursors on mass transport and growth rate during sublimation deposition of AlN crystal, J. Appl. Phys. **100**, 84901.1–84901.6 (2006)

24.31 R. Schlesser, R. Dalmau, D. Zhuang, R. Collazo, Z. Sitar: Crucible materials for growth of aluminum nitride crystals, J. Cryst. Growth **281**, 75–80 (2005)

24.32 R. Schlesser, R. Dalmau, R. Yakimova, Z. Sitar: Growth of AlN bulk crystals from the vapor phase, Mater. Res. Soc. Symp. Proc. **693**, I.9.4.1–I.9.4.6 (2002)

24.33 B. Liu, J.H. Edgar, B. Raghothamachar, M. Dudley, J.Y. Lin, H.X. Jiang, A. Sarua, M. Kuball: Free nucleation of aluminum nitride single crystals in HPBN crucible by sublimation, Mater. Sci. Eng. B **117**, 99–104 (2005)

24.34 G.A. Slack, J. Whitlock, K. Morgan, L.J. Schowalter: Properties of crucible materials for bulk growth of AlN, Mater. Res. Soc. Symp. Proc. **798**, Y10.74.1–Y10.74.4 (2004)

24.35 B.M. Epelbaum, D. Hoffman, M. Bickermann, A. Winnacker: Sublimation growth of bulk AlN crystals: materials compatibility and crystal quality, Mater. Sci. Forum **389-393**, 1445–1448 (2002)

24.36 C.M. Balkas, Z. Sitar, T. Zheleva, L. Bergman, R. Nemanich, R.F. Davis: Sublimation growth and characterization of bulk aluminum nitride single crystals, J. Cryst. Growth **179**, 363–370 (1997)

24.37 R. Dalmau, B. Raghothamachar, M. Dudley, R. Schlesser, Z. Sitar: Crucible selection in AlN bulk crystal growth, Mater. Res. Soc. Symp. Proc. **798**, Y2.9.1–Y2.9.5 (2004)

24.38 B. Liu, J.H. Edgar, Z. Gu, D. Zhuang, B. Raghothamachar, M. Dudley, A. Sarua, M. Kuball, H.M. Meyer III: The durability of various crucible materials for aluminum nitride crystal growth by sublimation, MRS Internet J. Nitride Semicond. Res. **9**, 6.1–6.11 (2004)

24.39 E.N. Mokhov, O.V. Avdeev, I.S. Barash, T.Y. Chemekova, A.D. Roenkov, A.S. Segal, A.A. Wolfson, Y.N. Makharov, M.G. Ramm, H. Helava: Sublimation growth of AlN bulk crystals in Ta crucibles, J. Cryst. Growth **281**, 93–100 (2005)

24.40 C. Hartmann, J. Wollweber, M. Albrecht, I. Rasin: Preparation and characterization of tantalum carbide as an optional crucible material for bulk aluminum nitride crystal growth via physical vapour transport, Phys. Status Solidi (c) **3**, 1608–1612 (2006)

24.41 L.J. Schowalter, J.C. Rojo, N. Yakolev, Y. Shusterman, K. Dovidenko, R. Wang, I. Bhat, G.A. Slack: Preparation and characterization of single-crystal aluminum nitride substrates, MRS Internet J. Nitride Semicond. Res. **5S1**, W6.7.1–W6.7.6 (2000)

24.42 J.C. Rojo, G.A. Slack, K. Morgan, L.J. Schowalter, M. Dudley: Growth of self-seeded aluminum nitride by sublimation-recondensation and substrate preparation, Mater. Res. Soc. Symp. Proc. **639**, G1.10.1–G1.10.6 (2001)

24.43 J.C. Rojo, L.J. Schowalter, K. Morgan, D.I. Florescu, F.H. Pollack, B. Raghothamachar, M. Dudley: Single-crystal aluminum nitride substrate preparation from bulk crystals, Mater. Res. Soc. Symp. Proc. **680E**, E.2.1.1–E.2.1.7 (2001)

24.44 L.J. Schowalter, G.A. Slack, J.B. Whitlock, K. Morgan, S.B. Schujman, B. Raghothamachar, M. Dudley, K.R. Evans: Fabrication of native, single-crystal AlN substrates, Phys. Status Solidi (c) **0**, 1997–2000 (2003)

24.45 R. Schlesser, R. Dalmau, Z. Sitar: Seeded growth of AlN bulk single crystals by sublimation, J. Cryst. Growth **241**, 416–420 (2002)

24.46 B.M. Epelbaum, C. Seitz, A. Magerl, M. Bickermann, A. Winnacker: Natural growth habit of bulk AlN crystals, J. Cryst. Growth **265**, 577–581 (2004)

24.47 M. Bickermann, B.M. Epelbaum, A. Winnacker: Structural properties of AlN crystals grown by physical vapor transport, Phys. Status Solidi (c) **2**, 2044–2048 (2005)

24.48 M. Bickermann, B.M. Epelbaum, M. Kazan, Z. Herro, P. Masri, A. Winnacker: Growth and characterization of bulk AlN substrates grown by PVT, Phys. Status Solidi (a) **202**, 531–535 (2005)

24.49 M. Bickermann, B.M. Epelbaum, A. Winnacker: Structural, optical, and electrical properties of bulk AlN crystals grown by PVT, Mater. Sci. Forum **457-460**, 1541–1544 (2004)

24.50 R. Dalmau, R. Schlesser, B.J. Rodriguez, R.J. Nemanich, Z. Sitar: AlN bulk crystals grown on SiC seeds, J. Cryst. Growth **281**, 68–74 (2005)

24.51 V. Noveski, R. Schlesser, B. Raghothamachar, M. Dudley, S. Mahajan, S. Beaudoin, Z. Sitar: Seeded growth of bulk AlN crystals and grain evolution in polycrystalline boules, J. Cryst. Growth **279**, 13–19 (2005)

24.52 D. Zhuang, Z.G. Herro, R. Schlesser, B. Raghothamachar, M. Dudley, Z. Sitar: Seeded growth of AlN crystals on nonpolar seeds via physical vapor transport, J. Electron. Mater. **35**, 1513–1517 (2006)

24.53 Y. Shi, Z.Y. Xie, L.H. Liu, B. Liu, J.H. Edgar, M. Kuball: Influence of buffer layer and 6H-SiC substrate polarity on the nucleation of AlN grown by the sublimation sandwich technique, J. Cryst. Growth **233**, 177–186 (2001)

24.54 Y. Shi, B. Liu, L. Liu, J.H. Edgar, E.A. Payzant, J.M. Hayes, M. Kuball: New technique for sublimation growth of AlN single crystals, MRS Internet J. Nitride Semicond. Res. **6**, 5.1–5.6 (2001)

24.55 Y. Shi, B. Liu, L. Liu, J.H. Edgar, H.M. Meyer III, E.A. Payzant, L.R. Walker, N.D. Evans, J.G. Swadener, J. Chaudhuri, J. Chaudhuri: Initial nucleation study and new technique for sublimation growth of AlN on SiC substrate, Phys. Status Solidi (a) **188**, 757–762 (2001)

24.56 L. Liu, B. Liu, Y. Shi, J.H. Edgar: Growth mode and defects in aluminum nitride sublimed on (0001) 6H-SiC substrates, MRS Internet J. Nitride Semicond. Res. **6**, 7.1–7.5 (2001)

24.57 L. Liu, D. Zhuang, B. Liu, Y. Shi, J.H. Edgar, S. Rajasingam, M. Kuball: Characterization of aluminum nitride crystals grown by sublimation, Phys. Status Solidi (a) **188**, 769–773 (2001)

24.58 B. Liu, Y. Shi, L. Liu, J.H. Edgar, D.N. Braski: Surface morphology and composition characterization at the initial stages of AlN crystal growth, Mater. Res. Soc. Symp. Proc. **639**, G3.13.1–G3.13.6 (2001)

24.59 L. Liu, B. Liu, J.H. Edgar, S. Rajasingam, M. Kuball: Raman characterization and stress analysis of AlN grown on SiC by sublimation, J. Appl. Phys. **92**, 5183–5188 (2002)

24.60 B. Raghothamachar, R. Dalmau, M. Dudley, R. Schlesser, D. Zhuang, Z. Herro, Z. Sitar: Structural characterization of bulk AlN single crystals grown from self-seeding and seeding by SiC substrates, Mater. Sci. Forum **527-529**, 1521–1524 (2006)

24.61 R. Dalmau, R. Schlesser, Z. Sitar: Polarity and morphology in seeded growth of bulk AlN on SiC, Phys. Status Solidi (c) **2**, 2036–2039 (2005)

24.62 B. Raghothamachar, M. Dudley, R. Dalmau, R. Schlesser, Z. Sitar: Synchrotron white beam x-ray topography (SWBXT) and high resolution triple axis diffraction studies on AlN layers grown on 4H- and 6H-SiC seeds, Mater. Res. Soc. Symp. Proc. **831**, E8.24.1–E8.24.6 (2005)

24.63 W.L. Sarney, L. Salamanca-Riba, T. Hossain, P. Zhou, H.N. Jayatirtha, H.H. Kang, R.D. Vispute, M. Spencer, K.A. Jones: TEM study of bulk AlN grown by physical vapor transport, MRS Internet J. Nitride Semicond. Res. **5S1**, W5.5.1–W5.5.6 (1999)

24.64 B.M. Epelbaum, M. Bickermann, A. Winnacker: Seeded PVT growth of aluminum nitride on silicon carbide, Mater. Sci. Forum **433-436**, 983–986 (2003)

24.65 B.M. Epelbaum, P. Heimann, M. Bickermann, A. Winnacker: Comparative study of initial growth stage in PVT growth of AlN on SiC and on native AlN susbtrates, Phys. Status Solidi (c) **2**, 2070–2073 (2005)

24.66 P. Heimann, B.M. Epelbaum, M. Bickermann, S. Nagata, A. Winnacker: The initial stage in PVT growth of aluminum nitride, Phys. Status Solidi (c) **3**, 1575–1578 (2006)

24.67 K. Balakrishnan, M. Banno, K. Nakano, G. Narita, N. Tsuchiya, M. Imura, M. Iwaya, S. Kamiyama, K. Shimono, T. Noro, T. Takagi, H. Amano, I. Akasaki: Sublimation growth of AlN bulk crystals by seeded and spontaneous nucleation methods, Mater. Res. Soc. Symp. Proc. **831**, E11.3.1–E11.3.7 (2005)

24.68 S. Wang, B. Raghothamachar, M. Dudley, A.G. Timmerman: Crystal growth and defect characterization of AlN single crystals, Mater. Res. Soc. Symp. Proc. **892**, FF30.06.1–FF30.06.6 (2006)

24.69 V. Noveski, R. Schlesser, J. Freitas Jr., S. Mahajan, S. Beaudoin, Z. Sitar: Vapor phase transport of AlN in an RF heated reactor: Low and high temperature studies, Mater. Res. Soc. Symp. Proc. **798**, Y2.8.1–Y2.8.6 (2004)

24.70 D. Zhuang, Z.G. Herro, R. Schlesser, Z. Sitar: Seeded growth of AlN single crystals by physical vapor transport, J. Cryst. Growth **287**, 372–375 (2006)

24.71 Z.G. Herro, D. Zhuang, R. Schlesser, R. Collazo, Z. Sitar: Growth of large AlN single crystals along the [0001] directions, Mater. Res. Soc. Symp. Proc. **892**, FF21.01.1–FF21.01.6 (2006)

24.72 Z.G. Herro, D. Zhuang, R. Schlesser, R. Collazo, Z. Sitar: Seeded growth of AlN on N- and Al- polar ⟨0001⟩ AlN seeds by physical vapor transport, J. Cryst. Growth **286**, 205–208 (2006)

24.73 B. Raghothamachar, J. Bai, M. Dudley, R. Dalmau, D. Zhuang, Z. Herro, R. Schlesser, Z. Sitar, B. Wang, M. Callahan, K. Rakes, P. Konkapaka, M. Spencer: Characterization of bulk grown GaN and AlN single crystal materials, J. Cryst. Growth **287**, 349–353 (2006)

24.74 M. Dudley, X. Huang: X-ray topography. In: *Encyclopedia of Materials: Science and Technology*, ed. by K.H.J. Buschow, R.W. Cahn, M.C. Flemings, B. Ilschner, E.J. Kramer, S. Mahajan, E. Veyssière (Elsevier, Oxford 2001) pp. 9813–9825

24.75 B. Raghothamachar, W.M. Vetter, M. Dudley, R. Dalmau, R. Schlesser, Z. Sitar, E. Michaels, J.W. Kolis: Synchrotron white beam topography characterization of physical vapor transport grown AlN and ammonothermal GaN, J. Cryst. Growth **246**, 271–280 (2002)

24.76 B. Raghothamachar, M. Dudley, J.C. Rojo, K. Morgan, L.J. Schowalter: X-ray characterization of bulk AlN single crystals grown by the sublimation technique, J. Cryst. Growth **250**, 244–250 (2003)

24.77 L.J. Schowalter, S.B. Schujman, W. Liu, M. Goorsky, M.C. Wood, J. Grandusky, F. Shahedipour-Sandvik: Development of native, single crystal AlN substrates for device applications, Phys. Status Solidi (a) **203**, 1667–1671 (2006)

24.78 R.T. Bondokov, K.E. Morgan, R. Shetty, W. Liu, G.A. Slack, M. Goorsky, L.J. Schowalter: Defect content evaluation in single-crystal AlN wafers, Mater. Res. Soc. Symp. Proc. **892**, FF30.03.1–FF30.03.6 (2006)

24.79 S.B. Schujman, W. Liu, N. Meyer, J.A. Smart, L.J. Schowalter: Atomic force microscope studies on native AlN substrates, Mater. Res. Soc. Symp. Proc. **892**, FF30.05.1–FF30.05.6 (2006)

24.80 J. Pastrňák, L. Roskovcová: Optical absorption edge of AlN single crystals, Phys. Status Solidi (b) **26**, 591–597 (1968)

24.81 W.M. Yim, E.J. Stofko, P.J. Zanzucchi, J.I. Pankove, M. Ettenberg, S.L. Gilbert: Epitaxially grown AlN and its optical band gap, J. Appl. Phys. **44**, 292–296 (1973)

24.82 G.A. Slack, T.F. McNelly: AlN single crystals, J. Cryst. Growth **42**, 560–563 (1977)

24.83 T. Wethkamp, K. Wilmers, C. Cobet, N. Esser, W. Richter, O. Ambacher, M. Stutzmann, M. Cardona: Dielectric function of hexagonal AlN films determined by spectroscopic ellipsometry in the vacuum-UV range, Phys. Rev. B **59**, 1845–1849 (1999)

24.84 E. Silveira, J.A. Freitas Jr., M. Kneissel, D.W. Treat, N.M. Johnson, G.A. Slack, L.J. Schowalter: Near-bandedge cathodoluminescence of an AlN homoepitaxial film, Appl. Phys. Lett. **84**, 3501–3503 (2004)

24.85 E. Kuokstis, J. Zhang, Q. Fareed, J.W. Yang, G. Simin, M.A. Khan, R. Gaska, M. Shur, C. Rojo, L. Schowalter: Near-band-edge photoluminescence of wurtzite-type AlN, Appl. Phys. Lett. **81**, 2755–2757 (2002)

24.86 I. Vurgaftman, J.R. Meyer: Band parameters for nitrogen-containing semiconductors, J. Appl. Phys. **94**, 3675–3696 (2003)

24.87 M. Suzuki, T. Uenoyama, A. Yanase: First-principles calculations of effective-mass parameters of AlN and GaN, Phys. Rev. B **52**, 8132–8139 (1995)

24.88 J.M. Wagner, F. Bechstedt: Properties of strained wurtzite GaN and AlN: Ab initio studies, Phys. Rev. B **66**, 115202.1–115202.20 (2002)

24.89 J. Li, K.B. Nam, L. Nakarmi, J.Y. Lin, H.X. Jiang, P. Carrier, S.H. Wei: Band structure and fundamental optical transitions in wurtzite AlN, Appl. Phys. Lett. **83**, 5163–5165 (2003)

24.90 E. Silveira, J.A. Freitas Jr., O.J. Glembocki, G.A. Slack, L.J. Schowalter: Excitonic structure of bulk AlN from optical reflectivity and cathodoluminescence measurements, Phys. Rev. B **71**, 41201 (2005)

24.91 M. Strassburg, J. Senawiratne, N. Dietz, U. Haboeck, A. Hoffmann, V. Noveski, R. Dalmau, R. Schlesser, Z. Sitar: The growth and optical properties of large, high-quality AlN single crystals, J. Appl. Phys. **96**, 5870–5876 (2004)

24.92 J. Senawiratne, M. Strassburg, N. Dietz, U. Haboeck, A. Hoffmann, V. Noveski, R. Dalmau, R. Schlesser, Z. Sitar: Raman, photoluminescence and absorption studies on high quality AlN single crystals, Phys. Status Solidi (c) **2**, 2774–2778 (2005)

24.93 M. Bickermann, B.M. Epelbaum, P. Heimann, Z.G. Herro, A. Winnacker: Orientation-dependent phonon observation in single-crystalline aluminum nitride, Appl. Phys. Lett. **86**, 131904 (2005)

24.94 M. Bickermann, P. Heimann, B.M. Epelbaum: Orientation-dependent properties of aluminum nitride single crystals, Phys. Status Solidi (c) **3**, 1902–1906 (2006)

24.95 E. Silveira, J.A. Freitas Jr., G.A. Slack, L.J. Schowalter: Cathodoluminescence studies of large bulk AlN crystals, Phys. Status Solidi (c) **0**, 2618–2622 (2003)

24.96 E. Silveira, J.A. Freitas, G.A. Slack, L.J. Schowalter, M. Kneissl, D.W. Treat, N.M. Johnson: Depth-resolved cathodoluminescence of a homoepitaxial AlN thin film, J. Cryst. Growth **281**, 188–193 (2005)

24.97 D.R. Lide (Ed): *CRC Handbook of Chemistry and Physics*, 74th edn (CRC Press, Boca Raton 1993) Chap. 5, p. 5

24.98 M. Bickermann, B.M. Epelbaum, A. Winnacker: Characterization of bulk AlN with low oxygen content, J. Cryst. Growth **269**, 432–442 (2004)

24.99 G.A. Slack, L.J. Schowalter, D. Morelli, J.A. Freitas Jr: Some effects of oxygen impurites on AlN and GaN, J. Cryst. Growth **246**, 287–298 (2002)

844

25. Growth of Single-Crystal Organic Semiconductors

Christian Kloc, Theo Siegrist, Jens Pflaum

Organic semiconductor crystal growth presents a very different set of challenges than their inorganic counterparts. Although single crystals of organic semiconductors can be grown by the same techniques used for inorganic semiconductors, the weak intermolecular bonds, low melting temperatures, and high vapor pressures and solvent solubilities require specific modifications to crystal growth techniques of these materials. Bulk crystals of only a handful of different materials have been grown from the melt. The Czochralski, Bridgman, and general melt growth techniques are hampered by the high vapor pressure, which causes fast evaporation of the material during the growth process. However, the significant vapor pressure of organic semiconductors makes gas-phase growth methods suitable for most of them. In general, multistep synthesis of organic molecules produces impure materials which need extensive purification. Small crystals, mostly for structure determinations, have been grown from organic solvents. Zone melting has been used for a few materials, but many organic molecules decompose before reaching the melting temperature. The crystal growth of volatile molecules in a stream of flowing gas is therefore a widely used method that combines purification and crystal growth. Although gas-phase grown crystals tend to be small in size, they have high structural quality and superior purity and are therefore preferred for physical property measurements.

25.1	**Basics**	845
25.2	**Theory of Nucleation and Crystal Growth**	847
	25.2.1 Stability Criteria for Nuclei	847
	25.2.2 Thermodynamic Considerations of Crystal Growth	847
	25.2.3 Growth Morphology in Relation to Symmetry	848
	25.2.4 Structural Defects	848
25.3	**Organic Materials of Interest for Semiconducting Single Crystals**	848
25.4	**Pregrowth Purification**	850
	25.4.1 Zone Refinement	851
	25.4.2 Sublimation and Its Modifications	852
25.5	**Crystal Growth**	854
	25.5.1 Melt Growth	855
	25.5.2 Growth from the Gas Phase	857
	25.5.3 Solvent-Based Growth Methods	862
25.6	**Quality of Organic Semiconducting Single Crystals**	862
25.7	**Organic Single-Crystalline Field-Effect Transistors**	863
25.8	**Conclusions**	864
References		865

25.1 Basics

Most organic molecular crystals are transparent materials that behave like insulators, with electrical resistivities in the range of 10^{13}–10^{18} Ω cm at room temperature. However, some organic molecules with delocalized π-electrons such as conjugated hydrocarbons, phthalocyanines, oligothiophenes, etc. may form colored crystals indicating a small energy gap between the highest occupied (HOMO) and the lowest unoccupied (LUMO) molecular orbitals. Crystals made from such substances often show significant absorption and emission of visible light, and photoconductivity and electrical conductivity, and in general behave more like semiconductors than insulators. Photoconductivity of anthracene, with its three conjugated benzene rings, intrigued researchers from the beginning of the 20th century [25.1]. The feature that light is emitted from anthracene crystals under the absorption of high-energy radiation such as neutrons or x-rays has been widely used in scintillation detectors [25.2]. Later, research began to explore whether organic semiconductors may be suitable for applications where inorganic semiconductors were already established. Such research required organic semiconductors in thin-film form for applications, and in single-crystal form mostly for basic research. Studies that focus on physical properties resulting from weak intramolecular bonds, high symmetry of molecules, and anisotropy of physical properties require measurements on single crystals. Therefore, many well-established crystal growth methods for inorganic materials have been adapted to the growth of organic semiconductor crystals and are discussed in this chapter.

Semiconductors composed from van der Waals bonded molecules, especially those comprising small organic molecules, may be soluble in numerous organic solvents, even at room temperature. This fact allows for low-temperature processing such as printing, spraying, casting, laminating, etc. to mention just a few. Low-temperature solubility distinguishes significantly organic semiconductors from covalently bonded inorganic semiconductors, which often can only be dissolved in high-temperature fluxes. Organic semiconductors therefore seem to be the perfect components for devices where the active materials, as well as the electrodes and conduction paths, may simply be printed by various techniques. Printing semiconducting elements on plastic foils promises large-scale production of inexpensive electronic devices such as bendable displays, field-effect transistor based circuits or large solar cell arrays [25.3, 4].

The power of organic molecular synthesis and crystal engineering allows the formation of new semiconductor materials through systematic variation of crystalline and molecular structure for structure–properties investigations. Optimization of structural, electrical or optical characteristics through chemical design of molecules may then result in improved electronic devices. Furthermore, new properties may give rise to novel applications and devices. However, understanding the relationship between the molecular composition, the packing of molecules into crystals, and the physical properties of the resulting organic semiconductors still presents a challenge.

Often, studies of physical properties of organic semiconductors were performed on polycrystalline or amorphous thin-film devices due to their ease of fabrication and good control over physical dimensions. However, grain boundaries, inhomogeneous distribution of impurities, and high defect densities make thin-film devices unattractive for studies of intrinsic properties [25.5, 6].

In contrast, single crystals contain only small amounts of impurities and defects, due to the inherent purification during crystallization. Therefore, properties measured on single crystals tend to be more reproducible than those measured on amorphous or polycrystalline thin-film samples.

Historically, properties of inorganic semiconductors, such as silicon, germanium or III–V compounds, have been studied on single crystals or single-crystalline epitaxial thin films. It is therefore reasonable to assume that the intrinsic properties of organic semiconductors can also be studied following the same path. Contrary to inorganic semiconductors, were the growth units are atoms, the growth units in organic crystals are molecules. Such extended moieties, with their internal structure, require reorientation and alignment before incorporation into a crystalline lattice. These processes need to be considered by using a specific crystal growth technique for the growth of a particular organic semiconductor.

In general, organic semiconductors crystallize in low-symmetry unit cells. Therefore, all physical properties are tensorial properties with often large anisotropies and can only be properly evaluated on single crystals. The availability of high-purity and high-quality single crystals of organic semiconductors is thus crucial in exploring the physics of such materials and is very helpful in new material design and in building devices with novel functionality and high performance [25.7, 8].

Single crystals of a series of organic semiconductors have been grown using well-described crystal growth techniques. Examples of monographs focused on the theory and practice of crystal growth are cited at the end of this chapter [25.9, 10]. The weak van der Waals bonds between the growth units (in particular molecules) in organic crystals limit the methods applicable to the preparation of single-crystalline specimens [25.11, 12]. In additional the growth units may decompose relatively easily, thus contaminating the crystals. Since the properties of organic semiconductors, like their inorganic counterparts, are sensitive to low levels of impurities and defects, the growth of high-quality organic semiconductors needs to consider purification, contamination, and defect formation in every step of the preparation process.

In this chapter, we will therefore focus on practical approaches for purification and growth of high-quality molecular semiconductor single crystals. In the past, reviews concentrated either on melt growth [25.12] or gas-phase techniques [25.13, 14]. Here, we attempt to compare different crystal growth techniques in order to suggest their advantages and disadvantages as well as practical criteria for the use of these techniques for organic semiconductors. We focus on technological differences between the growth of inorganic and organic semiconductors. We will further discuss defects and contaminants resulting from the use of particular crystal growth methods.

25.2 Theory of Nucleation and Crystal Growth

The growth kinetics of inorganic as well as organic materials are determined by their inherent parameters, such as the interaction between the individual species/units forming the crystals, and external parameters, such as the growth temperature or adsorption rate. Especially the set of external conditions such as the interaction between a substrate and the initial nucleation layer are often difficult to control, if at all. These parameters are critical for the initial step of nucleation and might even change during growth as a function of temperature.

Furthermore, a fundamental property of organics is the anisotropic extension of the electronic wavefunctions, meaning that the electron distribution shows strong spatial anisotropy of intermolecular overlap, in contrast to inorganic crystals with covalent or ionic bonding. The intention is to correlate the spatial anisotropy of the bonding strength with the tensorial properties, e.g., higher mobility values along certain crystallographic directions.

To characterize the influence of such external parameters on crystal preparation, it is helpful to highlight first the kinetics of crystal formation for the case of an ideal system. In this case it is appropriate to divide the crystal formation into two steps: the initial nucleation and the subsequent growth of the individual nuclei.

25.2.1 Stability Criteria for Nuclei

To initiate volume growth of the intended organic crystal, the formation of the initial nuclei on a given support is the essential step. Beside the optimization of nuclei density in the beginning of crystal formation, an important criterion is the smallest number of organic molecules forming a stable nucleus. As for their inorganic counterparts, stability is governed by the balance between loss in surface energy versus gain in volume energy upon incorporation of an additional molecular building block.

However, besides the van der Waals interaction, which was interpreted on the basis of temporarily fluctuating dipoles by London, the occurring internal degrees of freedom, e.g., the torsion of extended phenyl groups as in the case of 9,10-diphenylanthracene [25.15], might play a considerable role at the early stage of nucleation. Even thermally induced changes of the molecular shape in the gas phase can significantly influence the crystallinity of the molecular systems. For instance, the thermal energy during evaporation is not sufficient to planarize the tetracene backbone of rubrene, which is twisted due to pairwise repulsion of the attached phenyl rings. As a result, thin films of rubrene have proven to be amorphous without any indication of crystallinity [25.16,17]. In contrast, growth from gradient sublimation under a constant gas flow provides sufficient energy to overcome the energy barrier of 210 meV and to planarize the molecule averaged over time [25.16]. In the latter case, rubrene shows defined crystal growth with spatial extensions of its growth planes in the centimeter range.

25.2.2 Thermodynamic Considerations of Crystal Growth

The thermodynamic quantities describing the energetics of crystal growth and especially of nucleation are the sublimation enthalpy and the entropy. Neglecting permanent dipoles, the bonding strength between molecular entities is governed by the weak van der Waals force, resulting in a distribution of sublimation enthalpies from 45 kJ/mol for benzene up to 215 kJ/mol for ovalene [25.18]. Normalizing the heat of sublimation divided by the number of intermolecular contacts reveals an average energy per contact of 50 J/mol for various polyaromatic hydrocarbon crystals, independent of the molecular packing [25.18]. However, for molecules offering additional degrees of freedom, e.g., due to twisted functional groups, the gain in entropy might add a significant amount of energy, determining the final crystalline structure.

25.2.3 Growth Morphology in Relation to Symmetry

In view of the anisotropic physical properties such as the tensor of charge carrier mobility or the optical indicatrix, one of the central aspects of crystal growth is the formation and stability of certain crystallographic facets, planes or axes. Similar to their inorganic equivalents, high-indices surfaces are unstable over time due to their higher growth velocities compared with low-index planes. However, for organic materials the crystal habit, defined as the relative size of the stable faces, proves to be more sensitive to the external growth conditions as a consequence of the weak van der Waals binding energies. Tetracene, and many other oligoacenes grown by plate sublimation, shows a distinct stable crystal habit depending on the environmental conditions. Under vacuum, a preferential growth along the (ab)-plane is observed. This mode is determined by the long-range diffusion and the gain of energy by incorporating the individual molecules in this plane under almost equilibrium conditions. In an atmosphere of 400 Torr N_2, however, the mean free path of the sublimed molecule is shorter than the traveling distance in the volume, and the crystal growth is governed by the material transport rather than by the diffusion atop the (ab)-crystal surface. The crystals emanating from this growth regime are needle shaped, preferentially oriented with the needle axis along the c^*-direction.

25.2.4 Structural Defects

The growth of organic molecular crystals, occurring in most cases in a temperature range between room temperature up to about 700 K, is accompanied by the formation of structural defects, e.g., vacancies, dislocations, and growth sectors. According to Boltzmann's law, the density of vacancies n in a crystal is a function of the temperature T and the energy, E_v, required to form a vacancy: $n \sim \exp(-E_V/k_B T)$. Possible defects can be characterized by their dimensionality as vacancies (zero-dimensional, 0-D), dislocation lines or screw dislocations (one-dimensional, 1-D), or grain boundaries (two, dimensional 2-D). Even for high-quality anthracene crystals, thermal treatment might result in vacancy densities at room temperature of 10^{14}–10^{16} cm^{-3} after growth. Furthermore, due to the respective packing along the various crystallographic directions, an anisotropic distribution of defects has been observed. Especially in crystals with herringbone-type packing in the (ab)-plane and a (001) glide plane, enhanced formation of dislocation lines is observed along the (100)||[010] Burgers vector. The quadratic scaling of the elastic energy with the respective Burgers vector makes the b-direction (the shortest distance between adjacent molecules) energetically favorable for the formation of dislocation lines.

In addition, the external conditions during crystal growth may have a significant impact on the defect density [25.19]. In melt-grown crystalline samples, the interaction of the crystal with the walls of the ampoule, especially during the cooling process, might induce dislocation densities three orders of magnitude higher compared with similar crystals grown from gas phase. However, subsequent thermal treatments might decrease the defect density by adding thermal energy to the system during annealing.

25.3 Organic Materials of Interest for Semiconducting Single Crystals

Organic semiconductors represent a large group of solids comprised of organic π-conjugated small molecules, oligomers or polymers. Since polymers are rarely available in single-crystalline form, the materials discussed in this chapter are limited to crystals of small molecules and oligomers.

Organic semiconductors are mostly used as active materials in field-effect transistors, light-emitting diodes or solar cells. These applications determine the selection of molecules of interest for research and production. Figure 25.1 presents examples of molecules with semiconducting properties used for research and applications.

Linear or planar fused-ring compounds, heterocyclic oligomers, and fullerenes are among the most studied molecules with semiconducting properties.

Single crystals of polycyclic aromatic hydrocarbons were among the first reported organic materials exhibiting small but measurable conductivity and photoconductivity. This suggested that these materials contain delocalized π-electrons and belong to the family of semiconducting compounds [25.20, 21].

These early studies required large single crystals for physical measurements. Therefore, the semiconductors selected for these measurements included molecules available in relative large quantity, stable at tempera-

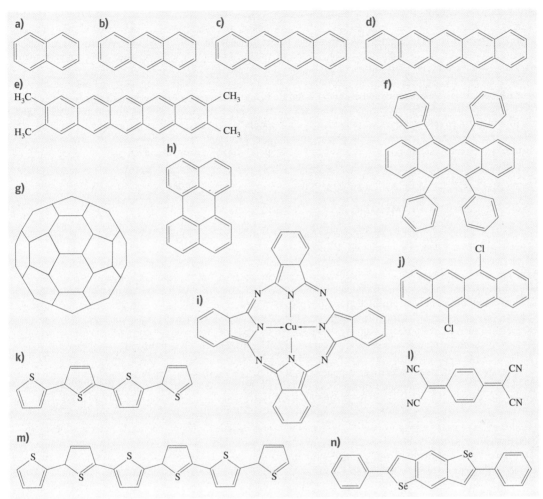

Fig. 25.1a–n Molecules of examples of the most studied organic semiconductors: (**a**) naphthalene, (**b**) anthracene, (**c**) tetracene, (**d**) pentacene, (**e**) 2,3,9,10-tetramethylpentacene (Me$_4$Pent), (**f**) rubrene, (**g**) fullerene (C$_{60}$), (**h**) perylene, (**i**) copper phthalocyanine, (**j**) 5,11-dichlorotetracene (DCT), (**k**) α-quaterthiophene (α-4T), (**l**) tetracyanoquinodimethane (TCNQ), (**m**) α-hexathiophene or α-sexithienyl (α-6T), (**n**) 2,6-diphenylbenzo[1,2-b:4,5-b']diselenophene (DPH-BDS)

tures close to their melting point, and grown in the form of crystalline ingots from the melt. Some materials, such as naphthalene, anthracene, stilbene, terphenyl, diphenylacetylene or quaterphenyl, have been chosen for crystal growth due to their strong scintillation properties [25.22]. Others, such as naphthalene, anthracene, and perylene, were selected for melt growth and time-of-flight measurements [25.12].

Oligothiophenes decompose below their melting temperature and can only be grown from gas phase in the form of small single crystals or thin films. These compounds are widely used in both single-crystal and thin-film forms for field-effect transistors [25.23, 24]. Hydrocarbons, such as rubrene, pentacene, and tetracene, are used to study the physical properties of organic semiconductors and for high-mobility transistors [25.8].

Fullerenes, such as C$_{60}$, and C$_{70}$, have received considerable attention due to their spherically delocalized π-electrons, yielding measurable conductivity and even superconductivity in alkali metal-C_{60} compounds [25.25].

Another group showing promise for practical application are cofacially stacked phthalocyanine systems and, often with central 3d transition-metal atom, porphyrines [25.26]. These compounds are exceptional due to their broad chemical variability and excellent stability. These properties, and the fact that they are available in sufficient quantity and quality, make porphyrines and phthalocyanines prominent candidates for efficient photovoltaic energy conversion.

The observation of light emission from organic diodes and the field effect in transistor configurations stimulated the design and development of many new organic π-conjugated molecules with semiconductor properties [25.27].

Many were synthesized and used only in thin-film field-effect transistors by evaporation or solvent-growth techniques. However, only a limited number of these compounds were grown in single-crystal form and only a few examples are known where the electronic transport properties were optimized.

25.4 Pregrowth Purification

For research purposes, batches of organic semiconductor materials are typically purchased from chemical suppliers. Some special, commercially unavailable molecules are synthesized by individual chemists. In both cases, the syntheses involve multistep organic chemical reactions yielding the expected molecules. Because the yield of an organic synthesis may be low, the desired molecules are often mixed with numerous, not always well-characterized, contaminants. Contaminants may be the molecules introduced as starting material, molecules formed by side reactions or decomposition processes, as well as solvents used in the process. In comparison with classical inorganic semiconductors, such as silicon or GaAs, where 6 N purity or better is available, organic semiconductors are significantly contaminated, with purity in the range of 95–98%. The contaminants are often not well defined and may form by reduction, oxidation, decomposition or photoinduced processes during storage. Therefore, before crystals can be grown, the source materials need to be purified.

The purification methods may be the same as those used for inorganic compounds: zone refining, sublimation or distillation. Additional, due to the solubility of organic semiconductors in organic solvents, recrystallization from organic solvents, or purification by gas- or liquid-phase chromatography, is feasible. The product of such purification requires subsequent removal of traces of solvents. Moreover, some specific contaminants, which are difficult to remove by traditional purification methods, may be altered chemically for effective removal. For instance, removing β-methylnaphthalene and thionaphthalene from naphthalene by fusing naphthalene with molten potassium, which chemically modifies the solid/liquid and solid/gas distribution coefficients, may be an example of chemically altering contaminants prior to removing them via sublimation and zone refining [25.28].

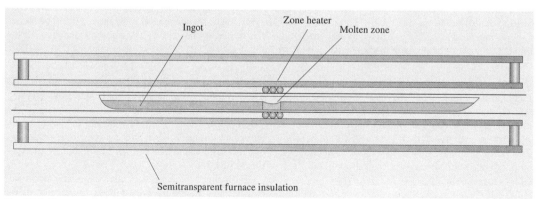

Fig. 25.2 Zone refining with one molten zone in a horizontal arrangement

Organic semiconductors, due to the peculiarity of the bonds, may not only decompose during storage but also dimerize and polymerize, especially if exposed to light. Anthracene is known to form dianthracene under illumination, but this dimer is not very stable and photolitically dissociates back to anthracene [25.29].

Pentacene, which has a mobility larger than $1\,\mathrm{cm}^2/(\mathrm{V\,s})$ in both thin-film and single-crystal field-effect transistors, easily oxidizes to 6,13-pentacenequinone [25.30], but also dimerizes and additionally reacts further to form a series of polycondensed aromatic hydrocarbons and undergoes a disproportionation reaction to 6,13-dihydropentacene [25.31].

25.4.1 Zone Refinement

Materials available in larger amounts, which can be melted without decomposition, such as anthracene, naphthalene or perylene, may be purified by repeated zone melting as the technique of choice. Established by *Pfann* for inorganic semiconducting materials [25.32], zone refinement was applied to anthracene [25.33, 34]. In this method, an ingot of about 20–50 cm length is formed and slowly dragged across a cylindrical heating zone of centimeter extension under an inert gas atmosphere. The inert gas on one hand protects the molten material from reaction with air and on the other hand reduces the evaporation of the material from the melted zone. The temperature of the heater is high enough to locally melt the material to form a short molten zone. Outside the heating zone the organic material remains in its solid phase (Figs. 25.2 and 25.3). A zone-refinement apparatus can be realized with an ingot fixed in either horizontal or vertical orientation. The heater moves along the fixed ingot, or alternatively, the ingot moves relative to fixed heater. To allow multiple melting and crystallization of zones along an ingot, a system with numerous heaters may be used (Fig. 25.4).

Due to the slow movement of the molten zone in one direction only, the purified material melts at one end of the zone and crystallizes at the other. Depending on the solubility of the respective impurities in the liquid and solid states, impurities with higher solubility in the melt than in the crystal are moved towards

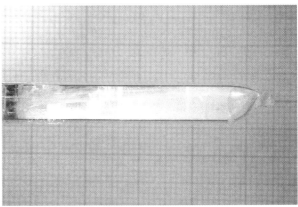

Fig. 25.3 Zone-refined anthracene. Contamination is moved to both ends of the ampoule, seen as a *dark* and a *light part* at each end of the anthracene ingot. Also, some cracks and crystal segments, formed due to crystal contraction during cooling, are seen

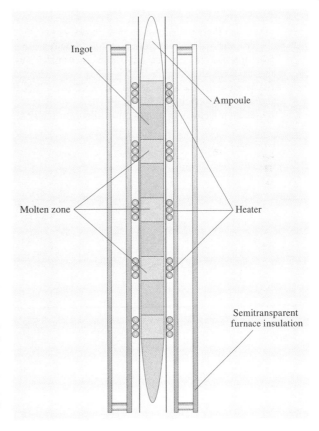

Fig. 25.4 Multizone refining system. To prevent evaporation of material from the melted zone, zone refining is run in vertical orientation. To decrease the time required for the process, a multizone furnace is used ▶

the molten interface, while impurities with lower solubility move to the crystallization interface. Therefore, the former impurities are pulled with the zone whereas the others are transferred to the opposite end of the ingot. Repetition of this process up to several hundred cycles will accumulate impurities on both ends of the ingot and pure material is collected from its central part (Fig. 25.3).

In the case that impurities have equal solubility in the melt and the solid, they cannot be separated via zone melting and need to be removed using other techniques such as sublimation or chemical precipitation. Some success has been reported by using an additional material forming an eutectic with the impurity. In this way the distribution coefficient is changed from an unfavorable value close to 1.0, to a smaller value which simplifies moving the impurities with the eutectic to the end of an ingot. For instance, tetracene, for eutectic zone melting purification, has been mixed with 2-naphthoic acid, while anthracene has been melted with benzoic acid. The zone refinement of such an eutectic mixture allows removing otherwise hard-to-remove impurities [25.35].

The purified material does not need to be removed from the zone-refining apparatus after each run of the molten zone along the ingot. Therefore, the purification process may be automatically repeated many times in the same ampoule [25.36]. Additionally, subsequent purification steps may be performed only on material collected from the central, purest part of the ingots. The purity of substances for which this method has been applied was superior, and low-temperature mobilities, evaluated by the time-of-flight method, were reported to be as high as a few hundred $cm^2/(V\,s)$ at low temperature. Because zone melting requires a large amount of material (on the gram scale) and a specific, noncommercial apparatus, this method is preferred for large-scale purification and has been used only for a limited number of organic semiconductors.

25.4.2 Sublimation and Its Modifications

For studies of new organic semiconductors, laboriously synthesized in only milligram amounts, standard vacuum sublimation (with various modifications) has been commonly used. In this method, material sublimes in vacuum in a temperature gradient between evaporation and deposition zones. If only a very small amount of the material is available, the deposition temperature needs to be significantly lower than the evaporation temperature. Under such conditions, easily evaporated molecules, such as most solvents, evaporate into the vacuum and most of the molecules deposit on some form of cold finger. Heavy molecules do not sublime at all and remain in the evaporation zone.

Vacuum sublimation in a temperature gradient is very efficient, but it often lacks selectivity and most

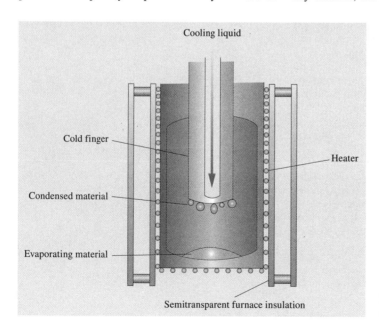

Fig. 25.5 Vacuum sublimation on a cold finger

molecules with comparable properties deposit simultaneously on cold surfaces.

Low-vapor-pressure molecules or solvents may be efficiently removed by sublimation. However, due to the weak van der Waals bonds between molecules, the volatility and evaporation enthalpies of organic molecules with comparable molecular mass are too close for efficient separation. It is worth mentioning at this point that this situation may become even more complex if the molecular species possess a permanent dipole moment, thereby significantly changing the sublimation enthalpies.

To improve the purification of organic semiconductors and change the mass transport kinetics, modified sublimation methods have been proposed. In one case, a carrier gas was introduced into the sublimation apparatus and a long deposition zone with a shallow temperature gradient from the evaporation zone to room temperature was used [25.13, 37]. In this way, the heavy impurity molecules, similar to vacuum sublimation, remain in the heating zone and the molecules with high vapor pressure (solvents, etc.) are removed from the deposition zone with the carrier gas stream. The carrier gas transports molecules with different molecular mass to different distances, allowing efficient separation of molecules with different mass.

This method is selective enough that it may even be used for separation of polymorphs or isomers. Polymorphs are crystalline substances formed from the same molecules but which crystallize in different structures and therefore possess different sublimation enthalpies.

Polymorphs of oligothiophenes, hexathiophenes, quaterthiophenes have been separated by this methods for crystal structure determinations.

Another modification of the sublimation technique is step sublimation. Here, the pristine material is placed in a boat at the end of a glass tube that is closed at

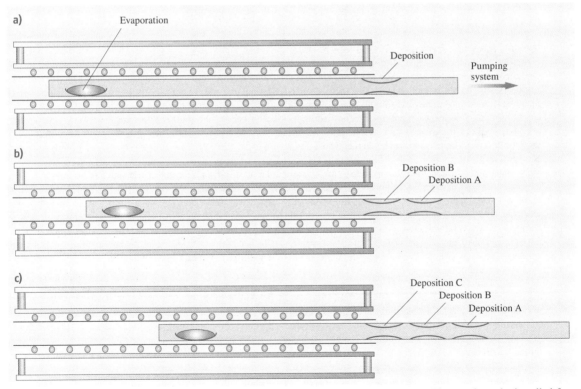

Fig. 25.6a-c The principle of step sublimation. The furnace is kept at constant temperature. The reaction tube is pulled from the furnace in such a way that material from the consecutive processes deposit at the positions marked A, B, C as depicted in (a-c)

one side. Connecting the open side to a pumping system, the tube is evacuated to pressures of 10^{-6} mbar and placed with the closed side into a transparent glass oven. Subsequent stepwise increases of the temperature, with the ingot partially translated out of the heated zone after each new temperature, allows for a detailed analysis of the respective contaminants subliming at different temperatures as well as the determination of the sublimation temperature of the intended material. Furthermore, the open arrangement of the setup allows for in situ optical investigations of the pristine material upon thermal treatment, e.g., for newly synthesized materials, changes between the solid and the liquid at a certain temperature can be observed. The technique can indicate changes in the chemical composition of the material as well as possible phase transitions. Due to its simplicity, this technique is well suited to estimate the thermal stability and sublimation conditions of newly synthesized molecules. Furthermore, this technique enables the purification of lager amounts of material compared with gradient sublimation and is therefore ideal as a prepurification step in combination with zone refinement.

In both of the above-described modifications the purified material needs to be removed after one purification step. For consecutive purification steps, the material needs to be reloaded into a new clean apparatus and the purification process may be repeated. To avoid this laborious and expensive procedure, a vapor zone-refining technique has been developed [25.35]. In this method, tetracene in the form of a long ingot was sublimed in an apparatus similar to a zone-refining system. The temperature, however, was kept below the melting temperature at which tetracene decomposes. Due to the high vapor pressure of tetracene, it evaporates from the short heated zone. Because some impurities may have lower and others higher vapor pressure than tetracene, the impurities are concentrated at both ends of the vapor zone-refining ingot. The efficiency of this purification method was confirmed through liquid-phase chromatography, but no other physical properties sensitive to small quantities of contaminants have been reported.

25.5 Crystal Growth

Organic compounds may be grown from the gas phase, solution or melt. Many well-established methods used for crystal growth of elements, (metals and nonmetals) or compounds (oxides, semiconductors or insulators) may be modified for growth of single-crystal organic semiconductors. The relative weak van der Waals bonds between molecules, the large dimensions of growth units, and the limited thermal stability of organic materials impose specific limitations on the use of these well-established techniques. In comparison with inorganic materials, melting temperatures of organic compounds are relatively low. Many small molecules appear even at room temperature as liquids and cannot be considered as practical semiconductors for room-temperature applications.

Thiophene (C_4H_4S), for example, has a melting temperature of only $-38\,°C$ and has not been reported as being used for electronic applications. Both 2,2′-bithiophene ($C_8H_6S_2$) and 2,2′:5′2″-terthiophene are solids at ambient temperatures, with melting temperatures of 32 and $93\,°C$, respectively. The larger oligothiophenes, i.e., quaterthiophene (four thiophene rings), quinquethiophene (five thiophene rings) [25.38], hexathiophene (sexithiophene, six thiophene rings) [25.24] or the longest synthesized octathiophene (eight thiophene rings), are all solid at room temperature and numerous authors report the use of these oligothiophene for field-effect transistors. The longest oligothiophene, α-8T is difficult to synthesize, as it needs high temperatures for evaporation but is not very stable around the gas-phase growth temperature of $340\,°C$ [25.39]. However the polymer polythiophene, and its derivatives, are soluble and are again very popular as thin-film field-effect-transistor materials.

Similar relations exist in the polycyclic aromatic hydrocarbons. Benzene is a liquid at room temperature; naphthalene is a volatile solid, anthracene melts at $210\,°C$, but photodimerizes under illumination. The longer hydrocarbons from the acene family, tetracene (naphthacene) and pentacene, are relatively stable solids with excellent semiconducting properties [25.40]. The next in the series, with six (hexacene) and seven (heptacene) conjugated benzene rings, are hard to synthesize and not very stable [25.41]. The polymer, an infinite linear chain of acenes or cyclic polyacenes, has not been synthesized [25.42].

The most frequently used single-crystal growth method for organic semiconductors are the numerous modifications of the vapor growth technique. Most organic molecules possess sufficient vapor pressure that the vapor growth method can be efficiently used. Some materials, such as anthracene, tetracene, pentacene, and rubrene, have been crystallized in sealed ampoules in the form of needles, platelets or even large bulk single crystals. However, most organic semiconductors have been grown by evaporation and condensation in a carrier gas stream. Oligothiophenes and acenes, for example, have been grown from the vapor phase, and resulting single crystals have been used for physical measurements as well as device preparation. Many organic molecules, designed and synthesized for exploration of their semiconducting properties, have been evaporated during heating and crystallized in the form of small (only a few to tens of micrometers long) crystals. Such small crystals may have sufficient quality that the crystal structure can be determined by diffraction methods (where dimensions of the order of $50 \times 50 \times 5\,\mu m^3$ are needed).

For some applications, such as scintillation detectors, where large crystals of a few cubic centimeters are required, the melt growth techniques are more practical. However, for other applications such as field-effect transistors, large crystals need to be cut and polished before preparing devices on their surfaces. Time-of-flight measurements revealed that the bulk quality of large organic crystals may be superior compared with at their surfaces. Unfortunately, due to the requirement for stability of the material up to the melting point, very few organic semiconductors have been grown from the melt.

Though the large molecules of oligothiophenes, acenes, and many others are excellent organic semiconductors, they have low vapor pressure, making it difficult to use them in gas-phase transport setups, and they often decompose before melting, excluding melt growth techniques. Additionally they show only limited solubility in solvents. These large molecules are sometimes functionalized with long alkane chains to improve solubility, but the substitution creates new molecules and therefore changes the physical properties of the semiconductor. Thus, due to their low solubility, crystals can only be grown from solvents by using high pressures and temperatures in a solvothermal method. However, to date, the quality of solvothermal-grown crystals was not optimized and the quality was insufficient for practical applications.

25.5.1 Melt Growth

Generally, materials that may be purified by zone melting may be crystallized from a melt by either the Bridgman or Czochralski method. In the Bridgman method, the material is melted in a container in a temperature gradient in such a way that crystallization starts at the lowest point of a narrowed ampoule. In most practical Bridgman processes the ampoule is sealed; as a result, the vapor cannot escape from the system. The growth of crystals may be initiated spontaneously or on a small oriented single-crystalline seed.

In the Czochralski method, the material is melted in a crucible at a temperature slightly above the melting point. A seed crystal, which is just below the melting temperature, is pulled from the melt. The crystallization begins on an oriented monocrystalline seed. During the crystal growth process, the melt evaporates from the crucible during the crystal pulling and the vapors continuously escape from the crucible. There-

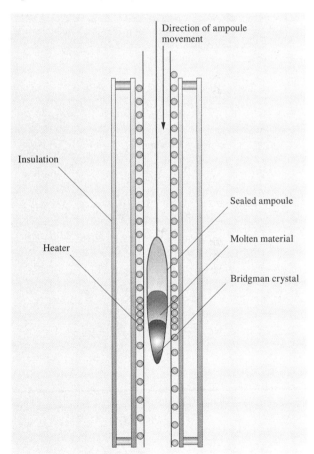

Fig. 25.7 The principle of the Bridgman crystal growth system in a sealed ampoule. A glass ampoule with molten material is moved through a temperature gradient. The crystal nucleates at the tip of the ampoule and grows from the melt when the lower part of the ampoule moves to the colder part of the furnace

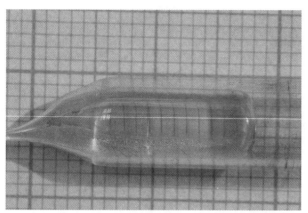

Fig. 25.8 Anthracene crystal grown by the Bridgman method inside a glass ampoule; the crystal sticks to the ampoule wall and is shown before removal from the ampoule

fore this method is suitable only for low-vapor-pressure materials.

Organic Semiconductors Grown by the Bridgman Technique

Due to their thermal stability upon melting, low-weight acenes such as naphthalene, anthracene [25.43, 44], 2,3-dimethylnaphthalene [25.45], perylene [25.46], and phenanthrene [25.47] have been grown by the Bridgman technique. Similar to zone refinement, the material is placed in an ampoule sealed under vacuum or at a pressure slightly lower than ambient; for example, 400 Torr of N_2, at room temperature, may be used. In a vertical oven with two different temperature zones, the organic compound is melted in the upper, hotter zone and then slowly lowered into the cooler region for crystallization. Depending on the organic materials, the growth speed can vary over a range from 0.1 mm/h to 0.1 mm/min.

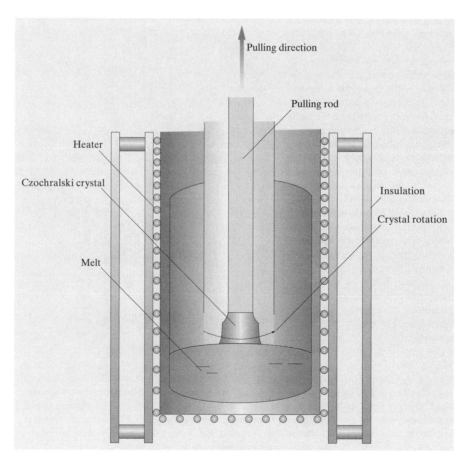

Fig. 25.9 A sketch of the Czochralski apparatus used for the growth of organic material. The pulling rod is slightly cooler than the melt. The heat is removed from the growing crystal through the pulling rod. The crystal is pulled from the melt. The vapor can escape during the growth process through the opening between pulling rod and the ampoule

To avoid possible formation of structural defects caused by mechanical vibrations during the process, the oven is moved instead of the ampoule. The resulting crystals are on the centimeter scale and provide a degree of purity that is as high as the pristine material, and sometimes even better due to the additional refinement cycle during growth.

A schematic diagram of the Bridgman crystal growth apparatus is shown in Fig. 25.7, and an anthracene single crystal grown by this method is shown in Fig. 25.8. A Pyrex or quartz glass ampoule containing a prepurified organic semiconductor is evacuated under vacuum, purged with inert gas, and sealed under vacuum or under reduced inert-gas pressure. The ampoule is heated slightly above the melting point of the material and then lowered through the temperature gradient.

Organic Semiconductors Grown by the Czochralski Technique

Crystals grown by the Bridgman method as well as ingots crystallized during zone refining suffer from cracks and stresses caused by contact with the walls of a container. To avoid this effect, the Czochralski technique was used for benzile, benzophenone, and stearic acid. The scheme of this method is presented in Fig. 25.9. A small seed crystal is slightly immersed into the melt and slowly rotated and pulled from the melt. The melt crystallizes on this seed, forming single crystals with the same orientation as the seed. The high vapor pressure of most organic compounds limits the usability of this method to a few organic semiconductors even though the quality of these crystals compares favorably with the perfection obtained by Bridgman or solution-grown crystals [25.48].

25.5.2 Growth from the Gas Phase

The most popular method of growing single crystals of many organic semiconductors is growth by gas-phase transport. The significant vapor pressure of van der Waals bonded materials [25.49] at moderate temperatures and the possible high purity and quality of vapor-grown crystals favor this method for the growth of single crystals for research purposes.

Even though the gas-phase growth technique is rather simple, the processes involved in evaporation and crystal formation are quite complex. The growth of crystals may be limited by processes occurring in the evaporating material, the gas-phase transport between evaporation and growth zone or on the surfaces of the crystal itself. Crystals may grow close to thermodynamic equilibrium within their own vapor, or far from equilibrium when the gas-phase transported molecules condense very fast or in the presence of an inert gas where molecules diffuse through the surface layer of adsorbed gas.

The growth unit of a molecular crystal is a single molecule, much larger than the growth unit of most inorganic crystals. Therefore, large molecules require additional energy to reorient, or even reconstruct, before incorporation into kinks or steps on the crystal surface.

For some molecules, the required rearrangement has been experimentally observed. Rubrene, which shows the highest charge carrier mobility at room temperature, crystallizes very poorly or as an amorphous layer if the growth temperature is low. The molecules strike a surface and stick to it where they hit, forming an amorphous conglomerate [25.16]. However, by increasing the growth temperature, the crystalline quality of thin films significantly improves [25.16]. To achieve a high enough growth temperature during condensation of rubrene thin film, a semiclosed method, i.e., hot-wall epitaxy, allowing growth at high vapor pressure, was used. Even higher growth temperatures are practical for rubrene and large well-formed rubrene single crystals are routinely obtained.

The presence of inert gas, in either an open or sealed system, radically changes the growth modus. *Horowitz*

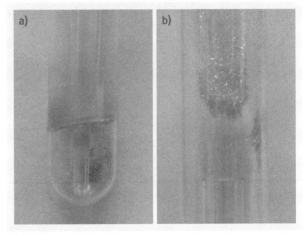

Fig. 25.10a,b α-Hexathiophene grown from the vapor phase. (a) Powder evaporated from the lower part of the ampoule, deposited in vacuum on the wall in the form of a fine crystalline powder. No crystals are observed. (b) When the ampoule was filled with 1 atm of an inert gas, up to centimeter-size crystals grew at the colder part of the ampoule

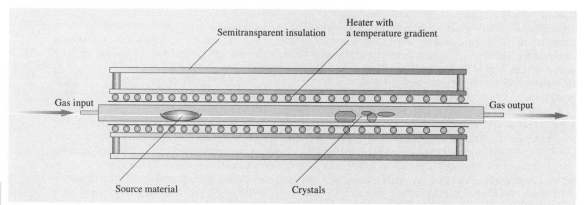

Fig. 25.11 The principle of a vapor-phase transport crystal growth system. The inert gas flows through the reactor, causing transport of molecules from high temperature, where the material evaporates, to the low temperature, where the material deposits. The high vapor pressure impurities are transported with the gas stream out of the furnace and the heavy contaminants remain at the evaporation zone. The presence of an inert gas modifies the surface processes, resulting in spontaneous formation of single crystals

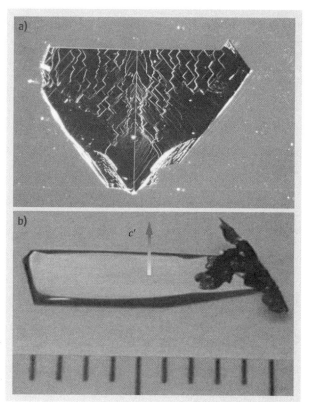

et al. reported α-hexathiophene single crystals grown at a few mbar of argon [25.50].

However, α-hexathiophene, during evaporation in vacuum, deposits as a powder, independent of the temperature gradient. At higher temperature, the vacuum evaporation is very fast and molecules are not able to rearrange before attaching to the growth sides and thus form a powder rather than crystals. After introducing an inert gas into the ampoule or applying a gas flow in an open system, the evaporation rate was significantly reduced, allowing even higher temperatures for evaporation and crystal growth. The molecules gain additional thermal energy, allowing surface diffusion to the nearest growth steps. There the molecules rearrange before forming single-crystalline platelets with well developed faces [25.37]. The same improvement in crystal growth was observed for quaterthiophene [25.51, 52].

At temperatures only slightly lower than the melting point, the occurrence of a rough or even quasiliquid surface layer is assumed. This layer influences the

Fig. 25.12a,b Examples of crystals grown from vapor phase. (**a**) Tetramethylpentacene grown in an inert gas stream: growth steps are visualized due to the use of illumination at shallow angle of incidence. (**b**) Tetracene grown in a closed ampoule: monolayer-high growth steps occur on the crystal surface (steps not visualized) ◄

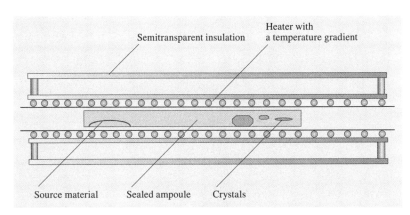

Fig. 25.13 Vapor-phase growth in a sealed glass ampoule. The ampoule may be filled with an inert gas under controlled pressure. The impurities cannot be removed from the ampoule during the growth process

adsorption, migration, surface diffusion, surface rearrangement, and nucleation of molecules and helps in the attachment of molecules to the crystal surface [25.53].

Various modifications of the gas-phase crystal growth technique have been described. They may be divided into processes run in sealed ampoules or in open tubes. Both processes, in a sealed or open tube, may run either in vacuum or in an inert gas atmosphere. In a sealed ampoule, the starting material is enclosed in a glass ampoule and the amount of the material is not changed during the process. However, due to chemical reactions, such as decompositions, photoreactions, polymerization, and others, the chemical composition of the material in a sealed ampoule may change. In an open tube system, a carrier gas is injected through one end of the tube and escapes from the other. The material for crystallization is located in the heated part of the tube, evaporates from there at elevated temperature, and condenses in the colder part of the tube [25.54]. The crystallizing material may undergo similar reactions as in the sealed ampoule. It may further react with the carrier gas and escape with it from the growth tube. The resulting composition and the amount of the material in the system may therefore change during the growth process.

Both variations of gas-phase crystal growth require only small quantities of the starting material. Therefore, many organic compounds have been grown in the form of small needles or platelets. These crystals, while often not large enough for device preparation, are large enough for x-ray diffraction analysis and structure determination. However to grow single crystals large enough for device fabrication, the process parameters need to be further optimized. To keep the temperature gradient between the evaporating and growing material small, sealed ampoules have frequently been used. To lower the evaporation rate and run the growth processes at higher temperatures, an inert gas has been introduced in both sealed and open systems. To purify the material during growth, the open system is preferred, as impurities are carried away from the growing crystal by a stream of inert gas.

Sublimation in sealed ampoules filled with an inert gas has been efficiently used for growth of anthracene, chrysene, diphenyl, p-terphenyl, acridine, pyrene, naphthalene, phenanthrene, and benzanthracene. Fluorescence measurements have been performed on such crystals [25.55].

Rubrene and pentacene crystals have been grown in a small temperature gradient formed in a vacuum sealed glass ampoule [25.31, 56]. In both cases growth temperatures were lower than the corresponding temperatures for growth in the gas stream. Also, the crystals produced by vacuum growth were much thicker because the surface migration of adsorbed molecules was limited at lower temperatures. However, the same

Fig. 25.14 C_{60} crystal grown in a sealed quartz-glass ampoule

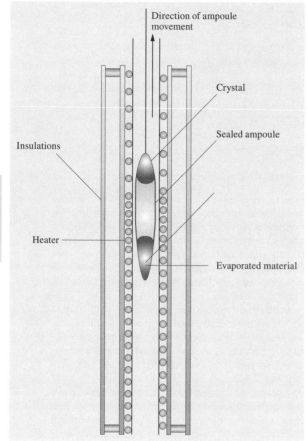

Fig. 25.15 Schematic diagram of the gas-phase growth of crystals while pulling the ampoule through a temperature gradient. Crystal nucleates in the narrow tip of the ampoule during pulling of the ampoule from the high-temperature zone in the lower part of the furnace to the colder, upper part of the furnace

Fig. 25.16 Tetracene crystal formed in the tip of an ampoule while pulling the ampoule through a temperature gradient

growth technique was absolutely inadequate for oligothiophenes, where a gas flow is required to keep a high temperature and to remove impurities from the system.

Larger crystals may be obtained from the gas phase by pulling a sealed glass ampoule up through a temperature gradient (the inverse of the Bridgman technique). In this technique, the material evaporates from the lower, hotter part of the ampoule and condenses at a narrow tip of an ampoule pulled into the colder part of the furnace. The conical tip of the ampoule is filled by the growing crystal. Tetracene single crystals have been successfully grown by pulling the ampoule with a speed of 5 mm/day [25.57].

During the growth in a sealed ampoule, even if pre-purified materials are used, some impurity molecules may be formed by decomposition, and some impurity molecules intercalated in the source material may be discharged during heating. To avoid contamination of the growing crystals, a part of the vapor may be removed through a small effusion hole placed close to the conical tip of the ampoule. This method, known as semi-open physical vapor transport, has been successfully used to grow nonlinear optical crystals [25.58, 59], but has not been explored for organic semiconductors.

One of the striking features of sublimation-grown crystals is their morphological habit. Many of the polyacenes, rubrene, and perylene derivatives show a preferred two-dimensional growth with aspect ratio between lateral extension and height of the order of 10^2-10^4. Such crystals are most suitable for transport studies carried out in field-effect transistor (FET) geometry by preparing metal contacts by thermal evaporation after growth. Since these measurements are governed by the charge carrier transport along the topmost layers of the crystal surface, contamination after growth by, for instance, photo-oxidation, has to be minimized by sample handling in a glovebox, under controlled lighting, etc.

The results of some field-effect transistor studies on organic single crystals with mobility values equal to or larger than $1 \text{ cm}^2/(\text{V s})$ are presented in Table 25.1.

Table 25.1 Organic semiconductors with a single-crystal field-effect transistor mobility above $1\ \text{cm}^2/(\text{V s})$

Structure	Mobility [$\text{cm}^2/(\text{V s})$]	Reference
Pentacene	2.2	*Roberson* et al. [25.31]
Rubrene	13	*Zeis* et al. [25.56]
Copper phthalocyanine	1.0	*Zeis* et al. [25.60]
2,6-Diphenylbenzo[1,2-*b*:4,5-*b'*]diselenophene	1.3	*Zeis* et al. [25.61]
Dichlorotetracene	1.6	*Moon* et al. [25.62]

25.5.3 Solvent-Based Growth Methods

The solubility, vapor pressure, and stability of molecules declines with increasing mass of the organic semiconductor. Therefore melt or gas-phase crystal growth methods are unsuitable for large molecules, since the molecules tend to break apart before melting or evaporation. The only growth method available in such cases is solution growth close to room temperature. In analogy to inorganic materials, the solubility may be increased by high pressure and moderate temperatures below the melting point, similar to quartz crystals, which are grown by the hydrothermal method a few hundred degrees below the melting point of SiO_2.

This concept has been used for organic materials [25.63], where a pressurized solvent (instead of water), was used. With this solvothermal method, it was shown that the practically insoluble material, hexathiophene, is well soluble in benzyl phenyl sulfide sealed in a glass ampoule. During cooling, hexathiophene precipitates to form small platelets. The crystal growth process has not been optimized, and resulting crystals were only a few hundred micrometers in dimensions, insufficient for transistor fabrication or transport measurement.

Small or polar molecules were successfully grown from organic solvents. During solvent evaporation or during cooling of saturated solutions, the material precipitates in the form of crystalline powders or, sometimes, small crystals. This is the method of choice for growth of small crystals for x-ray crystal structure determination. However, the solvent may intercalate or even form ordered mixed crystals. Larger crystals of organic semiconductors, such as anthracene and pyrene, were also grown from solvents by solvent evaporation or by temperature lowering and have been used for semiconducting parameter evaluation. Anthracene and pyrene crystals of a few millimeters, formed by slow evaporation of a solvent, were successfully used for dark conductivity and photoconductivity measurements [25.46].

Anthracene crystals with up to $2\,cm^2$ in area have been grown while suspended in a solvent. The essential feature of this method is the choice of the solvent, ethylene dichloride, with a density that is slightly greater than the density of anthracene at room temperature. Since the density of the solvent decreases faster than the density of anthracene upon heating, it is possible to find a temperature at which both solvent and anthracene density are equal and the crystal can be suspended and grown in solution. Additionally, a layer of xylene (a poor solvent for anthracene) was placed on the top of the anthracene/ethylene chloride solution. Xylene prevents ethylene dichloride from evaporating and further diffuses slowly into the ethylene dichloride and causes a decrease in anthracene solubility, thus creating supersaturation. Under such conditions, crystals can grow undisturbed while suspended in the solvent [25.64].

Further, large crystals of organic materials for nonlinear optics applications have been grown from solvents [25.65]. However, these crystals are transparent large-gap insulators and do not show properties typical of organic semiconductors; they are therefore not discussed in this chapter.

25.6 Quality of Organic Semiconducting Single Crystals

To understand the electronic properties of organic semiconductors, high-quality single crystals are required. In the past the structural quality of gas-phase-grown anthracene and Bridgman-grown 2,3-dimethyl naphthalene were studied with x-ray topography [25.19]. Because anthracene grew in the form of very thin platelets that were strongly warped, x-ray topography was severely hampered. Large, ultrapure Bridgman-grown anthracene as well as 2,3-dimethyl naphthalene single crystals were used for time-of-flight measurement of charge carrier mobilities. The x-ray topographs of these crystals revealed that defects were formed after the crystal growth process, during cooling of the crystals from the growth temperature to room temperature, during removal from ampoules, and during sample preparation for measurements. However, if properly and carefully handled, the quality of cleaved lamellae prepared from Bridgman-grown ingots displayed a highly perfect mosaic with only a few dislocations appearing, mostly at the side where the cleavage process caused stress and glide dislocations [25.19].

Double-crystal x-ray topography was performed on rubrene single crystals grown from the vapor phase [25.66]. In comparison with other organic semiconductors that crystallize in lower symmetries (monoclinic or triclinic), rubrene crystallizes in the orthorhombic system. Besides thin platelets, rubrene forms also thicker crystals suitable for structure analysis. The thermal expansion is not very anisotropic. The rocking curve width of a few crystals (selected under polarized light from a large batch grown in numerous growth runs) revealed a full-width at half-maximum (FWHM) of the (12 0 0) reflection of 0.013° for the best samples. Such a narrow rocking curve was also observed for the (004) reflection, but the FWHM was twice as large for the (020) reflection. Such a low mosaic spread is among the best exhibited by organic crystals. However, many other measured rubrene crystals displayed larger FWHMs of the order of 0.15°. The mechanism of forming such mosaic spreads in these crystals is not known. Furthermore, it is assumed that both the growth process and, more probably, any temperature gradient formed during the crystal cooling are responsible for defect formation in these crystals, often with an inhomogeneous distribution of dislocations.

Rubrene single crystals show high charge carrier mobility, close to ten times larger than values seen in other organic semiconductors such as tetracene or pentacene. However, the mechanisms by which defects, impurities, and intra- and intermolecular structure impact on charge carrier mobility is not clear as of today.

25.7 Organic Single-Crystalline Field-Effect Transistors

At the foundation of organic semiconductors as an active element of plastic electronics is the ability to use assemblies of molecules in FETs. Charge carrier transport has attracted attention for decades and was often studied using thin-film FET structures. In contrast, only a limited number of groups have focused on intrinsic electronic properties of single-crystal organic semiconductors. While the measurements on thin films, which are the configurations used for applications, are useful and indicate the potential of a particular organic compound, the presence of grain boundaries, impurities, and defects obscures the intrinsic performance that one can expect to achieve from a specific material. The alternative approach for evaluation and selection is the fabrication of an FET structure directly on a single crystal. In this way, the most efficient π-conjugated molecules can be selected from among the many commercially available organic compounds. The latter method allows better evaluation of intrinsic properties of organic semiconductors because the optical, electrical, and other properties are directly dependent on molecular properties of individual molecules that self-assemble (crystallize) on a nanoscale into well-ordered oriented grains, macroscopically defining a single crystal.

FET structures prepared on the natural surfaces of single crystals were described previously [25.8]. Two general variations are used. In the first, the source and drain electrodes, dielectric layer, and gate electrodes are fabricated by different techniques on a silicon wafer or plastic foil, and the single crystal is *stamped* (placed) onto such a device structure. In the second variation, the source and drain contacts are evaporated or printed onto the natural surface of a single crystal, then the dielectric is spun on or evaporated onto the crystal, and finally the gate electrode is fabricated on the dielectric layer. In the first method a strongly doped silicon wafer covered by an insulating layer of silicon oxide (SiO_2) with deposited metallic (mostly gold) electrodes

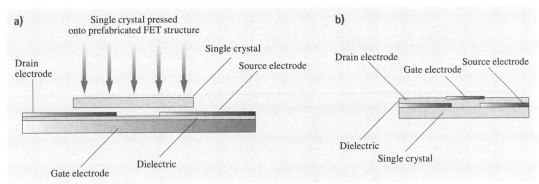

Fig. 25.17a,b Schematic diagram of two types of single-crystal field-effect transistor fabrication. (**a**) The FET electrodes and dielectric are prefabricated and a single crystals is attached by adhesion to the FET structure. (**b**) Electrodes and dielectric are built up step by step on the natural surface of a single crystal

is used as a support for crystals, which are gently affixed by adhesion to form the channel region between the source and gate electrodes. Good adhesion between the prefabricated FET support and crystal is crucial for proper operation of such a device. A similar concept of a prefabricated FET structure has been realized by using either a transparent indium oxide gate on glass or flexible foil and spun-on polymers as a gate dielectric. Again, strong adhesion between the electrodes, polymers, and crystals is crucial and may be controlled through chemical treatment of the polymer. An interesting modification of the last structure is an elastomeric stamp where the gate dielectric is replaced by an air gap. Since no direct contact between the dielectric (here air, inert gas or vacuum) and semiconductor surface occurs, the natural surface of the semiconductor may be studied and effects of the atmosphere on the FET properties may be observed.

For the second method, contact metals for p-type semiconductors are preferably gold or silver or low-work-function metals such as magnesium in the case of n-type materials. The lowest contact resistance on rubrene was obtained with painted contacts from a water-based solution of colloidal graphite. The gate insulation layer may consists of $0.5-2\,\mu\text{m}$ of parylene, prepared from the dimer Parylene N or C. The dimer is evaporated at $160\,°\text{C}$ and cracked to a monomer at $680\,°\text{C}$ that subsequently polymerizes on the surface of a crystal kept at room temperature. The gate electrode may either be evaporated or painted (colloidal graphite) directly on the parylene layer over the channel area between source and drain contacts.

In both FET structures, the semiconductor properties are tested only in the very thin surface layer of the single crystals (or thin films in a thin-film FET). However, this surface layer may be easily affected by intercalation/adsorption of gases present in the transistor environment. In this way, chemical sensors may be constructed, where the molecules are introduced into the channel area of the FET. Additionally, with the exception of the air-gap structure, the channel area may be modified/affected by the direct contact with a gate dielectric.

The highest FET mobility of $20\,\text{cm}^2/(\text{V s})$ in p-type rubrene or $1.6\,\text{cm}^2/(\text{V s})$ in n-type tetracyanoquinodimethane (TCNQ) [25.67] has been obtained in such devices. However, since FET structures probe only the top surface layers, the charge carrier mobility obtained in this way is not necessarily identical to the mobilities measured in the bulk.

The surface charge transport mobility generally decreases with decreasing temperature, an effect attributed to trapping states present at the surface. The bulk charge transport values of electrons and holes may be much higher, as was demonstrated by time-of-flight measurements in naphthalene, anthracene or perylene at low temperatures. In these cases the superior transport properties were attributed to meticulous purification and bulk crystal growth optimization as well as the insensitivity of the bulk properties on chemically poorly defined surface states.

In any case, electronic transport measurements are indispensable tools in evaluating the electronic traps affecting the charge transport in organic semiconductors. Together with optical methods and x-ray topography and diffraction to measure structural quality and elemental/molecular purity, the overall crystal quality of organic semiconductors can be efficiently assessed.

25.8 Conclusions

Single crystals of organic semiconductors have been grown mostly for basic research studies, for structure evaluation as well as for evaluation of maximum performance of organic semiconductors devices. Many purification and crystal growth methods have been applied to organic single crystals. Gas-phase transport is often used for growth of crystals used for field-effect transistors and x-ray structure determinations, whereas the Bridgman method dominates the growth of large crystals. So far, the highest electron and hole mobilities measured by the time-of-flight method have been achieved on crystals purified by extensive zone refining and grown from the melt by the Bridgman method.

However, the highest field-effect mobilities have been achieved on crystals purified by sublimation and grown from the vapor phase. The difference in mobilities between the bulk and surface measurements is attributed to different types of defects and contamination in the bulk and on the surfaces. These measurement values still reflect extrinsic effects and do not yet measure the intrinsic properties of organic semiconductor.

As of today, there are not enough studies of quality, purity, defect concentration, and optical and electrical properties on the same material grown by different methods. On the one hand this is a result of the instability of organic compounds and associated changes of their properties with time, while on the other hand it is challenging to use specific analysis methods for precise characterization of brittle crystals where defects alter with aging. Organic semiconductors need defect and impurity evaluation at part-per-million levels, similar to inorganic semiconductors. Even though organic semiconductors are much less studied than their inorganic counterparts, it seems that organic semiconductors will be an important component of future microelectronics. Organic light-emitting diodes (OLEDs) and displays are already commercially available. However, progress in organic material characterization is needed to guide improvements in purification and crystal growth that will consequently lead to the development of new, better organic semiconductors and higher-performance devices.

References

25.1 A. Byk, H. Borck: Photoelektrische Versuche mit Anthracen, Verh. Dtsch. Phys. Ges. **12**(15), 621–651 (1910)

25.2 J.B. Birks: *Scintillation Counters* (Pergamon, London 1953)

25.3 S.R. Forrest: The path to ubiquitous and low-cost organic electronic applications on plastic, Nature **428**, 911–918 (2004)

25.4 E. Reichmanis, H. Katz, C. Kloc, A. Maliakal: Plastic electronic devices: from materials design to devices applications, Bell Labs Tech. J. **10**, 87–105 (2005)

25.5 G. Horowitz, M. Hajlaoui: Mobility in polycrystalline oligothiophene field-effect transistors dependent on grain size, Adv. Mater. **12**, 1046–1050 (2000)

25.6 T.W. Kelley, C.D. Frisbie: Gate voltage dependent resistance of a single organic semiconductor grain boundary, J. Phys. Chem. B **105**, 4538–4540 (2001)

25.7 N. Karl: Charge-carrier mobility in organic crystals. In: *Organic Electronic Materials, Springer Series in Material Science*, Vol. 41, ed. by R. Farchioni, G. Grosso (Springer, Berlin Heidelberg 2001) pp. 283–326, Part II: Low Molecular Weight Organic Solids

25.8 M.E. Gershenson, V. Podzorov, A.F. Morpurgo: Colloquium: Electronic transport in single-crystal organic transistor, Rev. Mod. Phys. **78**, 973–989 (2006)

25.9 D.T.J. Hurle (Ed.): *Handbook of Crystal Growth* (Elsevier, Amsterdam 1994)

25.10 R.A. Laudise: *The Growth of Single Crystals* (Prentice-Hall, Englewood Cliffs 1970)

25.11 F.R. Lipsett: On the production of single crystals of naphthalene and anthracene, Can. J. Phys. **35**, 284–298 (1957)

25.12 N. Karl: High purity organic molecular crystals. In: *Crystals Growth, Properties and Applications*, Vol. 4, ed. by H.C. Freyhardt (Springer, Berlin Heidelberg 1980) pp. 1–100

25.13 R.A. Laudise, C. Kloc, P.G. Simpkins, T. Siegrist: Physical vapour growth of organic semiconductors, J. Cryst. Growth **187**, 449–454 (1998)

25.14 R.W.I. de Boer, M.E. Gershenson, A.F. Morpurgo, V. Podzorov: Organic single-crystal field-effect transistors, Phys. Status Solidi (a) **201**, 1302–1331 (2004)

25.15 A.T. Tripathi, M. Heinrich, T. Siegrist, J. Pflaum: Growth and electronic transport in 9,10-diphenyl-anthracene single crystals – An organic semiconductor of high electron and hole mobility, Adv. Mater. **19**(16), 2097–2101 (2007)

25.16 D. Käfer, G. Witte: Growth of crystalline rubrene films with enhanced stability, PhysChemChemPhys **7**, 2850–2853 (2005)

25.17 M.-C. Blüm, E. Ćavar, M. Pivetta, F. Patthey, W.-D. Schneider: Conservation of chirality in a hierarchical supramolecular self-assembled structure with pentagonal symmetry, Angew. Chem. Int. Ed. **44**, 5334–5337 (2005)

25.18 A. Gavezotti, G.R. Desiraju: A systematic analysis of packing energies and other packing parameters for fused-ring aromatic hydrocarbons, Acta Cryst. B **44**, 427–434 (1988)

25.19 H. Klapper: X-ray topography of organic crystals, Growth Cryst. **13**, 109–162 (1991)

25.20 O.H. LeBlanc: Hole and electron drift mobilities in anthracene, J. Chem. Phys. **33**, 626 (1960)

25.21 R.G. Kepler: Charge carrier production and mobility in anthracene crystals, Phys. Rev. **119**, 1226–1229 (1960)

25.22 J.B. Birks: *Scintillation Counters* (Pergamon, London 1953)

25.23 G. Horowitz, D. Fichou, X.Z. Peng, Z.G. Xu, F. Garnier: A field-effect transistor based on conjugated alpha-sexithienyl, Solid State Commun. **72**, 381–384 (1989)

25.24 G. Horowitz, F. Garnier, A. Yassar, R. Hajlaoui, F. Kouki: Field-effect transistor made with a sexithiophene single crystal, Adv. Mater. **8**, 52–54 (1996)

25.25 G.S. Hammond, V.J. Kuck (Eds.): *Fullerenes, Synthesis, Properties, and Chemistry of Large Carbon Clusters* (ACS, Washington 1992)

25.26 F.H. Moser, A.L. Thomas (Eds.): *The Phthalocyanines, Properties* (CRC, Boca Raton 1983)

25.27 A. Facchetti, H.E. Katz, T.J. Marks, J. Veinot: Organic semiconductor materials. In: *Printed Organic and Molecular Electronics*, ed. by D. Gamota, P. Brazis, K. Kalyanasundaram (Kluwer, Dordrecht 2004)

25.28 W. Warta, R. Stehle, N. Karl: Ultrapure, high mobility organic photoconductors, Appl. Phys. A **36**, 163–170 (1985)

25.29 E.A. Chandross, J. Ferguson: Photodimerization of crystalline anthracene. The photolytic disscociation of crystalline dianthracene, J. Chem. Phys. **45**, 3564–3567 (1966)

25.30 O.D. Jurchescu, J. Baas, T.M. Palstra: Effect of impurities on the mobility of single-crystal pentacene, Appl. Phys. Lett. **84**, 3061–3063 (2004)

25.31 L.B. Roberson, J. Kowalik, L.M. Tolbert, C. Kloc, R. Zeis, X. Chi, R. Fleming, C. Wilkins: Pentacene disproportionation during sublimation for field-effect transistors, J. Am. Chem. Soc. **127**, 3069–3075 (2005)

25.32 G. Pfann: Principles of zone-melting, J. Metals **4**, 747–753 (1952)

25.33 K.H. Probst, N. Karl: Energy levels of electron and hole traps in the band gap of doped anthracene crystals, Phys. Status Solidi (a) **27**, 499–508 (1975)

25.34 K.H. Probst, N. Karl: Energy levels of electron and hole traps in the band gap of doped anthracene crystals, Erratum, Phys. Status Solidi (a) **31**, 793 (1975)

25.35 G.J. Sloan, A.R. McGhie: Purification of tetracene: vapor zone refining and eutectic zone melting, Mol. Cryst. Liq. Cryst. **18**, 17–37 (1972)

25.36 M. Brissaud, C. Dolin, J. Le Duigou, B.S. McArdle, J.N. Sherwood: The purification and growth of large ultra-pure plastic single crystals of trimethylacetic acid, J. Cryst. Growth **38**, 134–138 (1977)

25.37 C. Kloc, P.G. Simpkins, T. Siegrist, R.A. Laudise: Physical vapor growth of centimeter-sized crystals of α-hexathiophene, J. Cryst. Growth **182**, 416–427 (1997)

25.38 M. Melucci, M. Gazzano, G. Barbarella, M. Cavallini, F. Biscarini, P. Maccagnani, P. Ostoja: Multiscale self-organization of the organic semiconductor α-quinquethiophene, J. Am. Chem. Soc. **125**, 10266–10274 (2003)

25.39 D. Fichou, B. Bachet, F. Demanze, I. Billy, G. Horowitz, F. Garnier: Growth and structural characterization of the quasi-2-D single crystal of α-octithiophene, Adv. Mater. **8**, 500–504 (1996)

25.40 J. Cornil, J.P. Calbert, J.L. Bredas: Electronic structure of pentacene single crystal: relation to transport properties, J. Am. Chem. Soc. **123**, 1250–1251 (2001)

25.41 W.J. Bailey, C.-W. Liao: Cyclic dienes XI. New syntheses of hexacene and heptacene, J. Am. Chem. Soc. **77**, 992–993 (1955)

25.42 S. Kivelson, O.L. Chapman: Polyacene and a new class of quasi-one-dimensional conductors, Phys. Rev. B **28**, 7236–7243 (1983)

25.43 H. Mette, H. Pick: Elektronenleitfähigkeit von Anthracen-Einkristallen, Z. Phys. **134**, 566–575 (1953), in German

25.44 H. Pick, W. Wissman: Elektronenleitung von Naphthalin-Einkristallen, Z. Phys. **138**, 436–440 (1954), in German

25.45 M. Tachibana, K. Kono, M. Shimizu, K. Kolima: Growth and dislocation characteristics of organic molecular crystals: 2,3-Dimethylnaphthalene, J. Cryst. Growth **198/199**, 665–669 (1999)

25.46 H. Inokuchi: Semi- and photo-conductivity of molecular single-crystals. Anthracene and pyrene, Bull. Chem. Soc. Jpn. **29**, 131–133 (1956)

25.47 B.J. McArdle, J.N. Sherwood: The growth and perfection of phenantrene single crystals, J. Cryst. Growth **22**, 193–200 (1974)

25.48 J. Bleay, R.M. Hooper, R.S. Narang, J.N. Sherwood: The growth of single crystals of some organic compounds by the Czochralski technique and the assessment of their perfection, J. Cryst. Growth **43**, 589–596 (1978)

25.49 C. Kloc, R.A. Laudise: Vapor pressures of organic semiconductors: α-hexathiophene α-quaterthiophene, J. Cryst. Growth **193**, 563–571 (1998)

25.50 G. Horowitz, B. Bachet, A. Yassar, P. Lang, F. Demanze, J. Fave, F. Garnier: Growth and characterization of sexithiophene single crystals, Chem. Mater. **7**, 1337–1341 (1995)

25.51 T. Siegrist, C. Kloc, R.A. Laudise, H.E. Katz, R.C. Haddon: Crystal growth, structure, and electronic band structure of α-4t polymorphs, Adv. Mater. **10**, 379–382 (1998)

25.52 L. Antolini, G. Horowitz, F. Kouki, F. Garnier: Polymorphism in oligothiophenes with an even number of thiophene subunits, Adv. Mater. **10**, 382–385 (1998)

25.53 T. Kuroda: Vapor growth mechanism of a crystal surface covered with a quasi-liquid layer – Effect of self-diffusion coefficient of the quasi-liquid layer on the growth rate, J. Cryst. Growth **99**, 83–87 (1990)

25.54 H. Meng, M. Bendikov, G. Mitchell, R. Helgeson, F. Wudl, Z. Bao, T. Siegrist, C. Kloc, C. Chen: Tetramethylpentacene: Remarkable absence of steric effect on field effect mobility, Adv. Mater. **15**, 1090–1093 (2003)

25.55 L.E. Lyons, G.C. Morris: Photo- and semiconductance in organic crystals. Part III. Photoeffects in dry air with eleven organic compounds, J. Chem. Soc. August, 3648–3660 (1957)

25.56 R. Zeis, C. Besnard, T. Siegrist, C. Schlockermann, X. Chi, C. Kloc: Field-effect studies on rubrene and impurities of rubrene, Chem. Mater. **18**, 244–248 (2006)

25.57 J. Niemax, A.K. Tripathi, J. Pflaum: Comparison of the electronic properties of sublimation- and vapor-Bridgman-grown crystals of tetracene, Appl. Phys. Lett. **86**, 122105 (2005)

25.58 G. Zuccalli, M. Zha, L. Zanotti, C. Paorici: Vapour growth of organic crystals by a semi-open Piz-

25.59 zarello (SOP) technique, Mater. Sci. Forum **203**, 35–38 (1996)

25.59 R.S. Fiegelson, R.K. Route, T.-M. Kao: Growth of urea crystals by physical vapor transport, J. Cryst. Growth **72**, 585–594 (1985)

25.60 R. Zeis, T. Siegrist, C. Kloc: Single-crystal field-effect transistors based on copper phthalocyanine, Appl. Phys. Lett. **86**, 022103 (2005)

25.61 R. Zeis, C. Kloc, K. Takimiya, Y. Kunugi, Y. Konda, N. Niihara, T. Otsubo: Single-crystal field-effect transistors based on organic selenium-containing semiconductor, Jpn. J. Appl. Phys. **44**(6A), 3712–3714 (2005)

25.62 H. Moon, R. Zeis, E.-J. Borkent, C. Besnard, A.J. Lovinger, T. Siegrist, C. Kloc, Z. Bao: Synthesis, crystal structure, and transistor performance of tetracene derivatives, J. Am. Chem. Soc. **126**, 15322–15323 (2004)

25.63 A.R. Laudise, P.M. Bridenbaugh, C. Kloc, S.L. Jouppi: Organo-thermal crystal growth of α6 thiophene, J. Cryst. Growth **178**, 585–592 (1997)

25.64 H. Kallman, M. Pope: Preparation of thin anthracene single crystals, Rev. Sci. Instrum. **29**, 993–994 (1958)

25.65 L. Zeng, M. Zha, M. Curti, L. Zanotti, C. Paorici: Solution growth for high anisotropy organic crystals, Mater. Sci. Forum **203**, 39–42 (1996)

25.66 B.D. Chapman, A. Checco, R. Pindak, T. Siegrist, C. Kloc: Dislocations and grain boundaries in semiconducting rubrene single crystals, J. Cryst. Growth **290**, 479–484 (2006)

25.67 E. Menard, V. Podzorov, S.-H. Jur, A. Gaur, M.E. Gershenson, J.A. Rogers: High performance n- and p-type single-crystal organic transistors with free-space gate dielectrics, Adv. Mater. **16**(22/23), 2097–2101 (2004)

# 26. Growth of III-Nitrides with Halide Vapor Phase Epitaxy (HVPE)

Carl Hemmingsson, Bo Monemar, Yoshinao Kumagai, Akinori Koukitu

III-nitrides can be grown by employing several different techniques, such as molecular-beam epitaxy (MBE), metalorganic vapor-phase epitaxy (MOVPE), halide vapor-phase epitaxy (HVPE), high-pressure solution growth, and sputtering. Each of these are suited for a particular application; the specific property of HVPE is a much larger growth rate, which makes this technique the natural choice for growth of very thick layers that can be used as high-quality native substrates for subsequent growth of device structures using other techniques. Such substrates will be needed for certain devices with high current density or high voltage load, where the high defect density caused by growth on foreign substrates (heteroepitaxy) cannot be tolerated. The HVPE technology is still under development, and below we present the present situation with emphasis on GaN. The thermodynamic limitations of HVPE growth are discussed first, including the high-temperature chemistry in both the source zone and growth zone of a growth reactor. Examples of the design of growth systems are given; in particular, issues such as flow patterns, parasitic growth, and growth rates are discussed. Methods to reduce the defect density for growth on foreign substrates are discussed, as well as various lift-off techniques to prepare free-standing GaN wafers. Common characterization techniques are mentioned, and important physical properties of high-quality GaN wafers are given. The ongoing developments of HVPE growth for AlN and InN are also briefly summarized.

26.1 Growth Chemistry and Thermodynamics . 869
26.2 HVPE Growth Equipment 872
26.3 Substrates and Templates
 for Bulk GaN Growth............................ 875
 26.3.1 Sapphire 875
 26.3.2 Silicon Carbide......................... 876
 26.3.3 GaAs...................................... 876
 26.3.4 Lattice-Matched Substrates 877
 26.3.5 Growth on Templates 877
 26.3.6 Basic 1S-ELO Structures............... 877
 26.3.7 2S-ELO 878
26.4 Substrate Removal Techniques............... 879
 26.4.1 Laser Lift-Off 879
 26.4.2 Self-Separation......................... 880
 26.4.3 Mechanical Polishing................. 881
 26.4.4 Plasma Etching 881
 26.4.5 Chemical Etching
   and Spontaneous Self-Separation 881
26.5 Doping Techniques for GaN in HVPE........ 882
 26.5.1 n-Type Doping of GaN 882
 26.5.2 p-Type Doping of GaN 882
26.6 Defect Densities, Dislocations,
 and Residual Impurities 883
26.7 Some Important Properties
 of HVPE-Grown Bulk GaN Material.......... 887
26.8 Growth of AlN by HVPE:
 Some Preliminary Results 888
26.9 Growth of InN by HVPE:
 Some Preliminary Results 890
References 891

26.1 Growth Chemistry and Thermodynamics

Commonly, a HVPE reactor for growing group III nitrides (GaN, AlN or InN) consists of two main zones: the source zone for forming chloride gas of a group III metal (Ga, Al or In) and the growth zone where the chloride of a group III metal and NH_3 are mixed to grow the nitride film. Therefore, understanding of the chemistry at both source zone and growth zone is essential to grow nitrides by HVPE.

In the source zone of a reactor, the group III metal is placed and maintained at a certain temperature. HCl gas is commonly used to form chloride gas of the group III metal. Therefore, the gaseous species formed at the source zone and their equilibrium partial pressures when HCl gas is introduced over the group III metal are important issues. Principally, gaseous species at the source zone are $III \cdot Cl$, $III \cdot Cl_2$, $III \cdot Cl_3$, $(III \cdot Cl_3)_2$, HCl, H_2, and inert gas (IG: N_2, He, Ar, etc.). The following chemical reactions occur simultaneously

$$III(s \text{ or } l) + HCl(g) = III \cdot Cl(g) + \tfrac{1}{2}H_2(g), \quad (26.1)$$

$$III(s \text{ or } l) + 2HCl(g) = III \cdot Cl_2(g) + H_2(g), \quad (26.2)$$

$$III(s \text{ or } l) + 3HCl(g) = III \cdot Cl_3(g) + \tfrac{3}{2}H_2(g), \quad (26.3)$$

$$2III \cdot Cl_3(g) = (III \cdot Cl_3)_2(g). \quad (26.4)$$

Equilibrium partial pressures of the gaseous species in the source zone can be calculated by a thermodynamic calculation [26.1]. It should be noted that a low source-zone temperature, a high HCl input partial pressure or a small surface area of group III metal might cause insufficient reactions in the source zone, i. e., the reactions will be kinetically limited.

Figure 26.1 shows equilibrium partial pressures of the gaseous species P_i in the source zone under a total pressure of the source zone ($\sum P_i$) of 1.0 atm, an input partial pressure of HCl (P^0_{HCl}) of 6×10^{-3} atm, and a mole fraction of hydrogen in the carrier gas (F^0) of 0.0 (inert carrier gas), calculated for Ga, Al, and In source zones. The source-zone temperature ranges from 300 to 1000 °C. In the Ga source zone, the major gaseous species of Ga is GaCl and its equilibrium partial pressure is almost equal to P^0_{HCl} at temperatures above 500 °C. This means that the predominant reaction at the Ga source zone is (26.1), and almost all HCl introduced into the source zone reacts with Ga metal. The In source zone shows the same tendency as the Ga source zone. Only the Al source zone shows a change of the predominant reaction with increase of the source-zone temperature. At low temperatures, the major gaseous species of Al is $AlCl_3$, with an equilibrium partial pressure one-third of P^0_{HCl}, whereas it is AlCl at high temperatures, above 800 °C. Although the carrier gas used in Fig. 26.1 is IG, the equilibrium partial pressures of group III chlorides do not vary when H_2 is used as a carrier gas ($F^0 = 1.0$).

Figure 26.2 shows the values of equilibrium constants K of the growth reactions of GaN, AlN, and InN using $III \cdot Cl$ or $III \cdot Cl_3$ and NH_3 as a function of the reciprocal of the reaction (growth-zone) temperature.

The equilibrium constants can be calculated using thermochemical tables [26.2, 3]. As seen in Fig. 26.1, the group III chloride formed at the source zone is $III \cdot Cl$ (monochloride) when the source-zone temperature is sufficiently high. However, with decreasing

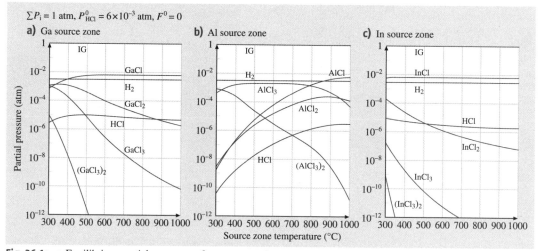

Fig. 26.1a–c Equilibrium partial pressures of gaseous species over group III metals placed in the source zone as a function of temperature calculated for (**a**) Ga, (**b**) Al, and (**c**) In source zones

source-zone temperature, the equilibrium partial pressure of III·Cl₃ usually increases. Also, GaCl₃, AlCl₃, and InCl₃ are available commercially in powder form, which makes it possible to sublimate them for growth of nitrides, as an alternative to the use of HCl. Therefore, equilibrium constants of growth reactions using trichlorides are also shown. In the figure, the following order of equilibrium constants K for growing nitrides is seen

$$K_{\text{AlN}} \gg K_{\text{GaN}} > K_{\text{InN}} \,. \tag{26.5}$$

In general, this order corresponds to the growth rate of nitrides using group III chlorides. For the growth reactions of GaN, the equilibrium constants are close to zero in both reactions using GaCl and GaCl₃. Therefore, growth of GaN is possible using GaCl or GaCl₃. In the AlN growth, the values of equilibrium constants are extremely large in both reactions using AlCl and AlCl₃, indicating high-speed growth of AlN. On the other hand, in the case of InN growth, the value of log K using InCl is significantly small and negative. Consequently, growth of InN is possible only using InCl₃ as an In source. Another point expected from the figure is the influence of H₂ on the growth of nitrides, since the growth reactions contain H₂ as a product when monochlorides are used. Therefore, thermodynamic analysis for the growth zone of a HVPE system is meaningful as well as that for source zone, since growth is commonly performed under mass-transportation-limited or thermodynamically controlled conditions.

Thermodynamic analysis of the HVPE growth zone has been performed and reported for GaN [26.4, 5], AlN [26.6], and InN [26.7]. In the literature the influence of various growth conditions on the driving force of growth is described. Here thermodynamic analysis of the conventional GaN growth zone is described as an example. When GaCl formed at the source zone and NH₃ are separately introduced into the growth zone with a carrier gas mixture of H₂ and IG, the following chemical reactions occur simultaneously

$$\text{GaCl(g)} + \text{NH}_3(\text{g}) = \text{GaN(s)} + \text{HCl(g)} + \text{H}_2(\text{g})\,, \tag{26.6}$$

$$\text{GaCl(g)} + \text{HCl(g)} = \text{GaCl}_2(\text{g}) + \tfrac{1}{2}\text{H}_2(\text{g})\,, \tag{26.7}$$

$$\text{GaCl(g)} + 2\text{HCl(g)} = \text{GaCl}_3(\text{g}) + \text{H}_2(\text{g})\,, \tag{26.8}$$

$$2\text{GaCl}_3(\text{g}) = (\text{GaCl}_3)_2(\text{g})\,. \tag{26.9}$$

Equation (26.6) is the main reaction producing GaN growth and the others (26.7–26.9) are secondary reactions. Then, gaseous species coexisting at the growth

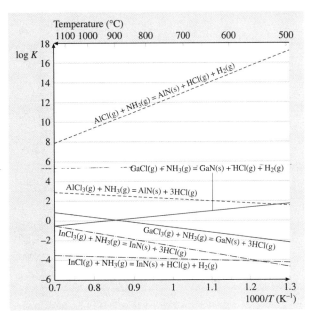

Fig. 26.2 Values of log K as a function of reciprocal of temperature calculated for growth reactions of GaN, AlN, and InN using mono- or trichloride of group III metals and NH₃

zone are GaCl, GaCl₂, GaCl₃, (GaCl₃)₂, NH₃, HCl, H₂, and IG. Figure 26.3 shows equilibrium partial pressures of gaseous species at the growth zone as a function of growth-zone temperature. Calculation of equilibrium partial pressures was performed after a procedure developed by *Koukitu* and *Seki* [26.8] for HVPE growth of arsenide and phosphide of Ga and In. The growth conditions are as follows: total pressure of the reactor ($\sum P_i$) of 1.0 atm, input partial pressure of GaCl (P^0_{GaCl}) of 6.0×10^{-3} atm, input V/III ratio ($P^0_{\text{NH}_3}/P^0_{\text{GaCl}}$) of 50, and mole fraction of hydrogen in the carrier gas (F^0) of 1.0 (H₂ carrier gas). An additional parameter α which denotes the mole fraction of decomposed NH₃ before GaN growth is introduced as follows

$$\text{NH}_3(\text{g}) \to (1-\alpha)\text{NH}_3(\text{g}) + \frac{\alpha}{2}\text{N}_2(\text{g}) + \frac{3\alpha}{2}\text{H}_2(\text{g})\,. \tag{26.10}$$

The value of α is difficult to determine exactly since it depends on the growth conditions, equipment, and temperature. In Fig. 26.3, the value is fixed at 0.03 according to the literature [26.9], and agreement between the calculated and experimental results [26.4–7]. It is seen that NH₃, GaCl, IG (N₂), and HCl are major gaseous species as well as H₂ used as the carrier gas.

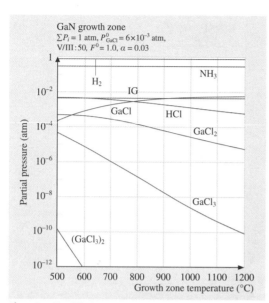
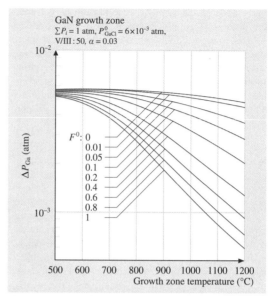

Fig. 26.3 Equilibrium partial pressures of gaseous species over GaN as a function of growth temperature

Fig. 26.4 Driving force for growth of GaN as a function of growth-zone temperature calculated for several values of F^0

The equilibrium partial pressure of HCl decreases with increase of growth-zone temperature, while that of GaCl increases with increase of growth-zone temperature and approaches a value nearly equal to P^0_{GaCl}. This means that the growth rate of GaN becomes smaller at high growth temperatures. It is of interest to calculate the driving force for growth as functions of various growth conditions. The driving force ΔP_{Ga} is given by

$$\Delta P_{Ga} = P^0_{GaCl} - \left(P_{GaCl} + P_{GaCl_2} + P_{GaCl_3} + 2 P_{(GaCl_3)_2} \right), \quad (26.11)$$

i.e., the difference between the number of Ga atoms put in and the amount of Ga atoms remaining in the vapor phase. Figure 26.4 shows the driving force for GaN growth using GaCl and NH_3 as a function of growth temperature calculated for several values of F^0. It is seen that the driving force decreases with increase of the growth temperature and F^0. Furthermore, the influence of H_2 on the decrease of the driving force is more significant at high growth temperatures. It has also been clarified that the driving force decreases with increase of H_2 in the carrier gas in GaN and InN growths using trichlorides [26.5, 7]. By contrast, growth of AlN using mono- or trichloride of Al does not depend on the presence of H_2, which is due to the extremely large equilibrium constants of the growth reactions, as can be seen in Fig. 26.2. Consequently, for high-speed and high-temperature growth of GaN by HVPE, use of inert carrier gas or mixed carrier gas of H_2 and IG with a small amount of H_2 is quite effective.

26.2 HVPE Growth Equipment

A schematic picture of a typical horizontal hot-walled HVPE growth system, which is the most commonly used geometry, is shown in Fig. 26.5. The reaction chamber is normally a quartz tube that is heated either with a multizone oven or by radiofrequency (RF) induction. The reactor has one low-temperature region where the halides are formed by letting high-purity HCl or Cl_2 gas flow over a melt of a group III metal. The temperature is normally kept about 800–900 °C in this section of the reactor. The precursors are reacting on the substrate, which is kept at a temperature about 1000–1100 °C. In order to minimize parasitic growth in the inlet of the growth zone and on the walls of the reactor, it is important that the ammonia and the halide are not mixed before they reach the substrate. Inert gases such as nitrogen (N_2), helium (He), argon (Ar) or hydrogen (H_2) and mixtures of these gases are used as carrier gases. In terms of the flow pattern, gases with lighter molecules, such as He or H_2, are more favorable, i. e., the laminar flow condition is obtained more easily.

In HVPE, a huge amount of ammonium chloride (NH_4Cl) is formed downstream of the growth zone due to the reaction between HCl and ammonia. Thus, care has to be taken in order to avoid clogging of the exhaust. The growth system has to be equipped with some type of trap at the outlet. The deposition of ammonium chloride can also be avoided by heating the outlet pipes above $\approx 150\,°C$. Growth of boules requires very long growth runs, thus it may be necessary to change the trap during the process.

The growth takes place under excess ammonia conditions and in a growth regime where the growth rate is mass transport limited by the vapor pressure of the halide precursor. Thus, in a horizontal system such as shown in Fig. 26.5, it is difficult to achieve a uniform deposition of high-quality material over a large-area substrate since the halide easily becomes depleted over the substrate. Using an inlet geometry where the halide

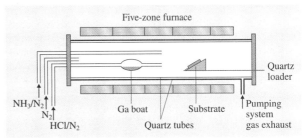

Fig. 26.5 Conventional horizontal HVPE system with a five-zone furnace

precursor is dispersed uniformly across the deposition zone by a showerhead can solve this problem. The proximity of the showerhead to the susceptor, on which the wafer is placed, minimizes premature reactions between the ammonia and the halide, which leads to good chemical efficiency. A showerhead in combination with rotation of the susceptor is suited to the deposition of uniform layers, allowing scalability to be combined with ease of operation. This type of reactor is normally used commercially in order to produce layers up to typically 300–500 μm. In order to grow thicker layers, this type of reactor is not suitable, mainly because of limitations due to parasitic deposition, which will create particle problems and changes of growth conditions during the growth run.

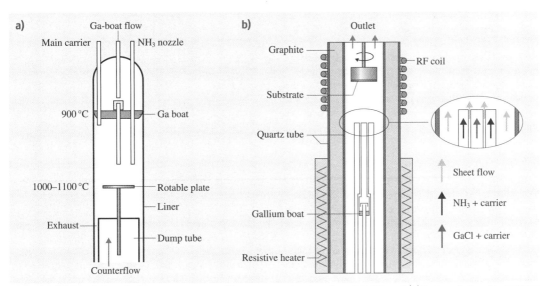

Fig. 26.6 (a) Top-fed vertical reactor geometry. The geometry is taken from [26.10] and **(b)** bottom-fed vertical HVPE reactor with a concentric inlet in order to minimize parasitical growth in the inlet

An alternative reactor geometry is a vertical setup, as shown in Fig. 26.6. The precursors can be introduced either from the top (Fig. 26.6a) or from the bottom (Fig. 26.6b). In the geometry where the gases are introduced from the top, the gas flow has to be controlled by adjusting the downward forced-convective flow to balance it with the upward buoyancy-driven convection. The problem with this geometry is a possible particle contamination over the substrate due to particles released from the parasitic growth upstream of the substrate. These particles may fall down and induce defect nucleation centers that degrade the quality of the grown crystal. By introducing the precursors from the bottom, this kind of problems can be minimized since the surface of the growing crystal is facing downwards. Figure 26.6b shows a schematic picture of a hot-walled vertical growth system where the precursors are introduced from the bottom. This system is designed for the growth of boule crystals (layers up to several millimeters). In order to avoid premature reactions between the GaCl and the ammonia, which will cause parasitic deposition at the inlet of the reactor, the inlet is designed with concentric tubes where the GaCl and the ammonia are separated using a sheet flow. In addition to the inlet sheet flow, a sheet flow close to the reactor walls prevents parasitic growth on the walls. The reactor is equipped with rotation of the sample holder, and the distance between the growing surface and the inlet can be kept constant by pulling of the holder.

In a vertical system one problem is to stabilize a high and constant conversion ratio of HCl to GaCl due to the small Ga area. This will affect the reactions (26.1–26.4). A reduction of the GaCl vapor pressure will reduce the growth rate, and the increase of the HCl vapor pressure increases the etching component in the growth process. Thus, it is important to design the Ga container in such a way that the HCl volume over the liquid Ga in the Ga container is kept approximately constant over time. Early work established that the GaCl conversion in a horizontal reactor with a large liquid Ga area was typically stabilized at a high value: the conversion efficiency to GaCl was above 95% [26.9]. By careful design of the Ga container, it has been shown that the conversion efficiency can be up to 98% in a vertical reactor [26.12], which is close to chemical equilibrium at 850 °C [26.13].

To achieve a desirable mass transport behavior, it is important to establish a stable, vortex-free flow field in the reactor. Any recirculating flow can create a nonuniform growth rate over the substrate, less control of the parasitic growth, and increased impurity incorporation.

Since the decay of ammonia is a slow process [26.9] which can be accelerated by catalytic effects, a recirculating flow where the gas may be in contact with the walls is especially severe in HVPE, since the parasitic growth on the walls can accelerate the decomposition of the ammonia and, consequently, change the growth conditions. It has been shown that, by minimizing the recirculation flows in the reactor, run-to-run reproducibility is improved drastically [26.14].

The flow pattern in a HVPE reactor is influenced by several mechanisms, such as reactor geometry, heat transfer characteristic, flow boundary conditions, and choice of carrier gases. With respect to recirculation flow, a bottom-fed vertical configuration where the colder gases are flowing upwards towards the normally hotter substrate is preferred because the buoyancy effect has a tendency to stabilize the flow. The use of a hot-walled system will also reduce the recirculation flows generated by natural convection since the precursors are efficiently heated and the thermal gradients are small in such a system. However, since the substrate is oriented perpendicular to the flow direction some care has to be taken in the choice of carrier gases and gas velocities. Figure 26.7 shows an example of modeling using different gas velocities and gas compositions in

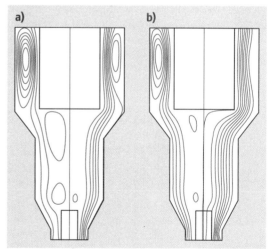

Fig. 26.7 (a) Streamline plots using N_2 as a carrier gas. The *left side* shows the streamlines using 0.5 l/min of carrier gas and the *right side* shows the situation when using 5 l/min of carrier gas and **(b)** shows the streamlines using 3 l/min of N_2 (*left side*) and 3 l/min of He (*right side*). The GaCl and NH_3 flow was 20 and 500 ml/min, respectively, in all cases (after [26.11])

a bottom-fed vertical configuration [26.11]. With a low total N_2 carrier gas flow (< 1 l/min), the growth is turbulent, with a vortex pattern in the middle of the substrate. At higher total flow the pattern evolves to an approximately laminar flow in the range $3-10$ l/min. N_2 is an attractive choice for carrier gas, since it is inert during the growth conditions, and does not affect the growth chemistry. However, in terms of the flow pattern, gases with lighter molecules, such as He or H_2, are more favorable, i.e., the laminar flow condition can be obtained using lower gas velocities, as clearly shown from Fig. 26.7b.

26.3 Substrates and Templates for Bulk GaN Growth

The ideal way of producing bulk GaN crystals is homoepitaxial growth on a GaN seed crystal. However, due to the lack of commercially available GaN wafers, the use of GaN substrates for HVPE bulk growth is not yet commonly employed. In the few reported experiments on homoepitaxial growth using HVPE, high-pressure grown platelets [26.15] have been used. The dislocation density is low ($< 100\,\text{cm}^{-2}$) in this type of substrate; however, the lateral size and thickness are limited to about 1 cm and 100 μm, respectively, and the carrier concentration is very high ($\approx 10^{19}\,\text{cm}^{-3}$). The high carrier concentration expands the lattice parameters compared with the relatively low-doped HVPE GaN, which results in stress and cracking problems for layers thicker than $\approx 100\,\mu\text{m}$ [26.16, 17]. Thus, due to the lack of commercial bulk GaN substrates, the growth of bulk crystals is mostly performed on foreign substrates (heteroepitaxy). Several different types of substrate are used or have been used for growth of thick GaN layers using HVPE, such as sapphire (Al_2O_3), SiC, GaAs, Si, MgO, ZnO, TiO_2 and $MgAl_2O_4$ [26.18], to mention a few.

Among these Al_2O_3 is the most commonly employed substrate material for bulk GaN growth, mainly due to the possibility of obtaining large-area sapphire wafers (2 inches) at relatively low cost. Most often, heteroepitaxial growth gives rise to numerous defects such as threading dislocations (TD), stacking faults, voids, and hexagonal pits due to the lattice mismatch and difference in thermal expansion between the substrate and the grown material. Table 26.1 summarizes these properties for some of the most important substrate materials used today. A nonpolar substrate can also give rise to inversion domains and mixed polarities in the grown layer. These defects are detrimental to device performance, causing high device leakage current, short minority-carrier lifetime, reduced thermal conductivity, etc. Other important properties of the substrate that have to be considered are those related to the surface, such as surface roughness, step height, terrace width, and wetting behavior.

26.3.1 Sapphire

Heteroepitaxial growth on sapphire involves several problems. The large difference in lattice constant between sapphire and GaN gives rise to a high density of TDs, and the larger thermal expansion coefficient of sapphire compared with GaN causes a biaxial compressive stress in the GaN layer as it is cooled from growth temperature. For thicker films ($> 50\,\mu\text{m}$), the stress can cause cracking of both the GaN epilayer and the substrate. GaN epitaxy on c-plane (0001) sapphire, which is the most common orientation, results in c-plane-oriented films, but with a 30° rotation of the in-plane GaN crystal directions with respect to the same directions in the sapphire. Without the in-plane rotation, the lattice mismatch between the GaN and the sapphire is

Table 26.1 Surface lattice parameters and thermal expansion coefficients of some technologically important substrate materials for GaN growth

Substrate	Crystal structure	Growth plane for wurtzite GaN	Surface lattice parameter (Å)	In-plane lattice mismatch to GaN (%)	Thermal expansion coefficient ($10^{-6}\,\text{K}^{-1}$)
GaN	Wurtzite	(0001)	3.189		5.6
Al_2O_3	Wurtzite	(0001)	4.758	16.1	7.5
6H-SiC	Wurtzite	(0001)	3.08	3.54	4.2
GaAs	Zincblende	(111)	2.54	20.2	6.03

about 30%. However, the in-plane rotation of the GaN minimizes the lattice difference between the GaN and the sapphire to about 16%. The good morphological, crystallographical, and electrical properties of c-plane-grown GaN is mostly responsible for the widespread use of the c-orientation.

Nevertheless, in some cases it is beneficial to use growth on planes other than the c-plane. In c-plane nitrides, we have a built-in electrostatic field in the [0001] direction due to the spontaneous and piezoelectric polarization [26.19]. This field is undesirable for optoelectronics devices since it results in an uncontrollable shift of the emission peak and/or a reduction of the emission efficiency. The influence of this field can be avoided by the use of GaN with other orientations. Thus, other orientations such as r-plane ($1\bar{1}02$), a-plane ($11\bar{2}0$), and m-plane ($10\bar{1}0$) sapphire have been used in order to grow nonpolar a-plane GaN [26.20] or to facilitate cleavage of the structure to form edge-emitting lasers [26.21].

By using a miscut sapphire wafer it has been shown that the GaN quality can be improved in MOVPE [26.22, 23]. In HVPE an improvement in the morphology has been observed. Using a slightly miscut c-plane substrate with an $\approx 0.3°$ off-orientation against the a-plane or the m-plane, a reduction in hillocks is observed [26.24]. It has been shown that a-plane off-oriented substrates suppress the hillock formation more efficiently than other directions.

Growth on sapphire requires a pretreatment of the sapphire to improve the wetting properties and the crystal quality. One pretreatment step is nitridation, which is done by exposing the sapphire substrate to ammonia at typical growth temperatures (1000–1100 °C). It is believed that a thin AlN or AlN–Al$_2$O$_3$ layer forms on the surface, which facilitates nucleation [26.25]. This type of pretreatment is also employed in MOVPE [26.26–29] and MBE [26.30] growth of GaN on sapphire substrates. A second step, which is commonly employed and has been shown to reduce the pit density and improve the crystal structure [26.31, 32] is GaCl pretreatment. The GaCl pretreatment is done just before the growth of GaN commences, by exposing the sapphire to GaCl for $\approx 10–20$ min at the growth temperature. It is believed that the GaCl pretreatment supplies nucleation centers for GaN islands and the promotion of the coalescence of the GaN layer in the early stages of growth.

The crystal quality has been shown to improve by the use of a low-temperature-grown GaN buffer layer (LT-GaN). This technique is used for MOVPE growth of GaN on sapphire and has also been shown to be useful for HVPE growth. This LT-GaN layer greatly improves the surface morphology and reduces the dislocation density of the subsequently grown high-temperature GaN layer. Initially, a thin GaN film on the order of 100 nm is grown at $\approx 550–650$ °C. This layer can serve as a starting layer for growth [26.33]. However, it has been shown that annealing of the LT-GaN layer in ammonia before commencing the high-temperature growth reduces the TD density further [26.34]. Upon annealing of the LT buffer, relatively large grains are formed that can serve as nucleation centers for the subsequent high-temperature growth.

26.3.2 Silicon Carbide

SiC is widely used for growth of device structures using MOVPE and MBE due to the fact that it has several advantages to sapphire such as a smaller lattice mismatch (3.1%), a higher thermal conductivity (3.8 W/cm K), it is conductive, SiC has a polarity which facilitates control of polarity of the GaN, and there is no in-plane rotation between the GaN and the SiC, which makes it possible to cleave the crystal for laser facet formation. However, it has also several disadvantages, i.e., it requires in most cases a buffer layer of AlN or AlGaN due to its poor wetting properties, and the surface roughness is one order of magnitude larger than that of sapphire. The cost of SiC is also much higher than that of sapphire. Most of the above-mentioned favorable properties are not very important for bulk growth and, consequently, SiC substrates are not commonly employed for bulk growth.

26.3.3 GaAs

GaAs is an interesting substrate material since it is brittle and can easily be mechanically removed after growth. It can also be easily removed by etching in aqua regia. Furthermore, it has a thermal expansion coefficient closer to that of GaN than any other foreign substrate material (Table 26.1). GaAs substrates can be used both for growth of zincblende GaN and wurtzite GaN depending on the orientation of the substrate. GaAs(001) gives zincblende GaN while GaAs(111) is used for growth of wurtzite GaN. In principle, zincblende GaN has better electronic properties for device applications, such as isotropic properties, higher mobility, and high optical gain; however, the quality of the zincblende GaN grown on GaAs is very poor due to coexistence between the two phases. Thus, hereafter we only consider growth of wurtzite GaN grown on GaAs(111).

One problem with growth of GaN on GaAs is the fact that GaAs is decomposed in ammonia ambient at a temperature where the growth of GaN normally takes place (above 1000 °C). This may lead to problems with a degraded GaAs surface before the growth starts and autodoping of As [26.35]. Thus, it is necessary to grow a buffer layer to protect the GaAs surface. One way to do that is to grow a thin protective GaN layer at lower temperatures [26.36, 37]. Using the epitaxial lateral overgrowth (ELO) technique in combination with a low-temperature GaN buffer, 2 inch free-standing GaN layers with a thickness of about 500 μm have been demonstrated [26.28].

26.3.4 Lattice-Matched Substrates

There exist several other potential substrate materials such as LiAlO$_2$, LiGaO$_2$, and NdGaO$_3$ with closer lattice matching to GaN than the substrate materials discussed above. Among these, γ-lithium aluminum oxide (LiAlO$_2$) has shown potential as a substrate material for growth of thick GaN using HVPE [26.38]. The orientation between the LiAlO$_2$ and the GaN is GaN(0001)||LiAlO$_2$(110) and GaN(0001)||LiAlO$_2$(001) with a lattice mismatch of −1.4% and −6.3%, respectively.

Lithium gallate (LiGaO$_2$) has the best lattice matching of all, with only 0.9% lattice mismatch in the basal plane, and has been used for growth of thick GaN [26.39]. However, the poor chemical and thermal stability of these substrate materials in the HVPE growth environment makes the growth conditions very critical. In order to avoid decomposition of the substrate surface before growth, elaborate starting procedures have to be applied, with pretreatment steps and low-temperature GaN buffer layers before growth can commence. However, due to a partial decomposition of the substrate during growth, the grown GaN layer is very often spontaneously separated from the substrate during cooling after growth.

26.3.5 Growth on Templates

A very common growth procedure in HVPE is to start the growth on a starting layer of GaN [26.40], AlN [26.41–43] or ZnO [26.25, 44–46] in order to facilitate the initial nucleation process. The starting layer is normally grown heteroepitaxially with MOVPE (GaN) or by sputtering (AlN, MgO). Among these different types of starting layer, MOVPE-grown GaN is the most frequently used. With the state of the art of the MOVPE GaN heteroepitaxial growth technique today, the starting layers have a TD density typically in the 10^9 cm^{-2} range. In order to reduce the TD density further, the ELO technique can be used. However, this reduction is obtained at the expense of several additional process steps.

26.3.6 Basic 1S-ELO Structures

As mention above, one way to reduce the TD density is the ELO process. This technique has been used in growth of GaAs and has also been successfully used for growth of GaN. A one-step ELO structure (1S-ELO) is fabricated as follows, as depicted in Fig. 26.8: A substrate is covered with a dielectric film (Fig. 26.8a), normally SiN or SiO$_2$, and patterned by conventional photolithography to form on its whole area a grating of mask-free seeding windows, as shown in Fig. 26.8b. The orientation of the stripes is normally in the $\{1\bar{1}00\}$

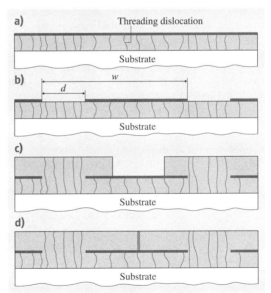

Fig. 26.8a–d Cross-sectional view perpendicular to the [1$\bar{1}$00] direction showing the different steps in the fabrication of a 1S-ELO structure. *Black thin vertical lines* represent threading dislocations. (**a**) Deposition of dielectric film, (**b**) after etching of the dielectric film, where d is the width of the stripes while w is the period of the stripes, (**c**) growth in the lateral direction over the film and (**d**) fully coalesced structure

direction of the GaN layer since it has been shown that lateral expansion is easier in this direction [26.47–50]. The width of the windows (d) in the mask and the periodicity of the lines (w) in the grating is typically about 3 and 10 μm, respectively, and the thickness of the dielectric film is typically about 100 nm. The dielectric is deposited using well-established techniques such as chemical vapor deposition (CVD) or plasma-enhanced CVD (PECVD). Then, an epitaxial layer is deposited on such a substrate. Initially, there is no nucleation on the dielectric stripes; however, as soon as the crystallization front passes the top layer of the mask, growth in the lateral direction over the film starts (Fig. 26.8c), which finally leads to coalescence of the layer, as shown in Fig. 26.8d, and a smooth (0001) surface suitable for further processing is formed.

The basic idea of the ELO technique is that the dielectric mask filters dislocations present in the substrate since defects cannot penetrate through the mask. Thus, in areas above the dielectric stripes where the GaN has grown laterally (wing region), the GaN should be dislocation free. However, in the region where the two laterally grown crystallization fronts meet on the mask, defects are created. Thus, using this technique, two regions with a higher defect density are formed, one in the region without mask and the other in the region where the two laterally growing parts meet on the mask. By using the 1S-ELO technique, the TD density can be reduced by about two orders of magnitude [26.47].

26.3.7 2S-ELO

To reduce the TD density further, double-layer ELO (or 2S-ELO) can be used, where a second dielectric mask layer is processed on a fully coalesced ELO structure. The idea is to cover the remaining regions where the TD is still high with a second dielectric mask. Thus, the remaining TDs are prevented from propagating to the second ELO layer. The mask can either be rotated by 60 or 90° [26.51] in order to keep the orientation along another equivalent $\langle 1\bar{1}00 \rangle$ orientation, or be displaced [26.52] from the first array of openings so that the second mask covers the coherent GaN over the window region of the first mask layer. From the theory of image dislocations and image forces [26.53], it is known that dislocations near a free surface bend towards the free surface. In the case of TDs in GaN, they can be bent 90° and start to propagate along the basal plane [26.47, 54, 55]. This phenomenon can be used to reduce the TDs over larger areas. A mask is formed in a similar way as for the conventional ELO

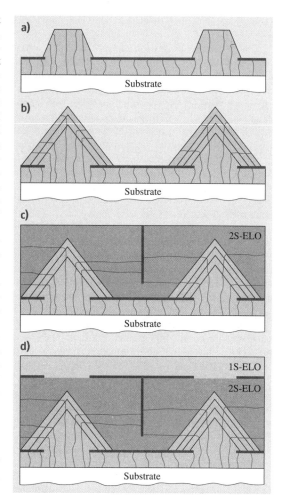

Fig. 26.9a–d Cross-sectional view perpendicular to the $[1\bar{1}00]$ direction showing the different steps in the fabrication of a 2S-ELOG structure (**a–c**) and a layered ELO structure (**d**). *Black thin vertical lines* represent threading dislocations. (**a,b**) The evolution of the formation of triangular stripes with $\{11\bar{2}2\}$ lateral facets, (**c**) fully coalesced 2S-ELOG structure and (**d**) additional 1S-ELO on 2S-ELO structure

process as described above. However, initially during the first step of growth, growth parameters that favor vertical expansion are selected, as shown in Fig. 26.9a. By tuning the process parameters, the stripes will adopt a triangular shape in cross section with $\{11\bar{2}2\}$ lateral facets. Thus, since the TDs are propagating perpendicular to the growing surface, they are bent 90°. This growth mode is kept until the (0001) top facet vanishes completely (Fig. 26.9b). In the second step, the growth condition is changed to promote lateral growth in order to fully coalesce the layer. This can be done either by adding Mg or increasing the growth temperature. The bent TDs from the neighboring window regions meet each other in the middle of the dielectric mask layer, as shown in the Fig. 26.9c. Thus, the TDs pile up above the middle of the mask and start to grow in the c-direction.

To further reduce the TDs, a 1S-ELO can be processed on a 2S-ELO with the mask located exactly above the first structure [26.56]. Thus, the TDs originating from the coalesced region on the 2S-ELO structure are efficiently prevented from propagating to the top GaN layer, as shown in Fig. 26.9d. Using this technique with several ELO structures stacked on each other, it has been shown that the TD density can be reduced down to or even below the $10^6\,\mathrm{cm}^{-2}$ range [26.57].

26.4 Substrate Removal Techniques

The development of advanced III-nitride devices, such as laser diodes (LDs) or some high-power lamps for general illumination, will require the growth of device structures on high-quality III-nitride substrates with a low TD density ($< 10^6\,\mathrm{cm}^{-2}$). These types of substrates are very difficult to produce by using heteroepitaxial growth on sapphire or SiC substrates, which is the common technique today. Thus, there is a strong need for free-standing GaN substrates with low TD density in order to fully exploit the III–V nitride materials.

Two major routes have appeared towards the production of bulk GaN wafers. One possibility is growing a thick boule followed by slicing and polishing. The alternative is growing a thick GaN layer on some foreign substrates, followed by substrate removal and proper polishing. Among these, the most attractive way to produce GaN substrates is directly by slicing of a GaN boule and polishing. However, despite many years of development and research, the process control and technical issues are not yet fully solved. Instead, several techniques using thick GaN layers grown by HVPE on foreign substrates followed by substrate removal have been developed.

26.4.1 Laser Lift-Off

The most used substrate material in HVPE is sapphire, which is very hard and chemical inert. This makes it very difficult to remove the substrate mechanically or chemically. Instead, a technique which relies on thermal decomposition of the GaN at the interface between the sapphire and the GaN film has attracted most attention in recent years. Using this so-called laser lift-off process (LLO) [26.58], large-area crack-free GaN substrates up to 2 inches in diameter with a thickness of about 300 μm have been demonstrated [26.59].

The idea behind the LLO process is to irradiate the GaN–sapphire substrate through the sapphire with a laser beam with an energy that is less than the bandgap of sapphire but larger then the bandgap of GaN. Thus, the laser beam will be absorbed at the GaN–sapphire interface, as shown in Fig. 26.10. If the power density of the laser beam is sufficient, a thin layer (of the order 100 nm) of GaN at the interface to the sapphire will decompose into liquid Ga and N_2 gas according to

$$2\mathrm{GaN} \rightarrow N_2(g) + 2\mathrm{Ga}(l) \,. \qquad (26.12)$$

The decomposition rate of GaN rapidly increases for elevated temperatures [26.60] and a high-power laser beam can easily obtain the onset of thermal decomposition. By scanning the beam, the GaN at the interface can be completely decomposed and, consequently, the GaN layer is delaminated from the sapphire.

Figure 26.11 shows a schematic picture of an LLO setup. It consists of a high-power laser with photon energy larger than the bandgap of GaN (3.43 eV). A suitable energy of 3.49 eV (355 nm) is obtained from the third harmonic of a neodymium-doped yttrium alu-

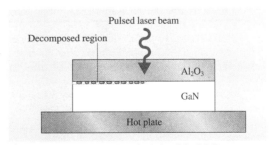

Fig. 26.10 Drawing of the principle of the LLO process

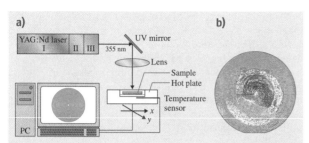
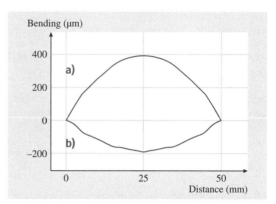

Fig. 26.11 (a) Schematic drawing of a laser lift-off setup. The movement of the wafer on the hot plate is done with a computer-controlled x–y stage tool. (b) Transmission light image of a 300 μm thick 2 inch free-standing GaN layer. On the backside of the wafer, traces from the laser lift-off process can be observed

Fig. 26.12a,b Experimentally observed bow for two GaN layers. (a) 250 μm thick layer residing on sapphire and (b) 300 μm thick free-standing GaN layer. A convex bow of the GaN layer is observed when the GaN layer is residing on sapphire. When the sapphire is removed, the bow switches to a concave bow (after [26.12], with permission from Elsevier)

minum garnet (Nd:YAG) laser. Another type of laser system that can be used is, for example, a KrF laser with an energy of 5 eV (248 nm). The laser beam is directed and focused on the backside of the sample. In order to reduce the stress due to the difference in thermal expansion coefficients between the substrate and the GaN layer and facilitate the thermal decomposition process, the sample is heated to about 700 °C. The entire sample holder is connected to an x–y stage, which is moved in a controlled fashion via a computer. The laser spot is moved in a spiral pattern from the perimeter to the center of the wafer. An example of a 300 μm thick 2 inch GaN wafer removed from the substrate by this technique is shown in Fig. 26.11b. The wafer is shown in transmitted light with the top surface up, and the spiral pattern from the scanning of the laser beam over the wafer can be observed as a circular pattern on the backside of the wafer. The process is very sensitive to the focusing of the laser beam and the laser output energy. Too high a power density generates cracks in the interface that can propagate through the layer and cause cracking of the wafer, while too small a power density might cause local islands where the layer has not separated. These islands create a strong nonuniform stress in the material during cool down after LLO, which may result in cracking of the layer. Another cause for incomplete separation is parasitic GaN growth on the backside of the substrate. The parasitic GaN absorbs the laser light and prevents the decomposition of the GaN at the interface.

The free-standing layers obtained after LLO are practically strain free. The main problem with free-standing GaN wafers is the substantial bowing of the wafers typically observed. Figure 26.12a shows an example of bowing of a 250 μm thick 2 inch layer residing on sapphire and a 300 μm thick free-standing GaN layer obtained by LLO. As can bee seen, the bowing can be quite large, in this example 400 μm before LLO and 200 μm after LLO. The sign of the bow is changed after LLO. Before LLO, the wafer is convex due to the larger thermal expansion coefficient of the sapphire, while afterwards the layer adopts a concave shape due to the defective interfacial region with large dislocation density and microcracks between the sapphire and the GaN, as shown in Fig. 26.12b. The layer shown in the example has too large a bow for further processing. Thus, in order to reduce the bowing, further processing such as annealing and/or removal of the highly defective region by etching is necessary.

26.4.2 Self-Separation

Many alternative techniques to LLO are under development since the LLO process normally has a relatively low yield. Several of these techniques use the stress that is built up in the GaN layer during cooling due to the difference in thermal expansion of the substrate and the GaN layer. By making a weak link that is provided by cavities between the substrate and the GaN layer, the separation of the substrate and the layer can be done spontaneously during cool down from the growth temperature.

Using ELO structures, voids can intentionally be created. By adjusting the growth process, the size and

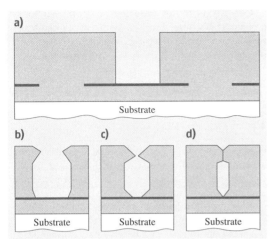

Fig. 26.13a–d Example of using ELO and mass transport in order to create voids at the substrate–GaN interface. (a) 1S-ELO structure with not fully coalesced wing region. (b–d) Evolution of the void under mass-transport conditions

shape of the voids can be engineered to facilitate separation during cooling [26.54, 61]. Figure 26.13 shows an example of a structure created by using mass transport. A conventional ELO structure with a SiO$_2$ or SiN mask is used as a starting template. During growth, in the stage when the lateral growth starts, the growth conditions are tuned such that the lateral facet becomes $\{11\bar{2}0\}$ (Fig. 26.13a). The growth is interrupted when the distance between the two lateral facets is a couple of microns. Then, by increasing the temperature to about 1100 °C under ammonia, mass transport will occur, as shown in Fig. 26.13b, c. At the end of the mass transport process, a vertical cavity has been created (Fig. 26.13b), which will facilitate the separation of the GaN layer from the substrate.

One of the more successful techniques that rely on self-separation is the so-called void-assisted separation (VAS) technique. The idea is basically the same as using ELO structures with cavities in the interface. Using this technique, high-quality free-standing substrates of both c-oriented [26.62] and a-oriented [26.63] GaN have been demonstrated. A thin layer of suitable metal (≈ 20 nm), generally Ti or Al, is deposited onto a thin MOVPE-grown GaN template prepared on a sapphire substrate. Subsequent annealing at ≈ 1050 °C of the metal-coated GaN template in a mixture of hydrogen and ammonia causes the metal to break up and form a nanonet-like metal nitride structure with many nanoscale holes with a typical diameter of about 20–30 nm. During the subsequent growth of the thick GaN layer, additional etching of the mask and the underlying GaN occurs, creating multiple voids at the substrate–GaN interface. If these voids exceed a critical size, it is found that the GaN layer is spontaneously separated from the substrate during cooling after growth. For a 250 μm thick GaN layer it was estimated that the critical void size is about 20 μm [26.63]. The nanonet-like metal nitride structure layer also serves as a growth mask, which efficiently blocks the propagation of threading dislocations along the growth axis.

26.4.3 Mechanical Polishing

Using mechanical polishing with diamond slurry, 30×30 mm^2 free-standing GaN substrates have been demonstrated from 400–450 μm thick GaN layers [26.64] grown on sapphire. However, this technique will be difficult to use for large-area wafers due to the strong convex bow of the layers that normally occurs after heteroepitaxial growth on foreign substrates. The bow makes it difficult to achieve uniform removal of the substrate by polishing.

26.4.4 Plasma Etching

Free-standing GaN wafers with a thickness of 200 μm and diameter of 30 mm have been demonstrated using reactive-ion etching (RIE) to remove the substrate. The GaN layer was grown on a SiC substrate and the substrate was removed by RIE in an SF$_6$-containing mixture gas [26.65, 66]. The main disadvantage with this technique is the long plasma processing times. Considering a typical etching rate of ≈ 100 nm/min for SiC, for a 300 μm thick SiC substrate an etching time of about 50 h is required.

26.4.5 Chemical Etching and Spontaneous Self-Separation

Sapphire and SiC are very hard and chemically inert and are difficult to etch chemically. However, for other types of substrates, which are softer and less chemically inert, the removal process can be done more easily. Growth of GaN on GaAs(111) has been shown to be a possible route to fabricate free-standing GaN. The substrate can most easily be removed by wet etching in aqua regia after growth. Using GaAs as the substrate, 2 inch free-standing GaN layers with thickness of about 500 μm have been demonstrated [26.28]. Chemical etching has also shown to be useful for separation of GaN grown on LiGaO$_2$. LiGaO$_2$ can be rapidly etched in a buffered basic solution at 50 °C [26.67] with an etch rate of about 0.25 µm/min.

GaN layers grown on softer and less chemically inert substrates have shown in many cases to be spontaneously separated from the substrate after growth. Examples are HVPE-grown GaN on NdGaO$_3$ [26.68, 69], LiGaO$_2$ [26.39] or LiAlO$_2$ [26.38].

26.5 Doping Techniques for GaN in HVPE

Doping in HVPE can be done either by introducing a diluted dopant gas, mixing of the dopant in the Ga melt or using a separate doping source exposed to HCl in order to form gaseous metal halides.

26.5.1 n-Type Doping of GaN

Nitride semiconductors can easily be n-type doped; in fact they often exhibit unintentional n-type conductivity. This background doping has mainly been attributed to unintentionally introduced impurities, such as oxygen or silicon, which can be introduced from quartz parts within the reactor or from the process gases. For intentional n-type doping, silicon is the most suitable doping element. Silicon can, in principle, act as both an acceptor and a donor, depending on whether it is substituting a gallium atom or a nitrogen atom. However, the silicon atom mainly replaces a gallium atom due to the low covalent radii difference between Si and Ga compared with the radii difference from nitrogen.

Both gaseous sources and solid silicon sources have been used in growth of n-type HVPE GaN. The most commonly used doping gas for n-type doping of MOVPE-grown GaN is silane (SiH$_4$) or disilane (Si$_2$H$_6$). Silane has also been reported to be a useful source for HVPE growth of GaN [26.70]. However, the entrance zone where the Ga boat is situated is rather far away in most HVPE systems, which may results in problems with decomposition of the silane (or disilane) before it reaches the substrate, since the decomposition of the silane (disilane) starts already at temperatures below the temperature of the Ga boat. This problem can be solved by using more stable gases such as dichlorosilane (SiH$_2$Cl$_2$) that decomposes at higher temperatures. Using dichlorosilane, doping levels up to 8×10^{18} cm^{-3} have been reported [26.71].

Solid Si can also be used as a Si source by exposing the Si to HCl to form SiCl$_x$. Using this technique, a free electron concentration in GaN up to 10^{18} cm^{-3} has been reported [26.72]. The temperature of the Si source can control the doping level. By changing the temperature in the range 200–400 °C the Si concentration was controlled in the range 1.0×10^{17}–2.5×10^{19} cm^{-3}. However, only 30–50% of the Si atoms were activated as donors.

Bulk growth requires a stable well-controlled doping source. This can be a problem with a solid Si source, as pointed out in [26.71]. These authors observed that controllability was poor using a solid Si source due to changes of the exposed Si area and morphology during growth.

Other n-type dopants that have been used in MOVPE growth of GaN are O, Ge, and Sn. However, there are no reports on intentional doping of HVPE-grown GaN using these dopants.

26.5.2 p-Type Doping of GaN

One of the major challenges associated with growth and development of III–V nitride-based materials is to understand and control the p-type doping. Today, magnesium (Mg) is the only element that has been shown to be a relatively efficient and controllable p-type impurity. Other elements such as Zn and Cd have too large activation energies to be useful as a p-type dopant for GaN due to a very low activation of these dopants at room temperature. Be introduces a shallow acceptor, but suffers from self-compensation problems, so p-type material is not obtained with Be doping [26.73]. However, despite Mg being the most suitable acceptor, it still has a relatively high thermal activation energy ($E_a \approx 0.17$ eV, after [26.74]). The high ionization energy results in only a few percent of the Mg acceptors being ionized at room temperature. One additional problem with Mg doping, which was for a long time the bottleneck for the development of III-nitride technology, is that hydrogen can passivate the Mg dopants by forming an Mg–H complex. The Mg dopant can be activated by removing the hydrogen either by postgrowth annealing in N$_2$ or an electron irradiation treatment using low-energy electron-beam irradiation (LEEBI) [26.75, 76]. For thick bulk-like material, thermal annealing is the only useful technique for activation of the acceptors since the penetration depth of low-energy electrons is small. This treatment is normally done in situ, directly after the growth. However, despite the thermal post-growth treatments, the acceptor thermal activation efficiency is still only a few percent. Thus, obtaining a suitably high hole concentration ($> 10^{18}$ cm^{-3} at 300 K) in p-doped GaN using Mg is not an easy task. The upper limit of Mg acceptors that can be incorporated into the lattice is about 10^{20} cm^{-3}, which gives a hole concentration only in the low 10^{18} cm^{-3} range in c-plane GaN. Increasing the Mg concentration does not improve the conductivity; in fact, it has been reported that it leads to an even lower hole concentration [26.77]. It has been suggested that the solubility of Mg is limited due to competing formation of Mg$_3$N$_2$ [26.78]. Increasing the Mg concentration in GaN beyond this limit may result in precipitation of Mg$_3$N$_2$, which leads to a reduction in the hole concentration and a decrease in crystal quality.

Doping of Mg in HVPE can be done by either using a separate boat for the doping species or mixing it with the Ga source. The HCl gas is reacted with the Mg, forming its halide MgCl, which is transported to the growth zone by some suitable carrier gas. A second technique, which has been used both for Zn and Mg doping [26.79], is to thermally evaporate the dopant and transport the species in a suitable carrier gas to the growth zone. To avoid passivation of the acceptors by formation of Mg–H complexes, the rationale would be to avoid using hydrogen as carrier gas. However, it has been suggested that hydrogen is actually beneficial for p-type doping [26.80–83] compared with the hydrogen-free case. It has been proposed that the hydrogen passivates the Mg acceptors during growth and, consequently, represses formation of native donors to compensate for the acceptors. However, despite the technological breakthroughs in growth and development of GaN-based devices, many properties of this material system are not well understood. Doping of bulk growth of GaN using HVPE is not well documented and only a few scientific reports have been presented.

26.6 Defect Densities, Dislocations, and Residual Impurities

Thick HVPE-grown GaN layers are typically developed for the purpose of later use as thick high-quality substrates for epitaxial growth of device structures. Such substrates need lapping and chemical polishing to obtain a suitable surface finish for later epitaxy with other growth techniques. Defects such as minor surface features will be removed during these processing steps, and will therefore not be discussed here. Important problems are wafer bowing, dislocations and stacking faults, impurities, and localized bulk defects.

A major problem already discussed above is the bowing of bulk GaN wafers produced by HVPE on a foreign substrate, which is removed by some lift-off techniques. It is important to reduce the original bowing before lift-off as much as possible; some techniques involving ELO structures with cavities are successful in this respect. Another technique to accomplish this is to grow under conditions where an O-rich defective layer is present close to the substrate. This layer can absorb the deformation during growth and cooling, and give a low-strain (and low-bowing) GaN layer prior to lift-off [26.84]. For substrates suitable for epitaxy the bowing should ideally not be larger than about 10 μm (measured as the elevation of the middle of a 2 inch wafer compared with the perimeter). If the grown freestanding wafer has a substantially larger bowing, the subsequent polishing may produce a variable lattice tilt across the final wafer, i.e., a variable miscut. This in turn might cause a variable growth morphology across the wafer in a later epitaxial growth process for a device structure. If the wafers are prepared from boules, these boules are usually grown with a length of about 10 mm. In this case the memory of a bowing is typically very much reduced, but these properties are not well documented in the literature to date.

A problem related to bowing is cracking. For thick layers on sapphire, cracking typically already occurs in situ during the first tens of microns of growth, due to the tensional strain developed in this initial growth process [26.85]. This cracking helps relax the strain during growth. These cracks typically overgrow (Fig. 26.14) and are generally not seen after growth of a 300 μm layer. Their presence in the interior of the material may well initiate more severe cracking during cool down after growth or during the polishing step. When GaN buffer layers are used for the growth it seems that this initial cracking behavior can be avoided, and the strain in the layer accumulates during growth, increasing the bowing of the wafer. Cracking during cool down is then a more critical issue, depending on the wafer thickness.

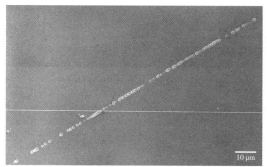

Fig. 26.14 Scanning electron microscopy (SEM) image of a HVPE-grown GaN layer on sapphire, showing an array of pits along a partly overgrown crack

Dislocations threading up through the wafer are severe defects in substrates, since they are typically replicated in the device structure epitaxially grown on top of the wafer. Dislocations are known to have a serious influence on carrier recombination in semiconductors; in GaN it seems that screw dislocations and mixed dislocations are most severe in terms of causing nonradiative recombination [26.86]. The threading dislocation density can be studied in several ways. Upon surface etching they typically form pits, which have a specific shape and size depending on the type of dislocations, and on the etch procedure used. This was studied by *Hino* et al. with HCl etching at 600 °C, where different pit shapes were produced by screw, mixed, and edge dislocations, respectively [26.86].

The dislocation types were established from separate transmission electron microscopy (TEM) studies. Complementary studies by cathodoluminescence (CL) topography reveal a different dark contrast, usually interpreted as due to nonradiative recombination, with strong contrast for screw and mixed dislocations, and quite a weak contrast for edge dislocations [26.86]. Similar results are observed from comparison of a different etching procedure (H_3PO_4 at 180 °C for 45 min) in Fig. 26.15. Here different sizes of pits observed by atomic force microscopy (AFM) (Fig. 26.15a) are observed [26.12]. The large pits are supposed to correspond to dislocations with a screw component, while the small pits correspond to edge dislocations. Figure 26.15b shows a CL topograph of another sample, showing mainly the high-contrast screw and mixed dislocations.

If the etching times are tuned to avoid overlap between different pits, the TD density of a layer can thus be determined by a simple etch pit count, a convenient method for a quick evaluation. The reliability of this technique can be tested by comparison of the dislocation count via etch pits and TEM studies of the same wafer, as reported in [26.86]. Obviously etch pit counting will omit dislocations that are running parallel to the surface, but these are usually not replicated in the overgrown epitaxial structure anyway.

The experience from HVPE growth of thick GaN layers is that the dislocation density is reduced with thickness in a manner that looks like a linear relation in a log–log plot, as seen in Fig. 26.16, where

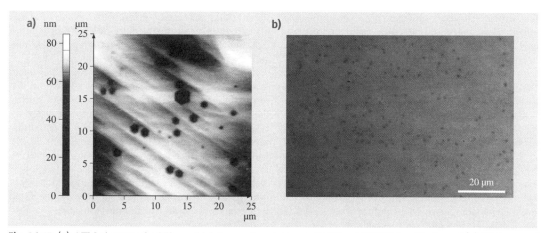

Fig. 26.15 (a) AFM pictures of a 250 μm thick layer after etching in H_3PO_4 at 180 °C for 45 min and (b) CL image of a 400 μm thick layer

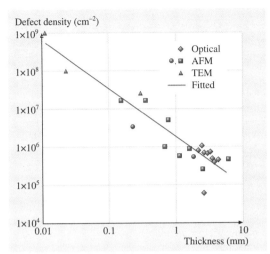

Fig. 26.17a,b Bright-field plan-view TEM images of an *a*-plane GaN film grown by HVPE with an MOVPE-GaN template layer taken close to the [11$\bar{2}$0] zone axis, slightly tilted: (**a**) along the [0001] direction, visualizing stacking faults formed in the basal plane of GaN by parallel lines, and (**b**) around the [0001] direction, visualizing partial dislocations having a nonzero [0001] component

Fig. 26.16 Threading dislocation density as a function of thickness of thick HVPE-grown GaN layers. *Points* labeled (◇), (■), and (▲) are taken from [26.89], while (●) from [26.87]

data from different groups are collected. For a thickness of 500 μm the TD density is typically at or below 10^7 cm^{-2}, while for a thickness of several millimeters a TD density of the order of 10^5 cm^{-2} is expected [26.87]. The mechanism for this has been investigated and suggested to be dislocation annihilation by recombination, driven by the dislocation strain fields and a certain mobility of dislocations at the growth temperature [26.88].

Stacking faults (SFs) are not abundant in GaN grown in the conventional polar (0001) direction. The situation is quite different in the case of growth of nonpolar wafers, such as *a*-plane and *m*-plane GaN on foreign substrates. *a*-Plane GaN can be grown on *r*-plane sapphire, and the defect density in such thick layers has been investigated [26.90]. Apart from dislocations a high density of stacking faults is observed, as shown in Fig. 26.17. These stacking faults thread up through the entire wafer, and are clearly undesirable defects, since they will propagate into an overgrown epitaxial structure [26.91]. The defects are found to be important recombination centers, and have a radiative emission characteristic for each type of SF (Fig. 26.18). This figure shows the low-temperature photoluminescence (PL) spectrum of an *a*-plane GaN layer grown by HVPE, with SF defects as shown in Fig. 26.17. The near-bandgap emission peak (NBE) is due to donor-

bound excitons, while the broad peak at 3.43 eV is related to basal-plane SFs [26.92].

Growth employing ELO techniques has been attempted to reduce the SF density in thick HVPE GaN layers [26.93]. In this way part of the wafer surface can be free from SFs. However, efficient methods to eliminate these defects over an entire nonpolar wafer have not yet been developed.

X-ray diffraction (XRD) is a standard technique to obtain an overall characteristic of the structural quality of a semiconductor sample. The presence of internal defects such as microcracks and wafer bowing strongly affects the XRD characteristics of thick GaN wafers produced in the way discussed above. The coherence length along the growth direction increases as the TD

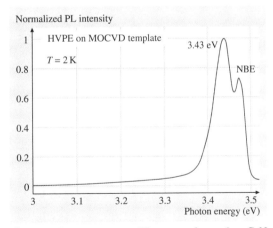

Fig. 26.18 Low-temperature PL spectra of an *a*-plane GaN layer grown by HVPE on an MOCVD-GaN template on *r*-plane sapphire

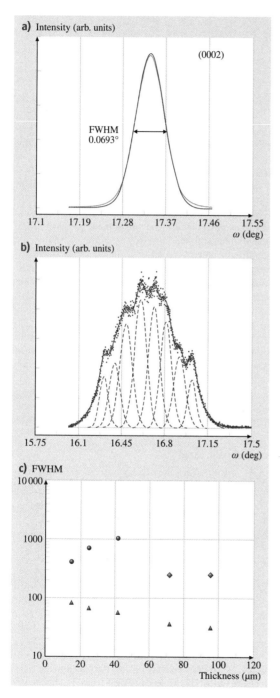

Fig. 26.19a–c Measured XRD line widths versus thickness for HVPE-grown GaN layers on sapphire. (**a**) Shows the line shape of a single rocking curve for a rather thin layer, while a thicker layer in (**b**) shows clear evidence of a mosaic substructure. (**c**) Shows a typical dependence of the line width on thickness for both rocking curve and the radial scan ◀

density is reduced. The full-width at half-maximum (FWHM) value of ω–2θ scans then decreases as the GaN film grows thicker, as seen in Fig. 26.19c for a set of moderately thick HVPE layers on sapphire. These data are consistent with those for another set of much thicker samples (about 300 μm), all having ω–2θ values of about 1 arcmin. The ω-scans show dramatically larger FWHM values, reflecting the presence of domain formation after a certain thickness (Fig. 26.19a,b), the residual bowing, and perhaps to some extent the internal cracks, and these effects are expected to contribute to the lattice tilt of the layers [26.94].

The characteristic XRD data for much thicker freestanding layers typically show much lower line widths. As an example, the FWHM for the ω-rocking curve of a 2 mm thick free-standing layer for the (002) reflection is 187 arcsec and the corresponding 2θ–ω peak has a FWHM of 36 arcsec, measured from the top Ga face [26.12].

Another class of defects of decisive importance for the quality of the material is impurities and point defects such as vacancies and interstitials. Residual impurities are introduced in the material during growth, and the density of such defects is critical mainly in the case of high-resistive material. The two common shallow donors in GaN are Si and O, both of which are easily introduced during growth, for different reasons. The Si contamination may strongly depend on the purity of the ammonia gas used; high-purity ammonia will minimize the Si contamination. There may also be some contribution from the attack on the quartz by parasitic growth upstream and in the growth zone. This process will depend on the proximity of the quartz walls to the substrate during growth. It has been shown that, with extended growth time, Si becomes the dominant residual impurity, while at the beginning of growth O is typically the dominant impurity incorporated. This is natural since O and water vapor are present, e.g., adsorbed or chemisorbed on quartz walls, or on the surface of the Ga source, and will gradually be released during a growth run, so that the amount of O contamination finally de-

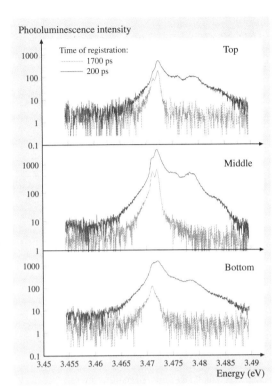

Fig. 26.20 Spatially resolved PL spectra at 2 K measured from different spots at the cross section of a 2 mm thick bulk GaN layer grown by HVPE, measured at two different delay times after the excitation pulse. The lower-energy O-related bound-exciton line (at about 3.471 eV) clearly dominates at the N face, while at the Ga face (grown last) Si is the dominant donor species (the peak at about 3.4725 eV) ◂

creases as the source of O is depleted. O can of course also be introduced via small air leaks in the system. An example is shown in Fig. 26.20, where the PL spectrum from the cross section of a 2 mm thick GaN wafer is dominated by the O bound exciton (BE) spectrum close to the N-side, but dominated by Si at the Ga side (grown

last). The ultimate value of Si contamination in a system not used for Si doping is presumably below 10^{15} cm^{-3}, but this has not been well documented to date.

Other important impurities are metal contaminants, in particular those from the iron group. These create deep levels in GaN, and thus act as compensation centers which decrease the efficiency of doping. The main source of these contaminants are the metal parts of the growth setup (tubing, manifolds, cylinders). Compounds including these metals may transport to the growth zone as particles, although these can be eliminated via particle filters. A more severe process is the transport of chloride molecules of these metals, promoted by the rather high vapor pressure they have already at room temperature. Such contaminants can easily be detected by optical spectroscopy, since these metals have characteristic spectra (typically sharp lines in the range 0.8–1.3 eV) related to the internal transitions in the ions [26.95]. Other important defects are vacancies. In n-type and undoped GaN the main vacancy defect is the V_{Ga}–O complex, with the so-called yellow luminescence (YL) as the optical signature [26.96]. These defects are deep acceptors, and can be eliminated by carefully excluding O from entering the material during growth.

26.7 Some Important Properties of HVPE-Grown Bulk GaN Material

At present, HVPE is the only method that can produce high-purity ($< 10^{16}$ cm^{-3}) strain-free GaN material with low dislocation density ($< 10^6$ cm^{-2}). Such material is therefore the prime choice for establishing the most relevant properties of bulk GaN. Such properties are lattice parameters, thermal expansion coefficients, thermal conductivity, and electron and hole mobilities. The lattice parameters obviously depend significantly on strain [26.97], but also on defect density and doping [26.98, 99], and values in the literature are therefore dependent on the material used. Recent studies on a 2 mm thick nominally undoped, free-standing GaN bulk layer with a residual doping of about 10^{16} cm^{-3} and a dislocation density of about 10^6 cm^{-2} probably represent the most accurate values of the lattice parameters for pure bulk GaN (Table 26.2). The lattice parameters as a function of temperature were also measured recently on a free-standing GaN layer, and thermal expansion coefficients were established with improved accuracy [26.100]. The thermal conductivity is another important property that depends strongly on the quality of the material. Early determinations of this quantity on rather defective HVPE GaN layers gave a value of 1.3 W/(cm K) [26.101].

Table 26.2 Some important material parameters for strain-free bulk GaN

Lattice parameter a (Å)	Lattice parameter c (Å) [26.105]	Thermal expansion coefficient α_a (10^{-6}/K) at 300 K	Thermal conductivity 300 K (W/(cm K))	Electron mobility 300 K (cm^2/(V s)) $n < 10^{16}$ cm^{-3}	Hole mobility 300 K (cm^2/(V s)) $p \approx 10^{18}$ cm^{-3}
3.18943 ±0.00015	5.18501 ±0.00015	4.3	2.3	1350	10

A more recent study on HVPE material with low dislocation density and low doping indicates that the phonon-scattering-limited thermal conductivity at room temperature is much higher, about 2.3 W/(cm K), i.e., considerably better than Si [26.102]. The best electron mobilities in GaN has been obtained in nominally undoped free-standing thick HVPE grown samples; a value of 1350 cm^2/(V s) has been reported [26.103]. The hole mobilities have so far been measured only in highly doped and highly defective GaN; no reliable data are available for low-doped HVPE-grown bulk p-GaN to our knowledge. Extrapolated values for such material indicate a limiting hole mobility above 200 cm^2/(V s) [26.104], to be confirmed in the future. In Mg-doped epitaxially grown device structures with a typical Mg concentration close to 10^{20} cm^{-3}, the hole mobility is at best about 10 cm^2/(V s) [26.104] (see also Table 26.2).

26.8 Growth of AlN by HVPE: Some Preliminary Results

High-quality and large-scale AlN wafers are recently in increasing demand as substrates for ultraviolet (UV) light-emitting devices and high-power high-frequency electronic devices. There have been several reports on growth of AlN by the sublimation–recondensation method [26.106, 107] and by solution growth [26.108]. Although AlN crystals with extremely low dislocation densities can be grown by these methods, expanding the size of the grown crystals is a challenging problem.

On the other hand, high-speed growth of thick AlN layers by HVPE followed by separation of the grown layers from the starting substrates is an interesting approach to prepare AlN wafers. As is widely known, GaN wafers of 2 inch diameter have been mass-produced by HVPE using GaAs [26.109] or sapphire [26.59, 62, 110, 111] as starting substrates. However, investigations concerning HVPE of AlN have been limited [26.1, 112–115] due to the fact that the molten Al or hot AlCl gas reacts violently with the quartz (SiO$_2$) reactor of the HVPE system, damaging the reactor. On the other hand, there are other Al chlorides such as AlCl$_2$, AlCl$_3$, and (AlCl$_3$)$_2$. Figure 26.21 shows the values of the equilibrium constants K of the thermodynamically feasible reactions between gaseous AlCl, AlCl$_2$, AlCl$_3$ or (AlCl$_3$)$_2$ and quartz as a function of the reciprocal of the reaction temperature [26.1]. The following order of the equilibrium constants K of the reactivity of Al chlorides is seen

$$K_{AlCl} > K_{AlCl_2} > K_{(AlCl_3)_2} > K_{AlCl_3} . \quad (26.13)$$

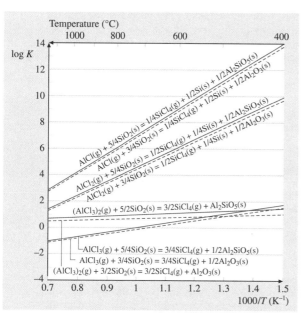

Fig. 26.21 Equilibrium constants (K) as a function of the reciprocal of reaction temperature for reactions between Al chlorides and quartz

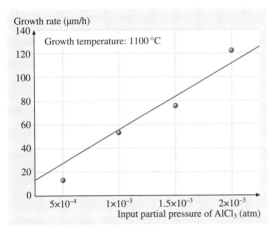

Fig. 26.22 Dependence of AlN growth rate on the $AlCl_3$ input partial pressure at $1100\,°C$

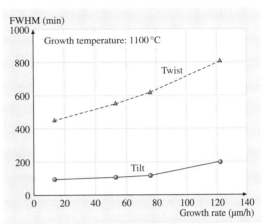

Fig. 26.23 FWHM of XRD rocking curves of the (0002) (tilt component) and the $(10\bar{1}0)$ (twist component) of AlN layers grown at $1100\,°C$ with various growth rates

The reactions between AlCl and quartz have extremely large values of K. On the contrary, reactions between $AlCl_3$ and quartz have small values of K, and negative values of log K at $700\,°C$ or above. Therefore, the reaction between $AlCl_3$ and the quartz reactor is negligible, and $AlCl_3$ is suitable as an Al source in AlN HVPE.

It has already been reported that preferential generation of $AlCl_3$ is possible [26.1] at the Al source zone of a conventional HVPE system by decreasing the source-zone temperature to about $500\,°C$ (Fig. 26.1b). Exploiting this fact, several groups have grown AlN using a conventional hot-walled quartz reactor. In Fig. 26.22, the dependence of AlN growth rate on the $AlCl_3$ input partial pressure is shown [26.114]. Al metal pellets (6 N grade) were placed in the source zone to enlarge the surface area and maintained at $500\,°C$. $AlCl_3$ was formed by the reaction between the Al metal and HCl gas and mixed with NH_3 in the growth zone where a (0001) sapphire substrate was placed. The input V/III ratio ($NH_3/AlCl_3$) was fixed at 2. Growth was performed at $1100\,°C$ under atmospheric pressure using H_2 as a carrier gas. A linear increase of the growth rate with increase of $AlCl_3$ input pressure is seen in the figure. With an $AlCl_3$ input partial pressure of 2×10^{-3} atm, the growth rate exceeds $100\,\mu m/h$. Therefore, high-speed growth of AlN by HVPE is possible just as in the HVPE growth of GaN.

Figure 26.23 shows the FWHM of the x-ray diffraction (XRD) rocking curves of (0002) (tilt component) and $(10\bar{1}0)$ (twist component) of AlN layers grown at $1100\,°C$ with various growth rates [26.114]. Although high-speed growth of AlN is possible at $1100\,°C$, the FWHM of the (0002) and the $(10\bar{1}0)$ increase with increasing growth rate. In order to grow high-quality AlN layers with a high growth rate, it is considered that a higher growth temperature will be required.

The result of absorption measurements for a $2\,\mu m$ thick AlN layer grown at $2\,\mu m/h$ on a sapphire substrate

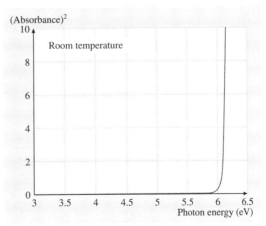

Fig. 26.24 Absorbance squared versus photon energy of an AlN layer measured at room temperature

is shown in Fig. 26.24, where absorbance squared is plotted as a function of photon energy. Extrapolation of the linear region to the horizontal axis gives the bandgap of AlN to be just above 6.0 eV (somewhat influenced by strain).

From the results shown above, the next challenges for AlN HVPE are thought to be high-temperature (> 1100 °C) growth and reduction of dislocation density in the grown layer using AlN templates, ELO structures, etc.

26.9 Growth of InN by HVPE: Some Preliminary Results

Recent progress in the growth of InN by MBE and MOVPE has clarified that the bandgap of InN is around 0.7 eV [26.116–118]. However, growth of thick InN layers remains difficult due to poor thermal stability, low growth rate of InN, and the lack of a suitable substrate material for epitaxy.

Although HVPE is a useful method for high-speed growth of GaN [26.47] and AlN [26.114] without formation of metal droplets, HVPE of InN is complicated: InN growth does not occur when InCl is used as an In source [26.119], while appreciable growth of InN occurs when $InCl_3$ powder is sublimated and transported by a carrier gas other than H_2 [26.119–121]. This is due to the fact that the equilibrium constant of the reaction between $InCl_3$ and NH_3 is larger than that between InCl and NH_3 (Fig. 26.2), and a suppression of InN growth in the presence of H_2 [26.7]. However, in order to grow high-quality InN layers without contamination, formation of $InCl_3$ gas in the source zone of a conventional HVPE system is needed, since commercially available $InCl_3$ powder inevitably contains water due to its hygroscopic nature. According to the thermodynamic analysis of the In source zone, where Cl_2 gas is introduced over the In metal, the equilibrium partial pressure of $InCl_3$ increases with decreasing source-zone temperature [26.122]. Here, it is essential to use Cl_2 instead of HCl to form $InCl_3$ because the reaction between In metal and HCl produces H_2, which suppresses the growth of InN at the growth zone [26.7].

We report preliminary results for successful HVPE growth of InN. Figure 26.25 shows the dependence of InN growth rate on the In source-zone temperature [26.122]. Cl_2 with an input partial pressure of 3.0×10^{-3} atm was introduced over the In metal using N_2 carrier gas, and NH_3 with an input partial pressure of 1.5×10^{-1} atm was separately introduced into the growth zone using N_2 carrier gas. Single-crystalline InN layers c-axis oriented without In droplets can be grown on a (0001) sapphire substrate at a growth zone temperature of 500 °C. It is seen that the growth rate increases with decreasing source-zone temperature above 450 °C. This is due to the increase of the $InCl_3$ equilibrium partial pressure at the In source zone. On the other

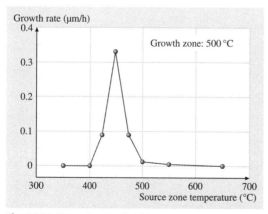

Fig. 26.25 Dependence of InN growth rate on the In source-zone temperature. Growth-zone temperature was fixed at 500 °C

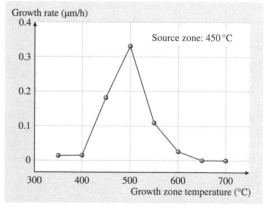

Fig. 26.26 Dependence of InN growth rate on the growth-zone temperature. In source-zone temperature was fixed at 450 °C

hand, with the source-zone temperature below 450 °C, the growth rate dropped rapidly because of insufficient reaction between Cl_2 and In metal.

As for the growth-zone temperature, InN growth is obtained within a narrow temperature range. Figure 26.26 shows the dependence of InN growth rate on the growth-zone temperature. The temperature of the In source zone was fixed at 450 °C. At growth temperature above 500 °C, the growth rate decreases as the growth-zone temperature increases. The growth rate also decreases with decrease of the growth-zone temperature below 500 °C due to insufficient reaction between $InCl_3$ and NH_3. Therefore, either an increase of Cl_2 input into the In source zone or an increase of the total gas flow rate in the reactor will be required to increase the growth rate of InN.

A room-temperature cathodoluminescence spectrum of the InN layer grown at 500 °C is shown in Fig. 26.27. An emission peak is seen at around 0.7 eV. Hall-effect measurements revealed that the InN layer had n-type conductivity with a carrier concentration of 3×10^{19} cm^{-3} and a carrier mobility of 890 cm^2/(V s).

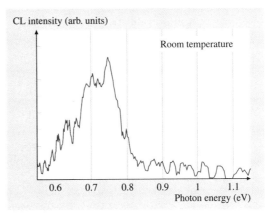

Fig. 26.27 Room-temperature CL spectrum of an InN layer grown at 500 °C by HVPE

The carrier concentration obtained is in agreement with that estimated from the dependence of the bandgap on electron concentration due to the Burstein–Moss effect [26.123].

References

26.1 Y. Kumagai, T. Yamane, T. Miyaji, H. Murakami, Y. Kangawa, A. Koukitu: Hydride vapor phase epitaxy of AlN: thermodynamic analysis of aluminum source and its application to growth, Phys. Status Solidi (c) **0**, 2498–2501 (2003)

26.2 M.W. Chase Jr. (Ed.): *NIST-JANAF Thermochemical Tables*, 4th edn. (National Institute of Standards and Technology, Gaithersburg 1998)

26.3 I. Barin (Ed.): *Thermochemical Data of Pure Substances* (Wiley, New York 1989)

26.4 A. Koukitu, S. Hama, T. Taki, H. Seki: Thermodynamic analysis of hydride vapor phase epitaxy of GaN, Jpn. J. Appl. Phys. **37**, 762–765 (1998)

26.5 Y. Kumagai, K. Takemoto, T. Hasegawa, A. Koukitu, H. Seki: Thermodynamics on tri-halide vapor-phase epitaxy of GaN and In$_x$Ga$_{1-y}$N using GaCl$_3$ and InCl$_3$, J. Cryst. Growth **231**, 57–67 (2001)

26.6 Y. Kumagai, K. Takemoto, J. Kikuchi, T. Hasegawa, H. Murakami, A. Koukitu: Thermodynamics on hydride vapor phase epitaxy of AlN using AlCl$_3$ and NH$_3$, Phys. Status Solidi (b) **243**, 1431–1435 (2006)

26.7 Y. Kumagai, K. Takemoto, A. Koukitu, H. Seki: Thermodynamics on halide vapor-phase epitaxy of InN using InCl and InCl$_3$, J. Cryst. Growth **222**, 118–124 (2001)

26.8 A. Koukitu, H. Seki: Thermodynamic analysis of the vapor growth of GaAs: the inert gas-hydrogen mixed carrier system, Jpn. J. Appl. Phys. **16**, 1967–1971 (1977)

26.9 V.S. Ban: Mass spectrometric studies of vapor-phase crystal growth. II. GaN, J. Electrochem. Soc. **119**, 761–765 (1972)

26.10 R.J. Molnar, W. Götz, L.T. Romano, J.M. Johnson: Growth of gallium nitride by hydride vapor-phase epitaxy, J. Cryst. Growth **178**, 147–156 (1997)

26.11 B. Monemar, H. Larsson, C. Hemmingsson, I.G. Ivanov, D. Gogova: Growth of thick GaN layers with hydride vapour phase epitaxy, J. Cryst. Growth **281**, 17–31 (2005)

26.12 C. Hemmingsson, P.P. Paskov, G. Pozina, M. Heuken, B. Schineller, B. Monemar: Growth of bulk GaN in a vertical hydride vapour phase epitaxy reactor, Superlattices Microstruct. **40**, 205–213 (2006)

26.13 V.S. Ban: Mass spectrometric and thermodynamics studies of the CVD of some III-V compounds, J. Cryst. Growth **17**, 19–30 (1972)

26.14 E. Richter, C. Hennig, M. Weyers, F. Habel, J.-D. Tsay, W.-Y. Liu, P. Brückner, F. Scholz, Y. Makarov, A. Segal, J. Kaeppeler: Reactor and growth process optimization for growth of thick GaN layers on sapphire substrates by HVPE, J. Cryst. Growth **277**, 6–12 (2005)

26.15 S. Porowski: High pressure growth of GaN – New prospects for blue lasers, J. Cryst. Growth **166**, 583–589 (1996)

26.16 I. Grzegory, B. Łucznik, M. Bočkowski, B. Pastuszka, G. Kamler, G. Nowak, M. Krysko,

26.17 S. Krukowski, S. Porowski: Crystallization of GaN by HVPE on pressure grown seeds, Phys. Status Solidi (a) **203**, 1654–1657 (2006)

26.17 B. Łucznik, B. Pastuszka, I. Grzegory, M. Boćkowski, G. Kamler, E. Litwin-Staszewska, S. Porowski: Deposition of thick GaN layers by HVPE on the pressure grown GaN substrates, J. Cryst. Growth **281**, 38–46 (2005)

26.18 L. Liu, J.H. Edgar: Substrates for gallium nitride epitaxy, Mater. Sci. Eng. Rep. **37**, 61–127 (2002)

26.19 O. Ambacher, J. Majewski, C. Miskys, A. Link, M. Hermann, M. Eickhoff, M. Stutzmann, F. Bernardini, V. Fiorentini, V. Tilak, B. Schaff, L.F. Eastman: Pyroelectric properties of Al(In)GaN/GaN hetero- and quantum well structures, J. Phys. Condens. Matter **14**, 3399–3434 (2002)

26.20 T. Paskova, P.P. Paskov, E. Valcheva, V. Darakchieva, J. Birch, A. Kasic, B. Arnaudov, S. Tungasmita, B. Monemar: Polar and nonpolar GaN grown by HVPE: Preferable substrates for nitride-based emitting devices, Phys. Status Solidi (a) **201**, 2265–2270 (2004)

26.21 W.A. Melton, J.I. Pankove: GaN growth on sapphire, J. Cryst. Growth **178**, 168–173 (1997)

26.22 B. Pécz, M.A. Di Forte-Poisson, F. Huet, G. Radnóczi, L. Tóth, V. Papaioannou, J. Stoemenos: Growth of GaN layers onto misoriented (0001) sapphire by metalorganic chemical vapor deposition, J. Appl. Phys. **86**, 6059–6067 (1999)

26.23 K. Hiramatsu, H. Amano, I. Akasaki, H. Kato, N. Koide, K. Manabe: MOVPE growth of GaN on a misoriented sapphire substrate, J. Cryst. Growth **107**, 509–512 (1991)

26.24 P. Brückner, M. Feneberg, K. Thonke, F. Habel, F. Scholz: High quality GaN layers grown on slightly miscut sapphire wafers, Mater. Res. Soc. Symp. Proc. **892**, 511–516 (2006)

26.25 R.J. Molnar, P. Maki, R. Aggarwal, Z.L. Liau, E.R. Brown, I. Melngailis, W. Götz, L.T. Romano, N.M. Johnson: Gallium nitride thick films grown by hydride vapor phase epitaxy, Mater. Res. Soc. Symp. **423**, 221–226 (1996)

26.26 S. Keller, B.P. Keller, Y.F. Wu, B. Heying, D. Kapolnek, J.S. Speck, U.K. Mishra, S.P. DenBaars: Influence of sapphire nitridation on properties of gallium nitride grown by metalorganic chemical vapor deposition, Appl. Phys. Lett. **68**, 1525–1527 (1996)

26.27 S. Fuke, H. Teshigawara, K. Kuwahara, Y. Takano, T. Ito, M. Yanagihara, K. Ohtsuka: Influences of initial nitridation and buffer layer deposition on the morphology of a (0001) GaN layer grown on sapphire substrates, J. Appl. Phys. **83**, 764–767 (1998)

26.28 K. Motoki, T. Okahisa, S. Nakahata, N. Matsumoto, H. Kimura, H. Kasai, K. Takemoto, K. Uematsu, M. Ueno, Y. Kumagai, A. Koukitu, H. Seki: Growth and characterization of freestanding GaN substrates, J. Cryst. Growth **237–239**, 912–921 (2002)

26.29 M. Ishida, T. Hashimoto, T. Takayama, O. Imafuji, M. Yuri, A. Yoshikama, K. Itoh, Y. Terakoshi, T. Sugino, J. Shirafuji: Growth of GaN thin films on sapphire substrate by low pressure MOCVD, Mater. Res. Soc. Symp. **468**, 69–74 (1997)

26.30 F. Widmann, G. Feuillet, B. Daudin, J.L. Rouviere: Low temperature sapphire nitridation: A clue to optimize GaN layers grown by molecular beam epitaxy, J. Appl. Phys. **85**, 1550–1555 (1999)

26.31 K. Naniwae, S. Itoh, H. Amano, K. Itoh, K. Hiramatsu, I. Akasaki: Growth of single crystal GaN substrate using hydride vapor phase epitaxy, J. Cryst. Growth **99**, 381–384 (1990)

26.32 R.J. Molnar, K.B. Nichols, P. Maki, E.R. Brown, I. Melngailis: The role of impurities in hydride vapor phase epitaxial grown gallium nitride, Mater. Res. Soc. Symp. Proc. **378**, 479–484 (1995)

26.33 J. Lee, H. Paek, J. Yoo, G. Kim, D. Kum: Low temperature buffer growth to improve hydride vapor phase epitaxy of GaN, Mater. Sci. Eng. B **59**, 12–15 (1999)

26.34 P.R. Tavernier, E.V. Etzkorn, Y. Wang, D.R. Clarke: Two-step growth of high-quality GaN by hydride vapor-phase epitaxy, Appl. Phys. Lett. **77**, 1804–1806 (2000)

26.35 H. Tsuchiya, K. Sunaba, M. Minami, T. Suemasu, F. Hasegawa: Influence of As autodoping from GaAs substrates on thick cubic GaN growth by halide vapor phase epitaxy, Jpn. J. Appl. Phys. **37**, L568–L570 (1998)

26.36 F. Hasegawa, M. Minami, K. Sunaba, T. Suemasu: Thick and smooth hexagonal GaN growth on GaAs (111) substrates at 1000 °C with halide vapor phase epitaxy, Jpn. J. Appl. Phys. **38**, L700–L702 (1999)

26.37 Y. Kumagai, H. Murakami, H. Seki, A. Koukitu: Thick and high-quality GaN growth on GaAs (111) substrates for preparation of freestanding GaN, J. Cryst. Growth **246**, 215–222 (2002)

26.38 E. Richter, C. Hennig, U. Zeimer, M. Weyers, G. Tränkle, P. Reiche, S. Ganschow, R. Uecker, K. Peters: Freestanding two inch c-plane GaN layers grown on (100) γ-lithium aluminium oxide by hydride vapour phase epitaxy, Phys. Status Solidi (c) **3**, 1439–1443 (2006)

26.39 O. Kryliouk, M. Reed, T. Dann, T. Anderson, B. Chai: Large area GaN substrates, Mater. Sci. Eng. B: Solid-State Mater. Adv. Technol. **66**, 26–29 (1999)

26.40 T. Paskova, S. Tungasmita, E. Valcheva, E. Svedberg, B. Arnaudov, S. Evtimova, P. Persson, A. Henry, R. Beccard, M. Heuken, B. Monemar: Hydride vapour phase homoepitaxlal growth of GaN on MOCVD-grown 'templates', MRS Internet J. Nitrid. Semicond. Res. **5**, W3.14 (2000)

26.41 H. Lee, M. Yuri, T. Ueda, J.S. Harris, K. Sin: Growth of thick GaN films on RF sputtered AlN buffer layer by hydride vapor phase epitaxy, J. Electron. Mater. **26**, 898–902 (1997)

26.42 E. Valcheva, T. Paskova, S. Tungasmita, P.O.Å. Persson, J. Birch, E.B. Svedberg, L. Hultman, B. Monemar: Interface structure of hydride vapor phase epitaxial GaN grown with high-temperature reactively sputtered AlN buffer, Appl. Phys. Lett. **76**, 1860–1862 (2000)

26.43 T. Paskova, E. Valcheva, J. Birch, S. Tungasmita, P.O.Å. Persson, P.P. Paskov, S. Evtimova, M. Abrashev, B. Monemar: Defect and stress relaxation in HVPE-GaN films using high temperature reactively sputtered AlN buffer, J. Cryst. Growth **230**, 381–386 (2001)

26.44 T. Detchprohm, K. Hiramatsu, H. Amano, I. Akasaki: Hydride vapor phase epitaxial growth of a high quality GaN film using a ZnO buffer layer, Appl. Phys. Lett. **61**, 2688–2690 (1992)

26.45 T. Ueda, T.-F. Huang, S. Spruytte, H. Lee, M. Yuri, K. Itoh, T. Baba, J.S. Harris Jr: Vapor phase epitaxy growth of GaN on pulsed laser deposited ZnO buffer layer, J. Cryst. Growth **187**, 340–346 (1998)

26.46 S. Gu, R. Zhang, J. Sun, L. Zhang, T.F. Kuech: The nature and impact of ZnO buffer layers on the initial stages of the hydride vapor phase epitaxy of GaN, MRS Internet J. Nitrid. Semicond. Res. **5**, W3.15 (2000)

26.47 A. Usui, H. Sunakawa, A. Sakai, A.A. Yamaguchi: Thick GaN epitaxial growth with low dislocation density by hydride vapor phase epitaxy, Jpn. J. Appl. Phys. **36**, L899–L902 (1997)

26.48 K. Hiramatsu, H. Matsushima, T. Shibata, N. Sawaki, K. Tadatomo, H. Okagawa, Y. Ohuchi, Y. Honda, T. Matsue: Selective area growth of GaN by MOVPE and HVPE, Mater. Res. Soc. Symp. Proc. **482**, 257–268 (1998)

26.49 G. Nataf, B. Beaumont, A. Bouillé, S. Haffouz, M. Vaille, P. Gibart: Lateral overgrowth of high quality GaN layers on GaN/Al$_2$O$_3$ patterned substrates by halide vapour-phase epitaxy, J. Cryst. Growth **192**, 73–78 (1998)

26.50 O. Parillaud, V. Wagner, H.J. Bühlmann, M. Ilegems: Localized Epitaxy of GaN by HVPE on patterned Substrates, MRS Internet J. Nitrid. Semicond. Res. **3**, 40 (1998)

26.51 Y. Zhonghai, M.A.L. Johnson, J.D. Brown, N.A. El-Masrys, J.F. Muth, J.W. Cook Jr, J.F. Schetzina, K.W. Haberern, H.S. Kong, J.A. Edmond: Epitaxial lateral overgrowth of GaN on SiC and sapphire substrates, MRS Internet J. Nitrid. Semicond. Res. **4S1**, G4.3 (1999)

26.52 R. F. Davis, O. H. Nam, T. Zheleva, M. Bremser: Methods of fabricating gallium nitride semiconductor layers by lateral overgrowth through masks, and gallium nitride semiconductor structures fabricated thereby, Patent W09944224 (1999)

26.53 J. Hirth, J. Lothe: The theory of straight dislocations. In: *Theory of Dislocations* (Wiley, New York 1982) pp. 55–95

26.54 P. Gibart, B. Beaumont, P. Vennéguès: Epitaxial lateral overgrowth of GaN. In: *Nitride Semiconductors, Handbook on Materials and Devices*, ed. by P. Ruterana, M. Albrecht, J. Neugebauer (Wiley, New York 2003) pp. 45–106

26.55 Z. Liliental-Weber, M. Benamara, W. Swider, J. Washburn, J. Park, P.A. Grudowski, C.J. Eiting, R.D. Dupuis: TEM study of defects in laterally overgrown GaN layers, MRS Internet J. Nitrid. Semicond. Res. **4S1**, G4.6 (2000)

26.56 H. Miyake, H. Mizutani, K. Hiramatsu, Y. Iyechika, Y. Honda, T. Maeda: Fabrication of GaN layer with low dislocation density using facet controlled ELO technique, Mater. Res. Symp. Proc. **639**, G5.3.1. (2001)

26.57 S.-I. Nagahama, N. Iwasa, M. Senoh, T. Matsushita, Y. Sugimoto, H. Kiyoku, T. Kozaki, M. Sano, H. Matsumura, H. Umemoto, K. Chocho, T. Mukai: High-power and long-lifetime InGaN multi-quantum-well laser diodes grown on low-dislocation-density GaN substrates, Jpn. J. Appl. Phys. **39**, L647–L650 (2000)

26.58 W.S. Wong, T. Sands, N.W. Cheung: Damage-free separation of GaN thin films from sapphire substrates, Appl. Phys. Lett. **72**, 599–601 (1998)

26.59 M.K. Kelly, R.P. Vaudo, V.M. Phanse, L. Gorgens, O. Ambacher, M. Stutzmann: Large free-standing GaN substrates by hydride vapor phase epitaxy and laser-induced liftoff, Jpn. J. Appl. Phys. **38**, L217–L219 (1999)

26.60 O. Ambacher, M.S. Brandt, R. Dimitrov, T. Metzger, M. Stutzmann, R.A. Fischer, A. Miehr, A. Bergmaier, G. Dollinger: Thermal stability and desorption of group III nitrides prepared by metal organic chemical vapor deposition, J. Vac. Sci. Technol. B **14**, 3532–3542 (1996)

26.61 D. Gogova, A. Kasic, H. Larsson, C. Hemmingsson, B. Monemar, F. Toumisto, K. Saarinen, L. Dobos, B. Pécz, P. Gibart, B. Beaumont: Strain-free bulk-like GaN grown by hydride-vapor-phase-epitaxy on two-step epitaxial lateral overgrown GaN template, J. Appl. Phys. **96**, 799–806 (2004)

26.62 Y. Oshima, T. Eri, M. Shibata, H. Sunakawa, K. Kobayashi, T. Ichihashi, A. Usui: Preparation of freestanding GaN wafers by hydride vapor phase epitaxy with void-assisted separation, Jpn. J. Appl. Phys. **42**, L1–L3 (2003)

26.63 P.R. Tavernier, B. Imer, S.P. DenBaars, D.R. Clarke: Growth of thick (11$\bar{2}$0) GaN using a metal interlayer, Appl. Phys. Lett. **85**, 4630–4632 (2004)

26.64 H.M. Kim, J.E. Oh, T.W. Kang, H.M. Kim, J.E. Oh, T.W. Kang: Preparation of large area free-standing GaN substrates by HVPE using mechanical polishing liftoff method, Mater. Lett. **47**, 276–280 (2001)

26.65 Y.V. Melnik, K.V. Vassilevski, I.P. Nikitina, A.I. Babanin, V.Y. Davydov, V.A. Dmitriev: Physical

26.66 Y. Melnik, A. Nikolaev, I. Nikitina, K.V. Vassilevski, V.A. Dmitriev: Properties of free-standing GaN bulk crystals grown by HVPE, Mater. Res. Soc. Symp. Proc. **482**, 269–274 (1998)

26.67 T.J. Kropewnicki, W.A. Doolittle, C. Carter-Coman, S. Kang, P.A. Kohl, N.M. Jokerst, A.S. Brown, S. April: Selective wet etching of lithium gallate, J. Electrochem. Soc. **145**, L88–L90 (1998)

26.68 O. Oda, T. Inoue, Y. Seki, K. Kainosho, S. Yaegashi, A. Wakahara, A. Yoshida, S. Kurai, Y. Yamada, T. Taguchi: GaN bulk substrates for GaN based LEDs and LDs, Phys. Status Solidi (a) **180**, 51–58 (2000)

26.69 A. Wakahara, T. Yamamoto, K. Ishio, A. Yoshida, Y. Seki, K. Kainosho, O. Oda: Hydride vapor phase epitaxy of GaN on $NdGaO_3$ substrate and realization of freestanding GaN wafers with 2-inch scale, Jpn. J. Appl. Phys. **39**, 2399–2401 (2000)

26.70 R. P. Vaudo, V. M. Phanse, J. Jayapalan, D. Wang, B.J. Skromme: Si-doping of GaN grown on sapphire by HVPE, abstract book, MRS fall meeting, (1997) D16.3

26.71 E. Richter, C. Hennig, U. Zeimer, L. Wang, M. Weyers, G. Tränkle: n-Type doping of HVPE-grown GaN using dichlorosilane, Phys. Status Solidi (a) **203**, 1658–1662 (2006)

26.72 A.V. Fomin, A.E. Nikolaev, I.P. Nikitina, A.S. Zubrilov, M.G. Mynbaeva, N.I. Kuznetsov, A.P. Kovarsky, B.J. Ber, D.V. Tsvetkov: Properties of Si-doped GaN layers grown by HVPE, Phys. Status Solidi (a) **188**, 433–437 (2001)

26.73 C. G. Van de Walle, S. Limpijumnong, J. Neugebauer: First-principles studies of beryllium doping of GaN, Phys. Rev. B **63**, 245205-1–245205-17 (2001)

26.74 W. Götz, N.M. Johnson, J. Walker, D.P. Bour, R.A. Street: Activation of acceptors in Mg-doped GaN grown by metalorganic chemical vapor deposition, Appl. Phys. Lett. **68**, 667–669 (1996)

26.75 H. Amano, M. Kito, K. Hiramatsu, I. Akasaki: p-Type conduction in Mg-doped GaN treated with low-energy electron beam irradiation (LEEBI), Jpn. J. Appl. Phys. **28**, L2112–L2114 (1989)

26.76 S. Nakamura, T. Mukai, M. Senoh, N. Iwasa: Thermal annealing effects on p-type Mg-doped GaN films, Jpn. J. Appl. Phys. **31**, L139–L142 (1992)

26.77 D.P. Bour, H.F. Chung, W. Götz, L. Romano, B.S. Krusor, D. Hofstetter, S. Rudaz, C.P. Kuo, F.A. Ponce, N.M. Johnson, M.G. Craford, R.D. Bringans: Characterization of OMVPE-grown AlGaInN heterostructures, Mater. Res. Soc. Symp. Proc. **449**, 509–518 (1997)

26.78 J. Neugebauer, C.G. Van de Walle: Theory of point defects and complexes in GaN, Mater. Res. Soc. Symp. Proc. **395**, 645–656 (1996)

26.79 A. Usikov, O. Kovalenkov, V. Ivantsov, V. Sukhoveev, V. Dmitriev, N. Shmidt, D. Poloskin, V. Petrov, V. Ratnikov: p-Type GaN epitaxial layers and AlGaN/GaN heterostructures with high hole concentration and mobility grown by HVPE, Mater. Res. Soc. Symp. Proc. **831**, 453–457 (2005)

26.80 J.A. Van Vechten, J.D. Zook, R.D. Horning, B. Goldenberg: Defeating compensation in wide gap semiconductors by growing in H that is removed by low temperature de-ionizing radiation, Jpn. J. Appl. Phys. **31**, 3662–3663 (1992)

26.81 J. Neugebauer, C.G. Van de Walle: Atomic geometry and electronic structure of native defects in GaN, Phys. Rev. B **50**, 8067–8070 (1994)

26.82 J. Neugebauer, C.G. Van de Walle: Hydrogen in GaN: Novel aspects of a common impurity, Phys. Rev. Lett. **75**, 4452–4455 (1995)

26.83 J. Neugebauer, C.G. Van de Walle: Gallium vacancies and the yellow luminescence in GaN, Appl. Phys. Lett. **69**, 503–505 (1996)

26.84 A. Kasic, D. Gogova, H. Larsson, I.G. Ivanov, C. Hemmingsson, R. Yakimova, B. Monemar, M. Heuken: Characterization of crack-free relaxed GaN grown on 2″ sapphire, J. Appl. Phys. **98**, 073525-1–073525-6 (2005)

26.85 E.V. Etzkorn, D.R. Clarke: Cracking of GaN films, J. Appl. Phys. **89**, 1025–1034 (2001)

26.86 T. Hino, S. Tomiya, T. Miyajima, K. Yanashima, S. Hashimoto, M. Ikeda: Characterization of threading dislocations in GaN epitaxial layers, Appl. Phys. Lett. **76**, 3421–3423 (2000)

26.87 R.P. Vaudo, X. Xu, C. Loria, A.D. Salant, J.S. Flynn, G.R. Brandes: GaN boule growth: A pathway to GaN wafers with improved material quality, Phys. Status Solidi (a) **194**, 494–497 (2002)

26.88 M. Albrecht, I.P. Nikitina, A.E. Nikolaev, V.Y. Melnik, V.A. Dmitriev, H.P. Strunk: Dislocation reduction in AlN and GaN bulk crystals grown by HVPE, Phys. Status Solidi (a) **176**, 453–458 (1999)

26.89 CREE homepage: www.cree.com/products/gan_tech.asp (24-Sep 2008)

26.90 T. Paskova, V. Darakchieva, P.P. Paskov, J. Birch, E. Valcheva, P.O.A. Persson, B. Arnaudov, S. Tungasmitta, B. Monemar: Properties of nonpolar a-plane GaN films grown by HVPE with AlN buffers, J. Cryst. Growth **281**, 55–61 (2005)

26.91 T. Paskova, R. Kroeger, P.P. Paskov, S. Figge, D. Hommel, B. Monemar, B. Haskell, P. Fini, J.S. Speck, S. Nakamura: Microscopic emission properties of nonpolar a-plane GaN grown by HVPE, Proc. SPIE **6121**, 47 (2006)

26.92 P.P. Paskov, T. Paskova, B. Monemar, S. Figge, D. Hommel, B.A. Haskell, P.T. Fini, J.S. Speck, S. Nakamura: Optical properties of nonpolar a-plane GaN layers, Superlattices Microstruct. **40**, 253–261 (2006)

26.93 B.A. Haskell, F. Wu, M.D. Craven, S. Matsuda, P.T. Fini, T. Fujii, K. Fujito, S.P. DenBaars, J.S. Speck, S. Nakamura: Defect reduction in ($11\bar{2}0$) a-plane gallium nitride via lateral epitaxial overgrowth by

26.94 B. Monemar, T. Paskova, C. Hemmingsson, H. Larsson, P.P. Paskov, I.G. Ivanov, A. Kasic: Growth of thick GaN layers by hydride vapor phase epitaxy, J. Ceram. Process. Res. **6**, 153–162 (2005)

26.95 J. Baur, U. Kaufmann, M. Kunzer, J. Schneider, H. Amano, I. Akasaki, T. Detchprohm, K. Hiramatsu: Characterization of residual transition metal ions in GaN and AlN, Mater. Sci. For. **196-201**, 55–60 (1995)

26.96 K. Saarinen, T. Laine, S. Kuisma, J. Nissilä, P. Hautojärvi, L. Dobrzynski, J.M. Baranowski, K. Pakula, R. Stepniewski, M. Wojdak, A. Wysmolek, T. Suski, M. Leszczynski, I. Grzegory, S. Porowski: Observation of native Ga vacancies in GaN by positron annihilation, Phys. Rev. Lett. **79**, 3030–3033 (1997)

26.97 M. Leszczynski, H. Teisseyre, T. Suski, I. Grzegory, M. Boćkowski, J. Jun, S. Porowski, K. Pakula, J.M. Baranowski, C.T. Foxon, T.S. Cheng: Lattice parameters of gallium nitride, Appl. Phys. Lett. **69**, 73–75 (1996)

26.98 V. Darakchieva, P.P. Paskov, T. Paskova, E. Valcheva, B. Monemar, M. Heuken: Lattice parameters of GaN layers grown on a-plane sapphire: Effect of in-plane strain anisotropy, Appl. Phys. Lett. **82**, 703–705 (2003)

26.99 C. G. Van de Walle: Effects of impurities on the lattice parameters of GaN, Phys. Rev. B **68**, 165209-1–165209-5 (2003)

26.100 C. Roder, S. Einfeldt, S. Figge, D. Hommel: Temperature dependence of the thermal expansion of GaN, Phys. Rev. B **72**, 085218-1–085218-6 (2005)

26.101 E.K. Sichel, J.I. Pankove: Thermal conductivity of GaN, 25–360 K, J. Phys. Chem. Solids. **38**, 330 (1977)

26.102 A. Jezowski, P. Stachowiak, T. Plackowski, T. Suski, S. Krukowski, M. Boćkowski, I. Grzegory, B. Danilchecnko, T. Paszkiewicz: Thermal conductivity of GaN crystals grown by high pressure method, Phys. Status Solidi (b) **240**, 447–450 (2003)

26.103 D.C. Look, J.R. Sizelove: Predicted maximum mobility in bulk GaN, Appl. Phys. Lett. **79**, 1133–1135 (2001)

26.104 J.W. Orton, C.T. Foxon: Group III nitride semiconductors for short wavelength light-emitting devices, Rep. Prog. Phys. **61**, 1–75 (1998)

26.105 V. Darakchieva, private communication, 2006

26.106 J.C. Rojo, G.A. Slack, K. Morgan, B. Raghothamachar, M. Dudley, L.J. Schowalter: Report on the growth of bulk aluminum nitride and subsequent substrate preparation, J. Cryst. Growth **231**, 317–321 (2001)

26.107 R. Schlesser, R. Dalmau, Z. Sitar: Seeded growth of AlN bulk single crystals by sublimation, J. Cryst. Growth **241**, 416–420 (2002)

26.108 M. Boćkowski, M. Wróblewski, B. Łucznik, I. Grzegory: Crystal growth of aluminum nitride under high pressure of nitrogen, Mater. Sci. Semicond. Process. **4**, 543–548 (2001)

26.109 K. Motoki, T. Okahisa, N. Matsumoto, M. Matsushima, H. Kimura, H. Kasai, K. Takemoto, K. Uematsu, T. Hirano, M. Nakayama, S. Nakahata, M. Ueno, D. Hara, Y. Kumagai, A. Koukitu, H. Seki: Preparation of large freestanding GaN substrates by hydride vapor phase epitaxy using GaAs as a starting substrate, Jpn. J. Appl. Phys. **40**, L140–L143 (2001)

26.110 S.S. Park, I.-W. Park, S.H. Choh: Free-standing GaN substrates by hydride vapor phase epitaxy, Jpn. J. Appl. Phys. **39**, L1141–L1142 (2000)

26.111 D. Gogova, H. Larsson, A. Kasic, G.R. Yazdi, I. Ivanov, R. Yakimova, B. Monemar, E. Aujol, E. Frayssinet, J.-P. Faurie, B. Beaumont, P. Gibart: High-quality 2″ bulk-like free-standing GaN grown by hydride vapor phase epitaxy on a Si-doped metal organic vapour phase epitaxial GaN template with an ultra low dislocation density, Jpn. J. Appl. Phys. **44**, 1181–1185 (2005)

26.112 A. Nikolaev, I. Nikitina, A. Zubrilov, M. Mynbaeva, Y. Melnik, V. Dmitriev: AlN wafers fabricated by hydride vapor phase epitaxy, Mater. Res. Soc. Symp. Proc. **595**, W6.5 (2000)

26.113 O.Y. Ledyaev, A.E. Cherenkov, A.E. Nikolaev, I.P. Nikitina, N.I. Kuznetsov, M.S. Dunaevski, A.N. Titkov, V.A. Dmitriev: Properties of AlN layers grown on SiC substrates in wide temperature range by HVPE, Phys. Status Solidi (c) **0**, 474–478 (2002)

26.114 Y. Kumagai, T. Yamane, A. Koukitu: Growth of thick AlN layers by hydride vapor-phase epitaxy, J. Cryst. Growth **281**, 62–67 (2005)

26.115 Y.-H. Liu, T. Tanabe, H. Miyake, K. Hiramatsu, T. Shibata, M. Tanaka, Y. Masa: Growth of thick AlN Layer by hydride vapor phase epitaxy, Jpn. J. Appl. Phys. **44**, L505–L507 (2005)

26.116 V.Y. Davydov, A.A. Klochikhin, R.P. Seisyan, V.V. Emtsev, S.V. Ivanov, F. Bechstedt, J. Furthmüller, H. Harima, A.V. Mudryi, J. Aderhold, O. Semchinova, J. Graul: Absorption and emission of hexagonal InN. Evidence of narrow fundamental band gap, Phys. Status Solidi (b) **229**, R1–R3 (2002)

26.117 J. Wu, W. Walukiewicz, K.M. Yu, J.W. Ager III, E.E. Haller, H. Lu, W.J. Schaff, Y. Saito, Y. Nanishi: Unusual properties of the fundamental band gap of InN, Appl. Phys. Lett. **80**, 3967–3969 (2002)

26.118 T. Matsuoka, H. Okamoto, M. Nakao, H. Harima, E. Kurimoto: Optical bandgap energy of wurtzite InN, Appl. Phys. Lett. **81**, 1246–1248 (2002)

26.119 N. Takahashi, J. Ogasawara, A. Koukitu: Vapor phase epitaxy of InN using InCl and $InCl_3$ sources, J. Cryst. Growth **172**, 298–302 (1997)

26.120 L.A. Marasina, I.G. Pichugin, M. Tlaczala: Preparation of InN epitaxial layers in $InCl_3$-NH_3 system, Krist. Tech. **12**, 541–545 (1977)

26.121 N. Takahashi, R. Matsumoto, A. Koukitu, H. Seki: Growth of InN at high temperature by halide vapor

26.122 Y. Kumagai, J. Kikuchi, Y. Nishizawa, H. Murakami, A. Koukitu: Hydride vapor phase epitaxy of InN by the formation of $InCl_3$ using In metal and Cl_2, J. Cryst. Growth **300**, 57–61 (2007)

phase epitaxy, Jpn. J. Appl. Phys. **36**, L743–L745 (1997)

26.123 J. Wu, W. Walukiewicz, S.X. Li, R. Armitage, J.C. Ho, E.R. Weber, E.E. Haller, H. Lu, W.J. Schaff, A. Barcz, R. Jakiela: Effects of electron concentration on the optical absorption edge of InN, Appl. Phys. Lett. **84**, 2805–2807 (2004)

27. Growth of Semiconductor Single Crystals from Vapor Phase

Ramasamy Dhanasekaran

Growth of single crystals from the vapor phase is considered to be an important method to obtain stoichiometric crystalline materials from inexpensive and readily available raw materials. Elements or compounds which are relatively volatile can be grown from vapor phase. Most II–VI, I–III–VI$_2$, and III–N compounds are high-melting-point materials which may be grown as single crystals by careful use of vapor phase. The chemical vapor transport (CVT) method has been widely used as an advantageous method to grow single crystals of different compounds at temperatures lower than their melting points. This method is quite useful for the growth of II–VI and I–III–VI$_2$ compounds, which generally have high melting point and large dissociation pressure at the melting point. In addition, they undergo solid-state phase transition during cooling or heating processes, which makes the growth of these compounds by some other methods, such as from the melt, difficult. In addition, the low growth temperature involved reduces defects produced by thermal strain, pollution from the crucible, and the cost of the growth equipment. II–VI compound semiconductors cover a very broad range of electronic and optical properties due to the large range of their energy gaps. These materials in the form of bulk single crystals or thin films are used in light emitters, detectors, linear and nonlinear optical devices, semiconductor electronics, and other devices. The development of growth technology for II–VI compound semiconductors from the vapor phase with the necessary theoretical background is important. I–III–VI$_2$ chalcopyrite compounds are of technological interest since they show promise for application in areas of visible and infrared light-emitting diodes, infrared detectors, optical parametric oscillators, upconverters, far-infrared generation, and solar energy conversion.

27.1 Classifications of Vapor Growth 899
27.2 Chemical Vapor Transport – Transport Kinetics 901
    27.2.1 Transport Models 901
    27.2.2 Physical Chemistry of Chemical Transport Reactions... 902
    27.2.3 Factors Affecting the CVT Reaction 903
    27.2.4 Choice of Transporting Agents...... 904
    27.2.5 Advantages and Limitations of CVT Method 904
27.3 Thermodynamic Considerations 905
    27.3.1 Estimation of Optimum Growth Parameters for the ZnSe–I$_2$ System by CVT 905
    27.3.2 Fluctuations in the Transport Rates................. 906
    27.3.3 Supersaturation Ratios in the ZnS$_x$Se$_{1-x}$ System 908
27.4 Growth of II–VI Compound Semiconductors by CVT 912
    27.4.1 Apparatus 912
    27.4.2 Preparation of Starting Materials.. 913
    27.4.3 Growth of ZnSe Single Crystals 914
    27.4.4 Growth of CdS Single Crystals 915
27.5 Growth of Nanomaterial from Vapor Phase 916
27.6 Growth of I–III–VI$_2$ Compounds.............. 917
    27.6.1 Growth of Undoped and Doped Crystals of CuAlS$_2$ 918
    27.6.2 Growth of Undoped and Doped Crystals of CuAlSe$_2$ 919
    27.6.3 Growth of CuGaS$_2$–Based Single Crystals ... 921
    27.6.4 Growth of AgGaS$_2$ and AgGaSe$_2$ Single Crystals 923
27.7 Growth of GaN by VPE 925
    27.7.1 Vapor-Phase Epitaxy (VPE)........... 925
    27.7.2 VPE GaN Film Growth 925
    27.7.3 Strength of HVPE Method............. 926
    27.7.4 Development of VPE System for the Growth of GaN 926

27.7.5 Growth of GaN by HVPE 927
27.7.6 Characterization
of GaN Films 928

27.8 **Conclusion** 929

References 930

A great deal of interest in these materials is generated by their chalcopyrite structure, which is noncentrosymmetric and makes them useful for second-harmonic generation. III-nitride semiconductors are of great interest to industry and the military due to their optoelectronic and mechanical properties, permitting the development of devices operating in the blue and ultraviolet regions of the spectrum and at high temperatures. Gallium nitride has particularly attracted considerable attention in this regard. Despite the progress recently achieved, some fundamental properties of GaN and its related compounds, such as InGaN and AlGaN, are still not fully understood. Epitaxial methods continue to lead the field of crystal growth by exploring new physics, materials science, and fabrication of novel devices. Good-quality GaN layers have been grown by employing vapor-phase epitaxy systems, the technical details of which are presented in detail herein.

During recent decades, ternary, quaternary, and more recently multinary semiconductor materials have been widely investigated because of their importance for solid-state device applications. However, their crystal growth technology is far from well understood. Due to practical difficulties involved in growing single crystals of most of these materials by melt techniques, regular production of small samples perfect enough and of good enough quality almost always depends upon the availability of vapor-phase (VP) chemical transport reactions (usually *iodine* transport) in closed tubes. One of the common drawbacks of these VP methods is the high level of supersaturation, and hence uncontrolled primary nucleation, which can give rise to many small-sized (often submillimeter) crystals. This high supersaturation is also responsible for various instability patterns (dendrites, overgrowths, etc.) that show up in the final crystals. Crystal growth from the vapor phase is in principle a flexible method of growing single crystals. This potential flexibility has occasionally been exploited, but more often the principles have been poorly understood [27.1]. Preparation of a crystal from the gas phase requires the general reaction

GAS (disordered phase) $\longrightarrow$
CRYSTAL (ordered phase) .

The gas phase may consist of molecules of the crystal substance, or of its separate constituents, if they are all volatile. Otherwise, the constituents may be reacted with a transporting agent to provide volatile species. In all cases, an inert gas is added to modify the transport kinetics. The basic reaction results in an increase in atomic order.

Important research efforts have been focused on II–VI compounds since the late 1950s. Because of their superiority over rival materials, these compounds are becoming of more importance day by day in research into electronic materials. Many experimental procedures have been improved, and theoretical models have been developed. However, in spite of their positive properties, the crystallographic properties of some II–VI semiconductor materials (e.g., ZnS and ZnSe) have hindered their application in the electronic industry due to polytypism.

Currently, I–III–VI$_2$ compounds (chalcopyrite ternary semiconductor), such as copper indium diselenide (CuInSe$_2$), copper indium disulfide (CuInS$_2$), and copper gallium diselenide (CuGaSe$_2$), are the most promising low-cost thin-film materials for solar cell applications due to their high absorption coefficient and excellent thermal stability. Photovoltaic (PV) cells based on these materials have potentials to be used on the Earth as well as in various space applications. Scientists from the US National Renewable Energy Laboratory (NREL), have achieved efficiency of about 19.2% with CuInSe$_2$ absorber in laboratory cells. The best efficiency of a thin-film solar cell up to now is 19.5%, by using Cu(In,Ga)Se$_2$ as absorber [27.2], which is comparable to that for polycrystalline Si solar cells. In view of their potential low cost and light weight, these solar cells have been expected to be useful in space applications in addition to the terrestrial domain. Thus, there is strong demand for improvement of the efficiency of PV cells based on these chalcopyrite materials.

The demonstration of blue light-emitting diodes (LEDs) and blue laser diodes (LDs) [27.3] has led to the emergence of III–V nitrides as a suitable material for high-power, high-temperature electronic and optoelectronic device production. The development of high-quality nitride-based devices is still restricted by the lack of suitable epitaxial substrates materials. Nitride-based structures are most commonly deposited on sapphire (Al_2O_3) substrates. The lattice mismatch of about 15% to GaN causes formation of threading dislocations (TDs) in the sapphire interface and device layers [27.4]. In order to improve GaN epitaxial layers, a two-step deposition process is often applied in both metalorganic chemical vapor deposition (MOCVD) [27.5, 6] and hydride vapor-phase epitaxy (HVPE) [27.7] systems. The application of HVPE technology assures high growth rates of about 500 nm/min for growth of GaN. Hence, sufficiently thick layers can be deposited in a relatively short time. This makes HVPE an advantageous fabrication option for freestanding GaN substrates.

In this chapter, growth of semiconductor single crystals from the vapor phase is explained in general, and the chemical vapor transport technique in particular. Growth of II–VI and I–III–VI_2 compounds is reviewed, and growth of GaN thin films using vapor-phase epitaxy is presented.

27.1 Classifications of Vapor Growth

The growth of crystal from vapor phase can be divided [27.7] into various categories:

1. *Sublimation* or *evaporation*, in which the vapor is obtained from the pure condensed phase at an appropriate temperature;
2. A compound may highly dissociate in the growth system, namely *dissociative sublimation*;
3. If a transporting reaction is used for one or more of the constituents of the crystal, the process is termed either *chemical vapor transport* (CVT) or *physical vapor transport* (PVT), distinguished on the basis of the conditions over the growing crystal. A comparison between PVT, physical vapor deposition (PVD), chemical vapor deposition (CVD), and CVT is shown in Table 27.1

PVT is a closed-tube technique that can be used to grow crystals if the vapor pressure of the material exceeds 10^{-2} Torr at some feasible temperature. Typically in these processes, the source is a solid which must be sublimed (dissociatively or similarly) to provide the vapor. The process by which vapor is obtained from the pure condensed phase at an appropriate temperature and transported to a deposition zone at a lower temperature is called physical vapor transport. The vapor atoms impinge on the substrate surface and become adsorbed, releasing part of their latent heat of condensation and migrating across the crystal surface. They become incorporated into the crystal lattice, releasing their remaining heat of condensation; otherwise they evaporate. This is a relatively high-temperature method.

In the CVT process, as shown in Table 27.1, vapor atoms or molecules are chemically different from those of the growing crystal. Chemical transport reactions are those in which a solid or liquid substance A reacts with a gas to form exclusively volatile products, which in turn undergo the reverse reaction at a different place in the system, resulting in the formation of A. The process appears to be sublimation; however, it does not possess appreciable vapor pressure at the temperatures applied. In addition to a reversible heterogeneous reaction, a concentration gradient must be established. The latter can be the result of a temperature difference, changes in the relative pressures, or the difference in free energy of formation of two substances.

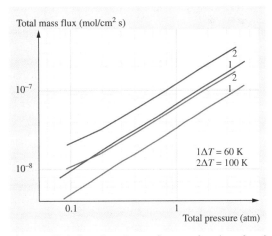

Fig. 27.1 Variation of total mass flux as a function of total pressure in the CdSe–I_2 system for the given two temperature differences between the source zone and growth zone and for the given diameter of 15 cm and length of 14 cm of the growth ampoule

Table 27.1 Comparison between PVT, CVT, PVD, and CVD

Parameters	Physical vapor transport (PVT)	Chemical vapor transport (CVT)	Physical vapor deposition (PVD)	Chemical vapor deposition (CVD)
Growth mechanism and bond energy	Monatomic, associative or dissociative sublimation Direction always from hot to cold ($T_s \to T_g$)	Reversible chemical reaction migration from hot to cold ($T_s \to T_g$) as well as from cold to hot ($T_g \to T_s$) possible	Direction always from hot to cold Nonconformal deposition Highly directional deposition Bonding energy is weak, less than 10 kcal/mol or 0.434 eV/atom	Conformal deposition Multidirectional deposition Bonding energy is strong compared with physical bond, up to 200 kcal/mol
Transport (carrier) gas	Without transport agent or inert gas	Needs a transporting agent such as I_2, Cl_2, etc.	Not necessary	Needs a transporting agent such as nitrogen, argon, etc.
Temperature difference (ΔT)	1–10 K	3–500 K	≈ 50 K to Several 100 K	≈ 50–100 K
Total system pressure (P)	High vacuum (or) inert gas of 0.01–0.5 atm	Up to 3 atm	1 atm	Atmospheric pressure to very low pressure
Thermophysical parameter	Constant	Varying	Constant	Varying
Compressible effect	Insignificant	Significant	Insignificant	Significant
Aspect ratio (l/d)	Typically 5	Typically 10	–	–
Possible flow pattern	Diffusion/advective and solutal convection	Diffusion/advective and thermal convection	Diffusion/advective and solutal convection	Diffusion/advective, thermal convection and laminar for HVPE GaN

The stationary temperature profile (STP) technique has been successfully applied for growing crystals since the growth of crystals by chemical vapor transport was initiated several decades ago. In the STP procedure, the source temperature (T_s) and growth temperature (T_g) are fixed at particular values after keeping a reverse temperature profile for the first few hours in order to clean the crystallization zone. As T_s and T_g are kept at constant values, the supersaturation remains unchanged during the whole growth period. Though this procedure can be adopted to obtain single crystals of many compounds, it may not be the most suitable technique for obtaining large-size single crystals. Also in most cases one ends up with too many crystals as the supersaturation is maintained constant throughout the growth experiment.

In the time-varying temperature profile (TVTP) method, the source temperature is increased linearly with time from a value below the growth temperature, while the growth temperature is kept constant. Hence the supersaturation increases form a negative value (i. e., at $T_s < T_g$) to positive values at which the formation of primary nuclei is possible. After the formation of primary nuclei, although T_s continues to increase, the supersaturation drops due to the decrease of the vaporized mass, which is consumed in producing the first crystal. Then, for a certain length of time depending on the particular vaporization kinetics of the charge, supersaturation remains sufficiently low that the first crystal can be further grown by secondary nucleation. After this, supersaturation increases again, enabling primary nucleation to take place. This TVTP procedure was adopted for growing ternary chalcopyrite crystals. It enables the growth of crystals much larger than those grown by the STP procedure. Also the quality of the crystals has been reported to be superior to those grown by adopting the STP procedure.

One of the drawbacks of the STP procedure is often the onset of constitutional undercooling from the very beginning of the growth process. In this respect, the TVTP procedure is an improvement because, in the first part of the growth period, this cause of instability is avoided. Constitutional undercooling appears as a result of both large supersaturation and the presence of mass convection in the vapor. Supersaturation can practically be reduced by working with small T_s and T_g differences, possibly at temperatures and pressures where thermodynamics predict low supersaturation. Mass convection, which can be reduced by working at suitably small values of the Grashof number, induces constitutional undercooling through the reduction of the diffusion boundary layer at the growing interface. It is concluded that the time-varying temperature profile method described above can be used to increase the size of the crystals of some ternary compounds. Enlargements of about two- to threefold were obtained by this procedure, as compared with the usual STP technique. Time-varying temperature profile methods have been applied by various researchers to improve the size of CdS [27.8], $CdIn_2S_4$, and $CuInS_2$ [27.9] single crystals grown by CVT technique.

In PVD there is a transfer of subliming molecules/atoms A by (saturated) carrier gas to a colder growth area.

In chemical vapor deposition gaseous compounds of the materials to be deposited are transported to a substrate surface, where a thermal reaction/deposition occurs. Reaction byproducts are then exhausted out of the system.

The other vapor growth methods are:

1. Metalorganic chemical vapor deposition (MOCVD)
2. Vapor-phase epitaxy (VPE)
3. Metalorganic vapor-phase epitaxy (MOVPE)
4. Organometallic vapor-phase epitaxy (OMVPE).

27.2 Chemical Vapor Transport – Transport Kinetics

Crystal growth from the melt using the classical methods of Czochralski, Kyropoulos or Bridgman becomes difficult if the compounds to be grown have high melting points. Additional difficulties arise if the compounds show appreciable dissociation at the melting point or melt only under elevated pressure. To obtain single crystals of such compounds, they are sublimed (either in vacuum or in a stream of carrier gas or vapor of its constituents) and allowed to react in the crystallization chamber [27.10–12]. The temperature required for these processes is high, in the vicinity of the sublimation point. Instead of directly vaporizing a solid at high temperature it may be vaporized at much lower temperature by forming highly volatile chemical intermediates and reacting back the resulting gas mixture at a different temperature, utilizing the temperature dependence of the chemical equilibrium involved [27.12–16]. By properly adjusting the two temperatures, the departure from chemical equilibrium in the vicinity of the growing seeds can be made small enough to avoid nucleation but large enough to make the seeds grow, i.e., appropriate supersaturation can be maintained.

27.2.1 Transport Models

In general transport models have been developed based on the following experimental conditions:

1. The system is one dimensional and the length of the growth ampoule is L units, with the source at $x = 0$, transported to a deposit at $x = L$.

2. The transport mechanism involves a chemical equilibrium which can be written as

$$S_{(s)} + aA_{(g)} = bB_{(g)} + cC_{(g)} + \ldots, \quad (27.1)$$

associated with an equilibrium constant K and an enthalpy change ΔH, where $S_{(s)}$ is the crystalline source materials, A is the transporting agent, and B and C are the gaseous products.

3. Diffusion is the only means of transport through the vapor phase; turbulent flow and thermal diffusion are ignored.

4. The mean free path is much smaller than the dimension of the system. As a consequence, a total pressure gradient cannot be considered. Impedance to the flow due to total pressure gradients is so small that we can ignore these gradients relative to partial pressure gradients.

5. The surface rate of the reaction is infinitely fast, so local equilibrium is essentially established in the neighborhoods of the source and the substrate.

6. The change of the equilibrium constant upon moving from $x = 0$ to $x = L$, leads to a difference in temperature between the source and growth zones ΔT, which is much smaller than T_g. An equivalent statement is $\Delta T = T_s - T_g \ll (RT_g T_s / \Delta H)$, according to the usual thermodynamic relationship between ΔH and ΔT.

7. $|\Delta H| \gg \text{RT} \cong \frac{1}{2} R(T_g + T_s)$. This also implies $\Delta T \ll T_g$.

8. All gases behave ideally.

Following *Schäfer*'s model [27.12], a transport reaction is defined as a reaction of a solid or liquid with a gaseous component in a temperature gradient to form exclusively gaseous products. For chemical transport caused only by diffusion, such as in a sealed silica ampoule, *Schäfer* [27.12] obtained a transport equation for the migration of a solid in a one-dimensional system. *Lever* [27.14] described the chemical transport of a solid involving several simultaneous heterogeneous and homogeneous equilibria [27.17, 18]. *Arizumi* and *Nishinaga* [27.19–21] proposed transport equations for the transport of materials (semiconductors) in closed tubes, assuming a one-dimensional system and considering diffusion and laminar flow as the fundamental transport mechanism (AN model).

Following the model of *Factor* and *Garrett* [27.22, 23] in a one-dimensional transport system, the flux rate J (mol/cm^2 s) of a component i caused by diffusion and Stefan flow can be described. The variation of total mass flux as a function of total pressure in the CdSe–I$_2$ system for various growth temperatures for given dimensions of the growth ampoule (diameter 15 cm and length 14 cm) is shown in Fig. 27.1.

Using the flux function method of *Richardson* and *Noläng*, it is possible to calculate the net fluxes J_k (mol/cm^2 s) based on an equilibrium model [27.24–26]. The flux function takes into account the Stefan flow, as well as the temperature and total pressure variation of the diffusion coefficient. Thus it becomes possible to compute the transport rates of the simultaneous migration of several condensed phases. A flux function Φ takes into account the temperature dependency of the diffusion coefficient. The advantage of this flux function is that different transport systems with different transporting agents, temperature gradients, and diffusion coefficients can be compared with each other [27.27–29].

27.2.2 Physical Chemistry of Chemical Transport Reactions

Partial Pressure

Consider a closed system containing $n+1$ components, all of which are in chemical equilibrium with one other [27.30]. The temperature is chosen such that only one component is solid, whereas the other n components are in gaseous state. For $n = 1$, the trivial case of sublimation of a solid via its gaseous atoms or molecules results. In the following, the case $n = 3$ is considered, which applies to the transport of metal chalcogenides by halogens. The equilibriums involved are [27.12]

$$\mathrm{MX_{(s)}} + \mathrm{I_{2(g)}} \Leftrightarrow \mathrm{MI_{2(g)}} + \tfrac{1}{2}\mathrm{X_{2(g)}}, \quad (27.2)$$

where M = Cd, Zn and X = S, Se.
A specific example is [27.13, 18, 31]

$$\mathrm{ZnS_{(s)}} + \mathrm{I_{2(g)}} \Leftrightarrow \mathrm{ZnI_{2(g)}} + \tfrac{1}{2}\mathrm{S_{2(g)}}. \quad (27.3)$$

In a CVT system, a solid substance MX reacts with a gaseous transporting agent I$_2$ (chalcogenides) at a higher temperature T_s to form exclusively vapor-phase products MX and $\tfrac{1}{2}$X at the source zone. The vapor products in turn undergo the reverse reaction at the growth zone at a lower temperature T_g

$$\mathrm{I_{2(g)}} \Leftrightarrow 2\mathrm{I_{(g)}}. \quad (27.4)$$

The corresponding equilibrium constants of (27.2) and (27.4) are

$$K_{(27.2)} = \frac{p_{\mathrm{MI_2}} \left(\tfrac{1}{2}p_{\mathrm{MI_2}}\right)^{\frac{1}{2}}}{p_{\mathrm{I_2}}}, \quad (27.5)$$

$$K_{(27.4)} = \frac{p_{\mathrm{I}}^2}{p_{\mathrm{I_2}}}, \quad (27.6)$$

respectively.
The total pressure of the system is given by

$$P = p_{\mathrm{I_2}} + p_{\mathrm{I}} + 1.5 p_{\mathrm{MI_2}}. \quad (27.7)$$

The equilibrium constant $K_i(T)$ ($i = 3.2, 3.4$) as a function of temperature T can be evaluated using the relations (27.2) and (27.4–27.6) [27.27–29]

$$\log K_i(T) = a_i - b_i T^{-1} - c_i T^{-2} - d_i \log(T) - e_i T. \quad (27.8)$$

If the temperature T and total pressure p are fixed and the value of $K_i(T)$ is known, then the partial pressure of species MI$_2$, X$_2$, I$_2$, and I can be determined by using (27.2) and (27.4–27.8). A graph showing the partial pressure of various gaseous species as a function of total pressure of the CdSe–I$_2$ system is shown in Fig. 27.2; for ZnSe–I$_2$ and other systems, refer to literature [27.13, 31–34].

Thermophysical Parameters

The diffusion coefficient of a binary gas mixture is based on the well-known Chapman–Enskog formula [27.29, 32, 35, 36]

$$D_{ij} = 1.858^{-27} \frac{\sqrt{T^3(1/\mathrm{M}_i + 1/\mathrm{M}_j)}}{P\sigma_{ij}^2 \Omega_{D_{ij}}}. \quad (27.9)$$

In (27.9) σ_{ij} is the characteristic length expressed in Å and Ω_D is the dimensionless diffusion collision integral. The binary collision integral was estimated from individual collision diameters σ_A and σ_B using the expression

$$\sigma_{ij} = \frac{\sigma_i + \sigma_j}{2}, \quad (27.10)$$

where Ω_D is a function of $k_B T/\varepsilon_{ij}$, where k_B is the Boltzmann's constant, $\varepsilon_{ij} = (\varepsilon_i \varepsilon_j)^{1/2}$, and ε_i and ε_j are the Lennard–Jones energy values for species i and j, respectively. The complete expression for Ω_D can be given as

$$\Omega_D = \frac{A}{T^{*B}} + \frac{C}{\exp(DT^*)}, \quad (27.11)$$

where $T^* = k_B T/\varepsilon_{ij}$ is the reduced temperature and the values of A, B, C, and D can be calculated from reported data. The diffusion collision integral and diffusion coefficients for various species are presented in Table 27.2.

The molecular viscosity μ and thermal conductivity k_T of the species can be calculated from Chapman–Enskog and Lennard–Jones potential theory [27.29, 37, 38]

$$\mu = 2.669 \times 10^{-26} \frac{\sqrt{MT}}{\sigma_{ij}^2 \Omega_\mu} \; [\text{kg}/(\text{m s})] \quad (27.12)$$

and

$$k_T = 8.324 \times 10^{-22} \frac{\sqrt{(T/M)}}{\sigma_{ij}^2 \Omega_\mu} \; [\text{W}/(\text{m K})], \quad (27.13)$$

where M is the molecular weight (kg/mol), T is the average temperature (K), σ_{ij} is the collision diameter, and Ω_μ is the viscosity collision integral.

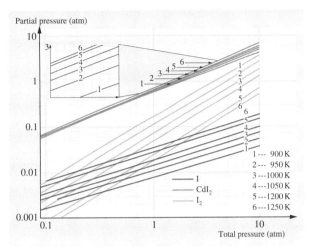

Fig. 27.2 Partial pressures of various gaseous species as a function of total pressure of the CdSe–I_2 system for different temperatures

27.2.3 Factors Affecting the CVT Reaction

Influence of Chemical Parameters on CVT Growth of Crystals

Generally chemical vapor transport reactions depend on chemical parameters such as the free energy change (ΔG), the concentration of the transporting agent (C_T), the temperatures at the source and growth zones (T_s and T_g), etc. [27.29, 39–42]. The free energy change for a reaction depends to a considerable extent on the partial pressures of the chemical species used in the reactions. The amount of material transportation from the source to the growth zone depends strongly on C_T. Also it has been observed by many authors that there is a strong dependence of the size and quality of the resulting crystals on C_T.

Table 27.2 Characteristic length and diffusion collision integral of different species

ij	σ_{ij} (Å)	Ω_{D-ij} at $T = 1173$ K	ij	σ_{ij} (Å)	Ω_{D-ij} at $T = 1173$ K
I_2-S_2	4.051	0.914	ZnI_2-Se_2	5.278	0.945
I-S_2	3.538	0.903	ZnI_2-I_2	4.955	0.954
I_2-Se_2	4.189	0.967	Cd-I_2	3.512	0.981
I-Se_2	4.409	0.935	Cd-S_2	3.422	0.914
I_2-I	4.253	0.938	Cd-Se_2	3.762	1.040
Zn-I_2	4.081	1.007	CdI_2-Cd	4.248	0.977
Zn-S_2	4.597	1.175	CdI_2-I_2	5.534	1.262
Zn-Se_2	4.323	0.946	CdI_2-Se_2	5.431	1.232
ZnI_2-S_2	4.673	1.189	CdI_2-S_2	5.072	1.129

If one is concerned only about the rate of transportation and not the quality of the crystals, the absolute values of the temperatures T_s and T_g can be varied over a relatively wide range.

Influence of Geometrical Parameters on CVT Growth of Crystals

The amount of material transported decreases linearly with the length of the tube due to the increase in flow resistance [27.13]. If several seeds are to be formed and to grow in one experiment then the crystallization chamber should not be too small, to avoid intergrowth between the crystals. The distribution of the temperature in the deposition zone must be as uniform as possible in order to avoid reevaporation of crystals from warmer parts during the experiment. The use of a tapered growth zone is advantageous for this purpose, and it induces nucleation at the corner of the tube where the temperature is low. The most important geometrical parameter is the tube cross-section q, which influences the transport rate decisively [27.12, 39, 43, 44].

27.2.4 Choice of Transporting Agents

The choice of transporting agent generally depends on the thermodynamics of the transport reaction and the volatility of the compounds involved [27.1, 22]. In selecting the transporting agents, the following conditions should be taken into consideration [27.40–42, 45–51]:

1. The transporting agent should be chosen such that all species that will be formed by reaction with the solid to be transported have adequate volatility. Moreover there should be good knowledge of the volatility and stability of all the vapor species formed in the reaction and their solubility in the growing crystal.
2. If it is assumed that element M is transported by halogen I through the transport reaction (27.2), the equilibrium constant for this reaction is given by (27.4). To obtain efficient transport, the equilibrium constant for the reactions must not be extreme and should have suitable magnitude at a convenient temperature.
3. The knowledge on the various components of the substances to be transported which have sizable vapor pressures in the temperature range selected for growth.
4. It should be known to the extent of the substance to be transported as undissociated, uncombined molecules in the vapor phase.

27.2.5 Advantages and Limitations of CVT Method

Advantages of Chemical Vapor Transport

1. The growth process is normally carried out at temperatures much lower than the melting point of the material, which is useful for the growth of materials with high melting points and those that exhibit appreciable dissociation at their stoichiometry melting points.
2. Decrease in contamination from the crucible due to the reduced growth temperature.
3. Allotropic crystalline forms can be grown.
4. Good stoichiometry control is possible.
5. Growth of epitaxial layers is possible.
6. In situ chemical vapor cleaning of the substrate/system is possible.
7. The solid–vapor interfaces exhibit higher interfacial morphological stability during growth than in the case of solid–liquid interfaces.

Limitations of Chemical Vapor Transport

1. Thermodynamics and kinetics are complex and poorly understood.
2. Deposition occurs on the substrate as well as on the walls of the container.
3. The reactive gases involved are dangerous in some cases and need special handling procedures.
4. Controlling nucleation is difficult, and growing several crystals due to primary nucleation will result in a decrease in their sizes for a given amount of charge materials.
5. Transporting agents will incorporate into the growing crystals during the growth process and alter the physical properties of the grown crystals.
6. There are difficulties incorporating some doping elements (e.g., Ga, In, Al) into the growing crystals because of their low vapor pressures.

27.3 Thermodynamic Considerations

27.3.1 Estimation of Optimum Growth Parameters for the ZnSe–I₂ System by CVT

The partial pressures inside the tube during growth, the temperature, the iodine concentration, and the tube dimensions have been taken into account for these calculations (Fig. 27.3). The vapor-phase chemical transport of ZnSe with iodine has been considered to consist of three equilibria

$$\mathrm{ZnSe} + \mathrm{I}_2 \Leftrightarrow \mathrm{ZnI}_2 + \tfrac{1}{2}\mathrm{Se}_2 \,, \quad (27.14)$$

$$\mathrm{ZnSe} + \mathrm{I}_2 \Leftrightarrow \mathrm{Zn} + 2\mathrm{I} + \tfrac{1}{2}\mathrm{Se}_2 \,, \quad (27.15)$$

$$\mathrm{ZnSe} + 2\mathrm{I} \Leftrightarrow \mathrm{ZnI}_2 + \tfrac{1}{2}\mathrm{Se}_2 \,. \quad (27.16)$$

Neglecting dissociation of ZnSe via

$$\mathrm{ZnSe} \Leftrightarrow \mathrm{Zn} + \tfrac{1}{2}\mathrm{Se}_2 \,, \quad (27.17)$$

and dissociation of ZnI₂ via

$$\mathrm{ZnI}_2 \Leftrightarrow \mathrm{Zn} + \mathrm{I}_2 \,, \quad (27.18)$$

it was assumed that, as a first approximation, the vapor phase contains only four components (I, I₂, ZnI₂, and Se₂) and also the law of mass conservation of iodine at equilibrium was considered

$$\mathrm{I}_2 \Leftrightarrow 2\mathrm{I} \,. \quad (27.19)$$

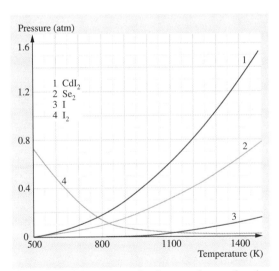

Fig. 27.3 Partial pressures of various species as a function of temperature for various iodine concentration

The partial pressures in terms of initial iodine concentration can be expressed by

$$p_{\mathrm{I}_2}^0 = p_{\mathrm{I}_2} + 0.5 p_{\mathrm{I}} + p_{\mathrm{ZnI}_2} \,. \quad (27.20)$$

The exact stoichiometry of the ZnSe source materials gives the equation

$$p_{\mathrm{ZnI}_2} = 2 p_{\mathrm{Se}_2} \,, \quad (27.21)$$

where $p_{\mathrm{I}_2}^0$ is the partial pressure of iodine concentration initially taken into account for the calculations, and the other p symbols are the partial pressures of the corresponding components at a given temperature. The equilibrium constants for the reactions (27.14) and (27.19) are given by

$$K_{(27.14)} = \frac{\left(p_{\mathrm{ZnI}_2} p_{\mathrm{Se}_2}^{0.5}\right)}{p_{\mathrm{I}_2}} \,, \quad (27.22)$$

$$K_{(27.19)} = \frac{p_{\mathrm{I}}^2}{p_{\mathrm{I}_2}} \,. \quad (27.23)$$

The values can be calculated as a function of temperature from [27.52]

$$K_{(27.14)} = 7.64 - 5849 T^{-1} - 4154 T^{-2}$$
$$- 0.83 \log T - 1.5 \times 10^{-4} T \,, \quad (27.24)$$

$$K_{(27.19)} = 4.34 - 7879 T^{-1} + 4264 T^{-2}$$
$$+ 0.33 \log T + 2 \times 10^{-5} T \,. \quad (27.25)$$

Using (27.20–27.25), the following equation can be derived

$$2 \frac{p_{\mathrm{I}}^2}{K_{2.6}} + \left(4 \frac{K_{(27.14)}}{K_{(27.19)}}\right)^{\tfrac{2}{3}} p_{\mathrm{I}}^{\tfrac{4}{3}} + p_{\mathrm{I}} - 2 \frac{CRT}{M} = 0 \,, \quad (27.26)$$

where p_{I} is the partial pressure of atomic iodine, C is the initial concentration of iodine, R is the universal gas constant, T is temperature, and M is the molar mass of iodine.

It was assumed that iodine behaves as an ideal gas in these calculations. From (27.26), the partial pressures of the components inside the tube during growth were calculated for iodine concentration in the range 0.5–10 mg/cm³ and for temperatures in the range 500–1200 °C. The partial pressures of I, I₂, ZnI₂, and Se₂ for temperatures of 500–1200 °C are plotted in Fig. 27.4 for iodine concentration of 2 mg/cm³.

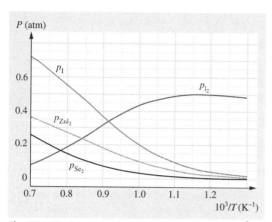

Fig. 27.4 Partial pressure of components for 2 mg/cm³ iodine

The amount of ZnI_2 formed with respect to the initial concentration of I_2 can be described by the ratio α

$$\alpha = \frac{p_{ZnI_2}}{p_{I_2}^0}. \qquad (27.27)$$

The variation in the ratio α for different iodine concentrations at different temperatures is shown in Fig. 27.5. The transport rate of ZnI_2 determines the size and quality of the crystal grown at the growth zone. It can be considered as the difference in α between the source and growth zones for given temperature. It can be approximated by

$$\Delta\alpha = \frac{\Delta p_{ZnI_2}}{p_{I_2}^0}. \qquad (27.28)$$

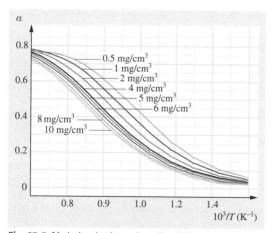

Fig. 27.5 Variation in the ratio α for different iodine concentrations at different temperatures

The temperature difference between two zones was fixed at 50 °C in the calculations. The number of moles of ZnSe transported from the source to the growth end as ZnI_2 in time t is given by

$$n_{ZnSe} \propto \Delta\alpha x t . \qquad (27.29)$$

The diffusion-limited transport rates (J_D) can be calculated using the formula [27.15, 53]

$$J_D = \frac{2D_{ZnI_2} p_T}{RTL} \ln\left(\frac{(2p_T - p_{ZnI_2(g)})}{(2p_T - p_{ZnI_2(s)})}\right), \qquad (27.30)$$

where D_{ZnI_2} is the diffusion coefficient of the component ZnI_2, p_T is the total pressure at the temperature T, $p_{ZnI_2(g)}$ is the partial pressure of ZnI_2 at the growth zone, $p_{ZnI_2(s)}$ is the partial pressure of ZnI_2 at the source zone, R is the universal gas constant, T is temperature, and L is the distance between the sublimed source and growing crystals in the ampoule.

27.3.2 Fluctuations in the Transport Rates

The iodine concentration of 2 mg/cm³ is considered for the present investigation. The partial pressure of the components I, ZnI_2, and Se_2 increases with increasing temperature. The partial pressure of I_2 decreases with increasing temperature. The ratio (α) between the partial pressure of ZnI_2 and that of initial iodine pressure is found to decrease with increasing iodine concentration at a given temperature. Also, at higher iodine concentrations, the change in partial pressures of ZnI_2 is found to be minimum. Hence it can be concluded that increase in iodine concentration has little effect on the partial pressure of ZnI_2. From (27.29), the transport rate of ZnI_2 is proportional to the difference in α values of source and deposition zones, i.e., $\Delta\alpha$. The variation of $\Delta\alpha$ with respect to temperature for the various iodine concentrations is shown in Fig. 27.6. For a given iodine concentration the value of $\Delta\alpha$ is found to increase gradually, reach a maximum, and then decrease with increasing temperature. As the iodine concentration increases, the maximum of the curve positions shifts to higher temperatures and also the curves become broadened. From the figure, it can be observed that, for each iodine concentration, there is an optimum temperature region near the peak position where the change in $\Delta\alpha$ is minimum. Hence the fluctuations in transport rates will be minimum in that region. The change in transport rate is proportional to Δp_{ZnI_2}, i.e., the difference between the partial pressures of ZnI_2

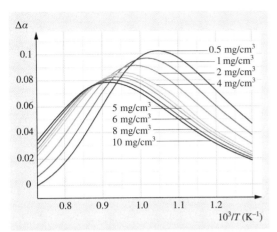

Fig. 27.6 Change in the value of $\Delta\alpha$ with respect to temperature for a given iodine concentration

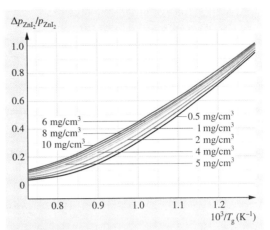

Fig. 27.8 Change in supersaturation ratio as a function of growth temperature for undercooling of 50 K and various iodine concentrations

in the source and deposition zones. So, it can be assumed that, without considering the type of migration along the ampoule, the growth rate is proportional to $\Delta\alpha$.

The fluctuation in transport rates affects the quality of the crystals grown. The fluctuation in transport rates should be minimum to achieve good-quality crystals. Hence it is essential to find the optimum conditions for which the fluctuation is minimum. However a slight drop in temperature at the deposition zone during the growth process causes multinucleation of crystals, which leads to intergrowth and changes in the composition of the crystals. The flow of the material is highly dependent on temperature, and control of temperature at the growth interface is always poorer than the accuracy shown by the temperature controllers in the furnace. Temperature variation up to $\pm 2\,\mathrm{K}$ can normally be expected inside the quartz ampoule during the growth process. The fluctuation in transport rates has been calculated for temperature fluctuation up to $\pm 2\,\mathrm{K}$ using (27.30), considering an iodine concentration of $2\,\mathrm{mg/cm^3}$ and a constant temperature difference between the two zones of 50 K; the results are shown in Fig. 27.7. It can be observed from the figure that the temperature range of 800–850 °C was found to have minor fluctuations in the transport rates. Somewhere in this temperature range, one can find a line nearly parallel to the x-axis and close to zero on the y-axis, for which the fluctuation in transport rate is expected to be minimum. It was therefore concluded that one could obtain good-quality crystals in the temperature range 800–850 °C.

The change in supersaturation ratio $\Delta p_{ZnI_2}/p_{ZnI_2}$ as a function of growth temperature for undercooling of 50 K and various iodine concentrations is shown in Fig. 27.8. It can be seen from the figure that an increase in iodine concentration produces only a small change in supersaturation at given temperature. It can be concluded that the iodine concentration has very little influence on the supersaturation ratio at a given temperature and thus the growth of the crystals.

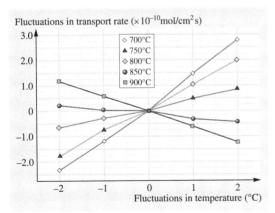

Fig. 27.7 Fluctuation in transport rates at different temperatures for $\Delta T = 50\,°C$ and for iodine concentration of $2\,\mathrm{mg/cm^3}$

Nucleation Calculations for the Growth of ZnS$_x$Se$_{1-x}$ Single Crystals by CVT

The binary nucleation theory was applied to find the critical radius and free energy for the formation of the critical nucleus. The formation and composition of ZnS$_x$Se$_{1-x}$ single crystals by CVT was found to be affected by the supersaturation ratios of S$_2$ and Se$_2$ and hence by the corresponding partial pressure values. The partial pressures of the components in the ZnS$_x$Se$_{1-x}$-I$_2$ system by CVT were found with the thermodynamic model as applied to the case of ZnSe in the previous section.

27.3.3 Supersaturation Ratios in the ZnS$_x$Se$_{1-x}$ System

The formation of ZnS$_x$Se$_{1-x}$ by iodine transport may take place by the following chemical equilibrium reactions.

At the source zone

$$\text{ZnS} + \text{ZnSe} + 2\text{I}_2 \Leftrightarrow 2\text{ZnI}_2 + \tfrac{1}{2}\text{S}_2 + \tfrac{1}{2}\text{Se}_2 \,. \quad (27.31)$$

At the growth zone

$$\text{ZnI}_2 + x\text{S}_2 + (1-x)\text{Se}_2 \Leftrightarrow \text{ZnS}_x\text{Se}_{1-x} \,. \quad (27.32)$$

It is assumed that the vapor phase contains only I, I$_2$, ZnI$_2$, S$_2$, and Se$_2$. The vapor-phase chemical transport of ZnS and ZnSe with I$_2$ in a closed tube follows the individual reaction steps as given below

$$\text{ZnS} + \text{I}_2 \Leftrightarrow \text{ZnI}_2 + \tfrac{1}{2}\text{S}_2 \,, \quad (27.33)$$

$$\text{ZnSe} + \text{I}_2 \Leftrightarrow \text{ZnI}_2 + \tfrac{1}{2}\text{Se}_2 \,. \quad (27.34)$$

Also, considering the mass conservation of iodine at the equilibrium

$$\text{I}_2 \Leftrightarrow 2\text{I} \,, \quad (27.35)$$

the partial pressures in terms of initial iodine concentration can be given by

$$p_{\text{I}_2}^0 = p_{\text{I}_2} + 0.5 p_\text{I} + p_{\text{ZnI}_2} \,. \quad (27.36)$$

The equilibrium constants for (27.33–27.35) are given by

$$K_{(27.33)} = 8.8 - 7539 T^{-1} + 8745 T^{-2}$$
$$\qquad - 1.19 \log T + 2.18 \times 10^{-6} T \,, \quad (27.37)$$

$$K_{(27.34)} = 7.64 - 5849 T^{-1} - 4154 T^{-2}$$
$$\qquad - 0.83 \log T - 1.5 \times 10^{-4} T \,, \quad (27.38)$$

$$K_{(27.35)} = 4.34 - 7879 T^{-1} + 4264 T^{-2}$$
$$\qquad + 0.33 \log T + 2 \times 10^{-5} T \,. \quad (27.39)$$

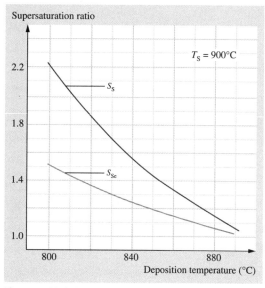

Fig. 27.9 Variation in supersaturation ratios of S$_2$ and Se$_2$ with deposition temperature

The partial pressures of the components inside the growth ampoule were calculated using these equations. The partial pressures of S$_2$ and Se$_2$ were found to affect the composition of ZnS$_x$Se$_{1-x}$ even when an ideal solid solution of ZnS and ZnSe (1 : 1) was assumed. Therefore the supersaturation ratios of S$_2$ and Se$_2$ have been obtained as a function of deposition temperature for a fixed source temperature and iodine concentration. The supersaturation ratio of the S$_2$ component in closed-tube vapor transport of ZnS$_x$Se$_{1-x}$ is given by

$$\frac{p_{\text{S}_2}}{p_{\text{S}_2}^*} = \frac{(p_{\text{S}_2})_\text{s}}{(p_{\text{S}_2})_\text{d}} \,, \quad (27.40)$$

where p_{S_2} is the partial pressure of sulphur at a given temperature, $p_{\text{S}_2}^*$ is the equilibrium partial pressure of sulphur, $(p_{\text{S}_2})_\text{s}$ is the partial pressure of sulphur at the source zone, and $(p_{\text{S}_2})_\text{d}$ is the partial pressure of sulfur at the deposition zone. Similarly, the supersaturation ratio of selenium can be calculated using

$$\frac{p_{\text{Se}_2}}{p_{\text{Se}_2}^*} = \frac{(p_{\text{Se}_2})_\text{s}}{(p_{\text{Se}_2})_\text{d}} \,. \quad (27.41)$$

Figure 27.9 shows the change in supersaturation ratios of S$_2$ and Se$_2$ with respect to varying growth zone temperature for fixed source temperature (900 °C).

Homogeneous Nucleation in the Growth of ZnS$_x$Se$_{1-x}$ Single Crystals

The spontaneous formation of a nucleus of another supersaturated phase is called homogeneous nucleation. The simple theory of homogeneous nucleation brings out two essential features of the nucleation process: (1) The vapor must be supersaturated with respect to the bulk condensate for small clusters to have any stability; the greater the supersaturation, the smaller the cluster which stands an even chance of increasing in size. (2) The rate of nucleation is strongly dependent on the size of the critical nucleus and hence on the supersaturation. In the present study, binary nucleation theory (capillarity approximation) was applied to the ZnS$_x$Se$_{1-x}$ system containing 1 : 1 ZnS and ZnSe

$$xZnS + (1-x)ZnSe \Leftrightarrow ZnS_xSe_{1-x} . \quad (27.42)$$

The free energy of formation of a nucleus is given by

$$\Delta G = -ak_BT \ln \frac{p_S}{xp_S^*} - bk_BT \ln \frac{p_{Se}}{(1-x)p_{Se}^*}$$
$$+ (aV_{ZnS} + bV_{ZnSe})^{2/3}(36\pi)^{1/3}\sigma , \quad (27.43)$$

where p_S/p_S^* is the supersaturation ratio of sulfur, p_{Se}/p_{Se}^* is the supersaturation ratio of selenium, a the number of ZnS molecules in the nucleus, b is the number of ZnSe molecules in the nucleus, x is the mole fraction of ZnS, $(1-x)$ is the mole fraction of ZnSe, V_{ZnS} is the molecular volume of ZnS, V_{ZnSe} is the molecular volume of ZnSe, σ is the interfacial tension (assumed to be constant at 1000 erg/cm^2), k_B is Boltzmann's constant, and T is temperature.

The free energy of formation of a nucleus corresponding to the saddle point can be obtained from

$$\left(\frac{\partial \Delta G}{\partial a}\right)_{T,P} = 0 , \quad \left(\frac{\partial \Delta G}{\partial b}\right)_{T,P} = 0 . \quad (27.44)$$

Applying this condition to (27.43) yields

$$-k_BT \ln \frac{p_S}{xp_S^*} + \tfrac{2}{3}V_{ZnS}(36\pi)^{1/3}$$
$$\times \sigma(a^*V_{ZnS} + b^*V_{ZnSe})^{-1/3} = 0 , \quad (27.45)$$

and

$$-k_BT \ln \frac{p_{Se}}{(1-x)p_{Se}^*} + \tfrac{2}{3}V_{ZnSe}(36\pi)^{1/3}$$
$$\times \sigma(a^*V_{ZnS} + b^*V_{ZnSe})^{-1/3} = 0 . \quad (27.46)$$

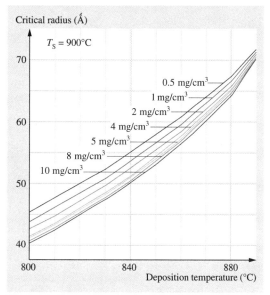

Fig. 27.10 Change in critical radius with deposition temperature

Assuming the shape of the nucleus to be spherical, one has

$$\tfrac{4}{3}\pi r^{*3} = (a^*V_{ZnS} + b^*V_{ZnSe}) ,$$

$$r^* = \left(\frac{3(a^*V_{ZnS} + b^*V_{ZnSe})}{4\pi}\right)^{\frac{1}{3}} , \quad (27.47)$$

solving (27.45) and (27.46) to

$$r^* = \frac{2\sigma V_{ZnS}}{k_BT \ln\left(\frac{p_S}{xp_S^*}\right)} . \quad (27.48)$$

The value of the critical radius has been calculated for iodine concentration of 0.5–10 mg/cm^3 and deposition temperature of 800–890 °C with a constant source temperature of 900 °C. The variation of the critical radius of the ZnS$_x$Se$_{1-x}$ nucleus is shown in Fig. 27.10 with respect to deposition temperature.

The free energy of formation of the critical nucleus is given by

$$\Delta G^* = \frac{16\pi\sigma^3}{3(\Delta G_v)^2} , \quad (27.49)$$

$$\Delta G^* = \frac{16\pi\sigma^3}{3k_B^2 T^2 \ln^2\left(\frac{p_S}{xp_S^*}\right)} . \quad (27.50)$$

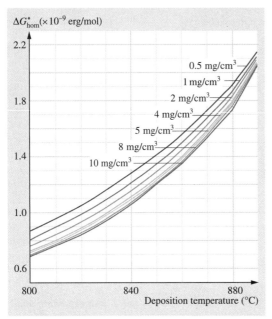

Fig. 27.11 Free energy change for the formation of a critical nucleus with deposition temperature under homogeneous nucleation condition

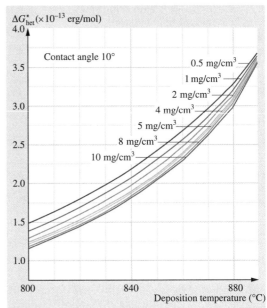

Fig. 27.12 Change in free energy for the formation of a critical heterogeneous nucleus (for contact angle 10°) with deposition temperature

The value of the free energy has been calculated as a function of deposition temperature for various iodine concentrations and is shown in Fig. 27.11.

Heterogeneous Nucleation of ZnS$_x$Se$_{1-x}$ Single Crystals

The nucleation of a condensed phase on the surface of a foreign body is called heterogeneous nucleation. The nucleation rate depends on the presence of available sites on the surface on which growth can take place to a considerable extent. For example, kinks, steps on a surface, inclusions, etc. facilitate the process of nucleation. In chemical vapor transport, deposition may occur on the ampoule wall so that transported vapor components react to form the crystal. Since the presence of a suitable surface induces nucleation at supersaturations lower than that required for spontaneous nucleation, the free energy associated with the formation of such a critical nucleus must be less than the corresponding free energy change. It is given by

$$\Delta G^*_{\text{het}} = \phi(\theta) \Delta G^*_{\text{hom}}, \quad (27.51)$$

where the factor $\phi(\theta)$ is less than 1. In treating such nucleation, a spherical cap-shaped embryo of radius r of solid phase S deposited on a substrate is considered. The contact angle between the cap and the substrate is θ. The ΔG^*_{het} for a critical nucleus is given by

$$\Delta G^*_{\text{het}} = \frac{16\pi \sigma^3_{13} \phi(\theta)}{3 \Delta G^2_v}, \quad (27.52)$$

where σ_{13} is the interfacial tension between the vapor and solid

$$\phi(\theta) = \frac{(2+\cos\theta)(1-\cos\theta)^2}{4}. \quad (27.53)$$

The contact angle θ varies between 0° and 180°. In this case, ΔG^*_{het} has been calculated for θ values of 10°, 30°, and 50° at different deposition zone temperatures, as shown in Figs. 27.12–27.14, respectively.

Variation in Composition with Growth Temperature

In vapor growth, the composition of the growing compound is controlled by the composition of the vapor over it. The undercooling or supersaturation plays the dominant role in determining the composition of the growing crystals. It has been found that the partial pressures of S_2 and Se_2 affect the formation and composition of

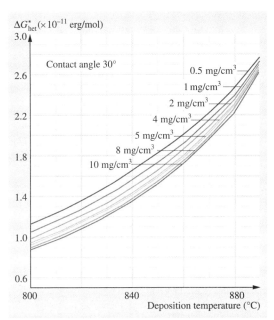

Fig. 27.13 Change in free energy for the formation of a critical heterogeneous nucleus (for contact angle 30°) with deposition temperature

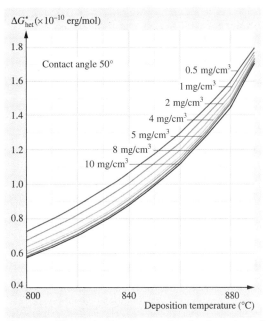

Fig. 27.14 Change in free energy for the formation of a critical heterogeneous nucleus (for contact angle 50°) with deposition temperature

ZnS_xSe_{1-x} even when the ideal solid solution of ZnS and ZnSe was assumed [27.54]. The mole fraction ratio of ZnS and ZnSe is proportional to corresponding partial pressure ratios, given by

$$\frac{x_{ZnS}}{x_{ZnSe}} = \frac{K_{ZnSe}}{K_{ZnS}} \left(\frac{p_{S_2}}{p_{Se_2}} \right)^{1/2}, \quad (27.54)$$

where K_{ZnS} is the dissociation constant of ZnS, K_{ZnSe} is the dissociation constant of ZnSe, x_{ZnS} is the mole fraction of ZnS, x_{ZnSe} is the mole fraction of ZnSe, and p_i is the partial pressure of the corresponding component ($i = S_2$ or Se_2). By considering this fact, in the present investigation, the supersaturations ratios of S_2 and Se_2 were taken into account for the calculation of the critical radius and free energy of formation. From Fig. 27.9, it was observed that the supersaturation decreases with decreasing undercooling (ΔT). Under conditions of smaller undercooling, the supersaturations of S_2 and Se_2 were found to be close enough; on the other hand, for large undercooling, the difference in the supersaturations of S_2 and Se_2 was large. So one could expect correct stoichiometry of the supersaturated vapor for the case of lower undercooling, resulting in crystal with equal x and $(1-x)$ values for an initial 1 : 1 solid

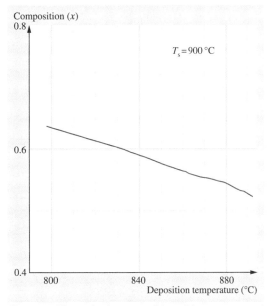

Fig. 27.15 Change in the composition of the grown crystal as a function of deposition temperature

source. For larger undercooling, the crystal would have an S-rich composition as the supersaturation of S will be high compared with that of Se. The calculated change of composition with deposition temperature is shown in Fig. 27.15. As the difference supersaturation ratio of S_2 and Se_2 increases, a crystal with sulfur-rich composition may be obtained.

The deposition temperature was varied from 800 °C to 890 °C for a fixed source zone temperature of 900 °C. The range of iodine concentrations used was $0.5-10\,\text{mg/cm}^3$. The critical radius was found to be high under conditions of smaller undercooling, and remained the same irrespective of iodine concentration. Under conditions of larger undercooling, the critical radius was found to decrease with increasing iodine concentration. The partial pressure of ZnI_2 increased with increasing iodine concentration (and hence supersaturation), resulting in the formation of large number of primary nuclei in the growth zone so that spurious nucleation occurred and hence the size of the critical nucleus decreased. The free energy of formation of a critical nucleus was calculated using (27.49) for different growth zone temperatures and iodine concentrations. As discussed earlier, ΔG^*_{het} was calculated for different contact angles between the substrate or ampoule wall and the crystal using classical nucleation theory. As the contact angle increased, the free energy also increased. The results are similar to that of ΔG^*_{hom} with respect to deposition temperature and iodine concentration.

27.4 Growth of II–VI Compound Semiconductors by CVT

This section deals in detail with the apparatus generally used for growth of II–VI compound single crystals by CVT technique and the preparation of starting materials for the growth process [27.54–68].

27.4.1 Apparatus

Double-zone or multizone furnaces are generally used for crystal growth by CVT technique [27.17, 54–59]. Depending on the required temperature profile and maximum operating temperature, the design and material used in fabrication of these furnaces will be different. Figure 27.16 shows a schematic of a double-zone horizontal furnace.

The muffles are made from ceramic materials or quartz tube of different lengths and diameters depending on the design of the furnace. The gap between the outer casing and muffle is packed with bricks and nonconducting materials such as asbestos powder in order to avoid heat loss. The ampoule is placed above the brick support. The two temperature zones must be considered as two different furnaces, and there is a gap between the two zones called the dead zone, which provides a smooth temperature gradient between the source and growth zone. The length of the dead zone is normally a few centimeters, depending on the required profile. The temperature of each zone is controlled by separate temperature controllers. Each zone has its own temperature sensors. For various regions, the temperature field in a typical cylindrical furnace is not radially symmetrical. If the axis of the furnace is horizontal, then the top part is usually a few degrees hotter then the bottom. If the furnace is set with its axial vertical, this effect is removed as for as the radial distributaries is concerned, but the heat field still depends on the configuration of the heating elements, etc.

In chemical vapor transport growth, the most common temperature measurement and control techniques are based on thermocouples. Chromel–alumel thermocouples can be used to measure up to 1273 K. The essential requirement for this is purity, which should be 99.999% of the individual metal or alloy. An alumina sheath has to be provided to protect the thermocouple to a considerable extent of its diameter. Generally these temperature sensors will give a very weak signal, on the order of mV. The weak signal will be amplified and used to control the power input of the furnace. The requirements of crystal grower are not fulfilled by

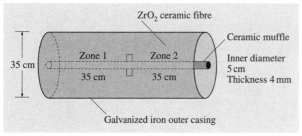

Fig. 27.16 Schematic diagram of a double-zone horizontal furnace for CVT growth

mechanical regulators and on–off controllers and nowadays they have been replaced by electronic controllers. To achieve a stable temperature without a setting error in real systems, proportional–integral–differential (PID) or Eurotherm controllers are used. Since the weak signals obtained from thermocouples are passed to the controllers, they must be carefully shielded. Otherwise stray voltages may be introduced by direct or capacitive coupling. Extension wires used to connect thermocouple outputs to the controller must be prepared from the same alloys with similar thermoelectric characteristics. Such cables are referred to as compensating cables. However these compensating cables can only minimize the error.

The ampoule may be made of any material which does not melt, soften or react appreciably with the species in the system at the operating temperature [27.13]. At temperature up to about 450 °C, Pyrex glass is suitable for many systems. At higher temperature silica is very commonly used, but aluminosilicates can be used as alternatives. Generally a cylindrically shape ampoule is used to grow the crystals, although different shapes of growth and ampoule have been studied. The effect of changing the shape of the growth end of the ampoule has been investigated by many researchers. A rounded end usually results in distributions of crystallites over several centimeters. A tapered, pointed end defines the cold region much better. The next stage was to extend the end of the ampoule by a narrow tube. Nucleation takes place inside the tube at several places, and after a time the crystallites grow together. The final stage of evolution was to provide a crystallite seed, rather than wait for nucleation. A convenient way of doing this is to trap a crystallite at the end of the ampoule by inserting a stopper above it. Generally it is considered that a rounded end to the ampoule gives the lowest possibility of developing unwanted nuclei, assuming that there is initially just one nucleus located centrally at the end of the ampoule.

Vacuum systems are used to evacuate the reaction tubes after filling with the charge materials and transporting agents. This is done for the following reasons:

1. It reduces the chance of explosion of the tube at operating temperature due to the increased vapor pressure of the elements.
2. It enables easy access to the growth zone as the mean free path of the gaseous particles becomes greater than or comparable to the size of the ampoule at lower total pressure.
3. It is advantageous for materials which exhibit a tendency for thermal decomposition and affinity for oxygen.

Oil rotary with diffusion pumps generally provide a vacuum level down to 10^{-6} Torr for loaded transporting agent and feed material within the ampoule.

Before loading the charge materials, the ampoules should be properly cleaned. Cleaning of the ampoule is one of the most important aspects of the growth of crystals by chemical vapor technique. If the ampoules are not cleaned properly, impurities may get into the growing crystal or serve as sites for heterogeneous nucleation and affect crystal quality. The general cleaning procedures include washing and etching (50 : 50 concentrated HNO_3 and 40% HF) and flame polishing. The polycrystalline of the material to be grown is taken in one tube (A) and the other tube (B) is loaded with transporting agent with the use of stopper D while it is cooled with Dewar flask to minimize sublimation of iodine. Then tube A will be cooled while B is heated gently. The iodine which has sublimed from tube B will be condensed in tube A, which will be sealed off in vacuum.

27.4.2 Preparation of Starting Materials

Purification
Whenever the impurity content of the starting elements is higher than desired, a careful check should be made to determine whether a purification step will effectively improve the situation. If one is dealing with an already high-purity semiconductor-grade material, one runs the risk that impurities will be introduced during handling or processing of the material. It is also possible that, even though multiple purification steps may sufficiently decrease certain impurity concentrations, one or two less desirable impurity elements may at the same time be incorporated into the material. Thus, it can occur that, after purification, the overall purity may be better, but the material may be of lesser quality with regards to a specific application. The above comments are not to be taken as a discouragement towards purification steps, but rather as a caution that, during the purification procedure and in the handling of high-purity material in general, great care must be taken to avoid unnecessary contamination. There are two types of purification steps which can often be incorporated during preparation and therefore deserve special mention. The removal of surface oxide is readily achieved by firing the metal compounds, i.e., Zn or Cd, in a stream of pure H_2 just

prior to use. The other step is distillation of the elemental compounds into apparatus.

Preparation of the Starting Compounds

If it is possible to prepare large quantities of the starting compounds in a single sealed system, the chance of obtaining high-purity or controlled impurity-doped material is greatly improved. There are several methods used to prepare ZnSe. In one procedure, ZnO, ZnS, and Se are reacted to yield ZnSe, which probably contains some residual sulfide and oxide.

The most direct method is rapid heating of Zn and Se under argon pressure of 200 atm in an autoclave to the melting point of ZnSe, followed by rapid quenching of the system. The reactor has been designed [27.69] along with a temperature profile for synthesizing ZnSe powder of ≈ 0.9 crystal density from the reaction of 5 N pure elemental zinc and H_2Se in the gas phase by means of H_2 carrier gas. The gas flow rates are: H_2 carrier at $100\,cm^3/min$, H_2 over zinc boat at $90\,cm^3/min$, and H_2 mixed with H_2Se at $20\,cm^3/min$ with H_2-to-H_2Se ratio of $40:1$. The zinc boat is loaded with 10 g zinc before each run, and used entirely to prevent cracking of the quartz upon cooling. ZnSe is controlled as a granular yellow powder both on the liner and attached in a solid chunk to the H_2/H_2Se nozzle. The rate of collection is $1.0\,g/h$ with 50% yield. This powder is then vapor transported under a dynamic vacuum of 10^{-5} Torr out of a furnace at $850\,°C$ using a 20 mm-diameter tube. Again a high-density ZnSe powder is collected at the mouth of the tube furnace while unused Zn and Se are transported further downstream. The approximate weight loss is 10%. All impurities usually detected by spectroscopic analysis are present at less than 0.1 ppm, including copper. This material is then collected and used for crystal growth.

27.4.3 Growth of ZnSe Single Crystals

Increasing attention has been focused on the growth of ZnSe single crystals for application as a substrate for blue light-emitting diodes [27.70, 71]. Growth of ZnSe single crystals was initiated in 1958, but no report of the size of the grown crystals was made in the early literature. The first report on the growth of ZnSe single crystals with mention of their size came from *Nitsche* [27.72], who used the CVT method for growth of these crystals with size of up to $4 \times 4 \times 3\,mm^3$. Since 1960 there have been a lot of reports on growth of ZnSe single crystals, mostly by vapor methods and in limited cases by other methods such as melt [27.69, 73], solution [27.74], and solid-state recrystallization [27.75]. The high melting point ($1520\,°C$), multitwin patterns produced during phase transitions from hexagonal to cubic at $1425\,°C$, and other defects produced at high temperatures have made growth of ZnSe from the melt difficult [27.76]. Attempts have been made by various researchers to use CVT with iodine as a transporting agent to grow ZnSe single crystals with larger size and improved properties suitable for device applications [27.77].

By using seed methods and applying smaller undercooling ($\Delta T = 7\,K$), ZnSe single crystals with dimensions of $20 \times 15 \times 15\,mm^3$ were grown [27.78]. Later it was shown that the grain size of the grown crystals depends strongly on the undercooling and growth ampoule geometry, and growth of ZnSe with dimensions of $24 \times 14 \times 14\,mm^3$ with undercooling as small as $7\,°C$ using an ampoule with a steep conical tip was reported [27.79]. Later there was no further improvement in the size of ZnSe single crystals grown by CVT, and attempts were oriented towards growth of these crystals with more attention to their morphological perfection and improved electrical and optical properties. *Kaldis* [27.80] discussed that the morphology of crystals grown from the vapor phase depends on the mechanism that is rate determining. Very recently, it has been shown that changes in experimental variables such as the system pressure, the temperature conditions, and the concentration of the transporting agent will affect the contribution of mass transport through the convection and diffusion mechanisms and affect the crystallographic perfection and morphological stability of ZnSe single crystals grown by the CVT method [27.81]. The growth aspect of ZnSe single crystals grown by CVT has been studied with special emphasis placed on the formation of different micromorphological patterns formed on the surface of crystals grown under different experimental conditions [27.58].

Pretreatment of starting material for the growth process was carried out by mixing a stoichiometric ratio of ZnSe elements with 5 N purity and loading them into a quartz ampoule. The mixture was heated in vacuum at a temperature of $850\,°C$ for 10 h in order to remove volatile impurities and obtain homogenized polycrystalline ZnSe. The formation of ZnSe was verified by x-ray diffractography. Ampoules with length of 23 cm and diameter of 1 cm were filled with 3 g heat-treated ZnSe polycrystalline powder along with iodine at concentration of $3\,mg/cm^3$ of the empty space of the ampoule. The ampoules, cooled by ice, were evacuated to 2×10^{-6} Torr and sealed off. The capsules were placed into a double-zone horizontal furnace controlled by Eurotherm controllers with accuracy of ± 0.1 K. PID controllers with accuracy of ± 5 K were also used in one of the growth runs. A reverse temperature profile was developed across the ampoule, with the growth zone at high temperature for 24 h to remove powders sticking to the deposition zone of the ampoules and diminish the active sites. Growth runs were carried out for different undercooling (ΔT) values using various compositions of starting materials and ampoule geometries while keeping the temperature of the growth zone constant at $890\,°C$. Each growth run was carried out for a week. At the end of each growth process, the furnace was slowly cooled to room temperature at a rate of $50\,°C/h$ to prevent thermal strain. The grown crystals have been characterized for their structure and morphology [27.80–85].

27.4.4 Growth of CdS Single Crystals

There has been a long-standing interest in CdS because of its optical and electrical properties. Due to its large bandgap (≈ 2.5 eV) and high quantum efficiency, CdS is used for applications from window material for solar cells to coloring plastics [27.86]. Single crystals of CdS have been grown largely by vapor and in limited cases by melt [27.87], solution, and flux methods [27.88]. Vapor growth of CdS at low temperatures was first achieved by *Nitsche* [27.72], who utilized the chemical vapor technique with iodine as the transporting agent. The growth of CdS by CVT technique was continued by various researchers, and crystals with different morphological habits and sizes were grown under different experimental conditions. *Paorici* [27.89] reported growth of CdS single crystals in the form of hollow prismatic rods in an iodine-rich atmosphere. *Kaldis* [27.90] has extensively studied the growth of CdS single crystals by CVT under different experimental conditions such as source temperature ($750\,°C < T_2 < 1120\,°C$), undercooling ($5\,°C < T < 200\,°C$), and iodine concentration ($0.1\,mg/cm^3 < c < 12\,mg/cm^3$) in order to find suitable conditions for control of nucleation and to improve the size of the grown crystals. These studies showed that, using the vertical pulling method with iodine transport, controlled nucleation was possible only at iodine concentration less than $0.2\,mg/cm^3$ and critical undercooling less than $15\,°C$. However, applying these conditions resulted in growth of crystals with cadmium deficiency. It was also reported that CdS grown under different experimental conditions exhibited different morphological habits (e.g., hexagonal platelets, hollow conical, and pyramidal) due to random nucleation and when using seed and undercooling of $5\,°C$. Change in morphology of CdS crystals with change in growth temperature and iodine concentration was also reported by other researchers. *Matsumoto* et al. [27.91] observed that CdS crystals grown at different temperatures and with different iodine concentrations showed different morphological habits such as needles, polyhedron, prisms, pyramids, rods, platelets, dendrites, and irregular shapes. *Attolini* et al. [27.8] improved the perfection of CdS single crystals grown by CVT by adopting the time-increasing supersaturation method, showing that crystals grown under small supersaturation do not exhibit skeletal and hollow forms. Later it was reported that using hydrogen as the transporting agent improved the size of the CdS single crystals and has several other advantages such as better control of nucleation, up to the point of obtaining large crystals and lack of contamination of grown crystals by the transporting agent [27.92]. *Paorici* and *Pelosi* [27.36] extensively studied closed-tube CVT techniques for the Cd:I_2 system from the thermodynamical as well as hydrodynamical point of view and calculated the mass transport rate as a function of overall pressure inside the tube, considering the contribution of thermal convection. *Attolini* et al. [27.93] applied the multidirectional productivity function for predicting the maximum transport efficiency as well as some other CVT features such as the relative importance of the various chemical reactions and the effect of inert gas on transport performance in the CdS:I_2 system.

CdS single crystals grown by CVT in closed tubes using iodine as the transporting agent show different morphological habits and degree of perfection when changing experimental conditions such as the growth temperature and iodine concentration. The growth temperature and iodine concentration have been changed over wide ranges to find empirically the most suitable growth conditions which yield CdS crystals of higher quality and large size. However the optimal growth temperature for a particular concentration of transporting agent for the growth of nearly perfect CdS single crystals remains to be determined. Hence the growth of CdS single crystals under different growth conditions has been studied to proof the agreement of theoretically predicted optimum conditions with experimental observations [27.94].

The experimental details for the growth of CdS single crystals followed at Crystal Growth Centre, Anna

University, are as follows: The starting material was cadmium sulfide of spectroscopic grade (5 N purity). Before use, the powder was heat-treated at 200–300 °C for several hours in vacuum. Three grams of heat-treated powder was filled into quartz ampoule having length of 20 cm and diameter of 1 cm, along with iodine. Due to the high vapor pressure of CdS, the ampoules were made from quartz tubes with thickness of 2 mm for safety reasons. The ampoule, cooled by ice, was evacuated to 2×10^{-6} Torr and sealed off. The capsule was placed into a double-zone horizontal furnace controlled by Eurotherm controllers. A reverse temperature profile was developed across the ampoule over several hours to remove the powder sticking to the tip or deposition zone of the ampoule. The growth temperature (T_g) was varied from 600–950 °C. The concentration of the transporter was varied from 1 to 10 mg/cm^3 and the difference in temperature of the source and deposition was kept constant at ($\Delta T = 50$ K) for all growth runs. Each growth run was carried out for 1 week. After that, the furnace was cooled at room temperature in 20 h.

Crystals grown at different growth temperatures with the same amount of iodine concentration showed different habits and optical quality. The crystals were orange, red, and pale yellow in color, with average dimensions of $3 \times 2 \times 1$ mm^3. The experimental results observed are in good agreement with the theoretical calculations. Crystals grown at growth temperature of 750 °C had complete faces, and their habits were mostly pyramidal and platelet. They were found to be more transparent compared with CdS crystal grown at other temperatures for the same transporter concentration. In general, the quality of the grown crystals was found to be good. This could be due to the stability of the flow of the materials transported from the source to the growth zone and the growth at the temperature of 750 °C predicted to be optimum for the iodine concentration of

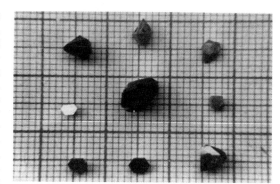

Fig. 27.17 CdS crystals grown at growth temperature of 750 °C

5 mg/cm^3. At the growth temperature of 600 °C, due to the higher partial supersaturation, more crystals, of smaller size, with needle and irregular habits, and of poor quality were grown. Crystals grown at 950 °C were found to have poorly developed faces in most cases and showed less transparency. The deficiency in the surface perfection and optical quality of CdS crystals grown at 600 °C and 950 °C compared with CdS crystals grown at 750 °C may be due to variations of the transfer rate caused by small fluctuations in the temperature of the deposition zone where the slope of the $\Delta \alpha$ curve is large. Since the optical transmission measurements were carried out for crystalline samples of the same thickness (2 mm) and equally polished surfaces, the difference in transmission could be due to scattering from defect centers. Because of the temperature fluctuation at the growth interface, defects such as precipitates, inclusions, and dislocations may get into the lattice of the growing crystals and affect their optical quality. Figure 27.17 shows some of the CdS single crystals grown at 750 °C for $c = 5$ mg/cm^3 and $\Delta T = 50$ K.

27.5 Growth of Nanomaterial from Vapor Phase

ZnSe can be used in optoelectronic devices such as LED and LDs in the blue wavelength region [27.95]. ZnS:Mn and ZnSe:Mn are known to be good luminescent materials [27.96, 97]. The recent interest in these semiconductors is due to the magnetic properties when transition metals are doped at lower concentrations, forming a separate class of research identified as diluted magnetic semiconductors [27.98]. The sp–d exchange interaction between ZnSe and Mn makes it possible to obtain magnetic character in nonmagnetic semiconductor, as applicable in magnetooptical devices [27.99]. Nanocrystalline growth of ZnS:Mn- and ZnSe:Mn-based semiconductors has attracted substantial interest in recent years [27.100, 101]. Nanocrystals and nanostructures of these materials were grown from aqueous solutions [27.102, 103]. Vapor growth methods such as MOVPE and MOCVD have also been used to synthesize nanostructures of

ZnSe [27.104, 105]. Magnetron sputtering has been used to deposit $SiO_2/CdTe/SiO_2$ nanocrystalline thin films [27.106]. A few II–VI compound nanocrystallites were grown on glassy matrix [27.107]. Sol–gel synthesis of ZnS was carried out in SiO_2 gels [27.108]. These nanostructures have usually been analyzed by characterizations techniques such as transmission electron microscopy (TEM), photoluminescence, Raman and ultraviolet–visible spectroscopy (UV-vis), etc. Electron spin resonance (ESR) has been used as a tool to analyze paramagnetic impurities in ZnS nanocrystals and quantum dots [27.109].

SiO_2 aerogels were obtained by sol–gel synthesis followed by annealing at high temperatures up to 400 °C. This matrix has been used to deposit nanocrystalline ZnSe:Mn on its pores by the chemical vapor transport method using iodine as transporting agent. ZnSe doped with 0.2 mol % Mn powder synthesized by vapor transport has been used in the experiment. The growth temperature was 800 °C and the temperature difference was 50 K. The experimental procedure has been discussed in detail [27.26]. The characterization details have been reported in the literature [27.104–115].

27.6 Growth of I–III–VI$_2$ Compounds

Nonvolatile solid substances can be transported through a vapor phase by chemical vapor transport (CVT) when suitable reactive gases are provided in the presence of a temperature gradient, such as when transforming solid substances into gaseous compounds via heterogeneous chemical reactions and vice versa. Vapor-phase chemical transport methods, as first described by *Schäfer* [27.12] and *Nitsche* et al. [27.116], are widely used in closed-tube arrangements for growing crystals. These vapor-grown crystals are often perfect enough and of good enough quality to be used in solid-state physics experiments.

Honeyman and *Wilkinson* [27.117] have grown $CuGaS_2$, $CuAlTe_2$, $AgGaS_2$, $AgAlS_2$, $AgAlSe_2$, and $AgAlTe_2$ single crystals from polycrystalline materials by chemical vapor transport using iodine as the transporting agent to form single crystals. Table 27.3 gives the details of the growth conditions used, such as temperatures, ampoule dimension, and transporting time, and the crystals obtained.

In all cases 5 mg/cm^3 iodine per unit volume of ampoule was used as the transporting agent. The silver compounds required much longer growth times than the copper compounds. This is probably due to the vapor pressure of silver iodide, which is approximately 30 times less than that of copper iodide at the same temperature. The tellurium compounds were also very difficult to grow and gave a low crystal yield. This may be due to the formation of $TeCl_4$, which would prevent the normal iodide vapor transport.

Paorici et al. [27.9] have grown $CuGaS_2$, $CuInS_2$, $CuAlS_2$, and $AgIn_5S_8$ single crystals using a temperature-variation method of the chemical vapor transport technique and iodine as the transporting agent. Larger

Table 27.3 Growth parameters of some I–III–VI$_2$ compounds by CVT method and their results

Compound	Ampoule diameter (mm)	Length of ampoule (cm)	Source–growth zone temperature (°C)	Transport time	Crystals obtained
CuGaS$_2$	18	20	800–700 850–750	3 days	High yield, yellow–green crystals stable in atmospheres
CuAlTe$_2$	18	20	780–650	5 months	Very low yield, red hexagonal platelets
AgGaS$_2$	18	20	840–740	2 weeks	Moderate yield straw-yellow transparent crystals
AgAlS$_2$	15	14	800–600	3 weeks	Colorless crystals, very unstable in air
AgAlSe$_2$	18	20	750–630	3 months	Low yield, yellow–black crystals
AgAlTe$_2$	18	20	830–630	2 weeks	No crystals

crystals were obtained by means of a time-varying temperature profile method. The principle of the method involves a gradually increase of the source temperature, which allows firstly reduced primary nucleation and secondly the avoidance of constitutional undercooling during the first stages of nucleus growth. The source temperature (at the hot end of the ampoule) T_s is slowly raised as a function of time, while the growth temperature T_g is kept constant. Initially, the ampoule is placed such that T_s is lower than T_g by about 100 °C (temperature inversion). This temperature inversion is a very important feature of the method, because of its cleaning effect on the quartz wall of the deposition zone. After 2–3 days of temperature inversion, T_s is adjusted to about 10 °C lower than T_g. Qualitatively, the growth mechanism can be described as follows. By increasing T_s, when T_s reaches some value, the supersaturation reaches a value for which primary nucleation has a high probability of occurring on some active sites of the quartz walls. If careful quartz treatment is performed, only a few primary nuclei are found. Since now the growth process of these primary nuclei will result in a more probable process, their development into large single crystals is expected. The two alternative methods, i. e., stationary temperature profile (STP) and time-varying temperature profile (TVTP), were used and the sizes of the crystal grown were compared.

27.6.1 Growth of Undoped and Doped Crystals of CuAlS$_2$

CuAlS$_2$ is a wide-bandgap (3.49 eV) member of the ternaries, which has been found to emit strong green and blue photoluminescence (PL) and therefore is considered a candidate material for blue and green light-emitting device realization [27.118, 119]. Single crystals were grown by chemical vapor transport technique from polycrystalline CuAlS$_2$, which was prepared by direct melting of the constituent elements in a BN crucible. The resulting crystals, which were typically plate-like with dimensions of $20 \times 10 \times 0.5\,\text{mm}^3$, were then annealed in evacuated and sealed quartz ampoules in the presence of 100 mg Zn metal placed at one end of the ampoule with the CuAlS$_2$ crystals placed at the other end. Thermal treatments were carried out for 50 h at different temperatures in the range 973–1173 K.

Shirakata et al. [27.120] have grown CuAlS$_2$ single crystals by chemical vapor transport using iodine as the transporting agent. Growth conditions of source and growth temperatures were 800–650 °C and iodine concentration was 5 mg/cm^3. PL measurements were carried out at low temperatures (77 and 10 K). Most of the crystals exhibited a strong, orange PL band peaking at 2.1 eV and weak peaks at 2.9 and 3.55 eV. These results can be explained by considering that a large concentration of antisite disorder defects should be expected in Cu- and Al-rich crystals, being acceptors such as Cu$_{Al}$ defects in Cu-rich samples and donors such as Al$_{Cu}$ defects in Al-rich samples. Thermal treatment in sulfur vapor will result in a decrease of the concentration of sulfur vacancies and an increase of the concentration of the cation vacancies V$_{Cu}$ and V$_{Al}$. Since the former defects are donor-like while the latter are acceptor-like the degree of compensation under sulfur annealing will be reduced, which was observed experimentally as an increase of p-type conductivity. The results for the sharp, near-bandgap PL lines are summarized in Table 27.4, where peak energies and assignment of the exciton PL lines in CuAlS$_2$ crystals are given. The crystals under consideration are as-grown ones of stoichiometric, Al-rich, and Cu-rich compositions as well as crystals annealed in sulfur and in vacuum. Strong lines are indicated by (++), weak lines by (+), and missing lines by (−).

Table 27.4 Summarized PL results on CVT-grown CuAlS$_2$ single crystals

Energy	Crystals			Annealed		Assignment
	As grown					
	Stoichiometric	Al-rich	Cu-rich	In sulfur	In vacuum	
3.376	+	−	+	−	−	−
3.567	+	Broad	+	+	+	−
3.550	++	−	++	++	++	Free exciton
3.540	++	−	++	+	++	Bound exciton
3.532	++	Broad	++	+	++	Bound exciton
3.500	+	−	−	++	−	Bound exciton
3.475	+	−	−	++	−	Bound exciton

Aksenov et al. [27.121] recorded PL spectra for $CuAlS_2$ single crystals doped with Zn by adding metallic zinc into the starting composition of the constituent elements prior to synthesis of the $CuAlS_2$ compound. Zn doping was carried out by using different techniques:

1. Zn dopant was added into the starting composition of constituent elements prior to synthesis of $CuAlS_2$ by direct melting. In this case the concentration of Zn in CVT-grown crystals was found to be equal to the nominal one.
2. Undoped CVT-grown crystals were annealed in evacuated and sealed quartz ampoules in the presence of ZnS and S at 973–1073 K for 60 h, the S pressure being 1–3 atm.
3. The undoped crystals were annealed at 873 K for 20–120 h in the presence of Zn, the Zn metal being placed at one end of the ampoule and the crystals to be doped at the other.

Aksenov et al. [27.122] have grown Cd-doped $CuAlS_2$ single crystals by CVT using iodine as the transporting agent, with grown crystal annealed under different atmospheres such as Cd. Annealing of the crystals in the presence of Cd at 700 °C did not result in any significant change in the PL properties of the crystals, the resulting PL spectrum being essentially the same for as-grown crystals. In the case of annealing at 900 °C for 50 h, an intense yellow–green photoluminescence band peaking at 565 nm was observed. This emission is interpreted as originating form donor–acceptor pair recombination, involving deep levels, formed by Cd-introduced defects.

27.6.2 Growth of Undoped and Doped Crystals of CuAlSe$_2$

$CuAlSe_2$ is one of the wide-bandgap ternary compounds with chalcopyrite-type structure and is promising as an optical application in the blue wavelength region. *Kurposhi* et al. [27.123] have grown single crystals of $CuAlSe_2$ by chemical vapor transport, which were plate-like and transparent with yellowish color. Relatively large single crystals with a well-developed (112) face and typical dimensions of $9 \times 5 \times 0.45$ mm^3 were obtained at source and growth temperatures of 1263 K and 1223 K and iodine concentration of 2–3 mg/cm^3. It was found that the light transmittance of the crystals decreased with during the days after growth. This may be due to surface oxidation of $CuAlSe_2$ crystals after growth, which is a serious problem for optical applications of this compound.

Chichibu et al. [27.124] have studied the resistivities, carrier concentrations, optical absorption, and PL of undoped and Cd- and Zn-doped $CuAlSe_2$ single crystals grown by CVT. The electrical and optical properties were almost unchanged after annealing under Se pressure. However, resistivity increased about seven orders of magnitude after annealing in vacuum. Resistivity also increased with Cd or Zn doping. The samples showed p-type conduction even with Cd or Zn doping. It was seen in all samples that the reduction of transmittance began from about 2.4 eV. A binding state such as a deep donor level which accompanies a lattice relaxation acts as an optical absorption center that needs more than 2.4 eV for its excitation. The assumed binding state appears to be related to a vacancy at the Se site because the absorption edge was almost unchanged after annealing in a Se atmosphere and shifted after annealing in vacuum. Therefore, taking into account that the absorption edge did not shift with Zn doping but did shift with Cd doping, the binding state may be a complex center consisting of the cation atoms and the nearest X_{Se}, where X is an atom or vacancy.

$CuAlSe_2$ single crystals were grown by CVT method using an ingot synthesized by a melt-grown technique in a rotating horizontal furnace using the elements Cu, Al, and Se with excess Se of 3 mol% form the stoichiometry. A concentration of iodine of 5 mg/cm^3 was used [27.125]. The crystal growth was carried out for 7 days at constant temperature of 850 °C for the source region and 700 °C for the growth zone. As-grown crystals had platelet and needle shape, with typically dimensions of about $9 \times 6 \times 0.5$ mm^3, and showed a greenish yellow color. In PL spectra measured at 80 K, two independent, broad emission bands with peak energies of 1.77 and 1.93 eV were observed.

Prabukanthan and *Dhanasekaran* [27.126] reported the growth of $CuInTe_2$ single crystal by CVT method. A polycrystalline ingot of stoichiometric $CuInTe_2$ was synthesized from copper, indium, and tellurium elements with 4 N purity. The stoichiometric ternary mixture was taken into a quartz ampoule and sealed in a quartz ampoule under vacuum of 2×10^{-6} Torr (0.3 mPa). The ternary mixture was gradually heated to 1323 K at a heating rate of 20 K/h. The ampoule was maintained at this temperature for 2 days until the reaction was complete. Then the furnace was cooled at a rate of 50 K/h. The ampoule was opened and the synthesized polycrystalline $CuInTe_2$ material was analyzed using powder x-ray diffraction (XRD). Single-phase polycrystalline $CuInTe_2$ powder showed a well-defined chalcopyrite structure. Two grams of synthesized CuInTe$_2$ polycrystalline material and 5 mg/cm^3 of high-purity iodine were taken into a quartz ampoule. The ampoule, cooled by ice, was evacuated to 2×10^{-6} Torr and then sealed off. The ampoule was placed into a double-zone horizontal electrical furnace controlled by Eurotherm temperature controller. During the first stage, the furnace was slowly heated. The temperatures of the source and the growth zones were allowed to reach 873 and 923 K, respectively, in order to remove the material and also to clean the growth zone of the ampoule. The duration was 20 h. After this, the temperatures of the source and growth zones were maintained at 923 and 873 K, respectively. After growth lasting 14 days, the furnace was slowly cooled at a rate of about 20 K/h. When the temperature of the ampoule reached room temperature it was opened to obtain CuInTe$_2$ crystals. The crystals were then cleaned in an ultrasonic bath containing a mixture of acetone and methanol, and then rinsed with deionized water. The CuInTe$_2$ single crystals obtained were black in color with mirror-like upper surface; the maximum dimensions of the crystals obtained was $15 \times 5 \times 3$ mm^3. Similarly, single crystals of CuInTe$_2$ were also grown by maintaining source and growth temperatures of 923–823 K and 923–773 K, respectively, for a period of 14 days. The dimensions of the crystals grown at 823 and 773 K were $7 \times 3 \times 5$ mm^3 and $4 \times 3 \times 3$ mm^3, respectively. The single crystals of CuInTe$_2$ grown at growth zone temperatures of 873, 823, and 773 K are shown in Fig. 27.18a–c. Single-crystal x-ray diffraction studies of CuInTe$_2$ single crystals carried out on crystals grown at different growth zone temperatures indicated tetragonal (chalcopyrite phase) structure.

Gombia et al. [27.127] have grown CuGaTe$_2$ and CuInTe$_2$ single crystals by chemical vapor transport technique and considered thermodynamically the growth parameters. The behavior of all gaseous species, the formation of spurious phases, i.e., liquid Te and solid TeI$_4$, and transport conditions were discussed. Polycrystalline CuGaTe$_2$ and CuInTe$_2$ ingots were prepared by fusion of the constituents, weighed in stoichiometric ratio. The mixture was sealed in a quartz ampoule under vacuum of 10^{-6} Torr and heated at a rate of about 10 °C/min up to 420 °C (below the Te melting point) for several hours, then kept at a temperature of 900 °C for CuInTe$_2$ and 950 °C for CuGaTe$_2$ for several hours to ensure complete reaction and homogenization. Finally the samples were cooled slowly to 650 °C, annealed at this temperature for 3 days, and quenched to room temperature.

A vapor transport method in a closed tube, using iodine as transport agent and polycrystalline CuGaTe$_2$ and CuInTe$_2$ as the starting source, was employed. To produce single crystals of significant size it is necessary to limit the number of potential sites for crystallization nuclei. Experimentally this can be done by thoroughly cleaning the quartz ampoule and using a low starting temperature gradient to assure a small supersaturation. To ensure the latter condition, a tubular eight-zone furnace permitting different gradients and profiles to establish conditions as close as possible to the theoretical ones was used. Inversion of the temperature profile between the source and crystallization zones reduced the number of nuclei. This process was performed automatically by means of an electronic programmer. The grown crystals were characterized and the resistivity,

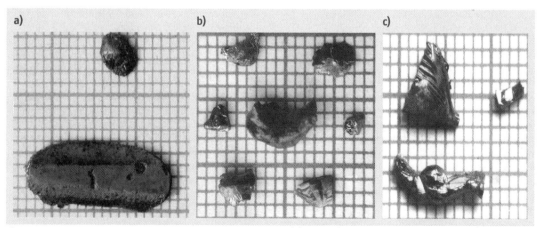

Fig. 27.18a–c CuInTe$_2$ crystals grown at growth temperature of (**a**) 873 K, (**b**) 823 K, and (**c**) 773 K

carrier mobility, and carrier concentration were determined in the temperature range between 80 and 300 K. CuInTe$_2$ crystal grown in this way and annealed under Cd atmosphere changes from p-type to n-type and also exhibits increased mobility.

27.6.3 Growth of CuGaS$_2$-Based Single Crystals

Yu et al. [27.128] have grown CuGaS$_2$ single crystals by two different methods: melt-growth and chemical vapor-phase transport. In the CVT technique iodine was used as a transporting agent. The crystals were synthesized from the constituent elements (Cu, Ga, and S of 6 N purity) in stoichiometric proportions in an evacuated quartz ampoule. The material was held in the temperature range 1050–1100 °C. It was cooled slowly at a rate of 2 °C/h to 700 °C and then cooled at a faster rate to room temperature. Crystals obtained from the melt varied in color from red to yellow–green. The crystals grown form the melt were crushed into powder and used as the initial charge for the vapor transport. This method involved placing the reacted material together with a small amount of iodine in an evacuated quartz ampoule, and heating it in a temperature gradient. The volatile iodine serves as the transporting agent. The temperature difference between the charge and crystallization zones was kept at 50 °C. The temperature of the charge zone was 1000 °C. The grown crystals were homogeneous and showed yellow–green color. The typical crystal dimensions obtained were $1 \times 3 \times 0.4$ mm^3.

Shirakata et al. [27.129] have grown CuGaS$_2$ single crystal by chemical vapor transport method using iodine as the transporting agent. Stoichiometric quantities of the elements Cu, Ga, and S (6 N) with total weight of 0.5 g were sealed into an evacuated quartz ampoule (10 mm inner diameter, 15 cm length) together with 3–25 mg/cm^3 iodine. The ampoule was placed in a two-zone horizontal furnace with temperature gradient of 900–700 °C and maintained for 4 days. With more than 10 mg/cm^3 iodine all source materials were transported to the other end of the ampoule, and plate-like crystals ($5 \times 5 \times 0.2$ mm^3 typical dimensions) with well-developed (112) plane were grown. With less than 5 mg/cm^3 iodine, some of the source materials remained and crystals were typically yellow–green, but the color became greenish as the quantity of iodine was increased. PL measurements have been carried out on CuGaS$_2$ single crystals at 4.2 and 77 K. At 4.2 K, high-quality single crystals grown by CVT method show sharp PL lines.

Prabukanthan and *Dhanasekaran* [27.130] have grown CuGaS$_2$ (CGS) single crystal by CVT method. Single crystals of CGS were grown by chemical vapor transport method using iodine as the transporting agent. The CGS crystals were grown at three different growth temperatures with 10 mg/cm^3 iodine concentration and temperature difference of 50, 100, and 150 K. The purity of the elements used for these experiments was 4 N. A mixture of the elements Cu, Ga, and S was taken into a quartz ampoule of length 18 cm and diameter 1 cm along with iodine at a concentration of 10 mg/cm^3. The ampoule, cooled by ice, was evacuated to around 2×10^{-6} Torr and sealed off. The ampoule was placed into a double-zone horizontal furnace controlled by temperature controllers with accuracy of ± 0.1 K. A reverse temperature profile was developed across the ampoule over several hours to clean the quartz walls of the growth zone. The duration was 20 h [27.131, 132]. After this, the temperatures of the source and growth zones were maintained at 1173 K and 1123 K, respectively. The growth duration was 7 days, after which the furnace was slowly cooled at a rate of about 10 K/h to 773 K and then at the rate of 60 K/h. The CGS single crystals obtained were yellow in color with maximum dimensions of $6 \times 4 \times 6$ mm^3. Similarly single crystals of CGS have been grown with the same iodine concentration and source zone temperature of 1173 K. Temperature differences of 100 and 150 K were maintained between source and growth zones, i.e., the temperature of the growth zone was maintained at 1073 and 1023 K, respectively. Growth was carried out for a period of 7 days in each case. The CGS single crystals obtained with growth zone temperature of 1073 and 1023 K were orange and green in color, respectively. The maximum dimensions of orange-colored CGS single crystals were $4 \times 2 \times 3$ mm^3 and of green-colored crystals were $15 \times 0.4 \times 1.2$ mm^3 (needle like) and $3 \times 2.5 \times 3$ mm^3.

Single crystals of CGS grown with growth zone temperature of 1123, 1073, and 1023 K are shown in Fig. 27.19. The temperature difference between the source and growth zones affects the quality and color of the resulting crystals. The crystal nucleation rate depends on the magnitude of supersaturation of the gas phase, which is proportional to the temperature difference between the source and growth zones. Normally the temperature difference between the source and growth zones is very low so that the formation of primary nucleation is controlled to form large-size crystals [27.17]. To initiate the crystallization processes, crystal nuclei have to be formed in the crystallization

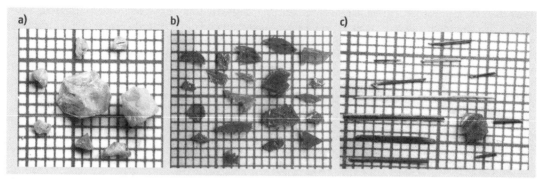

Fig. 27.19a–c As-grown single crystals of CuGaS$_2$ at different growth temperatures: (**a**) 1123 K, (**b**) 1073 K, and (**c**) 1023 K

zone. This is possible only if the gas phase is sufficiently supersaturated (i.e., the gas phase is in the unstable state). In the unstable state of high supersaturation, the rate of crystal nucleation is high and crystal nuclei are formed spontaneously in a short period of time. In the case of our experimental observations at growth temperature of 1023 K, the crystals grown were small in size due to high supersaturation ratio. However, at 1123 K, the crystals grown were larger in size due to the low supersaturation of the gas phase. Under chemical vapor transport conditions, the partial pressures of noble-metal iodides (CuI and GaI$_3$) are high compared with the partial pressure of sulfur. So controlling the stoichiometric composition is difficult. It is concluded from our experimental observations that, during the growth of CGS single crystal by CVT method with temperature difference between source and growth zones of 50 and 100 K, sulfur may play the main role in the transport process. The formation of other phases such as Cu$_2$S and Ga$_2$S$_3$ takes place during the growth of CGS single crystals at 1123 and 1073 K, respectively. However, when the temperature difference is maintained at 150 K, iodides such as CuI and GaI$_3$ may be the dominant gas species to from stoichiometric CGS single crystals.

Prabukanthan and *Dhanasekaran* [27.133] have grown Mn-doped CuGaS$_2$ single crystal by CVT method. Stoichiometric compositions of Cu, Ga, and S with 1 mol % Mn as precursor materials were taken into a quartz ampoule with iodine concentration of 10 mg/cm^3 (total weight 1 g, including Mn concentration). The quartz ampoule, cooled by ice, was evacuated to around 2×10^{-6} Torr and sealed off. The ampoule was placed into a double-zone horizontal electrical furnace. The furnace was controlled by Eurotherm temperature programmer and controller. The source and growth zones temperature were 1173 and 1023 K, respectively. The growth period was 7 days. The furnace was slowly cooled at a rate of 10 K/h to 873 K, after which it was cooled rapidly at a rate of 60 K/h. Similar growth conditions were used for growth of 2 mol % Mn-doped CuGaS$_2$ crystals. CuGaS$_2$ single crystals grown with 1 mol % and 2 mol % Mn doping were green and orange in color with dimensions of $2.5 \times 3 \times 4$ mm^3 and $4 \times 3 \times 3$ mm^3, respectively. Single crystals of 1 mol % and 2 mol % Mn-doped CuGaS$_2$ are shown in Fig. 27.20a,b. The Mn-doped CuGaS$_2$ single crystals were determined to be paramagnetic in nature. The increase of bulk conductivity of the Mn-doped CuGaS$_2$ single crystals at room temperature indicates an increase of hole concentration, and p-type conductivity was also found.

Tanaka et al. [27.134] have grown CuGaS$_{2-2x}$Se$_{2x}$ single crystal by CVT method. First, in order to prepare CuGaS$_2$ powder, Cu, Ga, and S were weighed in stoichiometric portions and sealed into evacuated silica tubes; then the tubes were heated in a furnace. The temperature was increased gradually to 1200 °C for about 24 h. CuGaSe$_2$ powders were also prepared in a similar way. Subsequently CuGaS$_{2-2x}$Se$_{2x}$ crystals were prepared by iodine transport method. CuGaS$_2$ and CuGaSe$_2$ powders were weighed in certain proportions, crushed, mixed, and then sealed in evacuated silica tubes (12 mm inner diameter, 10 cm length) with about 18 mg/cm^3 iodine. These ampoules were placed in a two-zone furnace. The source and growth zone temperatures were kept at 850–900 °C and 650–700 °C, respectively. Under these conditions all starting materials were transported by iodine within several days, and crystals of dimensions $3 \times 10 \times 0.5$ mm^3 were yielded.

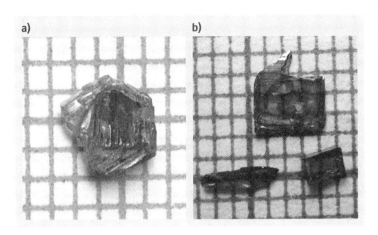

Fig. 27.20a,b As-grown single crystals of (**a**) 1 mol % Mn-doped and (**b**) 2 mol % Mn-doped CuGaS$_2$

The color of the crystals obtained varied continuously with composition: from dark green ($x = 0$), through red, to black ($x = 1$).

Sato et al. [27.135] have grown Fe-doped single crystals of CuAl$_{1-x}$Ga$_x$S$_2$ by CVT method using iodine as the transporting agent. Polycrystalline powders of CuAlS$_2$ and CuGaS$_2$ prepared by sintering were used as source materials. The source material and a transporting agent was sealed in vacuum in a silica ampoule with inner diameter of 13 mm and length of 20 cm. The iodine concentration was 25 mg/cm^3. The silica ampoule was placed in a double-zone electrical furnace. Then the temperature of the source and growth zones were kept at 900 °C and 700 °C, respectively. The dimensions of the crystals were $3 \times 4 \times 2$ mm^3. All of these undoped bulk crystals were black. Crystals with composition in the vicinity of CuAlS$_2$ were colored blue when polished to less than 0.5 mm, while those with the composition near CuGaS$_2$ were colored green, indicating the presence of trace impurity of iron. Crystals intentionally doped with Fe were also prepared with Fe concentrations of 0.1 mol % and 0.3 mol %. The grown crystals were characterized by PL. In addition, several researchers have grown CuGaS$_2$ single crystals, described in the literature [27.130, 132].

27.6.4 Growth of AgGaS$_2$ and AgGaSe$_2$ Single Crystals

Prabukanthan and *Dhanasekaran* [27.136] have reported the growth of AgGaS$_2$ single crystal by CVT method. The 1 : 1 : 2 mole ratio of Ag, Ga, and S with excess 0.5% sulfur as precursor materials, and 10 mg/cm^3 of high-purity iodine as the transporting agent, were taken into a quartz ampoule. In order to prevent deviation from stoichiometry resulting from possible volatile loss of sulfur during initial steps, it was found necessary to add excess sulfur. The starting materials were taken into a quartz ampoule (15 mm diameter, 180 mm length) evacuated to 2×10^{-6} Torr and then sealed off. The ampoule was placed into a double-zone (source and growth zones) horizontal electrical furnace. During the first stage the furnace was slowly heated at the rate of 10 K/h. The temperatures of the source and growth zones were allowed to reach 1023 and 1073 K, respectively, in order to clean the wall of the ampoule. The duration was 20 h. After this, the temperature of the source and growth zones were maintained at 1073 and 1023 K, respectively, for the next 15 days, after which the furnace was slowly cooled at a rate of 20 K/h. When the temperature of the ampoule reached room temperature it was opened to obtain yellow-colored AgGaS$_2$ crystals. Similarly single crystals of AgGaS$_2$ were grown with the same iodine concentration and source zone temperature of 1073 K. A temperature differences of 100 K was maintained between source and growth zones, so that the temperature of the growth zone was maintained at 973 K. The growth was carried out for a period of 15 days. The AgGaS$_2$ single crystals obtained were yellow in color. AgGaS$_2$ single crystals grown at growth zone temperatures of 1023 and 973 K are shown in Fig. 27.21a,b. Cut and polished AgGaS$_2$ single crystal grown at 1023 K is shown in Fig. 27.21c. Single-crystal XRD and powder XRD studies indicate that the as-grown AgGaS$_2$ crystals belong to the tetragonal (chalcopyrite) system with (112) plane as the dominant peak. The full width at half-maximum (FWHM) of the x-ray rocking curve

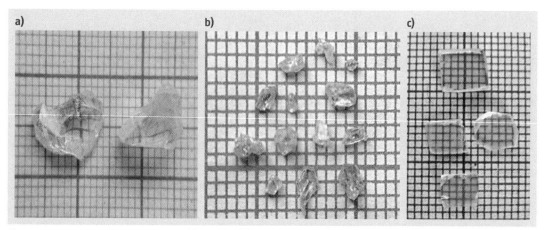

Fig. 27.21a–c AgGaS$_2$ crystals grown at (**a**) 1023 K, (**b**) 973 K, and (**c**) cut and polished AgGaS$_2$ crystals grown at 1023 K

for the as-grown AgGaS$_2$ single crystal was 5 arcsec. The quality of the crystals obtained with both temperature differences (100 and 50 K) between the source and growth zones was found to be good.

Noda et al. [27.137] have measured the transport rate for CVT growth of AgGaS$_2$ single crystal. The transport rate was measured by weighing the total amount of AgGaS$_2$ crystals grown in the growth zone. The important parameters affecting crystal growth were the amount of iodine, source temperature, and the temperature difference between the source and growth zones. The observed transport rate tends to increase with increasing ΔT at fixed T_s, but a higher transport rate does not necessarily mean larger single crystals. The other parameters affecting crystal growth is the presence of a maximum transport rate with increasing T_s at fixed ΔT. This indicates that condensation of the atoms at the growth zone was delayed with increasing growth temperature. All the as-grown crystals were transparent and yellowish in color and had rod and plate morphologies. The rod forms of single crystals were obtained when the growth temperature was more than 1173 K and the transport rate less than 10 mm/day. The biggest plate-type crystal ($7 \times 5 \times 2$ mm^3) was obtained under the conditions of $T_s = 1248$ K and $\Delta T = 75$ K for 1 week. The habit plane of the rod was (112) and the growth direction was nearly in the (112) direction. The PL spectra of the crystals were dependent on the amount of iodine used, and excitonic emission was observed, which means that good-quality crystals were obtained.

Polycrystalline AgGaS$_2$ was prepared by synthesis from its constituent elements Ag, Ga, and S with 5 N purity taken into a quartz tube. Polycrystalline powder (2 g) as a source, was sealed with iodine as the transporting agent at concentration of 5 mg/cm^3 into a quartz ampoule of length 20 cm and inner diameter 1 cm. The as-grown AgGaS$_2$ single crystals were transparent and yellowish in color.

Single crystals of AgGaSSe [27.56] have been successfully grown from 2.5 g polycrystalline powder AgGaSSe as source, sealed with iodine at concentration of 5 mg/cm^3 into a quartz ampoule of length 20 cm and inner diameter 1 cm.

Nigge et al. [27.138] have grown single crystals of AgGaSe$_2$ by CVT method using iodine as the transporting agent. Growth temperature of 770 °C and concentrations of the transporting agent of 1.6–1.7 mg I$_2$/cm^3 yielded compact single crystals with dimensions up to $8 \times 5 \times 5$ mm^3.

Tafreshi et al. [27.57] have grown single crystals of Cu$_{0.5}$Ag$_{0.5}$InSSe for the first time by chemical vapor transport technique using iodine as the transporting agent. A stoichiometric mixture of the constituent elements with 5 N purity was taken, and polycrystalline Cu$_{0.5}$Ag$_{0.5}$InSSe ingots were synthesized in a vertical furnace in vacuum at 900 °C. The polycrystalline synthesized powder of 2.5 g together with 5 mg/cm^3 5 N purity iodine was placed in a quartz ampoule with length 16 cm and diameter 1 cm. The ampoule, cooled by ice, was evacuated to 2×10^{-6} Torr and sealed off. The capsule was then placed into a double-zone horizontal furnace controlled by Eurotherm controllers. A reverse temperature profile was developed across the ampoule over several hours to remove the powder sticking at the tip or deposition zone of the ampoule. The source and growth zone temperatures were kept at 937 °C and 849 °C, respectively. The growth run was carried out for 1 week. After that, the furnace was cooled to room temperature in 20 h. The grown crystals were in the form of small platelets with dimensions of $1 \times 2 \times 0.5$ mm^3, joined together with mirror-like upper faces. X-ray analysis, surface analysis, and microindentation studies were carried out on the grown crystals to determine structure and lattice parameters, the growth mechanism, and mechanical properties, respectively. $Cu_{0.5}Ag_{0.5}InSSe$ was found to crystallize in tetragonal chalcopyrite structure.

27.7 Growth of GaN by VPE

Epitaxial growth can be achieved by solid-phase, liquid-phase, vapor-phase, and molecular-beam deposition. Vapor-phase growth is by far the most widely used technique for semiconductors. It consists of oriented crystal growth of a material transported from the gas phase onto a suitable solid substrate.

27.7.1 Vapor-Phase Epitaxy (VPE)

VPE systems are particularly employed in mass production of electronic devices because of their proven low cost and high throughput, in addition to their capability to grow advanced epitaxial structures. The fundamental reason for their success is due to the ease of dealing with low- and high-vapor-pressure elements. This is achieved by using specific chemical precursors, in the form of vapor, containing the desired elements. These precursors are brought into the reactor by a suitable carrier gas and normally mix shortly before reaching the substrate, giving rise to the nutrient phase for the crystal growth [27.139]. The release of the gas elements necessary for the construction of the crystalline layer may occur at the solid–gas interface or directly in the gas phase, depending on the type of precursors and the thermodynamic conditions.

In vapor-phase epitaxy with open flow systems a carrier gas containing the reactive species is forced to flow past the substrate crystal. At the crystal surface the species undergo a sequence of chemical reactions leading to extension of the substrate crystal lattice and formation of products which must leave the vicinity in order for the process to continue.

The sequence of steps which is generally assumed to occur is shown schematically in Fig. 27.22. The factors which influence the growth rate and material composition are the vapor pressure and the temperature, both of which can be precisely controlled. Depending upon the sources and the reactor type, one can distinguish between two special cases of VPE, namely hydride vapor-phase epitaxy (HVPE) and metalorganic vapor-phase epitaxy (MOVPE). The former uses inorganic sources and a hot-wall reactor, whereas the latter uses fully or partly organic sources and a cold-wall reactor.

27.7.2 VPE GaN Film Growth

Since the earliest pioneering work on growth of epitaxial films by VPE [27.140] until the early 1980s, VPE was a popular method for the growth of epitaxial layers of gallium nitride [27.141]. However, this technique was largely abandoned in the early 1980s because of its apparent inability to reduce the native defect concentration to nondegenerate levels and thus achieve p-type doping. This was presumed to be due to nitrogen vacancy defects, which would be thermodynamically favored at the high growth temperatures typically used in HVPE GaN growth. This interpretation predominated in spite of reports of nondegenerate films grown by HVPE [27.141] and careful growth stud-

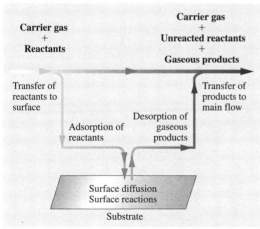

Fig. 27.22 The sequence of steps in a VPE process

ies which suggested that the incorporation behavior of this donor was inconsistent with a nitrogen vacancy defect [27.142]. With the advent of high-purity source materials and improved heteronucleation schemes, the growth of nondegenerate material has been reported by several groups.

The nature of the chemistry involved in GaN growth by HVPE technique differs from that of other III–V semiconductors. For instance, in GaAs growth by halide vapor-phase epitaxy, which uses the halide of the group V precursor $AsCl_3$ instead of its hydride AsH_3, thermal dissociation of the arsenic compounds results in the formation of As_4 and As_2 molecules, which typically remain volatile and chemically reactive and thus participate in film growth. In GaN HVPE, NH_3 is used as a source of nitrogen rather than nitrogen halide, NCl_3, which is highly explosive and highly unstable. In this process the thermal dissociation of NH_3 results in the formation of N_2 molecules, which are extremely stable and essentially unreactive at temperatures of interest [27.143–145].

Hydride VPE (HVPE) and chloride VPE (Cl-VPE) operate in a very similar manner and can be described by the final reaction as

$$GaCl + NH_3 \rightarrow GaN + HCl + H_2 \uparrow ,$$
$$GaCl_3 + NH_3 \rightarrow GaN + 3HCl . \qquad (27.55)$$

However, the precursor for generating gallium chloride is different in both techniques. In HVPE of GaN, the gallium source is gallium monochloride (GaCl), which is stable only at temperatures above 600 °C and is produced by the reaction of liquid gallium with HCl gas [27.146–149]. The supply of GaCl is controlled by the gallium cell temperature and the flow rates of HCl gas and H_2 carrier gas. Several researchers have chosen presynthesized $GaCl_3$ instead of HCl gas [27.134, 143–145]. In Cl-VPE, gallium trichloride ($GaCl_3$) is used as the Ga source, due to its high vapor pressure. The use of monohalide (GaCl) or trihalide is the main difference between HVPE and Cl-VPE, respectively.

27.7.3 Strength of HVPE Method

It is important to analyze the importance of HVPE with respect to the other techniques. *Shaw* [27.150] has demonstrated that the Cl-VPE technique is an equilibrium process. The other epitaxial techniques (MOVPE and MBE) operate far from equilibrium conditions [27.151–153]. The near-equilibrium nature arises because of the reversible processes occurring at the interface due to the volatility of group III at the operating temperatures; shift from equilibrium is caused by kinetic factors mainly due to ammonia decomposition. Based on this description it is easy to enumerate the strengths of VPE:

1. In VPE, being a near-equilibrium process, the growth rates are in principle uniquely determined by the mass input rate of the reactants; hence, unlike in nonequilibrium processes (MOVPE and MBE), very high growth rates ($> 10 \mu m/h$) can be easily achieved;
2. The lack of carbon incorporation into the film;
3. Cost effectiveness.

In the following, the details of the system and the process of chloride VPE for the growth of GaN layers are described. Furthermore, the optimum conditions for the growth of gallium nitride have been established. Based on our experimental investigations on the influence of the different growth parameters on the quality of the layer, it has been identified that the growth temperature plays a dominant role in the growth of GaN as compared with other parameters such as the flow rate, deposition time, etc.

27.7.4 Development of VPE System for the Growth of GaN

The development of experimental apparatus for the growth of GaN at our university is described [27.153]. The vapor-phase epitaxy (VPE) system mainly consists of a reaction chamber, single-zone furnace, $GaCl_3$ cell assembly, process control, and ammonia gas purifier assembly. The schematic diagram of the system is shown in Fig. 27.23.

As the film growth involves a high-temperature process, the reaction chamber is made out of a quartz tube of length 90 cm, wall thickness 2 mm, and diameter 70 mm. Both ends of the tube are made of stainless steel and the assembly consists of two ports on one end and three ports on the other. Integral structures are made to hold the quartz tube rigidly at either end. One end of the quartz is used for the transport of $GaCl_3$ vapor with nitrogen as carrier gas, and another inlet is used for another stream of nitrogen gas for diluting the reacting species, as the reaction is a dilute reaction. The other end of the reactor tube is used for an ammonia gas input port, exhaust line, and substrate holder setup.

The single-zone furnace consists of heating arrangements surrounded by metal boxes. A ceramic muffle of length 80 cm has been used for winding the heating

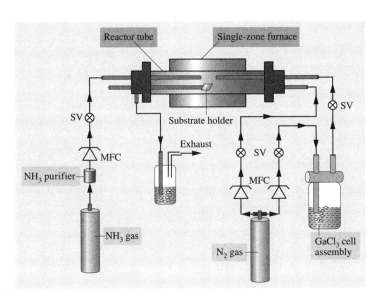

Fig. 27.23 Schematic diagram of the GaN vapor-phase epitaxy system

element (A1 Kanthal wire: SWG 18 type). The windings were done in such a way that the space between them gradually becomes narrow at the ends compared with at the center of the tube in order to compensate for heat losses at the ends of the tube to a certain extent. The windings were insulated by applying a paste of corundum cement and silica gel mixture (the silica gel provides a binding action to the refractory cement). The tube is then packed inside a stainless-steel container using a zirconia-grade fiber blanket of density $128\,\mathrm{kg/m^3}$, which acts as a heat shield. Two thermocouples are used: one to control the furnace temperature, placed at the middle of the ceramic tube, i.e., near the growth zone; and another mounted inside the quartz tube, very close to the boat, so that the growth temperature could be monitored correctly.

The $GaCl_3$ cell assembly is made of quartz tube and a special Teflon assembly. The crystalline $GaCl_3$ in the cell is melted by a liquid-paraffin bath in which it is immersed. An electric heater maintains the paraffin bath at a suitable temperature in the range $80-150\,°C$. $GaCl_3$ vapor was transported to the reactor by nitrogen carrier gas. The partial pressure of $GaCl_3$ could be controlled by both the $GaCl_3$ cell temperature and the flow rate of the N_2 gas. A mass flow controller and solenoid valves were used to control the flow rate of the carrier gas. A separate nitrogen gas line is used to dilute the ammonia gas to the required partial pressure. This nitrogen gas flow is measured by a float-type flow meter. The flow rate is varied by a needle valve. A computer program controls the temperature of both the furnace and the $GaCl_3$ bath. The entire growth scheme can be preprogrammed. The program also enables the heating and cooling cycle to be programmed and data to be acquired during the growth cycle.

The purifier system consists of molecular sieves, which is a palladium- and silica-based compound, to remove moisture, carbon monoxide, and carbon dioxide form the ammonia gas. The purified ammonia is expected to be of analytical grade. After this online purification of the ammonia gas, its flow is monitored by a gas flow (float-type) meter.

27.7.5 Growth of GaN by HVPE

The experimental conditions for growth of high-quality GaN layers using $GaCl_3$ precursor are discussed in this section. In view of the high-quality GaN growth, it is understood that careful cleaning of substrates prior to loading into the growth furnace is an important issue, because the quality of the grown layer (desired morphology and good optical properties) depends on this process, as does the reproducibility of the results. Sapphire ($Al_2O_3(0001)$) substrates were degreased in trichloroethylene (TCE), acetone, methanol, and deionized water in sequence for about 10 min. After completing the organic cleaning processes, in order to remove residual damage and scratches on the surface, substrates were chemically etched with $HCl + H_3PO_4$ (3 : 1) solution heated at $80\,°C$ for 15 min. Etched sapphire substrates were again rinsed thoroughly in deionized water.

A quartz plate of 8 cm² area was used as the substrate holder, located at an optimized distance from the $GaCl_3 : NH_3$ mixing zone. Substrates of 1 cm² area are used per run and are placed lengthwise along the deposition zone. Considerable variation of the amount and quantity of deposition was observed between layers grown in a single run with respect to the position of the substrate. It was observed that GaN layer always deposits in regions closer to the $GaCl_3 : NH_3$ mixing zone. Initially the substrates are placed close to the mixing zone to find the exact position of uniform deposition. The substrate position has been varied further downstream from the mixing zone. It is observed that the layer thickness decreases rapidly with distance from the mixing zone. Growth ceases completely at distances from 6 to 10 cm downstream of the mixing zone.

After loading the substrates, the reactor was purged with nitrogen gas. Subsequently, the substrate was heated to the desired growth temperature and $GaCl_3$ was supplied using nitrogen carrier gas. The nitrogen source, NH_3, is not stable at high temperature and will thermally decompose to nitrogen and hydrogen gases. At the typical growth temperature (1200 K) the equilibrium value is about 0.1. Hence most of the ammonia will decompose to nitrogen and hydrogen at that elevated temperature in thermodynamic equilibrium. However, it is not easy to reach thermodynamic equilibrium in such an open system. The actual value depends on the NH_3 temperature, partial pressure, residence time, and surface conditions. Both NH_3 partial pressure and residence time can be changed by varying the NH_3 and N_2 flow rates. The H_2 gas, produced from the thermal decomposition of ammonia, will promote the reduction of $GaCl_3$ to GaCl. If hydrogen reduction of $GaCl_3$ is not complete, the remaining $GaCl_3$ can react with ammonia via the following reaction:

$$GaCl_3 + NH_3 \rightarrow GaN + 3HCl. \quad (27.56)$$

No low-temperature buffer layer was used in these experiments. Typical N_2 flow rate for $GaCl_3$ transport was 0.1 l/min. During the process, 1.0–2.0 l/min NH_3 and 1.0–2.0 l/min of N_2 were introduced into the reactor zone. The growth temperature was varied in the range 925–1050 °C. Ambient pressure inside the reactor was maintained at 1 atm for all the experiments.

Under typical growth conditions, reaction (27.56) is thermodynamically favorable. A thick deposition of $GaCl_3$ is always observed on the exhaust side of the reactor, which implies that unreacted GaCl combines with HCl and forms $GaCl_3$, which is a more stable form of gallium chloride at lower temperature (< 600 °C).

27.7.6 Characterization of GaN Films

Following the development of the growth technique described previously, the next most important element aspect is characterization of the materials grown. Characterization of a material can be defined as complete description of its physical and chemical properties. A thorough and extensive characterization of a epilayer is very difficult because this would require a variety of tests using a number of sophisticated instruments. It is obvious that there are no ideal crystals in reality and all crystals grown by any technique contain some defects, impurities, and inhomogeneities. Most physical properties are therefore sensitive to deviation from ideality, and generally the characterization of the grown crystals is necessary.

The assessment of crystalline perfection is essential to interpret the structure-dependent properties. Postgrowth analysis of a epitaxial layer provides information on the processes that occurred during growth. Feedback from the analysis can be used to modify the growth process in order to improve the quality of the layers. Moreover, characterization of the grown epilayers forms an integral part of the growth studies performed by the crystal grower. The demand for layers of the highest quality is increasing, and only systematic characterization enables the crystal grower to optimize growth parameters in order to obtain better results.

An epitaxial layer may be characterized by a description of its chemical composition, its structure, its defects, and the spatial distribution of these three features. It is crucial to know the degree of purity and perfection of epilayers in order to interpret structure-dependent properties and determine whether the material can be successfully employed in experiments or for device fabrication.

GaN, being a technologically important material, has elicited a large number of characterization studies [27.154–159]; further optimization of device performance requires that the fundamental mechanisms upon which these devices operate be better understood. Optimization of GaN growth to reduce defect density is of paramount importance for achieving high-quality GaN layers. The grown epilayers were subjected to the following characterization studies that provide a basic understanding of the GaN material properties:

1. Crystal structure analysis by x-ray powder diffraction and high-resolution x-ray diffraction (HRXRD) techniques
2. Molecular structural analysis by Raman scattering studies
3. Determination of the bandgap of the film by UV–vis absorption spectra
4. Identification of excitonic transitions by both room- and low-temperature photoluminescence (PL) spectra
5. Surface morphology by scanning electron microscope (SEM) studies
6. Electrical transport properties and Hall-effect measurements.

X-ray diffraction was used to determine the crystallinity and orientation of the GaN films. GaN films grown under different experimental conditions were characterized for crystalline quality and phase formation by using powder x-ray diffractometer with Cu K_α radiation; the recorded XRD patterns are shown in Fig. 27.24. It is clear from this figure that films grown at 950 °C and 990 °C exhibit a high degree of c-axis orientation nature, which indicating that basal planes of the GaN film and sapphire substrate are parallel to each other. However, the XRD pattern for the film grown at 925 °C shows polycrystalline nature. When the temperature is increased to 950 °C and 990 °C, the polycrystalline nature of the film is suppressed and the single-crystalline nature dominates. It is observed from the XRD results that decomposition of NH_3 is very low at those temperatures and hence the amount of atomic nitrogen for the reaction is also very low.

From XRD analysis, it has been observed that GaN layers are single-phase wurtzite structure. For GaN grown at 925 °C and 950 °C, XRD spectra exhibit a sharp peak of wurtzite GaN (0002) at 34.25° with blunt peaks for GaN ($\bar{1}011$) and GaN ($\bar{1}100$) at $2\theta = 36.60°$ and $32.24°$, respectively. However the XRD spectrum for GaN grown at 990 °C shows diffraction only from the c-plane of GaN and the sapphire substrate. This indicates that preferentially oriented GaN layer is realized when the growth temperature is kept at 990 °C [27.152]. The lattice parameters have also been calculated from the XRD data as $a = 3.186$ Å and $c = 5.184$ Å, which are in very good agreement with reported values. The results of the other characterization studies are available in the literature [27.160–162]. Single-crystalline GaN nanowires on C-Al_2O_3 substrates have been synthesized by vapor-phase epitaxy process with the help of a Ni catalyst [27.163].

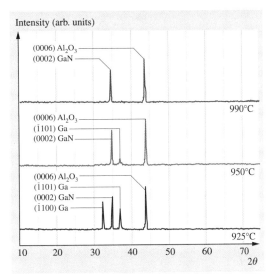

Fig. 27.24 XRD pattern of GaN grown on (0001) sapphire substrate at different growth temperatures

27.8 Conclusion

The growth of single crystals of II–VI and I–III–VI$_2$ compounds from vapor phase is a field of active research, with many researchers. The theoretical and thermodynamical background for the growth of these crystals have been given. The partial pressures of different components inside the growth ampoule during CVT growth of ZnSe single crystals were calculated using a thermodynamic model. Iodine was considered as the transporting agent. From the calculations, the optimum conditions for the growth of good-quality ZnSe single crystals were determined. The optimum temperature range was found to be 800–850 °C for an iodine concentration of 2 mg/cm^3. The undercooling and hence supersaturation ratios of S_2 and Se_2 were the factors determining the composition of ZnS_xSe_{1-x} single crystals grown in closed-tube vapor transport when ZnS and ZnSe were taken as source materials in 1 : 1 ratio. These values were calculated as a function of deposition temperature and iodine concentrations. The partial pressures were calculated using a thermodynamic model. The critical radius and free energy of formation were calculated using the supersaturation ratios by classical nucleation theory. Undercooling was found to have an influence on the supersaturations of S_2 and Se_2, and hence on the composition of the resulting crystals. Conditions of larger undercooling were found to favor S-rich composition whereas smaller undercooling was found to favor ZnS_xSe_{1-x} composition near to that of the source. ZnSe:Mn nanocrystals have been synthesized by chemical vapor transport method using iodine as the transporting agent on a matrix of SiO_2 aerogel. The cubic structure has been confirmed based on powder XRD results. The size of the nanocrystals calculated using Debye–Scherrer formula was ≈ 60 nm. This result is supported by TEM and electron diffraction pattern. A well-resolved ESR spectrum containing six lines corresponding to the Mn^{2+} ion in the ZnSe nanocrystal lattice was obtained and attributed to the presence of a low concentration of Mn. The luminescence spectrum shows the band-edge emission at around 466 nm and Mn-related emission at 570 nm. The growth of I–III–VI_2 compounds has been described. Our recent results on vapor-phase epitaxial growth of GaN have been presented.

References

27.1 M.M. Faktor, I. Garrett, R. Heckingbottom: Diffusional limitations in gas phase growth of crystals, J. Cryst. Growth **9**, 3–11 (1971)

27.2 M.A. Contreras, B. Egaas, K. Ramanathan, J. Hiltner, A. Swartzlander, F. Hasoon: Progress toward 20% efficiency in Cu(In,Ga)Se$_2$ polycrystalline thin-film solar cells, Prog. Photovolt. **7**, 311–316 (1999)

27.3 S. Nakamura, T. Mukai, M. Senon: Candela-class high-brightness InGaN/AlGaN double-heterostructure blue-light-emitting diodes, Appl. Phys. Lett. **64**, 687–1689 (1994)

27.4 T. Lang, M. Odnoblyudov, V. Bougrov, S. Suihkonen, M. Sopanen, H. Lipsanen: Validating gallium nitride growth kinetics using a precursor delivery showerhead as novel chemical reactor, J. Cryst. Growth **292**, 26–32 (2006)

27.5 I. Akasaki, H. Amano, Y. Koide, K. Hiramatsu, N. Sawaki: Effects of ain buffer layer on crystallographic structure and on electrical and optical properties of GaN and Ga$_{1-x}$Al$_x$N ($0 < x \leq 0.4$) films grown on sapphire substrate by MOVPE, J. Cryst. Growth **98**, 209–219 (1989)

27.6 M. Sumiya, N. Ogusu, Y. Yotsuda, M. Itoh, S. Fuke, S. Nakamura: Systematic analysis and control of low-temperature GaN buffer layers on sapphire substrates, J. Appl. Phys. **131**, 1311–1319 (2003)

27.7 G.H. Westphal: Convective transport in vapor growth systems, J. Cryst. Growth **65**, 105–123 (1983)

27.8 G. Attolini, C. Paorici, P. Ramasamy: Skeletal and hollow crystals of cadmium sulphide grown under time-increasing supersaturation, J. Cryst. Growth **78**, 181–184 (1986)

27.9 C. Paorici, L. Zanotti, G. Zuccalli: A temperature variation method for the growth of chalcopyrite crystals by iodine vapor transport, J. Cryst. Growth **43**, 705–710 (1978)

27.10 R.F. Lever, G. Mandel: Diffusion and the vapor transport of solids, J. Phys. Chem. Solids **23**, 599 (1962)

27.11 H. Hartmann: Vapor-phase epitaxy of II–VI compounds: A review, J. Cryst. Growth **31**, 323–332 (1975)

27.12 H. Schäfer: *Chemical Transport Reactions* (Academic, New York 1964)

27.13 R. Zuo, W. Wang: Theoretical study on chemical vapor transport of ZnS-I$_2$ system. Part II. Numerical modelling, J. Cryst. Growth **236**, 695–710 (2002)

27.14 R.F. Lever: Solid transport rate in the vapor-solvent growth system ZnS:I, J. Chem. Phys. **37**, 1078–1082 (1962)

27.15 G. Mandel: Diffusion and the vapor-transport of solids, J. Chem. Phys. **37**, 1177–1198 (1962)

27.16 G. Mandel: Vapor transport of solids by vapor phase reactions, J. Phys. Chem. Solids **23**, 587–598 (1962)

27.17 E. Lendvay: Growth of structurally pure cubic and hexagonal ZnS single crystals, J. Cryst. Growth **10**, 77–84 (1971)

27.18 R. Zuo, W. Wang: Theoretical study on chemical vapor transport of ZnS-I$_2$ system. Part I: Kinetic process and one-dimensional model, J. Cryst. Growth **236**, 687–694 (2002)

27.19 T. Arizumi, T. Nishinaga: Transport reaction in closed tube process, Jpn. J. Appl. Phys. **4**, 165–172 (1965)

27.20 T. Arizumi, T. Nishinaga: Thermodynamics of vapor growth of ZnSe-Ge-I$_2$ system in closed tube process, Jpn. J. Appl. Phys. **5**, 21–28 (1966)

27.21 H. Watanabe, T. Nishinaga, T. Arizumi: Vapor transport equations for III–V compound semiconductors, J. Cryst. Growth **17**, 183–188 (1972)

27.22 M.M. Faktor, I. Garrett: *Growth of Crystals from the Vapor* (Chapman and Hall, London 1974)

27.23 K. Noda, N. Mastsumura, S. Otsuka: Chemical-vapor-transport rate of ZnS in closed Tube, J. Electrochem. Soc. **137**, 1281–1294 (1990)

27.24 B.I. Noläng, M.W. Richardson: The transport flux function – A new method for predicting the rate of chemical transport in closed systems: I. Theory, J. Cryst. Growth **34**, 198–204 (1976)

27.25 B.I. Noläng, M.W. Richardson: The transport flux function – a new method for predicting the rate of chemical transport in flosed systems: II. A theoretical study of systems and experimental conditions for the chemical transport of SnO_2, J. Cryst. Growth **34**, 205–214 (1976)

27.26 M.W. Richardson, B.I. Noläng: Predicting the rate of chemical transport using the flux function method, J. Cryst. Growth **42**, 90–97 (1977)

27.27 K. Klosse: A new productivity function and stability criterion in chemical vapor transport processes, J. Solid State Chem. **15**, 105–116 (1975)

27.28 K. Klosse, P. Ullersma: Convection in a chemical vapor transport process, J. Cryst. Growth **18**, 167–174 (1973)

27.29 S.R. Brodkey, H.C. Hershey: *Transport Phenomena. A Unified Approach* (McGraw-Hill, New York 1988)

27.30 R. Nitsche, D.F. Sargent, P. Wild: Crystal growth of quaternary $1_2 2 4 6_4$ chalcogenides by iodine vapor transport, J. Cryst. Growth **1**, 52–53 (1967)

27.31 O. Senthil Kumar, S. Soundeswaran, R. Dhanasekaran: Thermodynamic calculations and growth of ZnSe single crystals by chemical vapor transport technique, Cryst. Growth Des. **2**, 585–589 (2002)

27.32 C. Paorici, C. Pelosi, G. Attolini, G. Zuccalli: Closed-tube chemical-transport mechanisms in the Cd:Te:H:Cl:N system, J. Cryst. Growth **28**, 358–364 (1975)

27.33 C. Paorici, G. Attolini, C. Pelosi, G. Zuccalli: Chemical transport mechanisms in the Cd:Te:H:I system, J. Cryst. Growth **21**, 227–234 (1974)

27.34 C. Paorici, L. Zecchina: The productivity function for multi-reaction chemical vapor transport in closed tubes, J. Cryst. Growth **97**, 267–272 (1989)

27.35 C. Paorici, L. Zecchina: Note on the productivity function in closed tube chemical vapor transport, J. Cryst. Growth **83**, 453–455 (1987)

27.36 C. Paorici, C. Pelosi: Kinetics of vapor growth in the system $CdS:I_2$, J. Cryst. Growth **35**, 65–72 (1976)

27.37 R.C. Reid, J.M. Prausnitz, T.K. Sherwood: *The Properties of Gases and Liquids*, 3rd edn. (McGraw-Hill, New York 1977)

27.38 K. Böttcher, H. Hartmann, D. Siche: Computational study on the CVT of the $ZnSe-I_2$ material system, J. Cryst. Growth **224**, 195–203 (2001)

27.39 B.I. Noläng, M.W. Richardson: The transport flux function – A new method for predicting the rate of chemical transport in closed systems: I. Theory, J. Cryst. Growth **34**, 198–204 (1976)

27.40 S. Fiechter, K. Eckert: Crystal growth of HfS_2 by chemical vapor transport with halogen (Cl, Br, I), J. Cryst. Growth **88**, 435–441 (1988)

27.41 O. Senthil Kumar, S. Soundeswaran, D. Kabiraj, D.K. Avasthi, R. Dhanasekaran: Effect of heat treatment and Si ion irradiation on ZnS_xSe_{1-x} single crystals grown by CVT method, J. Cryst. Growth **275**, e567–e570 (2005)

27.42 H. Ogawa, M. Nishio: Epitaxial growth and chemical vapor transport of ZnTe by closed-tube method, J. Cryst. Growth **52**, 263–268 (1981)

27.43 J.O. Hirschfelder, C.F. Curtiss, R.B. Bird: *Molecular Theory of Gases and Liquids* (Wiley, New York 1954)

27.44 F. Rosenberger, M.C. Delong, J.M. Olson: Heat transfer and temperature oscillations in chemical vapor transport crystal growth I, J. Cryst. Growth **19**, 317–328 (1973)

27.45 H. Watanabe, T. Nisi-iinaga, T. Arezume: Vapor transport equations for III–V compound semiconductors, J. Cryst. Growth **17**, 183–188 (1972)

27.46 M. Nishlo, H. Ogawa: Chemical vapor transport in the ZnTe-HCl closed-tube system and its thermodynamic analysis, J. Cryst. Growth **78**, 218–226 (1986)

27.47 M. Lenz, R. Gruehn: Developments in measuring and calculating chemical vapor transport phenomena demonstrated on Cr, Mo, W, and their compounds, Chem. Rev. **97**, 2967–2994 (1997)

27.48 Y.-G. Sha, C.-H. Su, F.R. Szofrari: Thermodynamic analysis and mass flux of the $HgZnTe-HgI_2$ chemical vapor transport system, J. Cryst. Growth **131**, 574–588 (1993)

27.49 D.W. Mackowski, V.R. Rao, R.W. Knight: Effect of solid phase heat transfer and wall deposition on crystal growth in physical vapor transport ampoules, J. Cryst. Growth **165**, 323–334 (1996)

27.50 A. Nadarajahk, F. Rosenberger, D. Alexander, J. Iwan: Effects of buoyancy-driven flow and thermal boundary conditions on physical vapor transport, J. Cryst. Growth **118**, 49–59 (1992)

27.51 M. Shiloh, J. Gutman: Growth of ZnO single crystals by chemical vapor transport, J. Cryst. Growth **11**, 105–109 (1971)

27.52 H. Hartmann, R. Mach, B. Selle: *Current Topics in Material Science*, Vol. 9, ed. by E. Kaldis (North-Holland, Amsterdam 1982)

27.53 K. Böttcher, H. Hartmann: Zinc selenide single crystal growth by chemical transport reactions, J. Cryst. Growth **146**, 53–58 (1995)

27.54 K. Mochizuki: Vapor growth and stoichiometry control of zinc sulfo-selenide, J. Cryst. Growth **58**, 87–94 (1982)

27.55 O. Senthil Kumar: Growth of pure and doped single crystals and nanocrystals of ZnSe and ZnS_xSe_{1-x} semiconductors by chemical vapor transport and their characterization, Ph.D. Thesis (Anna University, Chennai 2005)

27.56 M.J. Tafreshi: Growth of II–VI. I–III–VI$_2$ and CdIn$_2$S$_4$ single crystals by CVT method and their characterizations, Ph.D. Thesis (Anna University, Chennai 1996)

27.57 M.J. Tafreshi, K. Balakrishnan, R. Dhanasekaran: Growth and characterization of pentenary Cu$_{0.5}$Ag$_{0.5}$InSSe crystals grown by chemical vapor transport technique, Mater. Res. Bull. **30**, 1371–1377 (1995)

27.58 M.J. Tafreshi, K. Balakrishnan, R. Dhanasekaran: Micromorphological studies on the ZnSe single crystals grown by chemical vapor transport technique, J. Mater. Sci. **32**, 3517–3521 (1997)

27.59 W. Palosz: Vapor transport of ZnO in closed ampoules, J. Cryst. Growth **286**, 42–49 (2006)

27.60 D.W. Greenwell, B.L. Markham, F. Rosenberger: Numerical modeling of diffusive physical vapor transport in cylindrical ampoules, J. Cryst. Growth **51**, 413–425 (1981)

27.61 J.N. Butler, R.S. Brokaw: Thermal conductivity of gas mixtures in chemical equilibrium, J. Chem. Phys. **26**, 1636–1645 (1957)

27.62 G.T. Kim, J.T. Lin, O.C. Jones, M.E. Glicksman, W.M.B. Duval, N.B. Singh: Effects of convection during the physical vapor transport process: application of laser Doppler velocimetry, J. Cryst. Growth **165**, 429–437 (1996)

27.63 B.M. Bulakh, G.S. Pekar: The CdS crystal synthesis from vapors of the component elements, J. Cryst. Growth **8**, 99–103 (1971)

27.64 J.M. Ntep, S. Said Hassani, A. Lusson, A. Tromson-Carli, D. Ballutaud, G. Didier, R. Triboulet: ZnO growth by chemical vapor transport, J. Cryst. Growth **207**, 30–34 (1999)

27.65 G.R. Patzke, S. Locmelis, R. Wartchow, M. Binnewies: Chemical transport phenomena in the ZnO–Ga$_2$O$_3$ system, J. Cryst. Growth **203**, 141–148 (1999)

27.66 H. Wiedemeier, W. Palosz: Physical vapor transport of cadmium telluride in closed ampoules, J. Cryst. Growth **96**, 933–945 (1989)

27.67 T. Yamauchi, Y. Takahara, M. Naitoh, N. Narita: Growth mechanism of ZnSe single crystal by chemical vapor transport method, Physica B: Condens. Matter **376/377**, 778–781 (2006)

27.68 C.S. Fang, Q.T. Gu, J.Q. Wei, Q.W. Pan, W. Shi, J.Y. Wang: Growth of ZnSe single crystals, J. Cryst. Growth **209**, 542–546 (2000)

27.69 W.N. Holton, R.K. Watts, R.D. Stinedurf: Synthesis and melt growth of doped ZnSe, J. Cryst. Growth **6**, 97–100 (1969)

27.70 R. Nishizawa, K. Itoh, Y. Okuno, F. Sakurai: Blue light emission from ZnSe p–n junctions, J. Appl. Phys. **57**, 2210–2214 (1985)

27.71 N. Krasnov, Y. Purtov, F. Vaksman, V.V. Serdyuk: ZnSe blue-light-emitting diode, J. Cryst. Growth **125**, 373–374 (1995)

27.72 R. Nitsche: The growth of single crystals of binary and ternary chalcogenides by chemical transport reactions, J. Phys. Chem. Solids **17**, 163–165 (1960)

27.73 P. Rudolph, K. Umetsu, H.J. Koh, T. Fukuda: Growth of twin reduced ZnSe bulk crystals from the melt, J. Cryst. Growth **143**, 359–361 (1994)

27.74 H. Unuma, M. Higuchi, Y. Yamakawa, K. Kodaira, Y. Okano, K. Hoshikawa, T. Fukuda, T. Koyama: Liquid encapsulated flux growth of ZnSe single crystals from Se solvent, Jpn. J. Appl. Phys. **31**, L383–L384 (1992)

27.75 H.H. Woodbury, R.B. Hall: Diffusion of the chalcogens in the II–VI compounds, Phys. Rev. **157**, 641–645 (1967)

27.76 M. Shone, B. Greenberg, M. Kaczenski: Vertical zone growth and characterization of undoped and Na, P and Mn doped ZnSe, J. Cryst. Growth **86**, 132–137 (1988)

27.77 R.N. Bhargava: Materials growth and its impact on devices from wide band gap II–VI compounds, J. Cryst. Growth **86**, 873–879 (1990)

27.78 E. Kaldis: Crystal growth and growth rates of CdS by sublimation and chemical transport, J. Phys. Chem. Solids **26**, 1701–1732 (1965)

27.79 S. Fujita, H. Mimoto, H. Takebe, T. Noguchi: ZnO-based thin films synthesized by atmospheric pressure mist chemical vapor deposition, J. Cryst. Growth **47**, 326–334 (1979)

27.80 E. Kaldis: The chemistry of imperfect crystals, preparation, purification, crystal growth and phase theory by F.A. Kruger, J. Cryst. Growth **24**, 24–39 (1974)

27.81 K. Böttcher, H. Hartmann: Zinc selenide single crystal growth by chemical transport reactions, J. Cryst. Growth **146**, 53–58 (1995)

27.82 A.A. Simanovskii, N.N. Sheftal, E.I. Givargizov: Effect of crystallization condition on the morphology of ZnSe crystals, Growth Cryst. **9**, 225–229 (1975)

27.83 I. Nakada, E. Bauser: Origin of multiple steps in vapor growth of NbSe$_4$I$_{0.33}$, J. Cryst. Growth **96**, 243–257 (1989)

27.84 A.A. Simanovskii, N.N. Sheftal, E. Givargizov: Production of ZnSe single crystals via transport reaction, Growth Cryst. **7**, 224–229 (1969)

27.85 M.H.J. Hottenhuis, C.B. Lucasius: The influence of impurities on crystal growth: In situ observation of the (010) face of potassium hydrogen phthalate, J. Cryst. Growth **78**, 379–388 (1986)

27.86 Z. Sobiesierski, I.M. Dharmadasa, R.H. Williams: Photoluminescence as a probe of semiconductor surfaces: CdTe and CdS, J. Cryst. Growth **101**, 599–602 (1990)

27.87 W.E. Metcalf, R.H. Fahrig: High-pressure, high-temperature growth of cadmium sulfide crystals, J. Electrochem. Soc. **105**, 719–723 (1958)

27.88 M. Rubenstein: Solution growth of some II–VI compounds using tin as a solvent, J. Cryst. Growth **3**, 309–312 (1968)

27.89 C. Paorici: Iodine-doped hollow CdS crystals, J. Cryst. Growth **5**, 315–316 (1969)

27.90 E. Kaldis: Crystal growth and growth rates of CdS by sublimation and chemical transport, J. Cryst. Growth **5**, 376–390 (1969)

27.91 K. Matsumoto, K. Takagi, S. Kaneko: Kinetics of the cubic to hexagonal transformation of cadmium sulfide, J. Electrochem. Soc. **62**, 389–393 (1983)

27.92 G. Attolini, C. Paorici, L. Zanoti: Growth of cadmium sulphide single crystals by vapor-phase hydrogen transport, J. Cryst. Growth **56**, 254–258 (1982)

27.93 G. Attolini, C. Paorici, L. Zecchina: Productivity function for multireactional CVT and its application to iodine transport of cadmium sulphide, J. Cryst. Growth **99**, 731–736 (1990)

27.94 M.J. Tafreshi, K. Balakrishnan, R. Dhanasekaran: Growth, electrical conductivity and microindentation studies of $CuInS_2$ single crystals, J. Mater. Sci. Mater. Electron. **7**, 243–245 (1996)

27.95 S. Itoh, K. Nakano, A. Ishibashi: Raman studies of phosphorus-doped ZnSe, J. Cryst. Growth **214/215**, 1029–1034 (2000)

27.96 E. Chimczak, J.W. Allen: Energy transfer in the electroluminescence of ZnS:Mn and ZnSe:Mn driven by short voltage pulses, J. Phys. D: Appl. Phys. **18**, 951–957 (1985)

27.97 G. Jones, J. Woods: The luminescence of manganese-doped zinc selenide, J. Phys. D: Appl. Phys. **6**, 1640–1651 (1973)

27.98 C. Jin, B. Zhang, Z. Ling, J. Wang, X. Hou, Y. Segawa, X. Wang: Growth and optical characterization of diluted magnetic semiconductor $Zn_{1-x}Mn_xSe/ZnSe$ strained-layer superlattices, J. Appl. Phys. **81**, 5148–5153 (1997)

27.99 J.K. Furdyna, N. Samranth: Magnetic properties of diluted magnetic semiconductors: A review, J. Appl. Phys. **61**, 3526–3593 (1987)

27.100 J.F. Suyver, J.J. Kelly, A. Meijerink: Temperature-induced line broadening, line narrowing and line shift in the luminescence of nanocrystalline $ZnS:Mn^{2+}$, J. Lumin. **104**, 187–196 (2003)

27.101 H. Heulings IV, X. Huang, J. Li, T. Yuen, C.L. Lin: Mn-substituted inorganic-organic hybrid materials based on ZnSe: Nanostructures that may lead to magnetic semiconductors with a strong quantum confinement effect, Nano Lett. **1**, 521–525 (2001)

27.102 C. Wang, W.X. Zhang, X.F. Qian, X.M. Zhang, Y. Xie, Y.T. Qian: An aqueous approach to ZnSe and CdSe semiconductor nanocrystals, Mater. Chem. Phys. **60**, 99–102 (1999)

27.103 C. De Mello Donega, A.A. Bol, A. Meijerink: Time-resolved luminescence of $ZnS:Mn^{2+}$ nanocrystals, J. Lumin. **96**, 87–93 (2002)

27.104 D. Sarigiannis, J.D. Peck, G. Kioseoglou, A. Petrou, T.J. Mountziaris: Characterization of vapor-phase-grown ZnSe nanoparticles, Appl. Phys. Lett. **80**, 4024–4027 (2002)

27.105 X.T. Zhang, K.M. Ip, Z. Liu, Y.P. Leung, Q. Li, S.K. Hark: Structure and photoluminescence of ZnSe nanoribbons grown by metal organic chemical vapor deposition, Appl. Phys. Lett. **84**, 2641–2647 (2004)

27.106 S.K. Bera, S. Choudhuri, A.K. Pal: Electron transport properties of CdTe nanocrystals in $SiO_2/CdTe/SiO_2$ thin film structures, Thin Solid Films **415**, 68–77 (2002)

27.107 A.A. Lipovskii, E.V. Kolobkova, V.D. Petrikov: Formation of II–VI nanocrystals in a novel phosphate glass, J. Cryst. Growth **184/185**, 365–369 (1998)

27.108 P. Yang, M.K. Lu, C.F. Song, G.J. Zhou, Z.P. Ai, D. Xu, D.R. Yuan, X.F. Cheng: Strong visible-light emission of ZnS nanocrystals embedded in sol-gel silica xerogel, Mater. Sci. Eng. B **97**, 149–153 (2003)

27.109 P.A. Gonzalez Beerman, B.R. McGarvey, S. Muralidharan, R.C.W. Sung: EPR spectra of Mn^{2+}-doped ZnS quantum dots, Chem. Mater. **16**, 915–918 (2004)

27.110 N. Taghavinia, T. Yao: ZnS nanocrystals embedded in SiO_2 matrix, Physica E **21**, 96–102 (2004)

27.111 K. Manzoor, S.R. Vadera, N. Kumar, T.R.N. Kutty: Synthesis and photoluminescent properties of ZnS nanocrystals doped with copper and halogen, Mater. Chem. Phys. **82**, 718–725 (2003)

27.112 C. Liu, J. Liu, W. Xu: The g-factor-shift in $ZnS:Mn^{2+}$ nanocrystals/pyrex glasses composites, Mater. Sci. Eng. B **75**, 78–81 (2000)

27.113 K. Kawano, R. Nakata, M. Sumita: Electron spin resonance study of laser-annealed (Zn,Mn)O ceramics, Appl. Phys. Lett. **58**, 1742–1745 (1991)

27.114 M. Tanaka: Do triboluminescence spectra really show a spectral shift relative to photoluminescence spectra?, J. Lumin. **100**, 115–126 (2002)

27.115 V.J. Leppert, S. Mahamuni, N.R. Kumbhojkar, S.H. Risbud: Structural and optical characteristics of ZnSe nanocrystals synthesized in the presence of a polymer capping agent, Mater. Sci. Eng. B **52**, 89–92 (1998)

27.116 R. Nitsche, H.U. Bölsterli, M. Lichtensteiger: Crystal growth by chemical transport reactions I. Binary, ternary, and mixed-crystal chalcogenides, J. Phys. Chem. Solids **21**, 199–205 (1961)

27.117 W.N. Honeyman, K.H. Wilkinson: Growth and properties of single crystals of group I–III–VI_2 ternary semiconductors, J. Phys. D: Appl. Phys. **4**, 1182–1185 (1971)

27.118 I. Aksenov, K. Sato: Visible photoluminescence of Zn-doped $CuAlS_2$, Appl. Phys. Lett. **61**, 1063–1066 (1992)

27.119 O. Madelung, V. Rössler, M. Schulz: Optical properties of I–III–VI_2 compounds, Phys. Status Solidi **115**, K113–K118 (1989)

27.120 S. Shirakata, I. Aksenov, K. Sato, S. Isomura: Photoluminescence studies in $CuAlS_2$ crystals, Jpn. J. Appl. Phys **31**, L1071–L1074 (1992)

27.121 I. Aksenov, T. Yasuda, Y. Seawa, K. Sato: Violet photoluminescence in Zn-doped $CuAlS_2$, J. Appl. Phys **74**, 2106–2110 (1993)

27.122 I. Aksenov, T. Yasuda, T. Kai, N. Nishikawa, T. Ohgoh, K. Sato: Visible photoluminescence in undoped and Zn-doped $CuAlS_2$, Jpn. J. Appl. Phys **32**, 1068–1072 (1993)

27.123 N. Kuroishi, K. Mochizuki, K. Kimoto: Surface oxidation of CVT-grown $CuAlSe_2$, Mater. Lett. **57**, 1949–1954 (2003)

27.124 S. Chichibu, M. Shishikura, J. Ino, S. Matsumoto: Electrical and optical properties of $CuAlSe_2$ grown by iodine chemical vapor transport, J. Appl. Phys **70**, 1648–1654 (1991)

27.125 M.-S. Jin, C.-S. Yoon, H.-G. Kim, W.-T. Kim: Growth and characterization of $CuAlSe_2$ single crystals, J. Korean Phys. Soc. **26**, 628 (1993)

27.126 P. Prabukanthan, R. Dhanasekaran: Growth of $CuInTe_2$ single crystals by iodine transport and their characterization, Mater. Res. Bull. **43**, 1996–2004 (2008)

27.127 E. Gombia, F. Leccabue, C. Pelosi, D. Seuret: Vapor growth, thermodynamical study and characterization of $CuInTe_2$ and $CuGaTe_2$ single crystals, J. Cryst. Growth **65**, 391–396 (1983)

27.128 P.W. Yu, D.L. Downing, Y.S. Park: Electrical properties of $CuGaS_2$ single crystals, J. Appl. Phys. **45**, 5283 (1974)

27.129 S. Shirakata, K. Saiki, S. Isomura: Excitonic photoluminescence in $CuGaS_2$ crystals, J. Appl. Phys. **68**, 291–297 (1990)

27.130 P. Prabukanthan, R. Dhanasekaran: Growth of $CuGaS_2$ single crystal by chemical vapor transport and characterization, Cryst. Growth Des. **7**, 618–623 (2007)

27.131 S. Chichibu, S. Shirakata, A. Ogawa, R. Sudo, M. Uchida, Y. Harada, T. Wakiyama, M. Shishikura, S. Matsumoto, S. Isomura: Growth of $Cu(Al_xGa_{1-x})SSe$ pentenary alloy crystals by iodine chemical vapor transport method, J. Cryst. Growth **140**, 388–397 (1994)

27.132 K. Sugiyama, K. Mori, H. Miyake: Growth of epitaxy layer of $CuAlS_2$ on $CuGaS_2$ and characterization, J. Cryst. Growth **113**, 390–395 (1991)

27.133 P. Prabukanthan, R. Dhanasekaran: Influence of Mn doping on $CuGaS_2$ single crystals grown by CVT method and their characterization, J. Phys. D: Appl. Phys. **41**, 115102 (2008)

27.134 S. Tanaka, S. Kawami, H. Kobayashi, H. Sasakura: Luminescence in $CuGaS_{2-2x}Se_{2x}$ mixed crystals grown by chemical vapor transport, J. Phys. Chem. Solids **38**, 680–681 (1977)

27.135 K. Sato, K. Tanaka, K. Ishii, S. Matsuda: Crystal growth and photoluminescence studies in Fe-doped single crystals of $CuAl_{1-x}Ga_xS_2$, J. Cryst. Growth **99**, 772–775 (1990)

27.136 P. Prabukanthan, R. Dhanasekaran: Stoichiometric single crystal growth of $AgGaS_2$ by iodine transport method and characterization, Cryst. Res. Technol. **43**, 1292–1296 (2008)

27.137 Y. Noda, T. Kurasawa, N. Sugai, Y. Furukawa, K. Masumoto: Growth of $AgGaS_2$ single crystals by chemical transport reaction, J. Cryst. Growth **99**, 757–761 (1990)

27.138 K.M. Nigge, F.P. Baumgartner, E. Bucher: CVT-growth of $AgGaSe_2$ single crystals: Electrical and photoluminescence properties, Sol. Energy Mater. Sol. Cells **43**, 335–343 (1996)

27.139 R. Fornari: Vapor phase epitaxial growth and properties of III-nitride materials, Proc. Int. School Cryst. Growth Technol. Important Electron. Mater., ed. by K. Byrappa (2003) pp. 367–390

27.140 H.P. Maruska, J. Tietjen: The preparation and properties of vaporu deposited single crystalline GaN, J. Appl. Phys. Lett. **15**, 327–329 (1969)

27.141 M. Ilemgems: Vapor epitaxy of gallium nitride, J. Cryst. Growth **13/14**, 360–364 (1972)

27.142 W. Seifert, G. Fitzl, E. Butter: Study on the growth rate in VPE of GaN, J. Cryst. Growth **52**, 257–262 (1981)

27.143 R.J. Molnar, W. Gotz, L.T. Romano, N.M. Johnson: Growth of gallium nitride by hydride vapor-phase epitaxy, J. Cryst. Growth **178**, 147–156 (1997)

27.144 H. Lee, J.S. Harris: Observation of superstructure in high-quality pseudomorphic films of NiAl grown on GaAs, J. Cryst. Growth **169**, 689–693 (1996)

27.145 F. Dwikusuma, J. Mayer, T.F. Kuech: Nucleation and initial growth kinetics of GaN on sapphire substrate by hydride vapor-phase epitaxy, J. Cryst. Growth **258**, 65–74 (2003)

27.146 A. Shintani, S. Minagawa: Kinetics of the epitaxial growth of GaN using Ga, HCl and NH_3, J. Cryst. Growth **22**, 1–5 (1974)

27.147 G. Jacob, M. Boulou, M. Furtado: Effect of growth parameters on the properties of GaN:Zn epilayers, J. Cryst. Growth **42**, 136–143 (1977)

27.148 R. Fornari, M. Bosi, N. Armani, G. Attolini, C. Ferrari, C. Pelosi, G. Salviati: Hydride vapor phase epitaxy growth and characterisation of GaN layers, Mater. Sci. Eng. B **79**, 159–164 (2001)

27.149 X. Xu, R.P. Vaudo, C. Loria, A. Salant, G.R. Brandes, J. Chaudhuri: Growth and characterization of low defect GaN by hydride vapor-phase epitaxy, J. Cryst. Growth **246**, 223–229 (2002)

27.150 D.W. Shaw: *Mechanisms in Vapor Phase Epitaxy in Crystal Growth*, ed. by C.H.L. Goodman (Plenum Press, New York 1974) p. 25

27.151 G.B. Stringfellow: Fundamental aspects of vapor growth and epitaxy, J. Cryst. Growth **115**, 1–11 (1991)

27.152 G.B. Stringfellow: Fundamentals of thin film growth, J. Cryst. Growth **137**, 212–223 (1994)

27.153 E. Varadarajan, J. Kumar, R. Dhanasekaran: Fabrication of vapor-phase epitaxy system for the growth of gallium nitride, Proc. 6th Int. Conf. Optoelectron. Photonics **1**, 236–237 (2002)

27.154 I. Akasaki, H. Amano: Widegap column–III nitride semiconductors for UV/blue light emitting devices, J. Electrochem. Soc. **141**, 2266–2269 (1994)

27.155 S. Nakamura, M. Senoh, S. Nagahama, N. Iwasa, Y. Yamada, T. Matsushita, Y. Sugimoto, H. Kiyoku: Large conductance anisotropy in a novel two-dimensional electron system grown on vicinal (111) B GaAs with multiatomic steps, Appl. Phys. Lett. **69**, 3034–3041 (1996)

27.156 I. Akasaki, S. Sota, H. Sakai, T. Tanaka, M. Koike, H. Amano: Shortest wavelength semiconductor laser diode, Electron. Lett. **32**, 1105–1106 (1996)

27.157 S.J. Pearton, J.C. Zolper, R.J. Shul, F. Ren: GaN: Processing, defects, and devices, J. Appl. Phys. **86**, 1–6 (1999)

27.158 H. Morkoc: Comprehensive characterization of hydride VPE grown GaN layers and templates, Mater. Sci. Eng. R **33**, 135–207 (2001)

27.159 R. Fornari, M. Bosi, D. Bersani, G. Attolini, P.P. Lottici, C. Pelosi: Characterization of HVPE GaN layers by atomic force microscopy and Raman spectroscopy, Semicond. Sci. Technol. **16**, 776–782 (2001)

27.160 E. Varadarajan, P. Puviarasu, J. Kumar, R. Dhanasekaran: On the chloride vapor-phase epitaxy growth of GaN and its characterization, J. Cryst. Growth **260**, 43–49 (2004)

27.161 E. Varadarajan, J. Kumar, R. Dhanasekaran: Growth of GaN films by chloride vapor pahse epitaxy, J. Cryst. Growth **268**, 475–477 (2004)

27.162 E. Varadarajan, R. Dhanasekaran, D.K. Avasthi, J. Kumar: Structural, optical and electrical properties of high energy irradiated Cl-VPE grown gallium nitride, Mater. Sci. Eng. B **129**, 121–125 (2006)

27.163 T.I. Shin, H.J. Lee, W.Y. Song, H. Kim, S.-W. Kim, D.H. Yoon: High quality GaN nanowires synthesized from Ga_2O_3 with graphite powder using VPE method, Colloids. Surf. A **313/314**, 52–55 (2008)

936

Part E Epitaxial Growth and Thin Films

28 Epitaxial Growth of Silicon Carbide by Chemical Vapor Deposition
Ishwara B. Bhat, Troy, USA

29 Liquid-Phase Electroepitaxy of Semiconductors
Sadik Dost, Victoria, Canada

30 Epitaxial Lateral Overgrowth of Semiconductors
Zbigniew R. Zytkiewicz, Warszawa, Poland

31 Liquid-Phase Epitaxy of Advanced Materials
Christine F. Klemenz Rivenbark, Titusville, USA

32 Molecular-Beam Epitaxial Growth of HgCdTe
James W. Garland, Bolingbrook, USA
Sivalingam Sivananthan, Chicago, USA

33 Metalorganic Vapor-Phase Epitaxy of Diluted Nitrides and Arsenide Quantum Dots
Udo W. Pohl, Berlin, Germany

34 Formation of SiGe Heterostructures and Their Properties
Yasuhiro Shiraki, Tokyo, Japan
Akira Sakai, Osaka, Japan

35 Plasma Energetics in Pulsed Laser and Pulsed Electron Deposition
Mikhail D. Strikovski, Beltsville, USA
Jeonggoo Kim, Beltsville, USA
Solomon H. Kolagani, Beltsville, USA

28. Epitaxial Growth of Silicon Carbide by Chemical Vapor Deposition

Ishwara B. Bhat

The properties of silicon carbide materials are first reviewed, with special emphasis on properties related to power device applications. Epitaxial growth methods for SiC are then discussed with emphasis on recent results for epitaxial growth by the hot-wall chemical vapor deposition method. The growth mechanism for maintaining the polytype, namely *step-controlled epitaxy*, is discussed. Also described is the selective epitaxial growth carried out on SiC at the author's laboratory, including some unpublished work.

28.1 Polytypes of Silicon Carbide 941
28.2 Defects in SiC 942
    28.2.1 Micropipes 942
    28.2.2 Screw Dislocations..................... 942
    28.2.3 Growth Pits and Triangular Inclusions............ 943
28.3 Epitaxial Growth of Silicon Carbide 944
    28.3.1 Substrates for Silicon Carbide Growth........... 944

28.3.2 How to Control the Polytypes in SiC Homoepitaxy..................... 945
28.3.3 SiC Epitaxial Growth Techniques... 946
28.3.4 Chemical Vapor Deposition 946
28.4 Epitaxial Growth on Patterned Substrates 952
    28.4.1 Selective Epitaxial Growth 953
    28.4.2 Selective Epitaxial Growth of 4H-SiC Using TaC Mask............. 954
    28.4.3 Orientation Dependence of SiC Selective Growth................ 956
    28.4.4 Effects of Mask-to-Window Ratio ($M:W$) on SiC Selective Growth 957
    28.4.5 Effects of C/Si Ratio on SiC Selective Growth................ 959
    28.4.6 Mechanism of Selective Etching and Effect of Atomic Hydrogen..... 960
    28.4.7 Fabrication of 4H-SiC p–n Junction Diodes Using Selective Growth 960
28.5 Conclusions... 961
References 961

Advanced Si technology has brought about the very large-scale integrated circuits (VLSI) that are available today, and still further development of Si VLSI technology is expected. However, Si is approaching its performance limit due to intrinsic material properties, especially in applications related to high-power, high-temperature, and high-frequency devices. Thus, the development of new materials and technologies useful in this area is crucial.

SiC is a an extremely hard and inert IV–IV compound material having a lot of attractive features, in particular electrical properties which are suitable for advanced electronic devices that cannot be achieved using Si. An appreciation of the potential of SiC for electronics can be gained by examining Table 28.1, which compares the relevant material properties of SiC with Si and GaAs. As can be seen, SiC has large bandgaps (2.4–3.3 eV) [28.1–4], high breakdown fields ($\approx 3 \times 10^6$ V/cm) [28.5], high saturation electron velocities (2.7×10^6 cm/s) [28.6, 7], and high thermal conductivity (3.2–4.9 W/(cm K)) [28.8]. Owing to these excellent electrical and physical properties, SiC has been regarded as able to function well under high-temperature, high-power, and high-radiation conditions at which conventional semiconductors cannot perform adequately. This ability is expected to enable significant improvements for a wide variety of applications and systems. These applications range from greatly improved high-voltage switching for energy savings in public electric power distribution and electric vehicles, to more powerful microwave electronics for radar and communications, to sensors and controls for cleaner-

Table 28.1 Physical properties of 3C-, 6H-, 4H-SiC, Si, and GaAs at 300 K [28.1]

Properties	3C-SiC	6H-SiC	4H-SiC	Si	GaAs
Crystal structure	ZB	Hex.	Hex.	Dia.	ZB
Lattice constant (Å)	4.36	$a = 3.09$ $c = 15.12$	$a = 3.09$ $c = 10.08$	5.43	5.65
Band structure	Indirect	Indirect	Indirect	Indirect	Direct
Bandgap (eV)	2.4	3.0	3.3	1.11	1.43
Electron mobility (cm^2/(V s))	900	350^a 50^b	720^a 650^b	1400	8500
Breakdown field 10^6 (V/cm)	1.2	2.0	2.4	0.3	0.4
Thermal conductivity (W/(cm K))	3.2	4.9	3.7	1.5	0.5
Saturation drift velocity ×10^7 (cm/s)	2.0	2.0	2.0	1	2
Dielectric constant	9.6	9.7	10	11.8	12.8

^{a} mobility along a-axis, ^{b} mobility along c-axis

burning more fuel-efficient jet aircraft and automobile engines.

Theoretical simulations have indicated that SiC power metal-oxide semiconductor field-effect transistors (MOSFETs) and diode rectifiers would operate over higher voltage and temperature ranges, have superior switching characteristics, and yet have die sizes nearly 80% smaller than correspondingly rated silicon-based devices [28.9]. Hence, realization of practical SiC power devices will greatly impact the power electronics field.

The development of SiC for electronic applications has been a subject of study for more than 40 years. During the early years, a significant amount of fundamental research was performed, but the development of commercially viable SiC-based devices was limited by the low quality of bulk materials and inadequate epitaxial process. During the late 1980s, significant improvements in bulk and epitaxial process enabled commercial availability of device-quality wafers from Cree Research, Inc. Together, these factors have enabled the fabrication of higher quality device structures and have generated increased research activities in SiC electronic devices. A large number of devices, such as high-voltage Schottky rectifiers [28.10, 11], power metal-oxide semiconductor field-effect transistors [28.9], microwave and millimeter-wave devices [28.12], and high-temperature, radiation-resistant junction FETs (JFETs) [28.13], have been fabricated. However, SiC devices with quality sufficient for large-scale industrial applications were not available until recently. This is primarily due to the fact that the crystal growth and device fabrication technologies for SiC were not sufficiently developed to the degree required for reliable incorporation into electronic systems. Recent commercial availability of SiC Schottky diodes (up to 1200 V and 20 A) from Infineon Technology AG and Cree, Inc., should accelerate the introduction of SiC in commercial systems [28.14]. However, bipolar and power MOSFETs remain commercially elusive.

As for SiC epitaxial growth, homoepitaxial growth of SiC epitaxial layers on SiC substrates has been achieved by so-called *step-controlled growth* [28.15]. The quality of the epilayer and controllability of growth are improving with the improvement of the substrate quality and growth technology. However, abundant defects, including micropipes, screw dislocations, growth pits, step bunching, and 3C inclusions can still be found even in today's commercial wafers. Among these, growth pits, step bunching, and 3C inclusions can be caused by an unoptimized growth process while others are mainly due to defects propagated from the substrate. It is believed that these structural and surface morphological features are the major cause of poor device yield and premature device failure. Thus, these issues have to be addressed in order to advance the widespread commercial introduction of SiC-based systems, especially for high-power applications.

In this chapter, some basic properties of SiC that are relevant for good epitaxial growth are described. Recent progress on thick epitaxial growth as well as on selective growth of SiC will also be described.

28.1 Polytypes of Silicon Carbide

In this section, a concise review of the physical and electrical properties of SiC are presented. This is by no means complete, and further information can be found in the references of this chapter.

SiC has a large number of crystal structures (more than 200) [28.16], which are called polytypes. Considering close-packed structure, each layer of atoms along the c-axis can occupy three different positions, denoted by A, B, and C (Fig. 28.1). For SiC, a Si–C pair should be considered as one unit for atoms in close-packed structure. Variation of occupation sites along the c-axis brings about different crystal structures. They are usually represented by Ramsdell notation, which is denoted by the number of layers in the unit cell and a crystal system (C for cubic, H for hexagonal, R for rhombohedral). From ABCABCABC... stacking, we generate the 3C-SiC lattice, and from ABABAB... stacking, we generate the 2H-SiC lattice. Depending on the stacking order, the bonding between Si and C atoms in adjacent bilayer planes is of either zincblende (cubic) or wurtzite (hexagonal) nature. Zincblende bonds are rotated 60° with respect to nearest neighbors while hexagonal bonds are mirror images (Fig. 28.2). Each type of bond provides a slightly altered atomic environment, making some lattice sites inequivalent in polytypes. This also results in different ionization energy for dopants when substituted at these two different sites. Bonds in 3C-SiC are all cubic whereas bonds in 2H-SiC are all hexagonal. All of the other polytypes are mixtures of the fundamental zincblende and wurtzite bonds. Some common hexagonal polytypes with more

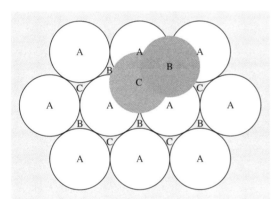

Fig. 28.1 Hexagonal close-packed structure

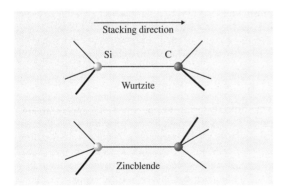

Fig. 28.2 Wurtzite bonding and zincblende bonding between Si and C atoms in adjacent planes. The three tetrahedral bonds are rotated 60° in the cubic case and mirror images in the hexagonal phase

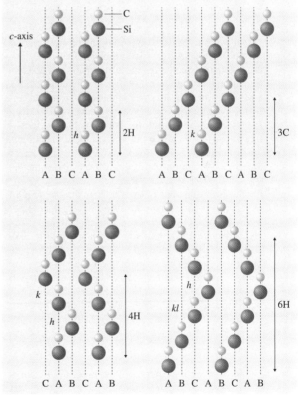

Fig. 28.3 Stacking sequences of different SiC polytypes projected on the [11$\bar{2}$0] plane. Examples of hexagon-stacked bilayers are labeled with h, while cubic-stacked bilayers are labeled with k

complex stacking sequences are 4H-SiC and 6H-SiC. 4H-SiC is composed equally of cubic and hexagonal bonds, whereas 6H-SiC is two-thirds cubic and one-third hexagonal. Despite the cubic elements, each has overall hexagonal crystal symmetry. The family of hexagonal and rhombohedral polytype is sometimes referred to as α-SiC, whereas 3C-SiC is referred to as β-SiC.

Figure 28.3 shows the stacking sequences for 3C-, 4H-, and 6H-SiC [28.16]. As can be seen, cubic SiC has a lattice constant of $a = 0.4349$ nm at room temperature. For hexagonal SiC, a is approximately 0.3080 nm for all polytypes whereas c varies with the number of layers in the unit cell, one single atomic layer being approximately 0.252 nm for all polytypes.

Why are there so many polytypes? The simplest level of understanding is that the formation energy difference between different polytypes is small. As calculated by *Park* et al. [28.17], the total energy difference for various polytypes is within 4.3 meV/atom. Hence, the temperature and other crystal growth conditions significantly affect the stability of the polytype. The occurrence probability of polytypes depends on the temperature. Generally, 3C-SiC is stable at temperature below 1900 °C, and 6H-SiC is stable at temperature above 1900 °C. A 3C-to-6H solid-phase transformation has been observed at temperature above 2150 °C [28.18, 19]. 4H-SiC also sometimes occurs in the high-temperature region, but the probability is low compared with 6H-SiC [28.20]. The 2H polytype is typically unstable because it can transform to a mixture of 3C and 6H polytypes at typical crystal growth temperatures [28.21]. *Powell* and *Will* have reported that the phase transformation away from 2H polytype can happen at temperatures as low as 400 °C [28.22].

28.2 Defects in SiC

Quality of materials is essential to the performance of the devices. As the development of devices proceed, demands on the quality of the bulk and epitaxial SiC wafers are increasing. Commercially available SiC wafers still contain a large number of various defects, including impurities, micropipes, growth pits, dislocations etc. Understanding the nature and properties of these defects is important to improve both the material quality and device performance. Since structural defects are currently limiting the introduction of commercial SiC devices, these defects are briefly reviewed.

SiC bulk and epitaxial wafers contain a large number of crystal structural defects, including micropipes, screw dislocations, growth pits, triangular inclusions, etc. Table 28.2 summarizes the properties of these defects and their impacts on device operation. More detailed descriptions of some of these defects are given below. Over the last few years, the concentrations of these structural defects have come down significantly, but their density is still high enough to cause problems in devices.

28.2.1 Micropipes

Among the various defects that exist in SiC crystals, screw dislocations lying along the [0001] axis are the most significant and are generally accepted to be one of the major factors limiting the extent of the successful applications of SiC [28.23]. These screw dislocations have been shown to have Burgers vector equal to nc (where c is the lattice parameter and n is an integer), with hollow cores becoming evident with $n \geq 2$ for 6H-SiC and $n \geq 3$ for 4H-SiC. These latter hollow-core dislocations are generally referred to as micropipes. The diameter of micropipes ranges from a few tens of nanometers to several tens of micrometers. Their density has come down in recent years, and micropipe-free wafers have been announced by commercial vendors [28.14]. Figure 28.4 shows an optical microscope picture of a micropipe in a typical 6H substrate used for our epitaxial growth. Using proper growth initiation procedure, it has been possible to close these micropipes during epitaxial growth [28.24]. However, this closure of the micropipe results in a group of elementary screw dislocations in its vicinity.

28.2.2 Screw Dislocations

When the Burgers vector of a SiC screw dislocation is small enough, hollow-core formation is avoided [28.25]. Nevertheless, the screw dislocation possesses many properties similar to micropipes, starting with an undesirable propensity to propagate through the entire thickness of bulk crystals and epilayers. These defects are present in average densities on the order of several thousands to around ten thousand per cm^2 in commercial wafers [28.21]. Because they have not declined in density over time as fast as micropipes, it

Table 28.2 Properties of SiC epilayer extended structural defects [28.29–31]

SiC epilayer defect	Density (/cm^2)	Observed defect source	Observed impact on high-field junction	Summary (comments)
Micropipe	< 30 (commercial) < 1 (best)	Substrate micropipes that propagate into epilayer	> 50% breakdown voltage reduction Microplasmas Increased leakage current	Improved in recent years such that the density is less than ≈ 1 cm^{-2} so, large-area power devices possible
Closed-core screw dislocation	≈ 3000 to $10\,000$	Substrate screw dislocations that propagate into epilayer	≈ 10–30% breakdown voltage reduction Softened breakdown Microplasmas	Density slowly improving, but will be present and affect ≈ 1 cm^2 power devices
Triangular 3C inclusions	< 5	Wafer preparation and epitaxial growth process	> 50% breakdown voltage reduction Increased leakage current	Density improving to where these may not be present in most ≈ 1 cm^2 power devices
Carrots and comet tails	< 5	Undetermined	Increased leakage current Nonsmooth surface seems likely to impact on Schottky rectifying properties	Density improving to where these may not be present in most ≈ 1 cm^2 power devices
Small growth pits	> 3000	Screw dislocations Wafer preparation and epitaxial growth	Nonsmooth surface seems likely to impact on Schottky rectifying properties	Density improving, but screw dislocation density may represent a limiting floor for these defects

would appear that all devices manufactured on mass-produced wafers will contain these defects for the foreseeable future. While not as detrimental to device characteristics as micropipes, experimental evidence is emerging that screw dislocations somewhat negatively impact on the electrical properties of high-field SiC junctions. One study that used x-ray topography to map screw dislocations demonstrated that elementary dislocations are detrimental to the reverse leakage and breakdown properties of low-voltage (< 250 V) 4H-SiC p$^+$n diodes [28.26]. Similar observations of increased reverse leakage, soft breakdown, and microplasma not associated with micropipes are reported in [28.27, 28]. The physical mechanisms and models for the electrical behavior of these dislocations remain to be further studied.

28.2.3 Growth Pits and Triangular Inclusions

Besides screw dislocations, other defects, such as growth pits, triangular inclusions, etc., have also been observed in SiC epilayers [28.32, 33]. Unlike micropipes and closed-core screw dislocations, which generally extend through the whole wafer, growth pits and triangular inclusions are primarily present in the epilayer. They are believed to be primarily caused by nonoptimized wafer preparation and epitaxial growth

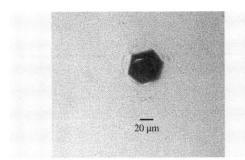

Fig. 28.4 Optical microscope image of a micropipe in a 6H-SiC substrate

processing. *Powell* and *Larkin* [28.33] also demonstrated that many morphological imperfections in the as-grown SiC epilayer surface are greatly impacted by surface polishing as well as the epitaxial growth initiation process. However, a recent study by *Schnabel* et al. [28.32], in which electron-beam-induced current (EBIC) measurements were correlated with x-ray screw dislocation mapping and surface growth pit mapping on a 6H-SiC epitaxial layer, revealed that all closed-core screw dislocations on the substrate mapped by x-ray resulted in a corresponding small growth pit on the as-grown epilayer. This suggested that epilayers process improvements may not be able to reduce small growth pit densities below bulk wafer screw dislocation densities.

The quality and smoothness of the semiconductor surface are well known to critically control the electrical properties of Schottky diode junctions. The presence of nonsmooth surface features, such as growth pits and triangular inclusions, could perturb the local electric field and thermionic carrier emission properties, which could result in locally increased current under both forward and reverse bias [28.21]. It is plausible that p–n junction devices are perhaps less affected by SiC surface defects, primarily due to the fact that the peak electric field occurs at the junction buried within the semiconductor. However, p–n junction characteristics can be expected to track surface imperfections in those cases where a surface defect arises due to an extended defect that actually runs through the thickness of the epilayer to intersect the p–n junction. For example, 3C triangular inclusions have been found to greatly increase leakage current and decrease breakdown voltage for kV-class epitaxial p–n junction diodes [28.34].

28.3 Epitaxial Growth of Silicon Carbide

To improve the quality of materials for use in devices and to produce complicated device structures, epitaxial growth techniques are necessary. This section contains a summary of several aspects of SiC epitaxial growth technology, including the SiC growth mechanism, commonly used reactors, SiC precursor sources, substrate materials, and dopant incorporation control. Also discussed are some recent advancements in SiC epitaxial growth.

28.3.1 Substrates for Silicon Carbide Growth

The majority of earlier chemical vapor deposition (CVD) crystal growth research was focused on determining the appropriate processes for obtaining low-defect-density single-crystal SiC films for the advancement of SiC-based devices. The first available substrates for SiC epitaxial growth were mainly small ($\approx 5\,\mathrm{mm}^2$) and irregularly shaped Lely and Acheson-derived SiC substrates that were only suitable for some epilayer physical property studies and limited prototype SiC-based devices. Therefore, dissimilar substrate materials of regular shapes and larger areas were usually chosen for CVD SiC epitaxial growth.

Motivated by the potential for Si–SiC device integration [28.35–37], many workers have attempted to grow SiC heteroepitaxially on Si. Since the growth temperature is limited by the melting point of the Si wafer ($< 1412\,°\mathrm{C}$), only cubic SiC can generally be grown on Si substrates, whether the substrate used is (100) or (111). This is because the more stable polytype at these temperatures is cubic. Large thermal and lattice mismatches between Si and SiC have limited the resultant material quality. Many defects such as antiphase boundaries (APBs), stacking-fault (SF) defects, and threading dislocations usually existed in these films and contributed to electrically leaky p–n junctions. APBs are the result of polar SiC growth on nonpolar Si substrate. In addition, other novel substrates such as silicon-on-insulator (SOI) [28.38], twist-bonded universal compliant (UC) substrates [28.38], porous silicon [28.39], and free-standing 3C-SiC (made by removing silicon substrates from thick SiC heteroepitaxial layers) [28.40] have been investigated. Epilayers with improved quality on these substrates have been reported, but p–n junctions built on these layers still show very high leakage current. Recently, a group in Japan has used *undulating* Si substrates to grow thick cubic SiC, followed by the removal of the Si substrates to obtain large-area 3C-SiC wafers [28.41, 42]. Using this undulating Si substrate, they succeeded in reducing the stacking fault density compared with films grown on a planar substrate. However, the large-scale commercial use of this 3C-SiC appears years away, as high-field device performance has remained relatively poor.

TiC has been used as an alternative to Si to grow 3C-SiC, since TiC is more closely lattice-matched to SiC [28.43]. Although 3C-SiC epitaxial layers on TiC were somewhat superior to those grown on Si, lack of high-quality TiC substrate material and high defect densities in the resulting SiC epitaxial layers have limited advancement of device research using 3C-SiC on TiC substrates.

As (0001)-oriented SiC substrates have become commercially available, more researchers started choosing SiC to grow SiC homoepitaxially. For epitaxial growth on (0001)SiC, typically the Si face has been the preferred surface, partly because of the superior epitaxial surface morphology and well-understood doping behavior compared with the C face. For 6H-SiC, the Si-face substrates are typically polished 3° off-axis from the basal plane, which increases the number of crystallographic steps on the surface to ensure SiC homoepitaxial growth (see the next section). In general, this homoepitaxial growth using off-axis SiC substrates proceeds by a step-controlled crystal growth process, as reported by *Itoh* and *Matsunami* [28.20]. Now the focus of much epitaxial crystal growth research has turned to 4H, mainly because of superior intrinsic properties such as higher bandgap, higher electron mobility, and lower dopant ionization energies. Similar to its off-axis 6H-SiC substrate predecessors, 4H-SiC substrates were initially polished 3° off-axis from the basal plane. However, researchers discovered that, compared with epitaxial growth on 6H-SiC substrates, 4H-SiC growth produces more triangular defects, which were later identified as 3C polytype inclusions [28.44, 45]. With the use of in situ pregrowth surface treatments, the occurrence of these 3C inclusions can be significantly reduced. Subsequently, the incidence of 3C inclusions was eliminated by increasing the 4H step density by using 8° off-axis substrates [28.46].

28.3.2 How to Control the Polytypes in SiC Homoepitaxy

As mentioned in the pervious section, the best way to obtain a high-quality epilayer is *homoepitaxy*, which is effectively free from lattice mismatch between the epilayers and the substrates. To ensure the growth of single polytypes, a method commonly called *step-controlled growth* has been employed in which the substrates are miscut a few degrees off from the (0001)Si face [28.20, 47]. Miscut SiC(0001) crystal surfaces consist of terraces and steps, and these steps reflect the substrate crystal structure, either 4H or 6H. According to the classical growth theory, adsorbed species migrate on the surface and are incorporated into the crystal at steps where the surface potential is low. Another competitive growth process of nucleation takes place on the terraces when the supersaturation is high enough. On the slightly misoriented (0001) plane, the step density is low and vast terraces exist. Then, crystal growth may occur on the terrace through two-dimensional nucleation due to the high supersaturation on the surface. In this case, the polytype of grown layers is determined by the growth conditions, mainly the growth temperature. At temperatures below 1800 °C, 3C-SiC is usually formed because it is stable at low temperature [28.48]. The growth of 3C-SiC can take two possible stacking orders, ABCABC... and ACBACB..., which leads to the formation antiphase boundaries (Fig. 28.5a). On off-oriented substrates, the step density is high and the terrace width is narrow enough for adsorbed species to reach steps. At a step, the incorporation site is uniquely determined by bonds from the step (Fig. 28.5). Hence, homoepitaxy can be achieved through lateral growth from the steps, inheriting the stacking order of the substrate. Hence, epitaxy is usually carried out on (0001) face of SiC (usually Si face) misoriented either 3.5° for 6H or 8° for 4H towards (11$\bar{2}$0) surface. In recent years, the emphasis has been to use reduced miscut wafers, such as 4°-off (0001) wafers, since they provide several advantages, such as reduction of crystal wastes while slicing wafers from the ingots and reduction of basal plane dislocations [28.49]. It is still possible to replicate substrate polytypes with SiC wafers with smaller offcut, but the optimum growth window for good growth

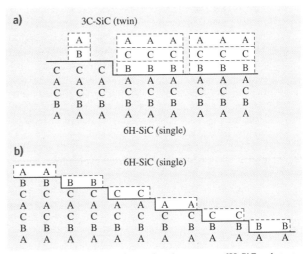

Fig. 28.5a,b Growth modes and polytypes on 6H-SiC substrate. (a) 3C-SiC growth by two-dimensional terrace nucleation, (b) homoepitaxy of 6H-SiC by step-flow growth (after [28.20])

is much narrower. In addition, increased step bunching is observed when the offcut angle is reduced, thus resulting in increased surface roughness.

28.3.3 SiC Epitaxial Growth Techniques

The purpose of epitaxy is to create a high-quality layer of desired thickness with controlled doping concentration, such as the abrupt p–n junction structure. Various epitaxial growth techniques have been used to achieve this goal. In this section, several important non-CVD epitaxial growth techniques that have been used for growing SiC are presented.

Liquid-phase epitaxy (LPE) is appealing since it is a growth process carried out at or close to equilibrium and thus produces high-quality materials [28.50]. During the LPE process, a SiC wafer, fixed on a graphite holder, is dipped into Si-based melt, which is heated to between 1000 and 2000 °C in a graphite crucible [28.50–54]. Growth takes place due to mass transfer from the crucible acting as the C source to the SiC wafer (seed) in the established temperature gradient. One of the most promising advantages of this method for SiC epitaxial growth is the potential to close micropipes emerging from the substrate. LPE epitaxial layer grows at low supersaturation compared with the physical vapor transport (PVT) growth technique, in which supersaturation is much higher and temperature fluctuations may easily result in local deviations from optimal growth conditions, leading to defect formation. LPE of SiC was shown to be an effective technique to overgrow micropipes in SiC wafers prepared by the physical vapor transport (PVT) technique [28.50–54]. Micropipe closing efficiency of > 80% has been demonstrated, and the growth rate ranges from several micrometer to millimeter per hour. LPE of SiC requires accurate control of thermal equilibrium conditions to avoid growth of different polytypes at the same time. It is difficult to control the doping concentration by LPE. Also the surface morphology of LPE SiC is rough, and terraces and steps are usually observed on the surface, which needs to be avoided for device fabrication.

Molecular-beam epitaxy (MBE) is an ideal technique for thin epitaxial layer deposition. Gas-source MBE (GSMBE) has been used for SiC homoepitaxial growth [28.55, 56]. It was found that low substrate temperature (< 1000 °C) resulted in deposition of 3C-SiC instead of replication of the stack sequences of the substrate. The growth rate is typically several nm/h. MBE has also been used for growing SiC on Si at relatively low temperature [28.57, 58] to avoid the formation of the voids that are usually observed in Si substrates at the SiC/Si interface when CVD techniques are used. Single-crystal 3C-SiC could be obtained by GSMBE at temperatures as low as 800–900 °C. MBE is suitable for low-dimensional modulation-doped structures or heterostructures of different SiC polytypes in the quantum regime due to its low growth temperature and growth rate. However, the thick epilayers required by high-voltage power device are unlikely to be obtained by MBE.

28.3.4 Chemical Vapor Deposition

The most common epitaxial growth technique for SiC is chemical vapor deposition (CVD), which has the advantages of precise control of the epitaxial layer thickness and impurity doping combined with a reasonable growth rate and good surface morphology. The growth temperatures for typical SiC CVD processes range from 1200 to 1800 °C, while the growth pressures vary from several tens of Torr to atmospheric pressure.

Precursors for SiC CVD Epitaxial Growth

Typical CVD techniques for SiC epitaxial layer growth employ separate sources of Si and C. Silane is the commonly used silicon source, while various choices of hydrocarbons have been used as carbon source. Both are introduced into the reactor with a high-purity H_2 carrier gas. Propane has been the most popular among the hydrocarbon choices, probably because of its association with the first successful demonstration of large-area epitaxial SiC on Si substrates [28.35], while methane, ethylene, and acetylene have been used less extensively. Methane is of interest because of the increased purity available from commercial sources compared with that of propane, although the larger thermal stability results in a smaller C source cracking efficiency [28.59].

Chlorine-based silicon sources, such as SiH_2Cl_2 and Si_2Cl_6, have been used to grow 3C-SiC on Si [28.60, 61]; one advantage is that Cl in the reactor could help to suppress silicon codeposition. Also, as reported by *Yagi* and *Nagasawa* [28.61], the supplied SiH_2Cl_2 is decomposed into $SiCl_2$ and H_2 at the growth temperature. The in situ generated $SiCl_2$ absorption is thought to be self-limited. Thus, atomic-layer epitaxial growth can be achieved by alternating supply of SiH_2Cl_2 and hydrocarbon. Using this method, *Yagi* and *Nagasawa* [28.61] have grown high-uniformity 3C-SiC on 6 inch Si substrates. The thickness deviation of the SiC film (4215 Å in thickness) was less than 1.3% over the substrate. Recently, *Miyanagi* and *Nishino* [28.62] used Si_2Cl_6 as a silicon source material to grow 4H-SiC. Si_2Cl_6 is a safer material than SiH_4, but large numbers of etching pits and scratch-like defects were seen on the resulting epilayers.

Single-source SiC precursors, such as hexamethyldisilane (HMDS) [28.63] and tetramethylsilane [28.64], have also been studied because they usually have much lower cracking temperature than that of multiple-source precursors, which could make lower-temperature ($< 1100\,°C$) SiC CVD possible. However, these chemicals are not available in sufficiently high purity, so their use will not result in layers with low background carrier concentration. Also it is impossible to adjust the C/Si ratio with a single-source precursor, which makes optimization of growth conditions difficult. One possible solution is to use an additional carbon source such as propane to control the C/Si ratio and hence the background doping concentration [28.65]. In addition, a few examples of their use in obtaining reproducible growth of high-quality SiC epitaxial layer have been reported. Hence, use of higher-purity chemicals such as silane and propane offers the best possible routes for growing layers for high-voltage applications. These chemicals can be further purified in situ in the reactor using resin-based purifiers [28.66].

CVD Reactor Configurations

The most commonly used simple reactor configuration for SiC CVD has been the quartz horizontal reactor operating at atmospheric or low pressure (Fig. 28.6). These reactors can be warm-walled or water-cooled. Researchers from the National Aeronautics and Space Administration (NASA) and Kyoto University have shown that high-quality SiC epilayers can be achieved using this type of reactor [28.33, 67]. The gases used are silane (SiH_4) and propane (C_3H_8) with a large amount of H_2 carrier gas. Figure 28.6 shows the schematic of the reactor. The growth rate is determined by the flow of silane gas, whereas the layer quality and the density of *defects* such as cubic SiC inclusions, dopant incorporations etc. are strongly dependent on the growth temperature and the flow rate of propane for a given growth rate. If the silane flow is increased beyond a certain value, the morphology of the grown layer deteriorates quickly. Growth rates of a few µm/h have been obtained using this type of system. *Nakamura* et al. [28.68] reported epitaxial growth in a horizontal cold-wall system with growth rates of up to $6\,µm/h$ at $1500\,°C$ by improving the initial growth conditions. Procedures to minimize doping and thickness nonuniformities include increasing the carrier gas flow (decreasing the residence time of the precursor), reducing the deposition pressure, and tilting the susceptor with respect to the gas flow. More importantly, the limited growth rate and parasitic deposits falling from the ceiling of the reactor wall make it difficult to grow thick ($> 50\,µm$) SiC epilayers with good surface morphology.

Vertical cold-wall reactors have also been built and used to grow SiC with high purity and crystal quality [28.69–71]. Precursor gases were introduced from the top of the reactor and the particles falling from the ceiling in the horizontal cold-wall reactor were avoided. The susceptor and attached sample were rotated at up to 1500 rpm to increase uniformity and prevent recirculation of the gases in the chamber. This fast rotation provides a pumping action for the gas flow. A background impurity level below $10^{14}\,cm^{-3}$ was obtained with a growth rate of $5-6\,µm/h$. Although a similar growth rate was obtained to that achieved in the horizontal cold-wall reactor, longer growth times could be

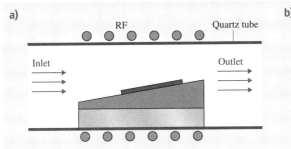

Fig. 28.6a,b Schematic drawing of a typical horizontal cold-wall SiC CVD reactor (**a**) and a horizontal water-cooled cold-wall reactor in operation at Rensselaer Polytechnic Institute (RPI) (**b**) (RF – radiofrequency)

used with the vertical reactor. Thick epilayers up to 50 μm have been obtained using this type of reactor.

In order to reproducibly grow thick, high-quality epilayers for power device applications, horizontal hot-wall reactors were proposed by *Kordina* and coworkers [28.72]. The idea of the hot wall was to obtain very good heating efficiency together with a high cracking efficiency of the precursors since gases could be much more efficiently heated in a hot-wall reactor. Also, since the system is heated more uniformly, Si droplet formation just above the sample surface may be suppressed, and hence a much higher silane flow can be used for growing SiC. This has been discussed as one of the problems that limit the growth rate in cold-wall reactors. Since the reactor walls are all heated (Fig. 28.7), the parasitic deposits on the reactor walls are polycrystalline SiC, which tends to stick to the wall more strongly than the amorphous SiC or byproducts that stick to the reactor wall in cold-wall reactors. Due to all these reasons, it is possible to grow SiC in a hot-wall reactor at a much higher rate and also for longer time without creating the *dusting* and surface morphology degradation that usually occur in cold-wall reactors. Several researchers have adapted this hot-wall process to grow SiC, and currently this appears to be the commonest process to grow SiC epitaxial films. For example, *Kimoto* et al. [28.73] have improved on this process, and have reported very high-purity SiC growth using this type of reactor. Epilayers with 50 μm thickness and n-type background doping in the low 10^{13} cm^{-3} range have been achieved reproducibly at 1550 °C and 80 Torr. With additional mechanical rotation, excellent thickness uniformity (1% standard deviation over mean value) and doping uniformity (6% standard deviation over mean value) have been achieved on 2 inch wafers.

using a horizontal hot-wall reactor [28.74]. Recently, *Thomas* and *Hecht* [28.75] reported a planetary horizontal hot-wall reactor with handling capacity of seven 2 inch or five 3 inch wafers. High-quality epilayers were grown by optimizing the process conditions. Excellent intrawafer homogeneity (thickness uniformity standard deviation of 2% of mean, doping uniformity standard deviation of 9% of mean) as well as wafer-to-wafer (thickness uniformity standard deviation of 1.6% of mean, doping uniformity standard deviation of 3.3% of mean) on 3 inch wafers have been achieved using this system. These horizontal hot-wall reactors with multiple-wafer capability are becoming industry-standard for growing epitaxial films for commercial production.

The vertical hot-wall reactor (also called the *chimney* CVD reactor) has also been developed as a high-temperature CVD technique in order to increase growth rates further [28.76–78]. The epitaxial growth of SiC in this geometry is investigated mainly for the purpose of growing thick epitaxial layers for high-power applications. In the chimney reactor, the high-temperature (1600–1900 °C) growth process enables epitaxial rates of 10–50 μm/h. Under optimum conditions, mirror-like SiC epilayers with low background doping of low 10^{13} cm^{-3} can be obtained. Recently, *Danno* et al. reported fast epitaxial growth (14–19 μm/h) on 4H-SiC(000$\bar{1}$) substrate using a chimney-type reactor [28.79]. High-quality epilayers with 100% micropipe closing have been grown at 1750 °C at C/Si ratio of 0.6. *Fujiwara* et al. reported the reduction of stacking faults in fast epitaxial growth by optimizing initial growth conditions during the heating process [28.80]. Stacking fault density could be dramatically reduced from 1000–9000 cm^{-2} to 0–5 cm^{-2}. The vertical hot-wall reactor has also been used in the author's own laboratory [28.81] for growing thick epitaxial films with high growth rate (> 20 μm/h).

Another type of epitaxial growth, termed *sublimation epitaxy*, was developed to grow bulk-like thick SiC, and was one of the first techniques to produce SiC epitaxial layers. This growth process is based on heating polycrystalline SiC source material to 1700–2000 °C under conditions at which it sublimes into the vapor phase and subsequently condenses onto a cooler SiC seed crystal. This is similar to the bulk growth process. It is possible to achieve high epitaxial growth rate (> 2 mm/h) with a stable mechanism, resulting in high-quality SiC. Thick homoepitaxial growth of 4H-SiC (50–100 μm) [28.82] and 3C-SiC (300 μm) [28.83] with high growth rates of 100 μm/h

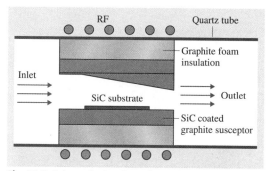

Fig. 28.7 Schematic drawing of a typical horizontal hot-wall SiC CVD reactor. The cross section of graphite wall is rectangular in shape

and 50 μm/h, respectively, have been demonstrated using this technique. Specular surface and better quality layers (compared with substrate) were obtained by this technique. Recently, *Sveinbjornsson* et al. reported higher field-effect channel mobility ($208\,\text{cm}^2/(\text{V s})$) in 4H-SiC MOSFETs fabricated on sublimation-grown epilayer compared with that ($125\,\text{cm}^2/\text{Vs}$) fabricated on a commercial CVD epilayer [28.84]. The mobility enhancement is attributed to better surface morphology of the sublimation epilayer, resulting in less surface scattering and lower density of interface states [28.85] as compared with the reference CVD epilayers. However, good results by sublimation epitaxy can only be obtained when the growth parameters are carefully controlled within a narrow window. One of the remaining challenges is to avoid the introduction of impurities from the growth environment into the epilayers. The purity of the solid polycrystalline SiC source material is the most critical issue in the growth of low-doped SiC by this method.

Recently, a new epitaxial growth process by introducing chlorinated silicon precursors or simply HCl addiction was proposed to achieve high growth rate of SiC [28.86–88]. A high growth rate of 90 μm/h has been obtained using methyltrichlorosilane (MTS) as a precursor [28.86], while growth rates of up to 49 μm/h have been reported using standard silane–propane–hydrogen growth chemistry with incorporation of HCl into the gas mixture [28.87]. No significant difference has been observed in terms of defects, doping uniformity or thickness uniformity compared with the standard process without HCl (or chlorinated silicon precursor). Although the growth mechanism is not fully understood, it is proposed that the addition of chlorine suppresses Si cluster formation in the gas phase, which is thought to be responsible for limiting the growth rate in horizontal cold-wall reactors. Therefore, epitaxial growth with both high growth rate and specular surface morphology is possible using this process. A complete investigation of this process is still ongoing to obtain high-quality SiC epilayers with even higher growth rates.

Doping Incorporation Technology

In order to ensure reproducible and reliable SiC semiconductor device characteristics, well-controlled dopant incorporation must be accomplished. Unlike the doping technology routinely used in the silicon semiconductor industry, SiC cannot be efficiently doped by diffusion at SiC growth temperature due to its low diffusion coefficient. Doping of SiC epitaxial layers of a device structure is accomplished either by flowing a dopant source into the reactor during growth or by implanting the dopant atoms.

Nitrogen is the most common n-type source for in situ doping. Since the atomic size of nitrogen (0.74 Å) is closer to that of carbon (0.77 Å), nitrogen tends to substitute at C sites to a first approximation, consistent with the reported experimental results [28.20]. As discussed previously, the bonding between Si and C atoms in adjacent bilayer planes can be of either zincblende (cubic) or wurtzite (hexagonal) nature. Each type of bond provides a slightly altered atomic environment, making some lattice sites inequivalent in polytypes. In 6H-SiC, three inequivalent sites are available (two cubic sites and one hexagonal site), whereas two inequivalent sites are available in 4H-SiC (one cubic site and one hexagonal site). The ionization energy will be different when a dopant is incorporated into different inequivalent sites. The ionization energies for nitrogen have been reported by several authors, but the values differ a lot. In general, measurements on epitaxial layers with higher donor concentration typically yield lower ionization energy due to Coulombic interactions with neighboring impurities [28.89]. Also, experimentally, it was found that the ionization energies of nitrogen in 4H-SiC are shallower than the corresponding values in 6H-SiC.

In situ nitrogen doping can be realized by passing nitrogen or ammonia into the reactor during the growth. However, control of net dopant incorporation by simply increasing or decreasing the flow of the dopant source is limited in terms of both reproducibility and doping range. This doping range is typically limited to 2×10^{16}–$5\times 10^{18}\,\text{cm}^{-3}$ for a nitrogen-doped n-type layer. If one increases the nitrogen flow further, either the growth rate is found to reduce or the surface morphology is found to deteriorate significantly [28.90]. So, an alternative method, termed *site-competition growth*, is used to increase the doping concentration [28.91]. It was found that nitrogen incorporation was very sensitive to the C/Si ratio employed during the growth and increased when C/Si decreased for growth on Si-face substrate (Fig. 28.8b). This was explained by site competition between nitrogen and carbon during the growth: since nitrogen and carbon occupy the same sites in SiC, decreasing C/Si will produce more C vacancies in SiC, which will in turn increase nitrogen incorporation. Use of this technique has enabled large expansion of the reproducible doping range, which as a result, now spans from 10^{14} to $5\times 10^{19}\,\text{cm}^{-3}$.

Phosphorous is another n-type dopant in SiC. However, phosphorous appears to have limited potential as a replacement for nitrogen, especially considering its larger atomic size and similar ionization energy to the nitrogen donor [28.91]. Understanding of the location of phosphorous in the SiC lattice as an isolated impurity, phosphorous-related complexes, and corresponding electronic levels is still poor. Hall measurements on phosphorous-implanted SiC epilayers have yielded two ionization energies: 80 ± 5 and 110 ± 5 meV for 6H-SiC [28.92] and 53 and 93 meV for 4H-SiC [28.93], which were assigned to the hexagonal and quasicubic sites of silicon, respectively. Recent research on phosphorous-implanted SiC indicates that the sheet resistance in high-dose phosphorous-implanted 4H-SiC is as much as an order of magnitude lower than those measured in nitrogen-implanted 4H-SiC with comparable doping and thermal processing [28.94]. Thus, there are some advantages in replacing nitrogen with phosphorous for applications where low sheet resistance is required.

Very little has been reported on in situ doping of phosphorous in CVD. The phosphorous donor ionization energy is also only available on phosphorous-implanted SiC, which generally contains lots of N or Al background. In one report [28.88] on the site-competition effect for phosphorous doping, *Larkin* reported that phosphorous substitutes at the Si site due to its similar atomic size (1.10 Å) to Si (1.17 Å). However, our own experimental results seem to be different from this conclusion, and it appears that phosphorous occupies both Si and C sites [28.95].

Aluminum serves as a shallow acceptor dopant in SiC and can be introduced during growth, or afterwards by implantation. In contrast to nitrogen, the aluminum acceptor resides at the Si lattice sites. Its ionization energy is only weakly dependent on polytype or to lattice sites h, k. The scatter of published data ranges from 190 meV [28.96] to 280 meV [28.97]. The in situ incorporation of Al strongly depends on the C/Si ratio during growth. Increasing C/Si will increase the Al doping concentration.

Boron is another acceptor dopant in SiC. During epitaxial growth, a considerable amount of hole-passivating hydrogen is simultaneously incorporated into the growing B-doped CVD epilayer. Postannealing at $1700\,°C$ in Ar is usually used to dissociate B–H and better activate the B acceptors [28.91]. Although the atomic size of boron (0.82 Å) is closer to that of C (0.77 Å), B–H has an atomic size (1.10 Å) more closely matched with that of Si (1.17 Å) and therefore should occupy Si sites. Experiments have also shown that boron incorporation increases with C/Si [28.91], which is in agreement with the site-competition theory. However, this does not imply that boron substitutes exclusively at Si sites. Experiments have shown that a small amount of boron can also occupy C sites and form an energetically deep complex termed a D-center. The energy level for the boron at the Si site is about 300 meV, as measured from the top of the valence band. The deeper center is about 600–700 meV above the valence band and manifests itself optically by producing donor–acceptor pair spectra at low temperature [28.97]. Its existence was also confirmed by deep-level transient

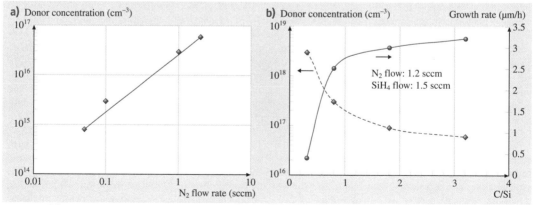

Fig. 28.8 (a) Carrier concentration versus N_2 flow for CVD of SiC. Increasing C/Si ratio can decrease the carrier concentration further as shown in (b) (after [28.95])

spectroscopy (DLTS) spectra [28.96]. The formation of shallow boron acceptors and deep D-centers can be adjusted during the doping process by the site-competition effect. Use of higher C/Si during growth will reduce the carbon vacancies, thus producing more shallow boron acceptors.

Epitaxial Growth on Nonstandard Crystal Orientations

After the success of epitaxial growth on off-axis (0001)Si-face substrate, growth on many other crystallographic faces of SiC has also been investigated. Most of the epitaxial growth studies on nonstandard crystal planes have focused on 4H-SiC due to its electronic properties advantages over 6H-SiC. Figure 28.9 shows schematics of several selected 4H-SiC lattice planes.

Epitaxial growth on $(000\bar{1})$C-face substrate has been carried out by CVD and compared with growth on (0001)Si face. SiC $(000\bar{1})$C face has a different surface atom configuration from (0001)Si face, and the electronegativity of Si and C atoms is different. This brings about various polarity dependences in chemical reactivity, growth kinetics, impurity incorporation efficiency, thermal oxidation rate, and oxide–SiC interface properties. SiC $(000\bar{1})$C face has shown superior properties of faster oxidation rate [28.98] and low surface roughness [28.96] compared with Si(0001) face. Therefore, it shortens the extremely time-consuming oxidation process on SiC(0001) and it is possible to obtain higher channel mobility of SiC MOSFETs due to reduced surface roughness scattering. Although it was found that the growth window on $(000\bar{1})$C face is narrower and the background doping level is higher compared with growth on (0001)Si face [28.97], high-quality epilayer with a background doping of mid-10^{14} cm^{-3} has been obtained following the development of the hot-wall CVD reactor [28.99]. The nitrogen donor incorporation efficiency has been found to be higher on $(000\bar{1})$C face, while the aluminum acceptor incorporation efficiency on $(000\bar{1})$C is lower compared with on (0001)Si face [28.100–102].

Epitaxial growth of 4H-SiC on $(1\bar{1}00)$ and $(11\bar{2}0)$ faces has also been investigated with the motivation to improve the SiC MOSFET channel mobility [28.103]. Both faces are perpendicular to the (0001) face, as shown in Fig. 28.9b,c, respectively, and no intentional off-angles are required for these faces to realize homoepitaxy of 4H-SiC, owing to the appearance of stacking information on the surface. The surface morphology of epitaxial layers grown on $(1\bar{1}00)$ face was found to be very poor, whereas the resulting SiC$(11\bar{2}0)$

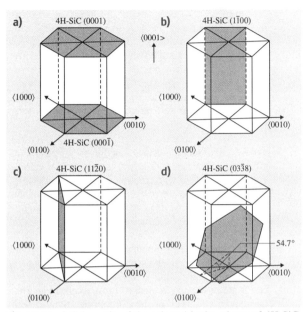

Fig. 28.9a–d Schematic of the selected lattice planes of 4H-SiC. (a) 4H-SiC(0001) and $(000\bar{1})$, (b) 4H-SiC$(1\bar{1}00)$, (c) 4H-SiC$(11\bar{2}0)$, and (d) 4H-SiC$(03\bar{3}8)$

epitaxial surface morphology was comparable to or even better than those grown on off-axis (0001)Si face. Although threading screw dislocations and edge dislocations parallel to the c-axis are present in $(11\bar{2}0)$ substrate, they are found not to propagate into epilayers. Extremely smooth and micropipe-free layers have been obtained by epitaxial growth on $(11\bar{2}0)$ substrate [28.104, 105]. Growth rates on $(1\bar{1}00)$ and $(11\bar{2}0)$ faces are almost the same as that on off-axis (0001) under the same conditions. The C/Si ratio dependence of dopant incorporation has been observed for epitaxial growth on $(11\bar{2}0)$ substrate [28.102, 105]. When nitrogen was introduced, epitaxial layers grown on $(11\bar{2}0)$ always showed higher donor concentration than on off-axis (0001). Although nitrogen incorporation was suppressed by increased C/Si ratio on $(11\bar{2}0)$, the C/Si dependence was weaker on $(11\bar{2}0)$ epilayers. A background doping concentration of low-10^{14} cm^{-3} for $(11\bar{2}0)$ epilayers can be achieved by increasing the C/Si ratio. In contrast to nitrogen doping, the acceptor (both Al and B) concentration of $(11\bar{2}0)$ epitaxial layers was lower than that of off-axis (0001) epitaxial layers under identical growth conditions.

To avoid propagation of defects along growth direction, SIXON Ltd., Japan, [28.106] developed crystal growth of 4H-SiC along the $\langle 03\bar{3}8 \rangle$ direction which is inclined to the c-axis by 54.7° as shown in Fig. 28.9d. During the bulk growth, defects along the $\langle 0001 \rangle$ direction propagate diagonally and terminate to the side of ingot. Epitaxial growth on $(03\bar{3}8)$4H-SiC has shown improved surface morphology [28.105, 107]. Microsteps and triangular defects have not been observed on the surface, and almost perfect micropipe closing has been realized in 4H-SiC$(03\bar{3}8)$ epitaxial growth. Growth rate on $(03\bar{3}8)$ is the same as on an off-axis (0001)Si face under the same conditions. Similar dopant incorporation efficiency as that on $(11\bar{2}0)$ has been found for epilayers on $(03\bar{3}8)$. The C/Si ratio dependence of impurity incorporation is intermediate between the case of the (0001)Si face and the $(000\bar{1})$C face. A background doping of low 10^{14} cm^{-3} can be achieved on $(03\bar{3}8)$, similar to that on $(11\bar{2}0)$.

Based on the experimental results, impurity incorporation in SiC growth is believed to be dominated mainly by surface bond configuration. Since both Al and B atoms occupy Si sites to form acceptors in SiC, the doping efficiency of these impurities may depend on the number of chemical bonds available from the surface C atoms. On (0001)Si plane, three bonds are available from neighboring C atoms on the surface, while two bonds are available from surface C atoms on $(11\bar{2}0)$ and $(03\bar{3}8)$. On $(000\bar{1})$C face, only one bond is available from the top C atom. This difference may be the reason why the doping efficiency of Al or B on $(11\bar{2}0)$ and $(03\bar{3}8)$ faces is lower than on (0001)Si face but higher than on $(000\bar{1})$C face. In a similar manner, N doping may be influenced by the number of chemical bonds from the surface Si atoms, as N atoms substitute at C sites. Therefore, nitrogen incorporation on $(11\bar{2}0)$ and $(03\bar{3}8)$ faces is more efficient than on (0001)Si face and less efficient than on $(000\bar{1})$C face.

The MOS interface properties of these nonstandard faces have also been studied to explore potential advantages for SiC power MOSFET applications. Although the interface state density (D_{it}) of the 4H-SiC(0001) MOS structure exhibits a rapid increase near the conduction-band edge, those on the other faces exhibit a rather flat D_{it} distribution. $(000\bar{1})$, $(11\bar{2}0)$, and $(03\bar{3}8)$ have about one order of magnitude lower interface density near the conduction-band edge than (0001) [28.108]. As known, the extremely high D_{it} near the band edge causes trapping of electrons induced in the inversion layer and significantly reduces the channel mobility of SiC MOSFETs [28.109]. Planar MOSFETs fabricated on 4H-SiC(0001), $(000\bar{1})$, $(11\bar{2}0)$, and $(03\bar{3}8)$ faces have shown that the effective channel mobility of those nonstandard faces is 2–3 times higher than that on (0001) face. Although the reason for this crystal face dependence is unclear, bond configurations and the polarity of nonstandard surfaces might lead to an intrinsic difference in chemical reactivity during oxidation and to a reduced number of electrically active states near the band edge. Recently, *Kimoto* et al. reported a new oxidation process using N_2O to reduce the interface state density [28.110]. The interface state density near the conduction-band edge was found to be much lower on 4H-SiC$(000\bar{1})$ and $(11\bar{2}0)$ ($\approx 3 \times 10^{11}$ (/cm^2 eV)) than on 4H-SiC(0001) ($\approx 1 \times 10^{12}$ (/cm^{2}eV)) with N_2O oxidation. A significant improvement of the channel mobility in SiC MOSFETs fabricated on $(000\bar{1})$ (> 30 cm^2/(V s)), and $(11\bar{2}0)$ (> 70 cm^2/(V s)) epilayers has been achieved [28.111, 112]. 4H-SiC$(000\bar{1})$, $(11\bar{2}0)$, and $(03\bar{3}8)$ faces are promising for MOS-based devices owing to the superior properties of the MOS interface. Note that research on $(03\bar{3}8)$ epitaxial growth and device applications is still limited due to the lack of substrates. All results have been reported by SIXON Ltd. and Kyoto University. Even though these orientations may have better device properties, the lack of availability of these wafers at lower cost may limit widespread use, and most epitaxial work will still be carried out on (0001)Si wafers.

28.4 Epitaxial Growth on Patterned Substrates

Epitaxial growth of SiC provides in situ doping of as-grown layers and better crystal quality as compared with ion implantation. Step-controlled epitaxy has been used at lower temperature (< 1800 °C) for high-quality crystal growth, avoiding SiC sublimation and surface roughening during growth. Therefore, if selective doping of SiC can be achieved via epitaxial growth, then it provides a promising alternative to ion implantation in SiC processing. There are two essential means to realize selective doping by epitaxial growth. One is to apply blanket growth on a trenched substrate, followed by a polishing process to remove excess epilayers on the mesas. The other is to use selective growth with a high-temperature mask that could be selectively re-

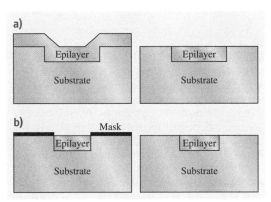

Fig. 28.10a,b Schematic of selective doping of SiC by epitaxial growth. (**a**) Blanket epitaxial growth followed by polishing to remove excess epilayers, (**b**) selective epitaxial growth followed by mask removal

moved after growth. These two processes are illustrated schematically in Fig. 28.10.

A few trials of homoepitaxial CVD growth of SiC on a trenched substrate have been reported. *Nordell* et al. [28.113] investigated the growth behavior at various C/Si ratios and temperatures. It was found that a low C/Si ratio gave smooth growth and small differences in growth rates between different lattice planes. A larger C/Si ratio gave more faceted growth, and the growth rate was lower in the $\langle 1\bar{1}00\rangle$ direction than in the $\langle 11\bar{2}0\rangle$ direction. *Chen* et al. [28.114] also carried out homoepitaxial CVD growth on trenched 4H-SiC substrate. Similar growth behavior was observed near trenches. It was found that the growth rate on trench side walls was usually lower than that on the bottom of a trench and on the top surface, and this difference became smaller with decreasing C/Si ratio. Growth on mesas oriented parallel to the substrate miscut direction showed clear step-flow growth, while growth on mesas oriented perpendicular to the miscut direction revealed the formation of (0001) plane. These studies focus more on crystal growth behavior; applications on device fabrication have hardly been reported.

Epitaxial growth on patterned substrates, specifically, 4H-SiC mesas, has been carried out at NASA [28.115–118]. 4H-SiC substrates were patterned with a series of mesas of various shapes and sizes (such as squares and hexagons) by dry etching techniques. When epitaxial growth is carried out on these substrates, step-free surfaces are obtained provided that the mesas do not contain any screw dislocations. Such techniques have been used to provide step-free large-size mesas up to $200\,\mu\text{m} \times 200\,\mu\text{m}$. Such mesas have been used for growth of cubic SiC or hexagonal GaN layers. The results from these studies have been reviewed before [28.117] and will not be repeated here.

28.4.1 Selective Epitaxial Growth

Selective epitaxial growth is a well-established technique for Si [28.119] and GaN [28.120, 121]. However, only a few works have been reported on selective growth of SiC. This is not surprising since high growth temperature above 1450 °C is generally required to grow high-quality epilayers of SiC by chemical vapor deposition. This high growth temperature makes it a challenge to identify an appropriate mask for selective epitaxial growth of SiC.

Several groups [28.122, 123] have demonstrated low-temperature ($\approx 1000-1200\,°\text{C}$) selective deposition of 3C-SiC on Si substrate using SiO_2 and Si_3N_4 as the mask. Two conflicting requirements became apparent: high growth temperatures were needed to produce the best crystal quality 3C-SiC films, but low growth temperatures were needed to minimize damage to the mask. Polycrystalline 3C-SiC was usually deposited in unmasked window when the temperature was lowered to avoid significant damage to the mask. In addition, delamination of dielectric mask was often observed. To improve selectivity, HCl was sometimes added to the carrier gas [28.124]. Lateral epitaxial overgrowth (LEO) has also been studied, and it was found that defect density in the LEO region was significantly reduced compared with films grown directly on Si substrate [28.125, 126].

Selective growth at higher temperature requires masks that can withstand the growth temperature of $> 1500\,°\text{C}$. One such mask is graphite. Recently, homoepitaxial selective growth of SiC has been reported using graphite mask [28.127–129]. *Chen* et al. investigated selective homoepitaxial growth of 4H-SiC on off-axis (0001) and $(11\bar{2}0)$ substrate at 1500 °C [28.128]. Polycrystalline 3C-SiC was observed on the graphite mask, and laterally overgrown polytype was identified as 3C-SiC in contrast to the 4H-SiC grown in window openings. Local epitaxy and lateral epitaxial overgrowth of SiC were even studied by using physical vapor transport (PVT) [28.129]. A lateral/vertical growth rate ratio of 6 was achieved. It was found that dislocations in the substrate propagated only into epilayers grown above the seed regions but not into the lateral overgrown region. The apparent disadvantage of the graphite mask is that the mask itself could act as a carbon source during high-temperature SiC epitaxial growth. This results in local variation in the C/Si ratio, which affects local crystal growth behavior and the doping concentration of grown layers. Also, polycrystalline SiC deposition on the mask interferes with the lateral growth. Because of these limitations, it is better to identify another mask that will withstand the high growth temperature but that will not cause contamination of the layer itself.

28.4.2 Selective Epitaxial Growth of 4H-SiC Using TaC Mask

At RPI, we have used TaC as the high-temperature mask for growing SiC selectively. Some salient features of this selective growth will now be presented, followed by some device results.

The mask to be used for selective epitaxial growth of SiC essentially needs to be stable under high temperature ($> 1400\,^\circ\text{C}$) in hydrogen ambient. Silicon oxide and silicon nitride are commonly used masks for selective growth of silicon and III–V compound semiconductors [28.119, 130, 131]. However, they are etched very quickly in hydrogen ambient under high temperatures and can be used only at relatively low temperature ($\approx 1000\,^\circ\text{C}$). TaC is a well-known refractory material with high melting point and high chemical resistance. TaC-coated components have been used with success in conventional vapor-phase epitaxy (VPE) of SiC [28.132]. In addition, TaC has been successfully used as a coating of the graphite susceptor in SiC epitaxy [28.95, 133].

A TaC mask was prepared by converting Ta to TaC in a CVD reactor. Specifically, high-purity Ta metal (99.999%) was first evaporated by electron-beam evaporation and patterned. Ta-coated SiC wafers were then placed in the CVD reactor, and the chamber was pumped down to 5×10^{-6} Torr using a turbo pump be-

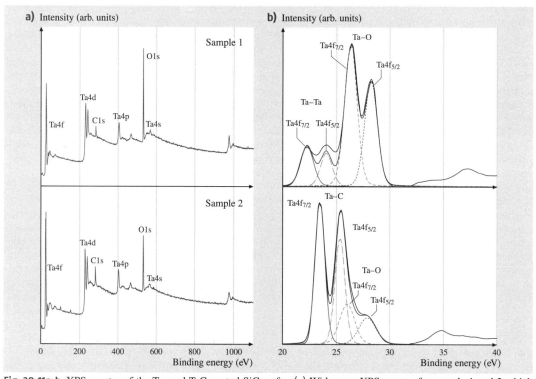

Fig. 28.11a,b XPS spectra of the Ta- and TaC-coated SiC wafer. (**a**) Wide-scan XPS spectra for sample 1 and 2 which have as-deposited 600–700 Å Ta film on the SiC. Sample 2 has been kept in C_3H_8/H_2 ambient at $1300\,^\circ\text{C}$ for 15 min. (**b**) XPS spectra from Ta4f for sample 1 and 2. They are fitted with Gaussian peaks which are assigned to Ta (Ta4f$_{7/2}$: 22.4 eV, Ta4f$_{5/2}$: 24.0 eV) Ta$_2$O$_5$ (Ta4f$_{7/2}$: 26.3 eV, Ta4f$_{5/2}$: 28.2 eV), TaC (Ta4f$_{7/2}$: 23.5 eV, Ta4f$_{5/2}$: 25.3 eV)

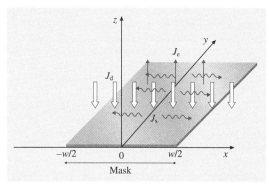

Fig. 28.12 Schematic of reactant species motion on the mask surface, taking into account surface diffusion

fore the susceptor was heated up. The wafers were then kept at 1300 °C in 180 ppm C_3H_8 diluted in H_2 ambient under 100 Torr pressure for 15 min. After the carburization process, the color of the mask changed from grey to golden yellow. X-ray photoelectron spectroscopy (XPS) has been used to confirm the formation of TaC, as show in Fig. 28.11. Sample 1 has as-deposited Ta on the surface whereas the carburization process was carried out on sample 2. Figure 28.11a shows wide-scan XPS spectra of the samples. The main peaks correspond to tantalum (Ta4s, Ta4p, Ta4d, and Ta4f), carbon (C1s), and oxygen (O1s). It is clear that the carbon fraction (18 and 31.5% for samples 1 and 2, respectively) increases after the carburization process. Also notice that some oxygen is present on the top surface of the films and that the amount of oxygen decreases after the carburization process. The same analysis was carried out for both samples at a depth of 100 Å below the surface (materials near the surface were removed by Ar^+ ion sputtering). The analysis shows that the amount of oxygen decreases significantly with depth. It is likely that the oxygen near the surface of the film is from natural oxidation.

Figure 28.11b shows high-resolution XPS spectra from Ta_{4f} for samples 1 and 2. They are fitted with six Gaussian peaks, which are assigned to Ta ($Ta4f_{7/2}$: 22.2 eV, $Ta4f_{5/2}$: 24.0 eV), Ta_2O_5 ($Ta4f_{7/2}$: 26.2 eV, $Ta4f_{5/2}$: 28.0 eV) and TaC ($Ta4f_{7/2}$: 23.4 eV, $Ta4f_{5/2}$: 25.4 eV). The binding energies are close to those reported elsewhere [28.134, 135]. Particularly, comparing the results with the work by *Gruzalski* and *Zehner* [28.134], who determined core-level binding energies of TaC_x over the range $0.5 < x < 1.0$, the carburized film here corresponds in composition to $TaC_{1.0}$.

The growth selectivity with a TaC mask is significantly dependent on the growth conditions. Polycrystalline film deposition occurs on the mask under nonoptimized conditions. The polycrystalline deposition on the masks is closely related to the surface concentration of the reactant species. Figure 28.12 shows a schematic diagram of the motion of reactant species on the mask region, taking account of surface diffusion. The reactant species arrive on the mask region with a flux density of J_d, migrate on the mask surface during a mean surface lifetime of τ_s, and then desorb from the mask surface and re-evaporate with a flux density of J_e. In general, the surface concentration of reactant species on the SiC surface could be assumed to be negligibly small due to the homoepitaxial incorporation of the species.

However, the surface concentration on the mask area cannot be neglected because the reactant species arriving on the mask surfaces are difficult to be incorporated into adsorption sites. Thus, the reactant species that arrive on the mask within a diffusion length (λ_s) of the edge of the window opening are incorporated into the substrate during epitaxial growth, and the surface concentration distribution across the mask is dependent

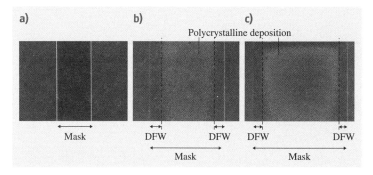

Fig. 28.13a–c SEM viewgraph of SiC polycrystal deposition on TaC mask at 1833 K. Flow rate is 2.24 sccm for both SiH_4 and C_3H_8. Growth pressure is 80 Torr. Width of the mask is (**a**) 100 μm, (**b**) 500 μm, (**c**) 1000 μm, while $M : W = 1$

on the surface diffusion length, which determines the profile of polycrystalline deposition.

Figure 28.13 shows the polycrystalline growth on a mask and the defect free width (DFW) when the mask widths are very large; this experiment gives an approximate surface diffusion length of the growth species on the TaC mask.

28.4.3 Orientation Dependence of SiC Selective Growth

A circular pattern was first used to study the orientation dependence of the selective growth of SiC. The circular pattern consists of 5 μm wide mask windows spaced every half degree as spokes of a wheel. The patterned sample was then carburized and SiC was grown in the CVD reactor. The flow rates were 1.2 and 1.2 sccm for C_3H_8 and SiH_4, respectively, using H_2 as the carrier gas. The epilayers were grown at 1550 °C under 80 Torr pressure, which would result in nominal planar growth rate of 3–4 μm/h. The growth lasted for 1 h.

Figure 28.14 shows tilted top views of selectively grown SiC window stripes along different directions. When the window stripes are parallel to the $\langle 11\bar{2}0 \rangle$ miscut direction, the growth on the exposed area followed the substrate orientation, and the top surface was smooth and specular (Fig. 28.14a). However, when the window stripes were aligned along $\langle 1\bar{1}00 \rangle$, the growth on the window stripes developed (0001) facet (Fig. 28.14c). For window stripes along a direction between $\langle 11\bar{2}0 \rangle$ and $\langle 1\bar{1}00 \rangle$, (0001) facet intersected the 8° off (0001) fronts and the extent of (0001) facet depended on the angle between the stripe orientation and the $\langle 11\bar{2}0 \rangle$ miscut direction.

To explain the formation of the (0001) facets, the effect of substrate offcut orientation has to be taken into account. As mentioned above, off-axis 4H-SiC(0001) 8° miscut towards $\langle 11\bar{2}0 \rangle$ was used in this work. When the stripe direction is along $\langle 11\bar{2}0 \rangle$ miscut, there is no restriction on the step-flow growth. New steps are continuously available and facet develops starting from the edges of the selective epitaxial layer. However, when the selective growth opening is along the $\langle 1\bar{1}00 \rangle$ direction, step flow was restricted within the opening window and pure step-flow growth makes all steps to move out of the edge, as depicted in Fig. 28.15. Cross-sectional scanning electron microscopy (SEM) of selectively grown films in $\langle 1\bar{1}00 \rangle$ strip openings indicated good agreement of w/h ratio with the schematic shown in Fig. 28.15. When the growth opening window is aligned between $\langle 1\bar{1}00 \rangle$ and $\langle 11\bar{2}0 \rangle$ with an angular displacement larger than 20° from these two orientations, step-flow growth is dominant, although partially restricted due to the angu-

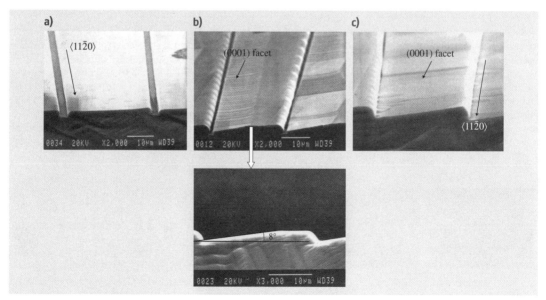

Fig. 28.14a–c Top view of the selective growth of 4H-SiC when window openings are along: (a) $\langle 11\bar{2}0 \rangle$ miscut direction, (b) direction between $\langle 11\bar{2}0 \rangle$ and $\langle 1\bar{1}00 \rangle$, (c) $\langle 1\bar{1}00 \rangle$ direction

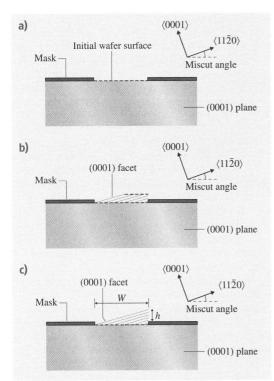

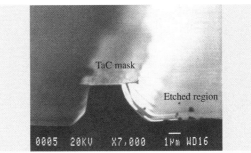

Fig. 28.16 In situ etching, instead of growth, occurs if SiH$_4$ flow is reduced by half compared with that used for Fig. 28.14. Note that the mask is intact after etching. SiH$_4$ flow rate: 0.6 sccm

Fig. 28.15a–c Schematic images of the (0001) facet formation on 4H-SiC off (0001) axis towards $\langle 11\bar{2}0 \rangle$ substrate: (a) after selective growth, (b) after selective growth for a short time, (c) after selective growth for a longer time

lar displacement between step orientation and opening window orientation. Therefore, (0001) facet was formed within the growth window, whereas the percentage of the window occupied by (0001) facet does not increase significantly with increasing angular displacement from $\langle 11\bar{2}0 \rangle$.

The morphology of growth on (0001) facet was rougher, whereas smoother surface was observed on 8° off-axis growth front (Fig. 28.14a–c). It is likely that some 3C-SiC may have also nucleated on (0001) facet at relatively low temperature ($\approx 1550\,^\circ$C). Similar facet formation was also reported by *Chen* et al. [28.136] during SiC deposition on mesa structures of 4H-SiC(0001) substrate. Since step-flow growth dominated on the off-axis SiC surface, the flat and specular epilayers without 3C-SiC were grown on the off-axis surface. Therefore, when selective epitaxial growth on planar off-axis substrate is designed for device fabrication, growth window openings should be aligned along the miscut orientation.

It is worth mentioning that the TaC mask can also act as an in situ etching mask instead of a selective growth mask. When the silane flow rate was reduced to half that used in Fig. 28.14 under otherwise identical conditions, 3 µm/h etching was obtained in unmasked region (Fig. 28.16). The mask itself was not etched, and etching of SiC under the mask left the mask hanging over the edge of the mesa. The mask peeled off during dicing for SEM observation, indicating that it was intact after etching. In situ etching studies using various gases have been carried out previously [28.95], and it was pointed out that the etch rate increases with increasing temperature under otherwise identical conditions. On the other hand, reactive atomic hydrogen is concluded to play a key role in etching of SiC. Therefore, the most plausible reasons for selective etching of SiC are: (1) the local temperature of the window opening region is higher than the nominal measured temperature of the susceptor since the conditions resulting in the etching of TaC-coated wafer are still within the growth window for a bare SiC substrate in our horizontal coldwall CVD system, and (2) Ta at high temperature may produce atomic hydrogen that may enhance the etching rate of SiC.

28.4.4 Effects of Mask-to-Window Ratio ($M : W$) on SiC Selective Growth

As mentioned previously, selective in situ etching occurred with lower SiH$_4$ flow, even though the same conditions result in growth when a bare SiC substrate is used. It is evident that the mask itself has an influence on the selective growth and/or etching. In order

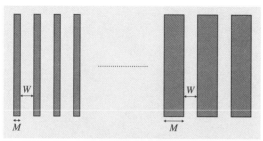

Fig. 28.17 Schematic of the linear pattern with varied mask (M) and window opening (W) width. M and W varied from 2 to 1000 μm. Pitch width (P) = mask width (M) + window opening width (W)

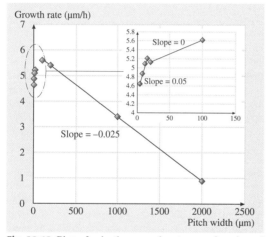

Fig. 28.18 Plot of selective growth rate as a function of pitch width. Pitch width (P) = mask width (M) + window width (W). $M:W = 1$. Flow rate is 2.24 sccm for both SiH_4 and C_3H_8. Growth temperature and pressure are 1833 K and 80 Torr, respectively. The *insert* shows an expanded plot for pitch width ≤ 100 μm

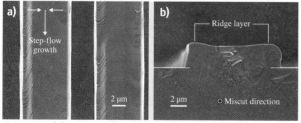

Fig. 28.19 (a) Top and (b) cross-sectional views of ridged layer grown an 4H-SiC(0001) substrate miscut towards $\langle 11\bar{2}0 \rangle$ direction. Stripes are along $\langle 11\bar{2}0 \rangle$ miscut direction

to identify the effects of the mask during the process, a linear pattern with varied mask and window opening width (Fig. 28.17) was made on the same substrate. This assures that the local conditions, such as temperature, precursor flow, pressure etc., for each window opening are the same. Therefore, different growth and/or etching behaviors, if any, are basically due to the existence of the mask. The growth was carried out on such a patterned substrate at temperature of 1560 °C under 80 Torr while the flow rate was kept at 2.24 sccm for both SiH_4 and C_3H_8. Figure 28.18 shows a plot of selective growth rate versus pitch width (pitch width = mask width (M) + window width (W)) when the mask width-to-window width ratio ($M:W$) was kept at 1. As can be seen, the selective growth rate increases with increasing pitch width when the pitch width is below 100 μm, indicating that the growth process rather than etching is dominant under these conditions and is enhanced by diffusion of species from the mask.

The enhanced growth due to the supply of the reactants from the mask is also illustrated by the ridge growth at the edges of the mask, as shown in Fig. 28.19. Since the window stripes were patterned along the $\langle 11\bar{2}0 \rangle$ miscut direction, miscut steps were provided continuously. Excess species supplied laterally from the mask could be easily incorporated at step sites near the edge. Therefore, the step-flow growth velocity became higher than that at the center area of the window opening. Figure 28.20 shows a schematic diagram of the vicinal surface and ridge growth due to the diffusion of species from the mask to the window. Step-flow growth occurs not only along but also perpendicular to the miscut direction.

However, the selective growth rate decreases with further increasing of the pitch width (≥ 200 μm). The growth rate reduces from 5.5 to 0.9 μm/h when the pitch width increases from 200 to 2000 μm. As is well known, both growth and etching process exist during SiC epitaxy and the etching process could be dominant under certain circumstances (e.g., higher temperature or lower precursor flow). In this experiment, the growth conditions were identical for all window openings since they were patterned on the same substrate and growth was carried out in the same run. It is believed that the decreased growth rate with increasing mask width is due to the enhanced etching by the mask. The enhanced etching process is probably caused by the production of atomic hydrogen by tantalum at high temperature, similar to the well-known hydrogenation catalysis by metals such as palladium [28.137]. Therefore, a larger area of the mask results in more atomic hydrogen and thus more

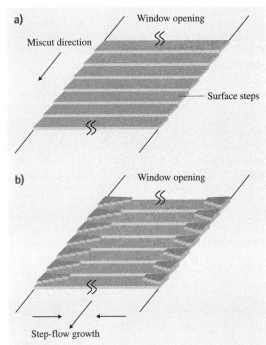

Fig. 28.20a,b Schematic of (**a**) (0001) vicinal surface and (**b**) ridge growth of the edge of the mask. Note that step-flow growth is along the directions parallel and perpendicular to miscut direction due to excess supply of species from the mask to the edge of window openings

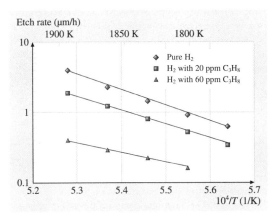

Fig. 28.21 Arrhenius plot of the etch rate of 4H-SiC as a function of temperature at different C_3H_8 flows. Reactor pressure is 100 Torr

etching of SiC in the vicinity. It is worth mentioning that significant etching of about 30 μm in 30 min has been obtained when growth was carried out on a patterned substrate with > 90% of the surface covered by the mask [28.90].

Based on the studies of growth rate as a function of mask layout, it is evident that the growth process is dominant and that growth is enhanced by reactant diffusion from the mask when the mask is narrow (< 100 μm in this study). This increase in growth rate is limited by the surface diffusion rate of reactant species on the mask. However, the etching process competes with the growth process when the mask area becomes larger, and the growth rate decreases (or even etching occurs) with further increases in mask width.

28.4.5 Effects of C/Si Ratio on SiC Selective Growth

It has been found from in situ etching study of SiC [28.95] that adding a small amount of C_3H_8 increases the hydrocarbon pressure and hence suppresses the etching of SiC. Figure 28.21 shows the variation of etch rate of a bare 4H-SiC substrate at different C_3H_8 flows, for etching carried out at temperature range of 1800–1900 K in hydrogen ambient. Therefore, changing the C_3H_8 flow during the growth with fixed SiH_4 flow would change the formation of hydrocarbon in the chamber and hence the etching of SiC. Figure 28.22 shows the dependence of selective growth rate on C/Si ratio at different pitch widths and $M : W$ ratios. When the C/Si ratio is low (= 1), the etching process is dominant and no mask-enhanced growth is observed for narrow mask width (≤ 100 μm), as observed previously. The selective growth rate decreases with increasing mask width, and almost no growth is obtained when the mask width was increased to 1000 μm. On a nonmasked wafer, such growth conditions would result in about 3 μm/h growth rate. With increasing of the C/Si ratio, as expected, the growth rate increases for all investigated pitch widths. Mask-enhanced growth for pitch width ≤ 100 μm is observed for C/Si ratio of both 3 and 6, and the growth rate saturates with further increasing of the mask width.

Decreased growth rate with increasing $M : W$ ratio for pitch width of 500 μm is obtained for a C/Si ratio of 1 and 3, as shown in Fig. 28.22. However, mask-enhanced growth, instead of etching, is achieved for pitch width of 500 μm for a C/Si ratio of 6, indicating

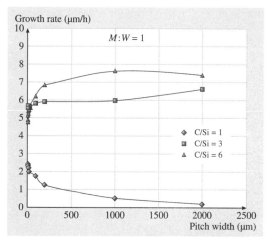

Fig. 28.22 Plot of selective growth rate as a function of pitch width at various C/Si ratios. Pitch width P = mask width (M) + window width (W). $M:W = 1$. Flow rate of SiH$_4$ is kept at 2.24 sccm. Growth temperature is 1813 K. Growth pressure is 80 Torr

that the etching process during epitaxy is suppressed by additional C$_3$H$_8$ flow.

28.4.6 Mechanism of Selective Etching and Effect of Atomic Hydrogen

It is worth pointing out that growth is obtained (Fig. 28.23) for extreme $M:W = 1000\,\mu\text{m}:5\,\mu\text{m}$ at a C/Si ratio of 6 with SiH$_4$ flow of 2.24 sccm, while selectivity is insignificantly deteriorated. It is evident from the experiment that in situ selective etching is enhanced by atomic hydrogen produced by the mask under high

Fig. 28.23 Cross-sectional view of selective growth with $M:W = 1000\,\mu\text{m}:5\,\mu\text{m}$ at C/Si = 6. Flow rate of SiH$_4$ is kept at 2.24 sccm. Growth temperature is 1813 K. Growth pressure is 80 Torr

temperature, which could be suppressed by increasing the C$_3$H$_8$ flow. Increasing the C$_3$H$_8$ flow is believed to be the most effective way to suppress in situ etching of SiC using a TaC mask without greatly deteriorating selectivity.

28.4.7 Fabrication of 4H-SiC p–n Junction Diodes Using Selective Growth

To illustrate the efficacy of selective growth for device fabrication, p–n junction diodes have been fabricated using selectively grown SiC. The starting wafer is commercial 8° off (0001)Si face, p-epitaxial layer grown on p$^+$ 4H-SiC substrate from CREE Research Inc. The nominal thickness and Al doping of the p-epilayer are 12 μm and $9.3 \times 10^{15}\,\text{cm}^{-3}$, respectively. Fabrication was started by cleaning the wafers using a standard solvent degrease, Piranha clean, and RCA clean. Alignment marks were then trenched using CHF$_3$/O$_2$ plasma through a titanium/nickel transfer mask while the metal mask was prepared using a lift-off process. Next, refill regions were formed by reactive ion etch (RIE) through an oxide mask. In this step, ≈ 600 Å tantalum was first deposited using an e-beam evaporator and a masking oxide was then deposited, patterned, and etched using a combination of dry and wet processes to open the refill window. Two etch steps followed to form refill trench. Tantalum and SiC trench were etched using CF$_4$/O$_2$ and CHF$_3$/O$_2$ plasma, respectively. The etched trenches were about 1.5 μm deep. After the trench etching, the masking oxide was stripped using buffered oxide etch (BOE). Since both tantalum and SiC are inert to BOE at room temperature, it is safe to use BOE for oxide removal. Next, the wafer was loaded into the horizontal cold-wall reactor and tantalum film was then converted to tantalum carbide by exposing the wafer in propane/hydrogen ambient at 1300 °C for 15 min as described previously. Epitaxial growth of SiC was carried out at temperature in the range of 1500–1600 °C. Silane, propane, and nitrogen were used as the precursors with hydrogen as the carrier gas. The flow rates of silane, propane, nitrogen, and hydrogen were kept at 2.2, 3.7, 8 sccm, and 12.5 slm, respectively, during the growth. The RIE-etched trenches were refilled by nitrogen-doped n-type epitaxial film grown by selective epitaxy. The doping concentration was measured by mercury probe capacitance–voltage (C–V) using an epilayer grown on planar substrate under the same conditions. The TaC mask was then removed after selective growth using wet etchant. Subsequently, diodes were fabricated using standard contact procedure.

The forward and reverse current–voltage (I–V) characteristics of a square-trench (250 µm × 250 µm) refilled diode were first compared with those of a mesa etched PIN circular diode. Similar I–V characteristics were observed for both diodes, indicating that damage from the RIE etching might have been cured during epitaxial growth and that the junction interface is as good as that in a PIN diode fabricated by planar epitaxial growth. The difference in the forward I–V characteristic at higher current is possibly due to the higher series resistance of the p-type base of the trench refilled diode.

28.5 Conclusions

In this chapter, some basic properties of SiC and its epitaxial growth process have been described. Most of the early epitaxial growth was carried out using propane and silane in a horizontal cold-wall reactor. In recent years, hot-wall reactors have been developed for SiC growth, allowing much higher growth rate with very good surface morphology and very low background doping concentration. High-voltage ($>10\,\text{kV}$) power devices have been demonstrated. Epitaxial growth of SiC on bulk SiC substrates has advanced significantly during the last decade, and commercial SiC devices have been introduced by several manufacturers. Further refinement of the basic reactor geometry and the growth process will be necessary to fully exploit the advantages of SiC over Si in devices. One such improvement may be the introduction of chlorine-containing precursors in addition to the more conventional silane and propane precursors. Some aspects of selective growth of SiC using TaC masks developed at the author's laboratory (mostly unpublished) are also reviewed.

References

28.1 T.P. Chow: SiC and GaN high voltage power switching devices, Mater. Sci. Forum **338-342**, 1155–1160 (2000)

28.2 R.G. Humphreys, D. Bimberg, W.J. Choyke: Wavelength modulated absorption in SiC, Solid State Comm. **39**, 163–167 (1981)

28.3 W.J. Choyke, L. Patrick: Exciton recombination radiation and phonon spectrum of 6H-SiC, Phys. Rev. **127**, 1868–1877 (1962)

28.4 W.V. Münch, I. Pfaffeneder: Breakdown field in vapor grown silicon carbide p-n junctions, J. Appl. Phys. **48**, 4831–4833 (1977)

28.5 D.K. Ferry: High field transport in wide band gap semiconductors, Phys. Rev. B1 **12**, 2361–2369 (1975)

28.6 R.P. Joshi: Monte Carlo calculations of the temperataure and field dependent electron transport parameters for 4H-SiC, J. Appl. Phys. **78**, 5518–5521 (1995)

28.7 E. Moruan, O. Noblanc, C. Dua, C. Brylinski: SiC microwave power devices, Mater. Sci. Forum **353-356**, 669–674 (2001)

28.8 G.A. Slack: Thermal conductivity of pure and impure silicon, silicon carbide, and diamond, J. Appl. Phys. **35**, 3460 (1964)

28.9 M. Bhatnagar, B.J. Baliga: Comparison of 6H-SiC, 3C-SiC and Si power devices, IEEE Trans. Electron. Dev. **40**(3), 645–655 (1993)

28.10 T. Kimoto, T. Urushidani, S. Kobayashi, H. Matsunami: High voltage SiC Schottky barrier diodes with low on-resistances, IEEE Electron. Dev. Lett. **14**, 548–550 (1993)

28.11 D. Alok, B.J. Baliga, P.K. Mclarty: A simple edge termination for silicon carbide devices with nearly ideal breakdown voltages, IEEE Electron. Dev. Lett. **15**, 394–395 (1994)

28.12 R.J. Trew, J.-B. Yan, P.M. Mock: The potential of diamond and SiC electronic devices for microwave and millimeter wave power applications, Proc. IEEE **79**(5), 598–620 (1991)

28.13 J.M. McGarrity, F.B. McLean, W.M. DeLancey, J. Palmour, C. Carter, J. Edmond, R.R. Oakley: SiC JFET radiation response, IEEE Trans. Nucl. Sci. **39**(6), 1974–1981 (1992)

28.14 See for example, http://www.cree.com/ and http://www.infineon.com/

28.15 H. Matsunami, T. Kimoto: Step controlled epitaxial growth of SiC: High quality homoepitaxy, Mater. Sci. Eng. **R20**(3), 125–166 (1997)

28.16 W.J. Choyke, G. Pensl: Physical properties of SiC, MRS Bull. **22**(3), 25–29 (1997)

28.17 C.H. Park, B.H. Cheong, K.H. Lee, K.J. Chang: Structural and electronic propertires of cubic, 2H, 4H and 6H SiC, Phys. Rev. B **49**(7), 4485–4493 (1994)

28.18 W.S. Yoo, H. Matsunami: Polytype-controlled single crystal growth of silicon carbide using 3C-6H solid state phase transformation, J. Appl. Phys. **70**(11), 7124–7131 (1991)

28.19 W.S. Yoo, H. Matsunami: Solid state phase transformation in cubic silicon carbide, Jpn. J. Appl. Phys. Part I (regular papers and short notes) **30**, 545–553 (1991)

28.20 A. Itoh, H. Matsunami: Single crystal growth of SiC and electronic devices, Crit. Rev. Solid State Mater. Sci. **22**(2), 111–197 (1997)

28.21 P.G. Neudeck: Electrical impact of SiC structural defects on high electric field devices, Mater. Sci. Forum **338–342**, 1161–1166 (2000)

28.22 J.A. Powell, H.A. Will: Low temperature solid state phase transformations in 2H SiC, J. Appl. Phys. **43**(4), 1400–1408 (1972)

28.23 M. Dudley, X. Huang: Characterization of SiC using synchrotron white beam x-ray topography, Mater. Sci. Forum **338–342**, 431–436 (2000)

28.24 I. Kamata, H. Tsuchida, T. Jikimoto, K. Izumi: Structural transformation of screw dislocation via thick 4H-SiC epitaxial growth, Jpn. J. Appl. Phys. **39**, 6496–6500 (2000)

28.25 P. Pirouz: On micropipes and nanopipes in SiC and GaN, Philos. Mag. A **78**, 727–736 (1998)

28.26 P.G. Neudeck, H. Wei, M. Dudley: Study of bulk and elementary screw dislocation assisted reverse breakdown in low-voltage (<250 V) 4H-SiC pn junction diodes: DC properties, IEEE Trans. Electron. Dev. **46**(3), 478–484 (1999)

28.27 U. Zimmermann, A. Hallen, A.O. Konstantinov, B. Breitholtz: Investigation of microplasma breakdown in 4H SiC, Mater. Res. Soc. Symp. Proc. **512**, 151–156 (1998)

28.28 A.O. Konstantinov, Q. Wahab, N. Nordell, U. Lindefelt: Study of Avalanche breakdown and impact ionization in 4H SiC, J. Electron. Mater. **27**(4), 335–341 (1998)

28.29 P.G. Neudeck, J.A. Powell: Performance limiting micropipe defects in SiC wafers, IEEE Electron. Dev. Lett. **15**, 63–65 (1994)

28.30 W. Si, M. Dudley, R. Glass, V. Tsvetkov, C. Carter Jr: Hollow core screw dislocations in 6H-SiC single crystals: A test of Frank's theory, J. Electon. Mater. **26**, 128–133 (1997)

28.31 W. Si, M. Dudley: Experimental studies of hollow core screw dislocations in 6H- and 4H-SiC single crystals, Mater. Sci. Forum **264–268**, 429–432 (1998)

28.32 C.M. Schnabel, M. Tabib-Azar, P.G. Neudeck, S.G. Bailey, H.B. Su, M. Dudle, R.P. Raffaelle: Correlation of EBIC and SWBXT imaged defects and epilayer growth pits in 6H-SiC Schottky diodes, Mater. Sci. Forum **338–342**, 489–492 (2000)

28.33 J.A. Powell, D.J. Larkin: Process induced morphological defects in epitaxial CVD silicon carbide, Phys. Status Solidi (b) **202**, 529–548 (1997)

28.34 T. Kimoto, N. Miyamoto, H. Matsunami: Performance limiting surface defects in SiC epitaxial pn junction diodes, IEEE Trans. Electron. Dev. **46**(3), 471–477 (1999)

28.35 S. Nishino, J.A. Powell, W. Will: Production of large area single crystal of 3C-SiC for semiconductor devices, Appl. Phys. Lett. **42**, 460 (1983)

28.36 S. Nishino, K. Matsumoto, Y. Chen, Y. Nishio: Epitaxial growth of 4H-SiC by sublimation close space technique, Mater. Sci. Eng. B **61/62**, 121–124 (1999)

28.37 H. Nakazawa, M. Suemitsu, S. Asami: Formation of high quality SiC on Si(001) at 900 °C using monomethylsilane gas source MBE, Mater. Sci. Forum **338–342**, 269–272 (2000)

28.38 M.E. Okhuysen, M.S. Mazzola, Y.-H. Lo: Low temperature growth of 3C-SiC on silicon for advanced substrate development, Mater. Sci. Forum **338–342**, 305–308 (2000)

28.39 M.-A. Hasan, A. Faik, D. Purser, D. Lieu: Heteroepitaxy of 3C-SiC on Si (100) using porous Si as a compliant seed crystal, Tech. Dig. Int. Conf. SiC Relat. Mater. ICSCRM2001 (Tsukuba 2001) pp. 492–493

28.40 Y. Ishida, K. Kushibe, T. Takahashi, H. Okumura, S. Yoshida: 3C-SiC homoepitaxial growth by chemical vapor deposition and Schottky barrier junction characteristics, Mater. Sci. Forum **389–393**, 275–278 (2002)

28.41 H. Nagasawa, K. Yagi, T. Kawahara: 3C-SiC heteroeptaxial growth on (001) Si undulant substrates, J. Cryst. Growth **237–239**, 1244–1249 (2002)

28.42 H. Nagasawa, K. Yagi, T. Kawahara, N. Hatta, G. Pensl, W.J. Choyke, T. Yamada, K.M. Itoh, A. Schoner: *Silicon Carbide: Recent Major Advances* (Springer, Berlin, Heidelberg 2004) p. 207

28.43 F.R. Chien, S.R. Nutt, J.M. Carulli Jr., N. Bunchan, C.P. Beetz Jr., W.S. Yoo: Heteroepitaxial growth of beta SiC films on TiC substrates: Interface structure and defects, J. Mater. Res. **9**(8), 2086–2095 (1994)

28.44 C. Hallin, A.O. Konstantinov, O. Kordina, E. Janzen: Mechanism of cubic SiC nucleation on off-axis substrates, Proc. 6th Int. Conf. SiC Relat. Mater. 1995, Inst. Phys. Conf. Ser. **142**, 85–88 (1996)

28.45 J.A. Powell, D.J. Larkin, P.B. Abel, L. Zhou, P. Pirouz: Effect if tilt angle on the morphology of SIC epitaxial films grown on vicinal (0001) SiC substrates. In: Silicon Carbide and Related Materials, Inst. Phys. Conf. Ser., Vol. 142 (1995) pp. 77–80

28.46 V.F. Tsvetkov, S.T. Allen, H.S. Kong, C.H. Carter Jr.: Recent progress in SiC crystal growth, Proc. 6th Int. Conf. SiC Relat. Mater. 1995, Inst. Phys. Conf. Ser. **142**, 17–22 (1996)

28.47 D.J. Larkin: An overview of SiC epitaxial growth, MRS Bulletin **22**(3), 36–41 (1997)

28.48 V. Heine, C. Cheng, R.J. Needs: The preference of SiC for growth in the metastable cubic form, J. Am. Ceram. Soc. **74**, 2630–2633 (1991)

28.49 K. Wada, T. Kimoto, K. Nishikawa, H. Matsunami: Epitaxial growth of 4H-SiC on 4° off-axis (0001) and (000-1) substrates by hot wall CVD, Mater. Sci. Forum **527–529**, 219–222 (2006)

28.50 S. Rendakova, V. Ivantsov, V. Dmitriev: High quality 6H- and 4H-SiC pn structures with stable elelctric breakdown grown by liquid phase epitaxy, Mater. Sci. Forum **264–268**, 163–166 (1998)

28.51 D.H. Hofmann, M.H. Muller: Prospects of the use of liquid phase techniques for the growth of bulk silicon carbide crystals, Mater. Sci. Eng. B **61/62**, 29–39 (1999)

28.52 R. Yakimova, M. Syväjärvi, S. Redankova, V.A. Dimitriev, A. Henry, E. Janzen: Micropipe healing in liquid phase epitaxial growth of SiC, Mater. Sci. Forum **338–342**, 237–240 (2000)

28.53 A. Tanaka, T. Ataka, E. Ohkura, H. Katsuno: Growth modes of silicon carbide in low-temperature liquid phase epitaxy, Jpn. J. Appl. Phys. **43**(11A), 7670–7671 (2004)

28.54 O. Filip, B. Epelbaum, M. Bickermann, A. Winnacker: Micropipe healing in SiC wafers by liquid-phase epitaxy in Si–Ge melts, J. Cryst. Growth **271**, 142–150 (2004)

28.55 T. Hatayama, S. Nakamura, K. Kurobe, T. Kimoto, T. Fuyuki, H. Matsunami: High-temperature surface structure transitions and growth of alpha-SiC (0001) in ultrahigh vacuum, Mater. Sci. Eng. B **61/62**, 135–138 (1999)

28.56 S. Nakamura, T. Hatayama, T. Kimoto, T. Fuyuki, H. Matsunami: Growth of SiC on 6H-SiC {01$\bar{1}$4} substrates by gas source molecular beam epitaxy, Mater. Sci. Forum **338–342**, 201–204 (2000)

28.57 T. Sugii, T. Aoyama, T. Ito: Low-temperature growth of beta-SiC on Si by gas-source MBE, J. Electrochem. Soc. **137**(3), 989–992 (1990)

28.58 H. Nakazawa, M. Suemitsu, S. Asami: Formation of high quality SiC on Si (001) at 900 °C using monomethylsilane gas-source MBE, Mater. Sci. Forum **338–342**, 269–272 (2000)

28.59 C. Hallin, I.G. Ivanov, T. Egillson, A. Henry, O. Kordina, E. Jansen: The material quality of CVD grown SiC using various precursors, J. Cryst. Growth **183**, 163 (1998)

28.60 Y. Gao, J.H. Edgar: Selective epitaxial growth of SiC: Thermodyanamic analysis of the Si–C–Cl–H and Si–C–Cl–H–O systems, J. Electrochem. Soc. **144**(5), 1875–1880 (1997)

28.61 K. Yagi, H. Nagasawa: 3C-SiC growth by alternate supply of SiH_2Cl_2 and C_2H_2, J. Cryst. Growth **174**, 653–657 (1997)

28.62 T. Miyanagi, S. Nishino: Hotwall CVD growth of 4H-SiC using Si2Cl6-C3H8-H2 system, Mater. Sci. Forum **389–393**, 199–202 (2002)

28.63 C. Sartel, V. Souliere, Y. Monteil, H. El-Harrouni, J.M. Bluet, G. Guillot: Epitaxial growth of 4H-SiC with hexamethyldisilane, Mater. Sci. Forum **389–393**, 263–266 (2002)

28.64 R. Rodriguez-Clemente, A. Figueras, S. Garelik, B. Armas, C. Combescure: Influence of temperature and tetramethylsilane partial pressure on the beta SiC depositin by cold wall chemical vapor deposition, J. Cryst. Growth **125**, 532–542 (1992)

28.65 T. Hatayama, H. Yano, Y. Uraoka, T. Fuyuki: High purity SiC epitaxial growth by chemical vapor deposition using CH_3SiH_3 and C_3H_3 sources, Mater. Sci. Forum **527–529**, 203–206 (2006)

28.66 See for example, http://www.saespuregas.com/

28.67 H. Matsunami, T. Kimoto: Step controlled epitaxial growth of SiC: High quality homoepitaxy, Mater. Sci. Eng. **R20**(3), 125–166 (1997)

28.68 S. Nakamura, T. Kimoto, H. Matsunami: Fast growth and doping characteristics of SiC in a horizontal cold wall CVD, Mater. Sci. Forum **389–393**, 183–186 (2002)

28.69 R. Rupp, A. Wiedenhofer, P. Friedrichs, D. Peters, R. Schorner, D. Stephani: Growth of SiC epitaxial layers in a vertical cold wall reactor suited for high voltage applications, Mater. Sci. Forum **264–268**, 89–96 (1998)

28.70 C. Sartel, J.M. Bluet, V. Souliere, I. El-Harrouni, Y. Monteil, M. Mermoux, G. Guillot: Characterization of homoepitaxial 4H-SiC layer grown from silane/propane system, Mater. Sci. Forum **433–436**, 165–168 (2003)

28.71 B. Thomas, W. Bartsch, R. Stein, R. Schorner, D. Stephani: Properties and suitability of 4H-SiC epitaxial layers grown at different CVD systems for high voltage applications, Mater. Sci. Forum **457–460**, 181–184 (2004)

28.72 O. Kordina, C. Hallin, A. Henry, J.P. Bergman, I. Ivanov, A. Ellison, N.T. Son, E. Janzen: Growth of SiC by hot-wall, CVD and HTCVD, Phys. Status Solidi (b) **202**, 321–334 (1996)

28.73 T. Kimoto, S. Nakazawa, K. Fujihira, T. Hirao, S. Nakamura, Y. Chen, H. Matsunami: Recent achievement and future challenges in SiC homoepitaxial growth, Mater. Sci. Forum **389–393**, 165–170 (2002)

28.74 A. Shoner, A. Konstantinov, S. Karlsson, R. Berge: Highly uniform epitaxial SiC-layer growth in a hot wall CVD reactor with mechanical rotation, Mater. Sci. Forum **389–393**, 187–190 (2002)

28.75 B. Thomas, C. Hecht: Epitaxial growth of n-type 4H-SiC on 3, wafers for power devices, Mater. Sci. Forum **483–485**, 141–146 (2005)

28.76 A. Ellison, J. Zhang, J. Peterson, A. Henry, Q. Wahab, J.P. Bergman, Y.N. Makarov, A. Vorob'ev, A. Vehanen, E. Janzen: High temperature CVD growth of SiC, Mater. Sci. Eng. B **61/62**, 113–120 (1999)

28.77 H. Fujiwara, K. Danno, T. Kimoto, T. Tojo, H. Matsunami: Fast epitaixal growth of thick 4H-SiC with specular surface by chimney-type vertical hot-wall chemical vapor depositon, Mater. Sci. Forum **457–460**, 205–208 (2004)

28.78 E. Janzén, J.P. Bergman, Ö. Danielsson, U. Forsberg, C. Hallin, J. ul Hassan, A. Henry, I.G. Ivanov, A. Kakanakova-Georgieva, P. Persson, Q. ul Wa-

28.79 K. Danno, T. Kimoto, K. Asano, Y. Sugawara, H. Matsunami: Fast epitaxial growth of high-purity 4H-SiC(0001) in a vertical hot-wall chemical vapor deposition, J. Electron. Mater. **34**(4), 324–329 (2005)

28.80 H. Fujiwara, T. Kimoto, T. Tojo, H. Matsunami: Reduction of stacking faults in fast epitaxial growth of 4H-SiC and its impacts on high-voltage Schottky diodes, Mater. Sci. Forum **483–485**, 151–154 (2005)

28.81 I. Bhat, Canhua Li: High growth rate epitaxy of SiC in a vertical hotwall reactor, unpublished

28.82 M. Syvajarvi, R. Yakimova, H. Jacobsson, M.K. Linnarsson, A. Henry, E. Janzén: High growth rate epitaxy of thick 4H-SiC layers, Mater. Sci. Forum **338–342**, 165–168 (2000)

28.83 T. Furusho, T. Miyanagi, Y. Okui, S. Ohshima, S. Nishino: Homoepitaxial growth of cubic silicon carbide by sublimation epitaxy, Mater. Sci. Forum **389–393**, 279–282 (2002)

28.84 E.O. Sveinbjornsson, H.O. Olafsson, G. Gudjonsson, F. Allerstam, P.A. Nilsson, M. Syvajarvi, R. Yakimova, C. Hallin, T. Rodle, R. Jos: High field effect mobility in Si face 4H-SiC MOSFET made on sublimation grown epitaxial material, Mater. Sci. Forum **483–485**, 841–844 (2005)

28.85 D. Ziane, J.M. Bluet, G. Guillot, P. Godignon, J. Monserrat, R. Ciechonski, M. Syvajarvi, R. Yakimova, L. Chen, P. Mawby: Characterizations of SiC/SiO$_2$ interface quality toward high power MOSFETs realization, Mater. Sci. Forum **457–460**, 1281–1286 (2004)

28.86 R.C. Glass, P. Lu, J.H. Edgar, O.J. Glembocki, P.B. Klein, E.R. Glaser, J. Perrin, J. Chaudhuri: High-speed homoepitaxy of SiC from methyltrichloro-silane by CVD, Int. Conf. Silicon Carbide Relat. Mater. (Pittsburgh 2005)

28.87 R. Myers, O. Kordina, Z. Shishkin, F. Yan, R.P. Devaty, S.E. Saddow: Effects of HCl additive on the growth rate of 4H-SiC in a hot wall CVD reactor, Int. Conf. Silicon Carbide Relat. Mater. (Pittsburgh 2005)

28.88 A. Veneroni, F. Omarini, M. Masi: Silicon carbide growth mechanisms from SiH$_4$, SiHCl$_3$ and nC$_3$H$_8$, Cryst. Growth Technol. **40**(10/11), 967–971 (2005)

28.89 G. Pensl, W.J. Choyke: Electrical and optical characterization of SiC, Physica B **185**, 264–283 (1993)

28.90 R. Wang, I. Bhat, unpublished results

28.91 D.J. Larkin: SiC dopant incorporation control by site competetion CVD, Phys. Status Solidi (b) **202**, 305–320 (1997)

28.92 T. Troffer, C. Peppermuller, G. Pensi, K. Rottner, A. Schoner: Phosphorus-related donors in 6H-SiC generated by ion implantation, J. Appl. Phys. **80**(7), 3739–3743 (1996)

28.93 M.A. Capano, J.A. Cooper Jr., M.R. Melloch, A. Saxler, W.C. Mitchel: Ionization energies and electron mobililties in phosphorous and nitrogen implanted SiC, J. Appl. Phys. **87**(12), 8773–8777 (2000)

28.94 S. Rao, T.P. Chow, I. Bhat: Dependence of the ionization energy of phosphorous donor in 4H-SiC on doping concentration, Mater. Sci. Forum **527–529**, 597–600 (2006)

28.95 R. Wang: SiC epitaxial growth for power device applications. Ph.D. Thesis (Rensselaer Polytechnic Institute, Troy 2002)

28.96 M.H. Anikin, A.A. Lebedev, A.L. Syrkin, A.V. Suvorov: Investigation of deep levels in SiC by capacitance spectroscopy methods, Sov. Phys. Semicond. **19**, 69–71 (1985)

28.97 W. Suttrop, G. Pensi, P. Lanig: Boron related deep centers in SiC, Appl. Phys. A **51**, 231–237 (1990)

28.98 A. Golz, G. Horstmann, E. Stein von Kamienski, H. Kurz: Oxidation kinetics of 3C, 4H and 6H silicon carbide, Proc. Sixth Int. Conf. Silicon Carbide Relat. Mater. (1996) pp. 633–636

28.99 K. Wada, T. Kimoto, K. Nishikawa, H. Matsunami: Improved surface morphology and background doping concentration in 4H-SiC (000$\bar{1}$) epitaxial growth by hot-wall CVD, Mater. Sci. Forum **483–485**, 85–88 (2005)

28.100 T. Yamamoto, T. Kimoto, H. Matsunami: Impurity incorporation mechanism in step-controlled epitaxy growth temperature and substrate off-angle dependence, Mater. Sci. Forum **264–268**, 111–116 (1998)

28.101 U. Forsberg, Ö. Danielsson, A. Henry, M.K. Linnarsson, E. Janzén: Nitrogen doping of epitaxial silicon carbide, J. Cryst. Growth **236**(1–3), 101–112 (2002)

28.102 U. Forsberg, Ö. Danielsson, A. Henry, M.K. Linnarsson, E. Janzén: Aluminium doping of epitaxial silicon carbide, J. Cryst. Growth **253**(1–4), 340–350 (2003)

28.103 T. Kimoto, T. Yamamoto, Z.Y. Chen, H. Matsunami: 4H-SiC (11$\bar{2}$0) epitaxial growth, Mater. Sci. Forum **338–342**, 189–192 (2000)

28.104 T. Kimoto, K. Hashimoto, K. Fujihira, K. Danno, S. Nakamura, Y. Negoro, H. Matsunami: Epitaxial growth and characterization of 4H-SiC(11$\bar{2}$0) and (03$\bar{3}$8), Mater. Res. Soc. Symp. Proc. **742**, 3–13 (2003)

28.105 Z. Zhang, Y. Gao, A.C. Arjunan, E.Y. Toupitsyn, P. Sadagopan, R. Kennedy, T.S. Sudarshan: CVD growth and characterization of 4H-SiC epitaxial film on (11$\bar{2}$0) as-cut substrates, Mater. Sci. Forum **483–485**, 113–116 (2005)

28.106 K. Nakayama, Y. Miyanagi, H. Shiomi, S. Nishino, T. Kimoto, H. Matsunami: The development of 4H-SiC {03$\bar{3}$8} wafers, Mater. Sci. Forum **389–393**, 123–127 (2002)

28.107 T. Kimoto, K. Danno, K. Fujihira, H. Shiomi, H. Matsunami: Complete micropipe dissociation in 4H-SiC (03-38) epitaxial growth and its impact on reverse characteristics of Schottky barrier diodes, Mater. Sci. Forum **433–436**, 197–200 (2003)

28.108 T. Kimoto, K. Danno, K. Fujihira, H. Shiomi, H. Matsunami: SiC epitaxy on non-standard surfaces, Mater. Sci. Forum **433–436**, 125–130 (2003)

28.109 E. Arnold, D. Alok: Effect of interface states on electron transport in 4H–SiC inversion layers, IEEE Trans. Electron. Dev. **48**, 1870–1877 (2001)

28.110 T. Kimoto, Y. Kanzaki, M. Noborio, H. Kawano, H. Matsunami: Interface properties of metal-oxide–semiconductor structures on 4H–SiC {0001} and (11$\bar{2}$0) formed by N_2O oxidation, Jpn. J. Appl. Phys. **44**(3), 1213–1218 (2005)

28.111 K. Fukuda, J. Senzaki, K. Kojima, T. Suzuki: High inversion channel mobility of MOSFET fabricated on 4H–SiC C(000$\bar{1}$) face using H_2 post-oxidation annealing, Mater. Sci. Forum **433–436**, 567–570 (2003)

28.112 T. Kimoto, H. Kawano, J. Suda: 1200V-class 4H–SiC RESURF MOSFETs with low on-resistances, Proc. 17th Int. Symp. Power Semicond. Devices IC's (2005) pp. 159–162

28.113 N. Nordell, S. Karlsson, A.O. Kenstantinov: Growth of 4H and 6H SiC in trenches and around stripe mesas, Mater. Sci. Forum **264–268**, 131–134 (1998)

28.114 Y. Chen, T. Kimoto, Y. Takeuchi, R.K. Malhan, H. Matsunami: Homoepitaxy of 4H–SiC on trenched (0001) Si face substrates by chemical vapor deposition, Jpn. J. Appl. Phys. **43**(7A), 4105–4109 (2004)

28.115 P.G. Neudeck, J.A. Powell, G.M. Beheim, E.L. Benavage, P.B. Abel, A.J. Trunek, D.J. Spry, M. Dudley, W.M. Vetter: Enlargement of step-free SiC surfaces by homoepitaxial web growth of thin SiC cantilevers, J. Appl. Phys. **92**, 2391–2400 (2002)

28.116 P.G. Neudeck, A.J. Trunek, D.J. Spry, J.A. Powell, H. Du, M. Skowronski, N.D. Bassim, M.A. Mastro, M.E. Twigg, R.T. Holm, R.L. Henry, C.R. Eddy Jr.: Recent results from epitaxial growth on step free 4H–SiC mesas, Mater. Res. Soc. Symp. Proc. **911**, B08–03 (2006)

28.117 P.G. Neudeck, A.J. Powell: Homoepitaxial and heteroepitaxial growth on step-free SiC mesas. In: *Silicon Carbide: Recent Major Advances*, ed. by W.J. Choyke, H. Matsunami, G. Pensi (Springer, New York 2003) p. 179

28.118 N.D. Bassima, M.E. Twigg, M.A. Mastro, C.R. Eddy Jr., T.J. Zega, R.L. Henry, J.C. Culbertson, R.T. Holm, P. Neudeck, J.A. Powell, A.J. Trunek: Dislocations in III-nitride films grown on 4H–SiC mesas with and without surface steps, J. Cryst. Growth **304**, 103–107 (2007)

28.119 M.R. Goulding: Selective epitaxial growth of silicon, Mater. Sci. Eng. B **17**(1–3), 47–67 (1993)

28.120 D. Kapolnek, S. Keller, R. Vetury, R.D. Underwood, P. Kozodoy, S.P. Den Baars, U.K. Mishra: Anisotropic epitaxial lateral growth in GaN selective area epitaxy, Appl. Phys. Lett. **71**(9), 1204–1206 (1997)

28.121 O. Nam, T.S. Zheleva, M.D. Bremser, R.F. Davis: Lateral epitaxial overgrowth of GaN films on SiO_2 areas via metalorganic vapor phase epitaxy, J. Electron. Mater. **27**(4), 233–237 (1998)

28.122 Y. Ohshita: Low temperature and selective growth of β-SiC using the $SiH_2Cl_2/i-C_4H_{10}/HCl/H_2$, Appl. Phys. Lett. **57**(6), 605–607 (1990)

28.123 J.H. Edgar, Y. Gao, J. Chaudhuri, S. Cheema, P.W. Yip, M.V. Sidorov: Selective epitaxial growth of silicon carbide on SiO_2 masked Si(100): The effects of temperature, J. Appl. Phys. **84**(1), 201–204 (1998)

28.124 K. Teker: Selective epitaxial growth of 3C–SiC on patterned Si using hexamethyldisilane by APCVD, J. Cryst. Growth **257**(3/4), 245–254 (2003)

28.125 S. Nishino, C. Jacob, Y. Okui, S. Ohshima, Y. Masuda: Lateral over-growth of 3C–SiC on patterned Si(111) substrates, J. Cryst. Growth **237–239**(2), 1250–1253 (2002)

28.126 A.R. Bushroa, C. Jacob, H. Saijo, S. Nishino: Lateral epitaxial overgrowth and reduction in defect density of 3C–SiC on patterned Si substrates, J. Cryst. Growth **271**(1/2), 200–206 (2004)

28.127 E. Eshun, C. Taylor, M.G. Spencer, K. Kornegay, I. Ferguson, A. Gurray, R. Stall: Homoepitaxial and selective area growth of 4H and 6H silicon carbide using a resistively heated vertical reactor, Mater. Res. Soc. Symp. **572**, 173–178 (1999)

28.128 Y. Chen, T. Kimoto, Y. Takeuchi, H. Matsunami: Homoepitaxial mesa structures on 4H–SiC (0001) and (11$\bar{2}$0) substrates by chemical vapor deposition, J. Cryst. Growth, **254**, 115–122 (2003)

28.129 Y. Khlebnikov, I. Khlebnikov, M. Parker, T.S. Sudarshan: Local epitaxy and lateral epitaxial overgrowth of SiC, J. Cryst. Growth **233**, 112–120 (2001)

28.130 R. Zhang, I. Bhat: Atomic force microscopy studies of CdTe films grown by epitaxial lateral overgrowth, J. Electron. Mater. **30**(11), 1370–1375 (2001)

28.131 B.A. Haskell, T.J. Baker, M.B. McLaurin, F. Wu, P.T. Fini, S.P. DenBaars, J.S. Speck, S. Nakamura: Defect reduction in *m*-plane gallium nitride via lateral epitaxial overgrowth by hydride phase epitaxy, Appl. Phys. Lett. **86**(11), 111917-1–111917-3 (2005)

28.132 A.A. Burk Jr., M.J. O'Loughlin, H.D. Nordby Jr.: SiC epitaxial layer growth in a novel multi-wafer vapor-phase epitaxial (VPE) reactor, J. Cryst. Growth **200**, 458–466 (1999)

28.133 L.B. Rowland, G.T. Dunne, J.A. Freitas Jr.: Initial results on thick 4H–SiC epitaxial layers grown using vapor phase epitaxy, Mater. Sci. Forum **338–342**, 161–164 (2000)

28.134 G.R. Gruzalski, D.M. Zehner: Defect states in substoichiometric tantalum carbide, Phys. Rev. B **34**(6), 3841–3848 (1986)

28.135 A.K. Dua, V.C. George: TaC coatings prepared by hot filament chemical vapour deposition: Characterization and properties, Thin Solid Films **247**, 34–38 (1994)

28.136 Y. Chen, T. Kimoto, Y. Takeuchi, H. Matsunami: Selective homoepitaxy of 4H–SiC on (0001) and (11$\bar{2}$0) masked substrates, J. Cryst. Growth **237–239**, 1224–1229 (2002)

28.137 Y. Fukai: *The Metal-Hydrogen System*, 2nd edn. (Springer, Berlin, Heidelberg 2005)

29. Liquid-Phase Electroepitaxy of Semiconductors

Sadik Dost

The chapter presents a review of the growth of single-crystal bulk semiconductors by liquid-phase electroepitaxy (LPEE). Following a short introduction, early modeling and theoretical studies on LPEE are briefly introduced. Recent experimental results on LPEE growth of GaAs/GaInAs single crystals under a static applied magnetic field are discussed in detail. The results of three-dimensional numerical simulations carried out for LPEE growth of GaAs under various electric and magnetic field levels are presented. The effect of magnetic field nonuniformities is numerically examined. Crystal growth experiments show that the application of a static magnetic field in LPEE growth of GaAs increases the growth rate very significantly. A continuum model to predict such high growth rates is also presented. The introduction of a new electric mobility in the model, i.e., the *electromagnetic* mobility, allows accurate predictions of both the growth rate and the growth interface shape. Space limitation required the citation of a limited number of references related to LPEE [29.1–73]. For details of many aspects of the LPEE growth process and its historical developments, the reader is referred to these references and also others cited therein.

29.1	Background	967
	29.1.1 Liquid-Phase Electroepitaxy	968
	29.1.2 Natural Convection	970
	29.1.3 Applied Magnetic Fields	970
	29.1.4 Observation of Growth Rate	970
29.2	Early Theoretical and Modeling Studies ..	971
	29.2.1 Peltier-Induced Growth Kinetics: Electromigration Mechanism	971
	29.2.2 A One-Dimensional Model	973
	29.2.3 Source-Current-Controlled (SCC) Growth	975
29.3	Two-Dimensional Continuum Models	977
29.4	LPEE Growth Under a Stationary Magnetic Field	978
	29.4.1 Experiments	979
29.5	Three-Dimensional Simulations	981
	29.5.1 Simulation Model	982
	29.5.2 Numerical Method	983
	29.5.3 Effect of Magnetic Field Strength ..	984
	29.5.4 Evolution of Interfaces	989
	29.5.5 Effect of High Electric and Magnetic Field Levels	990
29.6	High Growth Rates in LPEE: Electromagnetic Mobility	992
	29.6.1 Estimation of the Electromagnetic Mobility Value	993
	29.6.2 Simulations of High Growth Rates in a GaAs System	994
References	....	996

29.1 Background

Liquid-phase electroepitaxy (LPEE) is a solution growth technique by which layers of single crystals are grown at relatively low temperatures. It has great potential for producing high-quality, thick crystals of compound and alloy semiconductors. The LPEE growth technique was developed through the use of electric current for dopant modulation in LPE [29.1], and thereafter became a solution growth technique for growth of binary and ternary semiconductor crystals [29.1–11].

Growth in LPEE is initiated and sustained by passing an electric current through the substrate–solution–source system while the overall furnace temperature is kept constant. Since growth takes place at constant furnace temperature, LPEE has a number of advantages

Table 29.1 Potential applications of alloy semiconductors [29.12]

Semiconductor alloy	Applications
$Ga_{0.96}In_{0.04}As$	Substrates lattice matched to blue diodes and lasers (ZnSe)
$Ga_xIn_{1-x}As$, $InAs_{1-x}P_x$	Substrates suitable for OEICs operating in the 1.3–2 μm region
$Ga_xIn_{1-x}P$, $GaAs_{1-x}P_x$, $Al_xGa_{1-x}As$	Substrates for diodes and lasers operating in the visible range
$Hg_{1-x}Cd_xTe$, $Ga_{1-x}In_xSb$, $InAs_xSb_{1-x}$	Substrates for mid- and far-infrared detectors and lasers
$Si_{1-x}Ge_x$	Substrates for n-channel field-effect transistors (FETs)
	and efficient optoelectronic devices, solar cells, photodetectors
$Cd_xZn_{1-x}Te$	Substrates for γ- and x-ray detectors

such as steady and controlled growth rate, controlled doping, improved surface morphology and defect structure, low dislocation density, and improved electronic characteristics. In addition, the method is suitable for growing ternary and quaternary alloy crystals with desired compositions. This feature of LPEE has attracted interest in the growth of high-quality semiconductor crystals since the availability of such thick alloy substrates may solve the problems arising from lattice mismatch encountered in the integration of different material layers, and may open new horizons in the fabrication technology of optoelectronic devices and integrated circuits (OEICs). Due to the technological importance of LPEE, a number of experimental and modeling studies have been carried out in recent years.

Alloy semiconductors, such as GaInAs, GaInSb, GaInP, and CdZnTe, grown on commercially available GaAs, GaSb, GaP, and CdTe substrates, are of interest as lattice-matched substrates for novel semiconductor devices in optoelectronics (see [29.12] for more information). For instance, $Ga_{0.47}In_{0.53}As$ ternary alloy grown epitaxially on the lattice-matched InP substrate has been used as an active layer in lasers and photodetectors in optical communication systems [29.12–14]. It is a very good candidate for high-speed transistors because of its high carrier mobility. GaInAs epitaxial layers grown on GaAs substrates have also been used for high-electron-mobility transistors (HEMT) structures with significantly improved performance, and for strained-layer lasers, modulators, and detectors operating in the near-infrared region [29.15]. For this ternary material, however, only thin layers can be grown due to the lattice mismatch. As a result, many problems have been observed in lasers fabricated on such substrates, which are therefore restricted to the 0.8–1.1 μm region [29.16].

High-performance semiconductor lasers operating in the 2–5 μm range are highly desirable in optical-fiber communication systems employing low-loss fluoride-based fibers, laser radar, remote sensing of atmospheric gases, and molecular spectroscopy [29.17–19]. The availability of InAsP, GaInAs, and GaInSb substrates with desired thicknesses and quality would overcome this difficulty. These materials and many other desired alloy semiconductors either cannot be grown commercially, or are grown with inadequate thickness and quality, or cannot be grown reproducibly. LPEE has proven to have the potential for growing such crystals with the desired properties. The alloy materials that can be grown by LPEE are summarized in Table 29.1.

Semiconductors grown by LPEE show advantages over crystals grown by melt growth techniques, namely, lack of detectable electron traps [29.20], low vacancy densities [29.21], low dislocation densities [29.22], and high luminescence efficiency [29.23]. The distinct feature of the LPEE growth process is its capability to grow crystals with uniform crystal compositions. For example, millimeter-thick ingots of GaInAs [29.23–27], AlGaSb [29.28], and AlGaAs [29.29–32] exhibit remarkable compositional uniformity.

The above-mentioned features along with its low hardware cost make LPEE quite attractive for growth of high-quality alloy semiconductors in the form of both bulk crystals and buffer layers. However, reproducible growth of such crystals requires good understanding and control of the key mechanisms governing this process.

29.1.1 Liquid-Phase Electroepitaxy

In a typical LPEE growth system, graphite electrodes are placed at the top and bottom of the growth cell (Fig. 29.1). The substrate is placed at the bottom of the solution and the source material is placed between the solution and the upper electrode. The liquid contact zone located below the substrate provides a uniform, low-resistance electrical contact between the lower face of the substrate and the lower electrode, which is essential for satisfactory growth. The boron nitride *jacket* around the horizontal sandwich layers forms the cell

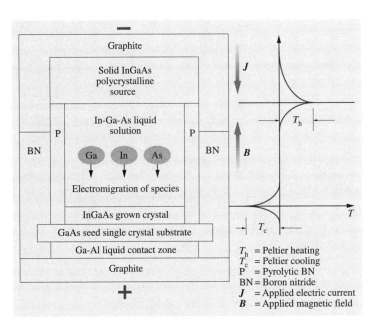

Fig. 29.1 Schematic view of an LPEE crucible for a GaInAs system. Note that the temperature distribution shown is not the actual one; it describes symbolically the Peltier cooling/heating at the interfaces

and acts as both a heat conductor and an electrical insulator. The growth temperature is typically in the range of 650–900 °C, depending on the material to be grown. After the system reaches thermodynamic equilibrium, the electric current is turned on and growth is initiated. During the growth process, the furnace temperature is kept constant. The applied electric current is the sole external driving force and the controlling element of the growth, and makes it possible to achieve a high growth rate and precise control of the process. The electric current passes through the lower electrode, contact zone, and substrate, but may bypass the source material into the upper electrode. A static magnetic field may also be used in LPEE. We discuss below some of the factors playing roles in the LPEE growth process.

Thermoelectric Effects

In a well-designed LPEE apparatus, vertical and horizontal external temperature gradients are effectively minimized. Thermal effects that may lead to temperature gradients in the solution may then be assumed to be solely due to the Peltier cooling/heating at the solid–liquid interfaces and the Joule heating in the grown crystal.

Peltier Cooling/Heating is a thermoelectric effect caused by the electric current passing across the solution–substrate (growth interface) and the substrate–contact zone interfaces. The electric current causes heat absorption or evolution at the interfaces, depending on the direction of the electric current and type of electrical conductivity of the crystal (n-type or p-type). In an equilibrated LPEE system with a positive polarity of the n-type lower electrode crystal, Peltier cooling occurs at the growth interface and is accompanied by Peltier heating at the substrate–contact zone interface. Thus, heat transport across the substrate affects the amount of cooling at the growth interface. Indeed, the amount of cooling at the growth interface increases with increasing substrate thickness. If the current passes through the solution–n-type source (dissolution) interface, Peltier heating occurs at this interface.

Being a semiconductor, the substrate has high electrical resistivity. The electric current passing through the substrate induces Joule heating proportional to the square of the current density and electrical resistivity (Joule heating produced in the solution and graphite electrodes is at least an order of magnitude lower because of the low electrical resistivity). The effect of Joule heating in the substrate increases with increasing substrate thickness and becomes significant for bulk crystals. A one-dimensional model presented in [29.32] suggests that the Joule heating may present itself as a thermal limiting factor (barrier) in the growth of very thick crystals, particularly in the growth of high-resistivity materials.

Growth Mechanisms

The main growth mechanism of LPEE is the transport mechanism known as *electromigration*. In the growth of compound and alloy semiconductors, the solutions are metallic conductors. In such solutions, electromigration takes place due to electron momentum exchange and electrostatic field forces [29.5, 9]. Under the influence of the electric field induced by the applied electric current, solute species migrate towards the anode with a velocity proportional to the solute mobility and electric field. Thus, when the substrate has positive polarity, the solution becomes supersaturated with solute near the substrate–solution interface, resulting in epitaxial growth.

The combined effect of Peltier cooling/heating and Joule heating results in an axial temperature gradient in the solution. This temperature gradient supersaturates the solution in the vicinity of the growth interface, leading to a further contribution to epitaxial growth. This is the second main growth mechanism of LPEE. Either electromigration or Peltier cooling can become dominant, depending on the particular growth conditions [29.8, 9, 33]. However, these contributions can be affected by the presence of natural convection in the solutions, as shown numerically in [29.34].

A typical growth rate in the LPEE growth of GaAs at a $3\,\text{A/cm}^2$ electric current density is about $0.5\,\text{mm/day}$. For the growth of thick crystals (several millimeters), mass transport in the liquid solution is mainly due to electromigration. The contribution of molecular diffusion is very small, as shown experimentally in [29.26, 27], and also numerically in [29.35, 36]. The growth rate increases with increasing electric current density. However, at higher electric current densities, for instance $10\,\text{A/cm}^2$ or higher, the growth becomes unstable [29.26, 27].

29.1.2 Natural Convection

The effect of natural convection has been observed in various experiments [29.26, 27, 37, 38]. It enhances the overall transport processes, and thus increases the growth rate, which is desirable. However, convection often has an adverse influence on growth kinetics, and on the structure and quality of grown crystals [29.37]. It has been observed that convective flow, resulting from both thermal and solutal gradients, leads to the growth of GaAs/GaInAs layers with nonuniform thickness profiles [29.8, 10, 26, 27]. Furthermore, convection has been found to limit the maximum achievable thickness in bulk crystal growth experiments [29.23], due to the loss of surface quality caused by unstable growth conditions [29.26, 27].

In the growth of alloy semiconductors, convection adds another dimension to the difficulty of the problem. In most alloys, the densities of the components are significantly different. This difference gives rise to inhomogeneity in the composition of the liquid solution during growth. In other words, gravity makes it difficult to maintain the solution with a uniform liquid composition. Less dense components move upwards, leading to the depletion of the required component(s) in the vicinity of the growth interface. This results in unsatisfactory growth. For example, in the case of GaInP, phosphorus, with the smallest density, tends to float.

29.1.3 Applied Magnetic Fields

In order to suppress convection, a static, external applied magnetic field has also been used in LPEE [29.26, 27]. As we will see later, the application of a vertical static magnetic field (perfectly aligned with the growth direction and the applied electric field) indeed suppresses convection significantly. However, it was also observed, unexpectedly, that the applied magnetic field increases growth rate very significantly. For instance, a field of 4.5 kG increases the growth rate about tenfold. Experimental [29.26, 27] and modeling [29.35, 36, 40, 41] studies have shown that the growth rate is also proportional to the intensity of the applied magnetic field; however its contribution to mass transport is about twice that of the applied electric current.

29.1.4 Observation of Growth Rate

One of the most significant advantages of the LPEE growth technique is the possibility of monitoring in situ the evolution of the growth interface. Although the subject of this chapter is the modeling and numerical simulations of the LPEE growth process, due to its significance we wanted to mention a great experimental technique developed in [29.74]. The technique is based on resistance measurements of the growth cell during LPEE growth. When scaled with the use of the technique of dopant modulation by current pulses (time markers) [29.1], the in situ monitoring allows the study of growth rate (averaged over the cross-section area of the crystal).

The literature on the experimental studies carried out on LPEE growth of various semiconductors is rich (for a detailed account of the subject, see [29.41]). In the following sections, we first present a short review of the early modeling and theoretical studies. Then, due to its significance, we give a short summary of the results of the recent experimental study of [29.26], carried out for LPEE growth of GaInAs under a strong static magnetic field. We also present a summary of the three-dimensional simulations carried our for LPEE growth of binary systems. Finally, a continuum model developed to predict the high growth rates observed in the LPEE growth of GaAs is introduced.

29.2 Early Theoretical and Modeling Studies

A number of conceptual/modeling studies have been carried out to have a better understanding of the relative contributions of electromigration, Peltier cooling, diffusion, and convection. These early studies contributed significantly to understanding of the LPEE growth process. Here, we present a summary of some of those early contributions.

29.2.1 Peltier-Induced Growth Kinetics: Electromigration Mechanism

An analytical treatment of the Peltier-induced growth kinetics in the LPEE growth of GaAs was presented in the absence of the effect of natural convection [29.10]. In this model, the growth rate was examined by considering diffusion and electromigration of As in a Ga-rich solution (Fig. 29.2). The mass balance for the solution and the grown layer is written as

$$\rho_L C_L(T_0) L = \rho_S C_S R(t) + \rho_L \int_R^L C(x,t) \, dx \, . \quad (29.1)$$

The distribution of solute $C(x,t)$ in the solution is expressed by writing the conservation of mass for the ionized species as

$$\frac{\partial C}{\partial t} = D_C \frac{\partial^2 C}{\partial x^2} - Z_{\text{eff}} \mu_E \frac{\partial}{\partial x}(EC) \, , \quad (29.2)$$

where

$$\mu_E = \frac{F D_C}{RT} \quad (29.3a)$$

and

$$Z_{\text{eff}} = Z - Z_0 \frac{\sigma^{-1}(C) - \sigma^{-1}(0)}{C \sigma^{-1}(0)} \, , \quad (29.3b)$$

and μ_E, F, and σ_E are the electric mobility, the Faraday number, and the conductivity of the solution, respectively, Z and Z_0 represent the valances of solute and solvent ions, Z_{eff} is the effective charge of solute species, and D_C is the diffusion coefficient. Differentiating (29.1) with respect to time and using (29.2), we obtain the growth rate as

$$V(t) = \rho_L \frac{D_C \, (\partial C/\partial x)|_{x=R} - \mu_E Z_{\text{eff}} E(R) C_L(T_1)}{\rho_S C_S - \rho_L C_L(T_1)} \, . \quad (29.4)$$

As can be seen, the calculation of the growth rate requires the electric field intensity E to be known, defined by the charge distribution in the solution and the solute concentration $C(x,t)$. Since the Ga (or Ga-rich) solution exhibits metallic properties, the participation of free electrons in the electric conductivity is much greater than that of the ionized solute (arsenic in this case) atoms. One may then assume that the electric field intensity E is constant across the solution and its value can be estimated from the expression $J = \sigma_E E$, which is simply the constitutive equation used for the electric

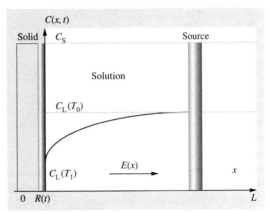

Fig. 29.2 Distribution of the solute concentration $C(x,t)$ in the solution (liquid) and solid: C_L = equilibrium composition of the solution, R = thickness of the grown layer, L = solution height, t = time, E = electric field intensity. In this setup electric current bypasses the source material (distribution of arsenic concentration in the solution after a very long time; after [29.10])

current with the conductivity σ_E of the solution [29.41]. In this case, the growth expression reduces to that given in [29.41, 63] with the assumption $E = Z_{\text{eff}} E$. Then, (29.2) reduces to the one-dimensional mass transport equation, which can be deduced from that given in [29.41, 63] as

$$\frac{\partial C}{\partial t} = D_C \frac{\partial^2 C}{\partial x^2} - \mu_E E \frac{\partial C}{\partial x}. \quad (29.5)$$

Equation (29.2) can be solved numerically under the following conditions (in the absence of Peltier effect at the source)

$$C(x, 0) = C_L(T_0),$$
$$C(R, 0) = C_L(T_1),$$
$$C(L, t) = C_L(T_0),$$
$$\left.\frac{\partial C}{\partial x}\right|_{x=L} = 0. \quad (29.6)$$

One can see that, in order to calculate the growth rate in (29.4), one must know the values (or functions) of $D_C(T)$, $C_L(T)$, $\Delta T \equiv T_1 - T_0$, $\sigma_E(C)$, and also Z_{eff}. Some of these values can be determined experimentally (such as $D_C(T)$, $\sigma_E(C)$, and Z_{eff}) and some of them (such as $C_L(T)$ and $\Delta T \equiv T_1 - T_0$) can be estimated through calculations (or numerical simulations). These values were estimated in [29.10] as follows.

The temperature difference $\Delta T \equiv T_1 - T_0$ was estimated directly by taking into account the fact that the growth of an epitaxial layer without the source on the solution surface occurs until the following condition is

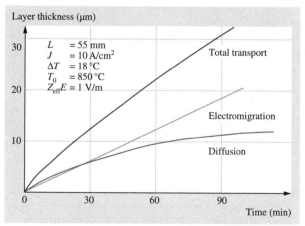

Fig. 29.4 Layer thickness versus time, and the relative contributions of electromigration and diffusion (after [29.10])

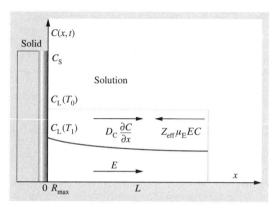

Fig. 29.3 Distribution of arsenic concentration in the solution after a very long time (after [29.10])

fulfilled (Fig. 29.3)

$$D_C \frac{\partial C}{\partial x} = Z_{\text{eff}} \mu_E E C, \quad (29.7)$$

which means that the diffusion, electromigration, and electrotransport streams balance each other. Solving (29.7) and substituting into (29.2) yields (see [29.10] for details)

$$C_L(T_1) = \frac{C_L(T_0) - (C_S \rho_S R_{\text{max}})/(\rho_L L)}{1 - (\mu_E L Z_{\text{eff}} E)/(2 D_C)}, \quad (29.8)$$

where R_{max} is the maximum thickness of the layer. Equation (29.8) and the exact knowledge of the liquidus curve $C_L(T)$ will determine $\Delta T \equiv T_1 - T_0$.

The electric conductivity of the Ga solution and the parameter Z_{eff} were estimated in [29.10] as follows. The product $E Z_{\text{eff}}$, which is linearly dependent on current density, was treated as the parameter to fit the experimental data of the growth rates obtained (see [29.10] for the experimental data) as

$$\frac{J}{Z_{\text{eff}} E} = \text{const.}, \quad \text{at } T_0 = \text{const.} \quad (29.9)$$

It follows that the fitting parameter $Z_{\text{eff}} E$ has to be proportional to the current density J. Agreement within $\pm 20\%$ between the calculated and measured values can be achieved, as suggested in [29.10] (Fig. 29.4). Actually in the growth of very thick crystals (during prolonged growth periods), as will be seen later [29.26], this agreement is very close, within a very small margin.

Once the values of $Z_{\text{eff}} E$ are known from the fitting procedure, the electrical conductivity σ_E, the effective charge Z_{eff}, and the electric field intensity E can be estimated. Using $J = \sigma_E E$, and (29.3b) we obtain

$$\sigma^{-1}(C) = \sigma^{-1}(0)\left(1 + \frac{ZC}{Z_0}\right)\frac{1+(1-\Pi)^{1/2}}{2} \quad (29.10)$$

and

$$Z_{\text{eff}} = Z - Z_0\frac{\left(1 + \frac{ZC}{Z_0}\right)\left[1+(1-\Pi)^{1/2}\right]-2}{2C} \quad (29.11a)$$

with

$$\Pi \equiv \frac{4Z_{\text{eff}}EC}{Z_0\sigma^{-1}(0)J\left(1+\frac{ZC}{Z_0}\right)^2}. \quad (29.11b)$$

Using (29.10) and (29.11a,b) the calculated values of Z_{eff} are tabulated in Table 29.2.

These results showed that mass transport in the LPEE growth of GaAs is due to both diffusion and electromigration of solute (As) in the liquid towards the seed substrate. The high value of the effective charge, estimated by fitting the theoretical values to those measured ones, justifies the conclusion that *the migration of arsenic species in the liquid solution is realized mainly due to collisions with electrons flowing across the solution.*

Indeed, this conclusion given in [29.10] for the mechanism of electromigration established the understanding of the electromigration process. As we will discuss it later, the high growth rates achieved under an applied magnetic field in the LPEE growth of GaAs [29.26] can only be explained by such a mechanism, as suggested in [29.61–63]:

The resistance during the collision of electrons with the species of the solution determines the mobility of the species and in turn the growth rate. This resistance is very much reduced under an applied static magnetic field due to the possibility that the charged species are aligned along the magnetic field lines

Table 29.2 Estimated values of electrical conductivity and effective charge [29.10]

T_0	$C_L(T_0)$	$(Z_{\text{eff}}E/J)$ $\times 10^3$	$\sigma^{-1}(C)$ $\times 10^5$	Z_{eff} (As^{3-})
(°C)	(at. %)	(Ω cm)	(Ω cm)	
750	1.38	-1.2	5	-24
800	2.23	-1.2	5.2	-23
850	2.75	-1.2	5.5	-22

(which are almost uniform along the growth direction in the growth system used in [29.26]), and leads to a very high mobility.

Thus, the mass transport due to electromigration increases tremendously and the total mobility of species depends not only on the electric mobility coefficient which is measured in the absence of an applied magnetic field, but also on the magnetic field intensity and a new material coefficient that is called the *magnetic mobility* in [29.61–63]. This understanding supports the earlier definition of electromigration in LPEE.

29.2.2 A One-Dimensional Model

Based on the conservation of species mass, a one-dimensional model for the LPEE growth process was given in [29.8, 9], where the contributions of the Peltier effect at the solid–solution interfaces and that of solute electromigration in the solution to the overall growth process were defined. According to this one-dimensional model, the contribution of electromigration to growth is dominant in the absence of convection, and the contribution of the Peltier effect can be dominant in the presence of convection. (The relative contributions of electromigration and the Peltier effect have been determined more clearly through two-dimensional numerical simulation models developed by the author; the reader is referred to [29.34, 48–50] for details.)

The one-dimensional mass transport equation was written as

$$\frac{\partial C}{\partial t} + V^g\frac{\partial C}{\partial x} = D_C\frac{\partial^2 C}{\partial x^2} - \mu_E E\frac{\partial C}{\partial x}. \quad (29.12)$$

The following boundary conditions were considered

$$-D_C\left.\frac{\partial C}{\partial x}\right|_0 + \mu_E E C_L = (C_S - C_L)V^g, \quad (29.13)$$

$C = C_0$ at $t = 0$ for all x, and
    at $t > 0$ for $x = \infty$ (absence of convection), or
$C = C_0$ at $t > 0$ for $x > \delta$ (presence of convection),
$C = C_L$ at $t > 0$ and $x = 0$ (growth follows the liquidus line).

Equation (29.12) was solved analytically with the assumption of a small growth velocity so that the second term on the left-hand side of (29.12) was neglected. The growth velocity was obtained for two cases: (i) for an infinitely long boundary layer ($\delta = \infty$, no convection), and (ii) for a finite boundary layer (with convection), which are given respectively as

$$V_T^g = \frac{\Delta T_p}{C_S - C_L} \frac{dC}{dT}\bigg|_L \left(\frac{D_C}{\pi t}\right)^{1/2} + \mu_E E \frac{C_L}{C_S - C_L} \quad (29.14)$$

and

$$V_T^g = \frac{\Delta T_p}{C_S - C_L} \frac{dC}{dT}\bigg|_L \frac{D_C}{\delta} + \mu_E E \frac{C_L}{C_S - C_L}. \quad (29.15)$$

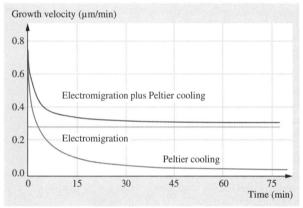

Fig. 29.5 Growth velocity of GaAs calculated from (29.14) (after [29.8, 9])

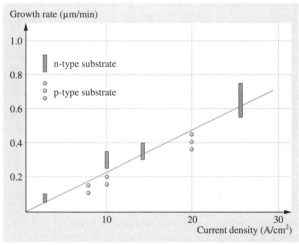

Fig. 29.6 Growth rate versus electric current density: 800 °C with a substrate thickness of 300 μm (after [29.8, 9])

In (29.14) and (29.15) the first terms represent the contribution of the Peltier effect and the second terms represent the contribution of electromigration to the growth velocity. Both terms depend linearly on the electric current through ΔT_p and E, respectively. Equation (29.14) was solved for LPEE growth of GaAs from a Ga/As solution at 800 °C growth temperature and using $J = 10\,\text{A}/\text{cm}^2$, $\Delta T_p = -3\,°\text{C}$, $D_C = 6.0 \times 10^{-5}\,\text{cm}^2/\text{s}$, and $\mu_E E = 10^{-5}\,\text{cm}/\text{s}$. The results are plotted in Fig. 29.5.

Equation (29.15) was also solved for various boundary layer thicknesses. It was shown that, as the boundary layer gets thinner, the contribution of Peltier cooling to the growth rate increases. The contribution of Peltier cooling to the growth rate also increases with increasing Peltier cooling at the growth interface (see [29.8, 9] for details).

According to the model presented above the relative contributions of Peltier cooling at the growth interface and electromigration of solute species in the liquid towards the interface depend on the experimental conditions. On the basis of this model, a quantitative criterion was given in [29.8, 9] to determine the relative contributions of the Peltier effect and electromigration. A critical substrate thickness d_c was defined, for which the contributions of the Peltier effect and electromigration are equal. When the substrate thickness is greater than the critical thickness, the Peltier effect dominates, and when the thickness is smaller than the critical thickness, electromigration dominates. Their conclusion then was that, in the growth of thick crystals and in the presence of significant convection, electromigration is dominated by Peltier cooling. On the other hand, with thin substrates and in the absence of significant convection, electromigration dominates electroepitaxial growth.

In order to verify the predictions of the one-dimensional model, specific LPEE experiments were devised in [29.8, 9]. Under specific growth conditions, using n- and p-type seed substrates, it was shown that the growth in the case of the p-type substrate was smaller than in the case of the n-type substrate. From the difference in growth rates, it was concluded that the contribution of Peltier cooling to the growth rate is less than 15%. The almost linear relationship between growth rate and electric current density observed experimentally is consistent with theoretical predictions (Fig. 29.6).

It was shown experimentally that the growth rate is also proportional (almost linearly) to current density for all growth temperatures used, and that, for a given current density, the growth rate increases with temperature. Results are summarized in Fig. 29.7 for LPEE growth of GaAs on a 0.3 mm thick n-type substrate.

Another important issue is the selection of a proper solution height when an LPEE crucible is designed since it influences the growth rate, among of course other growth parameters. LPEE experiments designed for the growth of GaAs at 800 °C using an n-type substrate showed that growth rate is linearly proportional to solution height up to a certain height (about 10 mm), above which the growth rate remains constant. The results of experimental measurements in [29.8, 9] are presented in Fig. 29.8. Naturally, the relative contribution of the Peltier effect to the growth rate varies with solution height due to the contribution of convection in the solution. Experiments carried out in [29.8, 9] (at 25 A/cm^2 electric current density and 900 °C) show that the contribution of Peltier cooling increases with solution height (Fig. 29.9).

29.2.3 Source-Current-Controlled (SCC) Growth

The source-current-controlled (SCC) method is a version of LPEE with the difference that the furnace temperature is lowered gradually during growth similar to in liquid-phase epitaxy (LPE) while the electric current is simultaneously used to generate a temperature gradient and supply additional solute to the

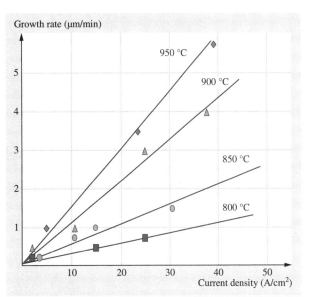

Fig. 29.7 Growth rate versus current density at various growth temperatures (after [29.8, 9])

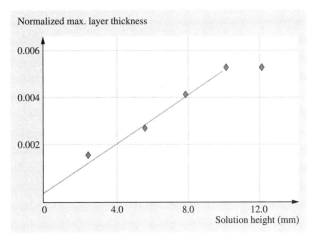

Fig. 29.8 Normalized maximum GaAs layer thickness as a function of solution height under the experimental conditions of 800 °C growth temperature, 0.3 mm thick n-type substrate (after [29.8, 9])

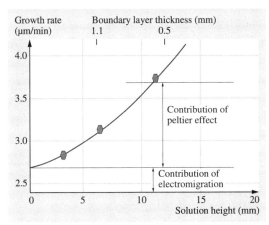

Fig. 29.9 Growth rate of GaAs from a Ga/As solution (at 900 °C and 25 A/cm^2) as a function of solution height. Substrates were Cr-doped and 0.3 mm thick. The estimated boundary layer thicknesses are also indicted at the *top* (after [29.8, 9])

liquid [29.64–67]. In this setup the applied electric current passes through the source, the solution, and then through the graphite on top, but bypasses the substrate (Fig. 29.10). Due to the passage of electric current through the source the temperature T_1 just above the

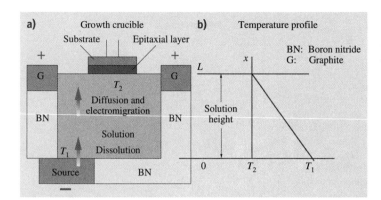

Fig. 29.10a,b Schematic view of the SCC (LPEE) growth crucible (a) domain and temperature profile (b) used in the model (after [29.64])

source is higher than that (T_2) just below the substrate. This temperature difference generates a temperature gradient between the source and the substrate. For this system, the thickness of the grown layer is calculated by solving the one-dimensional mass transport equation given in (29.5) (where we assume the mobility to be positive for current flow from source to substrate, opposite to the sign convection used in [29.64]: $\mu_E \equiv -\mu$) under the boundary conditions

$$C(0,t) = C_0 - (R_0/m_0)t,$$
$$C(L,t) = C_L - (R_L/m_L)t,$$
$$C(x,0) = C_0 + (C_L - C_0)(x/L), \quad (29.16)$$

where C_0 and C_L are the initial solution concentration in the vicinity of the growth interface and the source, respectively, and R_0 and R_L represent, respectively, the cooling rates at the growth and dissolution interfaces. The solution height is denoted by L, and m_0 and m_L are the slopes of the liquidus curves near the growth and dissolution interfaces, given by

$$m_0 \equiv \frac{1}{w_L} \left.\frac{dT}{dX^L}\right|_{x=0},$$

and

$$m_L \equiv \frac{1}{w_L} \left.\frac{dT}{dX^L}\right|_{x=L}, \quad (29.17)$$

where X^L represents the atomic fraction of the solute, and w_L is the solution density. Through an approximate analytical solution to (29.5), an expression for the layer thickness was given as

$$L = M_t/C_S, \quad (29.18)$$

where

$$M_t = D_C \left(\frac{\mu E}{2D_C} - \frac{1}{L}\right)\left(C_0 - \frac{R_0 t}{2m_0}\right)t$$
$$+ \frac{D_C}{L} \exp\left(\frac{\mu E}{2D_C}\right)\left(C_L - \frac{R_L t}{2m_L}\right)t$$
$$- \left[\frac{\pi}{L}\sum_{n=1}^{\infty}\frac{nT_n(0)}{A} + \frac{2}{L}\sum_{n=1}^{\infty}\frac{BC_L - C_0}{A}\right]$$
$$\times \left[\exp(-D_C A t) - 1\right]$$
$$+ \left(\frac{\pi}{L^3}\sum_{n=1}^{\infty}\frac{2n^2\pi}{A^2}\right)\left(\frac{R_0}{m_0} - B\frac{R_L}{m_L}\right)$$
$$\times \left[t + \frac{1}{D_C A}\exp(-D_C A t) - \frac{1}{D_C A}\right] \quad (29.19)$$

and C_S is the concentration of C atoms per unit volume of the grown layer. For an $A_xB_{1-x}C$ system, C_S is given by $C_S = 4/d^3$, where d is the lattice constant of the ternary compound. The layer thickness for diffusion- and electromigration-limited growth in a temperature-graded solution can be calculated using (29.18) and (29.19) [29.64].

The thickness of the grown layers of an $In_{0.53}Ga_{0.47}As$ system was calculated in [29.64] and various numerical data were given for the dependence of concentration on growth time, layer thickness as a function of temperature and cooling rate, the relative contributions of electromigration and diffusion to growth rate, and the composition of the grown crystal along the growth direction. These calculations were made using the value $\mu E = 0.1 - 0.001$ cm/min for the mobility of As in the Ga/As solution. It was stated in [29.64] that *the real value of μ can be determined through a comparison of experimental layer thicknesses with computed ones*. Indeed, as shown in [29.26, 62, 63] the mobility was calculated using experimental results of LPEE of a GaInAs system with and without the application of an external magnetic field.

The above analysis was extended in [29.68, 69] to a $Ga_xIn_{1-x}As$ system to determine the composition variation in the grown crystals. In this model, for the first time, phase equilibrium between the crystal and the solution is maintained while having consistency between the transported and incorporated mass or solute atoms at the growth interface. The comparison of experiments with computed results revealed that in the Ga/In/As solution the diffusion coefficient of Ga is about twice that of As, and the electric mobility of Ga is larger than that of As. The above results were instrumental in the development of future two-dimensional (2-D) and three-dimensional (3-D) numerical models and simulations for such systems.

29.3 Two-Dimensional Continuum Models

As mentioned earlier, the growth process of LPEE is quite complex and involves the interactions of various thermomechanical and electromagnetic fields. These include fluid flow, heat and mass transfer, electric and magnetic fields, various thermoelectric effects and their interactions in the liquid phase, and heat and electric conduction with various thermoelectric effects in the solid phase. In addition, the moving growth and dissolution interfaces with possible finite mass transport rates complicate the process further. One-dimensional and simple models are not sufficient for a full understanding of the various aspects of the LPEE growth process. To gain a better understanding of the growth process of LPEE, over the past 10 years a number of numerical simulation modeling studies have been carried out by the author's research group [29.34–36, 39–41, 46, 48–50, 61–63, 70–73].

For the first time, a two-dimensional computer simulation model for the LPEE growth process of GaAs was introduced in [29.70]. This simulation model was based on the rational mathematical model given for a binary system in [29.71]. The objective of this diffusion model was to examine the relative contributions of electromigration and Peltier cooling without introducing the additional complexity of natural convection. The model includes heat transfer, diffusive mass transport, electromigration, and Peltier and Joule effects. The governing equations are solved numerically using a finite volume method. Simulations are presented for three different growth cell configurations to investigate: (i) temperature and concentration distribution in the growth cell, (ii) the effect of applied electric current density and substrate thickness, and (iii) the contribution of electromigration and Peltier cooling to the overall growth rate.

Based on the simplifying assumptions used in this work, the simulation results showed that the magnitude of the relative temperature at the growing interface is controlled mainly by Peltier cooling for thin substrates (less than 2 mm) and small electric current densities (less than 20A/cm^2). As expected, Joule heating becomes significant only for thick substrates and high electric current densities. For all configurations investigated, electromigration is found to be the dominant growth mechanism. In critical regions of the growth cell, relatively small changes in the configuration are found to have a significant impact on the process, and on the degree of nonuniformity of the grown crystal.

This diffusion model has been followed by a number of numerical simulation models specific to particular growth cell configuration. As an extension of the diffusion model in [29.70], the effect of natural convection in LPEE was introduced for the first time in [29.34]. In this work, the effect of thermosolutal convection in LPEE growth of GaAs was investigated through a two-dimensional numerical simulation model that accounts for heat transfer and electric current distribution with Peltier and Joule effects, diffusive and convective mass transport including the effect of electromigration, and fluid flow coupled with temperature and concentration fields. Simulations were performed for two growth cell configurations and the results are analyzed to determine growth rates, substrate shape evolution, and relative contributions of Peltier cooling and electromigration. The simulations predicted and helped explain a number of experimentally observed features, which previous diffusion-based models failed to reproduce. As we will see later in detail, in general, electromigration is found to be the dominant growth mechanism, but the contribution of Peltier cooling to the overall growth rate is found to be significantly enhanced by thermosolutal convection in the solution, and Peltier cooling can in fact become the dominant growth mechanism for certain growth conditions and growth cell configuration. The overall growth rate is found to increase with increasing furnace temperature and applied electric current density. This thermosolutal convection model predicts an increased nonuniformity of the grown layers compared with the pure diffusion model in [29.70]. The shape of the grown layers is also shown to be very sensitive to changes in growth cell configuration.

The LPEE growth of ternary alloys was numerically simulated in [29.50] using a diffusion model for the growth of AlGaAs, and in [29.73] including the effect of convection in growth of GaInAs. The growth of $Al_xGa_{1-x}As$ is considered from a Ga-rich solution. The solution is dilute, with 0.345% Al and 2.445% As. The growth temperature is 850 °C, the electric current density is $10 A/cm^2$, and the solid composition is up to $x = 0.4$. Simulations are carried out in the absence of convection [29.50].

In the simulations carried out for LPEE growth of $Ga_xIn_{1-x}As$ ($x = 0.94$), the computational domain covered one-half of the crucible geometry, assuming symmetry of the geometric and boundary conditions with respect to the y-axis [29.73]. The initial temperature was chosen as 780 °C, corresponding to the experimental growth conditions. The solution compositions, determined from the Ga/In/As phase diagram for growth at 780 °C, were assumed to be uniformly distributed in the liquid zone. In this experimental setup the Ga/In/As solution is nondilute, with 29.85% Ga and 7.07% As.

Two-dimensional simulation models have also been developed for LPEE growth of crystals under magnetic fields [29.72]. The feasibility of using a magnetic field for suppressing convection in LPEE was first studied through a model for the growth of GaAs in [29.49]. The LPEE growth process of GaAs was numerically simulated under an applied static magnetic field using the finite element method based on the penalty function formulation [29.49]. It was found that this formulation is more robust and efficient than a mixed velocity–pressure formulation. The results of these early simulations have shown that the effect of natural convection can be reduced significantly by the application of an external static magnetic field. For the crucible selected, the uniformity of grown layers improves with increasing magnitude of the applied magnetic field. Results also showed that complete elimination of natural convection in the liquid phase requires the application of a very large magnetic field (about and more than 20 kG). However, this prediction is numerical, based on the model assumptions and simplifications made, and as we will see later in growth of bulk crystals by LPEE, the strong interaction between the applied magnetic and electric fields does not allow the use of a magnetic field intensity above a critical (maximum) value. For instance, this value was about 4.5 kG in LPEE growth of GaAs in experiments [29.26], and growth at higher field intensities led to unstable growth.

However, this early simulation model in [29.49] shed light on a number of issues in LPEE growth under magnetic field and laid the foundation for further experimental and numerical simulation studies in LPEE. It also provided a valuable insight into the use of the finite element technique for future studies. (A comprehensive review of the two-dimensional simulations carried out for LPEE can be found in [29.41].)

29.4 LPEE Growth Under a Stationary Magnetic Field

In spite of many significant advantages, LPEE has suffered historically from mainly three *shortfalls* towards its commercialization. The first is the achievable crystal thickness, which is relatively small, on the order of a few millimeters. This is mainly due to the combined effect of Peltier and Joule heating in the system, leading to higher temperature gradients and relatively strong natural convection in the liquid solution zone that cause unsatisfactory and unstable growth. This puts a limit on the achievable crystal thickness, particularly in the growth of bulk crystals, and providing less useful material. The second issue in LPEE has been its low growth rate. The growth rate in LPEE is almost linearly proportional to the applied electric current density, and increases with increasing electric current density. For instance, it is about 0.5 mm/day at $3 A/cm^2$. For higher electric current density levels, the growth rate will increase, but in the growth of thick (bulk) crystals the combined effect of temperature gradients and natural convection may lead to unstable growth. The third shortfall is the need for a single-crystal seed of the same composition as the crystal to be grown. Small compositional differences, on the order of 4% depending on the crystal lattice parameters, can be tolerated, but higher compositional differences may lead to unsatisfactory growth.

The first two shortfalls of LPEE have recently been addressed in [29.26]. By optimizing the growth parameters of LPEE, and also using a static external applied magnetic field, a number of thick, flat GaAs single crystals and $In_{0.04}Ga_{0.96}As$ single crystals of uniform compositions were grown, and the growth rate of LPEE was increased more than tenfold for a selected electric current density. The crystals grown under magnetic field or no magnetic field were all single crystals, and the results were reproducible in terms of crystal thickness, growth rate, and compositional uniformity. The addressing of the third shortfall has been attempted in [29.42–45] by utilizing the liquid-phase diffusion (LPD) technique.

A comprehensive experimental study of LPEE growth of GaAs and $Ga_{0.96}In_{0.04}As$ single crystals has been carried using the facility in the Crystal Growth Laboratory (CGL) of the University of Victoria. The LPEE experiments under no magnetic field have led to the growth of a large number of GaAs and $Ga_{0.96}In_{0.04}As$ single crystals of thicknesses up to 9 mm. It was possible to apply electric current densities of 3, 5, and 7 A/cm². The corresponding growth rates in these experiments were about 0.57, 0.75, and 1.25 mm/day, respectively. Growth interfaces were very flat, and the growth experiments were reproducible in terms of crystal thickness and growth rate. Experiments at higher electric current intensities were not successful.

Experiments at the 3, 5, and 7 A/cm² electric current density levels were repeated under various applied static magnetic field levels, starting at 3 A/cm² electric current density and 20 kG magnetic field level (based on an earlier initial numerical estimation in [29.49]). The LPEE experiments at the 4.5 kG and lower magnetic field levels were successful, but those under higher magnetic field levels were not. These experiments indicate that for the LPEE system used in [29.26] the 4.5 kG field level is the maximum (*critical*) field intensity above which the growth is not stable. The numerical simulations conducted under the same conditions yielded a lower *critical* magnetic field level of 2.0–3.0 kG [29.35]. In addition, the experimental LPEE growth rates under a magnetic field are much higher than those under no magnetic field. For instance, as we will see later, the growth rate at 4.5 kG magnetic field level was about ten times higher than that under no magnetic field (at $J = 3 A/cm^2$). The experiments performed at the $B = 1.0$ and 2.0 kG field levels (at $J = 3 A/cm^2$) were also successful, and the growth rates were also higher at 1.62 and 2.35 mm/day, respectively. Experiments showed that the application of an external magnetic field is very beneficial in increasing the growth rate in LPEE to a level competitive with other bulk crystal growth techniques.

At the higher magnetic field levels (even at the $J = 3 A/cm^2$ electric current density level), and the higher electric current density levels ($J = 10 A/cm^2$ or higher), the experiments did not lead to successful growth, but showed very interesting outcomes. Although very thick crystals were grown, even up to a 9 mm thickness, the growth processes were unstable and led to uneven grown crystals. From visual inspection of the grown crystals, the adverse effects of natural convection and strong electromagnetic interactions were obvious, causing either one-sided growth or, particularly in four experiments, to holes in the grown crystals [29.27]. It was considered that such growth (one-sided and with holes) is because of strong and unstable convection in the liquid zones (solution and contact zones) due to the strong interaction between the magnetic field and the applied electric current. Such predictions were confirmed qualitatively by numerical simulations [29.35, 36].

29.4.1 Experiments

In the initial LPEE experiments, a 20 kG magnetic field level was used based on the predictions in [29.49]. It was realized that this level of magnetic field was very strong, and the growth results were not successful. The applied magnetic field intensity was gradually lowered, and successful growth was first achieved at 4.5 kG.

Fig. 29.11a,b Two sample GaAs crystals grown without magnetic field. (**a**) All the material put into the solution well was depleted. (**b**) The growth has been stopped before the depletion of the solution in order to measure the growth rate accurately, and the upper part is the secondary growth during cooling (after [29.26])

A number of experiments were performed at this level. Experiments were also performed at 1.0 and 2.0 kG field levels to study the effect of the applied magnetic field on the growth rate. Details of the experimental results can be found in [29.26].

In the LPEE setup used, the grown crystal and the liquid solution could not be separated at the end of an experiment. Therefore, a secondary growth (like an LPE growth) occurs during the cooling period in the experiments that were stopped earlier (before depleting all the solution put in the growth well). Some experiments were stopped before depleting the solution for the purpose of determining the growth rate accurately. Most of the crystals were grown under low current density levels (3, 5, and 7 A/cm^2). Two sample GaAs crystals are shown in Fig. 29.11.

Attempts at these electric current density levels failed. This may be attributed to the strong convection

Fig. 29.13 Sample GaAs crystal grown at the $J = 3$ A/cm^2 electric current density and $B = 4.5$ kG magnetic field level (after [29.26])

in the solution at these levels of applied current during the prolonged growth periods in growth of *bulk* crystals. It must be mentioned that it is possible to use higher electric current densities (higher than 7 A/cm^2) in the growth of thin layers, as reported in the literature (e.g., [29.47]).

Experiments were conducted for various growth periods (from 1 to 8 days) at a temperature of 800 °C. The results are summarized in Fig. 29.12. As can be seen, the average thickness of the grown crystals is proportional to the applied electric current density. As expected, the thickness increases with time. It was concluded previously that the growth rate in LPEE is a linear function of the applied electric current density [29.12, 23]. The results shown in Fig. 29.12a indicate that this is approximately the case at the low electric current density levels, but that there is slight deviation from linearity (with a higher rate) at the higher electric current density levels. This deviation can be attributed to the enhanced natural convection in the solution. The increase of electric current density up to 7 A/cm^2 in the growth of *bulk* crystals by LPEE has been a significant contribution. This was possible due to the novel design features of the LPEE setup in [29.26].

Crystals were grown at 3 A/cm^2, without magnetic field, for various periods of time to determine the growth rate. Crystals were grown with thicknesses between 1.5 and 4.5 mm, with an average growth rate of 0.57 mm/day, as shown in Fig. 29.12b. In one of the experiments a thickness of 4.5 mm was achieved in 8 days. A number of crystals were also successfully grown under the electric current densities of 5 and 7 A/cm^2. The growth rates were 0.75 mm/day (2.25 mm thickness in 3 days) and 1.25 mm/day (3.75 mm thickness in 3 days), respectively. It must be mentioned that some of the growth thicknesses were not representative for calculating the growth rate since in such experiments the source material put into the well was completely depleted before stopping the growth (Fig. 29.12a). Those

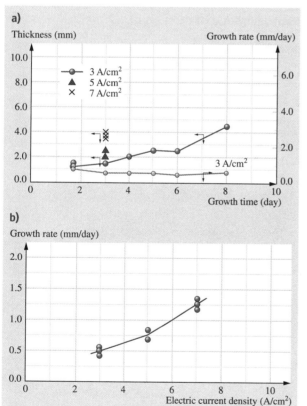

Fig. 29.12a,b Summary of the LPEE growth rates (b) and thickness (a) in the absence of applied magnetic field (after [29.26])

experiments were not included in the calculation of growth rates.

As mentioned earlier, in order to suppress the natural convection in the liquid solution for the purpose of prolonging and stabilizing the LPEE growth process for growing bulk single crystals (thicker crystals), an applied static magnetic field was used. A sample grown GaAs crystal is shown in Fig. 29.13.

The static applied magnetic field induces a magnetic body force acting on the moving particles of the liquid solution. The combined effect of the magnetic and gravitational body forces suppresses convection and prolongs growth. This beneficial effect of an applied magnetic field was the initial intention of the research program at UVic CGL. This goal was successfully achieved, by growing thick single crystals. However, the unexpected effect (a very *positive* effect of course) of the applied magnetic (at 4.5 kG and lower field levels) was the significant increase in the growth rate (about ten times at the 4.5 kG level, Fig. 29.14). The average growth rates were calculated at specific electric current density levels based on the selected representative experiments, those that were stopped deliberately before depleting the source material. This increase in the growth rate was almost the same at each of three electric current density levels, namely about 6.1, 7.8, and 10.5 mm/day at $J = 3$, 5, and 7 A/cm². Such a drastic increase in the growth rate elevates the LPEE growth process to the category of a bulk growth. Growth rates at the $B = 1.0$ and 2.0 kG levels were also higher, about 1.62 and 2.35 mm/day, respectively. Results show that the growth rate is also proportional to the applied magnetic field level (Fig. 29.15).

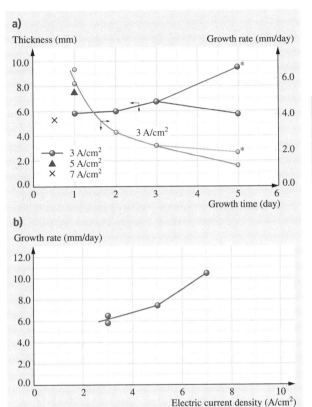

Fig. 29.14 Summary of the growth rates at the $B = 4.5$ kG magnetic field level (after [29.26])

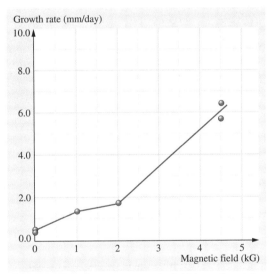

Fig. 29.15 Growth versus magnetic field intensity at $J = 3$ A/cm² (after [29.26])

29.5 Three-Dimensional Simulations

The earlier two-dimensional simulation results have shown that the level of applied magnetic field has a significant effect on natural convection in the solution, both in intensity and structure [29.46]. A stronger applied magnetic field leads to weaker convection in the solution, and more uniform interfaces. These two-dimensional numerical simulations have shed light on various aspects of the applied magnetic field in LPEE. However, these models, being two dimensional, have naturally neglected the contributions of the circumferential velocity and magnetic body force components. Therefore, the results presented earlier were qualitative and did not include information about three-dimensional effects such as deviations from axisymmetry, variations of the magnetic field components, and mixing in the solution.

In order to shed light on these three-dimensional effects, three-dimensional models were developed in [29.35, 36] for the LPEE growth of a binary system. To the best of our knowledge, these were the first three-dimensional simulation models developed for LPEE. Here, we present the key features of these models.

29.5.1 Simulation Model

The three-dimensional simulation studies carried out in [29.35, 36] were focused on the growth of a binary system (GaAs) for computational simplicity. In addition, the inclusion of the third component (In) in the analysis would not have a significant effect on the flow structures of the solution. The effect of magnetic field nonuniformity was also investigated. Below we present the model and the simulation results of [29.35, 36].

The growth cell selected for simulation is shown in Fig. 29.1. The governing equations are written explicitly in cylindrical coordinates for a binary system as follows

Continuity

$$\frac{1}{r}\frac{\partial}{\partial r}(ru) + \frac{1}{r}\frac{\partial v}{\partial \varphi} + \frac{\partial w}{\partial z} = 0. \tag{29.20}$$

Momentum

$$\frac{\partial u}{\partial t} + u\frac{\partial u}{\partial r} + \frac{v}{r}\frac{\partial u}{\partial \varphi} + w\frac{\partial u}{\partial z} - \frac{v^2}{r}$$
$$= \nu\left(\nabla^2 u - \frac{u}{r^2} - \frac{2}{r^2}\frac{\partial v}{\partial \varphi}\right) - \frac{1}{\rho_L}\frac{\partial p}{\partial r} - \frac{\sigma_E B^2 u}{\rho_L}, \tag{29.21}$$

$$\frac{\partial v}{\partial t} + u\frac{\partial v}{\partial r} + \frac{v}{r}\frac{\partial v}{\partial \varphi} + w\frac{\partial v}{\partial z} + \frac{uv}{r}$$
$$= \nu\left(\nabla^2 v - \frac{v}{r^2} + \frac{2}{r^2}\frac{\partial u}{\partial \varphi}\right) - \frac{1}{r\rho_L}\frac{\partial p}{\partial \varphi} - \frac{\sigma_E B^2 v}{\rho_L}, \tag{29.22}$$

$$\frac{\partial w}{\partial t} + u\frac{\partial w}{\partial r} + \frac{v}{r}\frac{\partial w}{\partial \varphi} + w\frac{\partial w}{\partial z}$$
$$= \nu\nabla^2 w - \frac{1}{\rho_L}\frac{\partial p}{\partial z} - g\beta_t(T-T_0) + g\beta_c(C-C_0). \tag{29.23}$$

Mass transport

$$\frac{\partial C}{\partial t} + (u+\mu_E E_r)\frac{\partial C}{\partial r} + \frac{v}{r}\frac{\partial C}{\partial \varphi}$$
$$+ (w+\mu_E E_z)\frac{\partial C}{\partial z} = D_C\nabla^2 C. \tag{29.24}$$

Energy

$$\frac{\partial T}{\partial t} + u\frac{\partial T}{\partial r} + \frac{v}{r}\frac{\partial T}{\partial \varphi} + w\frac{\partial T}{\partial z} = \alpha\nabla^2 T. \tag{29.25}$$

Electric charge balance

$$\frac{\partial^2 \phi}{\partial r^2} + \frac{1}{r}\frac{\partial \phi}{\partial r} + \frac{\partial^2 \phi}{\partial z} = 0, \tag{29.26}$$

where the gradient operator is defined as

$$\nabla^2 = \frac{1}{r}\frac{\partial}{\partial r}\left(r\frac{\partial}{\partial r}\right) + \frac{1}{r^2}\frac{\partial^2}{\partial \varphi^2} + \frac{\partial^2}{\partial z^2}. \tag{29.27}$$

The associated boundary and interface conditions are given as follows. Along the vertical wall

$$u=0, \quad v=0, \quad w=0, \quad T=T_g - \frac{z-z_0}{H}\Delta T,$$
$$\frac{\partial \phi}{\partial r}=0, \quad \frac{\partial C}{\partial r}=0. \tag{29.28}$$

Along the growth interface

$$u=0, \quad v=0, \quad w=0,$$
$$k_S\frac{\partial T}{\partial z} - k_L\frac{\partial T}{\partial z} = -\pi J,$$
$$-\sigma_E\frac{\partial \phi}{\partial r} = J, \quad C=C_1. \tag{29.29}$$

Along the dissolution interface

$$u=0, \quad v=0, \quad w=0,$$
$$k_S\frac{\partial T}{\partial z} - k_L\frac{\partial T}{\partial z} = +\pi J, \quad \phi=0, \quad C=C_2. \tag{29.30}$$

Initial conditions at $t=0$

$$C=C_0, \quad u=0, \quad v=0, \quad w=0, \quad T=T_g, \tag{29.31}$$

where u, v, and w are the velocity components in the r-, φ-, and z-directions, respectively, T is temperature, C is solute concentration, T_0 and C_0 are the reference values, H is solution height, and C_1 and C_2 are the concentrations on the growth and dissolution interfaces, respectively. Definitions of other symbols used here are given in Tables 29.3 and 29.4.

Table 29.3 Physical properties of the GaAs system [29.34]

Parameter	Value
Growth temperature T_g	800 °C
Current density J	10 A/cm^2
Peltier coefficient π	0.3 V
Solution electric conductivity σ_E	25 000 Ω^{-1} cm^{-1}
Solid (crystal and source) electric conductivity σ_S	40 Ω^{-1} cm^{-1}
Thermal diffusivity α	0.30 cm^2/s
Solutal diffusion coefficient D_C	4.0×10^{-5} cm^2/s
Solution kinematic viscosity ν	1.21×10^{-3} cm^2/s
Solution density ρ^L	5.63 g/cm^3
Solute electric mobility μ_E	0.027 cm^2/V s
Thermal expansion coefficient β_t	9.85×10^{-5} K^{-1}
Solutal expansion coefficient β_c	-8.4×10^{-2}
Thermal conductivities k	
Graphite	0.225 W/cm K
Ga-As solution and contact zone	0.526 W/cm K
GaAs substrate	0.082 W/cm K
Boron nitride: the r-direction	0.282 W/cm K
Boron nitride: the z-direction	0.440 W/cm K
Crystal radius	6.0 mm
Furnace radius	36.0 mm
Graphite height	19.0 mm
Solution height	6.0 mm
Substrate thickness	0.3 mm
Source thickness	5.0 mm
Contact zone height	2.0 mm

Table 29.4 Parameters of the GaAs system [29.35]

Parameter	Symbol	Value
Crystal radius	R_c	12.0 mm
Solution height	H	10.3 mm
Magnetic field intensity	B	0–12 kG
Nonuniformity coefficient	A	0–4
Thermal Grashof number	Gr_T	6.84×10^4
Solutal Grashof number	Gr_C	7.10×10^3
Hartmann number	Ha	0–871.5
Prandtl number	Pr	4.00×10^{-3}

flow field only, the evolution of the growth and dissolution interfaces is not included in the computations. The mass transport equation in the solution is solved simultaneously in order to take into account the influence of concentration field on the flow field (through the solutal Grashof number); however, the concentration field is not presented here for the sake of brevity. Transient terms are considered in the energy, mass transport, and momentum equations in order to account for possible unsteady flows and their influence. The simulation results are presented at $t = 1$ h since the flow field has fully developed by that time.

The required physical and growth parameters of the GaAs system are given in Tables 29.3 and 29.4. The Hartmann number is defined by

$$Ha = BH\sqrt{\sigma_E/\rho_L \nu} \,. \tag{29.32}$$

The simulations are carried out for half of the cylindrical cell domain for computational efficiency. However, to ensure that the half-domain solution represents fully the flow structure of the full domain, a full-domain solution was carried out for a case for which asymmetry was assured (under a large magnetic field, $B = 4$ kG). Results demonstrated that the half-plane treatment is reliable.

29.5.2 Numerical Method

The commercial CFX software was used to solve the field equations. The computation mesh in the liquid is $120 \times 40 \times 80$ in the r-, φ-, and z-directions, respectively, which was demonstrated to be sufficient for an accurate and stable solution. Since the focus is on the

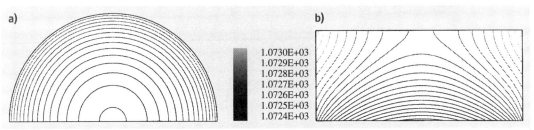

Fig. 29.16a,b Temperature distribution (K): (a) near the growth interface in the horizontal plane, (b) in the vertical plane at $\varphi = 0$ (after [29.35, 36])

Computed temperature distributions in the horizontal plane near the growth interface and also in the vertical plane at $\varphi = 0$ are given in Fig. 29.16. Temperature distributions agree with earlier 2-D solutions [29.34, 49, 50]. In addition, since the electric current is passing through the source in this setup, the computed isotherm patterns indicate that the shape of the growth interface will be single-humped (concave towards the crystal), as expected.

29.5.3 Effect of Magnetic Field Strength

For a better visualization, the simulation results for the flow field in the growth cell are presented in three distinct planes: the vertical (r–z) plane at $\varphi = 0$ that represents typical flow structures along the growth direction, the horizontal (r–φ) plane at the middle of the growth cell ($z = 5.15$ mm) where the changes in the flow field will be more prominent for radial and circum-

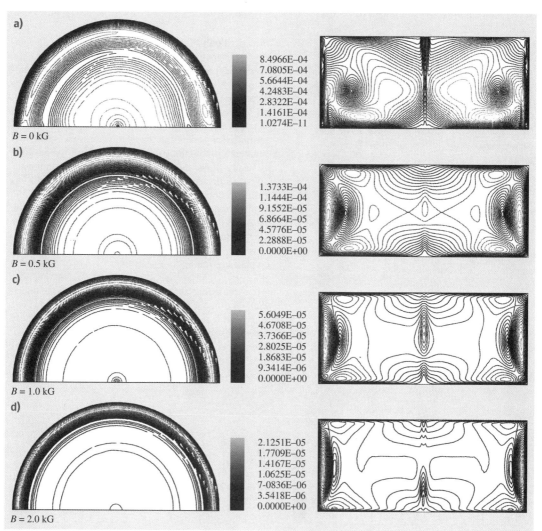

Fig. 29.17a–d Flow field in the horizontal plane in the middle of the solution zone (*left column*) and in the vertical plane at $\varphi = 0$ (*right column*) (after [29.35, 36])

ferential velocity components representing mixing, and finally the horizontal plane (r–φ) near the growth interface ($z = 0.4$ mm) where the flow field is often closely related to the quality of the grown crystals.

Figure 29.17 presents the simulation results for the flow field presented in the horizontal plane at the middle of the growth cell (left column) and in the vertical plane at $\varphi = 0$ (right column) for four levels of magnetic field strengths ($B = 0.0$, 0.5, 1.0, and 2.0 kG), while Fig. 29.18 shows the results for the flow field in the horizontal plane near the growth interface (left column) and in the vertical plane at $\varphi = 0$ (right column) for intermediate magnetic field intensities ($B = 2.5$ and 3.0 kG), and in the horizontal plane at the middle of the growth cell (left column) and in the vertical plane at $\varphi = 0$ (right column) for higher field levels ($B = 4.0$

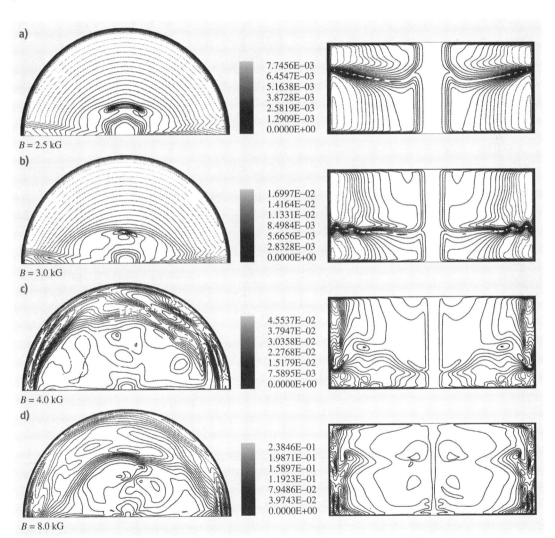

Fig. 29.18a–d Flow field in the horizontal plane near the growth interface at $z = 0.4$ mm (*left column*) and in the vertical plane at $\varphi = 0$ (*right column*) for (**a**) $B = 2.5$ kG, and (**b**) $B = 3.0$ kG, and in the horizontal plane at the middle of the growth cell (*left column*) and in the vertical plane (*right column*) for (**c**) $B = 4.0$ kG, and (**d**) $B = 8.0$ kG (after [29.35, 36])

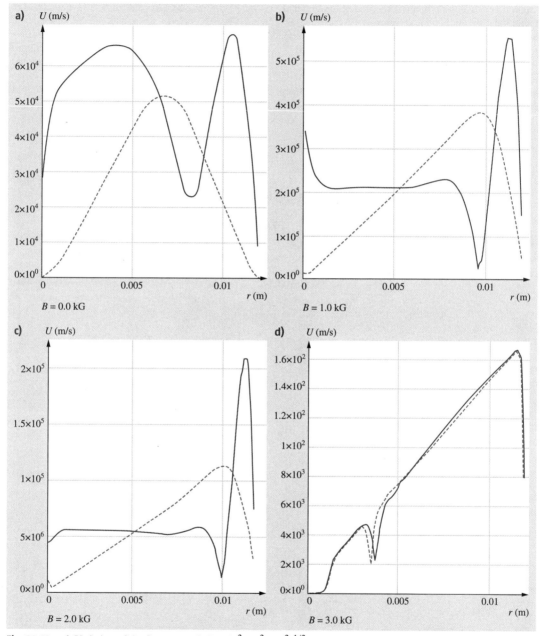

Fig. 29.19a–d Variation of the flow strength $U = (u^2 + v^2 + w^2)^{1/2}$ along the radial direction at $\varphi = \pi/2$ under various magnetic field strengths: (**a**) $B = 0.0$ kG, (**b**) $B = 1.0$ kG, (**c**) $B = 2.0$ kG, and (**d**) $B = 3.0$ kG. *Solid lines* at $z = 5.05$ mm (in the middle of the growth cell), and *dashed lines* at $z = 0.4$ mm (near the growth interface) (after [29.35, 36])

and 8.0 kG). Local magnitudes of the flow velocity $U = \sqrt{u^2 + v^2 + w^2}$ (which will be referred to as the *flow strength*) are computed in m/s, and the scales of flow strengths are shown in each figure.

The computed flow patterns in the vertical plane appear more complex than those of the 2-D simulations presented for the same LPEE growth system. This complexity can be attributed mainly to the inclusion of the contribution of the circumferential velocity component. In addition, the size of the growth crucible used in this study is twice that of the one used in [29.34, 49, 50]; this might have also contributed towards these differences. Such differences in the flow field of 2-D and 3-D models have also been observed in [29.39, 52, 53]. The strongest flow is seen in the lower part of the crucible cell, near the center of the half-domain along the r-direction. This is similar to what was observed in [29.34]. A weak flow cell forms just above the strongest flow cell. The flow in the horizontal planes is nearly homocentric. The flow in the horizontal plane near the growth interface is stronger in the middle region along the radial direction and becomes weaker and weaker near the growth cell wall or the axial center (Fig. 29.17a, left column). On the other hand, in the horizontal plane at the middle of the growth cell (Fig. 29.17a, right column) there are two maximum points for the flow intensity along the radial direction, with a relatively strong flow in the central region.

Note that the flow field in Fig. 29.17a is not strictly axisymmetric, especially in the middle of the growth cell, although the flow is stable. This result suggests that the present system is near the Grashof number that is a little bit lower than the critical Grashof number (2.5×10^5) under the same conditions (Prandtl number and geometry aspect ratio) given in [29.60] in the analysis of axisymmetry breaking of natural convection in a vertical Bridgman growth configuration.

Results are also summarized in Figs. 29.19 and 29.20 for various aspects of the flow field. Simulation results show three distinct characteristics depending on the level of applied magnetic field: (a) the *weak* magnetic field, with intensities from 0.0 to 2.0 kG, (b) the *intermediate* magnetic field, with levels from 2.0 to 3.0 kG, and (c) the *high* magnetic field, with intensities above 3.0 kG. Flow characteristics are quite different at each of these field levels.

Let us first focus on Fig. 29.17b–d, which represents results for magnetic field levels from 0.5 to 2.0 kG. In this category, the weak magnetic field level, an increase in the applied magnetic field strength results in significant reduction in the flow strength, which is, in general, desirable for a stable and controlled crystal growth. Flow cells are the strongest near the vertical wall and forms the so-called Hartmann layer [29.51–54]. At higher magnetic field strengths, the relative strength of these flow cells becomes increasingly stronger compared with the flow in the rest of the growth cell domain, and the Hartmann layer becomes thinner, which is in accordance with the scaling analysis given in [29.51, 52]. Furthermore, the strongest flow cells appearing in the lower part of the growth cell (Fig. 29.17a) move further down towards the growth interface with increasing strength, and new flow cells form near the dissolution interface. These strong flow cells form very visible boundary layers near the growth and dissolution interfaces (the so-called *end layers* [29.51, 52, 54]), and hence give rise to strong vertical velocity gradients in the vicinity of the growth interface, which may have an adverse effect on the crystal growth process.

One may state that the application of a magnetic field may not always be beneficial for the growth process [29.40, 54]. It was shown in [29.54] that the radial nonuniformity in vertical Bridgman is the most

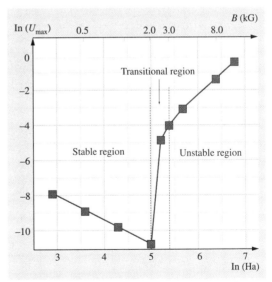

Fig. 29.20 Variation of the maximum velocity with magnetic field intensity: three distinct regions of stability of the flow field are obvious. The flow field is stable up to Ha = 150; in the region between Ha = 150 and 220 the flow is transitional; and above Ha = 220 the flow is unstable (after [29.35, 36])

significant at intermediate levels of magnetic field strength.

Flow fields become perfectly homocentric (and hence axisymmetric) with the increase of magnetic field strength, and hence the flow strength decreases. The flow velocities in the central region of the cell (in both the r- and z-directions) become more uniform at higher magnetic field levels, and form a core region of the uniform flow field, leading to a domain in which the flow is suppressed with the application of an applied magnetic field [29.51, 52, 54]. Indeed, an applied magnetic field at weak intensity levels can suppress fluid flows in a growth system, as shown in Fig. 29.17.

To the best of our knowledge, a magnetic field level higher than 8.0 kG is very strong for LPEE growth of thick crystals, due to the strong interaction between the applied magnetic and electric fields. Thus far, it was not possible to exceed the level of 4.5 kG in experiments, as higher magnetic field levels led to unsatisfactory growth [29.26]. The issue of LPEE growth under higher magnetic field levels and also the higher growth rates observed under an applied static magnetic field will be further discussed in the next section.

Figure 29.18 represents the simulation results for the intermediate magnetic intensity levels selected as $B = 2.5, 3.0, 4.0,$ and 8.0 kG. In this case, flow strength increases with magnetic field intensity. This result has not been widely reported in the relevant literature. It is possible that it is numerical in nature, and can be fixed by some innovative numerical treatments [29.56], but it is also possibly physical, as was observed experimentally in the LPEE growth of GaAs in [29.27].

Some experimental and numerical studies indeed demonstrated the enhancement of heat transfer (and hence flow strength) in a melt under a stationary magnetic field [29.57–59]. As can be seen from Fig. 29.18, the flow patterns show dramatic changes, as two flow cells were formed in each half of the vertical plane, with the upper cells getting larger and lower cells getting smaller with increasing magnetic field intensity. In the vertical plane, some strong unidirectional flows appear in the middle region along the r-direction, and some with very weak intensity in the middle, forming a small cylindrical region where the flow is nearly stationary. The flow fields are no longer axisymmetric and homocentric. Such an axisymmetry breaking may be caused by the unsteadiness of the flow field according to [29.60].

The flow stability was also examined at $B = 2.5$ and 3.0 kG levels. Comparing the results at different times, it was found that the flow field is essentially stable in spite of some small changes with time. One can then speculate that the reason for the axisymmetry breaking of the flow field at these magnetic field levels may be physical and due to a very delicate balance between the buoyancy and magnetic forces, even though the flow still remains stable. The variation of the flow strength along the radial direction for $B = 0.0, 1.0, 2.0,$ and 3.0 kG levels are shown in Fig. 29.19. Solid and dashed lines represent, respectively, the values at the middle and the lower regions of the solution zone. As can be seen, the flow strength fluctuates and shows differences in these regions. However, at higher magnetic field levels, the difference becomes less obvious. The variation of the flow strength in the radial direction near the growth interface is almost symmetric in the absence of a magnetic field. However, this symmetry is broken at higher magnetic field intensity levels.

Although higher magnetic field intensity levels, higher than 4.5 kG, appear not to be practical for LPEE growth of thick crystals [29.26], for the sake of completeness and also for the purpose of comparison with other studies, higher magnetic field intensity levels of $B = 4.0$ and 8.0 kG were also considered. The flow patterns become dramatically different and show large fluctuations with time, and temperature distributions show asymmetric behavior.

The maximum flow strength (the maximum magnitude of the local velocity vector) values U_{max} are presented in Fig. 29.20 for all magnetic field levels. As seen, the variation of the maximum flow strength shows three distinct regions of stability. The flow strength decreases with increasing magnetic field in the stable region (up to Ha = 150), but in the intermediate and unstable regions (between Ha = 150 and 220, and above Ha = 220, respectively) the flow strength increases. If one examines the logarithmic plot of U_{max} as a function of the Hartmann number Ha given in Fig. 29.20, we can see that within the stable region, the relationship between U_{max} and Ha obeys a power law of

$$U_{max} \propto \text{Ha}^{-5/4}, \tag{29.33}$$

which has been demonstrated by many authors [29.51–55], although the index of the power law is slightly different due to different system parameters and conditions. In the unstable region, this relationship becomes

$$U_{max} \propto \text{Ha}^{5/2}. \tag{29.34}$$

In the transitional (intermediate) region, on the other hand, as one expects, the change of the maximum strength (U_{max}) of the flow field with the Hartmann number is so dramatic that the power law is not suitable. The results given here for the intermediate and unstable regions were obtained for the first time in [29.35], and to the best of our knowledge, are not corroborated by anyone else in the literature.

29.5.4 Evolution of Interfaces

As mentioned earlier, the simulation results above were given at the end of a 1 h growth period ($t = 1.0$ h) and also for a *stationary* interface, for computational efficiency. This was sufficient for examining the flow field and the effect of a magnetic field on the flow structures. However, in order to draw meaningful conclusions for future experiments, and also examining the concentration fields in the solution and the evolution of interfaces, simulations were carried for a longer period of growth ($t = 20.0$ h) for three levels of magnetic fields, namely $B = 0.0$ kG (no applied magnetic field), $B = 1.0$ kG (the mid-field level in the stable region), and $B = 3.0$ kG (a field at the beginning of the unstable region), and also the evolution of growth and dissolution interfaces are included. Higher magnetic field levels are not considered since the reliability of computations for concentrations in the unstable region may be questionable.

Figure 29.21 summarizes the computed results for the flow field in the horizontal plane at the middle of the growth cell (left column) and in the vertical plane at $\varphi = 0$ (right column) for three levels of magnetic field intensities ($B = 0.0$, 1.0, and 3.0 kG). Flow strengths are computed in m/s, and the scales are shown in each figure. The inclusion of the evolution of growth and dissolution interfaces affected the flow patterns to a certain extent, although it is not significant in terms of magnitudes. The effect of the magnetic field in the stable region is obvious in suppressing the natural convection in the solution. At the $B = 3.0$ kG level, however, the flow patterns show signs of unstable flows.

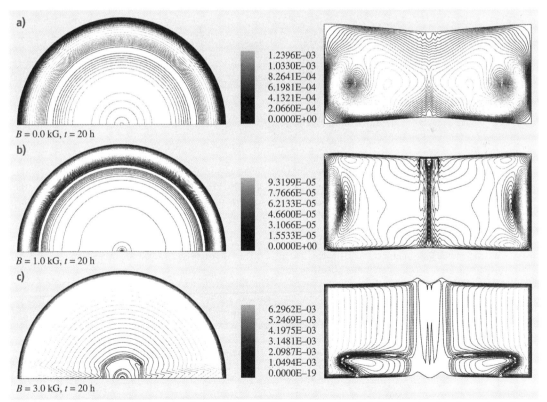

Fig. 29.21a–c Flow field in the horizontal plane in the middle of the growth cell (*left column*) and in the vertical plane at $\varphi = 0$ (*right column*) (after [29.35, 36])

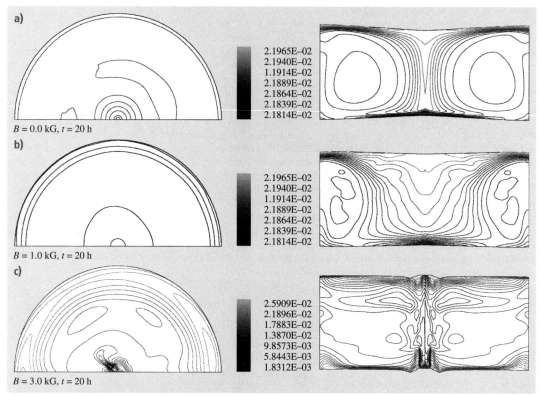

Fig. 29.22a–c Concentration distributions in the horizontal plane near the growth interface (*left column*) and in the vertical plane (*right column*) (after [29.35, 36])

The concentration distributions at the $B = 0.0$, 1.0, and 3.0 kG magnetic field levels are presented in Fig. 29.22 at $t = 20$ h. Comparing Fig. 29.22a,b one can see that the concentration gradients near the center of the growth interface and near the growth cell wall in the vicinity of the dissolution interface decrease with the increasing magnetic field levels. This is due to the reduction in flow strength, resulting in slower mass transfer towards the growth interface, consequently slowing down both the growth and dissolution rates. The concentration vortices in Fig. 29.22a nearly disappear in Fig. 29.22b. At the $B = 3.0$ kG level (Fig. 29.22c), concentration distributions exhibit significant changes compared with those of Fig. 29.22a,b. The concentration vortices near the center of the symmetric axis, in the vicinity of the growth interface, form strong concentration gradients and lead to fast growth rates in that region. Concentration layers are formed in the vicinity of both the growth and dissolution interfaces.

The computed growth rates obtained from simulations [29.35] under magnetic field do not predict the experimental growth rates [29.40]. We will discuss this issue later in detail.

29.5.5 Effect of High Electric and Magnetic Field Levels

The flow field was also numerically simulated under various electric current densities in [29.36]. The maximum flow strength shows the same trend at $J = 3$, 5, and 7 A/cm^2 electric field levels. The time evolution of the flow field was computed for various magnetic field intensity levels, and the one corresponding to $B = 4.0$ kG and $J = 7$ A/cm^2 is shown in Fig. 29.23. As can be seen, the flow patterns change with time and begin to become localized after 120 s of growth. This localization point is near the growth interface and is approximately at about a distance of a quarter of

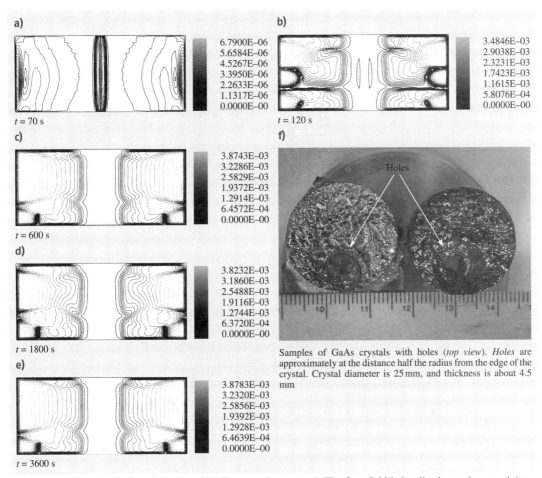

Fig. 29.23a–f Time evolution of the flow field (isostrength contours). The flow field is localized near the growth interface at about a distance of half of the crystal radius where approximately holes are observed (after [29.39]), samples from [29.27]

Samples of GaAs crystals with holes (*top view*). Holes are approximately at the distance half the radius from the edge of the crystal. Crystal diameter is 25 mm, and thickness is about 4.5 mm

the diameter of the crystal from the edge. This point is almost at the locations where the holes (or damages) were observed in crystals grown under high fields in [29.27]. Simulations are shown in Fig. 29.23 for the liquid solution zone, but it is possible that similar flow patterns can also be computed in the liquid contact zone below the seed crystal. The flow patterns computed for $B = 3.0$ kG and $J = 5$ A/cm^2 show similar patterns [29.36]. The numerical simulations are in qualitative agreement with experiments, but do not agree quantitatively on the critical value of the magnetic field. Numerical simulations predict a lower value (just over 2 kG) than the maximum experimental value of 4.5 kG.

Crystals were grown in [29.26] at field values up to $B = 4.5$ kG, and $J = 7$ A/cm^2. However, experiments at higher fields failed (two samples of such crystals are shown in Fig. 29.23f).

Three-dimensional numerical simulation results have shown that magnetic field intensities up to 2.0 kG suppress the flow structures in the solution, and the flow structures are stable and get weaker with the increasing magnetic field level. These levels of magnetic field are beneficial in suppressing the natural convection. However, field intensities higher than 2.0 kG change the flow patterns significantly, and at intensities higher than 3.0 kG the flow structures become unstable.

As mentioned earlier, the LPEE growth system developed in [29.26] allowed the growth of a large number of GaAs and Ga$_{0.96}$In$_{0.04}$As single crystals of thicknesses up to 9 mm. It was possible to apply electric current densities of 3, 5, and 7 A/cm^2, and the corresponding growth rates in these experiments with no magnetic field were, respectively, about 0.57, 0.75, and 1.25 mm/day. Growth interfaces were very flat, and the growth experiments were reproducible in terms of crystal thickness and growth rate. Experiments at higher electric current intensities were not successful. Experiments at 3, 5, and 7 A/cm^2 electric current densities were repeated under various applied static magnetic field levels. Results showed that LPEE experiments at the 4.5 kG and lower magnetic field levels were successful, but those under higher magnetic field levels were not.

In fact the experimental study of the LPEE growth of GaAs and Ga$_{0.96}$In$_{0.04}$As single crystals supported qualitatively the results of the three-dimensional numerical simulations [29.26]. It seems that the 4.5 kG field intensity level is a maximum (*critical*) value above which the growth is not stable. This experimental *critical* magnetic field level is higher than that predicted from the numerical simulations performed under the same growth conditions, which was somewhere between 2.0 and 3.0 kG [29.35]. Considering the complexity of the LPEE growth process, this is a good qualitative agreement.

29.6 High Growth Rates in LPEE: Electromagnetic Mobility

The LPEE experiments conducted in [29.26] yielded a very significant result that was not predicted from the modeling studies conducted earlier: the experimental LPEE growth rates under magnetic field were much higher than the predicted values. For instance, the growth rate at the 4.5 kG magnetic field level (at $J = 3$ A/cm^2) was about 6.1 mm/day, which is about 12 times higher than that with no magnetic field. Experiments performed at $B = 1.0$ and 2.0 kG field levels (at $J = 3$ A/cm^2) were also successful, and the growth rates were also higher: 1.62 and 2.35 mm/day, respectively. One more interesting observation of the LPEE experiments was that the direction of the applied magnetic field, either up or down, was not relevant. The growth rate was almost the same, being about 5–6% less when the magnetic field was in the direction of the applied electric field. As predicted from the three-dimensional models, at higher magnetic field levels (even with the $J = 3$ A/cm^2 electric current density level), and higher electric current density levels ($J = 10$ A/cm^2 or higher), experiments did not lead to successful growth, but showed very interesting outcomes [29.26, 36].

Although very thick crystals were grown, even up to 9 mm thickness, the growth processes were unstable, and led to uneven growth (Figs. 29.23 and 29.24). From visual inspection of the grown crystals, the adverse effect of natural convection was obvious, causing either one-sided growth or leading to holes in the grown crystals. It was considered that such growth (one-sided and with holes) is because of the strong and unstable convection in the liquid zones (solution and contact zones) due to the strong interaction between the applied magnetic field and the applied electric current. Such predictions were also confirmed qualitatively by the numerical simulations carried out by considering field nonuniformities in [29.35], and also by using a newly defined electromagnetic mobility in [29.40, 61, 62] and a new model in [29.63]. The simulated flow structures show the possibility of causing such nonuniform growth of crystals.

The contribution of electromigration under magnetic field was obtained through a nonlinear model in [29.63] (for the binary GaAs system, in the absence of the Soret effect) as

$$i = \rho_L D_C \nabla C + \rho_L (D_{EC} + D_{ECB} B) CE . \quad (29.35)$$

The second term in (29.35) represents the contribution of the applied electric current density to mass transport under the effect of a static external magnetic field. This term represents electromigration in the mass transport equation. Its coefficient, which is called the *total mobility* [29.61–63], is written in the following form for convenience

$$\mu_T \equiv D_{EC} + D_{ECB} B \equiv \mu_E + \mu_B B , \quad (29.36)$$

where the material constant μ_E (a second-order material coefficient) is the classical electric mobility of the solute (As) in the liquid solution (Ga/As solution) due to the applied electric current in the absence of an applied magnetic field. The constant μ_B is a third-order

Fig. 29.24a,b Samples of LPEE-grown GaAs crystals under high fields: grown under (**a**) $B = 20\,\mathrm{kG}$ and $J = 3\,\mathrm{A/cm^2}$, and (**b**) $B = 0$, and $J = 10\,\mathrm{A/cm^2}$ (after [29.26])

material coefficient that represents the contribution of the applied magnetic field intensity to the electromigration of species. It is zero (or insignificant) in the absence of an applied electric current. This term is called the electromagnetic mobility of solute [29.63]. Its value is determined using the data of [29.26]. The mass transport equation then becomes

$$(\mu_\mathrm{E} + \mu_\mathrm{B} B)(\boldsymbol{E} \cdot \nabla C) + D_\mathrm{C} \nabla^2 C = \frac{\partial C}{\partial t} + \boldsymbol{v} \cdot \nabla C \,. \tag{29.37}$$

29.6.1 Estimation of the Electromagnetic Mobility Value

Experiments show that the growth rate is proportional to the applied electric current density, and we have evaluated the value of μ_E in the Ga/As solution in the absence of an applied magnetic field. The numerical simulations based on this value verify the experimental growth rates at all three electric current levels ($J = 3$, 5, and 7 A/cm²). Of course, diffusion (the second term in (29.37)) and also natural convection (the last term on the right-hand side of (29.37)) contribute to the growth rate. However, as shown many times numerically, in LPEE the contribution of the first term (electromigration) is dominant, and the growth rate can be assumed approximately proportional to this term.

Experiments also show that the growth rate increases significantly in the presence of a static magnetic field, and is also proportional to the field intensity level as long as the field level is below a critical value, above which the growth is not stable [29.27]. Numerical values of the mobilities are calculated using the results of a large number of experiments in [29.26] in which the magnetic field vector B was applied both upward and downward. The growth rates in these experiments were almost the same whether B was up or down. In other words the mass transport due to electromigration was only dependent on the magnetic field intensity but not on its direction. This is also in compliance with the defined constitutive equations in [29.63]. Using the measured growth rates, the mobility values were computed (Table 29.5).

The dimensionless mobility is defined as

$$\mu = \frac{\mu_\mathrm{T}}{\mu_\mathrm{E}} = 1 + \frac{\mu_\mathrm{B}}{\mu_\mathrm{E}} \cong 1 + 2B \,, \tag{29.38}$$

Table 29.5 Numerical values of mobilities [29.26, 61, 63]

Experimental values				
Magnetic field (kG)	0.0	1.0	2.0	4.5
Electric current density (A/cm^2)	3.0	3.0	3.0	3.0
Growth rate (mm/day)	0.50	1.62	2.35	6.10
Computed values				
Electric mobility constant μ_E (m^2/V s)	0.7×10^{-5}	0.7×10^{-5}	0.7×10^{-5}	0.7×10^{-5}
Total mobility $\mu_T = \mu_E + \mu_B B$ (m^2/V s)	0.7×10^{-5}	2.3×10^{-5}	3.4×10^{-5}	7.1×10^{-5}
Electromagnetic mobility $\mu_B B$ (m^2/V s)	0.0	1.6×10^{-5}	2.7×10^{-5}	6.4×10^{-5}
Electromagnetic mobility constant μ_B (m^2/V s kG)	–	1.4×10^{-5}	1.4×10^{-5}	1.4×10^{-5}
Dimensionless mobility $\mu = \mu_T/\mu_B \cong 1 + 2B$	1	3	5	10

and is plotted in Fig. 29.25. As seen, the total mobility is almost linearly dependent on the magnetic field intensity, of course, within the limit of experimental measurements.

The first term in the mass transport equation reads explicitly

$$(\mu_E + \mu_B B)(\bm{E} + \bm{v} \times \bm{B}) \cdot \nabla C , \quad (29.39)$$

where the term $(\bm{v} \times \bm{B}) \cdot \nabla C$ is the contribution of the applied magnetic field due to the motion of the fluid particles (coupling term). Its contribution was found to be very small compared with that of $\bm{E} \cdot \nabla C$ (on the order of 3% based on a maximum velocity of 0.01 m/s and a 10 kG field level [29.35, 39]). Therefore, its contribution can be neglected in the model for computer simulations. Then, the electromigration term in the mass transport equation is written as

$$(\mu_E + \mu_B B)\bm{E} \cdot \nabla C = \mu_T \bm{E} \cdot \nabla C , \quad (29.40)$$

and the growth rate is computed by

$$V_n^g = \frac{\rho_L}{\rho_S} \left(D_C \frac{\partial C}{\partial n} + \mu_T C E_n \right) \frac{1}{C_S - C} , \quad (29.41)$$

or simply by, for the purpose of evaluating the mobility constants,

$$V^g = \frac{\rho_L}{\rho_S} \left(D_C \frac{\partial C}{\partial n} + \mu_T C E_z \right) \frac{1}{C_S - C} . \quad (29.42)$$

29.6.2 Simulations of High Growth Rates in a GaAs System

Simulations presented in the previous sections were repeated using the total mobility values in the mass transport equation (29.37). In earlier simulations only the electric mobility μ_E was used. A summary of the growth rates from these numerical simulations is presented in Fig. 29.26a. The values under no magnetic field are the experimental growth rates and are used to compute the value of μ_E. Naturally, they are coincident with the computed values. As seen, the growth rate decreases first with the magnetic field level and then increases with the magnetic field above the critical value. This pattern is similar to the pattern of experimental growth rates under various magnetic field levels, and also agrees with the numerical simulation results.

In order to predict the high experimental growth rates under an applied magnetic field, the mass transport equation in (29.24) was replaced with (29.37), and

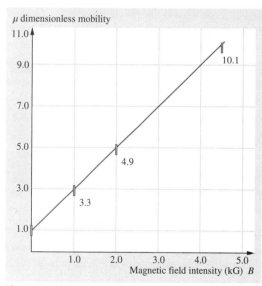

Fig. 29.25 Dependence of the total As mobility on magnetic field intensity (after [29.61, 63])

Fig. 29.26 (a) Computed growth rates are presented versus applied magnetic field with the use of constant electric mobility μ_E. *Squares* represent the values at $J = 3\,\text{A/cm}^2$ and *circles* denote for the values at $J = 7\,\text{A/cm}^2$. (b) Growth rates computed using the total mobility, $\mu_T = \mu_E + \mu_B B$, are represented by *full circles*. These values are in agreement with the experimental growth rates (*hollow circles*, invisible as coincident with the *full circles*). The growth rates under no magnetic field are also shown (*squares*) for comparison (after [29.36]) ▶

then the 3-D simulations were repeated using the total electromagnetic mobility $\mu_T = \mu_E + \mu_B B$ values given in Table 29.5. The growth rates computed using the total mobility μ_T are presented in Fig. 29.26b (full circles), and agree with those of experiments. The growth rates using only the electric mobility μ_E are also given in Fig. 29.26b for comparison (empty squares). For the sake of completeness and for comparison, the experimental growth rates under a magnetic field are also presented in Fig. 29.26b (note that the full and empty circles are coincident).

The above results show that the inclusion of the new total mobility that includes the contributions of both

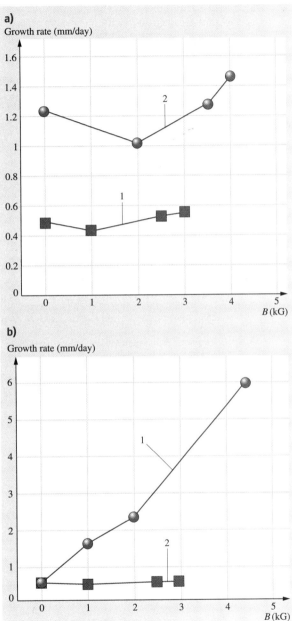

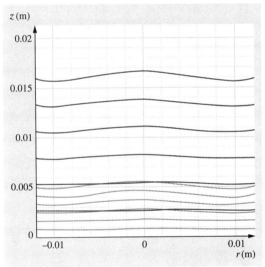

Fig. 29.27 Evolution of the computed growth interface using the total mobility $\mu_T = \mu_E + \mu_B B$ (*thick lines*) and using only the electric mobility μ_E (*thin lines*). Time increment between lines is 40 h. The shapes of the computed interfaces are in excellent agreement with those of experiments with and without magnetic field (see sample GaAs crystals in Fig. 29.11). It must be noted that the numerical simulations were let go until 15 mm growth thickness just to show that the interface remains flat. Thicknesses of the grown GaAs and GaInAs crystals were less than this value

the applied electric current and also its interaction with the applied magnetic field provide much more accurate predictions for growth rates in LPEE under a magnetic field.

When only the electric mobility is used in simulations, not only the growth rates but also the shapes of the evolving growth interface were not predicted accurately. However, when the total mobility was introduced, in addition to better predictions for growth rates, the growth interface shapes were also much closer to those of experiments.

In Fig. 29.27 the evolution of computed growth interface is presented. As seen, the computed interfaces are flatter and agree with the shapes of the interfaces of the crystals grown in [29.26], which are almost perfectly flat.

These results show that the introduction of a bulk constitutive coefficient representing the total electromagnetic mobility due to electromigration under magnetic fields in LPEE (the nonlinear model for LPEE under magnetic field in [29.63]) is a step in the right direction. It is by no means complete, since the interaction of electric and magnetic fields must also affect mass transport at the growth interface. Therefore, in addition to this model in the bulk (solution), a closer look may also be needed at various surface phenomena under the combined effect of applied electric and magnetic fields, in order to obtain better predictions from modeling.

References

29.1 M. Kumagawa, A.F. Witt, M. Lichtensteiger, H.C. Gatos: Current-controlled growth and dopant modulation in liquid-phase epitaxy, J. Electrochem. Soc. **120**, 583/584 (1973)

29.2 J.J. Daniele: Peltier-induced LPE and composition stabilization of GaAlAs, Appl. Phys. Lett. **27**, 373 (1975)

29.3 J.J. Daniele: Experiments showing absence of electromigration of As and Al in Peltier LPE of GaAs and $Ga_{1-x}Al_xAs$, J. Electrochem. Soc. **124**, 1143 (1977)

29.4 V.A. Gevorkyan, L.V. Golubev, S.G. Petrosyan, Y.A. Shik, Y.V. Shmatsev: Sov. Phys. Tech. Phys. **22**, 750–755 (1977)

29.5 L. Jastrzebski, H.C. Gatos, A.F. Witt: Electromigration in current-controlled LPE, J. Electrochem. Soc. **123**, 1121 (1976)

29.6 L. Jastrzebski, H.C. Gatos: Current-controlled growth, segregation and amphoteric behavior of Si in GaAs from Si-doped solutions, J. Cryst. Growth **42**, 309–314 (1977)

29.7 L. Jastrzebski, H.C. Gatos, A.F. Witt: Current-induced solution growth of garnet layers, J. Electrochem. Soc. **124**, 633 (1977)

29.8 L. Jastrzebski, Y. Imamura, H.C. Gatos: Thickness uniformity of GaAs layers grown by electroepitaxy, J. Electrochem. Soc. **125**, 1140 (1978)

29.9 L. Jastrzebski, J. Lagowski, H.C. Gatos, A.F. Witt: Determination of carrier concentration distribution in semiconductors by IR absorption – Si, J. Appl. Phys. **49**, 5909 (1978)

29.10 T. Bryskiewicz: Peltier-induced growth kinetics of liquid-phase epitaxial GaAs, J. Cryst. Growth **43**, 567–571 (1978)

29.11 A.F. Witt, H.C. Gatos, M. Lichtensteiger, C.J. Herman: Crystal-growth and segregation under zero gravity – Ge, J. Electrochem. Soc. **125**, 1832 (1978)

29.12 T. Bryskiewicz, A. Laferriere: Growth of alloy substrates by liquid-phase electroepitaxy - theoretical considerations, J. Cryst. Growth **129**, 429–442 (1993)

29.13 Y.H. Lo, R. Bhat, P.S.D. Lin, T.P. Lee: Long-wavelength optoelectronic integrated circuit transmitter, Proc. SPIE **1582**, 60–70 (1992)

29.14 W.P. Hong, G.K. Chang, R. Bhat, C. Nguyen, H.P. Lee, L. Wong, K.R. Runge: InP-based MSM-HEMT receiver OEICs for long-wavelength light wave systems, Proc. SPIE **1582**, 134–144 (1992)

29.15 W.J. Schaff, P.J. Tasker, M.C. Foisy, L.F. Eastman: Device applications of stained layer epitaxy. In: *Semiconductors and Semimetals*, Vol. 33, ed. by T.P. Pearsall (Academic, New York 1991) pp. 73–133

29.16 L.F. Eastman: In: *Optoelectronic Materials and Devices Concepts*, ed. by M. Razeghi (SPIE Optical Engineering Press, Bellingham 1991) p. 41

29.17 A. Andaspaeva, A.N. Baranov, A. Guseinov, A.N. Imenkov, L.M. Litvak: Sov. Tech. Phys. Lett. **14**, 377 (1988)

29.18 H.K. Choi, S.J. Eglash: High-efficiency high-power GaInAsSb-AlGaAsSb double-heterostructure lasers emitting at 2.3 μm, IEEE J. Quantum Electron. **27**, 1555 (1991)

29.19 S.J. Eglash, H.K. Choi: MBE growth, material properties, and performance of GaSb-based 2.2 μm diode-lasers, Inst. Phys. Conf. Ser. **120**(10), 487 (1992)

29.20 T. Bryskiewicz, C.F. Boucher Jr., J. Lagowski, H.C. Gatos: Bulk GaAs crystal-growth by liquid-phase electroepitaxy, J. Cryst. Growth **82**, 279–288 (1987)

29.21 S. Dannefear, P. Mascher, D. Kerr: Annealing of grown-in defects in GaAs, Proc. MRS **104**, 471 (1978)

29.22 C.F. Boucher Jr., O. Ueda, T. Bryskiewicz, J. Lagowski, H.C. Gatos: Elimination of disloca-

29.23 T. Bryskiewicz, M. Bugajski, J. Lagowski, H.C. Gatos: Growth and characterization of high quality LPEE GaAs bulk crystals, J. Cryst. Growth **85**, 136–141 (1987)
tions in bulk GaAs crystals grown by liquid-phase electroepitaxy, J. Appl. Phys. **61**, 359–364 (1987)

29.24 T. Bryskiewicz, M. Bugajski, B. Bryskiewicz, J. Lagowski, H.C. Gatos: LPEE growth an characterization of In(V)Ga/As(V) crystal, Proc. Inst. Phys. Ser. **91**(3), 259 (1988)

29.25 B. Bryskiewicz, T. Bryskiewicz, E. Jiran: Internal strain and dislocations in $In_xGa_{1-x}As$ crystals grown by liquid phase electroepitaxy, J. Electron. Mater. **24**, 203 (1995)

29.26 H. Sheibani, S. Dost, S. Sakai, B. Lent: Growth of bulk single crystals under applied magnetic field by liquid phase electroepitaxy, J. Cryst. Growth, **258**(3-4), 283–295 (2003)

29.27 H. Sheibani, Y.C. Liu, S. Sakai, B. Lent, S. Dost: The effect of applied magnetic field on the growth mechanisms of liquid phase electroepitaxy, Int. J. Eng. Sci. **41**, 401–415 (2003)

29.28 G. Bischopink, K.W. Benz: THM growth of $Al_xGa_{1-x}Sb$ bulk crystals, J. Cryst. Growth **128**, 466–470 (1993)

29.29 J.J. Daniele, A.J. Hebling: Peltier-induced liquid-phase epitaxy and compositional control of mm-thick layers of (Al,Ga)As, J. Appl. Phys. **52**, 4325–4327 (1981)

29.30 Z.R. Zytkiewicz: Influence of convection on the composition profiles of thick GaAlAs layers grown by liquid phase electroepitaxy, J. Cryst. Growth **131**, 426–430 (1993)

29.31 Z.R. Zytkiewicz: Liquid phase electroepitaxial growth of thick and compositionally uniform AlGaAs layers on GaAs substrates, J. Cryst. Growth **146**, 283–286 (1995)

29.32 Z.R. Zytkiewicz: Joule effect as a barrier for unrestricted growth of bulk crystals by liquid phase electroepitaxy, J. Cryst. Growth **172**, 259–268 (1997)

29.33 C. Takenaka, K. Nakajima: Effect of electric-current on the LPE growth of InP, J. Cryst. Growth **108**, 519 (1991)

29.34 N. Djilali, Z. Qin, S. Dost: Role of thermosolutal convection in liquid phase electroepitaxial growth of gallium arsenide, J. Cryst. Growth **149**, 153–166 (1995)

29.35 Y.C. Liu, Y. Okano, S. Dost: The effect of applied magnetic field on flow structures in liquid phase electroepitaxy – A three-dimensional simulation model, J. Cryst. Growth **244**, 12–26 (2002)

29.36 Y.C. Liu, S. Dost, H. Sheibani: A three dimensional numerical simulation for the transport structures in liquid phase electroepitaxy under applied magnetic field, Int. J. Trans. Phenom. **6**, 51–62 (2004)

29.37 R.W. Wilcox: Influence of convection on the growth of crystals from solution, J. Cryst. Growth **65**, 133 (1983)

29.38 S. Ostrach: Fluid mechanics in crystal growth – The 1982 Freeman scholar lecture, J. Fluids Eng. **105**, 5–20 (1983)

29.39 S. Dost, Y.C. Liu, B. Lent: A numerical simulation study for the effect of applied magnetic field in liquid phase electroepitaxy, J. Cryst. Growth **240**, 39–51 (2002)

29.40 S. Dost, B. Lent, H. Sheibani, Y.C. Liu: Recent developments in liquid phase electroepitaxial growth of bulk crystals under magnetic field, C. r. Mec. **332**(5/6), 413–428 (2004)

29.41 S. Dost, B. Lent: *Single Crystal Growth of Semiconductors from Metallic Solutions* (Elsevier, Amsterdam 2007)

29.42 M. Yildiz, S. Dost, B. Lent: Growth of bulk SiGe single crystals by liquid phase diffusion, J. Cryst. Growth **280**(1-2), 151–160 (2005)

29.43 M. Yildiz, S. Dost: A continuum model for the liquid phase diffusion growth of bulk SiGe single crystals, Int. J. Eng. Sci. **43**, 1059–1080 (2005)

29.44 E. Yildiz, S. Dost, M. Yildiz: A numerical simulation study for the effect of magnetic fields in liquid phase diffusion growth of SiGe single crystals, J. Cryst. Growth **291**, 497–511 (2006)

29.45 M. Yildiz, S. Dost, B. Lent: Evolution of the growth interface in liquid phase diffusion growth of bulk SiGe single crystals, Cryst. Res. Technol. **41**(3), 211–216 (2006)

29.46 S. Dost, Z. Qin: A model for liquid phase electroepitaxial growth of ternary alloy semiconductors – 1. Theory, Int. J. Electromagn. Mech. **7**(2), 109–128 (1996)

29.47 Y. Imamura, L. Jastrzebski, H.C. Gatos: Defect structure and electronic charateristics of GaAs layers grow by electroepitaxy and thermal LPE, J. Electrochem. **126**(8), 1381–1385 (1979)

29.48 Z. Qin, S. Dost, N. Djilali, B. Tabarrok: A finite element model for liquid phase electroepitaxy, Int. J. Numer. Methods Eng. **38**(23), 3949–3968 (1995)

29.49 Z. Qin, S. Dost, N. Djilali, B. Tabarrok: A model for liquid phase electroepitaxy under an external magnetic field – 2. Application, J. Cryst. Growth **153**, 131–139 (1995)

29.50 Z. Qin, S. Dost: A model for liquid phase electroepitaxial growth of ternary alloy semiconductors – 2. Application, Int. J. Appl. Electromagn. Mech. **7**(2), 129–142 (1996)

29.51 H. Ben Hadid, D. Henry: Numerical study of convection in the horizontal Bridgman configuration under the action of a constant magnetic field. Part 1. Two-dimensional flow, J. Fluid Mech. **333**, 23–56 (1996)

29.52 H. Ben Hadid, D. Henry: Numerical study of convection in the horizontal Bridgman configuration under the action of a constant magnetic field. Part 2. Three-dimensional flow, J. Fluid Mech. **333**, 57–83 (1996)

29.53 H. Ben Hadid, S. Vaux, S. Kaddeche: Three-dimensional flow transitions under a rotating magnetic field, J. Cryst. Growth **230**, 57–62 (2001)

29.54 D.H. Kim, P.M. Adornato, R.A. Brown: Effect of vertical magnetic field on convection and segregation in vertical Bridgman crystal growth, J. Cryst. Growth **89**, 339 (1988)

29.55 L. Davoust, M.D. Cowley, R. Moreau, R. Bolcato: Buoyancy-driven convection with a uniform magnetic field – Part 2. Experimental investigation, J. Fluid Mech. **400**, 59 (1999)

29.56 V. Kumar, S. Dost, F. Durst: Numerical modeling of crystal growth under strong magnetic fields: An application to the travelling heater method, Appl. Math. Modell. **31**(3), 589–605 (2006)

29.57 T. Tagawa, H. Ozoe: Enhancement of heat transfer rate by application of a static magnetic field during natural convection of liquid metal in a cube, J. Heat Transf. **119**, 265 (1997)

29.58 T. Tagawa, H. Ozoe: Enhanced heat transfer rate measured for natural convection in liquid gallium in a cubical enclosure under a static magnetic field, J. Heat Transf. **120**, 1027 (1998)

29.59 K. Terashima, J. Nishio, S. Washizuka, M. Watanabe: Magnetic field effect on residual impurity concentrations for LEC GaAs crystal growth, J. Cryst. Growth **84**, 247 (1987)

29.60 A.Y. Gelfgat, P.Z. Bar-Yoseph, A. Solan: Axisymmetry breaking instabilities of natural convection in a vertical Bridgman growth configuration, J. Cryst. Growth **220**, 316 (2000)

29.61 S. Dost, H. Sheibani, Y.C. Liu, B. Lent: Recent developments in modelling of liquid phase electroepitaxy under applied magnetic field, Cryst. Res. Technol. **40**(4/5), 313 (2005)

29.62 S. Dost, H. Sheibani, Y.C. Liu, B. Lent: On the high growth rates in electroepitaxial growth of bulk semiconductor crystals in magnetic field, J. Cryst. Growth **275**(1–2), e1–e6 (2005)

29.63 S. Dost, H. Sheibani: A mathematical model for solution growth of bulk crystals under magnetic field, Philos. Mag. **85**(33–35), 4331–4351 (2005)

29.64 K. Nakajima: Liquid phase epitaxial growth of very thick $In_{1-x}Ga_xAs$ layers with uniform composition by source-current-controlled method, J. Appl. Phys. **61**(9), 4626 (1987)

29.65 K. Nakajima, S. Yamazaki, I. Umebu: A new growth method using source current control to supply solute elements-demonstration of $In_{1-x}Ga_xAs$ case, Jpn. J. Appl. Phys. **23**(1), L26–L28 (1984)

29.66 K. Nakajima, S. Yamazaki: A new method to supply solute elements into growth solutions – Demonstration by liquid phase epitaxial growth of $In_{1-x}Ga_xAs$, J. Electrochem. Soc. **132**, 904 (1985)

29.67 K. Nakajima, S. Yamazaki: Growth of very thick $In_{1-x}Ga_xAs$ layers by source-current-controlled method, J. Cryst. Growth **74**, 39–47 (1986)

29.68 K. Nakajima: Calculation of composition variation of $In_{1-x}Ga_xAs$ ternary crystals for diffusion and electromigration limited growth from a temperature graded solution with source material, J. Cryst. Growth. **110**, 781–794 (1991)

29.69 K. Nakajima: Calculation of stresses in $In_{0.12}Ga_{0.88}As$ ternary bulk crystals with compositionally graded $In_{1-x}Ga_xAs$ layers on GaAs seeds, J. Cryst. Growth **113**, 477–484 (1991)

29.70 S. Dost, N. Djilali, Z. Qin: A two-dimensional diffusion model for liquid-phase electroepitaxial growth of GaAs, J. Cryst. Growth **143**(3/4), 141–154 (1994)

29.71 S. Dost, H.A. Erbay: A continuum model for liquid phase electroepitaxy, Int. J. Eng. Sci. **33**(10), 1385–1402 (1995)

29.72 S. Dost, Z. Qin: A model for liquid-phase electroepitaxy under an external magnetic field. Part 1. Theory, J. Cryst. Growth **153**, 123–130 (1995)

29.73 S. Dost, Z. Qin: A numerical simulation model for liquid phase electroepitaxial growth of GaInAs, J. Cryst. Growth **187**, 51–64 (1998)

29.74 A. Okamoto, S. Isozumi, J. Lagowski, H.C. Gatos: In situ monitoring of liquid phase electroepitaxial growth, J. Electrochem. Soc. **129**, 2095–2098 (1982)

30. Epitaxial Lateral Overgrowth of Semiconductors

Zbigniew R. Zytkiewicz

The state of the art and recent developments of lateral overgrowth of compound semiconductors are reviewed. First we focus on the mechanism of epitaxial lateral overgrowth (ELO) from the liquid phase, highlighting the phenomena that are crucial for growing high-quality layers with large aspect ratio. Epitaxy from the liquid phase has been chosen since the equilibrium growth techniques such as liquid-phase epitaxy (LPE) are the most suitable for lateral overgrowth. We then present numerous examples for which the defect filtration in the ELO procedure is very efficient and leads to significant progress in the development of high-performance semiconductor devices made of lattice-mismatched structures. Structural perfection of seams that appear when layers grown from neighboring seeds merge is also discussed. Next, we concentrate on strain commonly found in various ELO structures and arising due to the interaction of ELO layers with the mask. Its origin, and possible ways of its control, are presented. Then we show that the thermal strain in lattice-mismatched ELO structures can be relaxed by additional tilting of ELO wings while still preserving their high quality. Finally, recent progresses in the lateral overgrowth of semiconductors, including new mask materials and liquid-phase electroepitaxial growth on substrates coated by electrically conductive masks, are presented. New versions of the ELO technique from solution and from the vapor (growth from ridges and pendeo-epitaxy) are described and compared with standard ELO. A wide range of semiconductors, including III–V compounds grown from solution and vapor-grown GaN, are used to illustrate phenomena discussed. Very often, the similar behavior of various ELO structures reveals that the phenomena presented are not related to a specific group of compounds or their growth techniques, but have a much more general nature.

30.1 **Overview** 1000

30.2 **Mechanism of Epitaxial Lateral Overgrowth from the Liquid Phase** 1002
    30.2.1 Choice of Substrate Geometry for Growth of ELO Layers 1004
    30.2.2 Optimization of Liquid-Phase Lateral Overgrowth Procedure 1007

30.3 **Dislocations in ELO Layers** 1011
    30.3.1 Filtration of Substrate Dislocations in ELO 1011
    30.3.2 Structural Perfection of Coalescence Front in Fully Overgrown ELO Structures . 1014

30.4 **Strain in ELO Layers** 1016
    30.4.1 Mask-Induced Strain in Homoepitaxial ELO Layers 1017
    30.4.2 Thermal Strain in ELO Layers 1024

30.5 **Recent Progress in Lateral Overgrowth of Semiconductor Structures** 1026
    30.5.1 Developments in Liquid-Phase ELO Growth 1027
    30.5.2 New Concepts of ELO Growth 1030

30.6 **Concluding Remarks** 1034

References 1035

30.1 Overview

Modern micro- and optoelectronic devices consist of thin layers grown epitaxially on a substrate. These layers must be of high crystallographic quality. Very often however, there are no substrates available for lattice-matched epitaxy, and dislocations are generated at the layer–substrate interface to relax the lattice-mismatch strain. Segments of these dislocations thread to the surface of the epilayer, and then to the next-grown layers of the structure, deteriorating parameters of the device and leading to its fast degradation. To reduce the density of threading dislocations in lattice-mismatched heteroepitaxial structures a buffer layer with a graded or abrupt composition profile is usually deposited on an available substrate to obtain a layer with the required lattice parameter value. Strained layer superlattices are often inserted into the buffer to bend threading dislocations and partially prevent them from propagating to the surface [30.1]. In situ or ex situ annealing of buffers has also been employed to induce mutual reactions between threading dislocations that eventually lead to their annihilation [30.2]. In addition, for heterostructures with large lattice mismatch, buffer growth is usually initiated at relatively low temperature, and is then continued at a higher temperature (so-called two-step growth) [30.3]. All these sophisticated methods of buffer layer engineering have been proven to reduce significantly density of defects in lattice-mismatched heterostructures. However, despite the progress made, the best modern buffers still contain dislocations with a density of $\approx 10^6$ cm^{-2}, which is often too high for device applications. Therefore, the epitaxial lateral overgrowth (ELO) technique has been developed to block dislocations threading from the substrate or substrate/buffer structure to the next-grown epitaxial layers.

Figure 30.1 schematically illustrates the principle of epitaxial lateral overgrowth. First the substrate or substrate with a suitable buffer is covered by a thin amorphous masking film. Dielectrics, such as SiO$_2$ or Si$_3$N$_4$, are commonly used to mask the substrate, but other materials such as tungsten [30.4, 5], zirconium nitride [30.6] or graphite [30.7] have also been successfully applied for the ELO of III–V semiconductors. In the next step the substrate is patterned by conventional photolithography and etching to form on its whole area a grating of mask-free seeding windows of width W separated by masked areas of width M (Fig. 30.1a). The ELO growth starts by a selective-area epitaxy, i.e., nucleation takes place exclusively in narrow windows (Fig. 30.1b). When the layer becomes thicker than the masking film the growth proceeds also in a lateral direction over the mask. Finally, a new epitaxial layer fully covers the patterned substrate if the growth time is sufficiently long for coalescence of ELO stripes grown from neighboring seeds. The main advantage of the ELO process is that dislocations can thread from the substrate into the epitaxial layer through the very narrow windows, so respective etch pits should be visible only in the part of the ELO layer grown vertically from the seed (Fig. 30.1b). Since there is no continuity of crystal lattice at the interface between monocrystalline substrate and the amorphous mask, substrate dislocations cannot propagate into the mask and must vanish at the substrate surface. In this way the mask efficiently blocks dislocations and their density in the laterally grown parts (wings) of the layer should be significantly reduced. Indeed, there are many experimental evidences showing that the density of dislocations in the wings area is orders of magnitude lower than that observed in standard planar epilayers (Sect. 30.3.1). Thus the ELO process

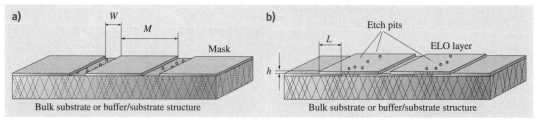

Fig. 30.1a,b Principle of epitaxial lateral overgrowth. (**a**) A substrate or substrate with a suitable buffer is covered by a thin amorphous masking film. Seeding windows of width W separated by masked areas of width M are opened up in the mask. (**b**) The ELO growth begins exclusively inside the seeding windows. Then the growth proceeds laterally over the mask. Note that substrate dislocations, marked by *dotted lines*, propagate to the ELO layer through the openings in the mask, only. L and h denote the width of the laterally grown part (wing) of the layer and its thickness, respectively

is a powerful method to grow epilayers with a low density of dislocations on heavily dislocated substrates such as the relaxed buffers deposited on lattice-mismatched substrates. When combined with well-developed methods of buffer layers engineering the ELO technique offers the possibility of producing high-quality substrates with adjustable value of lattice constant required by modern electronics. If ternary buffers are used the value of the lattice parameter might be controlled by choosing the proper composition of the buffer. Next, homoepitaxial lattice matched lateral overgrowth would be employed to remove the buffer dislocations.

A more detailed discussion on the dislocation filtration processes in ELO together with some examples will be presented in Sect. 30.3. However, it must be mentioned here that, according to the ELO concept presented in Fig. 30.1, the number of dislocations propagating into the layer should decrease as the width W of the seeding window decreases. In particular, the growth of a completely dislocation-free layer might be possible if the window width is smaller than the average distance between dislocations in the substrate. This means that for the density of dislocations of $\approx 10^{10}$ cm^{-2} the condition $W \ll 100$ nm should be fulfilled. In principle, windows a few nanometers wide should be sufficient for the transfer of information on the lattice parameter from the substrate to the epilayer, i.e., for epitaxial growth. Due to technical reasons, however, windows a few micrometers wide are usually applied. Note also that in many epitaxial systems dislocations threading through the seeding windows are inclined at some angle to the epilayer surface, so the width of dislocated area on the surface increases as the layer grows thicker [30.8]. Thus, the ELO layers should be as wide and as thin as possible in order to maximize dislocation-free wing area available for devices. In other words, the largest value of the aspect (width/thickness) ratio AR, defined in this work by the equation AR $= (2L + W)/h$, is required.

Besides the use for defect filtration during the heteroepitaxy of lattice-mismatched systems, the ELO technique in its homoepitaxial version has also found practical applications in the production of silicon-on-insulator structures [30.9], metal-oxide-semiconductor (MOS) transistors [30.10], field-effect transistors [30.11], solar cells [30.12], pressure sensors [30.13], and for three-dimensional device integration [30.14]. If metallic masks are applied they may be further used as buried electrical contacts to devices produced by ELO [30.11]. Finally, if the windows opened up in the insulating mask are so narrow that electrical conductance of ELO–substrate connections becomes small, the ELO process can be used for growing layers electrically separated from the substrate. This might be the way to obtain electrically insulated epilayers of semiconductors for which lattice-matched semi-insulating substrates are not available (e.g., GaSb).

To the best of our knowledge, the ELO process was initiated in 1980 when *McClelland* and his coworkers applied vapor-phase epitaxy (VPE) to grow GaAs epitaxial layers on masked-GaAs substrate [30.15]. Since the ELO layer was easily separable from its host substrate by cleaving, the technique was designed as an efficient way of producing GaAs epitaxial films on reusable substrates. *Tsaur* et al. were probably the first to report (in 1982) an efficient reduction of dislocation density in GaAs layers grown by VPE lateral overgrowth on Ge-coated Si substrates [30.16]. Then the ELO technique was used by *Jastrzebski*'s [30.9] and *Bauser*'s [30.17] groups to produce silicon-on-insulator structures by VPE and liquid-phase epitaxy (LPE), respectively. Finally, the efficient filtration of substrate dislocations has been reported in a series of papers on the growth of SiGe, GaAs, GaP, InP, InGaAs, and InGaP ELO layers on lattice-mismatched substrates [30.18–22]. However, the most spectacular recent achievement of the ELO technique was the breakthrough in the development of long-lifetime GaN/InGaN blue lasers, partly being due to the high efficiency of defects filtration during the lateral growth of GaN on sapphire [30.23].

Nowadays, a large body of ELO research concentrates on the lateral overgrowth of GaN epilayers on sapphire or SiC substrates using metalorganic VPE (MOVPE) or hydride VPE (HVPE) techniques. This is a result of the market demand for low-dislocation-density GaN substrates. GaN ELO activity is so dominant that sometimes the technique is called ELOG, which stands for epitaxial lateral overgrowth of gallium nitride. Apparently this is not correct since the principle of the ELO process is much more general and is not limited to either a specific group of materials or to a single epitaxial growth technique. In parallel, a great deal of increase in research on the lateral overgrowth of zincblende III–V epilayers on various substrates is observed.

The aim of this chapter is to present recent developments in the lateral overgrowth of compound semiconductors and review its present state. The chapter is organized as follows. In Sect. 30.2 we will focus on the mechanism of the ELO growth of zincblende III–V compounds from the liquid phase, highlighting the phenomena that are crucial for growing high-quality layers with large aspect ratio. Epitaxy from the liquid phase has been chosen since, as will be discussed in that section, the equilibrium growth techniques such as LPE are the most suitable for lateral overgrowth. The issue of defect filtration in the ELO procedure is widely addressed in Sect. 30.3.1. Numerous examples which are instrumental in showing that the defect filtration mechanism schematically shown in Fig. 30.1 is very efficient and leading to significant progresses in the development of high-performance semiconductor devices made of lattice mismatched structures, are presented. In Sect. 30.3.2 structural perfection of seams that appear when layers grown from neighboring seeds merge in fully overgrown ELO structures will be discussed. Strain in epitaxial layers may lead to the generation of new defects if it is too large. Therefore, the issue of strain in ELO layers will be extensively addressed in Sect. 30.4. First, we will concentrate on strain arising due to interaction of ELO layers with the mask underneath. This phenomenon leads to tilting of ELO wings and is commonly found in various ELO structures. The origin of mask-induced wing tilting and possible ways of its control will be presented. Then, the thermal strain in lattice-mismatched ELO structures will be discussed (Sect. 30.4.2). We will show that in ELO structures thermal strain can be relaxed by additional tilting of ELO wings while still preserving their high quality. Finally, in Sect. 30.5 the recent progress made in the lateral overgrowth of semiconductor structures is presented. New developments in the ELO growth from the liquid phase include new mask materials and liquid-phase electroepitaxial growth on substrates coated by electrically conductive masks. Then, the new versions of ELO technique from solution and from the vapor phase, namely growth from ridges and pendeo-epitaxy, will be discussed and compared with the standard ELO. Since the author's intention was to present the ELO growth of a wide range of compound semiconductors, both the zincblende III–V compounds grown from the liquid solution as well as the vapor-grown GaN will be used to illustrate the phenomena discussed. In particular, in Sects. 30.3 and 30.4 it will be shown that a quite similar behavior is observed in various ELO systems despite the large differences in their properties and the variety of growth techniques applied. This, in turn, clearly indicates that the phenomena presented are not related to a specific group of compounds or their growth techniques, but have much more general nature.

30.2 Mechanism of Epitaxial Lateral Overgrowth from the Liquid Phase

An efficient ELO procedure requires a large growth rate anisotropy, i.e., growth conditions are necessary at which the lateral growth of the epilayer is much faster than that in the direction normal to the substrate. This can be achieved by taking advantage of the natural growth anisotropy of various crystal faces.

Figure 30.2 schematically shows growth rate versus interface supersaturation for three types of crystal faces. On a perfect singular face, atoms can be incorporated into solid in the form of two-dimensional nuclei only. If the face is singular but imperfect, the surface imperfections (e.g., dislocations) supply steps necessary for its growth. Due to the limited rate of the surface processes involved, larger surface supersaturation is required in these two growth modes to get a notable growth velocity. On the other hand, atoms can be added to an atomically rough crystal face in a random way and the growth rate of such a face varies linearly with the interface supersaturation. As discussed by *Nishinaga* [30.18] the basic idea of ELO lies in fundamental dissimilarities between those growth modes. If a slowly grown facet covers the upper plane of ELO while the sidewalls are rough, then for low supersaturation, growth rates in ver-

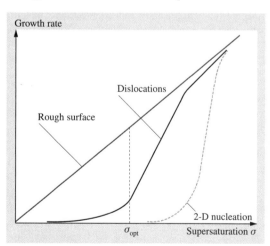

Fig. 30.2 Growth rate versus surface supersaturation for various crystal faces: perfect singular (*brown dashed line*), imperfect singular (*thin central line*), and atomically rough face (*solid line*). Dashed line schematically marks the surface supersaturation σ_{opt} that is optimal for large growth anisotropy in ELO

tical and horizontal directions differ significantly (see the vertical dashed line in Fig. 30.2 that schematically marks the optimal supersaturation σ_{opt}). Consequently, a large anisotropy of shape of ELO layers can be obtained.

Figure 30.2 indicates that there are two conditions necessary for successful lateral overgrowth. First, a proper geometry of the substrate must be chosen to keep the upper surface of the layer singular while the sidewalls of ELO remain atomically rough. Next, the growth conditions must be carefully adjusted for precise control of the surface supersaturation. These issues will be discussed in detail in Sects. 30.2.1 and 30.2.2. To illustrate that the mechanism of ELO growth presented above works in practice and to study the relative importance of various physical phenomena occurring during growth, we have conducted numerical simulations of ELO growth by LPE [30.24].

Figure 30.3 illustrates conceptually the solute movement in the liquid solution occurring during an ELO growth by LPE. Due to symmetry, only half of the substrate and liquid zone is shown. During LPE growth, the temperature of the system is slowly lowered to supersaturate the liquid solution. Since there is no nucleation on the mask, solute species diffuse exclusively towards the seeding area and are then incorporated into the growing ELO layer. Sidewall of the ELO layer is atomically rough, so there is no barrier there for incorporation of arriving species into the solid. Therefore, the solute concentration in the liquid zone near the side ELO face is equal to C_{eq}, the equilibrium concentration, a value that is determined by the phase diagram and actual temperature (Fig. 30.3). However, the upper ELO layer is faceted and the surface solute concentration there C_{in} is larger than the equilibrium concentration. This gives rise to a horizontal solute concentration gradient and so-called near-surface diffusion [30.25] in the liquid of solute species from upper ELO surface to its sidewall. It is obvious that the presence of near-surface diffusion significantly enhances lateral growth of ELO layers.

In the model [30.24], we consider a computational domain such as that shown in Fig. 30.3 and compute the change in the solute concentration field in the liquid zone as the system is cooled to force epitaxial growth on masked substrate. As discussed before, we assume a linear surface kinetic law to calculate the solute concentration C_{in} in the vicinity of the upper ELO surface while the solid–liquid phase equilibrium condition is used at the sidewall of the ELO layer. Moreover, the possible contribution of the Gibbs–Thomson effect is also considered to take into account the dependence of local equilibrium solute concentration on the surface curvature of the growing layer. Finally, the evolution of the solid–liquid interface shape with time is obtained using the growth velocity calculated from the solute concentration field in the vicinity of the growing layer. The results of calculations are discussed below, taking the GaAs/GaAs ELO system as an example due to its technological importance and the availability of the required input data.

Figure 30.4a shows the computed distribution of As concentration in the Ga/As solution after 2.5 h of ELO growth with an initial growth temperature of 650 °C and a cooling rate of 0.5 °C/min In the far field, isoconcentration contours are circular since at this length scale ELO crystal looks like a point located in the lower left corner of the computational domain. Close to the crystal, however, the As concentration distribution shows additional features. In particular, a solute concentration gradient appears along the upper ELO surface, leading to near-surface diffusion of As towards the ELO sidewall (compare Fig. 30.3) and enhancing lateral growth rate. Furthermore, the As concentration gradient close to sidewall of ELO is much larger than that at the upper face. This indicates that the layer grows faster laterally than in the vertical direction.

Figure 30.4b presents the cross section of GaAs ELO layer calculated for the same input parameters. As seen, the shape of the upper ELO surface is very flat, which closely agrees with experimental observa-

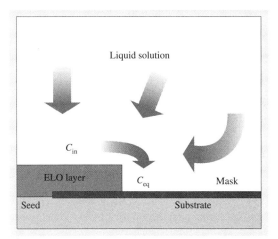

Fig. 30.3 Solute flow during LPE growth of ELO layer. The *arrow* from C_{in} to C_{eq} marks diffusion of solute in the liquid phase from upper surface of ELO to its sidewall (near-surface diffusion)

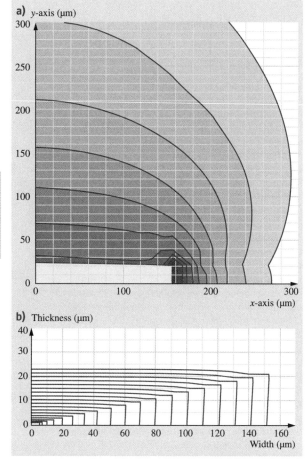

Fig. 30.4a,b Computed distribution of As concentration in the GaAs solution (**a**) and cross section of GaAs ELO layer (**b**) after 2.5 h LPE growth with initial growth temperature of 650 °C and cooling rate of 0.5 °C/min

rate [30.24]. Therefore, it can be successfully used to optimize the LPE growth procedure. In addition, we have shown that a spiky growth starts from the ELO corner, which is not observed in real experiments, if the dependence of solid–liquid equilibrium on surface curvature is not taken into account. This demonstrates the importance of Gibbs–Thomson effect in growth of ELO layers.

Having presented the mechanism of lateral overgrowth it is now worth to discuss which epitaxial growth technique is most suitable for ELO growth. As was mentioned earlier the surface supersaturation must be kept low to grow ELO layers with large aspect ratio. Otherwise, two-dimensional nucleation takes place on the upper ELO surface leading to higher vertical growth rate and consequently less growth anisotropy. Therefore the equilibrium growth techniques such as LPE should be chosen, if possible, for lateral overgrowth. Indeed, as will be shown later, ELO layers with aspect ratio as large as 130 can be grown from a liquid phase. Solution or melt growth of group III nitrides is extremely complicated due to low solubility of nitrogen in liquid metals (Sect. 30.5.2). Therefore, MOVPE or HVPE is commonly used nowadays to grow ELO structures of these compounds. Then, however, large supersaturation at the growing face makes control of growth anisotropy difficult, so lateral structures of GaN with aspect ratio up to 6 are usually obtained [30.27]. Molecular beam epitaxy (MBE) growth of ELO layers is even more complicated. The ELO technique requires that growth proceeds from the seeding areas only while deposition of polycrystalline material on the mask is hard to avoid during MBE [30.28, 29]. Sophisticated systems with molecular beams oriented at low angle to the substrate, low growth rate, and precise temperature control [30.30, 31] or special substrate preparation and growth procedures [30.32] are necessary to get the growth started selectively from the seeds. Even then, however, MBE-grown ELO layers are usually very narrow, which makes their application in devices production very difficult.

30.2.1 Choice of Substrate Geometry for Growth of ELO Layers

As noted above, large anisotropy of ELO growth rate can be obtained if the upper surface of the layer is facetted while the sidewalls of ELO remain atomically rough. In practice, for zincblende III–V semiconductors this can be achieved using (111)- or (100)-oriented substrate and by twisting the line openings in the mask by

tions. Moreover, the lateral growth rate is much larger than that in the vertical direction, resulting in a thin and wide layer. A similar shape of layer was obtained by *Yan* et al. in their two-dimensional simulations of InP ELO growth by LPE [30.26]. However, in our case the prediction of a flat top face is a *result* of surface kinetics and diffusion processes included in the model, while in their approach the vertical growth rate is *assumed* to be the same along the whole upper surface. It is also noteworthy to mention that our model provides a good prediction for the dependence of ELO aspect ratio on surface kinetic coefficient and cooling

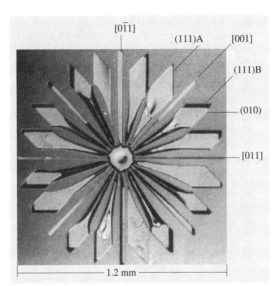

Fig. 30.5 Microphotograph of a GaAs ELO layer grown by LPE on a (100) GaAs substrate with a star-like pattern of seeds cut in the SiO$_2$ mask

some degrees off from the main crystallographic directions. Usually, optimal orientations of seeding lines in the substrate plane are determined experimentally by growing an ELO layer on a masked substrate with a star-like pattern of the seeds and by studying lateral growth rate versus seed orientation.

Figure 30.5 shows a plane view of a GaAs epilayer grown by LPE on a (100) GaAs substrate with a star-like pattern of openings cut in the SiO$_2$ mask [30.33]. The growth was forced by cooling the system from the temperature of 750 °C with a rate of 0.2 °C/min for 50 min. From Fig. 30.5 the dependence of lateral growth rate on the orientation of the seeds can be easily found. The width of the ELO is the largest when the stripe is ±15 and ±30° off-oriented from the [011] direction or its equivalent orientations, while it is the smallest for the [011] and [001] directions. From simple geometrical considerations the facets on the side walls of ELO layers can be identified as the slowly grown {111} and {100} planes. Their formation proceeded much faster when the stripe was oriented in the ⟨011⟩ and ⟨001⟩ directions, which slowed down the lateral development of the ELO layer. Therefore, on the (100) substrate the line seeds oriented 15, 30, 60 or 75° off from the ⟨011⟩ direction should be used to obtain a large value of the lateral-to-normal growth rate ratio. Note that the angular dependence of the lateral growth rate is centrosymmetric. As will be discussed later this is no longer true if slightly misoriented substrate is used. Note also that the width of ELO stripes shown in Fig. 30.5 increases with the distance r from the star center. As shown by *Zhang* and *Nishinaga* [30.34] this is due to the fact that the effective sink area for the growth units, and thus also the surface supersaturation, varies with r. For lateral overgrowth of GaSb on (100)-oriented GaSb substrates by LPE we have found the same dependence of ELO lateral growth rate on seed orientation. Results shown in Fig. 30.5 are in agreement with those published by *Zhang* and *Nishinaga* [30.34] and by *Naritsuka* and *Nishinaga* [30.35] for liquid-phase homoepitaxial lateral overgrowth of GaAs and InP, respectively. Similar angular dependence of lateral overgrowth has also been reported for ELO growth of GaAs by MOVPE [30.36] and for InP by VPE [30.37]. For (111)-oriented substrates analogous dependences can be found in [30.38, 39] for LPE ELO growth of GaAs and GaP, respectively.

In the case discussed thus far, the vertical growth of ELO was possible due to surface steps supplied by dislocations present in the substrates. However, sometimes dislocation-free substrates are used, for example, in silicon homoepitaxy. They must be slightly misoriented to provide steps necessary for crystal growth if uncontrolled two-dimensional nucleation is to be avoided. Then, the size and shape of ELO layer becomes dependent on the angle between seeding line and the substrate miscut directions. This issue has been discussed in detail by *Bergmann* et al. [30.40] and then by *Bergmann* [30.41] for lateral overgrowth of silicon on perfect Si substrates. On off-oriented imperfect substrates additional restriction applies for optimal direction of seed orientation. *Sakawa* and *Nishinaga* [30.42] have shown that, in such case, from many equivalent seed directions that are optimal on the (100) plane (Fig. 30.5), the one that should be chosen is that for which the density of misorientation steps inside the seeding area is the smallest.

The discussion above indicates that the presence of dislocations significantly affects the ELO growth mode. To present this issue in more detail Fig. 30.6 shows a sketch of the ELO layer grown from a line window on a misoriented, dislocation-free substrate. Direction of seeding line perpendicular to substrate miscut direction is assumed. As long as two-dimensional nucleation does not take place, the layer grows by flow of steps according to the misorientation direction (i. e., to the right side). Then the ELO growth is self-limited, i. e., it stops as soon as all the steps reach the edge of the layer

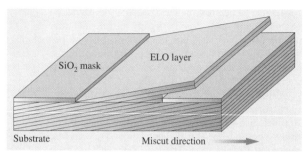

Fig. 30.6 Schematic illustration of ELO growth on misoriented, dislocation-free substrates; the seeding line is set perpendicular to the substrate miscut direction

Fig. 30.7 Cross section of a GaAs ELO layer grown on a GaAs substrate with the surface misoriented by 3° from the (100) plane; substrate is miscut in the right-hand side direction

Fig. 30.8 Scanning electron microscopy image of Si ELO layers grown by LPE on a SiO$_2$-masked Si substrate. Perfectly grown, defect-free layers are visible in the *foreground*; irregular-shaped defective layers are seen in the *background* (courtesy of Banhart)

in Fig. 30.6 that, for small substrate misorientation angles, substrate steps are partly in the shade of the mask edge and the ELO layer cannot get out of the seeding area if the mask is too thick. This prediction is also in agreement with experimental observations [30.40].

On the other hand, the situation is quite different if substrate dislocations contribute to the ELO growth. Figure 30.7 shows a cross section of a GaAs ELO layer grown on a GaAs substrate with the surface miscut by 3° from the (100) plane [30.43]. The seeds were oriented 15° off from the [0$\bar{1}$1] direction for fast lateral overgrowth, and were nearly perpendicular to the direction of substrate miscut. It is seen that, similar to the case shown in Fig. 30.6, the thickness of the ELO layer is not uniform. The reason for this is the same as before: substrate steps lead to the layer growth to the right-hand side. This time, however, growth to the left of the window (i.e., in the direction opposite to substrate miscut axis) is also observed. Such growth can be explained by considering dislocations (with a density of $1.5 \times 10^3 \, \text{cm}^{-2}$) present in our GaAs substrate. These dislocations provided steps that made an additional contribution to epitaxial growth and led to continuous growth of the layer to both sides of the seeding line. It is noteworthy that the surface of the layer is inclined to the substrate surface by an angle of $\approx 3°$ (Fig. 30.7). Simple geometrical considerations of sample geometry show that this is the angle at which the (100) plane intersects with the substrate plane in the (1$\bar{1}$0) cleavage section. This indicates that, despite substrate misorientation, the upper surface of the ELO layer forms the exact (100) plane, which is in agreement with our explanation.

As the final example of the essential role played by dislocations in the growth of ELO layers, Fig. 30.8 shows a scanning electron microscopy (SEM) image of silicon ELO layers grown by LPE on SiO$_2$-masked Si substrate [30.44]. Perfectly grown flat ELO layers of thickness around 2–3 μm are seen in the foreground. Studies by transmission electron microscopy (TEM) showed that they were entirely free of crystallographic defects. However, as shown in the background of the figure, some ELO layers of thickness exceeding 20 μm and quite different shape were also found on the same substrate wafer. They showed deep grooves and were about twice as wide as perfectly grown layers. TEM analysis revealed that those layers contained a regular arrangement of dislocations generated during ELO growth in the vicinity of SiO$_2$ mask edge. The origin of these dislocations is not fully clarified but it is believed that their formation is influenced by stresses caused

and the sidewall is covered by a slowly grown facet. In particular, this means that no growth should take place to the left of the seeding line. Indeed, such behavior has been observed during LPE growth of silicon ELO layers on perfect silicon substrates [30.41]. Note also by thermal oxidation of substrate wafer, the shape of the oxide edge, and nonuniform supersaturation of liquid solution volume. Once dislocations are generated they govern further development of the growth process. Instead of flow of steps supplied by substrate misorientation as illustrated in Fig. 30.6, dislocations become the dominant growth step source that leads to significant enhancement of vertical growth rate and worse surface morphology.

30.2.2 Optimization of Liquid-Phase Lateral Overgrowth Procedure

Having presented the effect of substrate geometry on the ELO growth process, in this section we will show examples of how parameters of an LPE procedure should be adjusted to optimize supersaturation near the crystal faces and obtain high growth anisotropy. In LPE, surface supersaturation reflects the relative magnitude of the solute supply from bulk of the liquid phase and solute consumption at the surface of growing crystal. Thus, the main parameters controlling these processes are the LPE growth temperature, the initial supercooling of the liquid phase, and the cooling rate.

Figure 30.9 shows lateral $V_{lat} = L/t$ and vertical $V_{ver} = h/t$ growth rates versus LPE growth temperature T_0 for nominally undoped GaAs ELO layers grown on SiO$_2$-masked GaAs substrates. Figure 30.10 presents values of the aspect ratio (AR) for the same layers [30.43]. As seen, for high LPE temperature the vertical ELO growth rate is large since some thermal roughening of the upper crystal face takes place. Moreover, surface kinetic processes are very fast, which leads to an additional increase in the vertical growth rate. This results in ELO layers with a small value of the aspect ratio. On the other hand, for very low growth temperatures the sidewalls cannot be considered as ideally rough. Most probably surface kinetic processes start to play a role there and lateral growth rate decreases, which leads again to a decrease of the aspect ratio. Thus, there is an optimum growth temperature T_{opt} at which ELO layers have the largest aspect ratio. As shown by Yan et al. [30.25] for InP ELO layers T_{opt} corresponds to the temperature at which the interface supersaturation at the upper ELO face is the smallest. For other growth temperatures surface supersaturation is larger than σ_{opt} (compare Fig. 30.2). It is obvious that the optimum ELO growth temperature depends on many parameters such as the slope of the liquidus curve of the phase diagram, the geometry of the LPE system, etc., so it must be determined experimen-

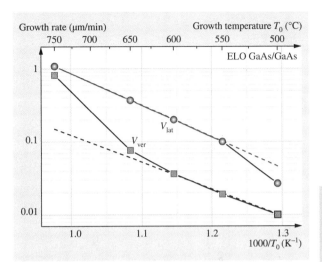

Fig. 30.9 Lateral V_{lat} (*dots*) and vertical V_{ver} (*squares*) growth rates versus LPE growth temperature T_0 for undoped GaAs ELO layers grown on GaAs substrates masked by SiO$_2$ film; seeding windows and masked area were 10 and 100 μm wide, respectively. The *dashed lines* guide the eye along the $V \propto \exp(-1/T_0)$ dependence

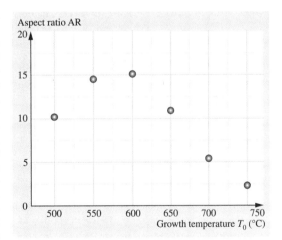

Fig. 30.10 Aspect ratio (AR) of undoped GaAs ELO layers versus LPE growth temperature T_0

tally for each particular case under study. Values of T_{opt} equal to 500, 580, and 530 °C have been reported for LPE growth of InP/InP [30.25], GaAs/GaAs [30.43], and GaAs/Si [30.45] ELO systems, respectively. For homoepitaxial GaSb ELO layers grown by LPE the optimal growth temperature $T_{opt} \leq 350$ °C has been found [30.46].

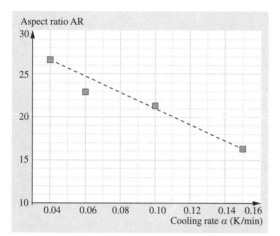

Fig. 30.11 Aspect ratio of Si-doped GaAs ELO layers grown at $T_0 = 575\,°C$ versus cooling rate; silicon concentration in the Ga-As solution was [Si] = 0.5 at. %

steps supplied by substrate miscut as shown in Fig. 30.6. Sidewall of the layer beginning to grow laterally is strongly curved, and due to the Gibbs–Thomson effect, it requires higher equilibrium solute concentration than the planar face. Thus, instead of growing laterally the layer is dissolved and cannot get out of the opening in the mask as long as the liquid phase is not supersaturated sufficiently [30.47].

The situation is different if dislocations enhancing vertical ELO growth are present in the substrate. To illustrate this point Fig. 30.12a shows a cross section of silicon-doped GaAs ELO layer grown on GaAs substrate by LPE without any initial supersaturation of the liquid solution [30.48]. As we have already reported [30.49], by proper chemical etching, boundaries between vertically and laterally grown parts of the layer can be revealed, allowing analysis of the temporal development of ELO shape. These boundaries are sketched in Fig. 30.12b by dashed black lines, while dashed brown lines schematically mark their shape that would be observed without the contribution of the Gibbs–Thomson effect. Note that the Gibbs–Thomson

To keep surface supersaturation at a low value, the cooling rate during LPE should be as low as possible. This is illustrated in Fig. 30.11, which shows the aspect ratio of Si-doped GaAs ELO layers versus cooling rate α [30.43]. The main idea behind that is to supply solute to the surface of the growing layer slowly enough that it can be transported by near-surface diffusion from upper to side walls of ELO and be incorporated there without any significant increase in surface supersaturation at the upper face. Similar dependence of aspect ratio on cooling rate has been found for InP [30.26] and GaSb [30.46] ELO systems. Also numerical simulations of ELO growth by LPE predict AR versus α dependence similar to that shown in Fig. 30.11 [30.24, 26].

Optimal choice of initial solution supercooling during ELO growth by LPE requires some additional effects to be taken into account. It is apparent that supercooling of the solution should be as small as possible. Then, the initial supersaturation of bulk of the liquid, and consequently that at the ELO surface, can be kept low. Indeed, we have found experimentally that, for the LPE growth of GaSb ELO structures, smaller values of initial melt supercooling lead to larger values of aspect ratio [30.46]. There are ELO systems, however, which require some initial supercooling of the liquid solution to initiate growth. Otherwise, the Gibbs–Thomson effect hinders the layer from growing out from the opening in the mask. In particular, this is the case for LPE growth of Si ELO layers on defect-free silicon substrates. Then the ELO layer grows only by flow of

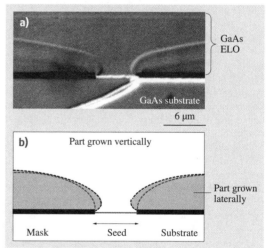

Fig. 30.12a,b Microphotograph (**a**) and schematic drawing (**b**) of the cross section of Si-doped GaAs ELO layer grown on GaAs substrate by LPE without initial supersaturation of the liquid solution. Boundaries between vertically and laterally grown parts of the layer are marked by *black dashed line* while their shape that would be observed without contribution of the Gibbs–Thomson effect are drawn by the *brown dashed lines* in (**b**). Note dissolution in the lateral direction induced by the Gibbs–Thomson effect at the beginning of growth

effect induces *dissolution* in the lateral direction instead of *growth* at the beginning of epitaxy. This is clearly visible as initial narrowing of the vertically grown part of ELO. However, during continuous cooling of the system, steps supplied by substrate dislocations still allow for vertical growth of the layer in the middle of the seed despite slow dissolution induced by the Gibbs–Thomson effect taking place in the direction parallel to the substrate. As soon as the layer grows thicker, curvature of the solid–liquid interface at the ELO sidewall decreases. Consequently, the contribution of the Gibbs–Thomson effect decreases and lateral overgrowth along the mask begins. We have proved that this explanation is correct by showing that the magnitude of lateral dissolution caused by the Gibbs–Thomson effect can be reduced if slightly supersaturated liquid solution is used to grow the layer [30.33]. This example again illustrates the important role of dislocations in the growth of ELO layers.

Doping has been found to be a useful way to reduce the vertical growth rate, leading to thin and wide ELO layers. Figure 30.13a shows an image of a GaAs ELO layer grown from a nominally undoped solution by LPE on SiO_2-masked GaAs substrate at 750 °C [30.49]. Since the growth temperature was much higher than the optimal temperature, the cross section of the layer has

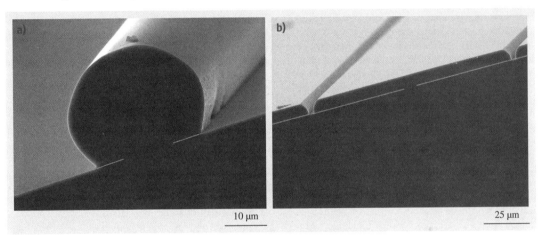

Fig. 30.13a,b Scanning electron microscopy images of undoped (**a**) and Si-doped (**b**) GaAs ELO layers grown at 750 °C on SiO_2-masked GaAs substrates. Besides doping, all the other growth parameters were the same for both layers

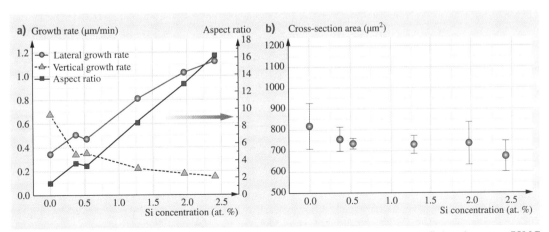

Fig. 30.14 (**a**) Lateral and vertical growth rates, and aspect ratio of GaAs ELO layers grown on GaAs substrates at 750 °C versus Si concentration [Si] added to the Ga-As liquid solution; (**b**) the cross sectional area of the same layers versus [Si]

a circular shape and the aspect ration is small (compare Fig. 30.10). A noticeable improvement of ELO shape has been achieved by adding 2.4 at. % Si to the Ga/As solution (Fig. 30.13b). As a result, the upper surface of the layer becomes very flat and the aspect ratio increases.

In order to study the effect in more detail, Fig. 30.14a shows a plot of lateral and vertical growth rates, and of the ELO aspect ratio of Si-doped GaAs layers versus the silicon concentration in the liquid phase. As seen, the vertical growth rate continuously decreases while the lateral one increases for higher dopant concentration in the liquid solution. As a result the ELO aspect ratio increases significantly with the doping level. It is important to emphasize that, whereas ELO layers become thinner and wider owing to the Si doping, within the limits of experimental error their cross sectional areas, being a measure of the amount of material deposited on the substrate, do not change (Fig. 30.14b). This means that the presence of Si has a negligible influence on the phase equilibrium, and a possible change of the liquid-zone supersaturation with doping cannot explain the effect presented. Thus, other processes affecting the redistribution of the growth units on the ELO surfaces should be considered.

The phenomenon of crystal habit modification by impurities is well known [30.50]. It might be attributed to hindering of the smooth step flow on the crystal face [30.51, 52]. The possible mechanism involves steps that are forced to squeeze between immobile impurity atoms tightly adsorbed on the surface. Thus, the steps have a local large curvature, which results in a decrease of their flow velocity. When applied to ELO growth this model leads us to the following interpretation of the data shown in Fig. 30.14: silicon atoms, when introduced to the liquid solution, are adsorbed on the upper ELO surface and render the steps there less operative, reducing in this way the vertical growth rate. On the other hand, the doping should have no direct influence on the random incorporation of growth units on atomically rough ELO faces. Since solute atoms cannot be easily incorporated onto the upper surface they are efficiently transported to the sidewalls by the near-surface diffusion. In that way the lateral growth rate increases at the expense of vertical one, while the total amount of material deposited on the substrate does not change.

It is worth mentioning that we have found a similar effect for Sn and Te dopants, although the reduction in the GaAs vertical growth rate caused by these impurities was much smaller than that for Si [30.49]. The reason is still unclear, but different values of the surface mobility of Si, Sn, and Te might be responsible. We have also observed the same increase of ELO aspect ratio by Si dopant during LPE growth of Si-doped GaSb ELO layers [30.46]. Change of ELO shape with doping similar to that shown in Fig. 30.13 has been observed during ELO growth of GaAs on GaAs-coated Si substrates by LPE [30.8]. *Kim* and *Lee* reported that selenium is even more efficient than silicon in reducing vertical growth rate of ELO GaAs [30.53], but our experiments do not confirm their conclusion. It is important to point out that, in principle, similar phenomenon should be expected in other epitaxial systems in which the upper ELO face grows by the flow of surface steps. Indeed, a strong influence of magnesium doping on the growth habit of the GaN ELO layers grown by MOVPE on sapphire was found recently [30.54]. The effect of Si doping was negligible in that case. This shows that the phenomenon of vertical growth rate reduction by doping is not a specific attribute of ELO layers grown by LPE and has much more general nature, although its magnitude depends on growth conditions and the most effective impurity must be found for each particular system under study. Basic research is still needed to obtain a deeper insight into the effect of impurities on the growing crystal and to indicate which best suits the growth of thin and wide ELO layers.

Finally it should be noted that in some systems the value of the ELO aspect ratio is much more sensitive to the presence of dopants in the liquid solution than to the LPE growth temperature. One may try to increase the aspect ratio by growing ELO layers at a relatively low optimal temperature T_{opt} as discussed earlier. However, the dopant solubility at this temperature might be too low to increase further the aspect ratio. Alternatively, some increase in the growth temperature over T_{opt} with a simultaneous increase of doping might be a reasonable way to grow thin and wide ELO layers. This procedure can give satisfactory results only if the aspect ratio increase due to higher doping is larger that its decrease caused by higher LPE growth temperature. We have found that this is the case for Si-doped GaSb ELO layers [30.46]. However, care is advised when trying this approach since heavy doping may lead to a worse surface morphology of the layers.

To summarize this section, Fig. 30.15 presents a scanning electron microscopy cross section image of a GaAs ELO layer grown in our laboratory by LPE on SiO_2-masked GaAs substrate. The layer was doped with silicon and the LPE growth conditions were carefully adjusted according to the rules discussed above to get

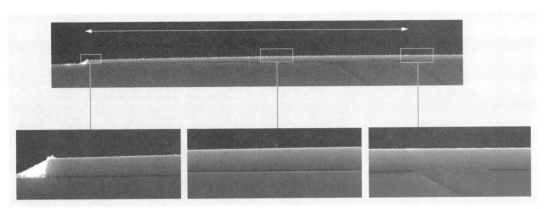

Fig. 30.15 Cross-section of Si-doped GaAs ELO layer grown by LPE on SiO_2-masked GaAs substrate. The thickness of the layer in its central part is 2.8 μm. The wing (marked by the *white arrow*) is 172 μm wide, yielding an aspect ratio of 126

the largest value of aspect ratio. The layer is 2.8 μm thick while its wings are 172 μm wide, which gives an aspect ratio as large as 126. Note that the thickness of the layer continuously decreases from the center to the edge. This is evidence of bending of the layer towards the mask as will be discussed in Sect. 30.4.1.

30.3 Dislocations in ELO Layers

As discussed in Sect. 30.1 the unique advantage of growing epitaxial layers by the ELO technique is that substrate dislocations are efficiently filtered during the growth, so low-dislocation-density epitaxial layers can be obtained on heavily dislocated substrates. In the following subsections this issue will be addressed in more detail. First, we will focus on the distribution of dislocations in lattice-mismatched ELO layers, showing examples where the process illustrated schematically in Fig. 30.1 is really observed in laboratory practice. In Sect. 30.3.2, the structural perfection of seams that appear when layers grown from neighboring seeds merge in fully overgrown ELO structure will be discussed.

30.3.1 Filtration of Substrate Dislocations in ELO

Figure 30.16a shows a cross section of a GaAs ELO layer grown by LPE on a silicon substrate. The structure consists of Si(100) substrate with 2 μm-thick GaAs buffer grown by MBE and coated with the SiO_2 masking film. Width and thickness of the layer are 85 and 11 μm, respectively. More growth details can be found elsewhere [30.55]. Figure 30.16b shows a plane view of the same layer with etch pits revealed by etching in molten KOH. It is noteworthy that the density of etch pits on the buffer surface is very high ($\approx 10^8 \, cm^{-2}$). On the contrary, wings of the layer are nearly dislocation free, and only the dislocations threading through the seeding window from the buffer are seen on the upper ELO surface. In agreement with the model presented in Fig. 30.1 these dislocations are confined in a narrow area above the seed. The same behavior is seen in Fig. 30.17, which presents a cross section (Fig. 30.17a) and plane view (Fig. 30.17b) of GaSb ELO layer grown by LPE on GaAs substrate coated by MBE-grown GaSb buffer [30.43]. These examples show that substrate defect filtration during ELO procedure is very efficient, and low-dislocation-density epilayers can be obtained by this technique despite high dislocation density in the seeding area.

Figure 30.18a presents a cross sectional TEM image of a GaAs ELO layer grown by LPE on a GaAs-coated Si substrate [30.56]. Note the large density of dislocations at the bottom of the figure, i.e., close to the GaAs/Si interface. Density of these defects decreases with thickness of the buffer and finally all of them are blocked by the SiO_2 film. Thus, no dislocation is

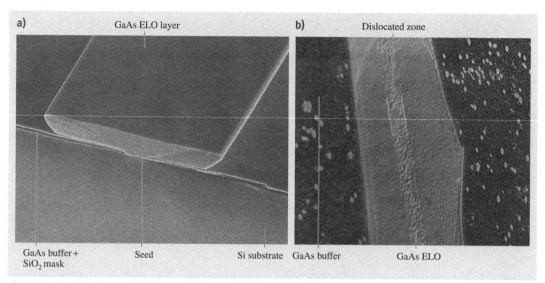

Fig. 30.16a,b SEM image of LPE-grown GaAs on Si ELO structure (**a**) and its plane view after etching in KOH to reveal the etch pit distribution (**b**). The layer is 85 μm wide

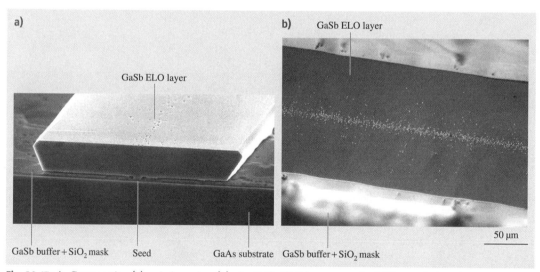

Fig. 30.17a,b Cross-section (**a**) and plane view (**b**) of a GaSb ELO layer grown by LPE on a GaAs substrate coated by 2 μm-thick GaSb buffer and SiO$_2$ mask; in (**b**) etch pits are revealed by chemical etching

seen in the GaAs wing area above the mask (compare Fig. 30.18b), which agrees with results shown in Figs. 30.16 and 30.17. Dislocations propagate to the ELO layer only through the opening in the mask. They are aligned on {111} planes, so the width of the defective area on the layer surface increases with the thickness of the layer. This is exactly the behavior postulated in Sect. 30.1 when explaining the ELO concept with the use of Fig. 30.1.

As the last example of an efficient filtration of substrate dislocations in ELO, Fig. 30.19 shows the distribution of dislocations analyzed by cross sectional

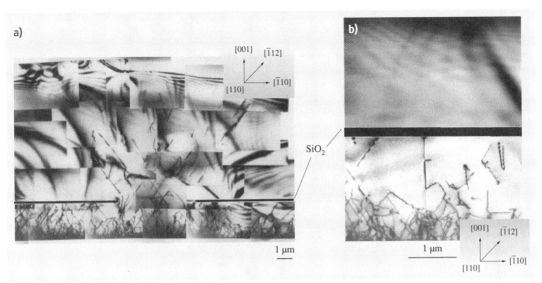

Fig. 30.18 (a) TEM cross sectional image of a GaAs ELO layer grown by LPE on a GaAs-coated Si substrate; (b) larger magnification of the ELO GaAs/SiO$_2$/GaAs buffer area; the mask position is marked by *solid black lines* for better visibility (courtesy of Tamura)

TEM in 140 μm-thick GaN ELO layers grown by HVPE on sapphire substrates [30.57]. The efficient blocking of buffer dislocations by the 1 μm-wide mask is clearly visible. Note also that some of the dislocations threading to the layer through the opening in the SiO$_2$ mask change their direction close to the mask edge and propagate parallel to the mask surface without reaching the upper ELO face. *Sakai* et al. claim that this phenomenon, instead of mechanical blocking of dislocations by the mask, gives the main contribution to the reduction of threading dislocation density in HVPE GaN films [30.58]. The mechanism of bending of threading dislocations in GaN ELO layers is not yet fully understood, although there are suggestions that it may be closely related to the change of the growth front direction and appearance of crystal facets at the very beginning of lateral overgrowth [30.58]. Probably, stress present during growth of the layers may play a role as well. It is worth mentioning that, to the best of our knowledge, there are no reports on similar behavior of dislocations in ELO structures of zincblende III–V materials. Actually, no evidence of dislocation bending in GaAs ELO layer grown on Si substrate is seen in Fig. 30.18. Some of the dislocations threading from the GaN buffer propagate through the GaN ELO layer without any change of direction, creating a defect zone on the surface above the seed, similar to those shown

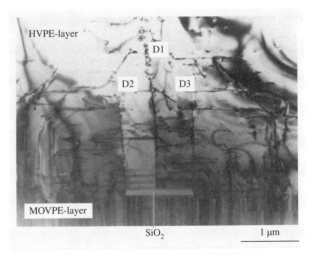

Fig. 30.19 TEM cross sectional image of a GaN ELO layer grown by HVPE on a GaN-coated sapphire substrate. Note that some dislocations threading from the buffer bend in the vicinity of the SiO$_2$ mask edge and propagate along the mask surface to ELO side-wall (courtesy of Sakai)

in Figs. 30.16b and 30.17b. In GaN these dislocations are aligned parallel to the *c*-axis, i.e., perpendicular to the substrate surface. Consequently, the width of the defect zone on the upper ELO surface does not change significantly with layer thickness. Again, this behavior is quite different from that observed in ELO structures of zincblende compounds (compare Fig. 30.18a). Note also that two additional types of defects, denoted by D1 and D2, aligned along the [0001] direction, are visible in Fig. 30.19: the D1 defect originates approximately from the center of the SiO_2 mask while the D2 defect originates from both edges of the mask. Their origin will be discussed in the following sections.

Finally, let us mention that high crystallographic perfection of ELO wings results in their better optical and electrical properties than those of planar buffers. In particular, we have shown that cathodoluminescence intensity is much higher in the wing area of the GaAs on Si ELO structure than that from the material grown vertically above the seed, and even higher than that from the MBE-grown GaAs buffer [30.20]. This should be expected, as dislocations are known to behave like centers of effective nonradiative recombination of excited carriers. A similar distribution of cathodoluminescence intensity has also been observed across surfaces of GaN on sapphire [30.59] and SiGe on Si [30.60] ELO structures. The advantage of using ELO layers for devices was first demonstrated by *Nakamura* et al., who reported in 1997 continuous-wave (CW) room-temperature operation of a blue, InGaN/GaN laser diode with a lifetime longer than 1000 h when deposited on GaN/sapphire ELO substrate [30.23]. Since then the lifetime of these diodes has been significantly increased, so they are commercially available at present. This remarkable improvement was due to a significant reduction of the threshold current density of the laser diodes fabricated on low-dislocation-density GaN ELO substrates. Similar behavior has been reported for GaAs-based lasers grown on GaAs/Si ELO substrate [30.61]. *Kozodoy* et al. have shown that the use of lateral overgrowth to eliminate dislocations leads to better electrical properties of GaN p-n junctions [30.62]. Reverse-bias leakage current was reduced by three orders of magnitude for diodes located on low-dislocation-density ELO wings. This has led to improved performance of light-emitting diodes [30.63,64]. Moreover, large reduction of dark current and sharper cutoff have been found for AlGaN solar-blind photodetectors fabricated on ELO substrates [30.65]. These examples provide the best evidence that the application of ELO technology led to significant progresses in the development of high-performance semiconductor devices made of lattice-mismatched epitaxial structures.

30.3.2 Structural Perfection of Coalescence Front in Fully Overgrown ELO Structures

In order to ensure high quality of ELO layers special care must be taken to avoid the generation of new defects during lateral overgrowth. The areas above the mask edges and fronts of coalescence of layers grown from neighboring seeds are the most sensitive since stress is the largest there. Therefore, factors influencing the structural perfection of ELO layers at these particular points will be briefly discussed.

Figure 30.20 shows a TEM image of the seeding area of the GaAs ELO layer grown by LPE on a SiO_2-masked GaAs substrate [30.66]. It is seen that the layer sticks closely to the SiO_2 mask. Contrast in the ELO image is very homogeneous, indicating that the overgrowth at the seed level is nearly stress free. Note that no dislocation is visible around the mask edge. As mentioned earlier, regular arrangements of dislocations generated during LPE growth in the vicinity of SiO_2 mask edge are sometimes observed in Si ELO layers [30.44]. Their creation leads to the growth of defective layers such as those shown in Fig. 30.8. It seems, however, that in this particular case defects originated due to specific LPE growth conditions used and that their generation can be completely avoided if the growth parameters are carefully adjusted. Indeed, the same group of authors has reported LPE growth of completely defect-free Si ELO layers [30.40]. A quite different behavior is observed in GaN ELO structures, in which dislocations above the mask edge are commonly found (see the D2 defect in Fig. 30.19). They are related to tilting of ELO wings as will be discussed in Sect. 30.4.1.

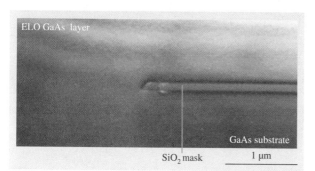

Fig. 30.20 TEM image of GaAs ELO/GaAs substrate interface at the edge of the SiO_2 mask. Note that no dislocation is seen above the mask edge

Coalescence fronts (seams) of ELO layers grown from adjacent seeding windows in fully overgrown ELO structures are other points where large mechanical stresses might be present. Thus, nucleation of dislocations is likely to occur there. Figure 30.21 presents a front of partial coalescence of two GaAs ELO layers grown from neighboring seeds by LPE on SiO_2-masked GaAs substrate [30.66]. From geometrical considerations the sidewalls of ELO layers were identified as close to the {232} planes. When approaching each other these planes separated mask area from the solution, thus the growth just above the mask terminated and a small void remained at the front of coalescence. Then the upper hollow filled up and the continuous ELO layer with a flat upper surface was finally obtained.

Creation of voids can significantly deteriorate the perfection of ELO layers. During their liquid-phase growth the solution is likely to be trapped inside the void. Since the thermal expansion coefficients of metallic inclusion and surrounding crystal are usually different this may lead to stresses that extend from the inclusion into the epitaxial layer and eventually to the formation of dislocations. This issue was extensively studied by *Nagel* et al. [30.67] and *Banhart* et al. [30.68] for LPE growth of Si ELO layers on thermally oxidized

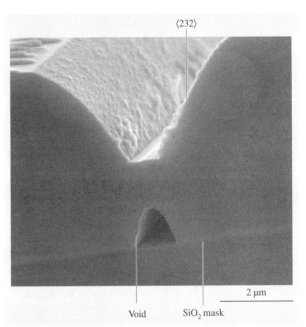

Fig. 30.21 SEM image of the front of partial coalescence of two GaAs ELO layers grown by LPE on SiO_2-masked GaAs substrate. Note the presence of inclined ⟨232⟩ sidewalls of ELO

Si substrates. They found that shape, size, orientation, and arrangements of seeding windows must be carefully designed for high quality of coalescence front. In particular, they have shown that 150 μm-long defect-free seams can be achieved in silicon lateral overgrowth by LPE if advancing layers start to join at only one point. Then the process should proceed in a zipper-like way in the direction of the seeding line. It is important that the growth solution could flow away along the window direction from the space between advancing growth fronts. When the merging layers join at two or more different points of the seam and coalescence proceeds inwards, solute inclusion and related dislocations are observed at the last point of coalescence. This usually happens if parallel seeding windows are used. Therefore, the application of nonparallel seeds has been advised to facilitate the zipper-like mechanism. This approach has been successfully employed

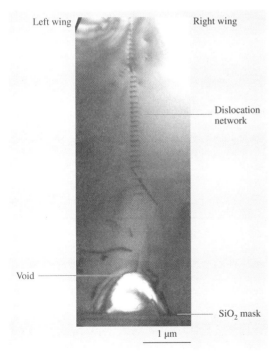

Fig. 30.22 TEM image of the front of coalescence of two GaAs ELO layers grown on SiO_2-masked GaAs substrate. Note the void at the mask surface and a well-organized network of dislocations separated by a distance of about 100 nm ◂

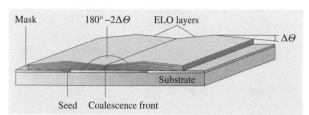

Fig. 30.23 Schematic drawing of merged ELO layers with their laterally grown parts tilted towards the mask. Since wings are tilted in opposite directions, the misorientation of the lattice planes at the seam is $2\Delta\Theta$, where $\Delta\Theta$ is the tilt angle of each wing

for defect-free coalescence of GaAs/GaAs [30.69] and InP/InP [30.70] ELO layers by LPE.

Figure 30.22 shows a TEM image of a coalescence front of two GaAs ELO layers grown by LPE on a SiO_2-masked GaAs substrate [30.66]. A void at the mask surface, similar to that seen in Fig. 30.21, is clearly visible. Probably, it was originally filled with gallium solvent that was later removed during the ion milling of the specimen. This might explain why some single dislocations surrounding the void are seen in its vicinity. The most important feature, however, is a well-organized set of dislocations starting at a distance of about 1 μm from the void and creating a low-angle grain boundary at the seam plane. These dislocations are parallel to the $\langle 010 \rangle$ direction with a Burgers vector of $\frac{1}{2}[110]$, both included in the (001) plane that is parallel to the surface of the substrate. Creation of the grain boundary at the seam can be explained as follows: very often laterally grown sections of ELO layers are tilted towards the mask with the tilt axis aligned parallel to the seeding window direction. This situation is schematically illustrated in Fig. 30.23 where the wing tilt angle is marked by $\Delta\Theta$. The phenomenon of wing tilting will be extensively discussed in Sect. 30.4.1. However, it must be noted here that it creates significant problems for coalescence of neighboring ELO layers as they tilt in opposite direction. Since the sense of bending must be reversed at the seam the crystal lattice is heavily stressed in this region and dislocations appear as soon as the critical shear stress is reached. As discussed by *Banhart* et al. [30.68] for LPE growth of Si ELO layers on thermally oxidized Si substrates the stress at the seam increases with the thickness of merging layers and the value of the tilt angle $\Delta\Theta$. They have found defect-free coalescence for Si ELO layers thinner than 3–4 μm and tilted less than 0.1°. For larger thicknesses and/or values of the tilt angle $\Delta\Theta$ generation of a low-angle grain boundary similar to that shown in Fig. 30.22 was observed.

The dislocation network created in our GaAs ELO layers is similar to the D1 defect found at the center of the mask in GaN ELO layers (Fig. 30.19). *Sakai* et al. claim that this grain boundary consists of dislocations that had propagated laterally in the GaN wing area and had changed their direction again in the seam area [30.57]. On the contrary, the density of threading dislocations in GaAs ELO layers is very low. Therefore, in our case new dislocations must have been generated and rearranged into the low-angle grain boundary at the seam to accommodate the mutual tilt of merging wings. A similar behavior has been reported by *Shih* et al. for fully overgrown Si ELO layers grown by chemical vapor deposition on SiO_2-masked Si substrates [30.71]. The result they found, i.e., that the number of seam defects increases with increasing buried oxide width, can be explained by larger tilt angles $\Delta\Theta$ for wider overgrowths.

The geometry of the dislocation network shown in Fig. 30.22 can be used to measure the relative misorientation of merging ELO lattice planes [30.66]. The value of the tilt angle $\Delta\Theta \approx 0.1°$ obtained in this way is nearly ten times smaller than that determined for the GaN on sapphire ELO layer shown in Fig. 30.19. Possible reasons and consequences of that will be discussed in the next section.

30.4 Strain in ELO Layers

All examples presented thus far prove that ELO layers are of much higher quality than the reference planar structures. Even if new dislocations are created at the fronts of coalescence the overall density of dislocations is still significantly reduced. However, ELO layers are not free of strain. In particular, there is a question about strain induced by the mask itself and/or its possible interaction with the overgrown epitaxial layer. Moreover, the lattice mismatch and thermal strain induced by different thermal expansion coefficients of various parts of ELO structure may result in large deformations of the layers. Although in real heteroepitaxial ELO structures all these phenomena take place simultaneously, for clarity of presentation we will first focus on the problem of interaction of ELO layers with the mask underneath (Sect. 30.4.1). Then, the issue of thermal strain in ELO structures will be addressed (Sect. 30.4.2). It is important to point out that ELO samples have a specific geometry. They consist of separated or coalesced monocrystalline stripes grown from line seeds cut in the mask, so different properties in the directions parallel and perpendicular to seeding lines may be expected. Therefore, before proceeding further a brief introduction will be given, presenting how the x-ray diffraction (XRD) technique is used to study a strain field in ELO layers.

Figure 30.24 shows the geometry commonly used for analysis of ELO layers by XRD. First, x-ray diffraction measurements are taken for the sample position in which the scattering plane (defined by the incident and diffracted wavevectors) is perpendicular to the seeding line direction. This corresponds to the axis of sample rotation during the ω scan being parallel to the seeds ($\varphi = 0°$). Next, the sample is rotated around the substrate normal and the measurements are repeated for the x-ray scattering plane being parallel to the seeds (i.e., for $\varphi = 90°$). Reciprocal space maps and/or x-ray diffraction curves (so-called *rocking curves*) are recorded for each sample positions.

Usually ELO layers are much narrower than the x-ray beam used in the standard x-ray diffraction experiments. Therefore, the diffraction pattern obtained contains information integrated over many stripes illuminated by the beam. To analyze strain distribution in a more detail we have developed a XRD technique in which the sample is moved in small steps and the x-ray beam, being much narrower than the ELO width, illuminates in sequence various parts of a single ELO stripe. For each position of the beam the diffraction pattern originating from a precisely defined area of the stripe is recorded. This allows us to analyze strain field *locally* inside a particular ELO stripe chosen. As will be shown later, this procedure allows the most characteristic features of the strain field in the samples to be determined.

30.4.1 Mask-Induced Strain in Homoepitaxial ELO Layers

Figure 30.25 shows x-ray rocking curves of GaAs ELO layer grown by LPE on SiO$_2$-masked GaAs substrate [30.7]. The layer was 10 μm thick and 90 μm wide. Both, the as-grown sample and that cut from the same wafer and etched to remove selectively the SiO$_2$ mask have been studied. The rocking curves presented

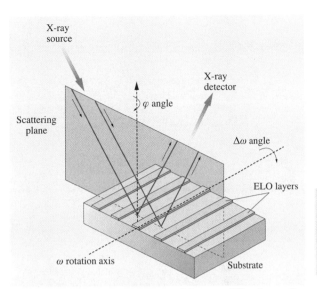

Fig. 30.24 Sketch of the geometry used for x-ray diffraction studies of ELO samples. The scattering plane is defined by the incident and diffracted vectors. φ is the angle between the direction of the seeds and the axis of sample rotation during the ω scan; $\varphi = 0°$ for the geometry shown

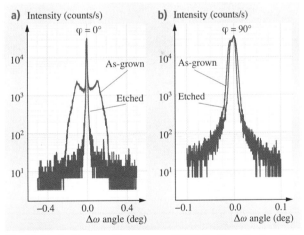

Fig. 30.25a,b X-ray rocking curves of (400) CuK_{α_1} reflection from the as-grown and etched GaAs ELO layer on the SiO$_2$-covered GaAs substrate measured with the ω-axis parallel (**a**) and perpendicular (**b**) to the seeding lines

in Fig. 30.25a,b have been measured with the $\varphi = 0°$ and $\varphi = 90°$ sample orientation, respectively. As can be seen from Fig. 30.25a the rocking curve of the as-grown

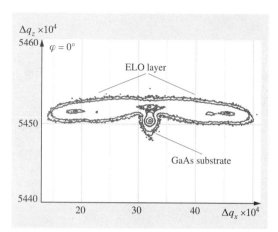

Fig. 30.26 Reciprocal-space map of (400) CuK$_{\alpha_1}$ reflection from the as-grown GaAs ELO layer on a SiO$_2$-covered GaAs substrate with the ω-axis parallel to the seeding lines. Both axes are marked in units of $\lambda/2d$, where λ is the wavelength of the x-ray radiation and d is the GaAs lattice spacing

sample is very broad. However, it becomes much narrower after the SiO$_2$ mask has been removed. On the contrary, for $\varphi = 90°$ the rocking curve is quite narrow and etching causes only a slight change of its shape (Fig. 30.25b). It is worth mentioning that during rocking curve measurements the x-ray diffraction is sensitive mainly to the distortion of the crystal planes in the scattering plane. Therefore, the large width of the rocking curve shown in Fig. 30.25a indicates that a significant deformation of the as-grown ELO takes place in the cross section plane perpendicular to the seeds.

Figure 30.26 shows the reciprocal space map of the same ELO layer. This map has been constructed for the same sample geometry as that used to measure the rocking curves presented in Fig. 30.25a. The distribution of the diffracted x-ray intensity proves that the reason for the large width of the rocking curve is the ELO stripe deformation in the direction perpendicular to the seeding line. This deformation is elastic since a narrow well-concentrated diffraction picture has been observed on the reciprocal lattice map if this is taken for the sample with the SiO$_2$ removed [30.72]. There are also slightly different positions of the ELO and substrate peaks visible in Fig. 30.26 that indicates some lattice misfit ($\approx 7 \times 10^{-5}$) between the heavily Si-doped ELO layer and undoped GaAs substrate [30.7].

When developing our ELO growth technology, occasionally we found GaAs ELO layers with an air gap between their laterally grown parts and the mask [30.73]. A cross section of such a layer is shown in Fig. 30.27a. Note the wings hanging over the substrate without any contact with the mask surface. Note also that the (400) CuK$_{\alpha_1}$ rocking curves of the layer are very narrow for both positions of the scattering plane (Fig. 30.27b), which means that deformation of the ELO wings, if any, is negligible. Comparison of the data shown in Figs. 30.25a and 30.27b proves that deformation of GaAs ELO wings must be due to their interaction with the mask surface.

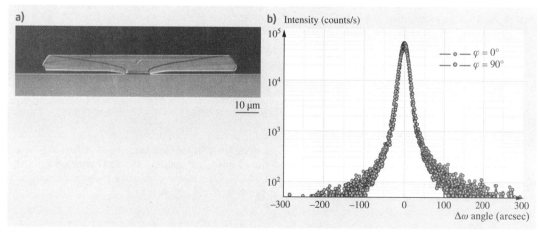

Fig. 30.27a,b Cross-section image (**a**) and x-ray rocking curves of (400) CuK$_{\alpha_1}$ reflection (**b**) of a GaAs ELO layer on a SiO$_2$-masked GaAs substrate. Note the air gap between the mask and the ELO wings

The results presented above allow us to present a picture of deformation of the ELO layer that is schematically illustrated in Fig. 30.28: During the growth the layer is bent towards the SiO$_2$ mask in the direction perpendicular to the seeding line (Fig. 30.28a). When the SiO$_2$ mask is removed by selective etching the ELO stripe behaves like a released spring and the wings tilt disappears (Fig. 30.28b). This corresponds to a narrowing of the curve shown in Fig. 30.25a. From a comparison of the rocking curves shown in Fig. 30.25b it can be concluded that similar deformation does not occur along the ELO stripe.

In the model presented in Fig. 30.28a it is postulated that the ELO wings are tilted towards the mask. Note, however, that the direction of tilt, upward or downward, cannot be directly determined from the shape of the rocking curve shown in Fig. 30.25a. Therefore, we have applied a spatially resolved XRD (SRXRD) technique in which the ELO stripe is moved in small steps, so the x-ray beam, being much narrower than the ELO width, illuminates in sequence various parts of the layer (see the geometry of the experiment sketched in Fig. 30.29c). For each position of the beam on the sample a local x-ray rocking curve originating from a precisely defined area of the stripe is recorded. The results of such a procedure are presented in Fig. 30.29b, which shows the set of x-ray rocking curves measured locally at various parts of a 304 μm-wide and 16.8 μm-thick GaAs ELO layer grown on SiO$_2$-masked GaAs substrate [30.74]. The standard rocking curve measured for the same sample with a 500 μm-wide x-ray beam is included in Fig. 30.29a for comparison. The diffraction plane was perpendicular to the seeds in these experiments. For the positions $x = -184$ μm and $x = 230$ μm the beam was out of the ELO area. Therefore, only narrow substrate peaks are visible. As soon as the beam approached the ELO stripe ($x = -138$ μm) a strong side-maximum, apparently due to radiation diffracted by the ELO edge, appeared in the curve. For $x = 0$ μm only the middle part of the stripe was illuminated by x-rays. As a result, the side-maxima are missing whereas the intensity of the substrate peak is significantly reduced. When the beam starts to move out of the stripe surface ($x = 138$ μm) the intensity in the central part of the curve decreased again, the side-maximum due to the edge reflection remained, while the intensity of the substrate peak increased. The data shown in Fig. 30.29b explains the origin of all the features visible in the standard rocking curve presented in Fig. 30.29a. In particular, this proves that the side-maxima are due to diffraction from the edges of the

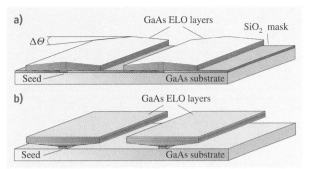

Fig. 30.28a,b Schematic drawing of the GaAs on GaAs ELO cross section. Bent layers are shown in the as-grown state (**a**), deformation of ELO disappears when the SiO$_2$ mask is removed by etching (**b**). $\Delta\Theta$ is the tilt of (100) crystal planes between the edge and the central part of ELO

ELO stripe. Therefore, the misorientation of ELO crystal planes must be the largest there. Consequently, half of the angular separation of the side-peaks on the standard rocking curve can be used as a good measure of the maximum tilt angle $\Delta\Theta$ of ELO lattice planes (compare Fig. 30.28a). Moreover, using a simple geometry of the diffraction system and analyzing the order in which the side-maxima appear while the beam moves across the stripe the direction of ELO tilt can be unambiguously determined. For example, the x-ray rocking curves of Fig. 30.29b show that the left wing diffracts x-rays at larger Bragg angle than the right one. This is clear evidence that the wings are tilted towards the mask [30.74]. It is worth mentioning that, improving the technique of spatially resolved x-ray diffraction presented above, we have employed it successfully for analysis of ELO layers using an x-ray beam as narrow as 20 μm [30.66]. Subsequently, even a 10 μm-wide beam moved in steps as small as 3 μm was applied, which allowed us to create x-ray rocking curve maps of bent ELO stripes [30.75]. Despite this progress the technique cannot be easily applied for GaN ELO layers. Since their widths are usually much smaller, electron diffraction [30.76] or synchrotron x-ray diffraction [30.77] must be used to determine the local magnitude and direction of the GaN wing tilt.

Additional evidence of the downward tilt of ELO wings comes from our studies of similar GaAs ELO samples by synchrotron x-ray topography (SXRT) [30.78, 79]. The basic idea of the experiment was to analyze how a white-beam synchrotron x-ray radiation collimated with the use of very narrow slits reflects back from the surface of the GaAs ELO sam-

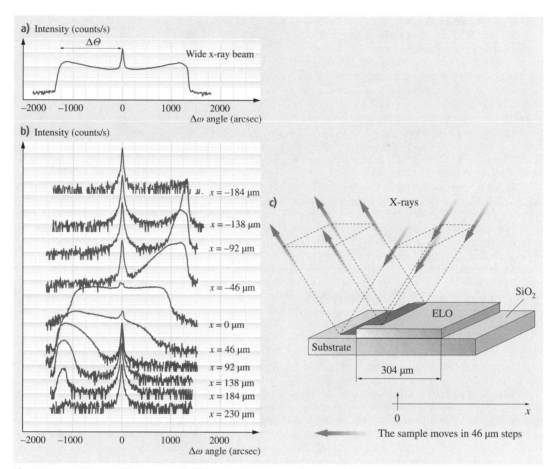

Fig. 30.29a–c X-ray rocking curves of a 304 μm-wide as-grown GaAs ELO layer on a GaAs substrate measured with the scattering plane perpendicular to the seeding lines with the use of a 500 μm-wide x-ray beam (**a**) and 90 μm-wide x-ray beam scanned across the ELO stripe in 46 μm steps (**b**). x denotes the position of the beam on the ELO. The $x = -184$ and 230 μm curves correspond to reflections from the substrate outside the ELO stripe. (**c**) Schematic geometry of the experiment

ple. More details on experimental conditions can be found in the source publications. The topograph obtained is shown in Fig. 30.30a where "w," "s," and "e" mark images of window seed, substrate, and ELO layer areas, respectively. The shape of the ELO layer image directly indicates a downward tilt of the GaAs ELO wings. Additionally, the image was simulated numerically (Fig. 30.30b) to gain information on the distribution of the (100) lattice plane tilt angle versus position across the single ELO stripe. This data allowed us to calculate the shape of (100) lattice planes in the sample as shown in Fig. 30.30c together with the experimental profile of ELO layer thickness determined by surface stylus profilometer. As seen, the numerical results agree very well with the experimental results.

It should be noted in Fig. 30.30c that the curvature of the (100) lattice planes is the largest in the central part of the ELO stripe. This finding has allowed us to postulate that bending of the ELO stripes starts at the very beginning of growth when the laterally overgrown parts are thin and flexible. Then, the bent crystal planes might be reproduced during subsequent growth, although retaining their shape [30.78]. As will be shown later, this is exactly the behavior observed experimentally.

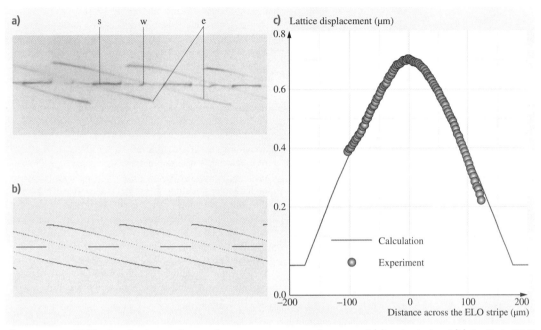

Fig. 30.30a–c 206 back-reflection section synchrotron x-ray topograph measured (**a**) and calculated (**b**) for GaAs ELO layer grown by LPE on SiO$_2$-masked GaAs substrate; "w," "s," and "e" mark topographic images of window seed, substrate and ELO layer, respectively; calculated (*solid line*) and measured (*dots*) displacements of ELO lattice planes across the single ELO stripe are shown in (**c**)

Note also that the data shown in Fig. 30.30c indicates a continuous decrease of layer thickness from the center of the layer to its edge, being due to the wing tilt. Indeed, this effect is observed experimentally as, for example, is clearly seen in the cross section image presented in Fig. 30.15. Its magnitude, namely the thickness change of $\approx 0.9\,\mu\mathrm{m}$ on a $\approx 170\,\mu\mathrm{m}$-wide wing, also agrees with data shown in Fig. 30.30c. A similar phenomenon has been reported for LPE-grown Si ELO layers. Since the upper ELO surface was a singular, atomically smooth face, tilting of the wings should lead to the presence of monoatomic steps at the Si/SiO$_2$ interface where silicon crystal planes terminate (see the sketch in Fig. 30.28a). Indeed, analysis of Si ELO layers by high-resolution transmission electron microscopy revealed the presence of such steps at the interface between the epitaxial layer and the SiO$_2$ film [30.17].

The tilting of ELO wings towards the mask has also been found for many other ELO systems, including Si layers grown by LPE on oxidized Si substrates [30.68] and GaN ELO layers grown by HVPE or MOVPE on sapphire or SiC substrates coated by SiO$_2$ or Si$_3$N$_4$ masks [30.57, 80–82]. *Fini* and coworkers used x-ray diffraction to observe the time evolution of lattice planes bending in GaN on sapphire ELO layers [30.83]. The scans in reciprocal space through the $(10\bar{1}3)$ diffraction peak of GaN obtained in situ, during growth of the structures by MOVPE, are shown for various growth times in Fig. 30.31a. Prior to epitaxy only the central peak, due to the diffraction from the GaN buffer, is visible. Emergence of the side-maximum is evident for the growth times $\approx 100\,\mathrm{s}$, and by $\approx 300\,\mathrm{s}$, distinct side-peaks have evolved, indicating wing tilt of $\approx 0.9°$. Postgrowth SEM studies of layer cross sections have shown that a significant lateral overgrowth started just at that moment. The wing peak narrows during subsequent growth while its position changes slightly, reaching a tilt of $\approx 1.19°$ after 3600 s of growth. The data collected during this elegant experiment are direct evidence that wing tilt must be due to interaction of the wing with the mask and that it starts at the very beginning of lateral overgrowth, in agreement with our earlier suggestion presented above when discussing the results shown in Fig. 30.30c [30.78].

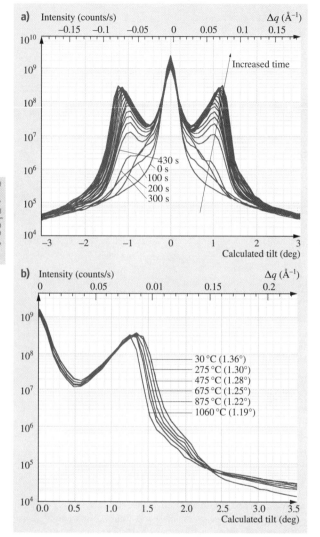

Fig. 30.31a,b Scans in reciprocal space trough the (10$\bar{1}$3) x-ray diffraction peak obtained for various growth times during MOVPE growth of GaN on sapphire ELO structures (**a**) and on cooling to room temperature (**b**). The scans in (**b**) are shown over half of the range measured (courtesy of Fini)

accommodated elastically. Generation of dislocations in the layer does not take place and deformation of the wings disappears completely when the mask is removed by postgrowth etching. Consequently, no defects above the mask edge (compare Fig. 30.20) and narrow x-ray rocking curve of the sample with the mask removed (Fig. 30.25a) are observed. On the contrary, tilt angles up to $\approx 1.5°$ are reported in literature for GaN ELO layers (Fig. 30.31a and [30.57, 80]). So large tilts cannot be accommodated elastically and the strain above the mask edge partially relaxes via the creation of a low-angle grain boundary. This relaxation process is energetically favorable if the ELO structure contains dislocations threading from the buffer. In such a case, the grain boundary can be easily formed from bent threading dislocations that pile up near the mask–seed edges. Indeed, as shown already in Fig. 30.19, cross sectional TEM studies of GaN on sapphire ELO structures have revealed additional low-angle tilt boundaries (the D2 defects) above the edges of the SiO_2 mask. Consequently, when the mask is etched away after the growth, similarly to our approach for GaAs/GaAs layers, only the elastic component of the bending strain is released. The plastic deformation of the crystal lattice above the mask edge remains, and as a result, only partial narrowing of the x-ray rocking curve takes place. This is exactly the behavior observed experimentally [30.81].

All the arguments presented up to now indicate that downward tilt of ELO wings originates from an interaction between laterally grown material and the mask. The nature of this interaction is still not fully clarified. *Kohler* with coworkers have suggested that an attractive van der Waals force starts to be active during LPE growth when the mutual distance between Si ELO wing and the upper surface of the SiO_2 mask falls below a certain value [30.84]. The surface tension of the solution should also be taken into account as a source of wing tilting during ELO growth from the liquid phase [30.47]. However, the tilt of GaN ELO wings seems to be too large to be related to adhesion or weak van der Waals forces. Therefore, a chemical reaction of laterally overgrown GaN with the SiO_2 underneath and/or densification of the oxide being due to a high temperature and chemically aggressive conditions (e.g., presence of ammonia) during vapor phase epitaxy of GaN have been considered as possible reasons of the tilt [30.83].

As discussed in Sect. 30.3.2, tilting of ELO wings towards the mask creates a significant problem for coalescence of neighboring stripes as they tilt in the opposite direction (Fig. 30.23). Therefore, there is no

The difference between the GaN and GaAs or Si ELO structures is the magnitude of wing tilt and the mechanism of its relaxation. As discussed earlier, tilt angles $\Delta\Theta \approx 0.1°$ are commonly found in LPE-grown GaAs ELO layers (Fig. 30.25a). Thus, the bending strain at the seed–mask edges is small enough to be doubt that for ELO layers of high crystallographic quality tilting of their wings should be reduced as much as possible.

We have proposed the following recipe for reduction of the mask-induced bending of the ELO layers: The initial vertical growth rate of ELO must be increased to start the lateral overgrowth at some microscopic distance from the upper surface of the mask [30.74]. Then the chance of ELO wings capture by attractive force and their interaction with the mask should be reduced. We have found that this recipe works efficiently in practice. In particular, tilting of the wings has been efficiently tailored by controlling the ratio of vertical to lateral growth rates at the beginning of ELO growth. This has been achieved by growing GaAs ELO layers on SiO_2-coated GaAs substrates with increasing density of dislocations. Then, the ratio of vertical to lateral growth rates at the beginning of the growth was increased due to the higher density of surface steps, which in turn led to reduction of the mask-induced tilt of ELO wings [30.73, 74]. In the limiting case of heavily dislocated GaAs substrates, namely on GaAs buffers grown by MBE on Si substrates, the vertical growth of GaAs ELO was so fast that air-bridged structures without any interaction with the mask were obtained [30.55]. Basically the same approach has been used by *Fini* et al. [30.86]. They have shown that the crystallographic quality of coalescence of neighboring ELO GaN stripes can be improved if vertical development of ELO stripes is forced at the beginning of the growth, followed by a change of growth conditions and fast lateral overgrowth of the structure.

If thermally activated processes during the growth itself are considered as origins of wing tilt, reduction of the ELO growth temperature might be helpful. Indeed, *Silier* with coworkers have reported negligible adhesion to the SiO_2 mask of Si ELO layers grown by LPE at 450 °C, in contrast to their previous growth experiments performed at high temperature of ≈ 900 °C [30.17]. Also replacement of commonly used dielectric films by other mask materials seems to be a promising way of reduction of mask-induced wing tilt. In particular, we have found that bending of GaAs ELO layers grown by LPE is negligible when the SiO_2 mask is replaced by a thin graphite film [30.7]. This finding has been explained by delayed beginning of lateral growth due to a change of the shape of the liquid solution in the vicinity of the ELO edge when the SiO_2 was replaced by graphite film not wetted by the gallium melt [30.73]. Similarly, tungsten has been found to be a promising mask material for MOVPE growth of GaN ELO structures with very low tilt of the wings [30.4]. In principle, however, it is difficult to predict a priori the mask material that would be the most suitable for each particular ELO case.

The most natural way to eliminate the mask-induced strain would be to grow the ELO structures with their wings having no contact with the mask surface. An example of such a GaAs structure grown by LPE was shown in Fig. 30.27. Another example is shown

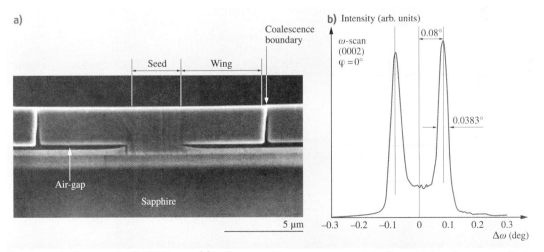

Fig. 30.32a,b Cross-sectional SEM image (**a**) and x-ray rocking curve measured for the scattering plane perpendicular to the seeding lines (**b**) of the air-bridged GaN on sapphire layers (after [30.85], © AIP 2000)

in Fig. 30.32a, which presents a cross section of air-bridged GaN layers grown by MOVPE on sapphire substrates coated by GaN buffer and silicon nitride mask [30.85]. Note that, due to profiling of the GaN buffer thickness, an air gap is formed during the growth above the mask and a free-standing laterally grown GaN is obtained. In this way, direct interaction of ELO wing with the mask is eliminated. Figure 30.32b shows the x-ray rocking curve of the layer measured with the x-ray scattering plane perpendicular to the seed direction. Two peaks originating from ELO wings are clearly visible. Each of them is much narrower than that obtained for standard GaN ELO layer (compare Fig. 30.31a), which means that the mask-induced bending of GaN wings is significantly suppressed. However, the different angular position of both peaks indicates that some residual tilt ($\approx 0.08°$) of the c-axis in the wing area relative to the underlying GaN is still present. Its origin is explained in the next section.

30.4.2 Thermal Strain in ELO Layers

Besides the mask-induced strain an additional deformation of ELO lattice planes may arise when the sample experiences a large stress upon cooling from the growth to room temperature due to the different thermal expansion coefficients of its components. This effect is commonly observed in planar heterostructures. In the following we will show that in the ELO structures thermal strain can be relaxed via additional tilting of the wings while still preserving their high quality.

Figure 30.33 shows a SEM image of a GaAs ELO layer grown by LPE on a silicon substrate coated by

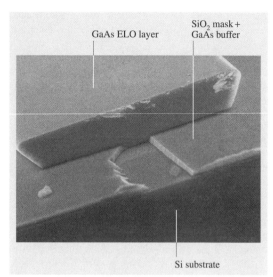

Fig. 30.33 Cross-sectional SEM image of the as-grown GaAs ELO layer on GaAs-coated Si substrate; the layer is 62 μm wide and 9.2 μm thick. Note that ELO wings overhang the mask

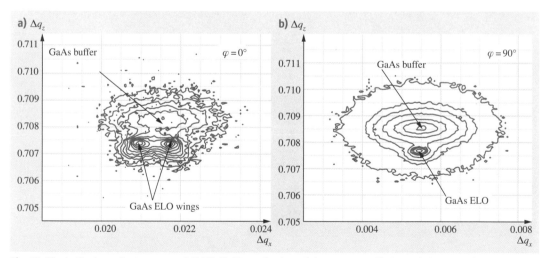

Fig. 30.34a,b Reciprocal-space map of (400) CuK_{α_1} reflection of the as-grown GaAs ELO layer on GaAs-coated Si substrate measured with the scattering plane perpendicular (**a**) and parallel (**b**) to the seeding lines. *Axes* are marked in units of $\lambda/2d$, where λ is the wavelength of the x-rays and d is the lattice spacing

a MBE-grown GaAs buffer layer and a SiO$_2$ mask. The layer was grown at 540 °C and was 62 µm wide and 9.2 µm thick. More details of the growth procedure can be found in the source publication [30.55]. Note that the layer does not stick to the substrate and its wings overhang the mask surface. This is the result of a fast vertical growth rate caused by buffer dislocations supplying surface steps as discussed in the previous section. Reciprocal-space maps measured for the same sample with the x-ray scattering plane perpendicular and parallel to the seeds are presented in Fig. 30.34a,b, respectively. The broad Bragg peak from the GaAs buffer as well as narrow, well-separated reflections originating from ELO wings are clearly visible. Note that, for $\varphi = 0°$, reflection from the layer consists of two peaks, whereas a single Bragg peak is visible only for $\varphi = 90°$. This is a quite similar behavior to that observed for our GaAs/GaAs ELO samples (compare Fig. 30.25) and indicates tilt of ELO wings in the plane perpendicular to the direction of the seeds.

Figure 30.35 shows x-ray rocking curves of the same sample before and after etching to remove selectively the SiO$_2$ mask. Note that the full-width at half-maximum of the reflection from the wing equals 94 arcsec, only. This is much less than that of the MBE-grown buffer (≈ 435 arcsec) and confirms the very high structural quality of laterally overgrown parts of the layer. However, the whole curve is so wide that tilt of ELO wings caused by their interaction with underlying

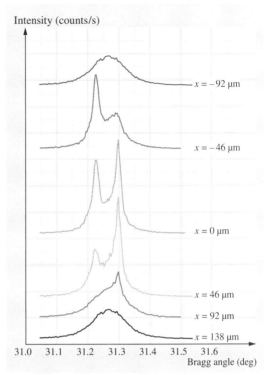

Fig. 30.36 (400) CuK_{α_1} rocking curves from as-grown GaAs on a Si ELO layer measured with a 90 µm-wide x-ray beam moved across the ELO stripe in 46 µm steps. The $x = -92$ and 138 µm positions correspond to reflections from the GaAs buffer outside the ELO stripe

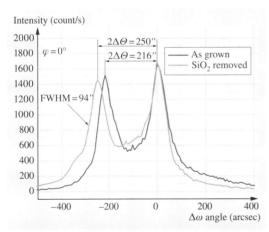

Fig. 30.35 X-ray rocking curves of (400) CuK_{α_1} reflection for the as-grown (*solid line*) and etched (*dashed line*) GaAs ELO layer on the SiO$_2$-coated Si substrate with GaAs buffer, with the x-ray scattering plane perpendicular to the direction of the seeding lines

SiO$_2$, similar to that presented for homoepitaxial GaAs ELO layers, might be suspected. It is worth noticing, however, that this time behavior of the system is quite different. The wing tilt angle increases when the mask is removed. Moreover, the wings hang over the mask, so the mask-induced strain, if any, should be negligible.

We have checked the direction of the tilt of the GaAs wings using the SRXRD technique described earlier (Fig. 30.29c). The set of x-ray rocking curves obtained from precisely defined areas of the GaAs stripe is shown in Fig. 30.36. The geometry of the experiment was exactly the same as before, i.e., the beam illuminated the left ELO wing first ($x = -46$ µm), then the central part of the ELO stripe ($x = 0$ µm), and finally left the sample on its right-hand side ($x = 92$ µm). Note the order in which ELO diffraction peaks appeared when the beam was illuminating subsequent parts of the stripe. The left ELO wing diffracted x-rays at smaller Bragg an-

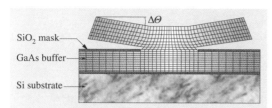

Fig. 30.37 Schematic drawing of the cross section of the as-grown GaAs ELO layer on a Si substrate. The tilt of the ELO wing is $\Delta\Theta$, where $2\Delta\Theta$ is the angular separation of the diffraction peaks, as marked in Fig. 30.35

gle than the right one. This is the opposite behavior to that observed for the GaAs/GaAs ELO system (compare Fig. 30.29) and indicates upward tilt of laterally overgrown parts of the layer [30.55]. All these findings indicate that processes other than wing adhesion to the mask must be involved.

We have explained the results presented above by taking into account the thermal strain that appears in the ELO structure during its postgrowth cooling to room temperature. This is illustrated in Fig. 30.37, which schematically shows deformation of lattice planes in the cross section plane of GaAs on Si ELO layer. A biaxial tensile strain caused by the different thermal contraction of epilayer and substrate is commonly observed in planar GaAs layers grown on Si substrates [30.87]. This strain disappears when the GaAs/Si structure is heated to $\approx 500\,°\text{C}$ [30.88]. Therefore, at the LPE growth temperature, the GaAs ELO layer grows essentially stress free. At room temperature, the basal plane of ELO layer should have the same tensile deformation in the direction perpendicular to the line seed as the upper surface of the buffer GaAs. We have shown that the ELO layer stands free and does not adhere to the mask. Therefore, unrestricted strain relaxation and free contraction of vertically grown volume of ELO should take place in its upper part, as shown in Fig. 30.37. This, in turn, must lead to the upward bending of ELO wings. Note, however, that this process does not affect the structural perfection of ELO wings, and the high crystallographic quality of laterally overgrown parts of the layer is preserved.

Finally, the shift of diffraction pattern seen in Fig. 30.35 for the etched sample should be explained. The thermal expansion coefficient of silicon oxide is much smaller than that for zincblende III–V semiconductors. Therefore, the SiO_2 film exerts a tensile stress on GaAs underneath and a compressive strain is present in oxide-free GaAs seeding windows [30.72]. In the GaAs/Si ELO system the thermal tensile strain in the GaAs seeding area exerted by the substrate is partially compensated by the mask-induced compression [30.55]. Therefore, deformation of ELO at its base, and consequently the ELO wings tilt angle, increase when the SiO_2 mask is etched away. This leads to increase of the angular separation of x-ray reflections, as shown in Fig. 30.35.

It is worth noting that our model of thermal strain relaxation in ELO layers predicts that the direction of wing tilt is correlated with the sign of the strain in the buffer. In particular, downward wings tilt should be observed if the buffer layer is under compressive thermal stress, which is usually the case for GaN ELO layers grown on SiC or sapphire substrates. This is exactly the behavior presented in Fig. 30.31 for the GaN on sapphire ELO system. At the growth temperature GaN ELO wings are tilted downward due to their interaction with the mask. Upon cooling to room temperature, the position of the wing peak increases from 1.19 to 1.36° (Fig. 30.31b), which means that an additional downward tilt induced by thermal stress has occurred. Qualitatively, the same phenomenon can be inferred from simulations via finite-element analysis of strain in ELO structures [30.89]. Most likely, the residual tilt found in GaN air-bridged structures illustrated in Fig. 30.32b might also be explained in a similar way by the thermal strain in the GaN buffer. All these examples prove that our model is correct. Moreover, they show that the phenomenon of thermal strain relaxation via tilting of ELO wings is not a specific attribute of GaAs/Si ELO structures grown by LPE but has much more general nature.

30.5 Recent Progress in Lateral Overgrowth of Semiconductor Structures

There is continuous progress in the application of the lateral overgrowth technology, and new versions of the ELO technique are being tested. In this section, we first present the recent progress made in liquid-phase ELO growth, including new mask materials and ELO growth by the liquid-phase electroepitaxy. Then, new concepts of ELO growth will be briefly reviewed and compared with the conventional ELO technique. Liquid-phase epitaxy of InGaAs bridge layers and pendeo-epitaxy of GaN layers from vapor and liquid phase will be used as examples.

30.5.1 Developments in Liquid-Phase ELO Growth

Recent developments in liquid-phase ELO growth include the application of novel mask materials. In particular, electrically conductive masking films are interesting since they can be used in photovoltaic and thermophotovoltaic devices as back mirrors for photon recycling and/or as buried electrical contacts. However, the main problem is to find mask material that adheres strongly to the substrate and is thermally and chemically stable to remain undamaged during the aggressive conditions of epitaxial growth. Moreover, it is of prime importance that no nucleation takes place on the mask during the growth since in ELO technique the layer should start growing exclusively from the opening in the mask.

We have applied tungsten for substrate masking in lateral overgrowth of GaAs by LPE [30.5]. Tungsten masks were already successfully used to grow GaN ELO layers on sapphire by MOVPE and dramatic reduction of strain in the layers was found when tungsten film instead of common dielectrics was used to mask the substrate [30.4]. However, MOVPE growth conditions for GaN are quite different from those for solution growth of zincblende III–V compounds, so applicability of a W mask for selective growth of GaAs by LPE has to be verified experimentally.

Figure 30.38 shows a SEM image of GaAs layer grown at 600 °C on a tungsten-masked GaAs substrate. As seen, the epitaxial growth is perfectly selective, i.e., the layer starts growing from the line seed without any nucleation of GaAs on the mask. However, the tungsten mask peels off from the substrate, which often leads to cracking of the ELO layer. This effect is attributed to a large grain size and the strain built into the metallic film sputtered onto the bare GaAs. It has been reported that the application of a AuZn wetting layer between the substrate and the tungsten film helps to reduce the tungsten grain size, and consequently leads to lower strain inside the mask [30.90]. Indeed, we have found that use of a AuZn interlayer significantly improves adhesion of the W mask to the GaAs substrate during LPE growth of GaAs and cracking of ELO layers by the mask peeling off from the substrate is avoided [30.5]. Unfortunately, sometimes a melt removal from W-masked substrate after growth was incomplete since tungsten is well wetted

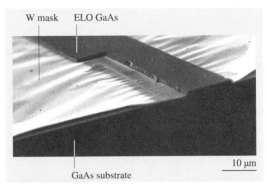

Fig. 30.38 SEM image of a GaAs ELO layer grown at 600 °C by LPE on a tungsten-masked GaAs substrate; mask deposited on bare GaAs substrate. Note that the mask peels off from the substrate, which leads to cracking of the layer

by gallium-rich solution. Consequently, local polycrystalline deposits from melt droplets left on the mask surface are often observed. This encouraged us to look for alternative mask materials.

We have found zirconium nitride films promising for electrically conductive masking in ELO technique [30.6]. This material has interesting properties resulting from its both metallic and covalent bonding characteristics. The covalent crystalline properties are: high melting point, extreme hardness and brittleness, and excellent thermal and chemical inertness. The metallic characteristics are high electrical conductivity and metallic reflectance [30.91]. Moreover, ZrN film when sputtered on a substrate usually has a fine-grained microstructure, thereby exhibiting a low strain level.

Figure 30.39 shows SEM image of a GaAs ELO layer grown at 750 °C by LPE on a ZrN-masked GaAs substrate. As seen from the figure, perfect growth selectivity is obtained. Moreover, the masking film is mechanically stable, adheres strongly to the substrate, and does not show any signs of mechanical degradation even at LPE growth temperature as high as 750 °C. A similar behavior can be seen in Fig. 30.40, which shows SEM image of a GaSb ELO layer grown at 480 °C on ZrN-masked GaSb substrate. This indicates that, due to the high stability of zirconium nitride, long time contact of the mask with Ga − As or Ga − Sb melts does not lead to degradation of its mechanical properties. Moreover, we have found that ZrN forms a low resistivity ohmic contact to n-type GaAs, which is important if the mask is to be further used as a buried electrical contact. Unfortunately, our experiments show

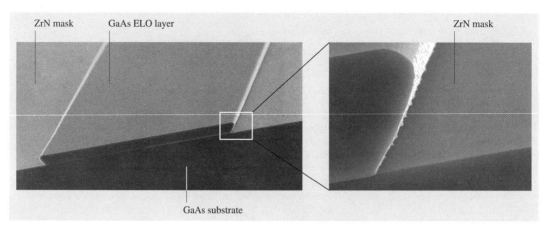

Fig. 30.39 SEM image of a GaAs ELO layer grown at 750 °C by LPE on a ZrN-masked GaAs substrate

that at temperatures higher than ≈ 580 °C the ZrN masks become highly resistive when they are heat-treated in hydrogen atmosphere during LPE growth. Most likely, ZrN is converted into insulating zirconium hydrides, though more work is needed to clarify this issue. This phenomenon sets an upper limit for LPE growth temperature if ZrN masks are to be used as electrical contacts. Thus, the ELO growth of GaAs becomes more difficult, though still possible. For ELO of GaSb layers however, typical LPE growth temperature is about 480 °C, which allowed us to grow high-quality GaSb ELO layers by LPE still preserving the high electrical conductivity of the ZrN mask.

As illustrated above, a proper choice of mask material is crucial for successful lateral overgrowth. However, masks that survive undamaged during a long growth process are difficult to find if extreme crystallization conditions are required. This is the case in solution growth of GaN. Since at typical conditions solubility of nitrogen in gallium is low, very high temperatures (≈ 1450 °C) and very high nitrogen pressure (≈ 1 GPa) are necessary to achieve a reasonable growth rate of GaN from Ga-N solution. *Bockowski* et al. have found that under such conditions silicon nitride, SiO_2, nickel, and molybdenum masks dissolve in the hot gallium while iridium reacts with Ga forming IrGa [30.92]. They have also found that tungsten platelets do not react with the Ga-N solution, so tungsten seemed to them a promising mask material for solution growth of GaN ELO layers. However this argument is not convincing enough since W films sputtered on a substrate have not been tested and other effects such as reduced chemical stability of thin films and cracking caused

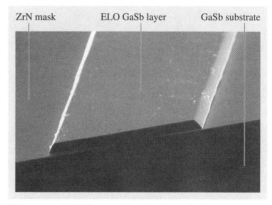

Fig. 30.40 SEM image of a GaSb ELO layer grown at 480 °C by LPE on a ZrN-masked GaSb substrate

by built in strain or weak adhesion to the substrate (compare Fig. 30.38) might cause problems during ELO growth.

Recently there has been increasing interest in lateral overgrowth of semiconductor layers by the liquid-phase electroepitaxy (LPEE) technique – a solution growth method in which layer crystallization is achieved by passing an electric current through the solid–liquid interface while the furnace temperature of the system is kept constant (see [30.93] for a review). The main mechanisms of solute transport toward the substrate during LPEE are diffusion due to the Peltier-effect-induced temperature gradient and electromigration of species (i.e., momentum exchange between free charge carriers and solute species). Technical realization of LPEE is slightly more complicated than that of LPE.

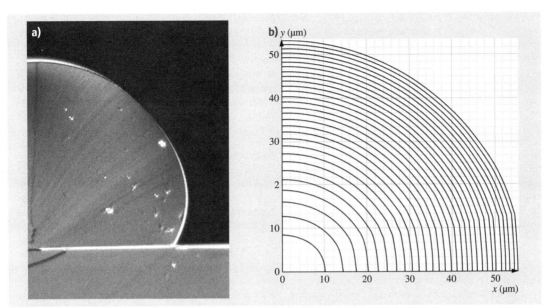

Fig. 30.41a,b SEM image of cross section (**a**) and calculated shape (**b**) of undoped GaAs ELO layer grown at 700 °C by LPEE on a SiO$_2$-masked GaAs substrate. The scale in both figures is the same for easier comparison. The time interval in (**b**) is 8 min

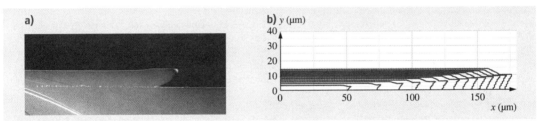

Fig. 30.42a,b SEM image of cross section (**a**) and calculated shape (**b**) of undoped GaAs ELO layer grown at 700 °C by LPEE on a GaAs substrate coated by electrically conductive (tungsten) mask. The scale in both figures is the same for easier comparison. The time interval in (**b**) is 16 min

However, electroepitaxy offers unique possibility to control the growth kinetics and properties of crystals produced. In particular, since the electric current flow affects the solute distribution in the liquid solution, by controlling the electrical current distribution areas with *local* supersaturation in the liquid zone can be produced, which is crucial for selective epitaxy techniques such as ELO. The progress achieved with the application of substrates coated by metallic masks has allowed a novel approach to ELO growth by LPEE.

Figures 30.41a and 30.42a show cross sections of undoped GaAs ELO layers grown at 700 °C by LPEE on (100) GaAs substrates. The difference between them is that they were grown on substrates masked by SiO$_2$ and tungsten films, respectively, while the growth parameters were kept the same in both cases (see [30.5] for more details). As seen, the ELO growth on the SiO$_2$-masked substrate is nearly isotropic while under the same conditions a wide and thin ELO layer is grown on the tungsten-masked substrate. This means that the shape and dimensions of the ELO layers grown by LPEE can be efficiently tailored by varying the electrical conductivity of the mask.

We have developed a simple mathematical model for LPEE growth of ELO layers, which predicts the electric current intensity and solute concentration distributions in the liquid zone under various LPEE growth conditions [30.94]. Moreover, the model allows for direct simulation of the evolution of the growth interface with time. Calculated shapes of the GaAs ELO layers grown by LPEE on GaAs substrates are plotted in Figs. 30.41b and 30.42b. They are shown on the same scale as the experimental images for easier comparison. As seen, the numerical results are in very good quantitative agreement with the experimental ones. The model allows us to postulate the following mechanism for ELO growth of semiconducting layers by LPEE: During growth on substrate masked by electrically insulating film electromigration of solute towards the seed dominates growth kinetics. Since the electric current distribution in the liquid zone around the seed is isotropic, this leads to a rounded shape of layers and to a low value of their aspect ratio despite surface kinetic processes on the upper ELO surface. On the other hand, if the substrate is coated by an electrically conductive mask and the resistivity of its contact to the substrate is low, after some initial ELO growth more electric current passes to the substrate directly through the melt–mask interface than through the ELO crystal and the seeding area. Here, the main role of electromigration is to supersaturate the solution just above the mask between the adjacent seeds. Then diffusion dominates ELO growth by supplying solute from these supersaturated areas directly to side-walls of growing layers. In agreement with our experiments, this mechanism significantly increases the lateral growth rate, thereby producing thin and wide ELO layers with flat upper facets. It is worth mentioning that a similar increase in the lateral growth rate in LPEE has been reported by *Mauk* and *Curran* for ELO of silicon from Si-Bi melt [30.95].

30.5.2 New Concepts of ELO Growth

The most characteristic feature of the ELO process presented so far is the change of the predominant growth direction – from vertical in the growth window to lateral in the regions over the masking film. In principle, however, it is possible to start lateral overgrowth from seeds oriented perpendicular to the substrate surface, so change of the growth direction is not necessary. If the substrate (bare substrate or substrate covered by a suitable buffer layer) is etched to create on its surface a pattern of ridges or elongated columns oriented similar to that of windows in ELO, conditions are provided such that the side-walls of the columns provide the crystallographic template for lateral growth.

As an example, Fig. 30.43 shows the geometry of a silicon substrate used for the LPE lateral overgrowth of Si from ridge seeds [30.17, 96]. A thin SiO_2 mask has been deposited in the trenches to avoid direct epitaxy on the substrate. During LPE, the growth starts from the sidewalls of the ridges and proceeds laterally over the mask. The top surface of the ridge is the dislocation-free Si(111) facet, so it contains no step sources. Therefore, under proper growth conditions the vertical growth is completely eliminated and the layer grows laterally without any change of its thickness. As the thickness of the layer is determined by the ridge height, it can be easily controlled. However, very thin layers are difficult to obtain. When the thickness of the layer decreases the radius of curvature of its side-wall decreases as well (Fig. 30.43). This leads to an increase in the equilibrium concentration of the solute around this wall due to the Gibbs–Thomson effect (compare discussion in Sect. 30.2.2). Therefore, a higher melt supersaturation must be used to grow thinner layers. Then, however, the vertical growth may be initiated if the supersaturation is large enough for two-dimensional nucleation at the top surface of the ridge. Using this technique 18.8 μm wide and as thin as 0.23 μm, i.e., with L/h ratio up to 82, Si layers have been grown laterally upon SiO_2-coated Si substrates by LPE [30.96].

In contrast with the case of homoepitaxy, in lattice-mismatched heterostructures the top surface of the columns is not dislocation free and must be covered by an additional mask to avoid excessive vertical growth. Figure 30.44a shows schematically how this approach has been applied by *Iida* et al. to grow by LPE InGaAs bridge layers on GaAs substrates [30.21]. First, the substrate covered by the SiN mask has been processed by photolithography and etching to fabricate on its surface a pattern of deep circular trenches. Then, an $In_xGa_{1-x}As$ ($x \approx 0.06$) layer has been grown on such substrate by LPE. It is worth noting that no InGaAs buffer is used in this case. The layer grew inwards from the side-wall of the trench, forming a bridge over the

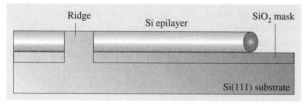

Fig. 30.43 Schematic illustration of the cross section of a Si epilayer grown by LPE on a SiO_2-masked Si substrate from a ridge seed

trench. Some growth over the mask outside the trench area was observed as well. Analysis of the layer surface revealed presence of dislocated zone with etch pits density $\approx 10^6\,\mathrm{cm}^{-2}$ only above the small area where the layer was in direct contact with the GaAs substrate (Fig. 30.44b). The rest of the layer, having diameter above 1 mm, was of very high quality with dislocation density below $10^4\,\mathrm{cm}^{-2}$ (Fig. 30.44c), so it could be used as substrate for further deposition of device layers. The same authors have tested the bridge growth from maskless trenches. Then the lateral growth from the trench wall shown in Fig. 30.44a is accompanied by additional growth of InGaAs from the trench bottom. As a result, new dislocations are found in the InGaAs bridge layer at the places where the heavily dislocated material growing from the trench bottom comes in contact with the bridge. If GaAs substrates are used the presented technique allows for growth of $\mathrm{In}_x\mathrm{Ga}_{1-x}\mathrm{As}$ layers with very low In concentrations ($x \approx 0.06$). For a higher indium content ($x \approx 0.20$), InAs wafer converted into InGaAs instead of GaAs has been used as the substrate [30.97]. The technology of LPE growth of bridge layers over the trenches seems to be very promising due to its simplicity, very high efficiency, and high quality of layers produced.

The method of lateral overgrowth that has received much attention recently is pendeo- (from the Latin: to *hang* or to be *suspended from*) epitaxy (PE). To grow PE GaN layers, a thin AlN wetting layer and GaN buffer are grown first on an available substrate as shown in Fig. 30.45a. Next, trenches deep enough to reach the substrate surface (or even slightly deeper) are etched in the structure. The bottoms of trenches and top surfaces of GaN columns are then covered by a suitable masking film, so the GaN column side-walls are the only surfaces exposed for subsequent GaN regrowth. During pendeo-epitaxy, the GaN layers overflow the trenches growing laterally from the column side-walls (Fig. 30.45b). Also some vertical growth takes place followed by lateral growth over the mask covering the top of GaN mesas. After sufficiently long growth time the PE layers starting from adjacent seeds merge at the fronts marked A and B in Fig. 30.45c, so a continuous PE GaN layer is formed. As mentioned earlier, dislocations thread in the GaN buffer mainly along the *c*-axis (i. e., perpendicular to the surface). Thus, only a very small part of them intersect the side-wall of GaN columns and propagate into the PE GaN layer. Therefore, the regrown material contains four to five orders of magnitude lower density of dislocations than that in the buffer [30.98]. Moreover, in contrast to the standard ELO case, also the disloca-

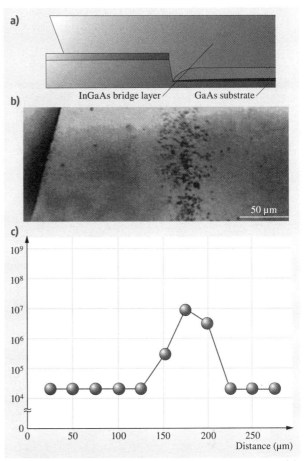

Fig. 30.44a–c Schematic illustrations of the layer cross section (**a**), photographs of the etched surface (**b**), and the profiles of etch pits density (**c**) in an InGaAs bridge layer grown by LPE on a SiO$_2$-masked (111) GaAs substrate (courtesy of Hayakawa)

tions originally present in the buffer (marked as TDs in Fig. 30.45) are prevented from reaching the PE layer if the GaN columns are capped by the mask. Thus, in principle, the density of dislocations should be reduced over the whole area of the substrate and the need to locate devices only in the laterally overgrown parts of the structures (as was in the case of ELO; see Sect. 30.3) could be eliminated. The PE technique in its version shown in Fig. 30.45 has been used by *Chen* et al. to grow by MOVPE GaN layers on SiO$_2$-masked sapphire substrates [30.99]. Studies of the layers with the use of transmission electron microscopy have proved that laterally overgrown parts of the layers are nearly dislo-

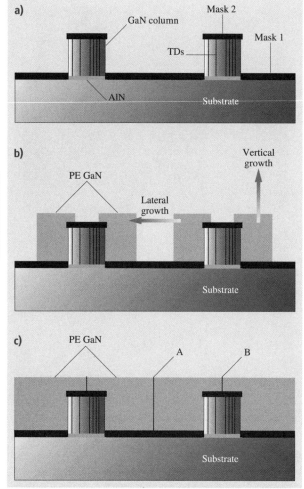

Fig. 30.45a–c Schematic of the process steps for pendeo-epitaxy of GaN: (**a**) etching of elongated columns in GaN buffer layer followed by masking of the substrate (mask 1) and the top column surfaces (mask 2); (**b**) epitaxial growth of GaN starts from exposed side-walls of GaN columns and proceeds laterally to fill the trenches; also some vertical growth occurs; (**c**) PE layers merge over the trenches (seam A) and over the capped GaN columns (seam B), so continuous PE GaN film is formed. TDs are the dislocations threading to the surface of the GaN column

cation free. However, new dislocations have been found at the fronts where the layers merge (marked A and B in Fig. 30.45c) as well as over the edges of masked GaN columns. In the light of what has been written in Sect. 30.4 above, this finding is not surprising. Most probably, interaction of the laterally overgrown parts of the layers with the SiO_2 mask has occurred, leading to their relative misorientation and defect creation above the mask edges and at the coalescence fronts. Thus, this is the same phenomenon as we have discussed already.

These results, as well as the whole discussion in Sect. 30.4, clearly show that properties of laterally overgrown layers are strongly influenced by the presence of a mask. In particular, the strain introduced by the mask and poisoning by impurities diffusing out from the mask at high temperature have negative impact on quality of the layers. Therefore, processes are needed in which layers can be grown laterally without any masking film. It is important to note that each mask requires additional technological steps. Thus, a maskless version of selective epitaxy would result in significant simplification of the whole procedure, which is especially important if the growth process under development is to be applied on industrial scale. *Zheleva* et al. have found that under proper MOVPE conditions neither GaN nor free Ga accumulate on 6H-SiC (0001) surface [30.98]. Thus, pendeo-epitaxy of GaN without a mask between the GaN columns (mask 1 in Fig. 30.45) became possible by replacement of sapphire by SiC. As before, PE growth started from the side-walls of the GaN columns and proceeded laterally at some distance from the SiC surface. Transmission electron microscopy showed that neither tilting nor low-angle tilt boundaries were present at the coalescence boundary over the trench (seam A in Fig. 30.45c). However, tilting of 0.2° and defects at seam B were still found in the portion of the coalesced GaN layer that interacted with the mask covering the GaN column [30.100]. This tilt can be further eliminated if a completely mask-free PE process (i.e., without masks 1 and 2) is employed, as has been reported for GaN [30.98] and AlGaN [30.101] on SiC systems. Then, defect-free coalescence can be obtained at both coalescence fronts [30.98]. The price paid for that, however, is enhanced vertical growth of the PE layer and free propagation into regrown GaN of threading dislocations originally present in the GaN buffer.

The next step towards commercialization of pendeo-epitaxy of GaN would be to replace SiC substrates, which are expensive and available with low diameters, by large-area silicon wafers. However, direct epitaxy of GaN on Si is difficult and usually results in polycrystalline films, most likely due to the prior formation of SiN_x film on the Si surface. Therefore, *Davis* et al. have employed a procedure in which Si(111) substrates are first covered by an epitaxial film of SiC, followed by their transfer to the MOVPE system for deposition of an AlN wetting layer and a GaN buffer [30.100]. Next, they could make use of the process route elaborated earlier for PE of GaN on bulk SiC, namely, substrate etching to define GaN columns and PE regrowth. The quality of the GaN layers they obtained was quite similar to that on bulk SiC substrates. The problem, however, was the thermal stress arising due to the large difference of thermal expansion coefficients between subsequent layers, which resulted in cracking of the structures on cooling if the SiC transition layer was too thin.

As the final example of new developments in lateral overgrowth, successful liquid-phase growth of GaN layers will be briefly presented. Solution growth is preferred for ELO of GaN since supersaturation at the crystal surface can be kept lower than that during vapor-phase epitaxy. Moreover, the growth velocity along the c-axis of GaN is usually much smaller than in perpendicular directions [30.102], so fast lateral growth is expected if (0001)-oriented substrates are used. Unfortunately, as discussed in the Sect. 30.5.1, mask materials that survive undamaged extreme conditions of GaN growth from liquid solution are difficult to find. Therefore, for ELO of GaN *Bockowski* with coworkers [30.103] have used unmasked sapphire substrate with GaN ridges cut from 3 μm-thick GaN buffer grown by MOVPE (Fig. 30.46a). Basically, this is the maskless version of the substrate drawn schematically in Fig. 30.45a. Figure 30.46b shows a plane view of the epilayer grown on such substrate from Ga-N solution at 1430 °C and nitrogen pressure of 1 GPa. The GaN stripes increased their width from 20 to 60 μm and grew ≈ 20 μm high during 5 h-long crystallization. Unfortunately, simultaneous direct crystallization of GaN on unprotected sapphire took place too, so small GaN grains are visible between the stripes. These grains disturbed lateral growth by consuming solute arriving from the bulk solution. Moreover, dislocations were generated in the layer when grains came into contact with laterally grown GaN, in the similar way to that observed in InGaAs bridge layers (compare Sect. 30.5.1). As the result, quite high dislocation density of ≈ 10^6 cm^{-2} was found in the GaN wing area, though it was much smaller than the ≈ 10^8 cm^{-2} dislocation density in the areas above the GaN ridges. It has been expected that parasitic GaN nucleation on bare sapphire might be reduced by decreasing the distance between neighboring GaN ridges [30.92]. It is not clear if this has really happened since decrease of ridges spacing from 300 to 60 μm did

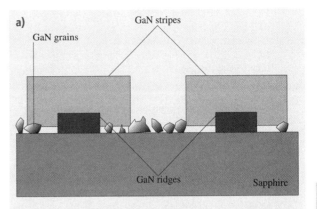

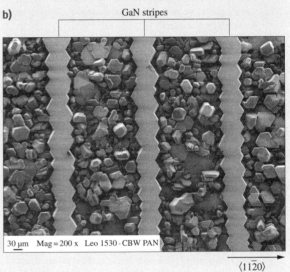

Fig. 30.46a,b Schematic cross section (not to scale) (**a**) and plane-view SEM image (**b**) of GaN layers grown from the Ga-N solution at high nitrogen pressure on sapphire substrate with 20 μm-wide 3 μm-high GaN ridges cut from 3 μm-thick MOVPE-grown GaN buffer. Spacing between the ridges was 300 μm (courtesy of Boćkowski)

not lead to a significant reduction of dislocation density in laterally grown GaN.

Another mode of lateral growth of GaN from Ga-N solution at high nitrogen pressure was found when ≈ 800 μm-thick free-standing GaN substrate grown by HVPE was patterned to create on its surface 5–10 μm-high GaN ridges [30.104]. During GaN growth at 1425 °C and nitrogen pressure of 1 GPa some growth took place directly from the unprotected HVPE sub-

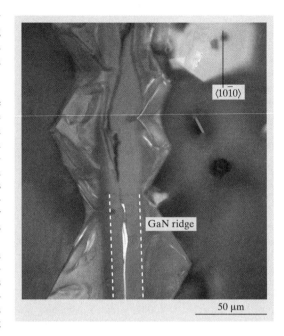

Fig. 30.47 SEM image of GaN layers grown from Ga-N solution at high nitrogen pressure on free-standing HVPE GaN substrate with 20 μm-wide 10 μm-high GaN ridges (*dashed lines*). Spacing between the ridges was 300 μm (courtesy of Boćkowski) ▶

strate as in the previous case. However, due to the specific distribution of solution supersaturation at the substrate surface, preferential lateral overgrowth from ridge edges was also observed. A plan view of such a sample is shown in Fig. 30.47. An important finding by Bockowski et al. was that laterally overgrown parts hang over the substrate if the ridge height was increased to 10 μm. This resulted in their better crystallographic quality. Indeed, analysis of the sample by defect-selective etching has shown that ELO wing areas were dislocation free.

These initial results obtained in lateral overgrowth of GaN from high-temperature Ga-N solution at high nitrogen pressure are very promising. Note that the widths of GaN wings are much larger than those commonly obtained by vapor growth techniques. However, it seems that for further progress application of a suitable mask will be needed to avoid direct crystallization on the substrate. If such a mask is found, much wider wings with lower dislocation density could be grown. Also, improvement of upper surface morphology is required for free-standing layers, similar to that shown in Fig. 30.47.

30.6 Concluding Remarks

A review of epitaxial lateral overgrowth is provided. The main idea behind lateral overgrowth techniques is selective epitaxy of a homoepitaxial layer on a relaxed buffer having a required lattice parameter. The seeds can be defined by patterning of the mask covering the buffer layer (ELO) or the growth starts from the exposed side-walls of ridges etched in the buffer (PE). Selective nucleation of the layer in the narrow seeds allows information on buffer lattice spacing to pass through, while blocking all (or the main part of) defects created due to lattice misfit between the buffer and its substrate. We have shown numerous examples in which this procedure works efficiently. Usually, practical realization of the idea above depends on the system considered. Quite often specific properties of the materials and growth technique used determine the way in which laterally overgrown structures are produced (*vide* maskless pendeo-epitaxy of GaN on SiC, solution growth of GaN, etc.). There are, however, some basic phenomena typical for all laterally overgrown structures. These include some general rules that should be followed to obtain wide and thin laterally overgrown layers (i.e., choice of optimal direction of the seeds, high quality of the buffer, low growth rate, use of doping to enhance lateral growth, etc.). Efficient filtration of dislocations originally present in the buffer is well evidenced in all laterally overgrown structures. Tilt of wings caused by their interaction with the mask underneath is commonly observed in both ELO and PE layers, independently of the growth technique used. If the tilt angle is small the accompanied strain can be accommodated elastically (e.g., in GaAs/GaAs ELOs). Otherwise, arrays of dislocations creating low-angle boundaries appear above the mask edges (e.g., in GaN ELO structures). Tilting of laterally grown parts of the layer leads to the generation of new defects at the coalescence front where two layers grown from neighboring seeds and having opposite tilt direction merge. Growing the layers standing free above the mask seems to be the best way to improve the quality of the coalescence front, although there are no general recipes for how to grow such structures. Thermal strain due to different thermal expansion coefficients of subsequent layers is also commonly found in laterally overgrown heterostructures. However, if the layer is attached to the buffer through the narrow seeds only, this strain is usually much smaller than in reference planar structures.

It is difficult to predict future developments in the ELO technology. From a scientific point of view there are still many questions to be answered. In particular, understanding of the microscopic interaction of ELO wings with the mask, identification of mechanisms leading to bending of dislocations threading in a seed area, and many other issues require further work. Basic crystal growth research including process modeling is needed to increase the width of laterally overgrown layers, improve the quality of coalescence fronts, develop technology for lateral overgrowth of ternary layers on binary substrates, find new mask materials, and simplify the whole growth procedure to meet industrial standards. Application of laterally overgrown layers will depend on progress in these areas. Market demands for novel semiconductor devices will decide if and which techniques of lateral epitaxy will find their permanent position in industrial practice. Nowadays there is great demand for group III nitride substrates, so lateral overgrowth of these compounds is most studied in many laboratories worldwide. It is believed that the importance of lateral overgrowth of *traditional* zincblende III–V ternary substrates will increase in the near future.

References

30.1 T. Kawai, H. Yonezu, Y. Ogasawara, D. Saito, K. Pak: Segregation and interdiffusion of In atoms in GaAs/InAs/GaAs heterostructures, J. Appl. Phys. **74**, 1770–1775 (1993)

30.2 N. Chand, R. People, F.A. Baiocchi, W.K. Wecht, A.Y. Cho: Significant improvement in crystalline quality of molecular beam epitaxially grown GaAs on Si (100) by rapid thermal annealing, Appl. Phys. Lett. **49**, 815–817 (1986)

30.3 N. Gopalakrishnan, K. Baskar, H. Kawanami, I. Sakata: Effects of the low temperature grown buffer layer thickness on the growth of GaAs on Si by MBE, J. Cryst. Growth **250**, 29–33 (2003)

30.4 Y. Honda, Y. Iyechika, T. Maeda, H. Miyake, K. Hiramatsu, H. Sone, N. Sawaki: Crystal orientation fluctuation of epitaxial-lateral-overgrown GaN with W mask and SiO_2 mask observed by transmission electron diffraction and x-ray rocking curves, Jpn. J. Appl. Phys. **38**, L1299–L1302 (1999)

30.5 D. Dobosz, Z.R. Zytkiewicz, T.T. Piotrowski, E. Kaminska, E. Papis, A. Piotrowska: Application of tungsten films for substrate masking in liquid phase epitaxial lateral overgrowth of GaAs, Cryst. Res. Technol. **38**, 297–301 (2003)

30.6 D. Dobosz, K. Golaszewska, Z.R. Zytkiewicz, E. Kaminska, A. Piotrowska, T.T. Piotrowski, A. Barcz, R. Jakiela: Properties of ZrN films as substrate masks in liquid phase epitaxial lateral overgrowth of compound semiconductors, Cryst. Res. Technol. **40**, 492–497 (2005)

30.7 Z.R. Zytkiewicz, J. Domagla, D. Dobosz, J. Bak-Misiuk: Microscopic bending of GaAs layers grown by epitaxial lateral overgrowth, J. Appl. Phys. **84**, 6937–6939 (1998)

30.8 S. Sakawa, T. Nishinaga: Effect of Si doping on epitaxial lateral overgrowth of GaAs on GaAs-coated Si substrate, Jpn. J. Appl. Phys. **31**, L359–L361 (1992)

30.9 L. Jastrzebski, J.F. Corboy, R. Soydan: Issues and problems involved in selective epitaxial growth of silicon for SOI fabrication, J. Electrochem. Soc. **136**, 3506–3513 (1989)

30.10 R. Bergmann, E. Czech, I. Silier, N. Nagel, E. Bauser, H.J. Queisser, R.P. Zingg, B. Hofflinger: MOS transistors with epitaxial Si, laterally grown over SiO_2 by liquid phase epitaxy, Appl. Phys. A **54**, 103–105 (1992)

30.11 H. Asai, S. Adachi, S. Ando, K. Oe: Lateral GaAs growth over tungsten gratings on (001) GaAs substrates by metalorganic chemical vapor deposition and applications to vertical field-effect transistors, J. Appl. Phys. **55**, 3868–3870 (1984)

30.12 J.M. Olchowik, A. Fave, B. Semmache, A. Laugier, A. Kaminski, W. Sadowski: Crystallisation of Si epitaxial lateral overgrowth layers for photovoltaic structures, Proc. 16th Eur. Photovolt. Sol. Energy Conf., Vol. 2 (2000) pp. 1286–1288

30.13 J.J. Pak, G.W. Neudeck, A.E. Kabir, D.W. DeRoo, S.E. Staller, J.H. Logsdon: A new method of forming a thin single-crystal silicon diaphragm using merged epitaxial lateral overgrowth for sensor applications, IEEE Electron. Dev. Lett. **12**, 614–616 (1991)

30.14 G.W. Neudeck, P.J. Schubert, J.L. Glenn Jr., J.A. Friedrich, W.A. Klaasen, R.P. Zingg, J.P. Denton: Three dimensional devices fabricated by silicon epitaxial lateral overgrowth, J. Electron. Mater. **19**, 1111–1117 (1990)

30.15 R.W. McClelland, C.O. Bozler, J.C.C. Fan: A technique for producing epitaxial films on reuseable substrates, Appl. Phys. Lett. **37**, 560–562 (1980)

30.16 B.-Y. Tsaur, R.W. McClelland, J.C.C. Fan, R.P. Gale, J.P. Salerno, B.A. Vojak, C.O. Bozler: Low-dislocation-density GaAs epilayers grown on Ge-coated Si substrates by means of lateral epi-

30.17 I. Silier, A. Gutjahr, N. Nagel, P.O. Hansson, E. Czech, M. Konuma, E. Bauser, F. Banhart, R. Kohler, H. Raidt, B. Jenichen: Solution growth of epitaxial semiconductor-on-insulator layers, J. Cryst. Growth **166**, 727–730 (1996)

30.18 T. Nishinaga: Epitaxial lateral overgrowth of III–V compounds for obtaining dislocation free layers, Cryst. Prop. Prep. **31**, 92–99 (1991)

30.19 P.O. Hansson, A. Gustafsson, M. Albrecht, R. Bergmann, H.P. Strunk, E. Bauser: High quality Ge_xSi_{1-x} by heteroepitaxial lateral overgrowth, J. Cryst. Growth **121**, 790–794 (1992)

30.20 Z.R. Zytkiewicz: Laterally overgrown structures as substrates for lattice mismatched epitaxy, Thin Solid Films **412**, 64–75 (2002)

30.21 Y. Hayakawa, S. Iida, T. Sakurai, H. Yanagida, M. Kikuzawa, T. Koyama, M. Kumagawa: Epitaxial lateral overgrowth of InGaAs on patterned GaAs substrates by liquid phase epitaxy, J. Cryst. Growth **169**, 613–620 (1996)

30.22 S. Nakayama, M. Kaneko, S. Aizawa, K. Kashiwa, N.S. Takahashi: InGaP lattice-mismatched LPE growth on GaAs substrates by epitaxial lateral overgrowth technique, J. Cryst. Growth **236**, 132–136 (2002)

30.23 S. Nakamura, M. Senoh, S. Nagahama, N. Iwasa, T. Yamada, T. Matsushita, H. Kiyoku, Y. Sugimoto, T. Kozaki, H. Umemoto, M. Sano, K. Chocho: InGaN/GaN/AlGaN-based laser diodes with modulation-doped strained-layer superlattices, Jpn. J. Appl. Phys. **36**, L1568–L1571 (1997)

30.24 Y.C. Liu, Z.R. Zytkiewicz, S. Dost: Computational analysis of lateral overgrowth of GaAs by liquid-phase epitaxy, J. Cryst. Growth **275**, e953–e957 (2005)

30.25 Z. Yan, S. Naritsuka, T. Nishinaga: Interface supersaturation in microchannel epitaxy of InP, J. Cryst. Growth **203**, 25–30 (1999)

30.26 Z. Yan, S. Naritsuka, T. Nishinaga: Two-dimensional numerical calculation of solute diffusion in microchannel epitaxy of InP, J. Cryst. Growth **209**, 1–7 (2000)

30.27 B. Beaumont, P. Vennegues, P. Gibart: Epitaxial lateral overgrowth of GaN, Phys. Status Solidi (b) **227**, 1–43 (2001)

30.28 R.J. Matyi, H. Shichijo, T.M. Moore, H.-L. Tsai: Microstructural characterization of patterned gallium arsenide grown on ⟨001⟩ silicon substrates, Appl. Phys. Lett. **51**, 18–20 (1987)

30.29 J.T. Torvik, J.I. Pankove, E. Iliopoulos, H.M. Ng, T.D. Moustakas: Optical properties of GaN grown over SiO_2 on SiC substrates by molecular beam epitaxy, Appl. Phys. Lett. **72**, 244–245 (1998)

30.30 G. Bacchin, T. Nishinaga: A new way to achieve both selective and lateral growth by molecular beam epitaxy: Low angle incidence microchannel epitaxy, J. Cryst. Growth **208**, 1–10 (2000)

30.31 S.C. Lee, K.J. Malloy, L.R. Dawson, S.R.J. Brueck: Selective growth and associated faceting and lateral overgrowth of GaAs on a nanoscale limited area bounded by a SiO_2 mask in molecular beam epitaxy, J. Appl. Phys. **92**, 6567–6571 (2002)

30.32 H. Tang, J.A. Bardwell, J.B. Webb, S. Moisa, J. Fraser, S. Rolfe: Selective growth of GaN on a SiC substrate patterned with an AlN seed layer by ammonia molecular-beam epitaxy, Appl. Phys. Lett. **79**, 2764–2766 (2001)

30.33 Z.R. Zytkiewicz: Epitaxial lateral overgrowth of GaAs: Principle and growth mechanism, Cryst. Res. Technol. **34**, 573–582 (1999)

30.34 S. Zhang, T. Nishinaga: Epitaxial lateral overgrowths of GaAs on (001) GaAs substrates by LPE: Growth behavior and mechanism, J. Cryst. Growth **99**, 292–296 (1990)

30.35 S. Naritsuka, T. Nishinaga: Epitaxial lateral overgrowth of InP by liquid phase epitaxy, J. Cryst. Growth **146**, 314–318 (1995)

30.36 R.P. Gale, R.W. McClelland, J.C.C. Fan, C.O. Bozler: Lateral epitaxial overgrowth of GaAs by organometallic chemical vapor deposition, Appl. Phys. Lett. **41**, 545–547 (1982)

30.37 J. Park, P.A. Barnes, C.C. Tin, A. Allerman: Lateral overgrowth and epitaxial lift-off of InP by halide vapor-phase epitaxy, J. Cryst. Growth **187**, 185–193 (1998)

30.38 T. Nishinaga, T. Nakano, S. Zhang: Epitaxial lateral overgrowth of GaAs by LPE, Jpn. J. Appl. Phys. **27**, L964–L967 (1988)

30.39 S. Zhang, T. Nishinaga: LPE lateral overgrowth of GaP, Jpn. J. Appl. Phys. **29**, 545–550 (1990)

30.40 R. Bergmann, E. Bauser, J.H. Werner: Defect-free epitaxial lateral overgrowth of oxidized (111) Si by liquid phase epitaxy, Appl. Phys. Lett. **57**, 351–353 (1990)

30.41 R. Bergmann: Model for defect-free epitaxial lateral overgrowth of Si over SiO_2 by liquid phase epitaxy, J. Cryst. Growth **110**, 823–834 (1991)

30.42 S. Sakawa, T. Nishinaga: Faceting of LPE GaAs grown on a misoriented Si(100) substrate, J. Cryst. Growth **115**, 145–149 (1991)

30.43 D. Dobosz, Z.R. Zytkiewicz: Epitaxial lateral overgrowth of semiconductor structures by liquid phase epitaxy, Int. J. Mater. Prod. Technol. **22**, 50–63 (2005)

30.44 F. Banhart, R. Bergmann, F. Phillipp, E. Bauser: Dislocation generation in silicon grown laterally over SiO_2 by liquid phase epitaxy, Appl. Phys. A **53**, 317–323 (1991)

30.45 Y.S. Chang, S. Naritsuka, T. Nishinaga: Optimization of growth condition for wide dislocation-free GaAs on Si substrate by microchannel epitaxy, J. Cryst. Growth **192**, 18–22 (1998)

30.46 D. Dobosz, Z.R. Zytkiewicz, E. Papis, E. Kaminska, A. Piotrowska: Epitaxial lateral overgrowth of GaSb layers by liquid phase epitaxy, J. Cryst. Growth **253**, 102–106 (2003)

30.47 H. Raidt, R. Kohler, F. Banhart, B. Jenichen, A. Gutjahr, M. Konuma, I. Silier, E. Bauser: Adhesion in growth of defect-free silicon over silicon oxide, J. Appl. Phys. **80**, 4101–4107 (1996)

30.48 Z.R. Zytkiewicz, D. Dobosz, Y.C. Liu, S. Dost: Recent progress in lateral overgrowth of semiconductor structures from the liquid phase, Cryst. Res. Technol. **40**, 321–328 (2005)

30.49 Z.R. Zytkiewicz, D. Dobosz, M. Pawlowska: Epitaxial lateral overgrowth of GaAs: Effect of doping on LPE growth behaviour, Semicond. Sci. Technol. **14**, 465–469 (1999)

30.50 J.C. Brice: *The Growth of Crystals from Liquids* (North-Holland, Amsterdam 1973)

30.51 M. Ohara, R.C. Reid: *Modelling Crystal Growth Rates from Solution* (Prentice-Hall, Englewood Cliffs 1973)

30.52 W.J.P. van Enckevort, A.C.J.F. van den Berg: Impurity blocking of crystal growth: a Monte Carlo study, J. Cryst. Growth **183**, 441–455 (1998)

30.53 D.-K. Kim, B.-T. Lee: Heteroepitaxial growth of GaAs on (100) GaAs and InP by selective liquid phase epitaxy, Jpn. J. Appl. Phys. **33**, 5870–5874 (1994)

30.54 B. Beaumont, S. Haffouz, P. Gibart: Magnesium induced changes in the selective growth of GaN by metalorganic vapor phase epitaxy, Appl. Phys. Lett. **72**, 921–923 (1998)

30.55 Z.R. Zytkiewicz, J. Domagala: Thermal strain in GaAs layers grown by epitaxial lateral overgrowth on Si substrates, Appl. Phys. Lett. **75**, 2749–2751 (1999)

30.56 T. Nishinaga: Microchannel epitaxy: an overview, J. Cryst. Growth **237–239**, 1410–1417 (2002)

30.57 A. Sakai, H. Sunakawa, A. Usui: Transmission electron microscopy of defects in GaN films formed by epitaxial lateral overgrowth, Appl. Phys. Lett. **73**, 481–483 (1998)

30.58 A. Sakai, H. Sunakawa, A. Usui: Defect structure in selectively grown GaN films with low threading dislocation density, Appl. Phys. Lett. **71**, 2259–2261 (1997)

30.59 Z. Yu, M.A.L. Johnson, T. McNulty, J.D. Brown, J.W. Cook, J.F. Schetzina: Study of the epitaxial lateral overgrowth (ELO) process for GaN on sapphire using scanning electron microscopy and monochromatic cathodoluminescence, MRS Internet J. Nitride Semicond. Res. **3**, 6 (1998)

30.60 A. Gustafsson, P.O. Hansson, E. Bauser: Cathodoluminescence from relaxed Ge_xSi_{1-x} grown by heteroepitaxial lateral overgrowth, J. Cryst. Growth **141**, 363–370 (1994)

30.61 Z.I. Kazi, P. Thilakan, T. Egawa, M. Umeno, T. Jimbo: Realization of GaAs/AlGaAs lasers on Si substrates using epitaxial lateral overgrowth by metalorganic chemical vapor deposition, Jpn. J. Appl. Phys., Part 1 **40**, 4903–4906 (2001)

30.62 P. Kozodoy, J.P. Ibbetson, H. Marchand, P.T. Fini, S. Keller, J.S. Speck, S.P. DenBaars, U.K. Mishra: Electrical characterization of GaN p-n junctions with and without threading dislocations, Appl. Phys. Lett. **73**, 975–977 (1998)

30.63 T. Mukai, K. Takekawa, S. Nakamura: InGaN-based blue light-emitting diodes grown on epitaxially laterally overgrown GaN substrates, Jpn. J. Appl. Phys., Part 2 **37**, L839–L841 (1998)

30.64 C. Sasaoka, H. Sunakawa, A. Kimura, M. Nido, A. Usui, A. Sakai: High-quality InGaN MQW on low-dislocation-density GaN substrate grown by hydride vapor-phase epitaxy, J. Cryst. Growth **189/190**, 61–66 (1998)

30.65 G. Parish, S. Keller, P. Kozodoy, J.P. Ibbetson, H. Marchand, P.T. Fini, S.B. Fleischer, S.P. DenBaars, U.K. Mishra, E.J. Tarsa: High-performance (Al,Ga)N-based solar-blind ultraviolet p-i-n detectors on laterally epitaxially overgrown GaN, Appl. Phys. Lett. **75**, 247–249 (1999)

30.66 Z.R. Zytkiewicz, J.Z. Domagala, D. Dobosz, L. Dobaczewski, A. Rocher, C. Clement, J. Crestou: Tilt and dislocations in epitaxial laterally overgrown GaAs layers, J. Appl. Phys. **101**, 013508 (2007)

30.67 N. Nagel, F. Banhart, E. Czech, I. Silier, F. Phillipp, E. Bauser: The coalescence of silicon layers grown over SiO_2 by liquid-phase epitaxy, Appl. Phys. A **57**, 249–254 (1993)

30.68 F. Banhart, N. Nagel, F. Phillipp, E. Czech, I. Silier, E. Bauser: The coalescence of silicon layers grown over SiO_2 by liquid-phase epitaxy, Appl. Phys. A **57**, 441–448 (1993)

30.69 W. Huang, T. Nishinaga, S. Naritsuka: Microchannel epitaxy of GaAs from parallel and nonparallel seeds, Jpn. J. Appl. Phys. **40**, 5373–5376 (2001)

30.70 Z. Yan, Y. Hamaoka, S. Naritsuka, T. Nishinaga: Coalescence in microchannel epitaxy of InP, J. Cryst. Growth **212**, 1–10 (2000)

30.71 Y.-C. Shih, J.-C. Lou, W.O. Oldham: Seam line defects in silicon-on-insulator by merged epitaxial lateral overgrowth, Appl. Phys. Lett. **65**, 1638–1640 (1994)

30.72 Z.R. Zytkiewicz, J. Domagala, D. Dobosz, J. Bak-Misiuk: Strain in GaAs layers grown by liquid phase epitaxial lateral overgrowth, J. Appl. Phys. **86**, 1965–1969 (1999)

30.73 Z.R. Zytkiewicz, J. Domagala, D. Dobosz: Control of adhesion to the mask of epitaxial laterally overgrown GaAs layers, J. Appl. Phys. **90**, 6140–6144 (2001)

30.74 Z.R. Zytkiewicz, J. Domagala, D. Dobosz: Mask-induced strain in GaAs layers grown by liquid phase epitaxial lateral overgrowth, Mater. Res. Soc. Symp. Proc. **570**, 273–278 (1999)

30.75 J.Z. Domagala, A. Czyzak, Z.R. Zytkiewicz: Imaging of strain in laterally overgrown GaAs layers by spatially resolved x-ray diffraction, Appl. Phys. Lett. **90**, 241904 (2007)

30.76 I.-H. Kim, C. Sone, O.-H. Nam, Y.-J. Park, T. Kim: Crystal tilting in GaN grown by pendoepitaxy method on sapphire substrate, Appl. Phys. Lett. **75**, 4109–4111 (1999)

30.77 D. Lubbert, T. Baumbach, P. Mikulik, P. Pernot, L. Helfen, R. Kohler, T.M. Katona, S. Keller, S.P. DenBaars: Local wing tilt analysis of laterally overgrown GaN by x-ray rocking curve imaging, J. Phys. D Appl. Phys. **38**, A50–A54 (2005)

30.78 R. Rantamaki, T. Tuomi, Z.R. Zytkiewicz, J. Domagala, P.J. McNally, A.N. Danilewsky: Synchrotron x-ray topographic and high-resolution diffraction analysis of mask-induced strain in epitaxial laterally overgrown GaAs layers, J. Appl. Phys. **86**, 4298–4303 (1999)

30.79 R. Rantamaki, T. Tuomi, Z.R. Zytkiewicz, J. Domagala, P.J. McNally, A.N. Danilewsky: Comparative analysis of synchrotron x-ray transmission and reflection topography techniques applied to epitaxial laterally overgrown GaAs layers, J. X-ray Sci. Technol. **8**, 277–288 (2000)

30.80 P. Fini, H. Marchand, J.P. Ibbetson, S.P. DenBaars, U.K. Mishra, J.S. Speck: Determination of tilt in the lateral epitaxial overgrowth of GaN using x-ray diffraction, J. Cryst. Growth **209**, 581–590 (2000)

30.81 M.H. Kim, Y. Choi, J. Yi, M. Yang, J. Jeon, S. Khym, S.J. Leem: Reduction in crystallographic tilting of lateral epitaxial overgrown GaN by removal of oxide mask, Appl. Phys. Lett. **79**, 1619–1621 (2001)

30.82 K. Hiramatsu, H. Matsushima, T. Shibata, Y. Kawagachi, N. Sawaki: Selective area growth and epitaxial lateral overgrowth of GaN by metalorganic vapor phase epitaxy and hydride vapor phase epitaxy, Mater. Sci. Eng. B **59**, 104–111 (1999)

30.83 P. Fini, A. Munkholm, C. Thompson, G.B. Stephenson, J.A. Eastman, M.V. Ramana Murthy, O. Auciello, L. Zhao, S.P. DenBaars, J.S. Speck: In situ, real-time measurement of wing tilt during lateral epitaxial overgrowth of GaN, Appl. Phys. Lett. **76**, 3893–3895 (2000)

30.84 R. Kohler, B. Jenichen, H. Raidt, E. Bauser, N. Nagel: Vertical stress in liquid-phase epitaxy Si layers on SiO_2/Si evaluated by x-ray double-crystal topography, J. Phys. D Appl. Phys. **28**, A50–A55 (1995)

30.85 I. Kidoguch, A. Ishibashi, G. Sugahara, Y. Ban: Air-bridged lateral epitaxial overgrowth of GaN thin films, Appl. Phys. Lett. **76**, 3768–3770 (2000)

30.86 P. Fini, L. Zhao, B. Morgan, M. Hansen, H. Marchand, J.P. Ibbetson, S.P. DenBaars, U.K. Mishra, J.S. Speck: High-quality coalescence of laterally overgrown GaN stripes on GaN/sapphire seed layers, Appl. Phys. Lett. **75**, 1706–1708 (1999)

30.87 S.F. Fang, K. Adomi, S. Iyer, H. Morkoc, H. Zabel, C. Choi, N. Otsuka: Gallium arsenide and other compound semiconductors on silicon, J. Appl. Phys. **68**, R31–R58 (1990), and references therein

30.88 N. Lucas, H. Zabel, H. Morkoc, H. Unlu: Anisotropy of thermal expansion of GaAs on Si(001), Appl. Phys. Lett. **52**, 2117–2119 (1988)

30.89 T.S. Zheleva, W.M. Asmawi, K.A. Jones: Pendeo-epitaxy versus lateral epitaxial overgrowth of GaN: A comparative study via finite element analysis, Phys. Status Solidi (a) **176**, 545–551 (1999)

30.90 E. Kaminska, A. Piotrowska, E. Mizera, M. Guziewicz, A. Barcz, E. Dynowska, S. Kwiatkowski: Rapid thermal nitridation of tungsten-capped shallow ohmic contacts to GaAs, Mater. Res. Soc. Symp. Proc. **337**, 349–354 (1994)

30.91 L.E. Toth: *Transition Metals Carbides and Nitrides* (Academic, New York 1971)

30.92 M. Boćkowski, I. Grzegory, G. Nowak, G. Kamler, B. Łucznik, M. Wróblewski, P. Kwiatkowski, K. Jasik, S. Krukowski, S. Porowski: Growth of GaN on patterned GaN/sapphire substrates with various metallic masks by high pressure solution method, Proc. SPIE **6121**, 612103-1–612103-9 (2006)

30.93 T. Bryskiewicz: Liquid phase electroepitaxy of semiconductor compounds, Prog. Cryst. Growth Charact. **12**, 29–43 (1986), and references therein

30.94 Y.C. Liu, Z.R. Zytkiewicz, S. Dost: A model for epitaxial lateral overgrowth of GaAs by liquid-phase electroepitaxy, J. Cryst. Growth **265**, 341–350 (2004)

30.95 M.G. Mauk, J.P. Curran: Electro-epitaxial lateral overgrowth of silicon from liquid-metal solutions, J. Cryst. Growth **225**, 348–353 (2001)

30.96 Y. Suzuki, T. Nishinaga: Si LPE lateral overgrowth from a ridge seed, Jpn. J. Appl. Phys. **29**, 2685–2689 (1990)

30.97 K. Balakrishnan, S. Iida, M. Kumagawa, Y. Hayakawa: A novel method to grow high quality $In_{1-x}Ga_xAs$ ELO and bridge layers with high indium compositions, J. Cryst. Growth **237-239**, 1525–1530 (2002)

30.98 T.S. Zheleva, S.A. Smith, D.B. Thomson, K.J. Linthicum, P. Rajagopal, R.F. Davis: Pendeoepitaxy: A new approach for lateral growth of gallium nitride films, J. Electron. Mater. **28**, L5–L8 (1999)

30.99 Y. Chen, R. Schneider, S.Y. Wang, R.S. Kern, C.H. Chen, C.P. Kou: Dislocation reduction in GaN thin films via lateral overgrowth from trenches, Appl. Phys. Lett. **75**, 2062–2063 (1999)

30.100 R.F. Davis, T. Gehrke, K.J. Linthicum, E. Preble, P. Rajagopal, C. Ronning, C. Zorman, M. Mehregany: Conventional and pendeo-epitaxial growth of GaN(0001) thin films on Si(111) substrates, J. Cryst. Growth **231**, 335–341 (2001)

30.101 R.F. Davis, T. Gehrke, K.J. Linthicum, T.S. Zheleva, E.A. Preble, P. Rajagopal, C.A. Zorman, M. Mehregany: Pendeo-epitaxial growth of thin films of gallium nitride and related materials and their characterization, J. Cryst. Growth **225**, 134–140 (2001)

30.102 I. Grzegory: High pressure growth of bulk GaN from solutions in gallium, J. Phys. Condens. Matter **13**, 6875–6892 (2001)

30.103 M. Boćkowski, I. Grzegory, J. Borysiuk, G. Kamler, B. Łucznik, M. Wróblewski, P. Kwiatkowski, K. Jasik, S. Krukowski, S. Porowski: Growth of GaN on patterned GaN/sapphire substrates by high pressure solution method, J. Cryst. Growth **281**, 11–16 (2005)

30.104 M. Boćkowski, I. Grzegory, G. Nowak, B. Łucznik, B. Pastuszka, G. Klamer, M. Wróblewski, P. Kwiatkowski, K. Jasik, W. Wawer, S. Krukowski, S. Porowski: Growth of GaN on patterned thick HVPE free standing GaN substrates by high pressure solution method, Phys. Status Solidi (c) **3**, 1487–1490 (2006)

1040

31. Liquid-Phase Epitaxy of Advanced Materials

Christine F. Klemenz Rivenbark

The performance of many electronic and optoelectronic devices critically depends on the structural quality and homogeneity of the base material, which is often an epitaxial film grown by either vapor-phase epitaxy (VPE) or liquid-phase epitaxy (LPE).

This chapter presents the state of the art in LPE growth of selected advanced materials:

1. High-temperature superconductors
2. Calcium gallium germanates (langasite-type materials)
3. III–V wide-bandgap nitrides.

It is not the aim, herein, to present LPE growth of more *traditional* III–V semiconductors (Si, Ge, GaAs, GaP, InP, GaP) and garnets, which have already been described extensively in the literature since about 1960. Instead, some of the most relevant literature references are given in the historical overview, which also provides a very good insight into the potential of LPE growth of newer materials. Despite the fact that LPE growth has gained less attention over the past decades, mainly due to the development of VPE growth techniques, there is a silver lining which clearly indicates that the highest-quality epitaxial films, for most efficient electronic and optoelectronic devices, will ultimately be achieved from liquid-assisted or LPE-grown films.

31.1 Historical Development of LPE 1042
31.2 Fundamentals of LPE and Solution Growth 1042
31.3 Requirements for Liquid-Phase Epitaxy .. 1044
31.4 Developing New Materials: On the Choice of the Epitaxial Deposition Method 1044
31.5 LPE of High-Temperature Superconductors ... 1046
   31.5.1 Phase Relations, Solvent System, and Solubility Curves 1046
   31.5.2 Heat of Solution 1048
   31.5.3 Supersaturation and Driving Force for LPE 1048
   31.5.4 Substrates and Epitaxial Relationship 1049
   31.5.5 LPE Growth System and Film Growth Procedure 1051
   31.5.6 Growth Mechanisms and Growth Parameters: Theory Versus Experiment 1052
31.6 LPE of Calcium Gallium Germanates 1055
   31.6.1 Solvent System 1055
   31.6.2 Substrates for Homoepitaxial LGT LPE Film Growth 1056
   31.6.3 LPE growth of LGS, LGT, and LGN 1057
   31.6.4 Structural and Chemical Characterization of Doped LGT LPE Films 1057
31.7 Liquid-Phase Epitaxy of Nitrides 1059
   31.7.1 Developments and Trends in LPE of GaN and AlN 1060
   31.7.2 Substrates for Epitaxy of Nitrides 1061
   31.7.3 Growth System and Optimization ... 1062
   31.7.4 Morphological Evolution of LPE-Grown Nitride Films 1062
31.8 Conclusions 1063
References 1064

Liquid-phase epitaxy (LPE) is a growth process where a film of crystalline material is deposited from a supersaturated solution onto a single-crystal substrate. This method was first described by *Nelson* in 1963 [31.1], who used a tipping LPE apparatus for the fabrication of GaAs laser diodes. Since then, the LPE technique has been widely investigated and applied to a variety of materials, among others II–V and III–V semiconductors, carbides, nitrides, and a variety of oxides.

31.1 Historical Development of LPE

Liquid-phase epitaxy has traditionally been the optimal method of choice for the synthesis of highest-quality material. However, this method present also challenges. Today, more *flexible* vapor growth techniques are generally preferred, especially when film defects are not a limiting factor for a given device application.

The potential of LPE to yield low-defect material has been documented in many instances, following the path of the development of technologically important materials. In LPE, the driving force (or supersaturation) for epitaxy is very low. Hence, growth occurs near to thermodynamic equilibrium. The atoms can migrate efficiently on the growing surface until they find the most energetically favorable position where they will incorporate. If supersaturation and other requirements are sufficiently met, the LPE growth process may result in single-crystalline films with very high homogeneity and purity. This was best demonstrated with the development of light-emitting diodes (LEDs), with compound semiconductors such as GaAs [31.2] and GaP [31.3]. Films grown by LPE contain inherently a very low concentration of point defects, which is often orders of magnitude lower than material grown by any other growth technique. Most relevant evidence is found in the research and development of green GaP LEDs. Due to an extremely low concentration of nonstoichiometric point defects in GaP LPE films, even at room temperature, the dominant emission transition mechanism is free exciton recombination. This enables a highly efficient nitrogen-free pure-green light emission (550 nm) [31.4] that cannot be obtained from GaP produced by other growth techniques.

As early as in 1974, studies on selective LPE growth of GaAlAs led to the discovery of the epitaxial lateral overgrowth (ELO) technique [31.5]. The combination of ELO and LPE [31.6] enabled dislocation-free films of many semiconductors, such as Si, GaAs, GaP, GaAs/Si, and InP [31.7–10]. In view of this, one may wonder why LPE growth is only marginally studied today. For example, research on wide-bandgap nitride semiconductors could highly profit from dislocation-free LPE films to be used as substrates. Often, the potential of a growth technique is evaluated from its commercialization perspective. Without contest, the broadest commercialization to report for films grown by LPE resides in the field of optoelectronic devices. Even today, the highest-efficiency red GaAs LEDs are commercially produced by LPE. Films grown by LPE are also used in other applications, such as γ-ray [31.11] and high-end infrared (IR) detectors based on CdHgTe/CdTe [31.12], for example, where low-defect films are required.

The LPE growth technique has been applied to oxides as early as 1968 by *Linares* [31.13], who was able to demonstrate high-quality yttrium iron garnet (YIG) films for the first time. Different magnetic garnet compositions were studied for their use in bubble domain devices, and this led to extensive studies on many fundamental and experimental aspects of LPE of oxides, including the development of different LPE growth techniques, which are summarized in excellent review papers [31.14–16]. The LPE growth method has been successfully applied for the growth of many oxides, among which one can also cite high-temperature superconductors (HTSCs), optical, nonlinear optical, and electrooptical materials, and ferroelectric and dielectric materials.

31.2 Fundamentals of LPE and Solution Growth

The fundamental mechanisms for LPE growth are similar to the growth of free-standing crystals from a solution (or flux), or to the growth of larger crystals by top-seeded solution growth (TSSG). Usually, the same solution can be used for the three methods. In growth from solutions, a supersaturation is created either by cooling or by solvent evaporation of the saturated solution. Another method that can be used is the transport of the growth species through a temperature gradient in the solution. In each case, the growth is controlled by the diffusion of growth species towards the growing interface and the counterdiffusion of solvent species towards the bulk of the solution.

In the initial stages of crystallization from solutions, some degree of supersaturation, or driving force, has to be established for nucleation to take place. A crystal can only develop once a nuclei has reached a critical size and thus is stable. In the phase diagram, the region between the liquidus temperature and the temperature at which nucleation proceeds spontaneously is called the metastable or Ostwald–Miers region. Within this region, heteronucleation is possible, for example, on the crucible, on a substrate or seed, or on an impurity. The width of the metastable region can vary greatly for different solvent–solute systems. It depends on many factors, such as the nature and complexity (number of components) of the crystallizing material, the viscosity of the solution, additives/dopants, and hydrodynamic flows. For oxides, undercoolings up to 150 °C are possible [31.17], whereas metallic semiconductor solutions allow only a few degrees or less of undercooling. When a seed or a substrate is used, the initial supersaturation required for crystallization is less than that required for spontaneous and free nucleation in the solution. This means that, in LPE, nucleation starts within a range between the liquidus temperature and the temperature for spontaneous homogeneous nucleation.

The relative supersaturation is expressed by

$$\sigma = \frac{x - x_e}{x_e}, \qquad (31.1)$$

where x and x_e represent the actual and equilibrium concentration of the solution.

The undercooling $\Delta T = T_e - T$ is related to the respective concentrations through the specific phase diagram of the system considered.

During the LPE film growth, a boundary layer of thickness δ is created at the growth front, in which a temperature and a concentration gradient exist. At the growth interface, two processes are determining: hydrodynamic flow and diffusion. For mass transport, flow is far more effective than diffusion. However, the growth is governed by the slowest process. In addition, the hydrodynamic flow decreases from the bulk of the solution to become negligible at the growth interface. Thus, at the growing interface, within the boundary layer, the process is solely controlled by diffusion, and the thickness of the boundary layer will be determined by the hydrodynamic flow. As a consequence, for a given system at a given supersaturation, the thickness of the boundary layer will ultimately determine the growth rate, which in turn can be adjusted through the substrate rotation rate.

In the case of LPE growth, for systems with relatively large Schmidt number (Sc = ν/D), the thickness of the diffusion boundary layer on the rotating substrate interface can be estimated from

$$\delta = 1.61 \left(\frac{\nu}{\varpi}\right)^{1/2} \left(\frac{D}{\nu}\right)^{-1/3}, \qquad (31.2)$$

where ν is the kinematic viscosity, ω is the rotation rate of the substrate, and D is the diffusion coefficient of the solute [31.18].

In LPE, the growth rate will not only be determined by the mass transport process of the growth species to the growing interface, but also by the incorporation of the solute atoms on the growing surface. Thus, supersaturation and surface kinetics should both be considered, and *Ghez* and *Giess* [31.19] developed a model for the growth of garnets that includes surface kinetic terms. However, in most growth rate studies, it is generally assumed that surface diffusion can be neglected [31.20]. For supersaturations σ below a critical value σ_c, the dependence between growth rate and supersaturation is quadratic, with $r \sim \sigma^2$ for $\sigma < \sigma_c$, and linear above, with $r \sim \sigma$ for $\sigma > \sigma_c$. High growth rates can be achieved when a continuous source of steps is present on the growing surface, for example, when growth spirals are present [31.20, 21].

In solution growth or LPE of complex crystals, the solute usually has a different composition than the solution, and the tendency for the incorporation of specific growth species in the growing crystals is expressed by the distribution coefficient k. The equilibrium distribution coefficient k_0 can be determined from the phase diagram. The effective distribution coefficient describes the concentration ratio between an element in the solution and in the crystal, as $k_{eff} = c_{crystal}/c_{solution}$. The relation describing the dependence of the effective segregation coefficient on the diffusion boundary layer δ and growth rate r has been derived by *Burton* et al. [31.22] as

$$k_{eff} = \frac{k_0}{k_0 + (1-k_0)\exp-\left(\frac{cr\delta}{D}\right)}. \qquad (31.3)$$

In this equation, c is a constant for the ratio of the mass density between the whole system and the element in the solution. The effective distribution coefficient is an extremely important parameter. If k_{eff} for a certain element is known, it is possible to adjust its concentration in a thin film or crystal, thereby opening the possibility to *engineer* the properties of a given material, which is important, for example, for the development of laser materials. For $k_{eff} < 1$, the concentration of the element considered will be smaller in the crystal than in the solution. Hence, this can be used in purification processes by crystallization.

31.3 Requirements for Liquid-Phase Epitaxy

For LPE growth of a new material, the greatest challenge is usually to find a suitable solvent system in which the material crystallizes first and as a single phase. This requires sufficient knowledge of the phase relations in the system considered. Unfortunately, complex phase diagrams are often not available, therefore the first choice is usually made empirically, based on experience with similar material systems.

Besides the solvent, the substrate has to fulfill several requirement, and not only for the epitaxial deposition itself, but there are also requirements depending on foreseen applications. If heteroepitaxy is considered, the growth of layers by LPE is very demanding. Since LPE is a near-equilibrium process, it is very sensitive to all parameters. Basic substrate requirements are low misorientation, low misfit at growth temperature, similar thermal expansion coefficients, and excellent chemical/thermal stability. An important factor for control of the growth mode is the misfit between substrate and film. In the case of LPE of garnets [31.13, 23] it has been demonstrated that the misfit $\Delta a (= a_{substrate} - a_{film})$, with a being the lattice constant, had to be less than 0.01Å in order to avoid cracking. In LPE, there are three major parameters that govern the film growth mode: the supersaturation, the substrate misfit, and the substrate misorientation. Extremely flat surfaces can only be obtained if all three values are extremely low; for example, in the case of c-oriented $YBa_2Cu_3O_{7-x}$ (YBCO) films grown by LPE on (110)$NdGaO_3$, it has been demonstrated that surfaces with interstep distances of $10\,\mu m$ between monosteps of 12 Å can only be achieved if the following conditions are jointly met: substrate misfit of $\leq 0.08\%$, substrate misorientation of $\leq 0.02\%$, and relative supersaturation of $\leq 0.18\%$ [31.24]. Even if values will vary for other material systems, these values can be taken as order-of-magnitude requirements.

There exist three major methods commonly used for LPE growth: the tipping, dipping, and sliding boat techniques. They have been extensively described in the literature [31.1, 14–16, 23, 25], and will not be reviewed herein, except for the specific materials discussed further below.

31.4 Developing New Materials: On the Choice of the Epitaxial Deposition Method

Deposition methods can be divided into two categories: growth from the vapor phase, and growth from the liquid phase. In vapor growth, or vapor-phase epitaxy (VPE), deposition occurs under vacuum conditions. VPE techniques include radiofrequency (RF) sputtering, laser ablation, thermal and electron-beam evaporation, ion-beam deposition, molecular-beam epitaxy (MBE), atomic layer epitaxy (ALE), chemical vapor deposition (CVD), metalorganic chemical vapor deposition (MOCVD), and plasma-assisted processes. These methods either create a vapor or molecular nutrient by thermal or electron beam evaporation, or the effusion of gas atoms from a Knudsen cell. The growth species are transported to the heated substrate, where nucleation and growth occur. Vapor-phase epitaxy is a very powerful method for the growth of semiconductors. In particular, VPE has the flexibility needed to grow and explore a variety of p/n-multilayer and quantum well structures, and allows the growth and and control of the deposition process down to atomic scale. Thus, over the past decades, VPE growth techniques have constantly improved and become widely applied, whereas interest in LPE growth declined.

The successes of VPE growth techniques, however, resides mainly in the field of conventional semiconductors, such as InGaAsSb and similar systems. Surprisingly, VPE growth of oxides and nitrides exhibits much lower performance, yielding rather defect-rich films that often contain secondary phases and numerous grain boundaries.

In oxides, difficulties typically arise due to the complexity of the material [31.25], the high melting point of constituent oxides, complex phase diagrams, and also often a limited temperature–pressure stability range which restricts the growth conditions that can be applied. Especially important is also the low mobility of growth species under vacuum conditions, which limits surface diffusion of adatoms. Thus, oxide films grown by VPE typically show two-dimensional nucleation of small islands that spread over the growing surface and coalesce. This *birth-and-spread* growth mechanism results in films with numerous grain boundaries, and surfaces with high density of steps with small interstep distances (typically < 50 nm) [31.26]. Surface nucleation can be suppressed by large misorientation angles of the substrates, thereby providing kinks and steps. However, films grown on misoriented substrates show a very high step density with very small interstep distances, and hence are rough. With increasing thickness of oxide VPE films, a columnar structure often develops, which is not desirable for most applications [31.27]. Similar behavior has been observed for nitride VPE films. In VPE, near-equilibrium conditions may exist very close to the growth steps, and kinetics is the dominant factor that governs the deposition. In contrast, in LPE, the low driving force is maintained over a much larger extensions, and when all substrate and growth parameters are sufficiently met, the layer-by-layer growth mechanism can proceed over macroscopic dimensions. In LPE, extremely flat surfaces may develop, and interstep distances of several microns are not uncommon [31.28, 29].

For the achievement of high structural film perfection, it is essential that growth and decomposition reactions coexist at the growing interface. It is postulated that the high stability of nitrides, combined with a low mobility of growth species under vacuum (VPE) conditions, limits the surface mobility of atoms and thereby also reversible reactions associated with surface diffusion. This could also tentatively explain the experimental limit to the minimum dislocation density observed in vapor-grown nitride films, where dislocations can only be reduced to about 10^5 cm^{-2} despite the use of ELO techniques [31.30–33]. The nucleation and growth kinetic processes are extremely complex, and the reasons for this experimentally observed limit are not well understood.

The achievable surface flatness, density of defects (point defects, dislocations), and film homogeneity/perfection are three aspects where the LPE technique may surpass VPE, and the choice of the deposition method should also be based on such consideration.

The growth system required for LPE growth (in any configuration) is very inexpensive compared with the equipment required for VPE growth. However, the achievement of high-quality LPE layers of complex oxides, nitrides or carbides can be very challenging, because this requires a multidisciplinary approach and high degree of expertise, especially in solution chemistry and thermodynamics, phase diagrams, and an excellent knowledge of the fundamentals of crystal growth and epitaxy. Like any other growth method, LPE presents advantages and disadvantages. Among the advantages, we can mention:

1. Films with highest structural perfection (low dislocation density, low concentration of point defects)
2. Automatic stoichiometry control
3. Dopant incorporation can be extremely homogeneous
4. Extremely flat surfaces (facets) can develop
5. Upscaling and mass production are possible.

Among the disadvantages, we can mention:

1. Usually thick films: minimum achievable is about 500 nm (atomic layer epitaxy not possible)
2. Not practical for the growth of heterostructures/quantum/multilayers
3. Rough interfaces frequent, due to back-dissolution
4. Very sensitive to substrate surface defects and parameters (misfit, misorientation)
5. Three major conditions have to be satisfied for the highest-quality LPE films:
   – Very small misfit
   – Misorientation
   – Supersaturation.

Only a few aspects were discussed in this introductory section. In the following sections, this paper discusses these challenges more precisely, with examples from recent studies on LPE growth of three different advanced materials, developed for different applications:

1. High-temperature superconductors, for flat (Josephson-junction) tunnel-device technology
2. Nitrides, for optoelectronic and high-power electronic devices
3. Novel frequency-agile piezoelectrics, for high-precision resonators and oscillators.

31.5 LPE of High-Temperature Superconductors

In 1986, the discovery of superconductivity at 35 K in a La-Ba cuprate [31.35], above the 23 K of Nb_3Ge, initiated the search for other cuprates with a higher critical temperature T_c. One year later, *Wu* et al. [31.36] found $YBa_2Cu_3O_7$ (YBCO) with a T_c of 90 K. This superconductivity above the boiling point of liquid nitrogen (77 K) stimulated worldwide research on this new group of superconducting materials (*ceramics*), and numerous applications based on tunneling were envisaged assuming that a timely solution of the material and crystal growth problems could be found. A new technological era seemed to be born, and this discovery initiated a race for the identification of new ceramic compounds with even higher critical temperatures. Among all high-temperature superconductors (HTSC), the so-called 123-cuprates became the most investigated. In particular, $YBa_2Cu_3O_{7-x}$ (YBCO) and $NdBa_2Cu_3O_{7-x}$ (NdBCO) attracted significant attention.

Due to their complexity and limited thermal and chemical stability, the growth process parameters that can be used for the synthesis of HTSCs are limited; for example, YBCO and NdBCO compounds melt incongruently below 1100 °C, and their oxygen content will depend on the growth conditions. In most cases, YBCO is grown as tetragonal $YBa_2Cu_3O_{6.2}$, which has to be oxidized at high temperature to be transformed to the superconducting orthorhombic $YBa_2Cu_3O_{6.93}$ phase with the highest T_c. This structural phase transition results in twinning, which in epitaxial layers depends on the substrate misfit and on the film thickness. Most thin-film growth efforts were concentrated on vapor-phase epitaxy (VPE), with more than 500 groups worldwide exploring chemical and physical vapor deposition of HTSCs. These VPE films typically show two-dimensional (2-D) nucleation and localized step flow, or spiral-island formation [31.37, 38]. Thus, step densities were very high, with interstep distances typically between 10 and 30 nm. For planar tunnel device technology, due to the very short coherence length in HTSCs, extremely flat film surfaces and interfaces were required. Thus, despite the challenges, researchers started to investigate liquid-phase epitaxy (LPE) growth of HTSCs. Early attempts in LPE growth of YBCO, Nd-BCO, $Bi_2Sr_2CaCu_2O_y$ (2212), and Tl(1223)/(1324) can be found in [31.39–52].

31.5.1 Phase Relations, Solvent System, and Solubility Curves

Phase Relations

Knowledge of phase relations and solubilities is of fundamental importance for crystal growth and liquid-phase epitaxy of a given compound. In order to avoid the crystallization of competing secondary phases, the growth has to occur in the primary crystallization field (PCF) of the material to be grown, where it crystallizes

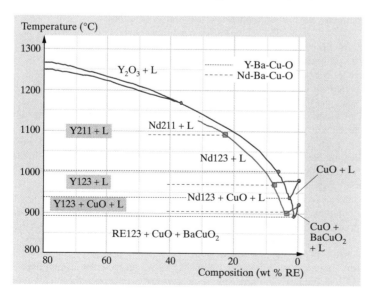

Fig. 31.1 Pseudobinary cut in the phase diagram of YBCO compared with NdBCO (after [31.34])

first, as a single phase, upon cooling. Due to the chemical and structural complexity of YBCO and NdBCO, this turned out to be very challenging [31.53, 54].

YBCO and NdBCO are incongruently melting in air, i.e., they decompose upon melting. Therefore, it is not possible to grow YBCO and NdBCO substrate crystals directly from their melt, for example, by Czochralski or Bridgman techniques. Therefore, they have to be grown from high-temperature solutions.

Solvent System

Finding a suitable solvent for a given crystal may sometimes represent a real challenge. The old concept *similia similibus solvuntur*, (Latin: similar is dissolved in similar) is the basic criterion to be followed. The optimum choice is a solvent which is chemically similar (in the type of bonding) to the solute, but which has sufficient crystal-chemical differences between the solvent and solute constituents in order to prevent the incorporation of solvent species into the solute structure. For example, for the growth of metals, metallic solutions can be used. For high-melting oxide compounds, oxides or fluorides (or a mixture of both) are generally used.

In the case of YBCO and NdBCO, attempts to use solvents traditionally applied to other oxides, for example, lead oxide and lead fluoride, were not successful [31.55]. Hence, a self-flux was chosen. Since the lowest possible growth temperature is desired, a solution composition near the binary eutectic between $BaCuO_2$ and CuO, which is at about 29 ± 1.5 mol % BaO at a temperature of $910 \pm 10\,°C$ in air [31.54], is used as solvent.

The two systems, NdBCO and YBCO, are not fundamentally different. Compared with YBCO, NdBCO has two advantages: a higher thermal stability reaching nearly $1100\,°C$ in air, and a much wider PCF allowing concentrations up to approximately 20 wt % for single-phase growth. Figure 31.1 shows a pseudobinary cut in the phase diagrams of YBCO compared with NdBCO [31.34].

Thus, YBCO LPE films (or crystals) can be grown from about 920 to $1000\,°C$ from the BaCuO/CuO eutectic solvent; for NdBCO, temperatures up to $1100\,°C$ can be applied. Crystals and films grown from this solvent are not superconducting, because they are depleted in oxygen. They have to be annealed in oxygen after the growth to become superconducting. This will be discussed in detail below.

Solubility Curves

In order to adjust the supersaturation, knowledge of the solubility of the solute in the solution is essential. In the case of YBCO and NdBCO, solubility studies were performed by observation of the formation of crystals on the top of the solution, and by crystallization experiments on nonpolished seeds dipped into the saturated solution [31.34]. Figure 31.2 shows the solubility data obtained (our work) compared with the data published by other authors.

There is a strong scatter in the published solubility data shown in Fig. 31.2, which can be tentatively explained by the difficulties of HTSCs and the quite different methods, solution compositions, chemicals, and crucibles used by the different authors [31.56, 57].

Observation of the formation and dissolution of crystals on top of a solution is usually not sufficiently accurate for the determination of the solubility curve. Indeed, due to the metastable (Ostwald–Miers) region,

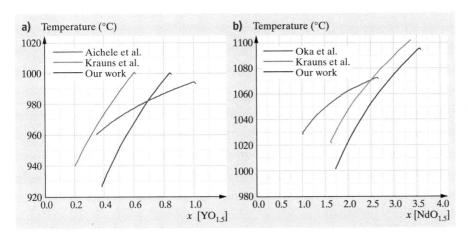

Fig. 31.2a,b Solubility of YBCO (**a**) and NdBCO (**b**) in the BaO/CuO flux at 31 mol %, compared with data published by other authors (after [31.34])

spontaneous crystallization does not necessarily occur at the liquidus temperature, because a stable nucleus of critical size needs to be formed first. Thus, the solution can be supersaturated by several degrees before a visible crystallite is observed. The use of a seed may allow the supersaturation required for nucleation and growth to be reduced, and may thus give a better estimate of the liquidus temperature. This became clear during our investigations, and our values shown in Fig. 31.2 were mainly obtained using seeds, and were further confirmed by the LPE experiments.

31.5.2 Heat of Solution

In an ideal solution there are no attraction forces between the chemical species in solution, and there is no change in internal energy upon mixing. Thus, all the chemical potentials can be expressed as

$$\mu_j = \mu_{0j} + RT \ln x_j \,, \tag{31.4}$$

where μ_{0j} is a function of temperature and pressure only, x_j is the mole fraction of the component j, and each component's chemical potential is a linear function of $\ln x_j$.

In the case of nonideal solutions, the mole fractions in (31.1) have to be replaced by the activities a_j, which are related to the concentrations by the activity coefficients γ_j according to

$$a_j = \gamma_j x_j \,. \tag{31.5}$$

For YBCO and NdBCO, there exist different possibilities to express the heat of solution. One may use a one-particle model by considering the reaction

$$\text{REBa}_2\text{Cu}_3\text{O}_{7-x}(s) \rightleftharpoons \text{REBa}_2\text{Cu}_3\text{O}_{7-y}(l) \,,$$

with the solid (s), liquid (l), and with RE = Y, Nd.

For an ideal solution, the relationship between the concentration, expressed in mole fraction, x_1 and x_2 at the temperatures T_1 and T_2, respectively, is then given by

$$\ln\left(\frac{x_2}{x_1}\right) = -\frac{\Delta H_{\text{fus}}}{R}\left(\frac{1}{T_2} - \frac{1}{T_1}\right), \tag{31.6}$$

where ΔH_{fus} represents the enthalpy of fusion. For an ideal solution, a linear plot of $\ln(x)$ versus $1/T$ is obtained. This Arrhenius-type relation allows the whole solubility curve to be extrapolated based on the melting point of the solute and a few solubility points. A deviation from linearity would indicate nonideal solution behavior. Since the volume change on melting is small, ΔH_{fus} may be approximated by the heat of solution L_{sol} [31.58, 59]. Then the heat of solution can be related to the saturation concentration of the solute in the solvent by the relationship

$$L_{\text{sol}} = RT^2 \frac{d \ln x_{\text{sat}}}{dT} \tag{31.7}$$

or

$$L_{\text{sol}} = 4.574 T_1 T_2 \frac{\log x_{\text{sat1}} - \log x_{\text{sat2}}}{T_1 - T_2}, \tag{31.8}$$

where x_{sat1} and x_{sat2} are the saturation concentrations at temperatures T_1 and T_2.

By using the above expressions and according to our refined experimental solubility curve, for YBa$_2$Cu$_3$O$_{7-x}$ crystals with $x \approx 0.8$, grown in air ($P_{\text{O}_2} \approx 0.2$ atm) in a flux with Ba-to-Cu ratio of 31 : 69, a value of $L_{\text{sol}} = 34.7$ kcal/mol at 1273 K is obtained [31.34], which is in good agreement with the calculations of *Tsagareisvili* et al. [31.60, 61], who obtained an enthalpy of melting of ΔH_{melt} of 36.44 and 41.03 kcal/mol for YBa$_2$Cu$_3$O$_6$ at 1446 K and YBa$_2$Cu$_3$O$_7$ at 1503 K, respectively.

For Nd$_{1.1}$Ba$_{1.9}$Cu$_3$O$_{7+x}$ crystals grown in air, at 1060 °C, from the same flux composition, a heat of solution of $L_{\text{sol}} = 28.1$ kcal/mol at 1333 K [31.34] is obtained.

The heat of solution can also be expressed by using a multiparticle model, similar to the eight-particle model used for garnets [31.62–64]. In this case, it is assumed that the REBCO molecule splits into several particles or molecules upon dissolution according to

$$\text{REBa}_2\text{Cu}_3\text{O}_{7-x}(s) \rightleftharpoons \text{RE}^{3+} + 2\text{Ba}^{2+} + 3\text{Cu}^{2+}$$
$$+ (7-x)\text{O}^{2-} \,,$$

or

$$\text{REBa}_2\text{Cu}_3\text{O}_{7-x}(s) \rightleftharpoons \text{REO}_{1.5}(l) + 2\text{BaO}(l)$$
$$+ 3\text{CuO}(l) \,.$$

31.5.3 Supersaturation and Driving Force for LPE

An important condition to achieve high-quality films and flat surfaces by LPE is the control of supersaturation. Knowledge of the solubility curve of a system allows one to estimate the supersaturation and the driving force for epitaxy for given growth conditions.

Crystals grow when the total free energy of the system can be decreased. Thus, the driving force for crystallization is the free energy difference

$$\Delta G = \Delta H - T\Delta S \qquad (31.9)$$

between the solid and the supersaturated solution. Here, ΔH and ΔS are the differences in enthalpy and entropy of the crystalline and fluid phase, respectively.

For near-equilibrium conditions, we can write

$$\Delta G = \Delta H \left(1 - \frac{T}{T_e}\right) \approx \Delta H \left(\frac{\Delta T}{T_e}\right), \qquad (31.10)$$

with $\Delta T = T_e - T$ being the difference between the equilibrium and actual temperature.

When we cool a solution that is at equilibrium, the concentration of solute x becomes greater than the equilibrium concentration x_e. Then, the free energy difference at the solid–liquid interface can be written as

$$\Delta G = \mathrm{RT} \ln\left(\frac{x}{x_e}\right) = \mathrm{RT} \ln\left(1 + \frac{x - x_e}{x_e}\right). \qquad (31.11)$$

For small values of supersaturation σ, we have

$$\ln\left(1 + \frac{x - x_e}{x_e}\right) \approx \frac{x - x_e}{x_e} = \sigma. \qquad (31.12)$$

Substitution into (31.11) yields

$$\Delta G = \mathrm{RT}\sigma. \qquad (31.13)$$

and by using (31.10), the following relation is obtained

$$\sigma = \frac{x - x_e}{x_e} = \Delta H \frac{\Delta T}{\mathrm{RT}\, T_e}, \qquad (31.14)$$

where σ is the relative supersaturation, x and x_e are the actual and equilibrium concentrations, and ΔH is the enthalpy of solution.

Besides σ, other expressions are often used

$$\alpha = \frac{x}{x_e} \quad \text{the supersaturation ratio},$$
$$\Delta T = T_e - T \quad \text{the undercooling}.$$

From the solubility curve of YBCO, if we assume an undercooling of $\Delta T = 0.5\,\mathrm{K}$, the Gibbs free energy difference (driving force) would then be $\Delta G(\Delta T = 0.5) = 13.6\,\mathrm{cal/mol}$, and for $\Delta T = 2.5\,\mathrm{K}$, $\Delta G(\Delta T = 2.5) = 68.1\,\mathrm{cal/mol}$.

31.5.4 Substrates and Epitaxial Relationship

For the growth of high-quality LPE films, the substrate has to fulfill several requirements: low misfit at the growth temperature, similar thermal expansion coefficients to film material, high chemical stability in the solvent, and excellent thermal stability, from the film growth to the application temperature. In the case of YBCO and NdBCO, the considered temperature range is from about $1100\,^\circ\mathrm{C}$ down to $77\,\mathrm{K}$ (liquid nitrogen). Furthermore, in LPE, a very low supersaturation has to be applied during film growth. Since all these requirements have to be satisfied for best results, it is evident that homoepitaxy would represent the best choice. In heteroepitaxy, there is generally no substrate material fulfilling all requirements, and compromises are necessary. Often, buffer layers (also called templates) of the same or a similar film material are used for better lattice matching. In the case of LPE of YBCO and NdBCO, the gallates $LaGaO_3$ (LGO) and $NdGaO_3$ (NGO) were extensively explored as substrate materials, as they present a better lattice match compared with other substrates such as $SrTiO_3$ or MgO. Both gallates and their solid solutions can be grown by Czochralski technique.

Due to the severe misfit requirements in LPE [31.65], precise knowledge of the lattice constants and thermal expansion coefficients of the substrate and HTSC film material is essential [31.66]. High-temperature lattice constant data available in the literature can vary significantly. This is not necessarily due to different determination methods or experimental inaccuracies, but can also be due to the crystal material itself. Indeed, the lattice constants of the same oxide material grown by Czochralski in different laboratories can vary, depending on the starting oxides and growth conditions applied. In addition, radial and longitudinal compositional variations are frequently observed in oxides grown by Czochralski technique. In the case of HTSCs, it was found more accurate to perform high-temperature x-ray measurements on film and specific substrate material used for LPE growth, using the same diffractometer/method [31.66]. For the measurements, single-crystalline film and substrate samples were ground into powder. A Philips Xpert diffractometer with CuK_α-radiation was used for the θ–2θ scans between 20 and $75°$ in the temperature range of about $35\text{–}1000\,^\circ\mathrm{C}$. The Bühler heating chamber HDK was equipped with a tightening tunable electrode for compensation of heat expansion of the Pt-strip heater, keeping the sample in focusing position. The samples were fixed with a Pt/Rh net with mesh size $200\,\mu\mathrm{m} \times 200\,\mu\mathrm{m}$, length 20 mm, and thickness $100\,\mu\mathrm{m}$, attached to the strip heater, reducing existing lateral temperature gradients over the sample. Lattice parameters were refined by Rietveld analysis. The most challenging aspect during such measurements (and also the greatest source of error) is usually lack of knowledge of the exact temperature of the sample. The use of an internal standard for calibration allows more accurate determinations. The thermal expansion behavior of alumina (Al_2O_3) and its chemical stability is well established. Therefore, Al_2O_3 reference powder was mixed with the gallates for calibration of the temperature [31.66].

The lattice parameters and lattice angles of the orthorhombic, tetragonal, and rhombohedral rare-earth gallates and cuprates (YBCO and NdBCO) can be transformed into a pseudocubic system. The pseudocubic cell of orthorhombic $REGaO_3$ (with RE = Y, Nd) consists of two [110]/2 axes, of which the distance is $0.5(a^2+b^2)^{1/2}$, and the orthogonal axis $c/2$. The pseudocubic angle $\gamma = 2\arctan(a/b) \neq 90°$ exists between the [110] directions. The substrate gallate planes (001) and (110) fit the film lattice parameters a, b, and $c/3$, depending on film orientation [31.65]. The epitaxial relations, in terms of pseudocubic lattice parameters, are shown schematically in Fig. 31.3, for c-oriented YBCO films on (001) and (110)NGO, on the left, compared with a-oriented films on (110)NGO, on the right. For epitaxy on the (001) gallate plane, both $d/2$ axes include the pseudocubic angle γ which slightly deviates from the cubic angle by $\Delta\gamma = 90° - \delta$. Thus, in-plane shear strains in the orthogonal film cell are introduced. For epitaxy on (110) planes the [110] direction pointing out of the (110) plane is inclined by $\Delta\gamma$, leading to *out-of-plane* shear strains in the film. The values of $\Delta\gamma$ at 1000 and 25 °C are 0.24 and 0.32° ($LaGaO_3$), and 0.34 and 0.74° ($NdGaO_3$).

Both compounds, YBCO and NdBCO, undergo a structural phase transition when they are cooled to room temperature after the growth. They are tetragonal ($a = b, c$) at LPE growth temperature of about

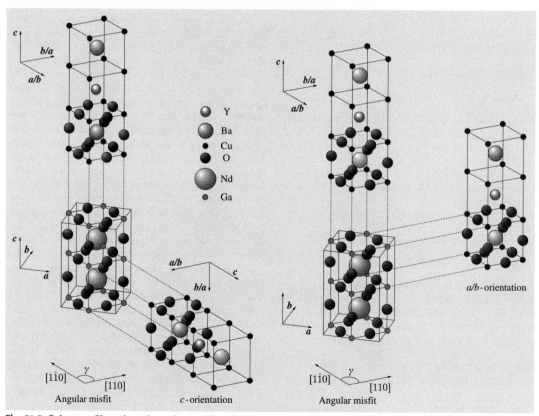

Fig. 31.3 Substrate–film orientations observed in epitaxial growth of YBCO on NGO

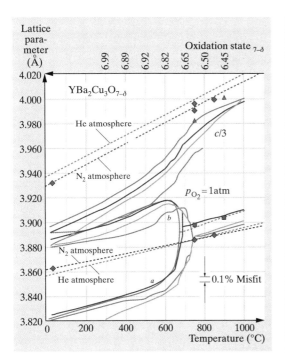

Fig. 31.4 Thermal expansion and oxidation behavior for YBCO according to different authors at $p_{O_2} = 1$ atm (after [31.66])

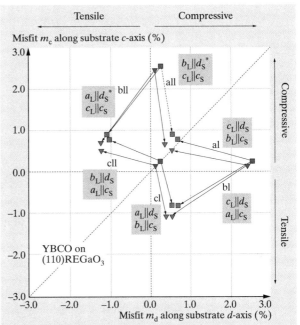

Fig. 31.5 Misfit relations of YBCO on (001)NdGaO$_3$ (□) and PrGaO$_3$ (∇) at 1000 °C (*full symbols*) and room temperature (*open symbols*). *Arrows* mark the path of cooling (and oxidation) after growth. Films are *a*-, *b*- or *c*-oriented (after [31.65])

1000 °C, and orthorhombic ($a \neq b, c$) once at room temperature. For YBCO, the splitting of the tetragonal *a*-axis into two *a*- and *b*-axes during this phase transition occurs at a temperature of about 600 °C, depending on the oxidation state of the material, as shown in Fig. 31.4. Since YBCO and NdBCO cannot directly be grown as orthorhombic phase, they have to be annealed in oxygen after the growth to become superconducting. This poses a formidable challenge for LPE growth of high-quality films, and will be discussed in detail in the film morphology section later.

The misfit relations between YBCO and (110) NdGaO$_3$ and PrGaO$_3$ at room temperature and at 1000 °C (approximate film growth temperature) [31.65, 66] are shown in Fig. 31.5. In this figure, assuming a rigid substrate, one can conclude that for *a*- and *b*-films at growth temperature the in-plane *c*-axis of the film, $c_L \parallel d_S$ (where "L" indicates film and "S" indicates substrate), is under high compressive strain of 2.5%, whereas the tetragonal *a*-axis (= *b*-axis) of the film shows only a small misfit of 0.15 and 0.2% for PrGaO$_3$ and NdGaO$_3$, respectively.

31.5.5 LPE Growth System and Film Growth Procedure

An important requirement for the achievement of high-quality LPE layers is precise control of the supersaturation σ. The supersaturation has to be small enough to prevent spontaneous three-dimensional nucleation, to prevent step bunching (described by the kinematic wave theory), and for layer-by-layer growth. The achievable growth mode depends also on substrate parameters, as discussed previously. Technically, precise control of supersaturation requires a corresponding precision of temperature control and programming, in combination with homogenization of the solution by forced convection. For preliminary LPE growth experiments, a chamber furnace can be sufficient. For highest-quality oxide films, low-gradient platinum-wound three-zone vertical LPE furnace systems allow

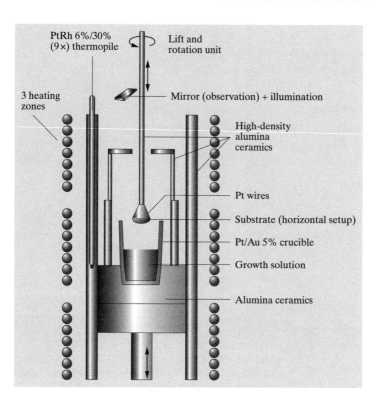

Fig. 31.6 Vertical three-zone LPE growth furnace system

the required control of temperature gradients [31.15, 16]. A typical three-heating-zone LPE growth furnace system is shown schematically in Fig. 31.6.

Such platinum LPE growth system can be used for LPE growth of a variety of oxides, up to 1200 °C. It is equipped with lift and rotation mechanisms. Optical-fiber illumination and mirror are used to observe the surface of the solution and for precise substrate dipping.

In a typical YBCO LPE growth experiment, high-purity starting chemicals (BaO_2, CuO, and Y_2O_3) are well mixed and introduced into the ceramic crucible. The crucible is placed in the furnace and heated to about 1010 °C during 9 h. After soaking for 7 h, the temperature is increased to 1040 °C and reduced to 1000 °C within 90 min. The equilibration to the liquidus temperature is done by the dipping of nonpolished test substrates into the solution. Then, several YBCO films are successively grown on (110)NGO by slow cooling of the solution. The substrates can be mounted in vertical or horizontal position, and the substrate rotation rate is typically between 15–30 and 100–120 rpm, for each configuration, respectively.

31.5.6 Growth Mechanisms and Growth Parameters: Theory Versus Experiment

The primary motivation for using the LPE growth technique for film growth has traditionally been the need for extremely homogeneous crystal material and/or flat surfaces for a given application. In the case of HTSCs, due to their short coherence length (≈ 4 and 12 Å, for a-oriented and c-oriented films, respectively), extremely flat film surfaces are required for planar tunnel-device applications. For c-oriented YBCO films, this represents one c-axis monostep height of 12 Å, separated from another monostep by 10 μm lateral distance. This formidable requirement can be better understood if we consider, for example, a step with a height of 1 m that would need to be separated from another step by a distance of 10 km. From this, it is obvious that only a growth mechanism providing a continuous source of steps, such as the layer-by-layer or spiral growth mode, would allow such surfaces to be obtained over macroscopic dimensions. This requirement leads to unprecedented challenges in film synthesis, and therefore precise evaluation of the tolerances on all substrate and growth parameters is required.

Besides substrate misfit and misorientation, knowledge of the supersaturation is especially critical. Supersaturation can be calculated from the phase diagram, for a given undercooling applied during the film growth. However, the values that are obtained are, at best, indicative only. Growth features, such as spirals, reflect the real conditions existing at the growing interface, and can thereby provide important fundamental data on growth parameters that cannot be measured. The approach discussed below can be applied to other materials as well, and the reader is encouraged to look into the original literature for full details.

In LPE of YBCO on (110)NdGaO$_3$ the film orientation depends on the supersaturation. It was found experimentally that the transition from a-oriented to mixed a/c to c-oriented films occurs with decreasing undercooling (supersaturation), with a threshold at about $\Delta T \approx 3$ K ($\sigma \approx 0.03$) [31.28].

Screw dislocations provide a continuous source of steps, which can propagate over macroscopic distances. It has been shown [31.20], that the distance between steps originating from a single dislocation is proportional to the size of the 2-D nucleus at a given supersaturation. For an Archimedean spiral, the relationship between supersaturation σ and interstep distance y_0 can be expressed by [31.67]

$$y_0 = 19\rho_c = \frac{19\gamma_m a}{k_B T \vartheta}, \quad (31.15)$$

with the relative supersaturation $\sigma = (n - n_e)/n_e$, where n and n_e represent the actual and equilibrium concentration of the solute, and a is the length of a growth unit. In the case of solution growth, the energy per growth unit (or molecule) γ_m on the edge of the critical nucleus is of the order of $1/6$ of the heat of solution per molecule. The anisotropy of γ_m, which will be higher for low-energy planes, is neglected in the following development.

In the case of YBCO, with $\gamma_m = 4.01 \times 10^{-20}$ J/molecule, $a = 3.9$ Å ($a = b$ for tetragonal YBCO at the growth temperature of 996 °C), and $T = 1269$ K, we can estimate the achievable interstep distances as a function of undercooling. This is shown in Fig. 31.7, for an Archimedean, monostep, single-source spiral. From this, the size of the critical nucleus was estimated to be about 23 nm for this surface. This is one order of magnitude larger than observed for YBCO films grown by VPE [31.68].

In this figure, we recognize that the undercooling has to be smaller than 0.17 °C in order to obtain interstep distances larger than 10 μm, as required for the specific application discussed herein. Such small under-

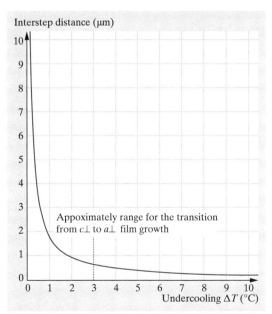

Fig. 31.7 Interstep distances as a function of undercooling (after [31.28])

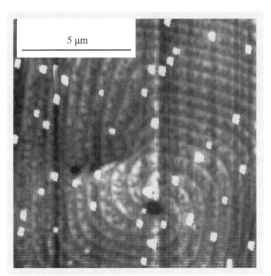

Fig. 31.8 AFM image of a complex YBCO spiral ($m = 2$ and $S = 6$) observed on a LPE film surface (after [31.29])

cooling sets stringent requirements on the temperature stability of the growth process.

Growth spirals can also emerge from groups of dislocations. In this case, the interstep distance is given by $y_0 = 4\pi r_c/\varepsilon$ [31.20], where r_c is the radius of the critical nucleus. For a group of dislocations of the same sign, ε can be as large as the number of dislocations contained in it. A group of S dislocations of the same sign, arranged on an array of length L, has an activity of ε times that of a single dislocation. Accordingly [31.20]

If $2\pi r_c \gg L$, then $\varepsilon = S$,

$2\pi r_c = L$, $\varepsilon = \dfrac{S}{2}$,

$2\pi r_c \ll L$, $\varepsilon = 2\pi r_c \dfrac{S}{L}$.

A comparison was made with complex growth spirals observed on c-oriented YBCO films grown on (110)NGO [31.29]. Figure 31.8 shows a typical YBCO spiral where six double steps (2×12 Å) emerge from the core region.

Interstep distances and step height of several complex growth spirals observed on single film surface of $1\,\text{mm} \times 1\,\text{mm}$, were measured by atomic force microscopy (AFM), and the following relation was found

$$y_0 = \frac{19\gamma_m am}{\varepsilon k_B T \vartheta}, \qquad (31.16)$$

with $2\pi r_c S/L < \varepsilon < S/2$, and where m represents the number of monosteps of the step source.

The high nucleus density in vapor-grown films results in high screw dislocation densities of typically $10^9\,\text{cm}^{-2}$ [31.68–70] and high grain boundary densities due to the coalescence of misaligned islands [31.71]. High critical current densities are generally observed for such defect-rich films. Island formation can be prevented and a pure step flow mode achieved when the substrate misorientation, which is 3–5° for physical vapor deposition (PVD) and chemical vapor deposition (CVD) [31.71], is adapted to the size of the critical nucleus. In contrast, in LPE, the nucleus density is lower.

With increasing radius of critical nuclei at low supersaturations, the substrate parameters misfit and misorientation become crucial for the initial growth stage and the growth mode of the films. Hence, in LPE, to understand the a/c-transition of YBCO on (110)NdGaO$_3$, the substrate strain energy contribution must be taken into account. Furthermore, the surface energy between a liquid and its solid is lower than the surface energy between a vapor and its solid. Therefore, in LPE, the strain energy is not negligible with respect to surface energy terms. In order to estimate the influence of substrate misfit and misorientation on the thermodynamically determined equilibrium size of the nucleus, characteristic substrate parameters were compared [31.24]. A relationship between misfit, relative supersaturation, radius of critical nucleus, and film orientation could be established, as shown in Fig. 31.9.

For the achievement of large interstep distances of $y_0 \approx 10\,\mu\text{m}$ between monosteps of 12 Å height for c-oriented YBCO layers on (110)NGO, the estimated corresponding undercooling is $\Delta T \approx 0.17\,\text{K}$. From Fig. 31.8, one can see that the maximum tolerable misfit with the substrate would then be 0.08%. For thin, strained c-YBCO films on (110)NGO with average misfit of 0.28%, it follows that a maximum interstep distance of $y_0 \approx 2.5\,\mu\text{m}$ between monosteps of 12 Å height can be obtained. It is also found that the substrate misorientation has to be smaller than 0.02° to avoid step bunching [31.24].

Only selected aspects were discussed in this section on HTSCs, to demonstrate the challenges and also

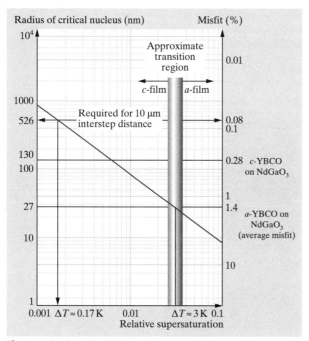

Fig. 31.9 Radius of critical nucleus as a function of misfit and supersaturation in LPE of YBCO (after [31.24])

the potential of this growth technique for the development of high-quality complex oxide films. Cracking and twinning of high-quality LPE REBCO films are further challenges to their application in devices. Whereas the T_c of LPE HTSC films was found comparable, after oxidation, to that of VPE-grown material, the critical current density was usually lower due the high quality of the LPE films (lack of pinning centers). The interested reader should refer to the specific literature for additional details.

31.6 LPE of Calcium Gallium Germanates

Langasite-type materials represent a family of trigonal acentric crystals whose structure was found to be similar to that of calcium gallium germanate, $Ca_3Ga_2Ge_4O_{14}$, (CGG, space group D_3^2-$P321$). They were first investigated as laser hosts [31.72], and rediscovered later for their outstanding piezoelectric properties [31.73]. So far, the most investigated ternary compositions have been $La_3Ga_5SiO_{14}$ (LGS), $La_3Ga_{5.5}Ta_{0.5}O_{14}$ (LGT), and $La_3Ga_{5.5}Nb_{0.5}O_{14}$ (LGN). Langasite-type materials are the first real competitors to quartz in high-precision resonators and high-temperature high-pressure sensors and displace a quartz technology that has been dominant over the past 50 years [31.74–76]. The advantages of langasites over quartz include lower acceleration sensitivity; higher piezoelectric coupling, which enables devices to be made smaller; and higher Q (quality factor), which reduces phase noise and enables higher-frequency operation. Unlike quartz, these materials have no phase transition below the melting point, which enables devices capable of high-temperature operation.

Ternary langasites LGS, LGT, and LGN, can be grown by Czochralski technique. However, due to inherent materials properties (phase relations/stability) and particularities of the growth process, these crystals and wafers show a defect structure and inhomogeneities [31.77] that lead to nonreproducibility in surface acoustic wave (SAW) device parameters [31.78] and discrepancies in fundamental measurements. Recently, high-quality 2 inch diameter LGT crystals have enabled the achievement of high Q-values [31.79] and overall better reproducibility in resonator parameters. It is also acknowledged that better values would probably by possible if the crystal quality could be further improved. In particular, better ordering of atoms in the LGT lattice and lower density of point defects could be beneficial.

Ternary langasites can be described as a mixed framework consisting of an octahedra (1a), and two types of tetrahedra, a small (2d) and a larger (3f). The holes of the Thomson cubes are occupied by large cations (3e) [31.80, 81]. Due to the presence of four sites for cations, the CGG-type crystal structure accommodates a wide spectrum of ions. Structure-forming ions, dopant ions, and their radii range (tolerance factor) for corresponding coordination number have been determined [31.81], and a stability diagram derived [31.82]. In ternary CGG-type materials, the four sites can be populated by more than one kind of cation, and this leads to a disordered structure of these materials. In LGT, the Ga and Ta ions share the same octahedral site, which leads to compositional inhomogeneities. Best ordering of atoms, and hence, highest Q-value, is expected from crystals grown near thermodynamic equilibrium. This was one of the reasons to investigate LPE growth of langasite materials. In addition, as we know from semiconductor studies (Sect. 31.1), the concentration of native intrinsic (point) defects is very low in crystal material grown by LPE or from solutions. From fundamental thermodynamics, for the same material, the point defect concentration will be lower if a growth process is used that allows a lower growth temperature. In Czochralski growth of ternary langasites, the growth temperature is about 1450 °C, whereas the same materials can be grown from solutions at about 950 °C [31.83–85]. This growth temperature difference results in a remarkable reduction in the concentration of intrinsic point defects. This may result in higher Q-values, and hence this was a further motivation to investigate LPE growth of langasites.

31.6.1 Solvent System

Finding a suitable solvent for the growth of such complex materials represents one of the major challenges. A first selection of possible fluxes is always done according to chemical and structural aspects, and through systematic studies of available respective or related phase diagrams. Compared with the solute, the solvent should be chemically similar in the type of bonding, but is should have components that present sufficient crystal-chemical differences to avoid incorporation of solvent species into the crystal.

In langasites, with decreasing temperature, the following crystallization sequence of possible phases is expected from respective melting points and considerations of the complexity of the corresponding oxide. Simple oxides (R_2O_3; $R = A, B$) have the highest melting temperature, followed by the perovskite (ABO_3) phase at a lower melting temperature, then the garnet ($A_3B_5O_{12}$) phase could be expected, and finally the langasite phase at the lowest temperature. However, when $A = La^{3+}$, the garnet phase does not exist, because La^{3+} is too large for the garnet crystal structure. Hence, in ternary langasites, $LaGaO_3$ is the phase likely to crystallize in a temperature region above the langasite liquidus. Hence, an ideal solvent should present sufficient solubility for the langasite phase below this temperature.

Alkali vandates, molybdates, and tungstates were widely used as fluxes for crystal growth of silicates and germanates. Li_2MoO_4:MoO_3 readily dissolves many oxides, and was used for the growth of BeO, GeO_2, SiO_2, TiO_2, Be_2SiO_4, Y_2SiO_5, as well as for emerald ($Be_3Al_2Si_6O_{18}$). Early attempts to crystallize the langasite phase from Li_2MoO_4:MoO_3 flux were not successful [31.85].

Another typical solvent for oxides is PbO. PbO-B_2O_3 fluxes as well as BaO-BaF_2-B_2O_3 ternary solvent systems were successfully applied for the flux and LPE growth of perovskites and garnets [31.16]. Bi_2O_3-based fluxes have similarities to PbO-based fluxes, and were explored as lead-free alternates for the growth of magnetic garnets. However, two major problems were reported: a flux difficult to remove after the growth, leaving usually a rough film surface after cleaning, and a growth-induced anisotropy of about one order of magnitude lower than with a PbO-B_2O_3 flux [31.16]. Due to the valence state of Bi^{3+}, bismuth-based solvents cannot be used for the growth of many oxides containing large rare-earth ions such as La^{3+}, because of possible substitution, therefore they were not investigated in those studies.

For the growth of LGS, LGT, and LGN, PbO-based solvent systems with various solute ion ratios and concentrations were tested by flux-growth experiments. In preliminary experiments performed between 950 and 1000 °C, lanthanum gallate was often obtained as secondary phase, and it became evident that the growth temperature had to be lower. However, PbO has a melting point of 886 °C and becomes very viscous around 900 °C. Thus, a eutectic-forming additive was sought to lower the growth temperature towards the expected stability field of the ternary langasites. The addition of MoO_3 was found to be beneficial, and LPE films could then be successfully grown in a lower temperature range from a solvent with PbO-to-MoO_3 ratio of 12 : 1 [31.85].

31.6.2 Substrates for Homoepitaxial LGT LPE Film Growth

In LPE, the substrate misfit is a determining factor in the achievable growth mode and film quality. The availability of Czochralski-grown LGS, LGN, and LGT substrates to be used as substrates for homoepitaxial LPE growth therefore represents an important advantage.

In case of homoepitaxy (LPE), the supersaturation required to initiate nucleation is extremely small. Hence, substrate surface defects, scratches, as well as residual strain due to the polishing process, will strongly affect the initial stages of film growth. Chemical etching is an efficient method that can be used to remove the damaged surface substrate layer prior to LPE. For LGT, similarly to perovskites and garnets, hot orthophosphoric acid was found to be particularly suitable, also to reveal substrate defects [31.77]. In langasites, striations can be revealed by etching the substrate in H_3PO_4 at 130 °C during 2–3 h, as shown in Fig. 31.10.

In ternary langasites, striations are likely due to the solid solubility between Si-Ga in LGS (tetrahedral site, CN IV), between Nb-Ga in LGN, and between Ta-Ga in LGT (octahedral site CN VI), associated with temperature fluctuations at the growth interface during Czochralski growth. These striations cause periodic (three-dimensional, 3-D) variations of lattice parameters and inhomogeneous dopant/impurity incorporation. In Czochralski growth of LGT, striations can be strongly reduced, below the detection limit of x-ray synchrotron topography, when all growth parameters are properly optimized [31.79].

When a film is grown by LPE on a striated substrate of same composition (homoepitaxy), the first monolayers of the film will try to adapt their composition to minimize the misfit strain. This phenomenon is well known, especially in semiconductor heteroepitaxy where it has been documented as the lattice pulling effect. A film grown on a striated substrate will show a variation of composition as a function of its thickness, which in turn may affect the film properties. It is therefore important to investigate such defects in the substrates used for epitaxial deposition.

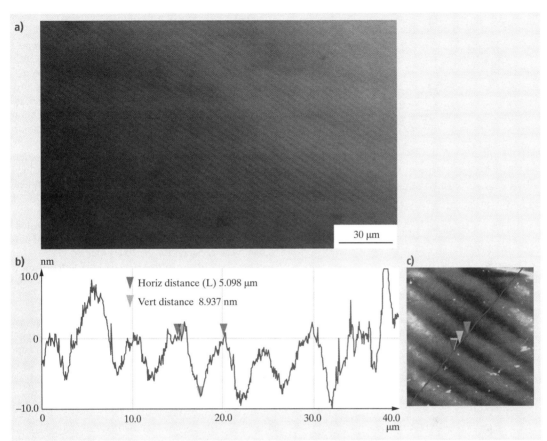

Fig. 31.10a–c Striations on *x*-cut LGT substrates after chemical etching in hot H_3PO_4. (**a**) Nomarski microphotography, (**b,c**) AFM profile

31.6.3 LPE growth of LGS, LGT, and LGN

Langasite LPE films are grown in a similar vertical furnace system to that previously described. In a typical experiment [31.84, 85], the starting oxides and dopants, consisting of 10–15 wt % stoichiometric LGT (or LGN) in the PbO solvent, are hand-mixed, introduced in a Pt crucible of about 45 cm^2, and covered with a lid. The crucible is placed in the furnace, and the furnace is heated to 980 °C in 5 h, followed by 24–48 h soaking at this temperature. Once during this time, the flux is gently mixed by using a fork made of Pt stripes mounted on an alumina rod. After soaking, the lid is removed and the temperature progressively is lowered to 900 °C and kept there for 5–10 h. The search for the liquidus is then carried out by slowly cooling the furnace while dipping unpolished LGT seeds and analyzing the phases that crystallize onto them by x-ray diffraction (XRD). Several LPE films are then grown in a temperature range of typically 900–850 °C. After the growth, the residual flux can easily be removed by cleaning the film in diluted nitric acid at room temperature.

31.6.4 Structural and Chemical Characterization of Doped LGT LPE Films

Various techniques have been used for the structural and chemical characterization of high-quality LPE films, and the best way to evaluate the films is usually to compare their quality with the substrates. In the following, only selected methods are discussed.

Rocking Curve Measurements

The structural characterization of high-quality crystals and films can be done be rocking curve measurements. During such measurements, the crystal is rotated (rocked) through the Bragg angle, and the reflected beam is measured by a fixed counter. Each potentially slightly misoriented subgrain comes into reflection as the crystal is rotated. Hence, the width of a rocking curve is a direct measure of the range of orientation present in a crystal. For a perfect crystal, the theoretical full-width at half-maximum (FWHM) is on the order of 0.003°. Very few natural crystals reach this value, and most single-crystals have FWHM 10–100 times larger.

In the case of LGT, films and substrates with different orientation were analyzed [31.86]. For x-oriented LGT substrates, typical value of about FWHM of 0.044° was obtained for the (004) rocking curve. The rocking curve of an x-oriented Al:Ti:LGT film is shown in Fig. 31.11. A FWHM of 0.035° was obtained, which indicates its very high structural perfection, better than that of the substrate.

The structural perfection of x-oriented LGT films was consistently better than that of LGT substrates. Also, the surface of x-LGT films is usually extremely

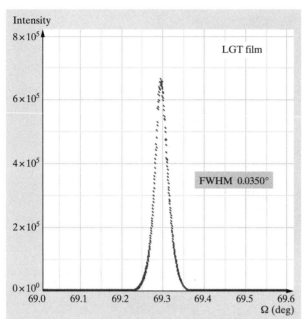

Fig. 31.11 Rocking curve of x-LGT film, (004) reflection (after [31.86])

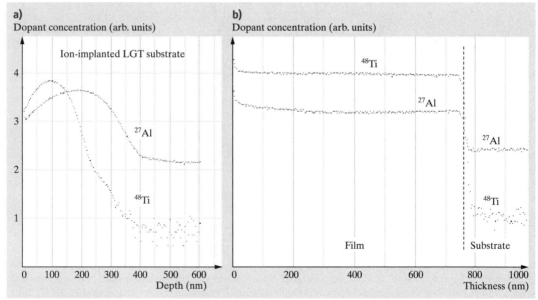

Fig. 31.12a,b Dopant concentration SIMS profile of: Al:Ti ion-co-implanted LGT substrate (**a**), and in situ codoped Al:Ti:LGT LPE film (**b**)

smooth, whereas *y*-oriented LGT films traditionally show a stepped surface [31.83]. Consequently, we obtained FWHM values slightly larger for *y*-oriented LGT LPE films, typically around 0.0465°. These values are quite remarkable for oxide films.

Quantitative Estimation of Dopant Concentration by SIMS

Secondary-ion mass spectroscopy (SIMS) is an analytical technique used for near-surface and small-area analysis. Depth profiling from about 1 nm to a few microns is possible. All elements and their isotopes can be detected by SIMS. Dynamic SIMS is used for bulk compositional analysis and depth profiles of trace elements. Several dopants were explored. Herein we focus on Al:Ti:LGT LPE films [31.86]. For quantitative measurement of the dopants concentration films by SIMS, standards are needed. Hence, Al and Ti ion-(co)implanted LGT substrates were used as standards. Figure 31.12a shows the SIMS profile of an Al:Ti ion-(co)implanted *x*-LGT substrate. The curved shape of the depth distribution is typical for ion-implanted samples, and demonstrate the inhomogeneous dopant concentration obtained by ion implantation. This should be compared with Fig. 31.12b, which is the SIMS profile obtained for an *x*-oriented Al:Ti:LGT LPE film.

One can see that the Al and Ti dopants are extremely homogeneously distributed through the whole thickness of the LGT LPE film, with almost constant value, and a very abrupt decrease at the film–substrate interface. The SIMS profile also allows a precise measurement of the thickness of the films, which is challenging to measure by other techniques in the case of almost perfect lattice match with the substrate (homoepitaxy).

The thickness of this film is about 770 nm. For a growth time of 30 min, these LGT films had typically thickness of 0.3–4 µm, depending on applied growth parameters. This also demonstrates that, in principle, relatively thin films can be grown by LPE technique. The mean dopant concentration values obtained for this film from the SIMS profile are: ^{48}Ti(100–700 nm) $\approx 2.72 \times 10^{19}$ atoms/cm^3, and ^{27}Al(200–700 nm) $\approx 5.67 \times 10^{18}$ atoms/cm^3 within the film depth given in brackets. If one assumes that the isotopic abundance is preserved, the corrected value for Ti is 3.69×10^{19} atoms/cm^3. These are equilibrium concentrations for the applied film growth temperature and conditions.

31.7 Liquid-Phase Epitaxy of Nitrides

Group III nitrides such as AlN, GaN, InN, and their alloys are the focus of intense research for their high potential in optoelectronic devices and high-power high-temperature electronics. These nitrides form ternary and quaternary solid solutions which enable the energy bandgap of the devices to be tuned from about 0.7 eV (InN) to 3.45 eV (GaN), to 6.2 eV (AlN). Hence, they are excellent candidates for engineering of materials with a specific bandgap. The III–V nitrides and their solid solutions are suitable for a wide rang of applications, for example, for high-temperature piezoelectric and pyroelectric applications, surface acoustic wave (SAW) devices, light-emitting diodes (LEDs), laser diodes (LDs), and ultraviolet (UV) detectors and sensors.

Wide-bandgap nitride semiconductors have been successfully commercialized as LEDs for solid-state lighting/illumination, compact laser sources for digital versatile disk (DVD) heads, and even high-power electronics in the form of high-electron-mobility transistors (HEMTs). However, doping issues, the lack of lattice-matched substrates, and high film defect density are currently major obstacles for the reliable development of high-performance devices.

The field of nitrides would highly profit from the availability of bulk GaN and AlN single crystals. However, their synthesis is faced with several difficulties. They cannot be grown from the melt by Czochralski or Bridgman technique due to their extremely high decomposition temperature of about 45 000 atm for GaN at a (theoretical) melting temperature of about 2500 °C. Various methods have been applied to synthesis of bulk nitrides, but none of them has yet provided substrates of sufficient quality and/or sufficient yield to be successfully implemented in device technology.

Interest in nitrides as semiconducting materials in the blue/UV range dates back to early 1970, with the first studies on epitaxial deposition of GaN with promising electrical and optical properties. However, the material showed strong n-type character, and p-type doping could not be achieved. Thus, interest in GaN faded until 1989, when p-type conduction could be demonstrated for the first time by low-energy electron-beam irradiation (LEEBI) of Mg-doped films [31.87]. The majority of GaN layers are grown from the vapor phase, by MOVPE, MBE, or HVPE, on mismatched substrates. On sapphire and SiC, the layers present a columnar structure consisting of many small hexagonal grains [31.88], which are tilted and rotated within the GaN film and which give rise to very high dislocation densities of about 10^9–10^{10} cm^{-2}. In spite of this, very bright LEDs and LDs have been demonstrated [31.89]. The highest brightness and lifetimes are reported for layers grown by the epitaxial lateral overgrowth (ELO) technique on sapphire [31.90] and SiC [31.91], which is due to the reduction of the dislocation density to about 10^5 cm^{-2} that can be obtained using ELO. This clearly demonstrates the impact of the crystalline perfection of the GaN layer on device performance.

Aluminum nitride has great potential in UV sensor devices. It also presents the required attributes for high-temperature high-power applications, where a high bandgap energy, high thermal and chemical stability, low leakage currents, high breakdown currents, high dielectric constant, and high resistance are required. In particular, AlN has emerged as a potential candidate for high-energy-density capacitors. Film defects such as low-angle grain boundaries (LAGBs), threading dislocations (TDs), and the development of a columnar structure are responsible for leakage currents. Hence, despite its potential, this material has not yet found broad application in such devices. Improving the quality of AlN films (and single crystals) is a critical challenge towards applications of AlN.

The case of InN and InGaN solid solutions is particularly interesting. This material received less attention than AlN and GaN until recently, when the energy gap of wurtzite InN was found to be about 0.7 eV [31.92–94] instead of the previously reported value of 1.9 eV [31.95]. This has extended the range of the energy gaps of group III nitride alloys from the deep-ultraviolet to the near-infrared spectral region. It has been shown that the bandgap of InGaN solid solutions can be varied continuously from 0.7 to 3.4 eV [31.96], providing a full-solar-spectrum material system for multijunction solar cells. In addition, a much greater radiation resistance has been reported for InGaN alloys compared with materials such as GaAs and GaInP. This makes them particularly suitable for radiation-hard high-efficiency solar cells for space exploration [31.97]. Here also, the highest photovoltaic cell efficiencies are expected from low-defect films and lattice-matched high-quality substrates, and this is a further area where LPE growth is particularly important to explore.

Today, the majority of AlInGaN devices rely on vapor-phase growth, often on highly mismatched substrates such as sapphire and SiC, leading to epilayers with large density of defects. As can be recognized from above, almost all applications of nitrides would greatly benefit from high-quality crystal/film nitride material. It is therefore surprising that LPE growth of nitrides is only marginally studied by a few groups.

31.7.1 Developments and Trends in LPE of GaN and AlN

LPE growth of GaN has been explored as early as 1972, by *Logan* and *Thurmond* [31.98], with the successful growth of GaN films on sapphire from Ga and Ga + Bi solvents, and ammonia gas as nitrogen source. A low solubility of $x_{GaN} \approx 3 \times 10^{-5}$ at 1150 °C at 1 atm pressure was reported. LPE growth of GaN on highly mismatched sapphire substrates (misfit $\approx 1.6\%$) turned out to be very challenging, and n-type material was obtained, similarly to vapor-grown films. GaN and InGaN films were successfully grown by LPE on a variety of substrates: sapphire, LiGaO$_2$ (LG), LiAlO$_2$ (LAO), and on MOVPE/HVPE buffer layers [31.99]. The kinetics of GaN formation from Ga solvent in ammonia atmosphere was investigated [31.100]. In 1964, *Glemser* [31.101] showed that Li$_3$N dissolves in melts based on lithium halides. Li$_3$N is one of the few nitrides that can be grown by Czochralski technique, and the solubility of nitrogen in molten Li$_3$N has been investigated [31.102]. Later, Li and Li$_3$N [31.103] fluxes were applied for the growth of GaN. Other solvent systems were investigated for the synthesis of nitride single crystals. Solutions based on Ga with Na have been successfully applied to the growth of GaN [31.104, 105], and dislocation densities as low as 2.3×10^5 cm^{-2} were reported [31.106]. Considering the fact that the LPE growth process that can be applied is very inexpensive compared with the multistep VPE growth technology necessary for similar film quality, there is significant potential in this approach. A remarkable aspect of LPE of GaN is that films with dislocation densities as low as 10^4 cm^{-2} can be grown on MOCVD GaN templates that have a much higher dislocation density (10^7–10^8 cm^{-2}) [31.107]. Most GaN LPE and flux-growth experiments were performed at pressure of about 40–50 atm in sealed crucibles [31.106, 107]. In such systems, not only the temperature but also the pressure becomes a variable. Besides challenges in nucleation control, growth stability is difficult to achieve in such systems. Hence, the use of solvents and LPE methods that allow the growth of GaN at atmospheric pressure may represent a better choice for the growth of thin high-quality LPE GaN films. From the low defect densities already obtained despite the use of nonideal templates [31.107] and nonoptimal LPE growth conditions, we can certainly predict that lowest dislocation (defects) densities will some day be reported for LPE-grown nitride films.

In the case of AlN, *Dugger* was the first to demonstrate the growth of AlN crystals of about 1 mm length from AlN-Ca_3N_2 solutions [31.108]. Recent studies with similar fluxes were only marginally successful, and the growth of AlN single crystals and LPE films from such solutions represents still a challenge.

31.7.2 Substrates for Epitaxy of Nitrides

Under near-equilibrium growth conditions, the substrate parameters misfit, misorientation, as well as the substrate surface structure after polishing/cleaning play a fundamental role. Only at practically zero misfit and low supersaturation can the layer-by-layer or Frank–van der Merwe growth mode, yielding films with highest structural perfection, be expected. Besides sapphire, MgO, SiC, $MgAl_2O_4$, and Si have been investigated as substrate material for nitrides, but all of them have too high a misfit for coherent overgrowth and layer-by-layer growth by LPE. Promising substrate materials such as $LiGaO_2$ (LGO) and $LiAlO_2$ (LAO) have emerged. However, their low chemical and thermal stability (about 900 °C) limits the growth parameters to be applied for any growth technique. Zinc oxide is actually the best lattice-matched substrate for nitrides, available at high yield [31.109]. The combination of ZnO and GaN in multilayer device structures would also be of particular interest to optoelectronic devices. However, ZnO has limited chemical stability with respect to halogenides at high temperature, and therefore cannot be used in chloride-HVPE nitride growth processes. This does not represent a major issue in LPE growth, where the substrate is immersed in the (metallic) solvent and the process is performed in such a way that the substrate never comes into direct contact with the reactive atmosphere.

Among all choices, homoepitaxial deposition provides the best possible conditions for highest-quality GaN LPE-grown films [31.99]. Hence, thick GaN films

or GaN substrate crystals would be of greatest interest, also as base structure for the fabrication of devices by VPE. Among all methods, the ammonothermal is actually the most promising method for obtaining large-size (≥ 2 inch) GaN at high yield. In 1997/1998 *Dwilinski* et al. demonstrated ammonothermal growth of GaN and AlN from supercritical ammonia and lithium and potassium amides [31.110]. The excellent properties reported led to intensive research in this area by different groups. It was found that GaN show a retrograde solubility when basic mineralizers are used, and quite high pressures (100–300 MPa) have to be applied [31.111, 112]. In acid mineralizers, the solubility behavior of GaN is conventional, and the process can be carried out at lower pressures (≤ 150 MPa) [31.113], which makes growth stability easier to achieve. The increasing availability of near-lattice-matched substrates will be very significant for major advances in the field.

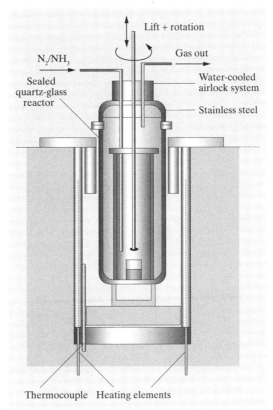

Fig. 31.13 Schematic view of the nitride LPE growth system (after [31.99])

31.7.3 Growth System and Optimization

In the following, a relatively simple growth system and method is presented, which was successfully used for the growth of GaN and InGaN films from Ga or Ga+Bi solutions, using NH_3 nitrogen source [31.99]. The film growth was performed in a sealed quartz glass reactor placed in a chamber furnace schematically shown in Fig. 31.13.

The crucible is placed in an appropriate temperature gradient in the middle of the reactor. A quartz liner is placed between the crucible and the reactor wall, with quartz glass plates placed on its top. This arrangement allows to reduce the convection of the gases above the crucible and is beneficial to prevent contamination of the solution by impurities/contaminants from the quartz reactor and/or airlock system.

Fig. 31.14 Typical surface morphology of GaN LPE film on sapphire (basal plane). A large lattice mismatch results in the Volmer–Weber growth mode with development of hexagonal GaN islands

Fig. 31.15 (a) Nomarski microphotography of the surface of a HVPE GaN template before LPE growth. (b) GaN film morphology at the early stage of LPE growth on the GaN HVPE template

In a typical experiment, the reactor with empty crucible is evacuated and cleaned several times with nitrogen gas and heated to about 850 °C. The liquid gallium or other (liquid/preheated) solvent can be introduced into the crucible from the top of the reactor, through the airlock system. The suitability of this method depend on the solvent used. If it is applied, this procedure has to be done under nitrogen overpressure, to prevent oxygen contamination. The temperature is then raised to the growth temperature. The substrate is fixed on an alumina or graphite substrate holder in horizontal, vertical or inclined position, and introduced into the reactor through the airlock system. The substrate is dipped into the solution, rotated, and the reactive gas mixture is then introduced into the reactor. This procedure allows the use of substrates that would not withstand the reactive gas atmosphere without degradation, for example, ZnO. In principle, various reactive gases can be applied. In the specific experiments described herein [31.99], a gas mixture of NH_3 and N_2 with ratio 1 : 4 was used. After the growth, before removing the substrates from the solution, pure nitrogen is introduced and the temperature reduced to about 800 °C while removing the film from the solution. Growth times of 10–72 h were applied, and several films were grown successively.

31.7.4 Morphological Evolution of LPE-Grown Nitride Films

Sapphire substrates (basal plane) have a lattice mismatch of -13.7% with GaN along the a direction, which leads to the Volmer–Weber (or islands) film growth mechanism. GaN LPE films grown directly on sapphire typically present hexagonal islands, which may spread over the entire substrate surface by coalescence, resulting in grain-boundary-rich films. Figure 31.14 shows a Nomarski microphotography of a GaN LPE on sapphire during the coalescence phase.

When GaN LPE films are grown on GaN templates, growth starts preferentially on existing defects. Etch back and regrowth mechanisms can easily take place, as growth proceeds near thermodynamic equilibrium, with very low supersaturation. Hence, surface mobility of atoms is high, which leads to the smoothing effect on growth hillocks present on GaN buffer layer. This is shown in the Nomarski microscopy images of Fig. 31.15. On the left, we can see the surface morphology of a GaN HVPE seed layer grown on sapphire before LPE. Conical islands and overall roughness along the macrosteps are clearly visible. On the right,

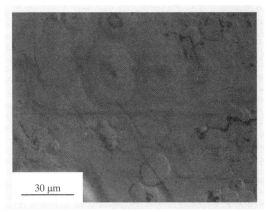

Fig. 31.16 Nomarski microphotograph of the surface of a GaN LPE film, grown on a HVPE MOVPE seed layer on sapphire (after [31.99])

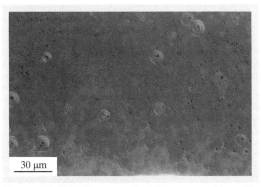

Fig. 31.17 InGaN film grown by LPE on c-sapphire, without a buffer layer. The lattice pulling effect can tentatively explain the (partial) suppression of the island growth mechanism that would be expected from the large lattice mismatch with the substrate

after LPE growth, we recognize the overgrown regions, with smooth regions that start to develop from the pre-existing islands. For longer growth time, such film may evolve into a smooth surface, with large flat areas.

Improved surface morphology can be obtained by using better-quality GaN templates, as shown in Fig. 31.16. The overgrown regions develop to rather flat areas already in the early stages of LPE film growth, when using a multilayer HVPE GaN template grown on a MOVPE GaN buffer layer on sapphire [31.99].

InGaN LPE films could be successfully grown [31.114] using the same growth system described above. However, the film morphologies that were obtained for films grown on sapphire, without a buffer layer, were quite surprising. In contrary to GaN LPE films, island growth was largely suppressed, and smooth surfaces did develop despite the lattice mismatch. Rather smooth, rounded hillocks developed and steps propagated over large surface area. This can be recognized in the Nomarski microphotographs of InGaN LPE film, which is shown in Fig. 31.17.

The InGaN film shown in Fig. 31.17 was grown from liquid Ga and ammonia reactive gas atmosphere by the procedure previously described. Growth conditions were similar to those applied for LPE of GaN [31.99], and film thickness was typically a few microns (GaN and InGaN). This significant difference in growth morphology can tentatively be explained by a possible lattice pulling effect that would minimize the lattice mismatch and hence reduce or suppress the island growth mechanism. Further investigations are needed to confirm this. However, this preliminary result is very promising, and LPE-grown high-quality InGaN solid-solution films may become a valuable approach in the fabrication of high-efficiency radiation-hard photovoltaic solar cells.

31.8 Conclusions

Liquid-phase epitaxy has been a major technique for the development of many materials, and for a variety of applications. The challenges of this technique, and especially the stringent requirements in all substrate and growth process parameters, have often been critical limiting factors. This has been shown especially in Sect. 31.5, where some aspects of LPE growth of high-temperature superconductors were presented. However, the structural perfection and homogeneity of the films that can potentially be achieved is the highest possible, when compared with any other melt or thin-film growth techniques. This is not only true for the case of semiconductors (for example, GaP; Sect. 31.1) but also in the case of complex oxides. There is a misconception that only thick films can be grown by LPE. Though this has been true for a majority of compounds investigated by LPE so far, the recent results presented in Sect. 31.6 on the LPE growth of langasites demonstrates that thin films of the order of a few hundred nanometers can also be grown by LPE.

The technologically important field of nitrides could greatly benefit from LPE-grown low-defect films, and this was discussed in Sect. 31.3. Several nitride applications actually hinge on the high density of film defects, and the lack of lattice-matched substrates. Despite the use of sophisticated multistep deposition methods and epitaxial lateral overgrowth, and the significant amount of research by several groups in this area over the past 10 years, the lower limit for the dislocation density achieved is about 10^5 cm^{-3}, which is two or three orders of magnitude too high for many applications. Especially here, the LPE technique could possibly lead to major advances, enabling material with lower dislocation density that could also be used as ideal substrates for growth of multilayer structures by VPE. In fact, LPE GaN films with low dislocation densities of the order of 10^2 cm^{-3} have already been demonstrated, and they can be grown without the need of any external patterning or the ELO technique. Besides GaN, AlN and InN could also benefit from LPE studies; for example, one would expect highest efficiencies for In-rich InGaN photovoltaic cells based on LPE-grown material.

Despite its challenges, the LPE technique offers the unique possibility to explore fundamental structure–property relations in a variety of doped and undoped materials grown under equilibrium conditions. This technique allows the study of incongruently melting materials with potentially interesting properties that are not investigated because they that cannot be grown by any other (melt) growth technique. In conclusion, there is significant margin for the discovery and study of novel materials using the LPE growth technique.

References

31.1 H. Nelson: Epitaxial growth from the liquid state and its application to the fabrication of tunnel and laser diodes, RCA Review **24**, 603 (1963)

31.2 L.R. Dawson: High-efficiency graded band-gap Ga$_{1-x}$Al$_x$As light-emitting- diodes, J. Appl. Phys. **48**, 2485–2492 (1977)

31.3 J.I. Nishizawa, Y. Okuno: Liquid-phase epitaxy of GaP by a temperature difference method under controlled vapor-pressure, IEEE Trans. Electron. Dev. **22**, 716–721 (1975)

31.4 J. Nishizawa, Y. Okuno, M. Koike, F. Sakurai: Bright pure green emission from N-free GaP LEDs, Jpn. J. Appl. Phys. **19**, 377–382 (1980)

31.5 M. Konagi, K. Takahashi: Formation of GaAs-(GaAl)As heterojunction transistors by liquid-phase epitaxy, Trans. IEEE J. **94-C**, 141 (1974)

31.6 E. Bauser, D. Kass, M. Warth, H.P. Strunk: Mater. Res. Soc. Symp. Proc. **54**, 267 (1986)

31.7 T. Nishinaga, T. Nakano, S. Zhang: Epitaxial lateral overgrowth of GaAs by LPE, Jpn. J. Appl. Phys. **27**, L964–L967 (1988)

31.8 Y. Suzuki, T. Nishinaga: Epitaxial lateral overgrowth of Si by LPE with Sn solution and its orientation dependence, Jpn. J. Appl. Phys. **28**, 440–445 (1989)

31.9 S. Kinoshita, Y. Suzuki, T. Nishinaga: Epitaxial lateral overgrowth of Si on non-planar substrate, J. Cryst. Growth **115**, 561–566 (1991)

31.10 T. Kochiya, Y. Oyama, T. Kimura, K. Suto, J.I. Nishizawa: Dislocation-free large area InP ELO layers by liquid-phase epitaxy, J. Cryst. Growth **281**, 263–274 (2005)

31.11 S.H. Shin, G.T. Niizawa, J.G. Pasko, G.L. Bostrup, F.J. Ryan, M. Khoshnevisan, C.I. Westmark, C. Fuller: P-I-N CdTe gamma-ray detectors by liquid phase epitaxy (LPE), Presented at the Nucl. Sci. Symp., Orlando (1984)

31.12 G. Bostrup, K.L. Hess, J. Ellsworth, D. Cooper, R. Haines: LPE HgCdTe on sapphire: Status and advancements, J. Electron. Mater. **30**, 560–566 (2001)

31.13 R.C. Linares: Epitaxial growth of narrow linewidth yttrium iron garnet films, J. Cryst. Growth **3/4**, 443–446 (1968)

31.14 H.J. Levinstein, S. Licht, R.W. Landorf, S. Blank: Growth of high-quality garnet thin films from supercooled melts, Appl. Phys. Lett. **19**, 486–488 (1971)

31.15 S.L. Blank: Crystal growth: Magnetic garnets by liquid-phase epitaxy, J. Educ. Modules Mater. Sci. Eng. **2**, 357–390 (1980)

31.16 M.H. Randles: Crystals for magnetic applications. In: *Crystals: Growth, Properties, and Applications*, Vol.1, ed. by C.J.M. Rooijmans (Springer, Berlin, Heidelberg 1978) pp. 71–96

31.17 D. Elwell, H.J. Scheel: *Crystal Growth from High-Temperature Solutions* (Academic, London 1975)

31.18 W. van Erk: The growth kinetics of garnet liquid-phase epitaxy using horizontal dipping, J. Cryst. Growth **43**, 446–456 (1978)

31.19 R. Ghez, E.A. Giess: Liquid phase epitaxial-growth kinetics of magnetic garnet films grown by isothermal dipping with axial rotation, Mater. Res. Bull. **8**, 31–42 (1973)

31.20 W.K. Burton, N. Cabrera, F.C. Frank: The growth of crystals and the equilibrium structure of their surface, Philos. Trans. R. Soc. Lond. Ser. A **243**, 299–358 (1951)

31.21 A.A. Chernov: Crystal growth from the solution and from the melt, Sov. Phys. Usp. **4**, 129 (1961)

31.22 J.A. Burton, R.C. Prim, W.P. Slichter: The distribution of solute in crystals grown from the melt, J. Chem. Phys. **21**, 1987–1991 (1953)

31.23 G. Winkler: *Magnetic Garnets* (Vieweg, Braunschweig 1981)

31.24 C. Klemenz, I. Utke, H.J. Scheel: Film orientation, growth parameters and growth modes in epitaxy of $YBa_2Cu_3O_x$, J. Cryst. Growth **204**, 62–68 (1999)

31.25 D.P. Norton: Science and technology of high-temperature superconducting films, Annu. Rev. Mater. Sci. **28**, 299–347 (1998)

31.26 H. Hilgenkamp, J. Mannhart: Grain boundaries in high-T_c superconducting films, Rev. Mod. Phys. **74**, 485–549 (2002)

31.27 K.H. Wu, S.P. Chen, J.Y. Juang, T.M. Uen, Y.S. Gou: Investigation of the evolution of $YBa_2Cu_3O_{7-d}$ films deposited by scanning pulsed laser deposition on different substrates, Physica C **289**, 230–242 (1997)

31.28 C. Klemenz, H.J. Scheel: Flat $YBa_2Cu_3O_{7-\delta}$ layers for planar tunnel-device technology, Physica C **265**, 126–134 (1996)

31.29 C. Klemenz: Hollow cores and step bunching effects on (001)YBCO surfaces grown by liquid-phase epitaxy, J. Cryst. Growth **187**, 221–227 (1998)

31.30 A. Chakraborty, K.C. Kim, F. Wu, J.S. Speck, S.P. DenBaars, U.K. Mishra: Defect reduction in nonpolar *a*-plane GaN films using in situ SiN_x nanomask, Appl. Phys. Lett. **89**, 041903 (2006)

31.31 K.A. Dunn, S.E. Babcock, D.S. Stone, R.J. Matyi, L. Zhang, T.F. Kuech: Dislocation arrangement in a thick LEO GaN film on sapphire, MRS Internet J. Nitride Semicond. Res. **5**, W2.11 (2000)

31.32 B.A. Haskell, T.J. Baker, M.B. McLaurin, F. Wu, P.T. Fini, S.P. DenBaars, J.S. Speck, S. Nakamura: Defect reduction in (1$\bar{1}$00) *m*-plane gallium nitride via lateral epitaxial overgrowth by hydride vapor phase epitaxy, Appl. Phys. Lett. **86**, L1197–L1199 (2005)

31.33 B.A. Haskell, F. Wu, M.D. Craven, S. Matsuda, P.T. Fini, T. Fujii, K. Fujito, S.P. DenBaars, J.S. Speck, S. Nakamura: Defect reduction in (1120) *a*-plane gallium nitride via lateral epitaxial overgrowth by hydride vapor-phase epitaxy, Appl. Phys. Lett. **83**, 644–646 (2003)

31.34 C. Klemenz, H.J. Scheel: Solubility of $YBa_2Cu_3O_{7-\delta}$ and $Nd_{1+x}Ba_{2-x}Cu_3O_{7+\delta}$ in the BaO/CuO flux, J. Cryst. Growth **200**, 435–440 (1999)

31.35 J.G. Bednorz, K.A. Müller: Possible high-T_c superconductivity in the Ba-La-Cu-O system, Z. Phys. B **64**, 189–193 (1986)

31.36 M.K. Wu, J.R. Ashburn, C.J. Torng, P.H. Hor, R.L. Meng, L. Gao, Z.J. Huang, Y.Q. Wang, C.W. Chu: Superconductivity at 93 K in a new mixed-phase Y-Ba-Cu-O compound system at ambient pressure, Phys. Rev. Lett. **58**, 908–910 (1987)

31.37 M. Hawley, I.D. Raistrick, J.G. Beery, R.J. Houlton: Growth mechanism of sputtered films of $YBa_2Cu_3O_7$ studied by scanning tunneling microscopy, Science **251**, 1587–1589 (1991)

31.38 C. Gerber, D. Anselmetti, J.G. Bednorz, J. Mannhart, D.G. Schlom: Screw dislocations in high-T_c films, Nature **350**, 279–280 (1991)

31.39 R.F. Belt, J. Ings, G. Diercks: Superconductor film growth on $LaGaO_3$ substrates by liquid-phase epitaxy, Appl. Phys. Lett. **56**, 1805–1807 (1990)

31.40 G. Balestrino, V. Foglietti, M. Marinelli, E. Milani, A. Paoletti, P. Paroli: Epitaxial films of BSCCO grown from liquid KCl solutions onto several substrates, IEEE Trans. Magn. **27**, 1589–1591 (1991)

31.41 G. Balestrino, V. Foglietti, M. Marinelli, E. Milani, A. Paoletti, P. Paroli, G. Luce: Structural and electrical properties of epitaxial films of BSCCO grown from liquid phase epitaxy, Solid State Commun. **76**, 503–505 (1990)

31.42 S. Narayanan, K.K. Raina, R.K. Pandey: Thin film growth of the 2122-phase of BCSCO superconductor with high degree of crystalline perfection, J. Mater. Res. **7**, 2303–2307 (1992)

31.43 H. Takeya, H. Takei: Preparation of high-T_c Bi-Sr-Ca-Cu-O films on MgO substrates by the liquid phase epitaxial (LPE) method, Jpn. J. Appl. Phys. **28**, L229–L232 (1989)

31.44 R.S. Liu, Y.T. Huang, P.T. Wu, J.J. Chu: Epitaxial growth of high-T_c Bi-Ca-Sr-Cu-O superconducting layer by LPE process, Jpn. J. Appl. Phys. **27**, L1470–L1472 (1988)

31.45 R.S. Liu, Y.T. Huang, J.M. Liang, P.T. Wu: Epitaxial growth of high-T_c superconducting Tl-Ca-Ba-Cu-O films by liquid phase epataxial process, Physica C **156**, 785–787 (1988)

31.46 A.S. Yue, W.S. Liao, H.J. Choi: LPE growth of high-T_c $Bi_2Sr_2Ca_1Cu_2O_x$ films, Cryogenics **32**, 596–599 (1992)

31.47 J.S. Shin, H. Ozaki: Superconducting Bi-Sr-Ca-Cu-O films prepared by the liquid phase epitaxial method, Physica C **173**, 93–98 (1991)

31.48 L.H. Perng, T.S. Chin, K.C. Chen, C.H. Lin: Y-Ba-Cu-O superconducting films grown on (100) magnesia and sapphire substrates by a melt growth method without crucible, Supercond. Sci. Technol. **3**, 233–237 (1990)

31.49 C. Klemenz, H.J. Scheel: Liquid-phase epitaxy of high-T_c superconductors, J. Cryst. Growth **129**, 421–428 (1993)

31.50 C. Dubs, K. Fischer, P. Görnert: Liquid phase epitaxy of $YBa_2Cu_3O_{7-x}$ on $NdGaO_3$ and $LaGaO_3$ substrates, J. Cryst. Growth **123**, 611–614 (1992)

31.51 H.J. Scheel, C. Klemenz, F.-K. Reinhart, H.P. Lang, H.-J. Güntherodt: Monosteps on extremely flat $YBa_2Cu_3O_{7-\delta}$ surfaces grown by liquid-phase epitaxy, Appl. Phys. Lett. **65**, 901–903 (1994)

31.52 T. Kitamura, M. Yoshida, Y. Yamada, Y. Shiohara, I. Hirabayashi, S. Tanaka: Crystalline orientation of $YBa_2Cu_3O_x$ film prepared by liquid-phase epitaxial growth on $NdGaO_3$ substrate, Appl. Phys. Lett. **66**, 1421–1423 (1995)

31.53 H.J. Scheel, F. Licci: Phase diagrams and crystal growth of oxide superconductors, Thermochim. Acta **174**, 115–130 (1991)

31.54 F. Licci, H.J. Scheel, P. Tissot: Determination of the eutectic composition by crystal growth and flux separation: example $BaCuO_2$-CuO_x, J. Cryst. Growth **112**, 600–605 (1991)

31.55 H.J. Scheel, F. Licci: Crystal growth of $YBa_2Cu_3O_{7-x}$, J. Cryst. Growth **85**, 607–614 (1987)

31.56 T. Aichele, S. Bornmann, C. Dubs, P. Görnert: Liquid-phase epitaxy (LPE) of $YBa_2Cu_3O_{7-x}$ high T_c superconductors, Cryst. Res. Technol. **32**, 1145–1154 (1997)

31.57 C. Krauns, M. Sumida, M. Tagami, Y. Yamada, Y. Shiohara: Solubility of RE elements into Ba-Cu-O melts and the enthalpy of dissolution, Z. Phys. B **96**, 207–212 (1994)

31.58 P. Bennema, J.P. van der Eerden: Crystal graphs, connected nets, roughening transition and the morphology of crystals. In: *Morphology of Crystals*, ed. by I. Sunagawa (Terra Scientific, Tokyo 1987), Chap. 1

31.59 G.H. Gilmer, K.A. Jackson: Computer simulation of crystal growth. In: *Crystal Growth and Materials*, ed. by E. Kaldis, H.J. Scheel (North Holland, Amsterdam 1977) pp. 80–114

31.60 D.S. Tsagareishvili, G.G. Gvelesiani, I.B. Baratashvili, G.K. Moiseev, N.A. Vatolin: Thermodynamic functions of $YBa_2Cu_3O_7$, $YBa_2Cu_3O_6$, Y_2BaCuO_5 and $BaCuO_2$, Russ. J. Phys. Chem. **64**, 1404–1406 (1990)

31.61 G.K. Moiseev, N.A. Vatolin, S.I. Zaizeva, N.I. Ilyinych, D.S. Tsagareishvili, G.G. Gvelesiani, I.B. Baratashvili, J. Sestàk: Calculation of thermodynamic properties of the phases in the Y-Ba-Cu-O system, Thermochim. Acta **198**, 267–278 (1992)

31.62 W. van Erk: The growth kinetics of garnet liquid-phase epitaxy using horizontal dipping, J. Cryst. Growth **43**, 446–456 (1979)

31.63 W. van Erk: A solubility model for rare-earth iron garnets in a PbO/B_2O_3 solution, J. Cryst. Growth **46**, 539–550 (1979)

31.64 W. van Erk: Growth of a mixed crystal from an ideal dilute solution, J. Cryst. Growth **57**, 71–83 (1982)

31.65 I. Utke, C. Klemenz, H.J. Scheel, M. Sasaura, S. Myazawa: Misfit problems in epitaxy of high-T_c superconductors, J. Cryst. Growth **174**, 806–812 (1997)

31.66 I. Utke, C. Klemenz, H.J. Scheel, P. Nüesch: High-temperature x-ray measurements of gallates and cuprates, J. Cryst. Growth **174**, 813–820 (1997)

31.67 N. Cabrera, M.M. Levine: On the dislocation theory of evaporation of crystals, Philos. Mag. **1**, 450–458 (1956)

31.68 H.P. Lang, H. Haefke, G. Leemann, H.J. Güntherodt: Scanning tunneling microscopy study of different growth stages of $YBa_2Cu_3O_{7-d}$ thin films, Physica C **194**, 81–91 (1992)

31.69 M. Hawley, I.D. Raistrick, J.G. Beery, R.J. Houlton: Growth mechanism of sputtered films of $YBa_2Cu_3O_7$ studied by scanning tunelling microscopy, Science **251**, 1587–1589 (1991)

31.70 C. Gerber, D. Anselmetti, J.G. Bednorz, J. Mannhart, D.G. Schlom: Screw dislocations in high-T_c films, Nature **350**, 279–280 (1991)

31.71 T. Nishinaga, H.J. Scheel: Crystal growth aspect of high-T_c superconductors. In: *Advances in Superconductivity VIII*, ed. by H. Hyakawa, Y. Enomoto (Springer, Tokyo 1996) pp. 33–38

31.72 A.A. Kaminskii, B.V. Mill, G.G. Khodzhabagyan, A.F. Konstantinova, A.I. Okorochkov, I.M. Silvestrova: Investigation of trigonal $(La_{1-x}Nd_x)_3Ga_5SiO_{14}$ crystals: I. Growth and optical properties, Phys. Status Solidi (a) **80**, 387–398 (1983)

31.73 Y.V. Pisarevski, P.A. Senushencov, P.A. Popov, B.V. Mill: New strong piezoelectric $La_3Ga_{5.5}Nb_{0.5}O_{14}$ with temperature compensation cuts, Proc. IEEE Int. Freq. Control Symp. (1995) pp. 653–656

31.74 S. Kück: Laser-related spectroscopy of ion-doped crystals for tunable solid-state lasers, Appl. Phys. B **72**, 515–562 (2001)

31.75 Y. Kim, A. Ballato: Force-frequency effects of Y-cut langanite and Y-cut langatate, IEEE Trans. Ultrason. Ferroelectr. Freq. Control **50**, 1678–1682 (2003)

31.76 J.A. Kosinski, R.A. Pastore Jr., X. Yang, J. Yang, J.A. Turner: Stress-induced frequency shifts in langasite thickness-mode resonators, Proc. IEEE Int. Freq. Control Symp. (2003) pp. 716–722

31.77 C. Klemenz, M. Berkowski, B. Deveaud-Pledran, D.C. Malocha: Defect structure of langasite-type crystals: a challenge for applications, Proc. IEEE Int. Freq. Control Symp. (2002) pp. 301–306

31.78 R. Fachberger, T. Holzheu, E. Riha, E. Born, P. Pongratz, H. Cerva: Langasite and langatate nonuniform material properties correlated to the performance of SAW-devices, Proc. IEEE Int. Freq. Control Symp. (2001) pp. 235–239

31.79 C.F. Klemenz, J. Luo, D. Shah: High-quality 2 inch $La_3Ga_{5.5}Ta_{0.5}O_{14}$ and $Ca_3TaGa_3Si_2O_{14}$ crystals for oscillators and resonators, Adv. Electron. Ceram. Mater. Ceram. Eng. Sci. Proc. **26**, 169–176 (2005)

31.80 B.V. Mill, Y.V. Pisarevshy, E.L. Belokoneva: Synthesis, growth, and some properties of single crystals with the $Ca_3Ga_2Ge_4O_{14}$ structure, Proc. Jt. Meet. Eur. Freq. Time Forum IEEE Int. Freq. Control Symp. (1999) pp. 829–834

31.81 B.V. Mill, M.V. Lomonosov: Two new lines of langasite family compositions, Proc. IEEE Int. Freq. Control Symp. (2001) pp. 255–262

31.82 V.I. Chani, K. Shimamura, Y.M. Yu, T. Fukuda: Design of new oxide crystals with improved structural stability, Mater. Sci. Eng. R **20**, 281–338 (1997)

31.83 C. Klemenz: Liquid-phase epitaxy of $La_3Ga_5SiO_{14}$, Proc. Eur. Freq. Time Forum (2001) pp. 42–45

31.84 C. Klemenz: High-quality langasite films grown by liquid-phase epitaxy, J. Cryst. Growth **237**, 714–719 (2002)

31.85 C. Klemenz: High-quality $La_3Ga_{5.5}Ta_{0.5}O_{14}$ and $La_3Ga_{5.5}Nb_{0.5}O_{14}$ LPE films for oscillators and resonators, J. Cryst. Growth **250**, 34–40 (2003)

31.86 C.F. Klemenz, A. Sayir: In-situ Al:Ti-co-doped $La_3Ga_{5.5}Ta_{0.5}O_{14}$ films for high-Q resonators, Proc. IEEE Int. Freq. Contol Symp. (2006) pp. 676–680

31.87 H. Amano, M. Kito, K. Hiramatsu, I. Akasaki: P-type conduction in Mg-doped GaN treated with low-energy electron beam irradiation (LEEBI), Jpn. J. Appl. Phys. **28**, L2112–L2114 (1989)

31.88 F.A. Ponce: Defects and interfaces in GaN epitaxy, MRS Bull. **22**, 51–57 (1997)

31.89 S. Nakamura, M. Senoh, S. Nagahama, N. Iwasa, T. Yamada, T. Matsushita, Y. Sugimoto, H. Kiyoku: Subband emissions of InGaN multi-quantum-well laser diodes under room-temperature continuous wave operation, Appl. Phys. Lett. **70**, 2753–2755 (1997)

31.90 T. Detchprohm, T. Kuroda, K. Hiramatsu, N. Sawaki, H. Goto: The selective growth in hydride vapor phase epitaxy of GaN, Inst. Phys. Conf. Ser. **142**, 859–862 (1996)

31.91 T.S. Zheleva, O. Nam, M.D. Bremser, R.F. Davis: Dislocation density reduction via lateral epitaxy in selectively grown GaN structures, Appl. Phys. Lett. **71**, 2472–2474 (1997)

31.92 V.Y. Davydov, A.A. Klochikhin, R.P. Seisyan, V.V. Emtsev, S.V. Ivanov, F. Bechstedt, J. Furthmüller, H. Harima, V. Mudryi, J. Aderhold, O. Semchinova, J. Graul: Absorption and emission of hexagonal InN. Evidence of narrow fundamental band gap, Phys. Status Solidi (b) **229**, R1–R3 (2002)

31.93 J. Wu, W. Walukiewicz, K.M. Yu, J.W. Ager, E.E. Haller, H. Lu, W.J. Schaff, Y. Saito, Y. Nanishi: Unusual properties of the fundamental bandgap of InN, Appl. Phys. Lett. **80**, 3967 (2002)

31.94 J. Wu, W. Walukiewicz, W. Shan, K.M. Yu, J.W. Ager, E.E. Haller, H. Lu, W.J. Schaff: Effects of the narrow band gap on the properties of InN, Phys. Rev. B **66**, 201403 (2002)

31.95 T.L. Tansley, C.P. Foley: Optical bandgap of indium nitride, J. Appl. Phys. **59**, 3241–3244 (1986)

31.96 J. Wu, W. Walukiewicz, K.M. Yu, J.W. Ager, E.E. Haller, H. Lu, W.J. Schaff: Small band gap bowing in $In_{1-x}Ga_xN$ alloys, Appl. Phys. Lett. **80**, 4741–4743 (2002)

31.97 J. Wu, W. Walukiewicz, K.M. Yu, W. Shan, J.W. Ager, E.E. Haller, H. Lu, W.J. Schaff, W.K. Metzger, S. Kurtz: Superior radiation resistance of $In_{1-x}Ga_xN$ alloys: Full-solar-spectrum photovoltaic material system, J. Appl. Phys. **94**, 6477–6482 (2003)

31.98 R.A. Logan, C.D. Thurmond: Heteroepitaxial thermal gradient solution growth of GaN, J. Electrochem. Soc. **119**, 1727–1734 (1972)

31.99 C. Klemenz, H.J. Scheel: Crystal growth and liquid-phase epitaxy of GaN, J. Cryst. Growth **211**, 62–67 (2000)

31.100 G. Sun, E. Meissner, P. Berwian, G. Müller, J. Friedrich: Study on the kinetics of the formation reaction of GaN from Ga-solutions under ammonia atmosphere, J. Cryst. Growth **305**, 326–334 (2007)

31.101 O. Glemser, M. Field, K. Kleine-Weischeide: Z. Anorg. Chem. **332**, 257–259 (1964)

31.102 R.M. Yonco, E. Veleckis, V.A. Maroni: Solubility of nitrogen in liquid lithium and thermal decomposition of solid Li_3N, J. Nucl. Mater. **57**, 317–324 (1975)

31.103 W.J. Wang, X.L. Chen, Y.T. Song, W.X. Yuan, Y.G. Cao, X. Wu: Assessment of Li-Ga-N ternary system and GaN single crystal growth, J. Cryst. Growth **264**, 13–16 (2004)

31.104 M. Yano, M. Okamoto, Y.K. Yap, M. Yoshimura, Y. Mori, T. Sasaki: Growth of nitride crystals, BN, AlN and GaN by using a Na flux, Diam. Relat. Mater. **9**, 512–515 (2000)

31.105 H. Yamane, M. Shimada, S.J. Clarke, F.J. DiSalvo: Preparation of GaN single crystals using a Na flux, Chem. Mater. **9**, 413–416 (1997)

31.106 F. Kawamura, T. Iwahashi, M. Morishita, K. Omae, M. Yoshimura, Y. Mori, T. Sasaki: Growth of transparent: Large size GaN single crystal with low dislocations using Ca-Na flux system, Jpn. J. Appl. Phys. **42**, L729–L731 (2003)

31.107 F. Kawamura, H. Umeda, M. Kawahara, M. Yoshimura, Y. Mori, T. Sasaki, H. Okado, K. Arakawa, H. Mori: Drastic decrease in dislocations during liquid-phase epitaxy growth of GaN single crystals using Na flux method without any artificial processes, Jpn. J. Appl. Phys. **45**, 2528–2530 (2006)

31.108 C.O. Dugger: The synthesis of aluminum nitride single crystals, Mater. Res. Bull. **9**, 331–336 (1974)

31.109 T. Fukuda, D. Ehrentraut: Prospects for the ammonothermal growth of large GaN crystal, J. Cryst. Growth **305**, 304–310 (2007)

31.110 R. Dwiliński, R. Doradziński, J. Garczyński, L. Sierzputowski, J.M. Baranowski, M. Kamińska: AMMONO method of GaN and AlN production, Diam. Relat. Mater. **7**, 1348–1350 (1998)

31.111 T. Hashimoto, K. Fujito, M. Saito, J.S. Speck, S. Nakamura: Ammonothermal growth of GaN on an over-1-inch seed crystal, Jpn. J. Appl. Phys. **44**, L1570–L1572 (2005)

31.112 B. Wang, M.J. Callahan, K.D. Rakes, L.O. Bouthillette, S.-Q. Wang, D.F. Bliss, J.W. Kolis: Ammonothermal growth of GaN crystals in alkaline solutions, J. Cryst. Growth **287**, 376–379 (2006)

31.113 D. Ehrentraut, N. Hoshino, Y. Kagamitani, A. Yoshikawa, T. Fukuda, H. Itoh, S. Kawabata: Temperature effect of ammonium halogenides as mineralizers on the phase stability of gallium nitride synthesized under acidic ammonothermal conditions, J. Mater. Chem. **17**, 886–893 (2007)

31.114 C. Klemenz: unpublished

32. Molecular-Beam Epitaxial Growth of HgCdTe

James W. Garland, Sivalingam Sivananthan

Epitaxial HgCdTe grown by molecular-beam epitaxy (MBE) is the material of choice for advanced infrared (IR) detection and imaging devices. Its bandgap is easily tunable over the entire IR range with only very small changes in lattice constant, offering the possibility of multilayer device structures and thus an unlimited choice of device designs, and it yields devices with quantum efficiencies as high as 0.99. Despite a number of unresolved challenges in achieving its ultimate promise for industrial application, the great achievements in the MBE growth of HgCdTe are made evident by its routine use in the industrial manufacture of focal-plane arrays (FPAs). MBE growth can be continuously monitored in situ by reflection high-energy electron diffraction, spectroscopic ellipsometry (SE), and other characterization tools, providing instantaneous feedback on the influence of growth conditions on film structure. This allows the growth of a large range of unique structures such as superlattices (SLs), quantum well devices, lasers, and advanced design devices such as multicolor and high-operating-temperature IR sensors and focal-plane arrays. This chapter considers the theory and practice of MBE growth of HgCdTe and HgTe/CdTe superlattices and the use of HgCdTe in IR devices, emphasizing such incompletely resolved issues as the choice and preparation of substrates, dislocation reduction, p-doping, and the uses of SE.

The theory of MBE growth is summarized briefly in Sect. 32.2, followed by a lengthy discussion of substrate-related issues in Sect. 32.3, including a summary of the relative merits and demerits of different substrate materials. The growth hardware is discussed very briefly in Sect. 32.4, followed by a discussion of the in situ characterization tools used for monitoring and control of the growth in Sect. 32.5 and of the growth procedure for HgCdTe in Sect. 32.6. A discussion of the doping of HgCdTe, including the serious issues still surrounding p-type doping, is given in Sect. 32.7. The properties achievable in MBE-grown HgCdTe are summarized in Sect. 32.8, with emphasis on the types of defects common in MBE-grown material, their effects on device performance, and possible methods to reduce the present defect densities. The use of MBE-grown HgTe/CdTe SLs for IR absorbers in lieu of HgCdTe alloy material is considered in Sect. 32.9. Finally, a brief discussion of the devices enabled by the MBE growth of HgCdTe and of their fabrication is given in Sects. 32.10 and 32.11, and a brief concluding summary of the chapter is given in Sect. 32.12.

32.1	**Overview**	1070
	32.1.1 Why HgCdTe Is Important	1071
	32.1.2 Why MBE Is the Preferred Method of Growth for HgCdTe IR Detectors and Imagers	1072
	32.1.3 General Description of the MBE Growth Technique	1072
32.2	**Theory of MBE Growth**	1073
	32.2.1 Pseudo-Equilibrium Theories	1074
	32.2.2 Kinetic Theories	1075
32.3	**Substrate Materials**	1076
	32.3.1 Substrate Orientation	1077
	32.3.2 CdZnTe Substrates	1077
	32.3.3 Si-Based Substrates	1084
	32.3.4 Other Substrates	1087
32.4	**Design of the Growth Hardware**	1088
	32.4.1 Mounting of the Substrate	1088
	32.4.2 Valving of the Effusion Cells	1089
32.5	**In situ Characterization Tools for Monitoring and Controlling the Growth**	1090
	32.5.1 Spectroscopic Ellipsometry (SE): Basic Theory and Experimental Setup	1090
	32.5.2 SE Data Analysis	1092

- 32.5.3 SE Study of Hg Absorption and Adsorption on CdTe 1098
- 32.5.4 Correlation Between the Quality of MBE-Grown HgCdTe and the Depolarization and Surface Roughness Coefficients Measured by in situ SE 1099
- 32.5.5 Surface Characterization by in situ RHEED 1100
- 32.5.6 Other in situ Tools for Controlling the Growth 1101

32.6 **Nucleation and Growth Procedure**.........1101
- 32.6.1 Nucleation and Growth of CdTe or ZnTe on Si.................. 1101
- 32.6.2 Substrate Preparation and Growth of HgCdTe 1102

32.7 **Dopants and Dopant Activation** 1104
- 32.7.1 Extrinsic n-Type Doping............. 1104
- 32.7.2 Extrinsic p-Type Doping............. 1105
- 32.7.3 In situ Group I Dopant Incorporation 1106
- 32.7.4 In situ Group V Dopant Incorporation 1106
- 32.7.5 As Activation............................ 1106

32.8 **Properties of HgCdTe Epilayers Grown by MBE**........1107
- 32.8.1 Electrical and Optical Properties .. 1107
- 32.8.2 Structural Properties.................. 1110
- 32.8.3 Surface Defects 1110

32.9 **HgTe/CdTe Superlattices**.......................1112
- 32.9.1 Theoretical Properties 1113
- 32.9.2 Growth 1114
- 32.9.3 Experimentally Observed Properties 1114

32.10 **Architectures of Advanced IR Detectors** ..1115
- 32.10.1 Reduction of Internal Detector Noise 1116
- 32.10.2 Increasing Detector Response 1116
- 32.10.3 High-Speed IR Detectors 1117
- 32.10.4 High-Operating-Temperature (HOT) IR Detectors 1118

32.11 **IR Focal-Plane Arrays (FPAs)**..................1118

32.12 **Conclusions**..1119

References 1121

Molecular-beam epitaxy (MBE) is a process for growing thin epitaxial films of a wide variety of materials, ranging from oxides to semiconductors to metals. Because of their high technological value, its first and still most common application is for the growth of compound semiconductors and their alloys. In MBE beams of atoms or molecules are incident in an ultrahigh vacuum upon a heated crystal processed to produce a nearly atomically clean, smooth surface. The arriving atoms form a crystalline layer in registry with the substrate, i.e., an epitaxial film. MBE allows doping levels and alloy compositions to be controlled precisely and changed rapidly, producing almost atomically abrupt homojunctions or heterojunctions. In 1969 Cho [32.1] published landmark results reporting the first in situ observations of the MBE growth process using high-energy electron diffraction. He demonstrated that MBE growth could produce atomically flat, ordered layers; thus these studies marked the beginning of the use of MBE for practical device fabrication. This structural analysis capability proved to be crucial for characterizing MBE epitaxy because it provided instantaneous feedback on the influence of growth conditions on film structure. That allowed the growth of a large range of unique structures such as superlattices (SLs), quantum well devices, lasers, and advanced design devices such as multicolor and high-operating-temperature IR sensors and focal-plane arrays.

MBE-grown HgCdTe is the material of choice for advanced infrared (IR) detection and imaging devices. Its bandgap is easily tunable over the entire IR range with only very small changes in lattice constant, it offers the possibility of multilayer device structures and thus an unlimited choice of device designs, and it yields devices with quantum efficiencies as high as 0.99. Despite a number of unresolved challenges in achieving its ultimate promise for industrial application, the great achievements in the MBE growth of HgCdTe are made evident by its routine use in the industrial manufacture of focal-plane arrays (FPAs).

32.1 Overview

The MBE growth technique for HgCdTe, the in situ characterization techniques used in its growth, the present status of achievable HgCdTe material quality, and its applications to IR detection are reviewed here.

Particular emphasis is placed on the issues remaining to be solved before MBE-grown HgCdTe can reach its full potential as an ideal material for IR devices. These issues include substrate choice and preparation, the p-doping of HgCdTe, and the reduction of material defect densities. Recent progress in addressing those issues is discussed, as are techniques still under investigation for addressing those issues. Except for points not documented elsewhere, detailed discussions are left to the references.

32.1.1 Why HgCdTe Is Important

Infrared detectors fall into two broad categories: photon and thermal detectors. Photon detectors offer much higher detection speeds and sensitivities as well as the ability to differentiate between objects at different temperatures using multicolor or hyperspectral detectors and to form high-resolution images using FPAs. These properties are essential for almost all military and space applications. Mercury cadmium telluride has been of great interest as a material for infrared (IR) detection since it was first grown almost 50 years ago and has dominated the market for more than two decades [32.2]. It is the only semiconductor system in which the optical absorption edge can be made to vary across the entire IR spectrum with an almost constant lattice parameter. The 300 K lattice constant changes only from 6.4614 Å for HgTe with a -0.14 eV bandgap to 6.4825 Å for CdTe with a 1.56 eV bandgap. This allows the high performance of photodiodes over a much larger range than InGaAs, for example, and allows the epitaxial growth of layers with different optical cutoffs on top of one another with very little strain, $\approx 0.1\%$, which in turn, along with the low-temperature MBE growth method, allows the fabrication of advanced detectors and FPAs, such as multicolor and high-operating-temperature (HOT) FPAs, and allows the growth of HgTe/CdTe SLs, either for very long-wavelength infrared (VLWIR) detection or as buffer layers. In addition to this great advantage, HgCdTe offers two other overwhelming advantages that explain its dominance in the IR marketplace. Both the thermal generation rate per unit volume of material and the quantum efficiency for photon absorption in the IR are higher in HgCdTe than in any competing material. From a fundamental point of view only the diffusion-limited operation of a photovoltaic diode need be considered. In that case only these two properties are important in determining the maximum useful operating temperature, as discussed by *Kinch* [32.3] and outlined below.

The following four different types of materials systems have been serious players in the IR systems marketplace:

1. Direct-bandgap semiconductors (minority-carrier transport):
   a) Binary III–V alloys such as InSb-based alloys
   b) Ternary II–VI alloys (tunable-bandgap HgCdTe)
   c) II–VI tunable-bandgap type I SLs (HgTe/CdTe)
   d) Type II and III SLs such as InAs/GaInSb.
2. Extrinsic semiconductors such as Si:Ga or Ge:Hg (majority-carrier transport)
3. Type I SL quantum-well infrared photodetectors (QWIPs) such as GaAs/AlGaAs QWIPs (majority-carrier transport)
4. Silicon Schottky barriers such as PtSi or IrSi (majority-carrier transport).

For the reasons discussed in the previous paragraph, HgCdTe is the dominant direct-bandgap semiconductor material for applications in all regions of the IR spectrum except the VLWIR. Even there, either HgCdTe or HgTe/CdTe SLs are expected to become strong or even dominant players.

In the long-wavelength IR (LWIR) spectral band the normalized detector dark current is four orders of magnitude smaller for direct-bandgap semiconductors such as HgCdTe than for any of the other three types of systems, so that direct-bandgap semiconductors such as HgCdTe give the highest background-limited performance (BLIP) operating temperature by far. This is extremely important because cryogenic cooling is expensive and greatly increases the weight and bulk of IR detection and imaging systems. The sensitivity of a detector can be expressed in a variety of ways, but the limiting factor in all of them is the fluctuation in the relevant carrier concentration, as this determines the minimum observable signal. Ideally, neglecting the background incident flux, the most commonly used measure of the detectivity D^* depends only on the properties of the IR material in the detector. It is two or more orders of magnitude higher for HgCdTe than for any of the other three types of systems. Although material growth issues have prevented HgCdTe from reaching its full potential, it is the material of choice for most detector systems. The only important advantage of III–V alloys and SL QWIPs is their lower cost arising from the preexisting large industrial infrastructure in III–V materials and device growth, processing, and packaging. For military systems, which dominate the market, the increasing requirements of succeeding generations of IR detection and imaging systems will further enhance the dominance of HgCdTe in the IR materials marketplace. No other material, even in principle, will be able to meet the requirements for higher operating temperatures (and ultimately totally uncooled operation), larger-area FPAs, faster frame rates, and multispectral operation with no increase, or even a reduction, in cost.

32.1.2 Why MBE Is the Preferred Method of Growth for HgCdTe IR Detectors and Imagers

Since the growth of HgCdTe by MBE was first reported in 1982 [32.4], more than 15 groups have been established worldwide for the growth of Hg-based alloys and SLs by MBE or metalorganic MBE (MOMBE) for IR device applications. Since that time enormous progress has been made in reducing defect densities, increasing growth areas, improving the control of alloy compositions and doping levels, and demonstrating increasingly sophisticated and powerful IR electrooptical devices, as discussed in later sections of this chapter. Molecular-beam epitaxy has many advantages over other methods for the epitaxial growth of HgCdTe [32.5]:

1. The MBE growth temperature for HgCdTe is only $\approx 185\,°C$, far below the liquid-phase epitaxy (LPE) and metalorganic vapor-phase epitaxy (MOVPE) (also called metalorganic chemical vapor deposition (MOCVD)) growth temperatures of $\approx 450\,°C$ and $300–400\,°C$, respectively. This offers several great advantages. It greatly reduces impurity outdiffusion from the substrate, it allows the formation of precisely controlled, atomically sharp or graded hetero- or homojunctions, and it allows the growth of SLs.
2. MBE is ideally suited for the deposition of multilayer structures; in addition to having only diffuse junctions, heterostructures grown by LPE usually are limited to two or three layers because different melts are required for the growth of layers having different compositions or dopings. MBE is the only technique suited to the growth of SLs and the only technique allowing in situ growth of CdTe and ZnS passivation layers for devices.
3. MBE growth allows the use of spectroscopic ellipsometry (SE), reflection high-energy electron diffraction (RHEED), and other in situ analytical characterization techniques to monitor and control the growth.
4. Unlike LPE, no aggressive medium is present in MBE growth. Combined with the low growth temperature and analytical facilities for surface control, this allows the use of alternative lower cost, more robust, larger size, more attainable substrates such as CdTe/Si with improved matching of the substrate and readout system thermal expansion coefficients.
5. In MBE no contamination of interfaces occurs because one can grow different compounds in the same growth run in the same system under ultrahigh vacuum.

In 1983 the first photovoltaic device fabricated from MBE-grown HgCdTe was reported [32.6]. By 1985 the quality of MBE-grown HgCdTe layers was comparable to that of the best layers grown by other techniques [32.7]. By 1997 the performance of IR FPAs made from MBE-grown material was comparable to that of those made from LPE materials [32.8]. Although materials-related HgCdTe problems remain, progress in MBE technology continues to outstrip that in more mature techniques such as LPE. Meanwhile, techniques closely related to conventional MBE and at first appearing to offer advantages have largely been abandoned: metalorganic MBE (MOMBE) [32.9, 10] because of its higher growth temperature and because of the ease of carbon contamination and void formation, and photo-assisted MBE (PAMBE) [32.11], which offered higher-quality growth and the as-grown activation of dopants with no self-compensation [32.12], because it was limited to extremely small-area growth.

32.1.3 General Description of the MBE Growth Technique

The reader is referred to [32.13, 14] for detailed and extensive reviews of the physical principles, equipment, and general procedures of MBE growth. Only a very brief general description is given here; specifics of the MBE growth of HgCdTe are discussed in Sects. 32.3–32.7. MBE uses localized beams of atoms or molecules in an ultrahigh-vacuum (UHV) environment as the source of the constituents to the growing surface of a substrate crystal. The UHV environment minimizes contamination of the growing surface. The beam atoms and molecules travel in nearly collision-free paths until arriving either at the substrate or at chilled walls of the chamber, where they condense and are effectively removed from the system. The substrate is kept at a moderately elevated temperature to provide sufficient thermal energy to the arriving atoms for them to migrate over the surface to lattice sites. When a shutter is interposed in a beam, the beam is effectively turned off almost instantly. These features make it possible to grow the films very slowly without contamination, and to change the composition of the arriving stream of atoms very abruptly; in fact, the composition of the flux can be changed in times much shorter than that needed to grow a single atomic layer of the film.

The vacuum environment surrounding the growing crystal must be kept as low as possible to avoid contamination that might affect surface morphology and the electrical properties of the epilayer grown. Because the growth rate is low, $\approx 1-3$ ML/s, to keep the arrival rate of a background species at 10^{-6} of the growth rate would require a partial pressure of $\approx 10^{-12}$ Torr; however, fortunately many of the background species have low reactivities (low sticking coefficients). For III–V growth, the best background pressures achievable are $\approx 10^{-11}$ Torr (except in brand-new growth chambers), and pressures commonly are $\approx 10^{-10}$ Torr; for HgCdTe growth the achievable pressures are ≈ 100 times higher because of Hg contamination. Typically substrates are loaded into the growth chamber via a load-lock chamber, where they are outgassed while the growth chamber remains under vacuum. Then, the vacuum is broken only in the small loading chamber, so only it needs to be evacuated. II–VI and III–V semiconductors should not be grown in the same chamber because the residual pressure from one would dope the other.

The growth chamber must include source ovens, beam shutters and their actuating mechanisms, a substrate holder and heater, in situ growth characterization tools, beam flux monitors, and cryopanels to act as cryopumps and to condense unused beam flux. All of these must be designed to minimize outgassing. The chamber must be designed to keep the substrate at a uniform well-controlled temperature, to provide uniform well-controlled fluxes to the substrate, to avoid temperature transients on the growth surface due to the shuttering and unshuttering of the atomic beams. Variations in the flux ratios and/or the temperature of the growing surface lead to compositional variations and even non-stoichiometry, which can destroy the crystalline quality of the growth. Typically, chambers are custom-designed for each type of material to be grown.

32.2 Theory of MBE Growth

Theories of MBE growth can be classified either as quasi-equilibrium thermodynamic theories or as kinetic growth theories. Although MBE growth typically is a far-from-equilibrium process, quasi-equilibrium theories have been found to yield valuable results even in cases in which desorption is negligible, so that the solid and vapor phases are far from equilibrium with one another. Only a partial equilibrium of the solid phase, not a global quasi-equilibrium, is required for the validity of most quasi-equilibrium theories. On the other hand kinetic theories are necessary to understand important aspects of MBE growth such as the surface morphology, as well as in determining the limits of validity of quasi-equilibrium thermodynamic theories. The structural quality of MBE-grown films depends largely on whether partial, or at least local, thermodynamic equilibrium is established on a time scale appropriate to the growth. If local thermodynamic equilibrium is established, one may use the usual notions of absolute rate theory and the resulting rate expressions for kinetic processes when describing MBE growth [32.15, 16]. However, whether the combined effect of individual kinetic processes is able to move the growing film toward a partial equilibrium state and to reach this state in a time scale relevant to the growth depends on the growth conditions and the specific nature of the material being grown. Under some conditions, local thermodynamic equilibrium may exist during the growth, but the kinetics may not be fast enough to ensure partial equilibrium across the entire growth surface or even macroscopic areas of that surface. The microstructural and even the chemical properties of a film grown under only local equilibrium conditions can be very different from those of the film if it were grown under partial equilibrium conditions.

The basic MBE growth models have been reviewed in detail by *Tsao* [32.14]. The driving forces for MBE growth are the difference $\Delta\mu^\alpha$ between the chemical potential μ^α of the vapor of the α element and its equilibrium chemical potential for sublimation μ^α_{eq}. For many elements desorption is negligible in MBE; on the other hand, the desorption rate of Hg is orders of magnitude higher than its absorption rate in the MBE growth of HgCdTe. This implies that $\Delta\mu^{\text{Hg}} \ll k_\text{B} T$ at the growth temperature. For growth on a planar surface there exist three basic MBE growth modes. First, there is the layer-by-layer, two-dimensional or Frank–van der Merwe growth mode, in which each atomic layer tends to be completed before the next layer is started. Second, there is the three-dimensional or Volmer–Weber mode, in which hillocks and cavities form on the surface. Third is Stranski–Krastanov growth, in which the growth begins as two dimensional and then becomes three dimensional. In fact, MBE growth of HgCdTe occurs on vicinal surfaces, with a well-defined miscut to a surface with low Miller indices. In this mode the growth surface is characterized by an array of steps between smooth terraces. The adatoms preferentially bond at the bottoms of the steps and the crystal grows as the steps advance across the growth surface with a velocity proportional to $\Delta\mu$ for the controlling chemical species (Te for HgCdTe) (step-flow growth).

A phenomenon of particular interest is the self-assembly of islands to form a regular array of quantum dots. To a first approximation, this can be understood as a form of growth in the Stranski–Krastanov regime. According to this point of view, based on general thermodynamic arguments, the deposited material starts to grow as a homogeneous wetting layer, because it has a lower surface energy than the substrate. After a critical thickness, the growth of three-dimensional islands that allow for partial elastic strain relaxation becomes energetically preferable to the two-dimensional film growth. These three-dimensional islands grow until they have reached an optimum size [32.17, 18]. Kinetics enters in this theory only as an external parameter, the island density, which is determined during the nucleation phase and assumed to remain constant during three-dimensional growth of the islands.

32.2.1 Pseudo-Equilibrium Theories

The general principles of the quasichemical thermodynamic theory were presented as early as the 1930s [32.19–24]. The formalism usually used was first given in 1956 [32.25]. The validity of the theory, which rests almost solely on the law of mass action, is unquestioned for thermodynamic equilibrium. However, by definition, thermodynamic equilibrium does not exist during crystal growth. Therefore, it is interesting to consider the past successes and failures of the theory in modeling MBE growth in terms of the degree of equilibrium of the growth. In the theory, one derives from any existing equilibrium condition a corresponding relationship between the probability of occurrence of some configuration of atoms, vacancies, etc. in the solid and the partial pressures of the gases with which the solid is approximately in equilibrium. For approximate global equilibrium during growth two requirements must be met. First, the sticking coefficient of that species must be very small, so that the ratio of the incident flux to the exeunt flux is very nearly

one. Second, any intermediate-state reactions must either be sufficiently fast not to hinder thermodynamic equilibrium or so slow that they take place only after absorption is complete, i.e., in the bulk below the growth surface. These two requirements suffice for the valid application of the theory to MBE growth. However, it has been shown that the application of the theory to MBE growth correctly describes the growth even in the absence of global equilibrium, provided that the second requirement holds.

As early as 1958, equilibrium thermodynamic arguments were used [32.26] to explain the different modes of epitaxial growth in terms of the surface energies at the epilayer–vacuum (σ), the epilayer–substrate (σ_i), and the substrate–vacuum (σ_s) layer interfaces. In partial equilibrium, layer-by-layer growth should be expected when $\sigma_s > \sigma + \sigma_i$, that is, when the change in the surface energy $\Delta\sigma = \sigma + \sigma_i - \sigma_s$ accompanying the deposition process is negative. When $\sigma_s < \sigma + \sigma_i$, or $\Delta\sigma > 0$, the film will grow as isolated islands. Stranski–Krastanov growth takes place when $\Delta\sigma$ changes sign from negative to positive because of strain energy after some characteristic thickness. This theory of course requires the atoms on the growing surface to have a high enough mobility to find the lowest-energy sites to occupy. Thus, it typically is valid only at the high end of the window of desirable growth temperatures.

Many authors have given first-principles thermodynamic treatments of the MBE growth of III–V compounds and alloys in agreement with experiment, even though their MBE growth is far from equilibrium for all chemical species involved. For example, the dependence of the desorption rate of Ga in $Ga_{1-x}Al_xAs$ on x and on the growth temperature was established [32.27], and it was shown [32.28–30] that dopant incorporation in the MBE growth of GaAs occurs in accord with the theory. Also, the growth rates and compositions of a series of III–V compounds and alloys have been calculated, and results in agreement with experiment were found [32.31]. Additionally, the In desorption rate in MBE-grown AlInAs alloys has been calculated as a function of the As beam flux and substrate induced strain, again in agreement with experiment [32.32, 33]. Finally, the surface segregation of base elements and impurities in MBE-grown II–V compounds has been described [32.34].

Of more interest for this review, several authors have applied equilibrium thermodynamic theory to the growth of HgCdTe with useful results. *Gailliard* [32.35] found that MBE growth of CdTe, HgTe, and HgCdTe is correctly described by the theory, and successfully calculated growth rates and compositions, as well as determining the range of acceptable growth conditions, all with no adjustable parameters. *Colin* and *Skauli* [32.36] used the theory to construct a general model for the optimization of the MBE growth conditions for HgCdTe. They calculated an optimal growth temperature as a function of Hg flux, Cd concentration, and growth rate. They also suggested that the growth of HgCdTe is improved by growing on a Te-rich surface, which is now common practice. *Vydyanath* et al. [32.37] used equilibrium thermodynamics to determine the Cd composition in In-doped HgCdTe as a function of the In and Te source temperatures. *Piquette* et al. [32.38] modified the results of Colin and Skauli by forcing the population of Hg atoms on the growth surface to be proportional to the HgTe fractional growth rate, finding a new formula for the optimized growth temperature that experimentally led to consistently lower defect densities. *Chang* et al. [32.39] have used that result to show that all thermodynamically equivalent growth conditions near optimized conditions lie on a straight line with slope -5385 in a graph of $\ln[p_{Hg}/v(1-x)]$ versus T^{-1}, where p_{Hg} is the pressure read by the Hg flux gauge, v is the growth rate, and x is the Cd concentration. Equilibrium thermodynamics has also been applied to study the density of Te-condensation-related voids as a function of Hg flux [32.10] and to determine that As dopants from an uncracked source are incorporated primarily as As_4 molecules [32.40], rather than being dissociated on the growth surface.

32.2.2 Kinetic Theories

The MBE growth of semiconductors is performed at relatively low temperatures and slow growth rates in comparison with LPE and vapor-phase epitaxy (VPE). Low growth temperatures result in growth mechanisms that are strongly dominated by surface kinetics and chemistry. As the source molecular beams are introduced into the MBE chamber some of these molecular species reach the substrate and become physisorbed or weakly chemisorbed to the surface. The adsorbed molecules can then migrate to energetically favorable lattice sites, step edges, nucleated islands, etc. and bond there. A thermodynamic redistribution near the surface layers then establishes the final configuration.

MBE growth depends on the sticking coefficients and surface kinetics of atomic or molecular beams impinging on a suitable substrate. The importance of kinetics in the MBE growth of HgCdTe is demonstrated by the strong dependence of the Hg sticking coefficient [32.41–43] and the quality of growth [32.43, 44] on substrate orientation. One can identify a number of parameters relating to the kinetics of growth; they include the flux F of the impinging species, the sticking coefficient S, the desorption rate R, substrate orientation, substrate temperature, etc. A variety of kinetic rate equation atomistic models [32.45] have been explored for growth from the vapor phase on a planar substrate. In the case of epitaxial HgCdTe MBE growth, surface kinetics place strict boundaries on optimal deposition conditions. It has been found [32.46] that surface kinetics limits the Hg flux, the substrate orientation, and the substrate temperature during MBE HgCdTe growth.

Various kinetics problems have been treated; a few examples follow. A model for the evolution of the profile of a growing interface and a Langevin equation (the Kardar–Parisi–Zhang (KPZ) equation) for the local growth of the profile have been proposed [32.47]. It has been argued [32.48] that the KPZ equation is not the relevant continuum equation for MBE growth at intermediate temperatures, where overhangs and bulk defects can be neglected. A new nonlinear growth equation was suggested as the relevant dynamical equation for ideal MBE growth at intermediate to high temperatures. When desorption and overhangs occur, the KPZ equation remains the relevant equation describing MBE growth.

The effect of finite size on smooth layer-by-layer MBE growth has been examined, and it was concluded that, for finite system sizes, growth can always be smooth as long as the diffusion length is comparable to the system size even though the asymptotic growth may be rough in the thermodynamic limit [32.49]. The heteroepitaxy of InAs on GaAs(001) has been discussed, and first-principles density-functional theory calculations for In diffusion on the strained GaAs substrate have been presented [32.50]. In particular, the effect of heteroepitaxial strain on the growth kinetics of coherently strained InAs islands was considered. The strain field around an island was found to cause a slowing down of material transport from the substrate towards the island, and thus to help achieve more homogeneous island sizes.

32.3 Substrate Materials

As with all epitaxial systems and techniques, the structural quality of HgCdTe epilayers deposited by MBE is inextricably linked to the quality of the substrate. The absence of any substrate material satisfactory in all respects for the MBE growth of HgCdTe is perhaps the primary factor limiting the performance of IR detectors and FPAs. In 1993 *Triboulet* et al. [32.51] compared the basic properties and the advantages and disadvantages of different substrate materials. Here, we summarize and update that discussion and consider the question of substrate orientation. Special attention is paid to growth on CdZnTe and CdTe/Si substrates. For CdZnTe emphasis is placed on the nature of substrate bulk and surface imperfections, on their effect on MBE-grown epilayers, and on methods to ameliorate those effects and screen out bad substrate wafers before growth. For CdTe/Si, emphasis is placed on possible methods to overcome or ameliorate the effects of the large lattice mismatch between Si and CdTe, including new methods being developed or under consideration. Several characteristics of the substrates in common use are compared in Table 32.1.

The first MBE growth of HgCdTe [32.4] was performed on a CdTe(111) substrate; early growths also were performed on CdTe(100) [32.6]. However, even the small lattice mismatch between CdTe and $Hg_{1-x}Cd_xTe$ with $x < 0.5$ caused threading dislocations and other crystalline defects. Also the growth on high-symmetry planes allowed the formation of a high density of microtwins in the HgCdTe [32.6]. The twinning can be eliminated by growth on higher-index planes or miscut surfaces, but bulk CdTe was abandoned as a substrate material because of a combination of factors, including the lattice mismatch, its low thermal conductivity, its poor crystalline quality, its brittleness, and its cost. $Cd_{0.96}Zn_{0.04}Te$ has become the substrate of choice for the growth of the highest-quality HgCdTe epilayers because of its in-principle

Table 32.1 Characteristics of the types of substrates currently used for MBE growth of HgCdTe

	$Cd_{1-y}Zn_yTe$ (y = 0.04)	GaAs	Si
Cost per cm^2	US$ 400 (high-pressure Bridgman growth)	US$ 0.71 (vertical gradient freeze growth) (US$ 28.00 w/CdTe buffer)	US$ 0.56 (float zone growth) (US$ 27.86 w/CdTe buffer)
Largest commercially available size (cm^2)	49	182.4	5228
Robustness of material	Brittle	Moderately robust	Strong
Vickers hardness (kg/mm^2 at 300 K)	60	360	1150–1330
Thermal conductivity (mW/(cm K))	55	500	1235
Surface preparation	Difficult, sometimes poor	Standardized by large III–V industry	Standardized by large Si industry
Substrate surface defect density (cm^{-2})	10^4	5×10^3 (10^5 on CdTe buffer)	10^2 (10^5 on CdTe buffer)
Lattice mismatch with HgCdTe (x = 0.2 at 300 K)	< 1%	13.6% (<1% w/CdTe)	19.47% (<1% w/CdTe)
Substrate etch pit density (cm^{-2})		< 500 on CdTe buffer	< 5000 on CdTe buffer
Thermal mismatch with HgCdTe (x = 0.2 at 300 K)	3.53%	27.04% (1.85% w/CdTe)	51.85% (1.85% w/CdTe)
Valence issues	None	Minor	Severe
Other	Impurities, crystalline defects, compositional uniformity problem, large thermal mismatch with Si ROIC	In situ surface preparation may cause contamination	Allows monolithic integration Requires buffer layer

perfect lattice match to IR HgCdTe, but is plagued with many problems. Composite substrates consisting of epilayers of CdTe or $Cd_{0.96}Zn_{0.04}Te$ grown on InSb, sapphire, Ge, GaAs, and Si have been used because of their greater robustness, much lower cost, higher thermal conductivity, greater lateral uniformity, and larger possible areas in comparison with $Cd_{0.96}Zn_{0.04}Te$ substrates. Of these substrates, two are still in common use: CdTe/GaAs and CdTe/Si. However, the very large lattice and thermal mismatches between the underlying substrate material and the top epilayer in these substrates lead to serious problems. Great progress has been made in solving these and other problems associated with each type of substrate still in common use, but for no substrate have the problems yet been completely solved.

32.3.1 Substrate Orientation

Substrate orientation is of primary importance in MBE growth, as was recognized very early for growth on CdTe substrates by noting the strong dependence of the Hg sticking coefficient on orientation [32.41–43] and as is seen from the strong dependence of the quality of growth on substrate orientation [32.43, 44]. It was recognized from the beginning that growth on a low-index CdTe plane such as the (100) or (111) planes led to microtwinning. On alternative substrates such as GaAs or Si growth on the (100) plane led to twinning and double domains rotated by 90° from one another due to the equivalence of different CdTe orientations on such planes. Two approaches were taken to reduce the symmetry of the substrate surface and thus eliminate this problem:

1. To slightly miscut the substrate away from a low-index surface plane [32.52–57]
2. To grow on higher-index planes containing steps between (100) or (111) flat terraces [32.58–61].

Today almost all growth is carried out on (211) substrates. However, the question of the preferred substrate orientation is not as clear for lattice-matched substrates as for alternative substrates. A brief review of the results for growth on substrates with various orientations is given here. In 1988 a study [32.46] of growth on (100), (111)B, (211)A, and (211)B substrates revealed that (211)B growth gives the best surface morphology, suppresses twinning, and gives higher Hg sticking coefficients and thus better control of composition than any other orientation studied except (111)B. In 1994 simultaneous growth was conducted [32.62] on (111)B substrates miscut by 4° towards (110) (labeled (776)) and on (211)B-oriented substrates. These orientations were described as (111)B surfaces tilted 19° towards and 4° away from (100), respectively. The primary difference between these two orientations was the width of the (111) terraces between steps: 53 and 10.6 Å, respectively. It was found that the (776) orientation with broader terraces was more susceptible to twinning at lower growth temperatures. In 2000 it was found [32.63] that growth on (211)B and (311)B surfaces under optimized conditions gives material of comparable quality. Recently, simultaneous MBE growth was conducted [32.44] in the (511)B, (311)B, (211)B, (111)B, and (255)B orientations tilted 15.8, 25.2, 35.3, 54.7, and 74.2° away from the (100) surface in the (011) direction. Over a nominal range of growth temperatures from 180 to 187 °C the x-ray rocking curve full-width at half-maximum (FWHM), and etch pit density (EPD) were measured for all samples, and surface defect densities were measured using Nomarski microscopy. Under the best growth conditions comparable results were obtainable for all orientations, although the (255) orientation allowed the lowest-temperature growth.

32.3.2 CdZnTe Substrates

A golden rule of epitaxial growth is to match the lattice parameter of the substrate to that of the epitaxial layer in order to avoid strain and dislocation formation in the epilayer. Although, because of its lattice match to HgCdTe, bulk CdZnTe is the substrate of choice for the MBE growth of the highest-quality material, especially for LWIR and VLWIR detection, it suffers from both intrinsic limitations and quality concerns, as listed in Table 32.1. Many of the intrinsic limitations – the very high cost, limited area, brittleness and large thermal mismatch with the Si readout integrated circuit (ROIC) in devices – do not affect the quality of MBE growth but do impact the practicality of use of the HgCdTe layers grown for device applications. The poor thermal conductivity and quality concerns – crystalline defects, impurities, nonuniform composition, surface roughness, imperfect surface flatness, and stoichiometry – all impact the quality of the HgCdTe grown on the substrates. Furthermore, all are amplified in importance by great variations in quality from substrate to substrate. These variations introduce great variations in HgCdTe material and thus both significantly lower yield in device production and make it difficult to identify optimal growth conditions; see, for example, [32.44]. For those reasons it is desirable to develop and use methods to screen substrates before the MBE growth of HgCdTe.

Effects of Poor Thermal Conductivity on HgCdTe Growth

Because the growth is highly noncongruent, the temperature window for the optimal growth of HgCdTe is very narrow, and the composition obtained for a given flux ratio is strongly temperature dependent. However, it is very important to have compositional and crystalline uniformity across HgCdTe epilayers to be used for FPAs. Thus, it is very important to keep the temperature as nearly uniform as possible across a growing HgCdTe layer. On the other hand, in MBE growth the outside edges, and especially the corners, of a growing epilayer become hotter than the central region. For that reason, it is highly desirable to use a substrate with a high thermal conductivity to minimize thermal gradients across a growing HgCdTe epilayer so as to minimize compositional gradients and pixel-to-pixel cutoff wavelength gradients in FPAs.

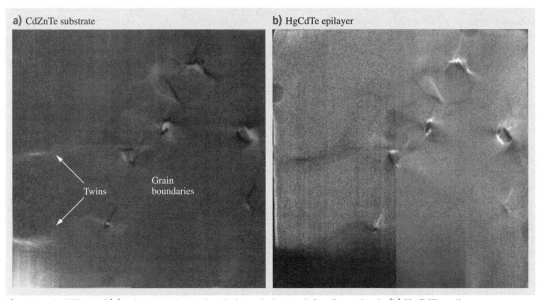

Fig. 32.1a,b Effects of (a) substrate twins and grain boundaries on defect formation in (b) HgCdTe epilayers

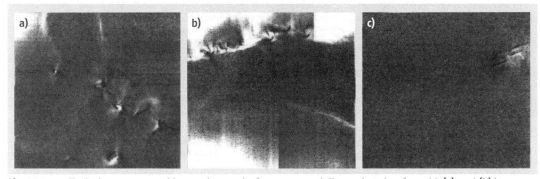

Fig. 32.2a–c Typical x-ray topographic mapping results from commercially purchased wafers with (a) and (b) large-area surface crystalline imperfections and (c) a moderate area of surface crystalline imperfection

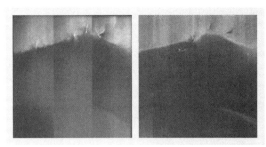

Fig. 32.3 X-ray topographic mapping results from HgCdTe epilayers grown on the substrates shown in Fig. 32.2a,b again showing that the defective features on the substrate are transferred to the HgCdTe epilayer

Effects of Substrate Crystalline Defects on HgCdTe Growth

Irrespective of the choice of substrate material, the crystalline quality of an MBE-grown HgCdTe epilayer depends on the crystalline quality of the substrate on which it is grown. The absence of a large quantity of substrate structural defects is a necessary, although not sufficient, condition for the MBE growth of a HgCdTe layer of high structural quality. The effects of substrate twins and grain boundaries on defect formation in HgCdTe epilayers is strikingly shown in Fig. 32.1; the HgCdTe surface defects form an exact map of the underlying CdZnTe defects. Of 12 commercially obtained substrates recently screened at the US Army Night Vision and Electronic Sensors Directorate, five showed large-area crystalline imperfections, as shown in Fig. 32.2a,b, three showed moderate areas of crystalline imperfections, as shown in Fig. 32.2c, and only four showed reasonably good surface crystalline perfection as revealed by x-ray topographic mapping, FWHM maps, and Fourier-transform IR (FTIR) transmittance maps. Results on others of these samples are given by *Carini* et al. [32.64]. When HgCdTe epilayers were grown on these substrates, the defective structures were transferred to the epilayers as shown in Fig. 32.3. Also, a high density of threading dislocations was found in the epilayers above the imperfections and very short carrier recombination lifetimes were measured [32.65]. These threading dislocations also can lead to microtwinning by allowing nucleation on the (111) terraces during growth as well as at the bottom of steps, thus giving three-dimensional rather than step-flow growth. This in turn leads to surface hillocks such as those shown in Fig. 32.4. More recently it was reported [32.44] that some substrates of a recently purchased batch had FWHM values of 15–25 arcsec, whereas others had values of 10–12 arcsec, with CdTe epilayers grown on the substrates having FWHM values limited by the substrate values.

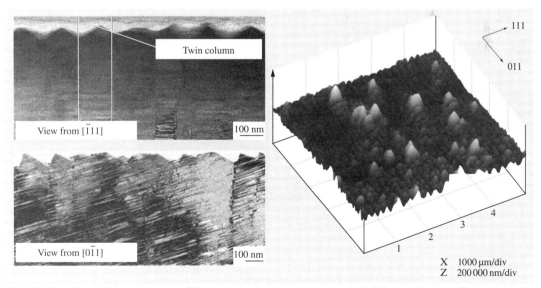

Fig. 32.4 Scanning electron microscopy (SEM) and atomic-force microscopy (AFM) images showing hillocks due to the formation of adjacent columns of twins

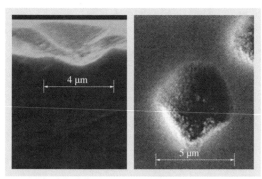

Fig. 32.5 SEM photographs of small voids

Another type of bulk defect found in CdZnTe substrates is Te precipitates. A large density of Te precipitates is unavoidable in as-grown CdZnTe boules because there is a large deviation from stoichiometry at the growth temperature. As the CdZnTe is cooled to room temperature, the excess Te forms precipitates from $<1\,\mu\text{m}$ to $\approx 25\,\mu\text{m}$ in size. Much attention has been paid to elimination of these defects [32.66, 67], yet they remain a problem. These substrate precipitates create dislocation clusters [32.68] and induce Te precipitates [32.66, 69] in epilayers grown on the substrates and crater or void defects on the epilayer surfaces [32.70]. These voids can short p–n junctions and hence blind pixels in FPAs. Scanning electron microscopy (SEM) images of typical voids are shown in Fig. 32.5. The Te precipitates also give rise to high hole concentrations in HgCdTe layers grown by LPE on the substrates [32.69].

Effects of Substrate Impurities

For best performance of LWIR HgCdTe FPAs, it is necessary to minimize low-temperature leakage currents. At temperatures such as 40 K the major sources of dark current crossing the p–n junction are believed to be tunneling and generation in the depletion region. These mechanisms involve trap sites that most likely are due to impurities. Some impurities also act as unwanted dopants. For these reasons, it is necessary to minimize impurities in the HgCdTe layers. For growth on CdZnTe substrates, one origin of impurities is the substrate, from which they diffuse during layer growth [32.71] and device fabrication. A well-known example is Cu [32.72, 73], which diffuses rapidly from substrate to epilayer, has been shown [32.74] to be preferentially attracted to HgCdTe over CdZnTe, and harms devices by shortening the carrier lifetime and converting weakly n-type material to p-type. Methods to extract mobile impurities from CdZnTe before MBE growth [32.75] and to getter and localize Cu at Te precipitates in CdZnTe [32.60] have been proposed, but Cu impurities remain a possible problem.

Effects of Nonuniform Composition

Zn segregation along the growth axis in the vertical Bridgman growth of CdZnTe leads to a gradual decrease in Zn composition from the bottom of a boule to the top, so that different wafers do not have the same lattice constant, allowing only part of a boule to be lattice-matched to a given composition of HgCdTe. More importantly, the Zn composition typically varies by about 1% from the center to the outside of a nominally 4% wafer, and that variation creates a variation in lattice constant equal to that caused by a 0.2% change in the composition of HgCdTe. Although small, that variation can cause a lateral variation in the density of threading dislocations in HgCdTe epilayers grown on CdZnTe substrates that is sufficient to cause significant lateral variations in the responsivity of LWIR and VLWIR FPAs. It also induces cross-hatch defects [32.76, 77] on the surface of MBE-grown HgCdTe epilayers, such as those shown in Fig. 32.6. Assuming CdZnTe and HgCdTe to relax with $a/2\langle 011\rangle$ Burger's vectors which propagate on the $\{111\}$ family of glide planes g, the resultant dislocation lines would be $\{u\} = \{[01\bar{1}], [\bar{2}13], [0\bar{1}1], [2\bar{3}\bar{1}]\}$. Misfit dislocations from the small lattice mismatch

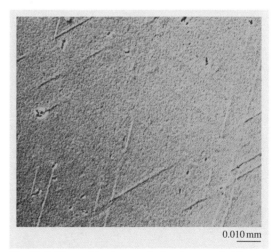

Fig. 32.6 Nomarski micrograph showing crosshatches on defect etched MBE-grown HgCdTe epilayers

and later from the thermal mismatch would allow threading dislocations to propagate on the {111} glide planes and close their dislocations along the *u* directions, resulting in the observed cross-hatch defects. Thus, these cross-hatch defects result from dislocations, but only when the dislocation density is low.

Effects of Substrate Roughness

Substrate surface roughness may be present on commercial substrates as received or may arise from imperfect surface preparation by the MBE grower. AFM surface roughness measurements on HgCdTe layers grown on substrates with smooth or rough surfaces show a direct correlation between substrate surface roughness and the surface roughness of the HgCdTe epilayer (Table 32.2). A high epilayer surface roughness on any lateral scale leads to imperfections in device fabrication and can lead to dead pixels in FPAs. High-frequency substrate roughness on a lateral scale of ≈ 100 Å or less inhibits the mobility of atoms on the epilayer growth surface and thus can lead to three-dimensional growth and poor crystalline quality. It has been observed that, for growth on moderately rough substrates (> 10 Å of roughness as measured by SE), the epilayer surface roughness increases in the initial growth stage [32.78] consistent with three-dimensional growth.

It also has been observed that high densities of elongated *needle* defects oriented in a preferred direction occur on the surfaces of HgCdTe epilayers grown on rough CdZnTe substrates. As shown in Fig. 32.7, the density of these defects was found to increase exponentially with increasing substrate surface roughness as measured by SE [32.78]. However, the MBE deposition of an 80 Å CdTe buffer layer before HgCdTe growth reduces the final needle defect density by an order of magnitude or more, as shown in Fig. 32.7. Needle defects occur only for growth on CdZnTe, their density does not correlate to the CdZnTe bulk qual-

Table 32.2 Correlation of substrate surface roughness and epilayer surface roughness

	Smooth substrate (Å)	Rough substrate (Å)
Substrate surface roughness	8	50
HgCdTe epilayer surface roughness	8	58

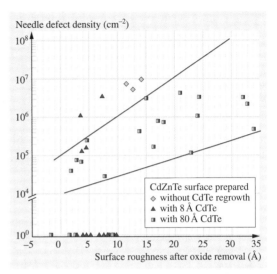

Fig. 32.7 Plot of the HgCdTe needle defect density versus surface roughness as measured by in situ SE with and without CdTe buffer layers. The *straight lines* are upper and lower limits of the defect density with an 80 Å CdTe buffer layer

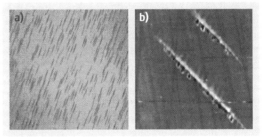

Fig. 32.8 (a) A Nomarski picture (1000×) of needle defects on a HgCdTe epilayer and (b) an AFM image of two needle defects (5 μm × 5 μm)

ity, and their connection to the surface roughness is not well understood. It also is not known how they affect device quality or how they correlate with surface preparation technique. It is possible that they could be eliminated by using plasma etching rather than wet chemical etching followed by thermal desorption of contaminants, the common practice. The length of the needle defects varies from a few micrometers to > 10 μm in different samples, their width is only 0.2–0.3 μm, and their height is ≈ 100 Å. Figure 32.8 shows (a) a Nomarski picture (1000×) of needle defects on a HgCdTe epilayer and (b) an AFM image of two needle defects. The HgCdTe epilayers grown on these substrates also were characterized by temperature-dependent, photoconductive-decay lifetime data. Fits to the data indicate the presence of mid-gap recombination centers, which were not removed by 250 K/24 h annealing under a Hg-rich atmosphere. The strong correlation observed between the substrate roughness measured by SE and the density of these centers leads us to believe that they originate from bulk defects related to the substrate roughness rather than from Hg vacancies.

Effects of Substrate Surface Curvature

Substrate surface curvature can have two deleterious effects. First, any deviation in surface flatness $\approx 0.5\,\mu$m or more can give problems in bonding to the flat Si ROIC surface for device fabrication. Second, the curvature will yield a surface orientation which is not constant across the wafer; that, in turn, may lead to epilayer MBE growth problems for a substrate which is only slightly miscut off a principal crystal axis, although it should have no effect on the usual growth on (211) surfaces.

Effects of Surface Nonstoichiometry and Contaminants

The surfaces of substrates must be prepared for MBE growth, i.e., polished and cleaned. The surfaces must be atomically clean, ordered, and have, at least approximately, the stoichiometry of the underlying wafer. The surfaces of commercially obtained wafers typically are covered with ≈ 1.5 ML of contaminants (mostly oxygen, carbon, and chlorine) [32.79]. The usual wet chemical preparation of the surface typically employs a 0.5% bromine in methanol solution that leaves behind an amorphous Te layer and residual carbon contamination, both of which must be removed for satisfactory nucleation of HgCdTe growth, although improved chemical polishing and cleaning procedures are under investigation [32.80, 81]. The substrate is then heated to 300–340 °C to remove the amorphous Te and residual contaminants such as carbon and oxygen to obtain a crystalline surface. This surface preparation procedure has been shown to leave $\approx 1/4$ of a monolayer of carbon as well as either excess Te or excess Zn [32.79, 82]. Incompletely removed amorphous Te leads to three-dimensional growth and microtwinning, and residual contaminants degrade the electrical properties of any HgCdTe epilayer grown on the substrate. Alternative wet chemical treatments lead to similar problems. An alternative electron cyclotron resonance Ar/H$_2$ plasma cleaning technique has been shown to leave the CdZnTe surface free of contaminants and with a roughness of only 4 Å as measured by interferometric microscopy [32.79]. However, HgCdTe epilayers grown by MBE on plasma-cleaned substrates were not found to be noticeably better than those grown on substrates cleaned by conventional wet chemical methods followed by annealing [32.79].

Characterization and Screening of CdZnTe Substrates

Many experimental methods have been used to characterize the crystalline quality, purity, and surface characteristics of CdZnTe substrate wafers. Poor crystalline quality – the presence of grain boundaries, low angle microcrystals, microtwinning, precipitates, and voids – can be examined in detail using cross-sectional SEM and TEM; however, those measurements are expensive, destructive, and can only be used to image microscopic sections of a substrate. X-ray topographic mapping measurements, which produce images such as those shown in Figs. 32.1 and 32.2, also require very expensive equipment and are not as detailed, but are nondestructive and capable of mapping an entire substrate wafer [32.64]. More detailed structural information, such as small variations in lattice spacing and the existence of regions of the substrate tilted only 0.011°, can be obtained by reciprocal-space mapping (RSM) and triple-axis x-ray diffractometry [32.83]. A simple gross quantitative measure of the substrate crystal structural perfection is given by the full-width at half-maximum (FWHM) of the x-ray double crystal rocking curve and by the existence or nonexistence of multiple peaks. The IR transmittance at wavelengths $< 6\,\mu$m [32.64] and χ, the ratio of channel counts to random counts in Rutherford backscattering spectroscopy [32.66], also give qualitative measures of the crystal structural perfection, with the IR transmittance being higher and χ being smaller for better crystals.

Individual Te precipitates and other extended defects can be seen by relatively simple nondestructive optical measurements: IR transmission microscopy [32.66, 69, 70], light-scattering IR tomography [32.84], and scanning photoluminescence (PL) [32.84]. Also, the presence of Te precipitates or excess Te atoms is revealed by an A(Te) peak in the Raman scattering spectra at $\approx 127\,\text{cm}^{-1}$ [32.67] and IR transmittance gives at least a qualitative measure of the density of Te precipitates, falling off rapidly with increasing wavelength for wafers having a high density of Te precipitates [32.67] A general rule of thumb is that, for wavelengths well above the cutoff wavelength, the transmittance should be > 63% in the absence of antireflection coatings. Reflection high-energy electron diffraction (RHEED), which is installed on all HgCdTe MBE growth chambers and is discussed in Sect. 32.5, gives yet another nondestructive qualitative description of the substrate crystalline perfection. Etch pit density (EPD) measurements reveal threading dislocations and Te precipitates at the substrate surface and are nondestructive except for the requirement of a surface etch, usually either an Everson [32.85] or a Nakagawa etch [32.86]; EPDs in the mid 10^4 cm^{-2} are desirable.

Quantitative measurements of substrate impurity dopant concentrations require destructive techniques. The most common are glow-discharge mass spectroscopy [32.73, 75, 87], sputter-initiated resonance ionization spectroscopy [32.70, 74], and secondary-ion mass spectroscopy (SIMS) [32.67, 88–90]. However, a qualitative measure can be obtained from the IR transmittance; for high impurity dopant levels the transmittance decreases by an amount proportional to $\ln(N_D - N_A)$ as the wavelength increases in the LWIR and VLWIR [32.84].

The Zn concentration and its lateral variations usually are measured by FTIR or spectral photometry, measuring the cutoff in the FTIR transmission in the near IR, because this is a very simple nondestructive measurement. However, the shape of the FTIR transmission curve is affected by crystalline imperfections, in particular by Te precipitates, in a way that can make the determination of the Zn concentration uncertain by as much as ± 0.015 [32.81]. It can be determined somewhat more accurately by SE using a high-resolution ellipsometer such as the M2000, as discussed in Sect. 32.5. A much more accurate method of determining the Zn concentration and its lateral variation is low-temperature PL mapping. Measurement of the reabsorption dip at the free exciton energy is capable of determining the Zn concentration within a precision of ± 0.0005 [32.81, 82]. The depth variation of the Zn concentration also can be measured, but only by destructive techniques: Auger electron spectroscopy (AES) or atomic absorption spectroscopy (AAS) on beveled edges or vertical slices of a substrate [32.88].

The surface roughness can be nondestructively and easily measured directly by Nomarski microscopy, Zygo interferometry, and AFM with increasing resolution but over decreasing areas, and indirectly by SE, as discussed in Sect. 32.5. Surface contamination also can be measured nondestructively by x-ray photoelectron spectroscopy (XPS), directly by measuring the photoelectron peak intensities corresponding to the different possible contaminants or indirectly by measuring the decrease in the intensity of the CdTe peak [32.82]. In screening substrates, the substrate should first be thermally cleaned in vacuum in the XPS chamber and then transferred in vacuum to the growth chamber after the XPS measurement. AES also can be used to verify the absence of contaminants on substrate surfaces [32.79] but is not as easily used to measure amounts of contamination.

Use of Buffer Layers on Substrates

The use of buffer layers to block the propagation of defects originating in substrates or at the substrate–epilayer interface has proven useful and is being optimized. The various commercial manufacturers of FPAs use buffer layers in the growth of HgCdTe, but the details of their growth procedures remain proprietary. Thin interfacial CdTe layers have proven effective in reducing the surface roughness, evening out variations in strain, and improving the quality of HgCdTe layers grown on the substrates [32.78]. However, in order to be fully effective in smoothing out the surface and reducing the dislocation density, these layers must be grown to a thickness exceeding the critical layer thickness, so that the lattice match to HgCdTe is lost. This suggests the use of HgTe/CdTe SLs as buffer layers. They are more effective at blocking dislocations and can be grown more nearly lattice matched to the HgCdTe epilayer to be grown. Cross-sectional TEM micrographs such as those shown in Fig. 32.9 show that the substrate microscopic roughness is smoothed out after several periods of SL growth (a) and that threading dislocations terminate (b) or bend (c) at the HgTe/CdTe interfaces in the SL, despite the small (0.32%) lattice mismatch between HgTe and CdTe which would normally be considered insufficient to induce the motion of threading dislocations [32.91, 92]. After growth, the SLs are interdiffused by annealing to produce a short-wavelength IR (SWIR) HgCdTe layer which serves as a window for backside-illuminated devices and typically has a lattice mismatch of $\leq 0.09\%$ with the final HgCdTe epilayer to be grown. It has been shown [32.91, 92] that the use of HgTe/CdTe SL buffer layers leads to reproducibly high-quality MBE-grown HgCdTe layers with EPDs $\approx 10^5$ cm^{-2} with values as low as 4×10^4 cm^{-2} and minority-carrier lifetimes close to the theoretical limits for perfect material. It also has been found that, even aside from the improvement of material characteristics, the presence of an SWIR buffer layer improves device quality by reducing the back-surface recombination velocity.

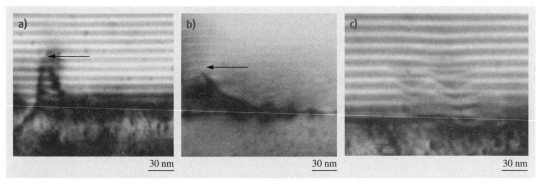

Fig. 32.9a–c Cross-sectional TEM micrographs showing the (**a**) bending and (**b**) termination of threading dislocations by HgTe/CdTe SL buffer layers, and (**c**) the smoothing of substrate roughness by SL buffer layers

32.3.3 Si-Based Substrates

There has been great interest in alternatives to CdZnTe as a substrate for the reasons outlined in Table 32.1 and detailed in Sect. 32.3.2 above. Of the many possible alternative substrate materials for the MBE growth of HgCdTe, Si has captured the greatest interest. HgCdTe was first grown on Si substrates in 1989 [32.93], and the growth of HgCdTe on Si having properties as good as those of HgCdTe grown on CdZnTe was reported in 1995 [32.94]. Although many factors favor Si over all other possible substrate materials, one factor in particular singles out Si uniquely – only Si offers the possibility of fabricating truly monolithic detectors and FPAs. Because the readout integrated circuit (ROIC) in all IR devices is based on Si, the use of any material other than Si for the substrate necessitates a hybrid-array technology wherein the detector and ROIC arrays are fabricated on separate manufacturing lines and subsequently indium-bump bonded to each other pixel by pixel. The functions of detection and charge readout would be monolithically integrated onto a single Si wafer in the ultimate IRFPA. This has been demonstrated by the epitaxial deposition of HgCdTe in selected areas of a Si ROIC [32.95].

Aside from the ultimate goal of monolithic integration, Si-based substrates offer many short-term benefits over all other possible substrate materials. First, Si wafers are commercially available in far larger sizes. This allows either much larger focal planes for greater resolution or greater numbers of smaller FPAs for reduced production costs. Second, in conjunction with that advantage, Si offers perfect thermal matching of its lattice to that of the ROIC, and thus no thermal mismatch limit on the size of FPAs. With other substrates, the thermal mismatch between substrate and Si ROIC creates a strain upon cycling down to cryogenic operating temperatures which could destroy the bump bonding for very large-area FPAs. Third, Si substrates allow the use of automated Si processing technology for increased efficiency in device fabrication. Fourth, Si is the most robust and durable possible substrate material, leading to less breakage and a higher device yield. Fifth, Si has the highest thermal conductivity, leading to the highest lateral uniformity of MBE-grown HgCdTe composition and thickness. Sixth, Si offers the lowest level of impurity migration into the HgCdTe. Finally, it has by far the lowest density of surface defects, although that advantage is overwhelmed by the problems associated with the large lattice and thermal mismatches between Si and HgCdTe. In summary, in every way that CdZnTe is a poor substrate, Si is clearly the best.

On the other hand, Si has the worst lattice and thermal mismatches with HgCdTe of any substrate material considered. The lattice parameter of HgCdTe ($\approx 6.46\,\text{Å}$) is $\approx 19\%$ larger than that of Si ($\approx 5.43\,\text{Å}$). The lattice mismatch is so extreme that HgCdTe does not even grow epitaxially on Si, so that a buffer layer is required. The buffer layers chosen have been $Cd_{0.96}Zn_{0.04}Te$ [32.96], CdSeTe [32.97], CdZnSeTe [32.98], and CdTe. Also, a first buffer layer of ZnTe, originally found to be necessary to preserve the substrate orientation [32.93], usually is grown before the deposition of the final buffer layer. Although $Cd_{0.96}Zn_{0.04}Te$ and CdZnSeTe buffer layers offer the advantage of an almost perfect lattice match to the HgCdTe to be grown, they have largely been abandoned because of the difficulty of obtaining epilayers of sufficient crystalline quality; the best values obtained for the x-ray diffraction (XRD) FWHM for lattice match to HgCdTe have been > 150 arcsec. Even CdTe buffer layers have a relatively poor crystalline quality because of their large lattice mismatch with Si. Even with special growth procedures involving the predeposition of ZnTe films on As-terminated Si surfaces [32.99, 100], CdTe buffer layers exhibit etch pit densities of $2 \times 10^5 - 5 \times 10^5$ cm^{-2}, indicative of threading dislocation densities of the order of $2 \times 10^5 - 5 \times 10^5$ cm^{-2}. This in turn leads to dislocation densities in the mid 10^6 cm^{-2} in the HgCdTe layers grown on these substrates by MBE. Although otherwise excellent material quality has been achieved [32.101–104], little success was achieved in lowering the dislocation density in HgCdTe grown on Si substrates without undesirable high-temperature cyclic annealing.

For dislocation densities $< 10^5$ cm^{-2} the dislocation density has little effect on the minority-carrier recombination time for $T > 77$ K, even for LWIR HgCdTe, because carrier recombination is dominated by intrinsic Auger mechanisms. However, for dislocation densities above the mid 10^5 cm^{-2} range and $T \leq 77$ K, carrier recombination is dominated by Shockley–Read–Hall (SRH) recombination due to the dislocations, and the minority-carrier recombination time for LWIR HgCdTe for is roughly inversely proportional to the dislocation density [32.105]. Also, dislocations can locally short out p–n junctions in devices, considerably increasing the dark current [32.106]. The MBE growth of HgCdTe on CdTe/Si substrates yields SWIR and mid-wavelength infrared (MWIR) photodiodes having performances similar to those grown on CdZnTe substrates at operating temperatures above 77 K [32.102, 103, 107, 108]. However, to date it has been incapable of yielding LWIR or VLWIR devices having similar performances (similar R_0A values) to the best devices grown on CdZnTe substrates [32.103, 109], except for devices with very small junction areas ($< 180\,\mu\text{m}^2$) [32.110] or for high background operation at temperatures ≥ 77 K [32.104]. LWIR FPAs fabricated from HgCdTe grown on Si give near-ideal performance even at 40 K if the junction sizes are sufficiently small to have no more than ≈ 5 threading dislocations passing through each junction area. This limits the ideal pixel pitch for LWIR FPAs to pixel sizes $< 25–30\,\mu$m [32.110].

There exist in principle two general ways to ameliorate the effects of the large lattice and thermal mismatches between Si and HgCdTe:

1. Reduce the as-grown dislocation density
2. Passivate the dislocations so as to reduce their effect on device properties.

Buffer Layers to Reduce the As-Grown Dislocation Density

If the as-grown dislocation density could be reduced by an order of magnitude, growth on composite Si-based substrates would be the clear choice for the MBE growth of HgCdTe. The use of improved buffer layers is expected to yield quantitative, but not qualitative, improvements. The authors have briefly explored the use of strained-layer CdTe/ZnTe SLs to lower the dislocation density and improve the crystal quality of the final CdTe buffer layer, but with only moderate success. Also, the growth of CdSeTe buffers on top of the usual CdTe buffers has been suggested. More recently, preliminary results [32.111] have shown that nanometer-thick perovskite oxide buffer layers tend to smooth the surface of the final CdTe buffer layer and may offer a promising approach to reducing the dislocation density. In particular, under certain growth conditions, an amorphous silicate layer forms between the perovskite and the Si, making the perovskite a potentially compliant buffer and thus potentially greatly reducing the dislocation density for growth on Si [32.112]. Another approach [32.113] would be the epitaxial growth of insulating Be chalcogenides and then BeCdTe or BeMgTe with a graded lattice parameter, initially lattice matched to Si (BeSe$_{0.45}$Te$_{0.55}$) and ending with a layer lattice matched to CdTe or HgCdTe. These layers could be grown in the same MBE reactor as a CdTe buffer layer, and the final HgCdTe epilayer would be grown in as part of one continuous growth run. They should grow with relatively few defects because the Be chalcogenides have very high stacking fault energies – more than ten times that of ZnS, which suppress the formation of dislocations. Also, the Be chalcogenides have a much higher thermal stability than CdTe or ZnTe. Alternatively, one could just use a single BeTe layer, which shows [32.114] good two-dimensional growth on Si:As and has an intermediate lattice constant. However, Be would be a very difficult contaminant to remove from the MBE growth chamber.

Thermal Cycling to Reduce the Dislocation Density

It has been shown that the thermal cycling of HgCdTe grown by MBE or MOCVD on CdTe/GaAs and CdTe/Si substrates can greatly reduce the HgCdTe dislocation density. Dislocation densities as low as 2.3×10^5, almost as low as those obtained for growth on CdZnTe, have been obtained by thermal cycling between 300 and 490 °C [32.115]. The reduction in the

HgCdTe dislocation density was accompanied by an increase in the CdTe dislocation density by two to three orders of magnitude. This effect was explained as due to the large thermally induced stresses between the CdTe and the Si or GaAs. The problem with this approach is that high-temperature annealing can cause substantial diffusion and destroy the sharpness of the hetero- and homojunctions required by many modern detector designs.

Growth on Si-Based Compliant Substrates

Growth on compliant substrates was first suggested by Lo [32.116]. Compliant substrates either take up most of the strain associated with a lattice mismatch or getter the dislocations formed at the epilayer–substrate interface. A compliant substrate is made much thinner than the epilayer grown on it and is free to slide against a thicker *handle* substrate. The most common example of a compliant substrate is a Si-on-insulator (SOI) structure with a Si layer ≈ 10 nm thick floating on an SiO_2 layer on a Si handle. Modification of the relaxation behavior in lattice-mismatched films due to growth on SOI compliant substrates is well documented. TEM measurements for SiGe grown on such films have shown dislocation densities reduced by more than five orders of magnitude [32.117]. AFM measurements have shown a factor of four reduction in surface roughness, and the XRD FWHM was shown to be reduced from 270 to 155 arcsec [32.118]. It has been shown that in this case the compliant thin Si layer does not deform elastically to absorb the strain due to the lattice mismatch, but rather facilitates a net downward image force on the dislocations so that they grow into the thin Si layer rather than the SiGe epilayer [32.117, 118]. In preliminary studies CdTe has been grown on an SOI substrate 20 nm thick. As opposed to the tensile strain found for growth on the usual Si substrates, the CdTe was found to be under compressive strain due to the difference between its thermal contraction and that of Si during the cooldown from its growth temperature to room temperature. However, the fact that CdTe has lower energies of formation for stacking faults and dislocations than does SiGe will make it more difficult to grow dislocation free CdTe on an SOI substrate. $Ge/SiO_2/Si$ would be an attractive alternative compliant substrate because Ge is more nearly lattice matched to CdTe than is Si and has even been used as a bulk substrate for the growth of HgCdTe. This substrate and a method for its growth are suggested in [32.113].

Selective-Area Growth on Patterned Substrates

There are two basic approaches to patterned heteroepitaxial growth. The first involves selective-area growth on a prepatterned substrate followed by lateral epitaxial overgrowth (LEO). The second involves the standard planar growth of a buffer layer followed by patterning of the buffer, annealing and then LEO. Both have been used with great success. The first approach has been used with MOCVD to grow GaAs on Si (4% lattice mismatch) [32.119] and GaN on Si (22% lattice mismatch) [32.119–121], as well as CdTe on Si (19% lattice mismatch) [32.122, 123], with dislocation densities much lower than those for planar growth in all cases. Patterned growth and LEO by MBE have been published only for GaAs on a patterned GaAs substrate [32.124] and for GaN on patterned Si [32.125] and has not been considered possible for CdTe. However, using nanopatterning with a CdTe seed layer, Bommena now has achieved the selective growth of CdTe on Si and LEO with MBE. Although much remains to be demonstrated, this work may open the door to finally overcoming the lattice mismatch between Si and CdTe. If nanopatterning does not sufficiently reduce the dislocation density at the CdTe surface, a further reduction could be achieved by patterning compliant SOI rather than Si.

The second approach has been used to remove almost all threading dislocations from patterned ZnSe epitaxial layers grown on GaAs by MOCVD by bending them over along glide planes to the pattern sidewalls [32.126, 127]. The activation energy for the glide process (≈ 0.7 eV) requires a high-temperature anneal, 400–600 °C, making this approach less desirable for MBE growth of CdTe.

Passivation of Dislocations

A possible alternative to achieving an order of magnitude reduction in the dislocation density would be to achieve a partial reduction in the dislocation density and attempt to passivate the remaining dislocations, i.e., eliminate the trap states normally introduced by dislocations. The hydrogen passivation of both shallow and deep levels has been well studied in many major semiconductors, particularly Si and GaAs. Perhaps, its most prominent application is in α-Si:H solar cells. Notably, it has been found to significantly reduce leakage currents through dislocation cores in GaN [32.128]. Atomic hydrogen has been found to passivate shallow donors and acceptors in virtually all semiconductors. The exact mechanisms by which hydrogen passivates deep levels are poorly understood, in part due to the lack of a clear understanding of the microscopic nature of many deep levels. The passivation of shallow and deep levels in CdTe and has received some attention; hydrogen was found to be an effective passivant when introduced in either molecular or atomic form, improving both electrical and optical properties [32.129–131].

The effects of hydrogen in HgCdTe still are not clear. Early research on the hydrogen passivation of HgCdTe involved electrochemical methods of introducing hydrogen into $Hg_{0.5}Cd_{0.5}Te$, followed by deep-level transient spectroscopy to determine its influence on deep levels [32.132, 133]. The dominant deep levels at approximately $E_g/2$ and $3E_g/4$ were significantly reduced in concentration after hydrogenation. Unfortunately, annealing at $70\,^\circ C$ restored their activity. Two more recent papers [32.134, 135] examine the effects of the hydrogenation of mercury-vacancy-doped MWIR and LWIR HgCdTe. Reference [32.134] concludes that electron cyclotron resonance (ECR) plasma hydrogenation is effective in passivating surface trap states, whereas [32.135] concludes that its only role is to passivate mercury vacancies.

Recently [32.136], attempts have been made to passivate the dislocations in HgCdTe grown on Si with hydrogen introduced via an ECR plasma. The dangling bonds formed along the dislocations give rise to states in the bandgap that act as Shockley–Read–Hall recombination centers. However, the attachment of hydrogen to the dangling bonds would be expected to eliminate the gap states, forming bonding states in the valence band and antibonding states in the conduction band. Hall mobility and minority-carrier lifetime data indicated that the incorporated hydrogen did passivate both scattering and recombination centers. It also was found that dislocations act as atomic diffusion channels both for outdiffusion, observed during annealing, and for indiffusion during ECR hydrogenation. Therefore it is reasonable to conclude that dislocations would be among the first type of defects to be passivated when samples are exposed to ECR plasmas, and that this passivation of dislocations was primarily responsible for the observed increases in mobilities and minority-carrier lifetimes. Preliminary stability studies indicated that the beneficial effects of hydrogenation remain after 3 months shelf storage and after heating to $80\,^\circ C$.

Te and As Monolayers on Si
The growth of an As monolayer on Si before the growth of a CdTe or ZnTe buffer layer has been shown to be crucial [32.137]. First-principles density-functional calculations of Te adsorption on Si(100) [32.138] and pseudopotential density-functional calculations of the electronic and atomic structure of monolayers of Te and As on Si(211) [32.139] have been performed. Te-covered surfaces were found not to have any definitive reconstruction, explaining conflicting experimental results for the reconstruction [32.138]. Both As and Te surfaces were found to be metallic [32.139]. More importantly, the effects of Te passivation and As passivation of the Si surface before CdTe growth have been compared experimentally [32.137]. The CdTe film grown on a substrate treated with a Te flux was found to exhibit a rough film–substrate interface and to have very poor crystalline quality with a $\approx (111)A$ orientation. In contrast, the CdTe film grown under identical conditions except that the Si substrate was treated with an As flux was found to have an atomically abrupt film–substrate interface and a single-domain structure in the technologically preferred (111)B orientation. On the As-passivated surface the CdTe growth is initiated with Cd atoms having one bond to an As atom and three bonds with Te atoms from the incoming Te flux, producing an initial smooth Te stabilized growth surface. Recently, the structure of the As and Te atoms on As-passivated Si(211) surfaces after the early stages of ZnTe growth has been investigated [32.140] by x-ray photoelectron spectroscopy (XPS) and ion-scattering spectroscopy (ISS). It was found that the Si(111) terraces are completely covered by a monolayer of As, but that the step edges were left free to be covered by the first layer of Te atoms, allowing optimum nucleation of either a ZnTe or a CdTe buffer.

32.3.4 Other Substrates

As seen in Table 32.1, the characteristics of GaAs are intermediate between those of CdZnTe and Si. GaAs was the first substrate commonly used as a substitute for CdTe, is still the most commonly used substrate for the LPE growth of HgCdTe, and is still used for the MOMBE and MBE growth of HgCdTe by some groups. Although the lattice mismatch between GaAs and HgCdTe is 13.6%, high-quality HgCdTe can be grown on CdTe/GaAs substrates. However, growth on GaAs does limit the growth area to $\approx 1/30$ of that allowed for growth on Si, does not allow monolithic integration, and allows the outdiffusion of Ga and As into the HgCdTe, which can seriously affect the electrical properties of the HgCdTe. To bring the diffused Ga and As in the HgCdTe down to background levels, one must grow either CdTe/ZnTe SL buffer layers [32.51] or thick (3–8 μm) CdTe buffer layers [32.51, 141] The dislocation density of HgCdTe grown on CdTe/GaAs is similar to that of HgCdTe grown on CdTe/Si : As.

Ge also is still used as a substrate for the MBE growth of HgCdTe with CdTe buffer layers. As in the case of Si, large-area Ge substrates are available. Two-color MWIR FPAs fabricated from HgCdTe grown on Ge substrates displayed responsivities, noise characteristics, and operabilities matching those of detectors fabricated from HgCdTe grown on CdZnTe substrates, although the HgCdTe grown on Ge substrates was of lower quality [32.142, 143]. Because Ge is a more homogeneous substrate than CdZnTe, the XRD FWHM of the HgCdTe grown on CdTe/Ge was more uniform; however, it was $\approx 50\%$ larger than that of the HgCdTe grown on the better CdZnTe substrates. Also, the dislocation density was two orders of magnitude higher in the HgCdTe grown on CdTe/Ge, indicating that CdTe/Ge substrates would not be suitable for LWIR or VLWIR material. In summary, Ge would appear to have most of the advantages and disadvantages of Si as a substrate material, with easier to prepare growth surfaces, but without quite as large a possible area, quite as high a crystal quality, or the great potential advantage of being used for monolithic detectors.

In the past, sapphire also has been extensively used as a substrate for the MBE, VPE, and LPE growth of HgCdTe and the fabrication of MWIR photodiodes [32.144], although it is no longer used for MBE growth. It has high crystalline quality, high electrical resistivity and thermal conductivity, and low cost, and is robust. However, like GaAs, sapphire substrates are available only with areas $\approx 1/30$ of that allowed for growth on Si. Also, CdTe buffers grown on sapphire exhibit a high density of dislocations, although that density is rather uniform, allowing the fabrication of highly uniform devices. Also, sapphire is not transparent in the LWIR and VLWIR, and thus cannot be used as a substrate for back-illuminated devices in those spectral ranges.

The final possible substrate material we consider is InSb. The physical properties of InSb suggest it as an almost ideal substrate for the epitaxial growth of HgCdTe. InSb and IR HgCdTe have almost identical lattice constants and coefficients of thermal expansion, and HgCdTe has been grown on InSb with XRD FWHM values as low as 18–22 arcsec [32.145]. Also, InSb wafers are available with diameters up to 4 inches. However, a number of difficult issues must be solved before InSb can be developed as a substrate material for the growth of HgCdTe. First, InSb has a low melting point (527 °C) and a tenacious native oxide, precluding thermal desorption as a wafer cleaning technique. The oxide can be removed in situ by Ar sputtering, and the induced damage partially removed by thermal annealing, but some surface and subsurface damage remains. Second, In_2Te_3 forms readily at the interface, creating structural defects that degrade any following CdTe or HgCdTe epitaxial growth, although this effect can be partially suppressed by using a Cd-enhanced flux during the initial CdTe growth. Finally, both In and Sb diffuse rapidly into CdTe and HgCdTe, where they strongly affect the electrical properties. Thus, appropriate diffusion blocking buffer layers would be required with the use of InSb substrates unless the substrates are removed before any annealing.

32.4 Design of the Growth Hardware

The reader is referred to [32.13] for a comprehensive review of the growth hardware for MBE growth. Here we discuss only two more recent growth hardware issues of importance for the MBE growth of HgCdTe: the mounting of the substrate on which the HgCdTe is to be grown, and the valving of the effusion cells which supply the growth materials.

32.4.1 Mounting of the Substrate

The standard mounting geometries for MBE substrates include contact and contactless configurations. In the contact configuration the substrate is bonded to a diffusion plate using a mounting or bonding medium such as In or Ga. Contactless configurations were introduced to reduce the risk of contamination from the mounting media such as In or Ga; thus they are referred to as free or In-free mounting configurations. Another advantage of contactless configurations is the smaller thermal mass, allowing more rapid substrate temperature adjustments. In contactless configurations, the substrate is sandwiched between a diffuser plate and either a Mo retaining spring plate for a square substrate such as CdZnTe or a Mo ring for round substrates such as Si. The diffusion plate can be just a Mo plate; however, to achieve a constant growth temperature with only a Mo plate one must severely ramp the nominal thermocouple temperature during growth [32.146]. Moreover, for a fixed thermocouple reading, the actual growth temperature depends strongly on the exact position of the thermocouple and on the mounting geometry, and its run-to-run variation was found to be $> 10\,°C$, as measured by optical pyrometry and by SE. As an improvement, the diffuser plate can be replaced by a Mo plate and a graphite disk in contact with the substrate. It also was found to be beneficial to put a sapphire disk between the Mo plate and the graphite disk because the flat sapphire disk offers better mechanical support to the soft graphite plate. The use of a diffuser plate with a graphite plate in contact with the sample increases thermal conduction, decreasing the importance of irradiation, thereby allowing much better temperature run-to-run reproducibility. Another mounting problem is that the substrate must be rotated at a constant rapid angular velocity to ensure a uniform beam flux across a growth wafer, and with a wobble as small as possible to avoid excessive noise in the SE data.

The temperature ramping recipe that has to be employed to keep the growth temperature constant depends on the substrate area compared with the area of the diffusion plate. For example, for the growth of large-area samples, the substrate can be either (i) directly exposed to the heater and heated by irradiation, or (ii) juxtaposed to a diffuser plate and heated mainly by conduction. Case (iii) is that of smaller-area substrates such as CdZnTe, which are mounted using a molybdenum spring plate that holds the substrate in contact with a diffuser plate facing the heater. The heating power is determined by a closed-loop feedback involving a thermocouple, which floats in the gap between the heater and the substrate. Its readings are related to both the heater and substrate temperatures, and in some cases even to the temperature of the cell walls. Thus, the grower of HgCdTe faces a serious temperature stabilization problem during growth, since it has a very small growth window in comparison with other MBE-grown materials. It has been experimentally observed that in case (i) the heater temperature must be ramped down [32.146], whereas in cases (ii) and (iii) (which have some similarity to mounting on a full molybdenum block) a temperature increase is necessary to keep the growth temperature constant [32.147]. Case (i) has been the most widely investigated, especially for III–V MBE-grown materials [32.148, 149]. Here the temperature transient is believed to stem from the fact that the substrate is largely transparent to infrared radiation, whereas the growing material is not. Thus the thermal absorptivity increases during growth, which results in an increase of the temperature. If the sample heating mechanism is conduction rather than irradiation, as in cases (ii) and (iii), material deposition results in a temperature decrease. This is because the emissivity of the front face increases without any substantial change in the heat conducted through the sample, as the substrate is much thicker than the epilayer being grown. Also radiative energy from the effusion cells can increase the substrate temperature, especially at nucleation, when temperature stability is essential to achieving good crystalline quality [32.150].

For these reasons, several optical methods have been used to measure the HgCdTe temperature in situ, such as (a) SE [32.151], (b) in situ FTIR spectroscopy [32.152], (c) dynamic reflectance spectroscopy (DRS) [32.153], and (d) absorption-edge spectroscopy (ABES) [32.154], on both II–VI and III–V materials. It has been shown that good reproducibility can be obtained from SE even though it gives no direct information on the growth temperature (Sect. 32.5.1). In the case of HgCdTe, DRS measurements must be carried out from the back face of the sample and generally require the deposition of a reflecting intermediate layer. For FTIR and ABES the HgCdTe must be optically thick to obtain reliable readings, so they are not applicable during the early stages of growth. However, information provided by RHEED spectra also can be used to indirectly analyze the growth surface temperature stability. In fact, RHEED and SE so far are the only measurements capable of characterizing growth nucleation in situ.

The employment of spring plates coated with materials with infrared emissivities similar to that of the HgCdTe to be grown allows the temperature ramping to be dramatically reduced or even eliminated for the growth of small-area samples. By employing a graphite plate in addition to a diffusion plate and coating the spring plates with an appropriate material, a dramatic improvement in growth yield was found. So long as substrates with similar characteristics were used, thermocouple run-to-run errors as low as $\pm(2-3)\,K$ were found for samples mounted with graphite plates and appropriately coated spring plates.

32.4.2 Valving of the Effusion Cells

Valved cells were introduced in the MBE growth of III–V materials as early as 1993 [32.155]. For the growth of HgCdTe a valved Hg source was introduced at the Microphysics Laboratory (MPL) of the University of Illinois at Chicago in 1998 and a valved Te source was introduced at the Rockwell Science Center in 1999 [32.156]. The use of a valved As cracker cell also recently has become common. The different valved cells share the same design concept, but with the Hg cell having a tube connected to the Hg reservoir to ease the Hg supply and the As cell having a high-temperature cracker to increase the ratio of As and As_2 to As_4 in the exeunt flux. The Hg and As fluxes can be precisely controlled and optimized by adjusting the computer-controlled needle valve for nearly continuous or abrupt flux changes with rapid flux stabilization, quick shutoff, and improved growth reproducibility. Also, the exit orifices of the valves can be customized to obtain excellent flux uniformity in any MBE systems. All these superior flux control capabilities are extremely desirable to maintain the optimized MBE growth of HgCdTe heterojunction material with complicated design architectures, which is impossible to achieve using conventional thermally controlled sources.

An As-valved cracker cell provides the convenience of a gas source with the safety of a solid source, which is beneficial to optimize flux profiles in time, reduce As consumption, and eliminate unintentional As incorporation into heterojunction structures. The source's cracking zone temperature can be adjusted to generate either As_4- or As_2-dominated flux beams. Such a valved cracker design also offers great flexibility to adjust the charge capacity and obtain optimized cracking efficiency.

32.5 In situ Characterization Tools for Monitoring and Controlling the Growth

Because the sticking coefficient of Hg is very much lower than those of Cd and Te, the growth window for MBE growth of HgCdTe is very small. The window in growth temperature for the growth of optimal epilayers has been reported to be as low as $\pm 0.5\,°C$, and that for the Hg flux $\pm 2.5\%$, which corresponds to $\pm 0.5\,°C$ in the temperature of the Hg cell [32.71]. This precise a degree of control is not possible, but it is very desirable to be able to control the growth temperature within $\pm 3\,°C$ and the Hg flux within $\pm 5\%$. Furthermore, to obtain a desired cutoff wavelength or cutoff energy $\pm 6\%$ in a device, one must control the Cd composition x within ± 0.002. Thus, excellent in situ monitoring and controlling of the growth are necessary, requiring multiple methods for monitoring and controlling the growth.

32.5.1 Spectroscopic Ellipsometry (SE): Basic Theory and Experimental Setup

Due to its simplicity, intrinsic accuracy and noninvasiveness, spectroscopic ellipsometry (SE) is widely used as an in situ tool to monitor, control, and characterize the MBE growth of $Hg_{1-x}Cd_xTe$ alloys. The growth temperature, alloy composition, and presence of defects, overlayers or roughness at the surface in principle all are measurable by in situ SE, if the appropriate data analysis technique is used. Here, greatest attention is given to:

1. The use of SE to determine the temperature and surface roughness of substrates and the composition and surface roughness of the HgCdTe during growth
2. The principal methods of analysis of SE data, pointing out the strengths and weaknesses of each method.

Various specialized applications of SE in the study of the MBE growth of $Hg_{1-x}Cd_xTe$ also are given.

Since shortly after its introduction [32.157] as an in situ characterization tool in 1991, ellipsometry has been applied primarily to find the $Hg_{1-x}Cd_xTe$ alloy composition x. Single-wavelength ellipsometers were first used for this purpose [32.158, 159]. In 1964 44-wavelength SE was used to determine x to within ± 0.01 during MOCVD growth of $Hg_{1-x}Cd_xTe$ and feedback control was established to force x to converge faster to its targeted value [32.160, 161]. Growth rate control and monitoring of the CdZnTe preparation procedure, its surface roughness, and its temperature were reported in 1996 [32.162]. By 1998 SE was mature enough to be introduced in a production environment, as the run-to-run precision in x had improved to ± 0.001 or ± 0.002 [32.163]. Now SE has entered the realm of standard use for compositional control, even in industry. In actuality, the capabilities of SE go well beyond the determination of x, although little work has been done to explore other applications.

Introduction to Ellipsometry

Ellipsometry is a polarimetric technique that uses the change of polarization of light upon reflection for the characterization of surfaces, interfaces, and thin films [32.164]. If the analyzed surface is homogeneous, isotropic, and not contaminated by the presence of overlayers, and neglecting the difference between the electronic structure at the surface, a single ellipsometric measurement with light of wavelength λ yields the complex optical dielectric function $\varepsilon_s(\lambda)$ of the material at that wavelength

$$\varepsilon_s(\lambda) = \varepsilon_a(\lambda) \left[\sin^2 \phi + \sin^2 \phi \tan^2 \phi \left(\frac{1-\rho}{1+\rho} \right)^2 \right], \tag{32.1}$$

where $\varepsilon_a(\lambda)$ is the dielectric function of the ambient, ϕ is the angle of incidence, and ρ is the complex reflectance ratio $r_s/r_p = \tan \Psi \, e^{i\Delta}$, where r_s and r_p are the Fresnel coefficients, i. e., the ratios between the incident and reflected electric fields for s- and p-polarized light, respectively. If $n > 1$ layers are optically accessible, $\varepsilon_s(\lambda)$ is replaced by a pseudodielectric function $\langle \varepsilon_s(\lambda) \rangle$ that depends on the dielectric function of each layer and the thickness of the first $n-1$ layers.

Multiwavelength ellipsometry (SE) provides the frequency-dependent dielectric function $\varepsilon_s(\omega)$, which is intimately connected to the semiconductor band structure and in particular to its critical-point (CP) energies. At these energies E_i there are discontinuities or infinite first derivatives in the joint density of states and thus, in the absence of line broadening, in the probability of photon absorption and hence in $\varepsilon_2(\omega)$, the imaginary part of $\varepsilon(\omega)$. Great effort has been devoted to the precise determination of the CP energies and line widths because they reflect any changes in temperature, composition, and other physical parameters. Given an ellipsometer with sufficient resolution and low noise, accurate determinations of the CP energies can be achieved by higher-order differentiation of the experimentally measured dielectric function [32.165], which greatly enhances the CP structure.

The actual measured quantities are the Fresnel coefficients of the sample, which usually are organized into Jones matrices of the form

$$J = \begin{pmatrix} r_{pp} & r_{ps} \\ r_{sp} & r_{ss} \end{pmatrix} = r_{ss} \begin{pmatrix} \rho_{pp} & \rho_{ps} \\ \rho_{sp} & 1 \end{pmatrix}. \tag{32.2}$$

The Jones matrix of a stack of layers is just the product of the Jones matrices of the layers in the stack. For isotropic materials, such as HgCdTe and its substrates, the Jones matrices reduce to diagonal form. Since no existing in situ ellipsometer can measure all four Fresnel coefficients in one reading at a single azimuthal angle, surface anisotropy due to oriented surface defects or the morphology of high-index surfaces can only be treated in an approximate fashion as a source of uncertainty.

The polarizing properties of any sample are given by its Mueller matrix, which relates the Stokes vector S of the exiting reflected, transmitted or scattered light to that of the incident light S_0. For a homogeneous, isotropic semiconductor, the Mueller matrix can be represented as the product of an ideal linear polarizer with transmission axis ψ and an ideal retarder of retardance Δ

$$\begin{aligned}
M &= \begin{pmatrix} 1 & -\cos 2\psi & 0 & 0 \\ -\cos 2\psi & 1 & 0 & 0 \\ 0 & 0 & \sin 2\psi \cos \Delta & \sin 2\psi \sin \Delta \\ 0 & 0 & -\sin 2\psi \sin \Delta & \sin 2\psi \cos \Delta \end{pmatrix} \\
&= \begin{pmatrix} 1 & -N & 0 & 0 \\ -N & 1 & 0 & 0 \\ 0 & 0 & C & S \\ 0 & 0 & -S & C \end{pmatrix}.
\end{aligned} \tag{32.3}$$

The final form in (32.3) holds even with depolarization; in general, $C^2 + S^2 + N^2 = p^2$, where $p = 1$ for nondepolarizing samples and $p = 0$ for completely depolarizing samples. C, S, and N completely characterize all isotropic, homogenous materials, even depolarizing ones. If the further assumption is made that a sample is nondepolarizing, then the knowledge of any two of C, S, and N suffices to calculate the pseudodielectric function.

Ellipsometer Designs

The primary ellipsometer designs in use for in situ SE are the rotating compensator ellipsometer (RCE) and the rotating analyzer ellipsometer (RAE). These designs differ in how fully they characterize the Mueller matrix. In the RAE [32.166], light generated by the source is first linearly polarized by a polarizer, then reflected by the sample, goes through an analyzer, and finally enters the photodetector. The presence of the first polarizer makes the measurement immune from polarization by the source optics. The analyzer is continuously rotated to generate a modulation of the detected signal, which is Fourier-analyzed to extract C and N. This design cannot measure S because that would require circularly polarized light. A simple calculation shows that the irradiance at the detector is given by

$$I = I_{dc} + I_s \sin 2\omega t + I_c \cos 2\omega t,$$

where

$$\frac{I_s}{I_{dc}} = \frac{C \sin 2\theta_p}{1 - N \cos 2\theta_p}, \quad \frac{I_c}{I_{dc}} = \frac{\cos 2\theta_p - N}{1 - N \cos 2\theta_p}. \quad (32.4)$$

The azimuthal angle of the polarizer θ_p can be determined through a calibration procedure [32.167, 168]. Because S does not appear in these equations, even in the absence of depolarization only S^2 is determined, and the sign of Δ is undetermined. Also, the sensitivity of the RAE ellipsometer is zero for $\varepsilon_2(\omega) = 0$, hence, at or below the optical energy gap of semiconductors. In particular, RAE ellipsometers designed for the characterization of HgCdTe in the region around the E_1 CP (2–3.5 eV) do not perform well for CdTe or Si in the same energy range.

The RCE was introduced to circumvent these limitations. A compensator is inserted between the polarizer and the sample. In the most common design, the compensator is rotated continuously, whereas the analyzer and the polarizer remain fixed. The output irradiance at the detector has extra Fourier components, of angular frequency 4ω,

$$I(t) = I_{dc} + I_c^{(2)} \cos(2\omega t) + I_s^{(2)} \sin(2\omega t) \\ + I_c^{(4)} \cos(4\omega t) + I_s^{(4)} \sin(4\omega t),$$

which allows C, S, and N to be calculated without ambiguity. Thus, the RCE determines the sign of Δ, gives accurate readings even for $\varepsilon_2(\omega) = 0$, and can measure depolarization. Several other ex situ designs which have been proposed can measure the anisotropic dielectric function and the depolarization at a single azimuthal angle, taking only two separate measurements, for example that by *Jellison* and *Modine* [32.169], but no applications to in situ measurements of HgCdTe have been proposed.

Retrofitting the MBE Chamber

The setup for in situ experiments is complicated by many constraints, and the data are subject to more sources of uncertainty than in an ex situ setup. Constraints are for example posed by the availability of ports for the windows, the window type, the intrinsic characteristics of the substrate manipulator, etc. The effect of windows on accuracy is very large for most ellipsometers [32.170]. Different ellipsometers measure different linear combinations of the Mueller matrix elements, and therefore differ in how fully they characterize and hence can correct for window imperfections. The RAE cannot measure even the out-of-plane component of birefringence during an experiment, while the RCE can. It is not possible with any instrument to separate the in-plane window retardation from that of the sample, even when the entire Mueller matrix is measured. To determine the in-plane window retardation one must resort to measurements with well-known samples, such as Si oxide of known thickness. Low-strain bakeable windows, guaranteed to have a birefringence of less than 1°, are commercially available, but even run-to-run variations of the order of 0.1°, which may occur when the windows are baked out, can be important. Also, the measured $\varepsilon_2(\omega)$ is extremely sensitive to changes in the angle and plane of incidence, so that much care must be taken to minimize wobble using wobble-free manipulators and a proper substrate mounting procedure. Unfortunately, substrate rotation cannot be avoided. Thus, it is common to equate the data acquisition time to the total rotation time of the substrate.

32.5.2 SE Data Analysis

To interpret ellipsometric results, the complex reflection coefficients must be computed from a model of the sample surface. In the general case, one constructs a model for the dielectric function which depends on the parameters to be determined. This is fitted to the experimental data using the standard Levenberg–Marquardt algorithm. The function itself and possibly its first, second or third derivatives, or a combination of them, are fitted. There are two principal approaches to parameterizing the optical constants of a bulk semiconductor, based either on the construction of calibrated libraries of optical dielectric functions or on the determination of their CP parameters. Library models are commonly used to find the composition and surface roughness of HgCdTe during growth, and the temperature and surface roughness of CdZnTe or CdTe before growth. Their main advantage is their ease of use. Unfortunately, when growth conditions are changed, such models can suffer from large run-to-run uncertainties, due to changes in surface morphology, small changes in temperature, and changes in window retardation and birefringence. The use of CP models for the dielectric function has been introduced to reduce these uncertainties.

Multilayer Models and the Virtual Interface Approximation

The most common multilayer model for HgCdTe, CdZnTe, and CdTe characterization is a three-layer model, comprised of epilayer or substrate, a surface roughness layer, and the ambient, normally vacuum. If the optical constants of the growing layer are unknown, the virtual interface approximation [32.171, 172] and dynamic algorithms can be used to extract them, the growth rates, and in some cases even the surface roughness of the growing layer at the beginning of growth. Exact expressions for the dielectric function ε_0 of a virtual surface located at the physical surface can be found as a function of the measured pseudodielectric function $\langle\varepsilon\rangle$. These expressions can be fitted to the experimental data to obtain estimates for both the near-surface composition and the growth rate, as shown [32.150] for MOCVD-grown HgCdTe. The virtual interface approximation has attained great importance for the characterization of HgCdTe, as it provides a way to monitor growth nucleation.

Surface Roughness

Surface roughness results in reflectance loss, polarization of the specularly reflected light, and the introduction of stray light reflected at near-specular angles. Most of the specularly reflected light is reflected coherently and stays polarized. Part is reflected incoherently and is depolarized, but the depolarization caused by a small degree of microscopic roughness is minimal and needs to be taken into account only for moderately to very rough surfaces. In that case, near-specular scattering may also play a role [32.173]. For roughness on a scale smaller than the wavelength of the reflected light, the changes in the coherent, polarized part of the reflected beam can be described by the Bruggeman effective medium theory (EMA) [32.174, 175]. That theory is derived from the Clausius–Mossotti expression [32.176] for the dielectric response of a heterogeneous medium of point-like microstructures embedded in a host for the case of an aggregate of two types of microstructures in which the volume fractions of the two types are comparable, so that the host must be chosen as some self-consistently chosen effective medium. In the EMA the dielectric function of the effective medium is chosen to be

$$\varepsilon = \frac{\varepsilon_a\varepsilon_b + \bar{\varepsilon}(f_a\varepsilon_a + f_b\varepsilon_b)}{\bar{\varepsilon} + (f_a\varepsilon_b + f_b\varepsilon_a)}, \qquad (32.5)$$

where ε_a and ε_b are the dielectric functions of the two types of microstructures, $\bar{\varepsilon} = (q^{-1} - 1)\varepsilon_b$, and q is a screening parameter between zero and one. The thickness t of the roughness layer is related to the mean-square roughness by $t = 2\sqrt{2}\sigma_z$ [32.177]. The void fraction in the roughness layer is commonly taken to be 50%, so that (32.5) reduces to the form $\varepsilon = [2\varepsilon_b + \bar{\varepsilon}(1+\varepsilon_b)]/[2\bar{\varepsilon} + (1+\varepsilon_b)]$.

Analysis Based on Dielectric Function Libraries

Libraries of dielectric functions are acquired at a set of values of one or more parameters, e.g., the temperature and/or the alloy composition, and interpolated between those values. In the simplest case, libraries are used in conjunction with multilayer models for the dielectric function. A typical model valid only for optically thick layers consists of a composition-dependent epilayer or temperature-dependent substrate, surface roughness, and ambient. The surface roughness is described in the EMA. Hence, a fit to the experimental data yields the thickness t of the roughness layer and either temperature or composition. In most in situ environments, it is also necessary to leave the angle of incidence ϕ free in the fit, thus having three independent fitting parameters.

To obtain a meaningful library, one's ellipsometer must be accurately calibrated, the windows must be free of birefringence, the angle of incidence must be known, and the sample composition and temperature must be known. Also the surface one is measuring must be exactly known, and no undesired overlayer or unknown roughness should be present. These conditions are very hard to attain in situ, but they can be attained at least approximately. A full calibration of the ellipsometer before data acquisition can be performed ex situ, with the angle of incidence as well as the sample type perfectly known. Then, standard calibration procedures can be performed. For most commercially available RAEs or RCEs, these calibration procedures are executed automatically by software algorithms. The window birefringence must be minimized using *strain-free* windows. Residual in-plane birefringence can be estimated in situ using a thermally grown SiO_2/Si calibration sample, the optical constants of which are well known. Since the window birefringence is approximately constant on any given MBE system, if low-strain windows are used, a precision of about $\pm 0.1°$ in the measured values of Δ is expected. The angle of incidence can be measured in situ in most cases, exploiting databases of room-temperature optical constants available in the literature.

Unfortunately, exact knowledge of a sample surface is very difficult to attain even in situ; submonolayer overlayers and roughness are always present, even in the cleanest in situ environment. Fortunately, they can be estimated by comparison with ex situ data at room temperature, and subtracted mathematically using the Jones formalism for multilayer analysis. The caveat here is that usually surface roughness and oxide layers are essentially indistinguishable in treating the effects of a thin overlayer of unknown thickness [32.178–180]. Usually an oxide overlayer is subtracted from the data to obtain the bulk dielectric function.

Since all in situ SE measurements of epilayer and substrate properties are based on the dielectric function library used, the measuring accuracy and precision largely depends on that library. Bearing in mind the caveats discussed above, *Johs* [32.181] developed a procedure to measure dielectric function libraries of growing materials, in particular of HgCdTe during MBE growth. He confronted with great success the problem of determining a library of dielectric functions for different Cd molar fractions, from $x = 0.2$ to 0.5, at the ideal growth temperature T_g. His paper discusses the difficulties that one encounters in measuring a compositional library $\langle\varepsilon(\omega; x, T_g)\rangle$ in great detail. Because $\varepsilon(\omega)$ depends strongly on T as well as x, the very narrow growth window for HgCdTe turns out to be very helpful in obtaining $\langle\varepsilon(\omega; x, T_g)\rangle$. Because of the very narrow growth window, the measured thickness of the surface roughness layer increases rapidly as the growth temperature departs from the ideal. This allows one to adjust the temperature to T_g even without knowing its value and thus to hold the uncertainty in the measured $\langle\varepsilon(\omega; x, T_g)\rangle$ to a minimum.

The compositional control of HgCdTe attained using the library developed by Johs has been excellent, as reported by a large number of authors [32.182–188]. Attention has not been focused on the real-time variations of the composition (which are inaccessible by routine FTIR measurements in any case), but only on its average value over the entire layer. This was obtained as follows: a set of 8–10 measured dielectric functions was suitably selected to span the entire growth process. A global fit was performed, treating the angle of incidence as a global parameter and leaving composition and surface roughness free to attain different values for each spectrum. Finally, the obtained x-values were averaged to get the average composition.

A fit to $\langle\varepsilon_2\rangle$ yields average compositional values that are in good agreement with those obtained by FTIR. The fact that a fit to the measured values of $\langle\varepsilon_2\rangle$ gives the best ellipsometric measurement of x is due to two factors:

1. $\langle\varepsilon_2\rangle$ contains much more direct information on the joint density of states (JDS) than either $\langle\varepsilon_1\rangle$ or Δ and Ψ
2. $\langle\varepsilon_2\rangle$ is much less affected by variations in the window birefringence.

These two factors also assure that $\langle\varepsilon_2\rangle$ gives a very accurate measurement of x when proper care is taken and when the library used is appropriate for the growth temperature and growth and SE measurement configurations used. However, it should be emphasized that the excellent results found for compositional monitoring and control require fully reproducible growth conditions. Changes in the geometry of the growth setup, the growth rate, the sample mounting, etc. have been shown to lead to substantial errors in the HgCdTe compositions measured using the library function of Johs. Also, in situ measurements of the composition y in CdZnTe as a function of temperature using a library of $Cd_{1-y}Zn_y Te$ dielectric functions have been performed for three samples with $z = 0.0$, 0.045, and 0.125, starting at room temperature, going up to $350\,°C$, and coming back down to $150\,°C$. He found that the composition measured upon coming back down in temperature from $350\,°C$ was very different from than that measured going up in temperature, ≈ 0.05 higher over the range $150–300\,°C$ for the CdTe sample. He also measured the temperatures using the same library, and found errors in the measured temperatures from -10 to $-30\,°C$ for the CdTe sample upon coming back down in temperature.

The acquisition of a two-dimensional (2-D) library $\langle\varepsilon(\omega; x, T)\rangle$ of HgCdTe pseudodielectric functions as a function of the two variables x and T is much more difficult than might be expected a priori. This is because high-quality HgCdTe only grows within a $\pm 5\,°C$ window, so that to get a temperature library one must interrupt the growth and ramp the temperature to different values. This must be done leaving the layer under Hg flux to avoid irreparable damage to the surface. We have found no systematic dependence of the CP energies on T, in data taken with this procedure, probably because of Hg evaporation, which alters the composition near the surface, producing random changes in the dielectric function and in any measured CP energies. No alternative procedure has been proposed to date.

These difficulties may be overcome in the near future, for example by a more accurate control of the Hg flux. However, even if a two-dimensional library were obtained, one must still conclude that one could not obtain x and T simultaneously from such a library without SE data greatly improved over that provided by the 88-wavelength RAE now commonly used, which does not provide the resolution necessary to allow the accurate determination of CP energies through multiple differentiation of the data. It has been shown that a fit to the undifferentiated data leaving four parameters free (x, T, t, and ϕ) yields a reasonably good estimate for T, but also substantially increases the uncertainty in x and t. A better approach would be based on the interplay of a 2-D library for $\langle\varepsilon(\omega; x, T)\rangle$ and a knowledge of the energy $E_1(x, T)$ obtained by multiply differentiating that library with respect to ω. The energy E_1 could be obtained very precisely and would be almost independent of any surface overlayer. One could find T as a function of x from a measured E_1, reducing a two-dimensional fitting problem to a one-dimensional problem using the undifferentiated library. One cannot simultaneously determine x and T by measuring two CP parameters without an ellipsometer capable of simultaneously measuring E_0 and E_1 accurately. No such ellipsometer is available at present. Measurements of the other E_i do not give sufficient information; $E_1 + \Delta_1$ and $E_0 + \Delta_0$ cannot be measured sufficiently precisely, and $E_1 + \Delta_1$ moves almost in parallel with E_1 with changes in x or T. Also, the line widths Γ_i depend strongly on material quality and cannot be measured as precisely as the E_i.

A new set of libraries of $\varepsilon(\omega; T)$ for $Cd_{1-z}Zn_zTe$ for $z = 0.0$, 0.045, 0.125, and 1.0 is available from MPL. That library ranges from 75 to 300 °C in increments of 25 °C with increments of 15 Å in wavelength from 1.2 to 5.0 eV and increments of 30 Å from 0.7 to 1.2 eV. The library was obtained by ramping the temperature of each CdZnTe sample up in 25 °C steps, pausing at each step to allow the sample to come to a steady state, and then measuring $\varepsilon(\omega)$ at that temperature. The following procedure [32.151] was used. Clean $Cd_{1-z}Zn_zTe$ was introduced in the MBE chamber after etching in a 0.5% bromine and methanol solution, and rinsing in methanol and deionized (DI) water. The surface oxide was desorbed at $\approx 250\,°C$, the substrate was ramped to different temperatures, and its pseudodielectric function $\langle\varepsilon(\omega; T, \phi, t)\rangle$ was measured at each temperature T. The temperature at the surface was measured with a thermocouple, and the temperature measurements were calibrated according to the melting points of Sn and In. A fit to room-temperature data gave the angle of incidence ϕ and the thickness t of the surface roughness overlayer, using calibrated data acquired ex situ. This procedure allowed us to obtain a set of spectra $\varepsilon_{CZT}(\omega; T_1, \ldots, T_n)$ with overlayer effects removed. The library was obtained using an M-2000 ellipsometer, which has 220 wavelengths from 1000 to 1700 nm, allowing the differentiation of the data to determine the E_0 and E_1 CP energies using a CP model, as discussed immediately below.

One can never ensure that one is measuring the optical constants of an abruptly terminated, clean, and smooth semiconductor surface. All that can be ensured is that our optical constants are reproducible, under the same conditions, on the same system. One cannot ensure the absolute accuracy of a library of measured optical constants, because of window birefringence and the presence of overlayers, but one can expect one's measurements to be reproducible from run to run on the same system. However, system-to-system portability of a library is never certain.

Analysis Based on Critical Point (CP) Models

An analysis of the dielectric function which includes a CP analysis in principle is superior to an analysis based on libraries, for the following reasons. First, much of the information obtainable from the derivatives of the pseudodielectric function is complementary to that obtainable from fits to the function itself and is not available from fits to the function itself. That information consists of the energies E_i and intrinsic line widths Γ_i of all band-structure CPs within the spectral range of the SE data. Second, the use of a parameterized theoretical model for the dielectric function allows us to measure directly the phase shift of the incoming light in SE as it passes though a surface overlayer and thus to measure directly the optical thickness of the overlayer, which is not otherwise possible. The E_i are virtually independent of surface overlayers and, given sufficiently good SE data, can be obtained very precisely.

High accuracy in the determination of the E_i and Γ_i is paramount in in situ applications. The problems with CP analyses are that the data fitting is more difficult and that low-noise, high-resolution SE data is required to perform the higher-order differentiation needed to obtain accurate values for the CP parameters. It has been found [32.189] that it is impossible to perform a useful CP analysis using only the 88 wavelengths available with an M-88 ellipsometer with the typical signal-to-noise ratio (SNR) of in situ data. However, the use of an M-2000 ellipsometer in situ has been found to yield sufficient resolution and a sufficiently small noise level to allow the profitable use of a CP model for data analysis.

Several CP models have been proposed, but only the Kim model [32.190] is capable of simultaneously fitting the optical dielectric function and its derivatives. The optical dielectric function is given by the fundamental expression [32.191]

$$\varepsilon(\omega) = 1 + \frac{8\pi^2 e^2}{m^2}$$
$$\times \sum_{c,v} \int W_{cv}(E)\,dE$$
$$\times \{[\Phi(\hbar\omega - E) - \Phi(\hbar\omega + E)]\}. \quad (32.6)$$

In this expression, $W_{cv}(E) = P_{cv}(E)^2 J_{cv}(E)$ represents the product of the Brillouin-zone-average transition probability at the energy E and a joint density of states (JDS), $J_{cv}(E)$. $\Phi(\hbar\omega \pm E)$ is the broadening function and is equal to $|\hbar\omega \pm E + i\Gamma|^{-1}$ for Lorentzian broadening; in general, the broadening is intermediate between Lorentzian and Gaussian [32.192]. The JDS is given by the formula

$$J_{cv}(E) = \int_{BZ} \delta[E_c(\mathbf{k}) - E_v(\mathbf{k}) - E]\,d\mathbf{k}, \quad (32.7)$$

and counts the number of states per unit energy between a given pair of conduction (c) and valence (v) bands that differ by an energy E. The JDS and hence the $W_{cv}(E)$ are smoothly varying functions of energy away from the critical points.

The Kim model parameterizes the fundamental expression (32.6) by approximating the $W_{cv}(E)$ by low-order polynomials in energy multiplied by functions of energy having the proper analytic behavior at the CPs. Thus the model has two parameters for each CP, E_i and Γ_i, along with amplitude parameters with respect to which the dielectric function is linear. $W_{cv}(E)$ is divided into seven contributions, one for each pair of connected CPs. The resultant JDS is shown schematically in Fig. 32.10. The expression for each region

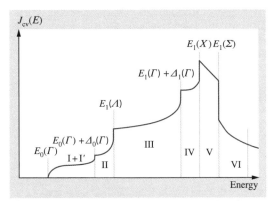

Fig. 32.10 Schematic diagram of the joint density of states of a zincblende semiconductor

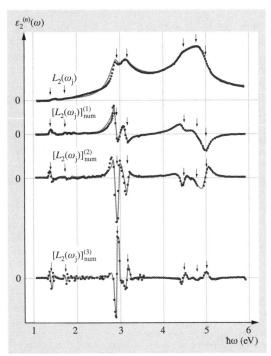

Fig. 32.11 Simultaneous fit of the Kim, et al. model to the experimental $\varepsilon_2(\omega)$ for GaAs and its first three numerical derivatives allowing a mixture of Lorentzian and Gaussian broadening for the four lowest-energy critical points. The *arrows* show the energies of the seven critical points used in the fit. The *black dots* show the experimental $\varepsilon_2(\omega)$ and its numerical derivatives; the *red curves* show the modeled $\varepsilon_2(\omega)$ and its numerical derivatives

in Fig. 32.10 is given in [32.190] and is fitted independently. The expressions for the seven contributions are substituted in (32.6) and integrated analytically. A proper harmonic oscillator is also added to account for out-of-range contributions to $\varepsilon_1(\omega)$.

The Kim model allows:

1. Predictions for the region below the fundamental gap, but above the onset of phonon transitions
2. The measurement of the true JDS
3. The measurement of the thickness of overlayers using the measured phase shifts at the CPs.

It was initially tested by simultaneously fitting $\varepsilon_2(\omega)$ for GaAs and its first three derivatives; the results are shown in Fig. 32.11; the fit to $\varepsilon_2(\omega)$ was good to 0.4%. It was later used to fit the dielectric functions of CdTe [32.193] and ZnSe [32.194] and their derivatives and to obtain functional forms for the temperature-dependent dielectric functions of CdTe [32.195] and HgCdTe [32.196] and the composition dependent dielectric function of AlGaAs [32.197]. It was found that, with a total of 75 free parameters, the measured $\varepsilon_2(\omega)$ for $Al_{1-x}Ga_xAs$ and its derivatives for all compositions could be fitted by functions of x and ω within a fractional root-mean-square (rms) error of only 1.2% in $\varepsilon_2(\omega; x)$. When measuring in situ, fewer parameters should be used because the data are affected by much larger uncertainties.

One value of using the Kim model is its failure to fit the measured $\varepsilon_2(\omega)$ for materials having poor crystalline quality or having overlayers not explicitly included in the fitting through the use of a multilayer model. This means that it cannot give incorrect CP energies to compensate for the effects of overlayers or poor crystalline quality, so the CP energies given are very reliable. This is illustrated by its failure to fit the alloy series $Hg_{1-y}Zn_yTe$ and $Cd_{1-y}Zn_yTe$. Fits to samples having different alloy compositions [32.198] failed to yield a systematic dependence of the room-temperature Γ_i on y over the compositional range $0 < y < 1$. The poor material quality of these alloy systems led to fluctuations in the CP line widths with alloy composition that made it impossible to obtain functional forms for their composition-dependent dielectric functions. That result highlights the sensitivity of this model to even small changes in CP parameters. To be able to monitor such changes opens new horizons for the in situ monitoring of HgCdTe growth.

Comparison of Results from Fitting to a Library and from CP Analysis Using the Kim Model

Consider a few cases in which both library functions and a CP analysis have been used in fitting data obtained from an M-2000 ellipsometer to measure composition or temperature. Abad found that the CP energies found for CdZnTe from the derivatives of the measured dielectric function are essentially unaffected by surface changes caused by annealing, even though those changes cause significant changes in the dielectric function itself. Also, *Badano* et al. [32.199] recently have studied the effect of surface roughness on the ellipsometric response from CdTe epilayers having up to $\approx$ 200 Å-thick surface roughness with a correlation length of several micrometers. SE data were taken ex situ at room temperature to reduce the noise in the data and the CP line widths, so as to improve the precision of measurement of the CP energies. Although the surface roughness introduced noticeable changes in the dielectric function, the CP energies found were unaffected by the roughness. Since the sample surface temperature can be defined solely by the CP energy E_1, this confirmed that the sample surface does not affect the temperature measured using the Kim model. These results, which argue in favor of using CP analysis, are not unexpected, because the dielectric function of a microscopically disordered surface does not have sharp derivatives.

Also, *Badano* et al. [32.200] recently have compared the use of library functions and the use of the Kim model for the measurement of the temperature of CdTe epilayers grown on Ge. The layers were ramped up and down in temperature as in the study by Abad, but only up to $\approx 250\,°C$, and were measured under a Te flux and under no flux. The reproducibility obtained using the two different data analysis methods, upon ramping the temperature up and then back down, were similar, except that the Te flux affected the values obtained for T using library functions but not those obtained using the Kim model. An error analysis of the CP fitting procedure suggested that noise in the data should introduce a random error of approximately $\pm 5\,K$ in the value of T found using the Kim mode. However, the values of T obtained using the Kim model were reproducible within $\pm 2\,K$, an outstanding result. The values for T found using a library were systematically higher than those found using the Kim model. Since values obtained using a library are much more subject to systematic error while those obtained from the Kim model are much more subject to random error, it would be reasonable to ascribe this systematic difference to systematic error arising from the use of library functions.

On the other hand, even with the use of an M-2000 ellipsometer, the noise in SE data induced random dispersions in y and T of the order of ± 0.005 in y and $\pm 4\,K$ in T for $Cd_{1-y}Zn_yTe$. Those uncertainties are somewhat too large, although substantially better than those found using a library of dielectric functions. On the other hand, it suggests that by measuring y ex situ at room temperature and then, knowing y, measuring T in situ, one could obtain more precise values than are now available. The uncertainties would be substantially reduced by not having to determine y and T simultaneously, and the uncertainty in y would be further reduced by measuring it ex situ. However, the large variations in y which sometimes exist from point to point on a CdZnTe substrate would make this method problematic, because the in situ and ex situ measurement points on the CdZnTe may not be identical. Also, Abad found the Kim model to be substantially more difficult to use than library fitting with the software provided with the ellipsometer. This result is discouraging with regard to the possibility of being able to measure x and T in situ simultaneously for HgCdTe, although the method proposed for HgCdTe is different, involving the determination of x from a library of dielectric functions and T from the determination of the CP energy E_1.

In conclusion, the use of the Kim model eliminates errors arising with the use of library functions due to changes in incident flux, surface damage and surface states, window coatings, etc., but introduces uncertainties due to the relatively large amount of noise in in situ SE data. Although for measuring temperature those uncertainties appear to be smaller than the errors which easily can occur with the use of library functions, they may need to be reduced further for SE to become a valuable tool for measuring T. Also, the use of the Kim model entails fits that are not as robust and may converge slowly, a serious problem for the in situ control of growth. The use of the Kim model for the in situ determination of substrate temperature with an M-2000 ellipsometer is promising, but would require a more robust fitting procedure to be justified in a well-controlled production environment, in which the use of

library functions determines T reproducibly even if not accurately.

However, the use of the Kim model would have two other advantages, namely the determination of the CP line widths and, unlike fitting to library functions, the determination of absolute values for the surface roughness or the thickness of other overlayers. Results from Badano et al. show that the use of a CP model can pinpoint the presence of overlayers on a sample, and does not fit SE data unless a proper model for such an overlayer is adopted. We envision the possibility of acquiring libraries of true bulk dielectric functions, by eliminating the roughness overlayer from the data, and possibly also the effect of other overlayers due to the presence of the impinging fluxes. More work is being done in this field.

32.5.3 SE Study of Hg Absorption and Adsorption on CdTe

SE has been shown [32.201, 202] to measure the relative abundance of chemisorbed and physisorbed monolayers of Hg on the CdTe substrate surface prior to growth nucleation. Data with submonolayer resolution were obtained. The substrate dielectric function was measured after the substrate was prepared using a standard procedure, but before subjecting it to Hg flux. Then, the Hg flux was opened and the dielectric function was remeasured. In the analysis of the change in the pseudodielectric function due to Hg deposition, two major approximations were made. First, the surface roughness and the chemisorbed and physisorbed Hg were treated as separate layers with the physisorbed Hg on top of the chemisorbed Hg, and, second, the three-dimensional (3-D) form of $\varepsilon(\omega; x)$ for $Hg_{1-x}Cd_xTe$ for large x was used to mimic the effect of chemisorbed Hg in the surface region. Thus, the data were interpreted using a five-layer model comprised of CdTe substrate, surface roughness, chemisorbed Hg, physisorbed Hg, and ambient. The quantities left free in the fit to the data were the thicknesses of the two Hg layers, that of the roughness layer, and the plasma frequency ω_p in the Drude form assumed for the dielectric function of the physisorbed Hg layer. The values found for the relative thicknesses of the chemisorbed and physisorbed layers are physically reasonable and show the expected dependence on the temperature and the Hg flux, as shown in Fig. 32.12. Although the absolute numbers provided by the analysis undoubtedly are not accurate, the results confirm our general ideas regarding the nature of the CdTe(211) surface under a Hg flux, and is valuable at least techno-

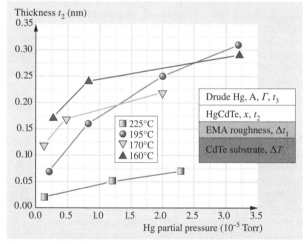

Fig. 32.12 Calculated thickness t_2 of the chemisorbed Hg layer, obtained as a function of the Hg flux for different substrate temperatures. The uncertainty in each thickness obtained from the correlation matrix is approximately ±0.4 Å. The values for the Hg partial pressure are uncorrected N_2 pressure readings. The data at 195 °C were acquired during a separate run than the data taken at 225, 160, and 170 °C. In the 195 °C run, the sample was exposed to Te flux down to a lower temperature, thus it was likely to have a higher Te coverage. The *inset* shows the assumed structure and the fitting parameters used in fitting the SE data

32.5.4 Correlation Between the Quality of MBE-Grown HgCdTe and the Depolarization and Surface Roughness Coefficients Measured by in situ SE

The SE surface roughness thickness parameter t measured using an effective medium approximation (EMA) for the roughness is routinely used to monitor growth. The surface roughness measured by SE during growth is a good indicator of the final quality of HgCdTe layers. Substantive increases in t are always associated with the onset of 3-D growth and the degeneration of material quality, mostly associated with the formation of twins. In addition, the substrate submicroscopic roughness as measured by SE correlates well with the qualitative roughness inferred from reflection high-energy electron diffraction (RHEED). Due to the nature of the EMA model, it is sensitive to roughness on an atomic scale and can give information about how small-scale defects form. However, it does not model roughness on a micrometer scale.

The evolution of surface roughness with increasing epilayer thickness for CdTe grown under nonideal conditions has been studied [32.199]. Both the use of the Kim model and the use of library functions gave surface roughness values which initially increased with increasing epilayer thickness and then decreased. It was found that the apparent decrease arose from the onset of macroscopic roughness, which could not be fitted by the EMA. For rough samples with roughness on much larger than a nanometer scale, the measured dielectric function departs significantly from the form

Table 32.3 Correlation between the depolarization coefficient $\bar{u}$, the surface morphology, the SRH contribution to the minority carrier recombination rate, the EMA surface roughness, and the etch pit density

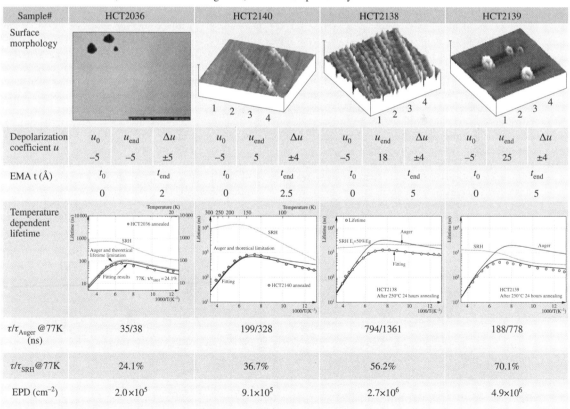

it takes with only atomic-scale roughness because the EMA is not suitable for describing larger-scale roughness with lower lateral spatial frequency. Because the RAE is very sensitive to depolarization effects, we have introduced a model for the analysis of RAE SE data, which contains a depolarization coefficient parameter.

The depolarization coefficient u for the RAE is defined by the equations

$$\cos 2\psi = (1+u)\cos 2\psi' \quad \text{and}$$
$$\cos \Delta = \cos \Delta' \left(1 + \frac{1}{\sin^2 \psi'} u\right), \tag{32.8}$$

where ψ' and Δ' are the values of ψ and Δ measured by SE in the presence of depolarization, and ψ and Δ are the values in the absence of depolarization [32.203]. The quantity u is the ratio of the intensity of the light scattered incoherently to that of the specularly reflected light [32.204]. Obviously, u is a function of wavelength or frequency, $u = u(\omega)$, but no measurements of $u(\omega)$ are available in the literature for rough HgCdTe or CdZnTe. We define $\bar{u}$ as the value of u averaged over the wavelength range of our SE data. It measures surface roughness on a scale larger than that of the EMA surface roughness, a lateral scale more nearly of order the wavelength of the incident polarized light.

We have discovered correlations between the minority-carrier recombination lifetime τ and $\bar{u}$. It correlates well enough with τ to be regarded as a useful predictor during growth. Typical results showing the correlation between $\bar{u}$ measured in situ, the surface morphology, the EPD, and the relative importance of Shockley–Read–Hall recombination processes are shown in Table 32.3. This suggests that the measurement of $\bar{u}$ can be developed as an in situ early warning of the degeneration of material quality, so that the MBE growth conditions can be reoptimized to stop the degeneration of layer quality, and hence improve the material growth yield.

32.5.5 Surface Characterization by in situ RHEED

In general, reflection high-energy electron diffraction (RHEED) has been the most important analytical tool available to the MBE grower, although SE has now taken over that role for the growth of HgCdTe. RHEED employs electron guns and display units such as fluorescent screens and/or recording devices such as charge-coupled device (CCD) cameras to obtain information on the structure and/or morphology of crystal surfaces. It is installed on virtually all MBE systems, and various authors have given detailed discussions of applications of RHEED in MBE growth [32.205]. It is particularly well suited for the in situ characterization of the growth surface during MBE growth. MBE provides the high vacuum and clean surface required for RHEED measurements; in return, RHEED does not interfere with the incoming particle beams during growth and is particularly sensitive to the top few layers or even the top monolayer. It uses relatively high electron energies (5–100 keV, typically 20 keV for HgCdTe) and very low impact angles ($< 5°$). The high energy of the electron beam makes the reflected image from an ordered surface sharp, and the low impact angle allows the electrons to pass only a few atomic layers into the crystal, giving an image that represents the structure of the surface instead of that of the whole crystal. RHEED signals from single-crystal surfaces that are flat on a nanometer scale produce a RHEED pattern with lines rather than dots because a 2-D array has a one-dimensional Fourier transform. Thus, the RHEED image of a smooth ordered surface shows sharp one-dimensional (1-D) rods, rather than zero-dimensional (0-D) dots. In contrast, an oxide substrate surface layer, being amorphous,

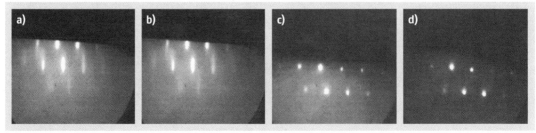

Fig. 32.13a–d RHEED patterns from CdZnTe substrates for different values of the surface roughness as measured by SE: (**a**) 3 Å, (**b**) 10 Å, (**c**) 20 Å, and (**d**) 30 Å

gives rise to a diffuse diffraction pattern, and relatively rough surfaces produce a RHEED pattern having dots rather than rods. The effect of CdZnTe substrate surface roughness is shown in Fig. 32.13. The HgCdTe (211)B MBE surface normally has several atomic layers of surface roughness, so the actual RHEED image normally has reflection and transmission contributions, which makes the quantitative analysis of the image very difficult. On the other hand, the analysis of RHEED data from 2-D growth on low-index surfaces has attracted great interest, with over 500 publications on oscillations in the RHEED intensity alone. Such oscillations, with a period equal to the growth period, were first observed in the early 1980s in the growth of GaAs [32.206–208], and in 1989 were observed in the (100) growth of HgTe/CdTe SLs [32.209]. They were first explained quantitatively in 1998 [32.210] and remain the subject of theoretical study even today [32.211].

32.5.6 Other in situ Tools for Controlling the Growth

Other commonly used in situ tools for controlling the growth include a heater and thermocouple behind the substrate to control the growth temperature and a thermocouple and/or ion flux gauge on each effusion cell to control the flux rates. The difficulty in determining the growth temperature from the thermocouple reading was discussed above in Sect. 32.4.1. Because of that difficulty, an optical pyrometer sometimes is used either to calibrate the thermocouple or to directly control the growth temperature. Because of the extreme sensitivity of the fluxes on effusion cell temperatures, the use of flux gages (and valved cells) is recommended for any fluxes which need to be well controlled. Finally, in situ FTIR, although not commonly used, can be used for temperature monitoring [32.152].

32.6 Nucleation and Growth Procedure

The nucleation of HgCdTe growth always occurs on either CdTe, CdZnTe with $\approx 4\%$ Zn or CdSeTe with $\approx 4\%$ Se. The more difficult nucleation process is the nucleation of CdTe, CdZnTe, CdSeTe or ZnTe on Si or another severely non-lattice-matched substrate material. We consider first the nucleation of CdTe or ZnTe on Si and then the nucleation and growth of HgCdTe.

32.6.1 Nucleation and Growth of CdTe or ZnTe on Si

The MBE growth of CdTe on Si has been studied for over a decade [32.54, 93, 97, 99, 212–215]. The primary issues to be faced when growing Te-based II–VI compounds on Si(211) are the high density of surface steps, which exacerbates the difficulty of removing oxygen and other contaminants from the surface, controlling the polarity of the II–VI layer to be grown, and finally reducing the dislocation density in that layer. Substrate preparation is a crucial step in the growth of Te-based II–VI compounds on Si. The two most important techniques for preparing silicon surfaces for low-defect epitaxy are oxidation and hydrogen passivation. Surface oxidation primarily serves to oxidize the adventitious surface contaminants to facilitate their low-temperature removal. Wet chemical techniques such as the RCA technique [32.216] or a modified RCA technique [32.217], which leave an oxide film $\approx 12\,\text{Å}$ thick on the surface, are effective in reducing contamination by metals and surface carbon contamination. One such technique is to clean the Si surface in a mixture of $NH_4OH : H_2O_2 : H_2O$ (1 : 1 : 5, 85 °C), etch the oxide in dilute HF (2%), and reoxidize the surface in a solution of $HCl : H_2O_2 : H_2O$ (1 : 1 : 5, 65 °C). However, these techniques require the oxide layer to be thermally desorbed in situ at temperatures $\geq 850\,°C$. Hydrogen passivation describes the formation of a self-protective layer on silicon surfaces upon exposure to HF acid. The hydrogen passivation layer has the important advantage of being volatile at $\approx 560\,°C$. A hydrogen passivation treatment has been developed [32.218] that greatly reduces the oxygen and carbon contamination of the Si surface without an extensive prebaking or the high-temperature thermal desorption required after RCA cleaning. This treatment allows the growth of epitaxial layers with reduced dislocation densities, but is substantially more difficult than the RCA treatment.

After thermal desorption of the passivant layer, the substrate is quickly cooled to $\approx 450–500\,°C$ under an As_4 or cracked arsenic flux and then cooled to the nucleation temperature of $\approx 300–340\,°C$ with either a Te_2 flux or no incident flux. This leaves a monolayer of As on the surface of the Si. The absorption of a precursor As layer has been shown to greatly reduce the sticking coefficient of Te on the Si surface and increase its surface mobility [32.219–221] and to be essential to the MBE growth of ZnTe [32.219] or CdTe [32.220] with good growth morphology. A model has been proposed [32.221] for the effect of the As, and density-functional calculations of the geometrical and electronic structures of As and Te on Si(211) have been performed [32.222].

Different groups have developed a variety of recipes for the remainder of the growth procedure. Either a thin ZnTe buffer layer ($\approx$ 60–500 Å thick) is grown by migration-enhanced epitaxy [32.96, 97, 99] at 230–300 °C or a thicker layer (0.1–1 μm) is grown by MBE at $\approx$ 220 °C and then annealed at $\approx$ 360–500 °C under a Te_2 flux before the final thick CdTe layer is grown at $\approx$ 300 °C. The growth of an As precursor and a ZnTe buffer ensures the growth of CdTe (211)B on Si (211)B. To further enhance the quality of the final CdTe epilayer, a ZnTe/CdTe strained-layer SL buffer could be grown before the CdTe growth. It also has been reported [32.97, 223] that the quality of the CdTe growth is enhanced by periodically interrupting the growth and flash annealing at $\approx$ 530 °C or periodically annealing $\approx$ 10 min at $\approx$ 360–380 °C under a Te_2 flux. Reported CdTe growth rates vary from 0.7 μm/h at 300 °C to 1.0 μm/h at 270 °C. Both the desorption of the Si passivation layer and the subsequent growth are monitored by RHEED and SE.

32.6.2 Substrate Preparation and Growth of HgCdTe

The following nucleation and growth procedures are those followed at MPL and at EPIR Technologies; similar procedures are followed at other laboratories and production facilities. A 12 keV RHEED system and an SE system are used routinely to monitor the substrate preparation and HgCdTe growth in situ. All substrates are first degreased in two separately heated trichloroethylene baths followed by rinses in two methanol baths. They are then prepared for HgCdTe growth by etching in a bromine/methanol solution, followed by several methanol and DI water rinses. This leaves a Te-rich surface. For CdZnTe substrates, the etch is performed for 20 s in a 0.5 vol. % solution to remove the top 1000–2000 Å of the CdZnTe, which may have polishing damage. For composite Si-based substrates, it is performed for 5 s in a 1 vol. % solution to remove 500–1000 Å of the CdTe, CdSeTe or CdZnSeTe top layer. Finally, the substrate is dried with nitrogen and loaded into the MBE system. Then it is slowly heated to 300 °C with no incident flux. At $\approx$ 135 °C the surface Te becomes mobile, decreasing

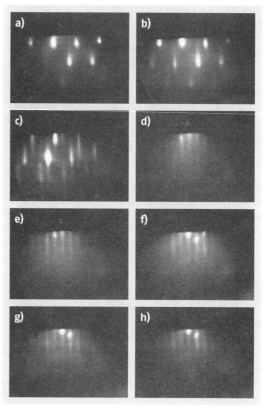

Fig. 32.14a–h In situ RHEED pictures of a HgCdTe sample grown on a CdZnTe substrate during MBE growth: (**a**) CdZnTe substrate as loaded, (**b**) after removal of Te and its oxide, (**c**) after CdTe regrowth and partial cooling towards the growth temperature, (**d**) at the moment HgCdTe growth is started, (**e**) 30 s of growth, (**f**) 1 h of growth, (**g**) about 2 h of growth, (**h**) after 3 h 45 min (end of growth)

the intensity of the rods in the RHEED pattern and causing dots to appear. At 195–205 °C the Te begins to evaporate, causing further changes in the RHEED pattern. At $\approx$ 220 °C the surface Te has evaporated, and the RHEED pattern again becomes streaky with rods. RHEED patterns taken during CdZnTe substrate preparation are shown in Fig. 32.14a–c. For the CdZnTe substrates, the Te shutter is opened at 300 °C, and the temperature is increased to $\approx$ 320 °C. The substrate is maintained at $\approx$ 320 °C for 5–10 min and then is slowly cooled down to 300 °C, where 200–500 Å of CdTe is grown to create a smooth surface. At 300 °C an additional thin CdTe layer is grown on the Si-based substrates and is annealed at $\approx 320\,°C$. This procedure ensures a smooth and clean initial surface prior to HgCdTe nucleation.

After the final CdTe growth, all substrates are cooled to the HgCdTe growth temperature, $185\pm5\,°C$, as measured by calibration of the thermocouple behind the substrate. Four temperatures can be used to calibrate the thermocouple readings:

1. The onset of Te surface mobility on CdTe at $\approx 135\,°C$
2. The onset of Te evaporation at $\approx 195\,°C$
3. The melting temperatures of In droplets on the surface of calibration substrates at $156\,°C$
4. The melting temperatures of Sn droplets on the surface of calibration substrates at $232\,°C$.

The growth of HgCdTe is then begun. A Hg flux is present even before the growth shutters are opened, so that the growth starts with a monolayer of Hg. The growth proceeds at a rate of $2-3\,\mu m/h$ with $\approx 3\times10^{-4}-3.5\times10^{-4}$ Torr Hg flux, 10^{-6} Torr Te_2 flux, and 2×10^{-7} Torr CdTe flux. The growth temperature is continuously adjusted to minimize the roughness of the growing surface as measured by SE, and the CdTe flux is continuously adjusted to maintain the SE-measured composition at the targeted value. The evolution of the RHEED pattern during the MBE growth of a HgCdTe on CdZnTe substrate is shown in Fig. 32.14d–h. If the growth temperature and/or growth rate are too high, Te precipitates and crater defects (also called voids) form due to the greatly reduced Hg sticking coefficient; if too low, microtwinning occurs and surface hillocks appear. To obtain lateral uniformity of the growth, the sample is continuously rotated and the growth fluxes are incident off center at an angle to the normal to the growth surface.

Due to initial island formation, the roughness of the growth initially increases up to a knee point corresponding to a critical growth thickness. Both the SE surface roughness and the RHEED intensity of a specular spot were studied at MPL during the early stages of growth at temperatures from 170 to $196\,°C$. The knee was found to occur in the first minute of growth. At the lowest temperatures the increase in roughness continued beyond the knee, but much more slowly, at intermediate temperatures a very slow recovery was observed beyond the knee, and at higher temperatures an exponential decay with a time constant $\approx 3\,min$ was observed. After the HgCdTe growth was paused, a recovery was observed under Hg flux at all temperatures. The critical thickness is plotted for two representative samples in Fig. 32.15, hct2352 having a higher growth rate than hct2387 at the same temperature. The critical thickness of layer hct2387 shows a much weaker dependence on temperature than layer hct2352. Their critical thicknesses are similar above $185\,°C$, but layer hct2352 has higher critical thicknesses below $185\,°C$. Those differences can be described in terms of the dimensionless ratio $\alpha = Jw^2/Dn_0$. This is the ratio of the input adatom flux J times the square of the (111) terrace width w to

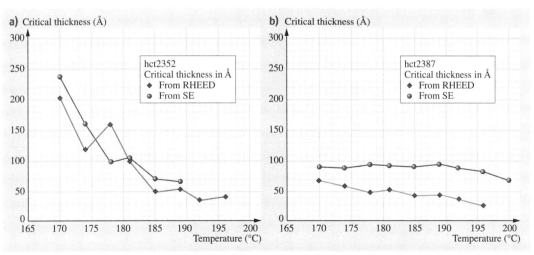

Fig. 32.15a,b Temperature dependence of the critical thickness during the initial stages of growth. The results shown in (a) are from a sample with a higher growth rate than that in (b)

the diffusion constant D times n_0, the concentration of lattice sites for island formation. The results shown in Fig. 32.15 can be explained as follows:

1. At higher temperatures (small α), the rate of clustering is low, and growth proceeds largely by the incorporation of mobile adatoms into step edges. Because most nucleation occurs at the bottoms of step edges, the activation energies for nucleation and island formation are predominantly mechanical, temperature-insensitive energies The critical thicknesses would be predicted to be of the same order as the step heights, $\approx 20\,\text{Å}$.
2. At lower temperatures (large α), the adatoms are less mobile, so the island concentration builds up until the islands largely overlap. The lower the temperature, the lower the adatom mobilities, and thus the steeper the islands and the greater the island heights before overlapping becomes important. Therefore, the critical thickness increases rapidly as the temperature is decreased. The islands and step edges merge, and less densely populated fresh roughened terraces are formed (Stranski–Krastanov growth). The build-up of islands is then repeated, but to a lesser extent, until a steady-state island concentration is reached when the nucleation rate is balanced by the capture of islands due to convection.
3. At very low temperatures (very large α), thermal diffusion is very low and hence the steepness of islands on the terraces is so large that the surface becomes extremely rough and continues to grow rougher even after reaching the critical thickness, eventually saturating and becoming time independent (3-D growth).

With a relatively small increase in temperature, 3-D growth turns into terraced hillock growth (modified S-K-M) and then into step-flow growth, as explained by the above discussion. It also explains why, as the input flux or the substrate roughness increases, step-flow growth requires a higher substrate temperature (lower α) due to a higher level of nucleation.

32.7 Dopants and Dopant Activation

The active components of both currently used and advanced IR photodetectors being developed, as well as other optoelectronic devices, use photovoltaic p–n junctions. In order to form sharp, precisely controlled interfaces and p–n junctions and to obtain well controlled doping levels, it is essential to grow such junctions by MBE with extrinsic n- and p-type doping. Although the stable, well-controlled n-type extrinsic doping of as-grown HgCdTe and HgTe/CdTe SLs has been demonstrated by many authors, the p-type extrinsic doping of these materials has remained a problem. In particular, the reproducible p-type activation of As, the best low-diffusivity p-type dopant in these materials, has repeatedly been reported to occur reproducibly only at annealing temperatures sufficiently high to vitiate many of the advantages of MBE as a growth technique. Even though almost all device fabricators have demonstrated the ability to fabricate high-performance HgCdTe-based photovoltaic photodiodes using various p-doping techniques, interdiffusion across interfaces, dopant diffusion, and the formation of Hg vacancies all become problems at the high annealing temperatures required for the reproducible p-type activation of group V dopants. The absence of a reproducible p-type doping technique which does not require high-temperature annealing remains a major stumbling block to successful implementation of high-yield fabrication of MBE-grown HgCdTe IR FPAs. This problem, attempts to solve it in the past, current progress, and alternative p-type dopants are reviewed briefly here.

32.7.1 Extrinsic n-Type Doping

Stable, well-controlled n-type extrinsic doping has not proven to be a major problem. The incorporation of In during MBE growth has been used to achieve this goal for as-grown HgCdTe [32.224–226] over a wide range of dopant concentrations (10^{14}–$10^{19}\,\text{cm}^{-3}$) and for HgTe/CdTe SLs [32.227–229] over the somewhat smaller range from 10^{15} to $10^{18}\,\text{cm}^{-3}$. A low-temperature postgrowth anneal can be used to optimize the structural and electronic properties of the doped material, but is not required for the n-type activation of the dopant In atoms. The only problem in achieving well-controlled n-type doping arises at very low doping levels from the intrinsic p-type doping which arises from Hg vacancies. For very small In concentrations, these vacancies can compensate the n-type conduction, lowering the mobility and destroying the otherwise precise control of the doping. However, Hg diffuses easily through interstitial sites in HgTe and IR-absorber HgCdTe material, even in perfect material, having an activation energy of only about 0.67 eV for diffusion via interstitial sites [32.230], and diffuses even more rapidly along dislocation cores [32.231]. This allows the Hg vacancy density to be reduced to values below 10^{13} cm^{-3} by low-temperature annealing under a high Hg vapor overpressure, essentially eliminating this problem, although it could reoccur during subsequent device processing steps or during long-term storage.

32.7.2 Extrinsic p-Type Doping

Extrinsic doping is necessary:

1. Because intrinsic p-type doping by Hg vacancies is not stable against diffusion [32.230, 232, 233] from p- to n-type regions
2. Because the density profile of Hg vacancies cannot be well controlled
3. Because Hg vacancies significantly degrade carrier mobilities [32.234] and act as SRH recombination centers for minority carriers, shortening the carrier lifetimes [32.225, 235–239] and hence decreasing the quantum efficiency of detectors having p-type absorption layers.

On the other hand, achieving the stable, well-controlled extrinsic p-type doping of MBE-grown HgCdTe and HgTe/CdTe SLs without negating the advantages of MBE growth offers difficult and fundamental challenges. The primary problem arises because the MBE growth of HgCdTe is optimally performed under Te-rich conditions [32.226, 233, 240–242], so that Hg vacancies are the dominant native point defects in as-grown HgCdTe. Although their density typically [32.239] is only $\approx 10^{14}$ cm^{-3}, it is many orders of magnitude higher than the density of Te vacancies in as-grown HgCdTe [32.230, 232, 243–245]. Therefore, in equilibrium, group V atoms in as-grown HgCdTe are less likely to be located on anion (Te) sites than on cation (Hg or Cd) sites, where they form neutral complexes or act as donors rather than acceptors. This has been confirmed by both theory [32.230, 232, 243–245] and experiment [32.224, 225, 233, 241, 242, 246–251]. (Of course, even for very high As concentrations, the number of cation vacancy sites remaining to be filled by As atoms is essentially independent of the number already filled by As atoms.) On the other hand, group I atoms, which also are located primarily on cation sites, and thus are p-type activated, suffer from high diffusivities.

Postgrowth As Incorporation

Extrinsic p-type doping can be achieved either by the ex situ implantation or diffusion of dopant ions into n-type layers or by their in situ incorporation during MBE growth. As is the preferred p-type dopant for implantation or in situ incorporation during MBE growth. As implantation [32.252–255] is the key procedure of the planar device fabrication process, and has been employed in most IR FPA fabrications. However, ion implantation induces considerable damage, which reduces both minority-carrier lifetimes and carrier mobilities, and therefore is detrimental to device performance. A high-temperature ($\geq 425\,°$C) anneal is usually performed after ion implantation. This anneal serves a dual purpose:

1. It activates the As as a p-type dopant
2. It reduces the residual implantation-induced damage.

This technology has been used successfully for the fabrication of mid-wave IR (MWIR) FPAs. However, such a high-temperature anneal itself is detrimental in several ways: it leads to the possibility of introducing a much higher density of Hg vacancies – at 500 °C the equilibrium density of Hg vacancies has been calculated to be above 10^{18} cm^{-3} even under a 1 atm Hg partial pressure [32.244, 245] – and both interdiffusion across interfaces and dopant diffusion become problems, vitiating primary advantages of the MBE growth technique. Furthermore, even the residual damage remaining after annealing is severely detrimental to the performance of long-wavelength IR (LWIR) devices, due to their narrower bandgaps – bandgap states induce significant tunneling in devices with very narrow bandgaps, resulting in large dark currents. An alternative method [32.256, 257] for the postgrowth incorporation of As is to diffuse it in from the surface. However, that requires an even more deleterious thermal treatment. In addition, ion implantation and As indiffusion can be performed only in-plane, which dramatically limits the possible detector structure designs for which these techniques are applicable. The next generation of IR detectors, including single-focus multicolor detectors and high-operating-temperature LWIR detectors, cannot be fabricated in-plane. They will require in situ incorporation of extrinsic p-dopants during MBE growth and their activation as-grown or after only low-temperature ($< 250\,°$C) annealing.

32.7.3 In situ Group I Dopant Incorporation

In principle, either group I or group V elements could act as acceptors in HgTe or HgCdTe – group I elements on cation (Hg or Cd) sites or group V elements on anion (Te) sites. However, unlike liquid-phase epitaxial growth, the MBE growth of HgTe and narrow-gap HgCdTe must be performed under Te-rich conditions [32.224, 233, 241, 242], so that during growth the number of cation vacancies is several orders of magnitude higher than the number of Te vacancies [32.230, 232]. Thus, group I dopant atoms in these materials reside almost entirely on cation sites and act as singly ionized p-dopants in the as-grown material, unlike group V dopant atoms. For this reason group I elements were the first to be tried as extrinsic p-dopants. The incorporation of Cu, Ag, Au, and Li as p-type dopants in HgCdTe epilayers and their diffusivities have been studied by many authors [32.224, 227, 233, 239, 258–269]. Even though the doped materials showed close to 100% activation and excellent transport properties and minority-carrier recombination times, all of these dopants proved to be unstable in HgCdTe, diffusing out of the doped region during further MBE growth or during the low-temperature postgrowth anneal. This behavior would appear to prevent their use in FPA technology and hence limit their usefulness as an acceptor dopant, especially for the abrupt junctions required for IR detectors. Thus, they suffered a period of relative neglect in the late 1980s. However, Au in particular is again being studied intensively as a p-type dopant, as its diffusivity may be able to be held within acceptable bounds. We have found no perceptible diffusion of Au dopants in a HgTe/CdTe SL after 3 years of storage at room temperature.

32.7.4 In situ Group V Dopant Incorporation

In recent years, because of the high diffusivities of group I elements and the damage caused by ion implantation, studies on the p-type doping of HgCdTe and of HgTe/CdTe SLs have focused on in situ doping with group V elements, especially with As [32.224, 229, 262, 270–286], because it is more easily activated than the other group V elements. The larger size of the group V atoms, and hence their lower diffusivity, allows the growth of stable, well-controlled p–n junctions, and thus sharp interfaces and/or precisely controlled dopant profiles. The incident As flux in most older experiments was uncracked and consisted almost entirely of As_4 tetramers, which are physisorbed as tetramers or clusters of tetramers, not as single atoms [32.40], and which calculations suggest [32.270] remain as tetramers in as-grown HgCdTe. The use of cracked As has become common, so that the As is mostly chemisorbed in atomic or diatomic form, as happens with the use of a CdAs flux. This gives an As sticking coefficient orders of magnitude higher but appears not to ease the problem of obtaining p-type activation of the As without a high-temperature anneal.

32.7.5 As Activation

Because the As atoms in as-grown MBE-grown HgCdTe reside primarily on cation sites rather than Te sites, annealing under a Hg overpressure is required to activate the As as a p-dopant by transferring it to Te sites. Several authors have proposed models for the activation process and performed calculations [32.242, 244, 245, 271, 272], all assuming As incorporation as isolated atoms. The standard activation anneal is 10 min at $\approx 425\,°C$ followed by 24 h at $\approx 250\,°C$ under a Hg overpressure to fill the Hg vacancies created during the previous anneal. The anneal at a temperature well above $350\,°C$ induces Te evaporation, which allows the As atoms to move to the resultant Te vacancies in the second-stage anneal. This annealing procedure yields almost complete activation of the As up to concentrations of order $10^{18}-10^{19}\,cm^{-3}$ depending on the temperature of the first-stage anneal [32.244, 245, 273, 274], with the doping abruptly saturating at that level, presumably due to all of the Te vacancies created during the first-stage anneal being filled.

Unfortunately, this annealing procedure introduces undesirable broadening of interfaces and dopant diffusion, and thus limits many of the advantages of MBE as a low-temperature growth technique. Therefore, a variety of approaches have been used in attempts to obtain the p-type activation of As either as-grown or after only a low-temperature anneal. Although each of these approaches has shown promise, none have consistently given full activation of the As. One approach [32.242, 247, 275, 276] is to use a two-stage anneal with $T \leq 300\,°C$ in both stages, with Te vacancy–Te interstitial pairs formed in the first stage and As atoms filling the Te vacancies in the second stage. Another approach [32.250] is to force As atoms out of cation vacancy sites by using a high Hg overpressure in both stages of a two-stage low-temperature anneal. The lack of consistently reproducible results obtained using these approaches is not understood. A rather thorough study of As incorporation, activation, and diffusion is given in [32.250].

Other approaches have involved either modified MBE growth methods or different methods of As incorporation. One approach [32.277–280] was to irradiate the growing HgCdTe with a laser beam (photo-assisted MBE, PAMBE), shifting the growth toward cation-rich conditions by causing Te desorption and thus allowing As incorporation on the vacant Te sites. This was combined with the idea of growing HgTe/CdTe SLs and then forming HgCdTe by annealing the SLs to interdiffuse the Hg and Cd. That allowed the As to be incorporated only during the CdTe growth, when the number of cation vacancies is much lower so that it is easier to incorporate the As on Te sites. This will yield almost complete p-activation of the As, but at a high price. PAMBE has been abandoned because it limits the growth to impractically small areas. Also, annealing the SL broadens any device interfaces. Another approach [32.281, 282] was to use a Te cracker cell so as to increase the Hg sticking coefficient and obtain As activation as-grown. Success was reported for (111) and (211) growth, but appears not to have been consistently reproducible.

An approach in which only the method of As incorporation is changed, not the growth technique, is that of planar doping. In planar doping the As flux is shut off most of the time, but periodically the CdTe and Te fluxes are shut off and the As flux is turned on, along with the Hg flux, which is never shut off. This approach was originated [32.283] in 1995, has been used many times [32.255, 283–286] (and in unpublished results of the authors) and has been observed to give full p-activation of the As in the as-grown material [32.283, 284]. Although full activation in the as-grown material is not observed consistently, at least partial activation is observed, and full activation appears to be consistently obtainable by combining planar doping with anneals at $\leq 300\,°C$. More consistent activation is obtainable in the planar doping of CdTe or of the CdTe layers of a HgTe/CdTe SL.

In conclusion, although progress is being made in solving the problem of obtaining well-controlled fully reproducible p-type doping of HgCdTe without the use of undesirable high-temperature anneals, this remains one of the major problems in the growth of material for advanced HgCdTe-based IR detectors and focal-plane arrays. The most promising avenue of attack on this problem appears to be planar doping with As coupled with the development of an optimal sequence of low-temperature anneals.

32.8 Properties of HgCdTe Epilayers Grown by MBE

The most important epilayer properties for detector and FPA performance are the minority-carrier lifetime, the carrier concentration and mobility, the perfection of the epilayer surface (to ensure the absence of dead pixels), and the lateral uniformity of the epilayer (to ensure uniformity of pixel response). Substrate imperfections, poor lattice matching and thermal matching between the substrate and the epilayer, and nonoptimal growth conditions all can degrade these properties. We consider first the electrical and optical properties.

32.8.1 Electrical and Optical Properties

Minority-carrier recombination processes in HgCdTe directly decrease photon detection and increase noise generation in photovoltaic infrared devices. Hence, it is especially important to maximize the minority-carrier lifetime τ. The three commonly recognized recombination mechanisms – Auger, radiative, and Shockley–Read–Hall (SRH) – are characterized by recombination lifetimes that are usually measured by fitting photoconductive decay data, which measure the total recombination lifetime $\tau = (\tau_{\text{Auger}}^{-1} + \tau_{\text{radiative}}^{-1} + \tau_{\text{SRH}}^{-1})^{-1}$. One can determine each recombination rate separately by fitting the measured values of τ as a function of T. Of these three mechanisms, two (Auger and radiative recombination) are determined by intrinsic material properties, such as the composition and doping level, which can be precisely measured by other techniques, e.g., the Hall effect and infrared absorption mapping [32.287]. The Auger and radiative lifetimes can be calculated from those measured quantities using sets of semi-empirical formulas developed by *Beattie*, *Landsberg*, and *Blackmore* [32.288, 289] based on Kane's $\mathbf{k}\cdot\mathbf{p}$ model with several key parameters determined experimentally [32.290]. These intrinsic mechanisms establish upper limits on the lifetime. The differences between the calculated and measured lifetime values are attributed to SRH recombination, which arises from impurities or defects and can be fitted within the theoretical framework given by *Shockley*, *Read*, and *Hall* [32.291, 292]. A primary criterion for a high-

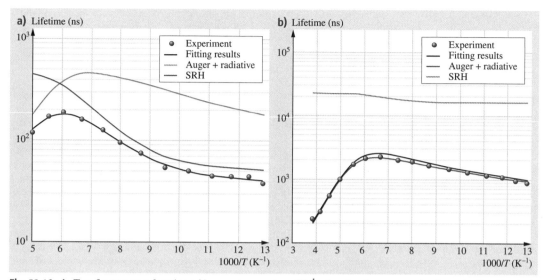

Fig. 32.16a,b Two fits to τ as a function of inverse temperature (T^{-1}) showing the measured values, the fitted intrinsic (*dotted*) and SRH (*solid*) lifetimes, and the fit to the measured values of τ; (**a**) shows a case in which SRH recombination dominates below 150 K, and (**b**) shows a case in which the SRH recombination rate is small and hence less well determined

quality layer is $\tau_{SRH} \gg \tau$. Two fits to τ are shown in Fig. 32.16: (a) with SRH processes dominant below ≈ 150 K and (b) with SRH processes negligible. Note the difference in the dependence of τ on T^{-1} in the two cases. Also, note that in Fig. 32.16b τ_{SRH} is very poorly determined because SRH recombination is so unimportant in that case.

In general, it is found that, for growth on CdZnTe substrates, the strong Auger recombination dominates the determination of τ and thus limits detector performance for p-type layers. However, intrinsic lifetimes are much longer in n-type materials, so it is more difficult to achieve the condition $\tau_{SRH} \gg \tau$. Others [32.293] have found that SRH recombination is dominant for low In doping levels in n-type MWIR HgCdTe samples, but that τ increases with increasing doping, suggesting that In may passivate the SRH defects for MWIR HgCdTe. On the other hand, we have found at MPL that lifetimes $\approx 1 \mu$s dominated by intrinsic recombination can be reproducibly achieved in n-type LWIR material with doping levels $\approx 10^{15}$ cm^{-3} after annealing under a Hg-rich atmosphere to reduce the concentration of Hg vacancies, which act as SRH recombination centers.

Two other electrical characteristics of a grown HgCdTe epilayer are important: the carrier concentration and the mobility. In the absence of compensation (the presence of only n- or only p-type dopants) both are routinely determined from Hall-effect measurements. However, in the presence of compensation their determination requires Hall-effect measurements as a function of magnetic field as well as temperature [32.293]. For LWIR $Hg_{1-x}Cd_xTe$ layers with $x = 0.23$ grown on Si-based substrates, at 77 K mobilities $\mu \approx 10^5$ cm^2/Vs are consistently obtainable for n-type layers with $n \approx 10^{15}$ cm^{-3}, $\mu \approx 500$ cm^2/Vs for p-type layers with $p \approx 1.3 \times 10^{16}$ cm^{-3}, and $\mu \approx 200$ cm^2/Vs for p-type layers with $p \approx 2.0 \times 10^{16}$ cm^{-3}. For layers of the same composition grown on CdZnTe substrates the electron mobilities reported are lower by a factor of ≈ 2–3.

The primary optical characteristic of HgCdTe epilayers is the IR transmittance or absorptivity, which is measured by FTIR spectroscopy. A general review of the IR optical characterization of HgCdTe and of FTIR spectroscopy in general is given elsewhere [32.294]; here we outline how FTIR measurements can be used to assess epilayer quality. Below the energy gap, impurity absorption can be measured and carrier concentrations can be measured by free carrier absorption. By scanning transmission/reflection mapping, the uniformity of composition and thickness across a wafer can be meas-

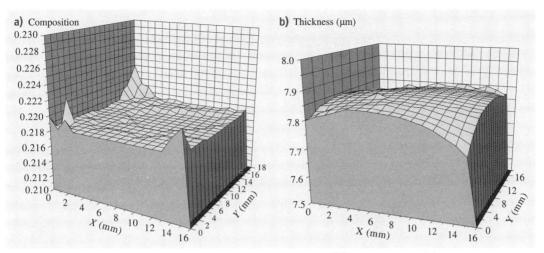

Fig. 32.17a,b Mapping results for a HgCdTe epilayer grown on CdZnTe: (**a**) composition and (**b**) thickness

ured. Typical results obtained at MPL for growth on square 20 mm × 20 mm CdZnTe substrates are shown in Fig. 32.17; the average HgCdTe composition was $x = 0.2182$ with a standard deviation of 0.0006, and the average thickness was $d = 7.84\,\mu\text{m}$ with a standard deviation of $0.03\,\mu\text{m}$ [32.287]. Higher compositions and thinner material was found in about $2 \times 2\,\text{mm}^2$ areas in each corner, presumably due to the low thermal conductivity resulting in a higher growth temperature in the corners and hence a lower Hg sticking coefficient there. This explanation is supported by the existence of a higher density of voids or crater defects near the corners; these are found primarily in Te-rich or higher-T growth. For growth on 3 inch-diameter Cd/Si substrates, there was no evidence of nonuniformity in T, and the nonuniformities in x and d over the central $20 \times 20\,\text{mm}^2$ area were much smaller than for growth on CdZnTe, although over the entire samples somewhat larger nonuniformities were observed, presumably due to flux nonuniformities.

Of greater importance, the Urbach tail energy, which arises primarily from electron–phonon interactions, alloy disorder, and structural disorder, has been shown to be useful in characterizing material quality [32.295]. The absorption coefficient below the energy gap can be expressed as [32.296, 297]

$$\alpha(\omega) = \alpha_0 \exp\left[\frac{\hbar\omega - E_0}{W(T)}\right], \tag{32.9}$$

where $W(T)$ is defined as the Urbach tail energy. It measures the steepness of the band tail and can be obtained by fitting the the absorption data below the energy gap. We obtained values for $W(T)$ by measuring $\alpha(\omega)$ from 4 K to room temperature and fitting the logarithmically linear part of $\alpha(\omega)$ below the bandgap. Those samples with surface microtwin defects arising from low growth temperatures or with high dislocation densities were seen to consistently show relatively high Urbach tail energies having large standard deviations over small areas, suggesting that nonuniformly distributed structural disorder dominates the nonuniformity of device performance figures of merit in HgCdTe IR FPAs. Moreover, Urbach tail energy mapping measurements we have performed correlate strongly with Cd composition mappings and with crystalline defects revealed by x-ray topograph mappings. The contribution of the electron–phonon interaction was calculated and found to be much smaller than the contribution of alloy and structural disorder and to give the observed small temperature dependence of $W(T)$ [32.295]. We also calculated the contribution of alloy disorder, $W_{\text{alloy}} = x(1-x)\text{VBO}^2 m^*_{\text{HH}} L^2 / 7.2\hbar^2$, where x is the alloy composition, VBO is the valence band offset, and L is the alloy disorder correlation length (estimated to be of the order of one interatomic spacing). It also was found to be much smaller than the contribution of structural disorder. $W(T)$ also was found to be only weakly dependent on T for CdZnTe substrates, for which the alloy contribution also is very small because of the small Zn concentration. Thus, $W(T)$ serves as a quantitative measure of structural disorder for both HgCdTe epilayers and CdZnTe substrates.

32.8.2 Structural Properties

The structural properties of HgCdTe depend greatly on the type of substrate used for its growth. For (211)B growth on CdZnTe, substrate defects are the primary cause of structural defects in the HgCdTe. For (211)B growth on CdTe/ZnTe/Si, the lattice and thermal mismatches at the ZnTe/Si and ZnTe/CdTe interfaces are primary causes of structural defects in the HgCdTe. The lattice mismatches are much greater than any mismatch resulting from thermal changes, but the thermal mismatches can cause strains much greater in magnitude than the residual lattice-mismatch strains remaining after relaxation by dislocation formation during growth. For simplicity consider a CdTe/Si interface. Following *Carmody* et al. [32.110], the equilibrium residual strain in a CdTe film of thickness h is $\varepsilon(h) = (h_c/h)\varepsilon_{\mathrm{misfit}} \approx 10^{-4} \times \varepsilon_{\mathrm{misfit}}$ for a film of thickness $5\,\mu\mathrm{m}$, where $h_c \approx 5\,\text{Å}$ is the equilibrium critical thickness and $\varepsilon_{\mathrm{misfit}} = -0.193$ is the lattice-mismatch strain, giving a residual stress $\sigma_{\mathrm{resid}} \approx 1.4\,\mathrm{MPa}$. Of course, the CdTe does not come to a complete equilibrium during growth, but even so the stress remaining at the growth temperature at the end of growth is much less than the thermal stress resulting from cooling to room temperature, which is $\approx 45\,\mathrm{MPa}$. Also, the relaxation of the strain induced during cooldown will be much less complete than the relaxation of the lattice-mismatch strain because of the much shorter time at an elevated temperature. Thus, incomplete relaxation upon cooldown can be a greater source of strain than the large lattice mismatch.

The most common measure of crystalline perfection is the XRD FWHM. Under optimal growth conditions the XRD FWHM of HgCdTe grown on CdZnTe substrates is limited only by the substrate FWHM, at least for (112)B, (113)B, and (552)B orientations, and is $\approx 20\,\mathrm{arcsec}$ [32.44]. However, the XRD FWHM is substantially larger for HgCdTe grown on CdTe/Si substrates. From 1991 to 1995 the best values obtained were reduced from 210 to 65 arcsec, but little progress has been made in achieving further reduction. In fact, typical values of the FWHM are still above 100 arcsec, and the best consistently obtainable value is $\approx 100 \pm 30\,\mathrm{arcsec}$, limited by the FWHM of the underlying CdTe layer [32.96]. One possible source of the 65 arcsec XRD line broadening is incomplete lattice relaxation giving rise to changes in the lattice constant in the growth direction as one moves away from the surface. This possibility is supported by an anticorrelation between the density of threading dislocations and the FWHM; if true, it would explain why no further significant decrease in the FWHM has been observed since 1995. The primary sources of the extra width observed before 1995 were microtwinning and low-angle grain boundaries due to growth at nonoptimal orientations and with nonoptimal buffer layers. The extra width above 65 arcsec still usually seen today probably arises in large part from low-angle grain boundaries.

Another measure is the etch pit density, which measures the density of threading dislocations at the HgCdTe surface. Dislocations adversely affect HgCdTe electrical properties and device performance in many ways [32.298]; the deleterious effects of dislocations were discussed briefly in Sect. 32.3.3. Threading dislocation densities below the mid $10^5\,\mathrm{cm}^{-2}$ are not a serious problem unless they form clusters or bend over to form misfit dislocations parallel to the surface under device contacts [32.298]. Growth on CdZnTe yields etch pit densities of order low 10^4 to low $10^6\,\mathrm{cm}^{-2}$, depending primarily on the substrate defect density. Growth on CdTe/ZnTe/Si typically yields etch pit densities of order mid 10^6 to mid $10^7\,\mathrm{cm}^{-2}$. As discussed in Sect. 32.3.3, this density of threading dislocations to date has prohibited the use of Si-based substrates for the growth of HgCdTe for LWIR and VLWIR detectors and FPAs.

Other structural defects commonly encountered before the last decade, such as two-phase growth, twinning and microtwinning, and low-angle grain boundaries, have been greatly reduced by choice of growth orientation, proper substrate preparation and the use of buffer layers, and improved control over growth conditions. However great care must be exercised to maintain the level of these defects at an acceptable level, especially twinning, which occurs at low growth temperatures T or high Hg flux, and can cause harmful surface defects. Under optimal growth conditions T is high enough that the surface diffusion length of incoming atoms is approximately the terrace width. Then, growth nucleation occurs at kinks or step edges, and step-flow growth occurs. If T is too low or in the presence of defects, growth can also nucleate on the terraces, so that growth from different points must coalesce, forming grain boundaries or laminar microtwins.

32.8.3 Surface Defects

The common surface defects are voids or crater defects, large triangle defects (like a type of void and treated here under the same heading), faceted microvoids or small triangle defects (similar to microvoids and in-

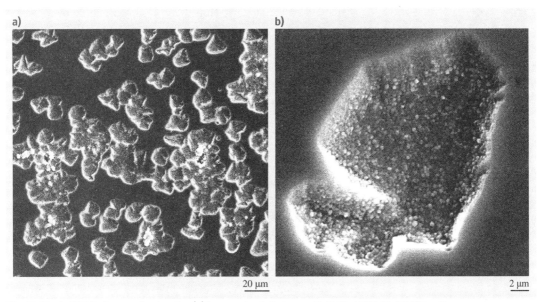

Fig. 32.18a,b SEM micrographs showing (**a**) large crater defects on a HgCdTe/CdZnTe (211)B epilayer grown on a poor substrate and (**b**) the typical morphology of a large crater defect

cluded here under the same heading), flakes, hillocks, needle defects, cross-hatching, and surface roughness. The reduction of the density of surface macroscopic defects (voids, microvoids of size 3–5 μm, and flakes) is one of the most serious challenges in MBE growth of HgCdTe. Voids or crater defects, discussed briefly in Sect. 32.3.2 and shown in Figs. 32.5 and 32.18, are the surface manifestations of threading defects associated with Te precipitates. Small and mid-size voids arise during growth from incomplete Te_2 dissociation (low T) or excess Te under Te-rich growth conditions (high T, low Hg flux or high growth rate) [32.39, 299–301]. For growth on CdZnTe substrates, large voids can arise from dust or other particulate matter on the substrate surface [32.183, 302] or Te precipitates in CdZnTe substrates [32.66, 69]. Detailed descriptions of their origin and morphology are given in [32.300] and [32.301]. They range in size from 2 to 40 μm [32.302] and can bypass p–n junctions and give premature reverse-bias breakdowns or even blind sensor elements [32.303]; hence, they are extremely harmful to FPA performance. They can be minimized by tight control of the growth conditions [32.39, 304] and by the proper preparation of CdZnTe substrates [32.66–70]. Typically, the void or crater density is mid to high $10^2\,cm^{-2}$ for growth on Si-based substrates and low 10^2 to high $10^3\,cm^{-2}$ for growth on CdZnTe substrates [32.38], depending strongly on both the substrate quality and the growth conditions. The minimum reproducible void density is high $10^2\,cm^{-2}$ for growth on Si-based substrates and $\approx 10^3\,cm^{-2}$ for growth on CdZnTe substrates.

Faceted microvoids have a sharp triangular shape and smooth, slightly raised surfaces. For growth on Si-based substrates, they are typically 3–5 μm long, their density is $\leq$ low $10^2\,cm^{-2}$, and they almost always result in electrical shorts at 140 K, blinding sensor elements [32.303]. At EPIR Technologies, 1 μm-long faceted microvoids have been found on the ZnTe and CdTe epilayers in CdTe/ZnTe/Si substrates and have been identified with microtwinning. To the best of the authors' knowledge, there has been no report of faceted microvoids for HgCdTe grown on CdZnTe. It is reasonable to assume that the microvoids observed on the surface of HgCdTe epilayers grown on CdTe/ZnTe/Si arise from microvoids on the substrate surface.

Like faceted microvoids, flakes have been reported only for growth on Si-based substrates. They have been investigated at the US Army Research Laboratory [32.96], the Night Vision Electronic Sensors Directorate [32.96], and EPIR Technologies. A typical flake defect is shown in Fig. 32.19. Flakes on the HgCdTe epilayers are known to result from correspond-

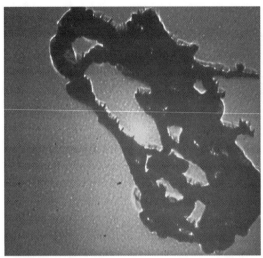

Fig. 32.19 An optical micrograph of a typical flake defect

ing flakes on the top CdTe layers of the substrates; they are completely a substrate problem. Unlike other defects, they have no known intrinsic origin. They arise primarily from sputtering from the CdTe cell during the substrate CdTe growth, but also from spitting from accumulated deposits in the growth chamber. They have irregular shapes, lateral dimensions ranging from several micrometers to > 1 mm and vertical dimensions ranging from $< 1\,\mu$m to $> 10\,\mu$m. Their density can be reduced to $\approx 2\,\text{cm}^{-2}$ by reducing the CdTe growth rate to $\leq 2\,\mu$m/h and taking care not to overheat the CdTe source, and has been reduced to $\approx 0.2\,\text{cm}^{-2}$ by EPIR Technologies by taking other measures. However, because of their very large sizes and 3-D nature, even one large flake is extremely deleterious to FPA performance. Large flakes must be completely eliminated for satisfactory FPA fabrication. They are the primary factor preventing the commercial success of HgCdTe grown on Si-based substrates, because the high dislocation density in HgCdTe grown on Si-based substrates is not harmful for most SWIR and MWIR applications.

Hillocks arise from structural defects – microtwinning or 3-D growth – and thus have largely been brought under control by the advent of (211)B growth and improved control of the growth conditions [32.305]. Needle defects were considered in Sect. 32.3.2. They occur only for growth on CdZnTe, and their density has been reduced to the range from mid $10^4\,\text{cm}^{-2}$ to mid $10^5\,\text{cm}^{-2}$ for better substrates and mid $10^6\,\text{cm}^{-2}$ for the worst substrates. The densities of needle defects and threading dislocations have been found to be correlated, but not in a cause-and-effect manner. This correlation makes it difficult to see any effect of the needle defects on electrical properties, and no effect has been seen. They appear not to be a problem, at least not below the mid $10^5\,\text{cm}^{-2}$ range. Cross-hatching is a sign of high-quality growth and is not detrimental to device performance. Surface roughness is an indicator of suboptimal growth, but also is not in itself detrimental to device performance.

32.9 HgTe/CdTe Superlattices

HgTe/CdTe SLs were first proposed in 1979 [32.304]. Their first actual growth was performed in 1982 [32.306]. They possess several potential advantages over HgCdTe alloy material for use in IR detectors, especially for very long wavelengths, in particular:

1. Their bands can be engineered to suppress Auger recombination relative to that in comparable bulk detectors, and SRH lifetimes up to $20\,\mu$s have been measured [32.307].
2. The band-to-band tunneling currents are lower than those in comparable bulk detectors due to greater effective masses in the growth-axis direction.
3. Their optical cutoff energies E_c or wavelengths λ_c are sharper than those in comparable bulk detectors for well and barrier widths ≤ 90 Å.
4. By doping only in the CdTe barrier layers, reproducible As activation can be achieved at lower temperatures than in MBE-grown HgCdTe alloys.
5. There is no alloy disorder scattering in HgTe wells such as that in HgCdTe alloys.
6. By doping only in the CdTe barrier layers, one can largely eliminate carrier scattering off dopant atoms. Also, their electrons and holes are in the same layers, resulting in strong optical absorption as compared with that in type II SLs.

However, the usefulness of HgTe/CdTe SLs as a narrow-bandgap optical material relies on the stability of the constituent Hg, Cd, and Te atoms to remain in place across the heterointerfaces of the SL structure. Interdiffusion in HgTe/CdTe SLs has been studied since shortly after their first growth [32.308–315]. Significant interdiffusion would drastically decrease the cutoff wavelength of a SL and greatly reduce the potential advantages that SLs have over conventional bulk HgCdTe devices. The FTIR spectra and E_c values of HgTe/CdTe SLs change rapidly at even moderate annealing temperatures $T_{an} < 300\,°C$ after growth, with the cutoff energies becoming higher. This has furthered serious concerns about the practicality of SLs for use as absorbers in the LWIR ($8-12\,\mu m$) and VLWIR ($>15\,\mu m$) due to the presumed smearing out and further instability of the HgTe/CdTe interfaces during annealing, device processing, and possibly even after long storage times. On the other hand, we have found that the interdiffusion which takes place during growth and initially thereafter results in a sharpening of the initially rough SL interfaces.

The novel electrical and optical properties of these structures have been reviewed by many authors [32.308, 316–320]. The MBE growth of HgTe/CdTe SLs was initiated at many industrial and other research laboratories in the mid to late 1980s – Rockwell in 1983, Honeywell in 1984, North Carolina State University in 1985, Texas Instruments and Bell Labs in 1986, Hughes Research Laboratory in 1987, and McDonnell Douglas in 1988. Research on HgTe/CdTe SLs peaked in the late 1980s and early 1990s, when MBE was still a relatively new technology and the growth of HgTe-based materials was in its infancy. Because even better control over the growth is required to grow a high-quality HgTe/CdTe SL than to grow high-quality HgCdTe alloy material, for practical technological reasons, the great advantages possessed by HgTe/CdTe SLs in theory were not realized in practice despite rather comprehensive research efforts. Among the factors that hindered the growth of high-quality HgTe/CdTe SLs at that time were the poor design of the Hg cells used, the absence of computer controlled shutters on the MBE system cells, and the unavailability of consistently high-quality substrates. Thus, the interest in HgTe/CdTe SLs peaked well ahead of its time.

32.9.1 Theoretical Properties

Either inverted-band or normal HgTe/CdTe SLs can be grown. With $50\,\text{Å}$ CdTe barrier thicknesses d_b, a SL has a normal band structure if the HgTe well thickness d_w is less than $90\,\text{Å}$ and an inverted band structure if $d_w > 90\,\text{Å}$. SLs having $d_w < 70\,\text{Å}$ are preferred because SLs with a direct band structure have much longer Auger lifetimes τ_A. For a p-type $60\,\text{Å}$

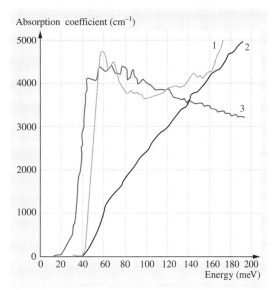

Fig. 32.20 Computed fundamental optical absorption spectra for two SLs (*curve 1* for a normal-band-structure $60\,\text{Å}$ HgTe/$50\,\text{Å}$ Hg$_{0.05}$Cd$_{0.95}$Te SL and *curve 3* for an inverted-band-structure $110\,\text{Å}$ HgTe/$50\,\text{Å}$ Hg$_{0.05}$Cd$_{0.95}$Te SL) and for bulk HgCdTe alloy (*curve 2*), all with optical bandgaps of approximately $40\,\text{meV}$ at $40\,\text{K}$

HgTe/$50\,\text{Å}$ CdTe SL having a $28.8\,\mu m$ optical cutoff and $p = 5\times 10^{15}\,\text{cm}^{-3}$, $\tau_A = 7\,\text{ns}$ at $40\,\text{K}$, whereas $\tau_A \ll 1\,\text{ns}$ at $40\,\text{K}$ for a HgCdTe alloy having the same optical cutoff and doping level. For the same doping level, a $50\,\text{Å}$ HgTe/$50\,\text{Å}$ CdTe SL having a $15\,\mu m$ cutoff has an intrinsic lifetime of $\approx 1\,\mu s$.

Another theoretical property of great interest is the sharpness of the optical absorption edge. Here, HgTe/CdTe SLs have a great advantage over HgCdTe alloys. As shown in Fig. 32.20 for a normal SL, an inverted-band SL, and bulk HgCdTe alloy, all having optical cutoffs of $\approx 29\,\mu m$ (optical bandgaps of $\approx 40\,\text{meV}$), both the normal and the inverted band SLs have much sharper absorption edges than the alloy. Note that interfacial roughness, as long as it is uniform over lateral distances of order $1\,\mu m$ and occupies no more than $\approx 1/2$ of the SL volume, does not significantly affect the sharpness of the SL absorption edges. Even a poorly grown SL will display a sharp absorption edge and other sharp structure from the SL subband structure.

For CdHgTe barrier layers containing $\le 20\%$ Hg with $d_b \ge 50\,\text{Å}$, the normal-band SL energy gap and optical cutoff are almost independent of d_b and the Hg concentration, and depend almost entirely on d_w and on the Cd concentration in the wells, with the dependence on Cd concentration being ≈ 0.9 times as large as that in the HgCdTe alloy with the same cutoff. If one could prevent Cd diffusion into the wells so that they would be pure HgTe, one could control d_w well enough to determine the optical cutoff more accurately in a SL than one could in the alloy by controlling the concentration. However, in reality the combination of interfacial roughness and diffusion make the determination of the optical cutoff less accurate in the SL than in the alloy.

32.9.2 Growth

The growth of HgTe/CdTe offers four challenges not present in the growth of HgCdTe alloy material:

1. The growth of CdTe is optimally performed at $\approx 270\,°C$, $\approx 80\,°C$ higher than the optimal growth temperature of HgTe or Hg-rich HgCdTe
2. The HgTe well widths must be precisely controlled within a fraction of a monolayer
3. The SL interfacial roughness must be minimized and made reproducible to obtain good control of the optical cutoff
4. p-Type doping must be activated as-grown or after only short anneals at temperatures $\leq 300\,°C$.

The substrate preparation and in situ growth monitoring are the same as for the growth of HgCdTe. The growth temperature must be that for HgTe, so the CdTe growth cannot be performed under optimum conditions. Also, the Hg flux must be present at all times, so the barrier layers grown are $Cd_{1-y}Hg_yTe$ with $y \approx 0.05$ rather than $y = 0$. This small amount of Hg in the barriers does not significantly affect the optical cutoff or the band structure.

For (211)B growth the control of the well widths must be achieved by precise control of the growth temperature and the Hg and Te fluxes or by using the virtual interface approximation [32.171, 172] and an ellipsometer capable of taking data below the SL optical energy gap. For (211)B growth, which is step-flow growth, one cannot measure the growth of each monolayer by monitoring RHEED oscillations as has been done in the past for (100) growth [32.209]. In step-flow growth the terraces remain unchanged during the growth except for moving laterally; there is no progression from an empty atomic layer to a layer randomly partially occupied to a full layer, and therefore no RHEED oscillations. To have RHEED oscillations one must grow on a surface conducive to the formation of grain boundaries and twinning.

Modulated doping with In in the barrier layers is used for in situ n-type doping, and planar doping with As in the barrier layers, as described in Sect. 32.7.5, is used for in situ p-type doping. The flux shutters are computer controlled and operate on the schedules shown in Fig. 32.21a,b for In and As doping, respectively.

32.9.3 Experimentally Observed Properties

The important observed properties of HgTe/CdTe SLs which do not correspond closely to HgCdTe alloy properties are the macroscopic and microscopic HgTe layer uniformities, the amount of interfacial roughness, and the amount of Cd interdiffused into the HgTe layers during growth. All other observed properties of HgTe/CdTe SLs match expectations based on the observed properties of MBE-grown HgCdTe al-

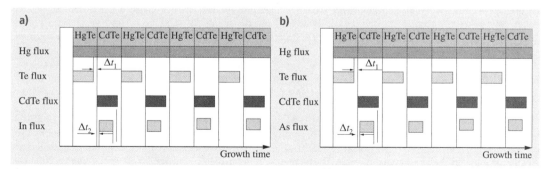

Fig. 32.21a,b MBE shutter sequences for HgTe/CdTe SL growth with: (a) In modulation doping and (b) As planar doping in the barrier layers

loys. Furthermore, As planar doping of the CdTe layers yields p-activation either as-grown or after only a short low-temperature anneal at $T \leq 300\,°C$.

In situ RHEED and SE measurements indicate high-quality growth, and ex situ XRD measurements yield results indicative of high structural quality. Also, scanning XRD and FTIR measurements have shown the layer uniformities to be excellent. However, both interfacial roughness and Cd interdiffusion during growth pose serious problems. For early (100) growth on $Hg_{0.73}Cd_{0.27}Te/Hg_{0.15}Cd_{0.85}Te$ SLs, it was found by quantitative chemical mapping that the FWHM of the as-grown interfacial roughness is only ≈ 2 monolayers near the surface of an as-grown $0.75\,\mu m$-thick SL [32.313]. More recently, we have found by TEM that for (211)B growth the FWHM of the as-grown interfacial roughness can be as large as ≈ 7 monolayers (ML) at the surface of a $10\,\mu m$-thick SL and ≈ 5.5 ML at the substrate, with no evidence during growth of poor growth morphology. Much of this observed roughness arose from the step-flow nature of (211)B growth and the existence of (100) steps several monolayers high between successive (111) terraces.

In agreement with earlier calculations [32.311, 314], our TEM results show that during growth and during subsequent annealing the CdTe layers become narrower and the HgTe layers wider. Cd atoms diffuse into the HgTe layers from the interfacial roughness region and later, more slowly, from the edges of the CdTe layers, with very little diffusion into the CdTe layers. We have shown that interdiffusion becomes unimportant once the Cd in the alloyed interfacial roughness layers diffuses throughout the HgTe well layers. This is because the diffusion coefficient for Cd in HgTe is much higher than that for Hg in CdTe and because the diffusion coefficient for Cd in HgCdTe increases rapidly with increasing Hg content [32.314]. For a SL of thickness $\approx 10\,\mu m$, the growth time is sufficiently long, $\approx 10\,h$, to induce considerable Cd diffusion far from the surface, but there is little diffusion near the surface, which was exposed to the growth temperature only briefly. Thus a SL of thickness $\approx 10\,\mu m$ displays a greatly broadened optical absorption edge, because the absorption edge moves to a much higher energy near the substrate than that at the surface, due to Cd diffusion during growth, which increases with increasing time exposed to the growth temperature. In order to avoid this problem, one must either:

1. Grow only thin SL epilayers, say $< 3\,\mu m$ thick
2. Anneal away the as-grown interfacial roughness throughout the epilayer.

It is not a practical option to greatly broaden the HgTe wells or to reduce the height or thickness of the CdTe barriers; either of those options would result in the loss of the primary SL advantages.

The problem with growing only thin SL epilayers is that thin layers give less absorption and thus lower quantum efficiencies. However, this could be overcome by stacking layers on top of one another in a device as is done by DRS in their three-color detector. The problem with annealing away the as-grown interfacial roughness throughout the epilayer is that the resultant Cd concentration in the well layers can easily become too high to obtain an absorption edge in the VLWIR. At $40\,K$ a (211)B $60\,Å$ HgTe/$50\,Å$ CdTe SL has a $24.57\,\mu m$ cutoff. Assuming an average Cd concentration of 0.5 in the roughness layers, to preserve a cutoff $\geq 14\,\mu m$ one must anneal away ≤ 3 ML of interfacial roughness; to preserve a cutoff $\geq 10\,\mu m$ one must anneal away < 6 ML of interfacial roughness.

To make SLs practical for use in the VLWIR, one must greatly reduce the amount of as-grown interfacial roughness, possibly by growing on a slightly miscut (100) or (111) plane. On the other hand, going back to (100) or (111) growth would reintroduce all of the problems solved by (211)B growth, such as microtwinning, surface hillocks, and 3-D growth. The achievement of well-controlled SL layer thicknesses with little interfacial roughness is essential to the practical use of SLs for VLWIR detection and possibly even for the LWIR.

32.10 Architectures of Advanced IR Detectors

The mechanisms that convert IR radiation to a device's electrical output and properties (current out, capacitance, etc.) are described by device physics. The reader is referred to [32.321] for a complete description of photon detector behavior. Here, we deal with photon detectors in which IR photons are absorbed by the detector material, creating excess carriers that form an output current proportional to the photon intensity.

An IR detector is a multilayer structure. These layers include contact metal, photon absorbing material, substrates, etc. The architecture describes how the layers are assembled to form the desired structure as well

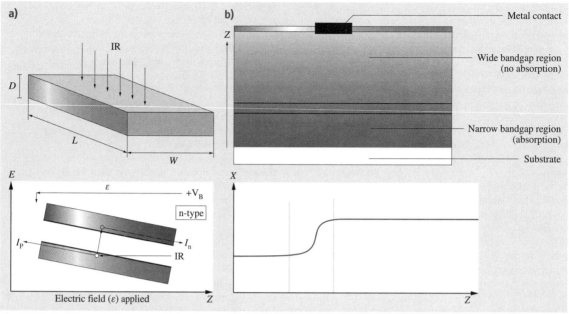

Fig. 32.22a,b Schematic diagrams of (**a**) a photoconductor and (**b**) a graded double-layer heterojunction

as the function of each layer. The two principal IR detectors used today are:

1. The photoconductor (PC), a resistor whose resistance changes under IR illumination, shown in Fig. 32.22a
2. The diode, a graded double-layer heterojunction (DLHJ) shown in Fig. 32.22b, which is a high-impedance minority-carrier device operating under reverse bias.

A detailed description of the performance of these two detector architectures is given in [32.321]. However, these two architectures are not adequate for all IR detector applications and under all external conditions. There are specific applications that require improvement in:

1. The performance at the upper end of the IR spectrum
2. The magnitude of the operating temperature
3. The speed of the detector response
4. The lowering of the detector internal noise.

In summary, there is a need to reduce power consumption, hence lowering cooling costs, to raise the operating temperature while keeping noise at a minimum, to increase the speed of detector response to rapidly changing targets, to increase the number of photoelectrons created per IR photon in and hence to increase the output signal, and/or to increase the detector signal-to-noise ratio. To resolve some of these needs, new detector architectures have been introduced and/or advanced material growth technologies have been exploited.

32.10.1 Reduction of Internal Detector Noise

One of the problems with the graded DLHJ is that optimal performance depends on placing the p–n junction extremely accurately within the graded bandgap region. Optimal p–n junction placement allows the minimization of diode leakage and g-r noise, hence improving the performance. Device physics calculations [32.322] show that the p–n junction must be placed within a few hundred Angstroms of the optimal position; pushing the junction too far into the wider gap region will cause barriers to form, reducing the minority photocurrent. To perform the precise p–n junction placement requires MBE layer growth.

32.10.2 Increasing Detector Response

The response of a detector is determined partly by the quantum efficiency (QE), a measure of the number of carriers generated per incident photon. For the standard PC or DLHJ, the QE is less than one, which results in an extremely low output current, of the order of nanoamperes. This means that an IR detector requires a preamplifier with a large amplification factor and extremely low noise. This is a tall order, especially if the preamplifier circuit must fit into an area roughly the size of current IR detectors, $\approx 30\text{--}50\,\mu\text{m}$. It is desirable then to increase the number of current carriers produced per incident IR photon. In order to do so, one must take advantage of a process called impact ionization. Here, an incoming photon creates a free electron with energy well above the conduction-band energy. As the electron moves through the crystal lattice, if the photon-generated free carrier has sufficient energy, when it collides with a bound electron on an atom, that electron becomes a free carrier. At this stage, one photon has created two free carriers. This process will continue until no electron has sufficient energy to free a new electron. To preserve charge neutrality, a free hole is also created. The multiplication process leads to a QE > 1. The probability that a photon-created carrier will create another carrier via impact ionization is called the impact ionization coefficient for that carrier.

The problem with this process is ionization noise. Impact ionization noise is at a maximum if the impact ionization coefficients for electrons and holes are equal. To reduce the noise, one must turn to a more complex device architecture such as the sawtooth variation in concentration shown in Fig. 32.23. Here the teeth created by varying the Cd composition in HgCdTe cause one of the impact ionization coefficients, either

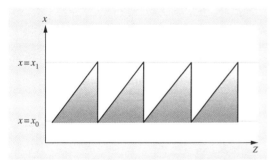

Fig. 32.23 Cd composition in HgCdTe versus distance from substrate for a potential avalanche photodiode (APD) design. This structure was proposed before it could be realized by MBE growth. The sawtooth significantly reduces one carrier's impact ionization coefficient compared with the other, which reduces the impact ionization coefficient

for electrons or for holes, to be small. Hence, the ratio of the two coefficients is large. This minimizes the impact ionization noise, resulting in a large multiplication factor (increased response) without increasing the noise, which would defeat the purpose. To create the desired sawtooth structure and maximize multiplication while minimizing impact ionization noise requires precise control of the MBE growth process.

32.10.3 High-Speed IR Detectors

A homojunction detector is a three-layer structure: a p-layer dominated by positively charged hole free carriers; a junction layer in which there is a large built-in electric field, hence the layer is depleted of free carriers; and an n layer dominated by negatively charged electrons. A photon-generated free carrier will diffuse to the edge of the junction and will drift at some rate to the opposite edge. In the absence of an external bias voltage, the junction crossing speed is not high for junction widths of $\approx 0.5\,\mu\text{m}$. One can increase the transit speed by applying an external bias whose field is in the same direction as the built-in field. The applied field is a reverse-bias field, which will increase the junction width. So, with the increased width of the region that will support an electric field and with the added external field, the increased total field exerting a force on a carrier will cause the carrier to transit the junction more rapidly.

The transit speed is too low, hence the response time is too long, for many applications such as detecting the wavefront of an electromagnetic wave. To increase the transit speed, one needs to significantly increase the external bias. However, there is a limit to the reverse bias one can apply to the junction before the junction breaks down, avalanches, and behaves like a resistor. To get around this, a three-layer device can be built with the outer layers doped n-type and p-type, respectively. The middle layer is an undoped (intrinsic) layer that contains few free carriers, and hence can support a large electric field. Because the width of the intrinsic layer is much greater than the junction width, say 10–20 times larger, the device can support a very large external voltage, hence a a very large external field without the device breaking down. This type of device is called a p-i-n diode, where i stands for intrinsic. To increase the speed of detector response, the effective width where there are few carriers, and hence which can support a large electric field junction, can be increased by at least a factor of 20 over the p–n junction width. Furthermore, the p- and n-side doping can be made much larger. Hence,

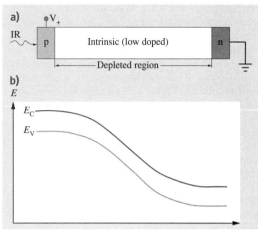

Fig. 32.24a,b Schematic diagram of a rapid-response p–i–n detector. (**a**) Device structure with a wide depleted region with no free carriers present. (**b**) Spatial band structure

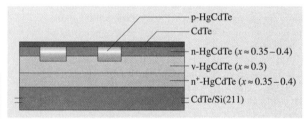

Fig. 32.25 Schematic diagram of a HOT detector architecture used at EPIR Technologies

the electric field in the junction is increased and acts over a longer distance. The carrier when it enters the field of the junction region will feel a very large force that will sweep the carrier to the opposite contact in an extremely short time. As a result, the p–i–n detector could fully respond to IR radiation in a time on the order of a nanosecond or less. The actual structure and the spatial band structure of a p–i–n diode are shown schematically in Fig. 32.24.

32.10.4 High-Operating-Temperature (HOT) IR Detectors

Normally, an IR detector must be cooled so that the near-field IR radiation from its surrounding structure does not interfere with the IR signal coming from a few meters to a few thousand kilometers away. The reason for this is that a body produces IR radiation with a power which increases with increasing temperature T. The near-field radiation is a source of noise because there are mechanisms that generate random numbers of electron–hole pairs, which in turn contribute random fluctuations in the output current and thus noise. Therefore, the higher the operating temperature of the detector, the lower the signal-to-noise ratio (SNR). Depending on the cutoff wavelength, IR photon detectors are operated at anywhere from 4 to 77 K. This lays a heavy burden on the system due to the cooler's weight, power needs, and lifetime.

The ideal, to avoid the cost and weight of cooling, is a detector that yields a good SNR at room temperature, that is, a HOT detector. To achieve this, one must suppress the mechanisms that yield the random (thermal) generation of free carriers. A HOT detector is one that via a specific multilayer architecture suppresses one or more of the principal thermal generation mechanisms. While still in a very early stage of development, these structures promise operation near room temperature or at least at thermoelectrically cooled temperatures. One HOT architecture is shown in Fig. 32.25. Note that the structure requires precise control of layer thickness and doping, and hence is an ideal application of MBE growth technology. Recent references on HOT detectors include [32.323–325].

32.11 IR Focal-Plane Arrays (FPAs)

An IR FPA consists of four layers involving different kinds of materials. The top layer is the detector array, consisting of an array of HgCdTe detectors arranged in a linear or checkerboard pattern. The detector array is grown on a substrate, the second layer, which usually is CdZnTe, GaAs or Si based. Given a sufficiently large substrate and reasonable sized arrays, several detector arrays can be grown on a single substrate. The output of an IR detector is very low, on the order of nA. Therefore, the detector array must be connected to a signal processor, which has a preamplifier for each detector plus various signal processing functions such as multiplexing and analogue-to-digital (A/D) conversion. This readout integrated circuit (ROIC) or multiplexor (MUX) is fabricated from silicon. The difference between the thermal coefficients of expansion of HgCdTe or CdTe and Si is quite large. Therefore, each detector must be connected to its preamplifier on the ROIC by an In bump. Therefore, a 256×256 detector array requires more than 65 000 separate In bumps. Detector sizes typically are $20\,\mu m$ on a side to $30\,\mu m$ on a side, making the In bump cross section extremely small, perhaps $10 \times 10\,\mu m^2$. In is used because it is a good conductor, is ductile, and can accommodate the strain introduced by the mismatched thermal coefficients of expansion. The array of In bump electrical contacts forms the third layer. The Si ROIC is the fourth layer.

IR FPAs designed for astronomical investigations can be as large as 2000×2000. This requires a very precise manufacturing process involving many steps of photolithography. Therefore, one becomes immediately concerned about yield versus cost versus quantity required for a given application.

32.12 Conclusions

The inherent properties of HgCdTe make this alloy the most preferred material for IR detection. During the past several decades HgCdTe alloys have been used as the primary material for IR detectors, which are most important for a variety of military and space applications. Significant advances in the MBE growth of HgCdTe alloys since the early 1980s have offered a new dimension in the ability to fabricate innovative detector structures, control multiple layer heterostructures, and use alternative substrates such as Si for large-format two-dimensional arrays. Today, the MBE of HgCdTe has emerged as a viable technology for research and development and the manufacture of high-performance IR FPAs.

Several factors have contributed to the growth and advancement of HgCdTe MBE technology. The large body of literature published by researchers studying various aspects of HgCdTe deposition experiments and theory has generated much useful information. Various advantages of the MBE technique such as the availability of SE, RHEED, and other in situ analytical characterization techniques to monitor and control the growth have enabled the verification of these experimental data and the related theories. The knowledge base has matured HgCdTe MBE technology in recent years to a point similar to that of III–V compound semiconductor MBE technology in the past. Unlike many III–V semiconductors, it took a while to understand the HgCdTe MBE process because of its stringent thermodynamic constraints. These constraints offer only a narrow window of parameter space for optimum epitaxy. These difficulties are further compounded by temperature changes in the growing HgCdTe layer due to its enhanced IR absorption. In situ tools, particularly SE and RHEED, helped to better understand MBE HgCdTe growth. In the past, and to a certain extent even today, LPE has been the preferred approach for HgCdTe growth. Today, however MBE technology with its various advantages and advancement is becoming more attractive and is required for detectors with advanced architectures.

Abrupt p–n junctions and the control of layer thickness were first demonstrated in GaAs III–V semiconductor MBE technology. Recently, the maturation of the HgCdTe MBE process has also enabled fabrication of device structures with various architectures such as p^+/n, n^+/p, p–i–n, etc. Fabrication of these structures in HgCdTe is now possible because of the thin-film process available in MBE wherein HgCdTe layers can be controlled within submicrometer thicknesses. MBE also has facilitated multiple-layer heterostructure growth with various compositions of stacked HgCdTe layers. Unlike LPE, MBE allows the control of multiple heterostructure growth and thus allows for new device designs and optimization. This very feature also makes it suitable for superlattice structures, and adds the ability to implement bandgap engineering in this alloy. Therefore, MBE has been used to demonstrate both ion-implanted and in situ doped p–n junction devices, and various device designs including superlattice and multicolor detectors have been reported.

The low-temperature and ultrahigh-vacuum nature of MBE adds to its many advantages. Paramount among these is the ability to achieve abrupt material junctions with minimum interdiffusion or interface contamination, and the possibility of depositing HgCdTe on foreign substrates such as Si, GaAs or Al_2O_3. Traditionally, bulk CdZnTe has been preferred as a substrate material for HgCdTe epitaxy. In the past, all three major growth techniques, LPE, MBE, and MOCVD, employed bulk CdZnTe substrates because of their close lattice and chemical match with HgCdTe alloys. During the past decade significant advances have occurred in the MBE of CdTe and HgCdTe on Si and

GaAs substrates. This has been possible due to the low temperature at which MBE occurs, where cross-contamination of the substrate material in the HgCdTe layer does not occur. The primary advantage of growing on Si or GaAs is the opportunity to produce very large-area HgCdTe epitaxial layers. Large-area growth is limited in bulk CdZnTe substrates because of the lack of high-quality large substrates and because of the large difference between the thermal coefficients of expansion of HgCdTe or CdTe and that of Si. The use of Si or GaAs offers more dies of IRFPAs per wafer, large-format IRFPAs, ease of device processing, and a low-cost option for manufacturing.

The challenge of HgCdTe MBE technology has been to produce material structures with structural, optical, and electrical characteristics better than or equivalent to the best LPE-grown materials. MBE technology has developed to the point at which MBE HgCdTe layers grown on bulk CdZnTe substrates have characteristics comparable to those of LPE material. The technology has reached the state at which the quality of MBE HgCdTe layers depends mostly on the characteristics of the substrate material. MBE is sensitive to even minute changes in substrate morphology, and tends to replicate those morphological defects in the growing HgCdTe layer. Various techniques have been developed to mitigate this problem so as to improve the crystalline quality of the HgCdTe layer. The use of buffer layers to block the propagation of defects originating in substrates or at the substrate–epilayer interface has proven useful and is being studied. A thin interfacial CdTe layer has proven effective in reducing substrate surface roughness, but has limitations. The most significant improvement is achieved by growing an interfacial HgTe/CdTe superlattice as a buffer layer. These superlattices are more effective at blocking dislocations and can be grown more nearly lattice matched to the HgCdTe epilayer to be grown.

Today the main challenge of HgCdTe MBE technology is to grow very high-crystalline-quality layers on Si. This epitaxial combination (HgCdTe/Si) creates an extreme 19.2% epitaxial mismatch. Thus growing HgCdTe layers with crystalline quality that is equivalent in characteristics to layers grown on bulk CdZnTe substrates is very difficult. Various techniques have been developed to reduce defect densities in HgCdTe layers grown on Si. Examples of these are variations of II–VI buffer layers such as CdTe and CdSeTe, and other techniques such as reduced-area growth, hydrogen passivation of defects, thermal anneal cycling, etc. Significant progress has been made in the MBE of CdTe and CdSeTe on Si, and extended defect densities as low as $2 \times 10^5\,\text{cm}^{-2}$ have been achieved. However, progress in reducing defects in the HgCdTe layers grown on buffer/Si substrates is lagging. The best MBE HgCdTe layers grown on buffer/Si substrates achieved thus far exhibit defect densities of $2-5 \times 10^6\,\text{cm}^{-2}$. Clearly, more research and development is necessary to reduce the defect density by at least an order of magnitude.

Another remaining challenge in the MBE of HgCdTe is to be able to reproducibly grow layers with the desired levels of p-type doping stable against diffusion. In order to form sharp, precisely controlled interfaces and p–n junctions and to obtain well-controlled doping levels, it is essential to grow such junctions by MBE with extrinsic n- and p-type doping. Although stable, well-controlled n-type extrinsic doping of as-grown HgCdTe and HgTe/CdTe SLs has been demonstrated by many authors, p-type extrinsic doping of these materials has remained a problem. Unfortunately, As, which has a low diffusion coefficient, can be reproducibly activated in HgCdTe only at annealing temperatures significantly higher than the growth temperature. Annealing at such high temperatures limits many of the advantages of MBE as a growth technique. Even though high-performance HgCdTe-based photovoltaic photodiodes using various p-doping techniques (in situ and implantation) have been demonstrated, interdiffusion across interfaces, dopant diffusion, and the formation of Hg vacancies all become problems at the high annealing temperatures required for reproducible p-type activation of the preferred group V dopants. The absence of a reproducible p-type doping technique which does not require high-temperature annealing remains a major stumbling block to successfully implementing high-yield fabrication of MBE-grown HgCdTe IRFPAs. Much progress is needed in developing a reproducible in situ extrinsic p-type doping process for MBE technology to advance further.

It is worth noting that, despite the various difficulties faced by the HgCdTe MBE technology, the state of IRFPA technology can still be advanced significantly by further maturation of the MBE process. Considering the current military engagements, advancement of IRFPA technology is needed more than ever. MBE has the potential to emerge as a leading technique to develop innovative IR device architectures and other detector array designs that have not been possible in the past. Already, multicolor IRFPAs have been fabricated from HgCdTe grown by MBE. With more research it is potentially possible to use the HgCdTe MBE process to develop very large-format IRFPAs. The advent of this technology will undoubtedly lead to important applications required to constantly monitor battlefields and borders. The progress of MBE in the area of alternative substrate technology would lead to more robust and low-cost IRFPAs. This will be most desirable in military applications, where more IR capability must be deployed.

References

32.1   A.Y. Cho: Morphology of epitaxial growth of GaAs by a molecular beam method: The observation of surface structures, J. Appl. Phys. **41**, 782–786 (1970)

32.2   S. Sivananthan, M.D. Lange, G. Monfroy, J.P. Faurie: New achievements in $Hg_{1-x}Cd_xTe$ grown by molecular-beam epitaxy, J. Vac. Sci. Technol. B **6**, 788–793 (1987)

32.3   M.A. Kinch: Fundamental physics of infrared detector materials, J. Electron. Mater. **29**, 809–817 (2000)

32.4   J.P. Faurie, A. Million: Molecular beam epitaxy of II–VI compounds $Cd_xHg_{1-x}Te$, J. Cryst. Growth **54**, 582–585 (1982)

32.5   Y.G. Sidorov, S.A. Dvoretsky, M.V. Yakushev, N.N. Mikhailov, V.S. Varavin, V.I. Liberman: Peculiarities of the MBE growth physics and technology of narrow-gap II–VI compounds, Thin Solid Films **306**, 253–265 (1997)

32.6   J.P. Faurie, A. Million, R. Boch, J.L. Tissot: Latest developments in the growth of $Cd_xHg_{1-x}Te$ and CdTe–HgTe superlattices by molecular beam epitaxy, J. Vac. Sci. Technol. A **1**, 1593–1597 (1983)

32.7   J.P. Faurie, M. Boukerche, J. Reno, S. Sivananthan, C. Hsu: Molecular beam epitaxy of alloys and superlattices involving mercury, J. Vac. Sci. Technol. A **3**, 55–59 (1985)

32.8   L. He, J.R. Yang, S.L. Wang, S.P. Guo, M.F. Yu, X.Q. Chen, W.Z. Fang, Y.M. Qiao, Q.Y. Zhang, R.J. Ding, T.L. Xin: A study of MBE growth and thermal annealing of p-type long wavelength HgCdTe, J. Cryst. Growth **175/176**, 677–681 (1997)

32.9   C.R. Abernethy: Compound semiconductor growth by metalorganic molecular beam epitaxy (MOMBE), Mater. Sci. Eng. R **14**, 203–254 (1995)

32.10 A. Parikh, S.D. Pearson, R.N. Bicknell-Tassius, L.H. Zhang, R. Benz, C.J. Summers: Optimization of the structural properties of $Hg_{1-x}Cd_xTe$ ($x = 0.18 - 0.30$) alloys: Growth and modeling, J. Electron. Mater. **26**, 524–528 (1997)

32.11 J.F. Schetzina: Photoassisted MBE growth of II–VI films and superlattices, Appl. Surf. Sci. **80**, 171–185 (1994)

32.12 R.N. Bicknell, N.C. Giles, J.F. Schetzina, C. Hitzman: Controlled substitutional doping of CdTe thin films grown by photoassisted molecular-beam epitaxy, J. Vac. Sci. Technol. A **5**, 3059–3063 (1987)

32.13 J.R. Arthur: Molecular beam epitaxy, Surf. Sci. **500**, 189–217 (2002)

32.14 J.Y. Tsao: *Materials Fundamentals of Molecular Beam Epitaxy* (Academic, Boston 1993)

32.15 A. Madhukar: Far from equilibrium vapour phase growth of lattice matched III–V compound semiconductor interfaces: some basic concepts and Monte-Carlo computer simulations, Surf. Sci. **132**, 344–374 (1983)

32.16 M.A. Herman: Approaches to understanding MBE growth phenomena, Thin Solid Films **267**, 1–14 (1995)

32.17 L.G. Wang, P. Kratzer, M. Scheffler, N. Moll: Formation and stability of self-assembled coherent islands in highly mismatched heteroepitaxy, Phys. Rev. Lett. **82**, 4042–4045 (1999)

32.18 L.G. Wang, P. Kratzer, N. Moll, M. Scheffler: Size, shape, and stability of InAs quantum dots on the GaAs(001) substrate, Phys. Rev. B **62**, 1897–1904 (2000)

32.19 C. Wagner, W. Schottky: Theory of controlled mixed phases, Z. Phys. Chem. B **11**, 163 (1931)

32.20 C. Wagner, W. Schottky: Theorie der geordneten Mischphasen II (Diffusionsvorgänge), Z. Phys. Chem. Bodenstein-Festband, 177 (1931), in German

32.21 C. Wagner: Theory of ordered mixture phases. III. Appearances of irregularity in polar compounds as a basis for ion conduction and electron conduction, Z. Phys. Chem. B **22**, 181 (1933)

32.22 C. Wagner: Errors in the classifications of crystalized polar compounds as basis for electron and ion conduction, Z. Elektrochem. **39**, 543 (1933)

32.23 W. Schottky: The mechanism of ion movement in solid electrolytes, Z. Phys. Chem. B **29**, 335 (1935)

32.24 W. Schottky: Statistics and thermodynamics of disorder states in crystals, especially in small disarranged states, Z. Elektrochem. **45**, 33 (1939)

32.25 F.A. Kröger, H.J. Vink: *Solid State Physics* Vol. 3, ed. by F. Seitz (Academic, New York 1956) p. 307

32.26 E. Bauer: Phänomenologische Theorie der Kristallabscheidung an Oberflächen I, Z. Kristallogr. **110**, 372–394 (1958), in German

32.27 R. Heckingbottom: Thermodynamic aspects of molecular beam epitaxy: high temperature growth in the $GaAs/Ga_{1-x}Al_xAs$ system, J. Vac. Sci. Technol. B **3**, 572–575 (1985)

32.28 R. Heckingbottom, C.J. Todd, G.J. Davies: The interplay of thermodynamics and kinetics in molecular beam epitaxy (MBE) of doped gallium arsenide, J. Electrochem. Soc. **127**, 444–450 (1980)

32.29 R. Heckingbottom, G.J. Davies: Germanium doping of gallium arsenide grown by molecular beam epitaxy – Some thermodynamic aspects, J. Cryst. Growth **50**, 644–647 (1980)

32.30 R. Heckingbottom, G.J. Davies, K.A. Prior: Growth and doping of gallium arsenide using molecular beam epitaxy (MBE): Thermodynamic and kinetic aspects, Surf. Sci. **132**, 375–389 (1983)

32.31 H. Seki, A. Koukitu: Thermodynamic analysis of molecular beam epitaxy of III–V semiconductors, J. Cryst. Growth **78**, 342–352 (1986)

32.32 F. Turco, J.C. Guillaume, J. Massies: Thermodynamic analysis of the molecular beam epitaxy of AlInAs alloys, J. Cryst. Growth **88**, 282–290 (1988)

32.33 J.Y. Shen, C. Chatillon: Thermodynamic analysis of molecular beam epitaxy of III–V compounds; application to the $Ga_yIn_{1-y}As$ multilayer epitaxy, J. Cryst. Growth **106**, 553–565 (1990)

32.34 S.V. Ivanov, P.S. Kop'ev, N.N. Ledentsov: Thermodynamic analysis of segregation effects in MBE of $A^{III}-B^V$ compounds, J. Cryst. Growth **111**, 151–161 (1991)

32.35 J.P. Gailliard: A thermodynamical model of molecular beam epitaxy, application to the growth of II–VI semiconductors, Rev. Phys. Appl. **22**, 457–463 (1987)

32.36 T. Colin, T. Skauli: Applications of thermodynamical modeling in molecular beam epitaxy of $Cd_xHg_{1-x}Te$, J. Electron. Mater. **26**, 688–696 (1997)

32.37 H.R. Vydyanath, F. Aqariden, P.S. Wijewarnasuriya, S. Sivananthan, G. Chambers, L. Becker: Analysis of the variation in the composition as a function of growth parameters in the MBE growth of indium doped $Hg_{1-x}Cd_xTe$, J. Electron. Mater. **27**, 504–506 (1998)

32.38 E.C. Piquette, M. Zandian, D.D. Edwall, J.M. Arias: MBE growth of HgCdTe epilayers with reduced visible defect densities: kinetics considerations and substrate limitations, J. Electron. Mater. **30**, 627–631 (2001)

32.39 Y. Chang, G. Badano, J. Zhao, C.H. Grein, S. Sivananthan, T. Aoki, D.J. Smith: Formation mechanism of crater defects on HgCdTe/CdZnTe (211)B epilayers grown by molecular beam epitaxy, Appl. Phys. Lett. **83**, 4785–4787 (2003)

32.40 J.W. Garland, C.H. Grein, B. Yang, P.S. Wijewarnasuriya, F. Aqariden: Evidence that arsenic is incorporated as As_4 molecules in the molecular beam epitaxial growth of $Hg_{1-x}Cd_xTe:As$, Appl. Phys. Lett. **74**, 1975–1977 (1999)

32.41 S. Sivananthan, X. Chu, J. Reno, J.P. Faurie: Relation between crystallographic orientation and the condensation coefficients of Hg, Cd and Te during molecular beam epitaxial growth of $Hg_{1-x}Cd_xTe$ and CdTe, J. Appl. Phys. **60**, 1359–1363 (1986)

32.42 S. Sivananthan, X. Chu, J.P. Faurie: Dependence of the condensation coefficient of Hg on the orientation and the stability of the Hg-Te bond for the growth of $Hg_{1-x}M_xTe$ (M = Cd, Mn, Zn), J. Vac. Sci. Technol. B **5**, 694–698 (1987)

32.43 J.P. Faurie: Developments and trends in MBE of II–VI Hg-based compounds, J. Cryst. Growth **81**, 483–488 (1987)

32.44 L.A. Almeida, M. Groenert, J.H. Dinan: Influence of substrate orientation on the growth of HgCdTe by molecular beam epitaxy, J. Electron. Mater. **35**, 1214–1218 (2006)

32.45 J.A. Venables: Kinetic studies of nucleation and growth at surfaces, Thin Solid Films **50**, 357–369 (1978)

32.46 R.J. Koestner, H.F. Schaake: Kinetics of molecular beam epitaxial HgCdTe growth, J. Vac. Sci. Technol. A **6**, 2834–2839 (1988)

32.47 M. Kardar, G. Parisi, Y. Zhang: Dynamic scaling of growing interfaces, Phys. Rev. Lett. **56**, 889–892 (1986)

32.48 P.I. Tamborenea, Z.-W. Lai, S. Das Sarma: Molecular beam epitaxial growth: Simulation and continuum theory, Surf. Sci. **267**, 1–4 (1992)

32.49 S. Das Sarma, Z.-W. Lai, P.I. Tamborenea: Crossover effects in models of kinetic growth with surface diffusion, Surf. Sci. Lett. **268**, L311–L318 (1992)

32.50 P. Kratzer, E. Penev, M. Scheffler: First-principles studies of kinetics in epitaxial growth of III–V semiconductors, Appl. Phys. A **75**, 79–88 (2002)

32.51 R. Triboulet, A. Tromson-Carli, D. Lorans, T. Nguyen Duy: Substrate issues for the growth of mercury cadmium telluride, J. Electron. Mater. **22**, 827–834 (1993)

32.52 R. Sporken, Y.P. Chen, S. Sivananthan, M.D. Lange, J.P. Faurie: Current status of direct growth of CdTe and HgCdTe on silicon by molecular-beam epitaxy, J. Vac. Sci. Technol. B **10**, 1405–1409 (1992)

32.53 R. Korenstein, P. Madison, J.P. Hallock: Growth of (111)CdTe on GaAs/Si and Si substrates for HgCdTe epitaxy, J. Vac. Sci. Technol. B **10**, 1370–1375 (1992)

32.54 Y.P. Chen, S. Sivananthan, J.P. Faurie: Structure of CdTe(111)B grown by MBE on misoriented Si(001), J. Electron. Mater. **22**, 951–957 (1993)

32.55 Y.P. Chen, J.P. Faurie, S. Sivananthan, G.C. Hua, N. Otsuka: Suppression of twin formation in CdTe(111)B epilayers grown by molecular beam epitaxy on misoriented Si(001), J. Electron. Mater. **24**, 475–481 (1995)

32.56 L.A. Almeida, Y.P. Chen, J.P. Faurie, S. Sivananthan: D.J. Smith, S.C.Y. Tsen: Growth of high quality CdTe on Si substrates by molecular beam epitaxy, J. Electron. Mater. **25**, 1402–1405 (1996)

32.57 M. Kawano, A. Ajisawa, N. Oda, M. Nagashima, H. Wada: HgCdTe and CdTe($\bar{1}\bar{1}3$)B growth on Si(112)5° off by molecular beam epitaxy, Appl. Phys. Lett. **69**, 2876–2879 (1996)

32.58 M.D. Lange, R. Sporken, K.K. Mahavadi, J.P. Faurie, Y. Nakamura, N. Otsuka: Molecular beam epi-

32.58 taxy and characterization of CdTe(211) and CdTe(133) films on GaAs(211)B substrates, Appl. Phys. Lett. **58**, 1988–1990 (1991)

32.59 J.P. Faurie, R. Sporken, Y.P. Chen, M.D. Lange, S. Sivananthan: Heteroepitaxy of CdTe on GaAs and silicon substrates, Mater. Sci. Eng. B **16**, 51–56 (1993)

32.60 V.S. Varavin, S.A. Dvoretsky, V.I. Liberman, N.N. Mikhailov, Y.G. Sidorov: The controlled growth of high-quality mercury cadmium telluride, Thin Solid Films **267**, 121–125 (1995)

32.61 T.J. de Lyon, D. Rajaval, S.M. Johnson, C.A. Cockrum: Molecular-beam epitaxial growth of CdTe(112) on Si(112) substrates, Appl. Phys. Lett. **66**, 2119–2121 (1995)

32.62 T. Colin, D. Minsas, S. Gjoen, R. Sizmann, S. Lovold: Influence of surface step density on the growth of mercury cadmium telluride by molecular beam epitaxy, Mater. Res. Soc. Symp. Proc. **340**, 575 (1994)

32.63 F. Aqariden, H.D. Shih, A.M. Turner, D. Chandra, P.K. Liao: Molecular beam epitaxial growth of HgCdTe on CdZnTe(311)B, J. Electron. Mater. **29**, 727–728 (2000)

32.64 G.A. Carini, C. Arnone, A.E. Bolotnikov, G.S. Camarda, R. de Wames, J.H. Dinan, J.K. Markunias, B. Raghothamachar, S. Sivananthan, R. Smith, J. Zhao, Z. Zhong, R.B. James: Material quality characterization of substrates for HgCdTe epitaxy, J. Electron. Mater. **35**, 1495–1502 (2006)

32.65 H. Abad, J. Zhao, G. Badano, Y. Chang, S. Sivananthan: Correlation of pre-growth surface morphology of substrates with the quality of HgCdTe epilayers, Mil. Sens. Symp. (Tucson 2004)

32.66 H.R. Vydyanath, J. Ellsworth, J.J. Kennedy, B. Dean, C.J. Johnson, G.T. Neugebauer, J. Sepich, P.K. Liao: Recipe to minimize Te precipitation in CdTe and (Cd,Zn)Te crystals, J. Vac. Sci. Technol. B **10**, 1476–1484 (1992)

32.67 B. Li, J. Zhu, X. Zhang, J. Chu: Effect of annealing on near-stoichiometric and non-stoichiometric wafers, J. Cryst. Growth **181**, 204–209 (1997)

32.68 S.J.C. Irvine, A. Stafford, M.U. Ahmed: Substrate/layer relationships in II-VIs, J. Cryst. Growth **197**, 616–625 (1999)

32.69 E. Weiss, O. Klin, E. Benory, E. Kedar, Y. Juravel: Substrate quality impact on the carrier concentration of undoped annealed HgCdTe LPE layers, J. Electron. Mater. **30**, 756–761 (2001)

32.70 S. Sen, C.S. Liang, D.R. Rhiger, J.E. Stannard, H.F. Arlinghaus: Reduction of defects and relation to epitaxial HgCdTe quality, J. Electron. Mater. **25**, 1188–1195 (1996)

32.71 J.P. Faurie, S. Sivananthan, P.S. Wijewarnasuriya: Current status of the growth of mercury cadmium telluride by molecular beam epitaxy on (211)B HgCdTe substrates, SPIE Proc. **1735**, 141–150 (1992)

32.72 R. Korenstein, R.J. Olsen, D. Lee, P.K. Liao, C.A. Castro: Copper outdiffusion from substrates and its effect on the properties of metalorganic chemical-vapor deposition-grown HgCdTe, J. Electron. Mater. **24**, 511–514 (1995)

32.73 J.P. Tower, S.P. Tobin, M. Kestigian, P.W. Norton, A.B. Bollong, H.F. Shaake, C.K. Ard: Substrate impurities and their effects on LPE HgCdTe, J. Electron. Mater. **24**, 497–504 (1995)

32.74 J.P. Tower, S.P. Tobin, P.W. Norton, A.B. Bollong, A. Socha, J.H. Tregilgas, C.K. Ard, H.F. Arlinghaus: Trace copper measurements and electrical effects in LPE HgCdTe, J. Electron. Mater. **25**, 1183–1187 (1996)

32.75 S. Sen, D.R. Rhiger, C.R. Curtis, P.R. Norton: Extraction of mobile impurities from CdZnTe, J. Electron. Mater. **29**, 775–780 (2000)

32.76 D.R. Rhiger, J.M. Peterson, R.M. Emerson, E.E. Gordon, S. Sen, Y. Chen, M. Dudley: Investigation of the cross-hatch pattern and localized defects in epitaxial HgCdTe, J. Electron. Mater. **27**, 615–623 (1998)

32.77 M. Martinka, L.A. Almeida, J.D. Benson, J.H. Dinan: Suppression of strain-induced cross-hatch on molecular beam epitaxy (211)B HgCdTe, J. Electron. Mater. **31**, 732–737 (2001)

32.78 J. Zhao, Y. Chang, G. Badano, S. Sivananthan, J. Markunas, S. Lewis, J.H. Dinan, P.S. Wijewarnasuriya, Y. Chen, G. Brill, N. Dhar: Correlation of (211)B substrate surface morphology and HgCdTe(211)B epilayer defects, J. Electron. Mater. **33**, 881–885 (2005)

32.79 J.N. Johnson, L.A. Almeida, M. Martinka, J.D. Benson, J.H. Dinan: Use of electron cyclotron resonance plasmas to prepare (211)B substrates for HgCdTe molecular beam epitaxy, J. Electron. Mater. **28**, 817–820 (1999)

32.80 R. Singh, S. Velicu, J. Crocco, Y. Chang, J. Zhao, L.A. Almeida, J. Markunas, A. Kaleczyc, J.H. Dinan: Molecular beam epitaxy growth of high-quality HgCdTe LWIR layers on polished and repolished substrates, J. Electron. Mater. **34**, 885–890 (2005)

32.81 P. Moravec, P. Höschl, J. Franc, E. Belas, R. Fesh, R. Grill, P. Horodyský, P. Praus: Chemical polishing of substrates fabricated from crystals grown by the vertical-gradient freezing method, J. Electron. Mater. **35**, 1206–1218 (2006)

32.82 Y.S. Wu, C.R. Becker, A. Waag, R.N. Bicknell-Tassius, G. Landwehr: Thermal effects on (100) substrates as studied by x-ray photoelectron spectroscopy and reflection high energy electron diffraction, Appl. Phys. Lett. **60**, 1878–1880 (1992)

32.83 R.H. Sewell, C.A. Musca, J.M. Dell, L. Faraone, B.F. Usher, T. Dieing: High-resolution x-ray diffraction studies of molecular beam epitaxy-grown HgCdTe heterostructures and substrates, J. Electron. Mater. **34**, 795–803 (2005)

32.84 R. Triboulet, A. Durand, P. Gall, J. Bonnafé, J.P. Fillard, S.K. Krawszyk: Qualification by optical means of CdTe substrates, J. Cryst. Growth **117**, 227–232 (1992)

32.85 W.J. Everson, C.K. Ard, J.L. Sepich, B.E. Dean, G.T. Neugebauer, H.F. Shaake: Etch pit characterization of CdTe and CdZnTe substrates for use in mercury cadmium telluride epitaxy, J. Electron. Mater. **24**, 505–510 (1995)

32.86 K. Nakagawa, K. Maeda, S. Takeuchi: Observation of dislocations in cadmium telluride by cathodoluminescence microscopy, Appl. Phys. Lett. **34**, 574–575 (1979)

32.87 M. Kestigian, A.B. Bollong, J.J. Derby, H.L. Glass, K. Harris, H.L. Hettich, P.K. Liao, P. Mitra, P.W. Norton, H. Wadley: Cadmium zinc telluride growth, characterization, and evaluation, J. Electron. Mater. **28**, 726–731 (1999)

32.88 P. Capper, E.S. O'Keefe, C. Maxey, D. Dutton, P. Mackett, C. Butler, I. Gale: Matrix and impurity element distributions in CdHgTe (CMT) and (Cd,Zn)(Te,Se) compounds by chemical analysis, J. Cryst. Growth **161**, 104–118 (1996)

32.89 B. Li, J. Zhu, X. Zhang, J. Chu: Effect of annealing on near-stoichiometric and non-stoichiometric wafers, J. Cryst. Growth **181**, 204–209 (1997)

32.90 S.L. Price, H.L. Hettich, S. Sen, M.C. Currie, D.R. Rhiger, E.O. McLean: Progress in substrate producibility and critical drivers of IRFPA yield originating with substrates, J. Electron. Mater. **27**, 564–572 (1998)

32.91 Y. Chang, J. Zhao, H. Abad, C.H. Grein, S. Sivananthan, T. Aoki, D.J. Smith: Performance and reproducibility enhancement of HgCdTe molecular beam epitaxy growth on substrates using interfacial HgTe/CdTe superlattice layers, Appl. Phys. Lett. **86**, 131924 (2005)

32.92 Y. Chang, C.H. Grein, J. Zhao, S. Sivananthan, C.Z. Wang, T. Aoki, D.J. Smith, P.S. Wijewarnasuriya, V. Nathan: Improve molecular beam epitaxy growth of HgCdTe on (211)B substrates using interfacial layers of HgTe/CdTe superlattices, J. Appl. Phys. **100**, 114316-1–114316-6 (2006)

32.93 R. Sporken, S. Sivananthan, K.K. Mohavadi, G. Monfroy, M. Boukerche, J.P. Faurie: Molecular beam epitaxial growth of CdTe and HgCdTe on Si(100), Appl. Phys. Lett. **55**, 1879–1881 (1989)

32.94 S. Sivananthan, Y.P. Chen, P.S. Wijewarnasuriya, J.P. Faurie, F.T. Smith, P.W. Norton: Properties of $Hg_{1-x}Cd_xTe$ grown on CdZnTe and Si substrates, Inst. Phys. Conf. Ser. **144**, 239–244 (1995)

32.95 S. Velicu, T.S. Lee, C.H. Grein, P. Boieriu, Y.P. Chen, N.K. Dhar, J. Dinan, D. Lianos: Monolithically integrated HgCdTe focal plane arrays, J. Electron. Mater. **34**, 820–831 (2005)

32.96 L.A. Almeida, L. Hirsch, M. Martinka, P.R. Boyd, J.H. Dinan: Improved morphology and crystalline quality of MBE CdZnTe/Si, J. Electron. Mater. **30**, 608–610 (2001)

32.97 Y.P. Chen, G. Brill, N.K. Dhar: MBE growth of CdSeTe/Si composite substrate for long-wavelength IR HgCdTe applications, J. Cryst. Growth **252**, 270–274 (2004)

32.98 Y.P. Chen, G. Brill, E.M. Campo, T. Hierl, J.C.M. Hwang, N.K. Dhar: Molecular beam epitaxial growth of $Cd_{1-y}Zn_ySe_xTe_{1-x}$ on Si(211), J. Electron. Mater. **33**, 498–502 (2004)

32.99 N.K. Dhar, C.E.C. Wood, A. Gray, H.-Y. Wei, L. Salamanca, J.H. Dinan: Heteroepitaxy of CdTe on {211} Si using crystallized amorphous ZnTe templates, J. Vac. Sci. Technol. B **14**, 2366–2370 (1996)

32.100 S. Rujirawat, L.A. Almeida, Y.P. Chen, S. Sivananthan, D.J. Smith: High quality large-area CdTe(211)B on Si(211) grown by molecular beam epitaxy, Appl. Phys. Lett. **71**, 1810–1812 (1998)

32.101 J.M. Peterson, J.A. Franklin, M. Reddy, S.M. Johnson, E. Smith, W.A. Radford, I. Kasai: High-quality large-area MBE HgCdTe/Si, J. Electron. Mater. **35**, 1283–1286 (2006)

32.102 M.F. Vilela, A.A. Buell, M.D. Newton, G.M. Venzor, A.C. Childs, J.M. Peterson, J.J. Franklin, R.E. Bornfreund, W.A. Radford, S.M. Johnson: Growth and control of middle wave infrared (MWIR) $Hg_{(1-x)}Cd_xTe$ on Si by molecular beam epitaxy, J. Electron. Mater. **34**, 898–904 (2005)

32.103 M. Carmody, J.G. Pasko, D. Edwall, R. Bailey, J. Arias, S. Cabelli, I. Bajaj, L.A. Almeida, J.H. Dinan, M. Groenert, A.J. Stoltz, Y. Chen, G. Brill, N.K. Dhar: Molecular beam epitaxy grown long wavelength infrared HgCdTe on Si detector performance, J. Electron. Mater. **34**, 832–838 (2005)

32.104 M. Carmody, J.G. Pasko, D. Edwall, M. Daraselia, L.A. Almeida, J. Molstad, J.H. Dinan, J.K. Markunas, Y. Chen, G. Brill, N.K. Dhar: Long wavelength infrared, molecular beam epitaxy, HgCdTe-on-Si diode performance, J. Electron. Mater. **33**, 531–537 (2004)

32.105 K. Jóźwikowski, A. Rogalski: Effect of dislocations on performance of LWIR HgCdTe photodiodes, J. Electron. Mater. **29**, 736–741 (2000)

32.106 S.M. Johnson, D.R. Rhiger, J.P. Rosbeck, J.M. Peterson, S.M. Taylor, M.E. Boyd: Effect of dislocations on the electrical and optical properties of long-wavelength infrared HgCdTe photovoltaic detectors, J. Vac. Sci. Technol. B **10**, 1499–1506 (1992)

32.107 T.J. de Lyon, R.D. Rajavel, J.A. Vigil, J.E. Jensen, O.K. Wu, C.A. Cockrum, S.M. Johnson, G.M. Venzor, S.L. Bailey, I. Kasai, W.L. Ahlgren, M.S. Smith: Molecular beam epitaxial growth of HgCdTe infrared focal-plane arrays on Si substrates for midwave infrared applications, J. Electron. Mater. **27**, 550–555 (1998)

32.108 T.J. de Lyon, J.E. Jensen, I. Kasai, G.M. Venzor, K. Kosai, J.B. de Bruin, W.L. Ahlgren: Molecular-beam epitaxial growth and high-temperature

32.108 ...performance of HgCdTe midwave infrared detectors, J. Electron. Mater. **31**, 220–226 (2002)

32.109 S.M. Johnson, A.A. Buell, M.F. Vilela, J.M. Peterson, J.B. Varesi, M.D. Newton, G.M. Venzor, R.E. Bornfreund, W.A. Radford, E.P.G. Smith, J.P. Rosbeck, T.J. de Lyon, J.E. Jensen, V. Nathan: HgCdTe/Si materials for long wavelength infrared detectors, J. Electron. Mater. **33**, 526–530 (2004)

32.110 M. Carmody, J.G. Pasko, D. Edwall, R. Bailey, J. Arias, M. Groenert, L.A. Almeida, J.H. Dinan, Y. Chen, G. Brill, N.K. Dhar: LWIR HgCdTe on Si detector performance and analysis, J. Electron. Mater. **35**, 1417–1422 (2006)

32.111 E.M. Campo, S. Nakahara, T. Hierl, J.C.M. Hwang, Y. Chen, G. Brill, N.K. Dhar, V. Vaithyanathan, D.G. Schlom, X.-M. Fang, J.M. Fastenau: Epitaxial growth of CdTe on Si through perovskite oxide buffers, J. Electron. Mater. **35**, 1219–1223 (2006)

32.112 Y. Liang, H. Li, J. Finder, C. Overgaard, J. Kulik, D. McCready, S. Shutthanandan: Mater. Res. Soc. Symp. (2004) p. 218

32.113 T.D. Golding, O.W. Holland, M.J. Kim, J.H. Dinan, L.A. Almeida, J.M. Arias, J. Bajaj, H.D. Shih, W.P. Kirk: HgCdTe on Si: Present status and novel buffer layer concepts, J. Electron. Mater. **32**, 882–889 (2003)

32.114 X. Zhou, S. Jiang, W.P. Kirk: Molecular beam epitaxy of BeTe on vicinal Si(100) surfaces, J. Cryst. Growth **175/176**, 624–631 (1997)

32.115 S.H. Shin, J.M. Arias, D.D. Edwall, M. Zandian, J.G. Pasko, R.E. DeWames: Dislocation reduction in HgCdTe on GaAs and Si, J. Vac. Sci. Technol. B **10**, 1492–1498 (1992)

32.116 Y. Lo: New approach to grow pseudomorphic structures over the critical thickness, Appl. Phys. Lett. **59**, 2311–2313 (1991)

32.117 Z. Yang, J. Alperin, W.I. Wang, S.S. Iyer, T.S. Kuan, F. Semendy: In situ relaxed $Si_{1-x}Ge_x$ epitaxial layers with low threading dislocation densities grown on compliant Si-on-insulator substrates, J. Vac. Sci. Technol. B **16**, 1489–1491 (1998)

32.118 P.D. Moran, D.M. Hansen, R.J. Matyi, L.J. Mawst, T.F. Kuech: Experimental test for elastic compliance during growth on glass-bonded compliant substrates, Appl. Phys. Lett. **76**, 2541–2543 (2000)

32.119 S.D. Hersee, D. Zubia, R. Bommena, X. Sun, M. Fairchild, S. Zhang, D. Burckel, A. Frauenglass, S.R.J. Brueck: Nanoheteroepitaxy for the integration of highly mismatched semiconductor materials, IEEE J. Quantum Electron. **QE-38**, 1017–1028 (2002)

32.120 S. Nakamura, M. Senoh, S.-I. Nagahama, N. Iwasa, T. Yamada, T. Matsushita, H. Kiyoku, Y. Sugimoto, T. Kozaki, H. Umemoto, M. Sano, K. Chocho: InGaN/GaN/AlGaN-based laser diodes with modulation-doped strained-layer superlattices grown on an epitaxially laterally overgrown GaN substrate, Appl. Phys. Lett. **72**, 211–213 (1998)

32.121 H. Marchand, X.H. Wu, J.P. Ibbetson, P.T. Fini, P. Kozodoy, S. Keller, J.S. Speck, S.P. DenBaars, U.K. Mishra: Microstructure of GaN laterally overgrown by metalorganic chemical vapor deposition, Appl. Phys. Lett. **73**, 747–749 (1998)

32.122 R. Zhang, I. Bhat: Selective growth of CdTe on Si and GaAs substrates using metalorganic vapor phase epitaxy, J. Electron. Mater. **29**, 765–769 (2000)

32.123 I. Bhat, R. Zhang: Anisotropy in selective metalorganic vapor phase epitaxy of CdTe on GaAs and Si substrates, J. Electron. Mater. **35**, 1293–1298 (2006)

32.124 S.C. Lee, K.J. Malloy, L.R. Dawson: Selected growth and associated faceting and lateral overgrowth of GaAs on a nanoscale limited area bounded by a SiO_2 mask in molecular beam epitaxy, J. Appl. Phys. **92**, 6567–6571 (2002)

32.125 R. Bommena, C. Fulk, J. Zhao, T.S. Lee, S. Sivananthan, S.R.J. Brueck, S.D. Hersee: Cadmium telluride growth on patterned substrates for mercury cadmium telluride infrared detectors, J. Electron. Mater. **34**, 704–709 (2005)

32.126 X.G. Zhang, P. Li, G. Zhao, D.W. Parent, F.C. Jain, J.E. Ayers: Removal of threading dislocations from patterned heteroepitaxial semiconductors by glide to sidewalls, J. Electron. Mater. **27**, 1248–1253 (1998)

32.127 X.G. Zhang, A. Rodriguez, P. Li, F.C. Jain, J.E. Ayers: Patterned heteroepitaxial processing applied to ZnSe and $ZnS_{0.02}Se_{0.98}$ on GaAs(001), J. Appl. Phys. **91**, 3912–3917 (2002)

32.128 Y. Dong, R.M. Feenstra, D.W. Greve, J.C. Moore, M.D. Sievert, A.A. Baski: Effects of hydrogen on the morphology and electrical properties of GaN grown by plasma-assisted molecular-beam epitaxy, Appl. Phys. Lett. **86**, 121914 (2005)

32.129 Y.F. Chen, C.S. Tsai, Y.H. Chang, Y.M. Chang, T.K. Chen, Y.M. Pang: Hydrogen passivation in $Cd_{1-x}Zn_xTe$ studied by photoluminescence, Appl. Phys. Lett. **58**, 493–495 (1991)

32.130 A.P. Jacobs, Q.X. Zhao, M. Willander, T. Baron, N. Magnea: Hydrogen passivation of nitrogen acceptors confined in CdZnTe quantum well structures, J. Appl. Phys. **90**, 2329–2332 (2001)

32.131 H.Y. Lee, T.W. Kang, T.W. Kim: Temperature dependence of the optical properties in $p-Cd_{0.96}Zn_{0.04}Te$ single crystals, J. Mater. Res. **16**, 2196–2199 (2001)

32.132 A.I. Evstigneev, V.F. Kuleshov, G.A. Lubochkova, M.V. Pashkovskii, E.B. Yakimov, N.A. Yarkin: Influence of hydrogen on the concentration of deep-level centers in $Cd_xHg_{1-x}Te$ crystals, Sov. Phys. Semicond. **19**, 562 (1985)

32.133 S.P. Komissarchuk, L.N. Limarenko, E.P. Lopatinskaya: *Narrow Gap Semiconductors and Semimetals* (LVOV, Moscow 1983) p. 126

32.134 H. Jung, H. Lee, C. Kim: Enhancement of the steady state minority carrier lifetime in HgCdTe photodiode using ECR plasma hydrogenation, J. Electron. Mater. **25**, 1266–1269 (1996)

32.135 Y. Kim, T. Kim, D. Redfern, C. Musca, H. Lee, C. Kim: Characteristics of gradually doped LWIR diodes by hydrogenation, J. Electron. Mater. **29**, 859–864 (2000)

32.136 P. Boieriu, C.H. Grein, S. Velicu, J. Garland, C. Fulk, A. Stoltz, W. Mason, L. Bubulac, R. DeWames, J.H. Dinan: Effects of hydrogen on majority carrier transport and minority carrier lifetimes in LWIR HgCdTe on Si, Appl. Phys. Lett. **88**, 62106 (2006)

32.137 Y. Xin, S. Rujirawat, N.D. Browning, R. Sporken, S. Sivananthan, S.J. Pennycook, N.K. Dhar: The effect of As passivation on the molecular beam epitaxial growth of high-quality single-domain CdTe(111)B on Si(111) substrates, Appl. Phys. Lett. **75**, 349–351 (1999)

32.138 P. Sen, S. Ciraci, I.P. Batra, C.H. Grein, S. Sivananthan: Finite temperature studies of Te adsorption on Si(001), Surf. Sci. **519**, 79–89 (2002)

32.139 P. Sen, I.P. Batra, S. Sivananthan, C.H. Grein, Nabir Dhar, S. Ciraci, Electronic structure of Te- and As-covered Si(211), Phys. Rev. B **68**, 045314 (2003)

32.140 M. Jaime-Vazquez, M. Martinka, R.N. Jacobs, M. Groenert: In-situ spectroscopic study of the As and Te on the Si(112) surface for high-quality epitaxial layers, J. Electron. Mater. **35**, 1455–1460 (2006)

32.141 S.D. Chen, L. Lin, X.Z. He, M.J. Ying, R.Q. Wu: High quality HgCdTe epilayers grown on (211)B GaAs by molecular beam epitaxy, J. Cryst. Growth **152**, 261–265 (1995)

32.142 P. Ballet, F. Noël, F. Pottier, S. Plissard, J.P. Zanatta, J. Baylet, O. Gravrand, E. De Borniol, S. Martin, P. Castelain, J.P. Chamonal, A. Million, G. Destefanis: Dual-band infrared detectors made on high-quality HgCdTe epilayers grown by molecular beam epitaxy on CdZnTe or CdTe/Ge substrates, J. Electron. Mater. **33**, 667–672 (2004)

32.143 J.P. Zanatta, G. Badano, P. Ballet, C. Largeron, J. Baylet, O. Gravrand, J. Rothman, P. Castelain, J.P. Chamonal, A. Million, G. Destefanis, S. Mibord, E. Brochier, P. Costa: Molecular beam epitaxy growth of HgCdTe on Ge for third-generation infrared detectors, J. Electron. Mater. **35**, 1231–1236 (2006)

32.144 A.I. D'Souza, J. Bajaj, R.E. de Wames, D.D. Edwall, P.S. Wijewarnasuriya, N. Nayar: MWIR DLPH photodiode performance dependence on substrate material, J. Electron. Mater. **27**, 727–732 (1998)

32.145 A.J. Norieka, R.F.C. Farrow, F.A. Shirland, W.J. Takai, J. Greggi Jr., S. Wood, W.J. Choyke: Characterization of molecular beam epitaxially grown HgCdTe on CdTe and InSb buffer layers, J. Vac. Sci. Technol. A **4**, 2081–2085 (1986)

32.146 G. Brill, S. Velicu, Y. Chen, N.K. Dhar, T.S. Lee, Y. Selamet, S. Sivananthan: MBE growth and device processing of MWIR HgCdTe on large area Si substrates, J. Electron. Mater. **30**, 717–722 (2001)

32.147 S. Sivananthan: Experimental study on the properties of HgCdTe grown by molecular beam epitxy. Ph.D. Thesis (Department of Physics. Univ. Illinois, Chicago 1987) p. 160

32.148 B.V. Shanabrook, J.R. Waterman, J.L. Davis, R.J. Wagner: Large temperature changes induced by molecular-beam epitaxial growth on radiatively heated substrates, Appl. Phys. Lett. **61**, 2338–2340 (1992)

32.149 B.V. Shanabrook, J.R. Waterman, J.L. Davis, R.J. Wagner, D.S. Katzer: Variations in substrate temperature induced by molecular-beam epitaxial growth on radiatively heated substrates, J. Vac. Sci. Technol. B **11**, 994–997 (1993)

32.150 P. Thompson, Y. Li, J.J. Zhou, D.L. Sato, L. Flanders, H.P. Lee: Diffuse reflectance spectroscopy measurement of substrate temperature and temperature transient during molecular beam epitaxy and implications for low-temperature III-V epitaxy, Appl. Phys. Lett. **70**, 1605–1607 (1997)

32.151 L.A. Almeida, N.K. Dhar, M. Martinka, J.H. Dinan: HgCdTe heteroepitaxy on three-inch (112) CdZnTe/Si: Ellipsometric control of substrate temperature, J. Electron. Mater. **29**, 754–759 (2000)

32.152 M. Daraselia, C.H. Grein, R. Rujirawat, B. Yang, S. Sivananthan, F. Aqariden, S. Shih: In-situ monitoring of temperature and alloy composition of $Hg_{1-x}Cd_xTe$ using FTIR spectroscopic techniques, J. Electron. Mater. **28**, 743–748 (1999)

32.153 T.P. Pearsall, S.R. Saban, J. Booth, B.T. Beard Jr., S.R. Johnson: Precision of noninvasive temperature measurement by diffuse reflectance spectroscopy, Rev. Sci. Instrum. **66**, 4977–4980 (1995)

32.154 T.J. de Lyon, J.A. Roth, D.H. Chow: Substrate temperature measurement by absorption-edge spectroscopy during molecular beam epitaxy of narrow-band gap semiconductor films, J. Vac. Sci. Technol. B **15**, 329–336 (1997)

32.155 F.G. Johnson, G.W. Wicks, R.E. Viturro, R. LaForce: Molecular-beam epitaxial growth of arsenide/phosphide heterostructures using valved, solid group V sources, J. Vac. Sci. Technol. B **11**, 823–825 (1993)

32.156 D.D. Edwall, D.B. Young, A.C. Chen, M. Zandian, J.M. Arias, B. Dlugosch, S. Priddy: Initial Evaluation of a valved Te source for MBE growth of HgCdTe, J. Electron. Mater. **28**, 740–742 (1999)

32.157 W.V. McLevige, J.M. Arias, D.D. Edwall, S.L. Johnson: Ellipsometric profiling of HgCdTe heterostructures, J. Vac. Sci. Technol. B **9**, 2483–2486 (1991)

32.158 D.R. Rhiger: Use of ellipsometry to characterize the surface of HgCdTe, J. Electron. Mater. **22**, 887–898 (1993)

32.159 K.K. Svitashev, S.A. Dvoretsky, Y.G. Sidorov, V.A. Shvets, A.S. Mardezhov, I.E. Nis, V.S. Varavin, V. Liberman, V.G. Remesnik: The growth of high-quality MCT films by MBE using in-situ ellipsometry, Cryst. Res. Technol. **29**, 931–937 (1994)

32.160 S.D. Murthy, I.B. Bhat, B. Johs, S. Pittal, P. He: Application of spectroscopic ellipsometry for real-time control of CdTe and HgCdTe growth in an OMCVD system, J. Electron. Mater. **24**, 445–449 (1995)

32.161 S. Pittal, B. Johs, P. He, J.A. Woollam, S.D. Murthy, I.B. Bath: In-situ monitoring and control of MOCVD growth using multiwavelength ellipsometry, Compound Semiconductors, Inst. Phys. Conf. Ser. **141**, 41–44 (1994)

32.162 J.D. Benson, A.B. Cornfeld, M. Martinka, K.M. Singley, Z. Derzko, P.J. Shorten, J.H. Dinan: In-situ spectroscopic ellipsometry of HgCdTe, J. Electron. Mater. **25**, 1406–1410 (1996)

32.163 B. Johs, C. Herzinger, J. Dinan, A. Cornfeld, J.D. Benson, D. Doctor, G. Olson, I. Ferguson, M. Pelcynski, P. Chow, C.H. Kuo, S. Johnson: Real-time monitoring and control of epitaxial semiconductor growth in a production environment by in situ spectroscopic ellipsometry, Thin Solid Films **313**, 490–495 (1998)

32.164 R.M.A. Azzam, N.M. Bashara: *Ellipsometry and Polarized Light* (Elsevier North-Holland, New York 1977)

32.165 J.W. Garland, C. Kim, H. Abad, P.M. Raccah: Determination of accurate critical-point energies, linewidths and line shapes from spectroscopic ellipsometry data, Thin Solid Films **233**, 148–152 (1993)

32.166 D.E. Aspnes, G.P. Shwartz, G.J. Gualtieri, A.A. Studna: Optical properties of gallium arsenide and its electrochemically grown anodic oxide from 1.5 to 6.0 eV, J. Electrochem. Soc. **128**, 590–597 (1981)

32.167 R.W. Collins: Automatic rotating element ellipsometers: calibration, operation, and real-time applications, Rev. Sci. Instrum. **61**, 2029–2062 (1990)

32.168 B. Johs: Regression calibration method for rotating element ellipsometers, Thin Solid Films **234**, 395–398 (1993)

32.169 G.E. Jellison, F.A. Modine: Two-modulator generalized ellipsometry: Theory, Appl. Opt. **36**, 8190–8198 (1997)

32.170 G.E. Jellison: Windows in ellipsometry measurements, Appl. Opt. **38**, 4784–4789 (1999)

32.171 D.E. Aspnes: Minimal-data approaches for determining outer-layer dielectric responses of films from kinetic reflectometric and ellipsometric measurements, J. Opt. Soc. Am. **10**, 974–983 (1993)

32.172 D.E. Aspnes: Optical approaches to determine near-surface compositions during epitaxy, J. Vac. Sci. Technol. A **14**, 960–966 (1996)

32.173 S.F. Nee: Polarization of specular reflection and near-specular scattering by a rough surface, Appl. Opt. **35**, 3570–3582 (1996)

32.174 D.A. Aspnes, J.B. Theeten, F. Hottier: Investigation of effective-medium models of microscopic surface roughness by spectroscopic ellipsometry, Phys. Rev. B **20**, 3292–3302 (1979)

32.175 H. Fujiwara, J. Koh, P.I. Rovira, R.W. Collins: Assessment of effective-medium theories in the analysis of nucleation and microscopic surface roughness evolution for semiconductor thin films, Phys. Rev. B **61**, 10832–10844 (2000)

32.176 J.D. Jackson: *Classical Electrodynamics* (Wiley, New York 1962), Sect. 6

32.177 S.F. Nee: Ellipsometric analysis for surface roughness and texture, Appl. Opt. **27**, 2819–2831 (1988)

32.178 D.E. Aspnes: *Optical Properties of Solids: New Developments*, ed. by O. Seraphin (North-Holland, Amsterdam 1976) p. 799

32.179 D.E. Aspnes, A.A. Studna: Optical detection and minimization of surface overlayers on semiconductors using spectroscopic ellipsometry, SPIE Proc. **276**, 227–232 (1981)

32.180 D.E. Aspnes, A.A. Studna: Chemical etching and cleaning procedures for silicon, germanium and some III–V compound semiconductors, Appl. Phys. Lett. **39**, 316–318 (1981)

32.181 B. Johs, C.M. Herzinger, J.H. Dinan, A. Cornfeld, J.D. Benson: Development of a parametric optical constant model for HgCdTe for control of composition by spectroscopic ellipsometry during MBE growth, Thin Solid Films **313/314**, 137–142 (1998)

32.182 J.D. Phillips, D.D. Edwall, D.L. Lee: Control of very-long-wavelength infrared HgCdTe detector-cutoff wavelength, J. Electron. Mater. **31**, 664–668 (2002)

32.183 L.A. Almeida, M. Thomas, W. Larsen, K. Spariosu, D.D. Edwall, J.D. Benson, W. Mason, A.J. Stolt, J.H. Dinan: Development and fabrication of two-color mid- and short-wavelength infrared simultaneous unipolar multispectral integrated technology focal-plane arrays, J. Electron. Mater. **31**, 669–676 (2002)

32.184 T.J. DeLyon, G.L. Olson, J.A. Roth, J.E. Jensen, A.T. Hunter, M.D. Jack, S.L. Bailey: HgCdTe composition determination using spectroscopic ellipsometry during molecular beam epitaxy growth of near-infrared avalanche photodiode device structures, J. Electron. Mater. **31**, 688–693 (2002)

32.185 J. Phillips, D. Edwall, D. Lee, J. Arias: Growth of HgCdTe for long-wavelength infrared detectors using automated control from spectroscopic ellipsometry measurements, J. Vac. Sci. Technol. B **19**, 1580–1584 (2001)

32.186 D. Edwall, J. Phillips, D. Lee, J. Arias: Composition control of long wavelength MBE HgCdTe using in-situ spectroscopic ellipsometry, J. Electron. Mater. **30**, 643–646 (2001)

32.187 L.A. Almeida, J.H. Dinan: In situ compositional control of advanced HgCdTe-based IR detectors, J. Cryst. Growth **202**, 22–25 (1999)

32.188 L.A. Almeida, J.N. Johnson, J.D. Benson, J.H. Dinan, B. Johs: Automated compositional control of

32.189 Hg$_{1-x}$Cd$_x$Te during MBE using in situ spectroscopic ellipsometry, J. Electron. Mater. **27**, 500–503 (1998)

32.189 G. Badano, J.W. Garland, S. Sivananthan: Accuracy of the in situ determination of the temperature by ellipsometry before the growth of HgCdTe by MBE, J. Cryst. Growth **251**, 571–575 (2003)

32.190 C.C. Kim, J.W. Garland, H. Abad, P.M. Raccah: Modeling the optical dielectric function of semiconductors – Extension of the critical-point parabolic-band approximation, Phys. Rev. B **45**, 11749–11767 (1992)

32.191 H. Ehrenreich, M.H. Cohen: Self-consistent field approach to the many-electron problem, Phys. Rev. **115**, 786–790 (1959)

32.192 J.W. Garland, H. Abad, M. Vaccaro, P.M. Raccah: Line shape of the optical dielectric function, Appl. Phys. Lett. **52**, 1176–1178 (1988)

32.193 C.C. Kim, S. Sivananthan: Modeling the optical dielectric function of II–VI compound CdTe, J. Appl. Phys. **78**, 4003–4010 (1995)

32.194 C.C. Kim, S. Sivananthan: Optical properties of ZnSe and its modeling, Phys. Rev. B **53**, 1475–1483 (1996)

32.195 C.C. Kim, M. Daraselia, J.W. Garland, S. Sivananthan: Temperature dependence of the optical properties of CdTe, Phys. Rev. B **56**, 4786–4797 (1997)

32.196 C.C. Kim, S. Sivananthan: Temperature dependence of the optical properties of Hg$_{1-x}$Cd$_x$Te, J. Electron. Mater. **26**, 561–566 (1997)

32.197 C.C. Kim, J.W. Garland, P.M. Raccah: Modeling the optical dielectric function of the aluminum gallium arsenide alloy system Al$_x$Ga$_{1-x}$As, Phys. Rev. B **47**, 1876–1888 (1993)

32.198 O. Castaing, J.T. Benhlal, R. Granger: An attempt to model the dielectric function in II–VI ternary compounds Hg$_{1-x}$Zn$_x$Te and Cd$_{1-x}$Zn$_x$Te, Eur. Phys. J. B **7**, 563–572 (1999)

32.199 G. Badano, P. Ballet, J.P. Zanatta, X. Baudry, A. Million: Ellipsometry of rough CdTe(211)B-Ge(211) surfaces grown by molecular beam epitaxy, J. Opt. Soc. Am. B **23**, 2089–2096 (2006)

32.200 G. Badano, P. Ballet, X. Baudry, J.P. Zanatta, A. Million, J.W. Garland: Molecular beam epitaxy of BeTe on vicinal Si(100) surfaces, J. Cryst. Growth **296**, 129–134 (2006)

32.201 G. Badano, Y. Chang, J.W. Garland, S. Sivananthan: In-situ ellipsometry studies of adsorption of Hg on CdTe(211)B/Si(211) and molecular beam epitaxy growth of HgCdTe(211)B, J. Electron. Mater. **33**, 583–589 (2004)

32.202 G. Badano, Y. Chang, J.W. Garland, S. Sivananthan: Temperature-dependent adsorption of Hg on CdTe(211)B studied by spectroscopic ellipsometry, Appl. Phys. Lett. **83**, 2324–2326 (2003)

32.203 D.G.M. Anderson, R. Barakat: Necessary and sufficient conditions for a Mueller matrix to be derivable from a Jones matrix, J. Opt. Soc. Am. A **11**, 2305–2319 (1994)

32.204 S.F. Nee: Polarization of specular reflection and near-specular scattering by a rough surface, Appl. Opt. **35**, 3570–3582 (1996)

32.205 W. Braun: *Applied RHEED, Reflection High-Energy Electron Diffraction During Crystal Growth*, Springer Tracts in Modern Physics, Vol. 154 (Springer, Berlin 1999)

32.206 J.J. Harris, B.A. Joyce, P.J. Dobson: Oscillations in the surface structure of Sn-doped GaAs during growth by MBE, Surf. Sci. **103**, L90–L96 (1981)

32.207 C.E.C. Wood: RED intensity oscillations during MBE of GaAs, Surf. Sci. **108**, L441–443 (1981)

32.208 J.M. van Hove, C.S. Lent, P.R. Pukite, P.I. Cohen: Damped oscillations in reflection high energy electron diffraction during GaAs MBE, J. Vac. Sci. Technol. B **1**, 741–746 (1983)

32.209 J. Arias, J. Singh: Use of cation-stabilized conditions to improve compatibility of CdTe and HgTe molecular beam epitaxy, Appl. Phys. Lett. **55**, 1561–1563 (1989)

32.210 W. Braun, L. Däweritz, K.H. Ploog: Origin of electron diffraction oscillations during crystal growth, Phys. Rev. Lett. **80**, 4935–4938 (1998)

32.211 M. Itoh, T. Ohno: Probing the submonolayer morphology change in epitaxial growth: A simulation study, Appl. Phys. Lett. **90**, 073111 (2007)

32.212 T.J. de Lyon, S.M. Johnson, C.A. Cockrum, O.K. Wu, W.J. Hamilton, G.S. Kamath: CdZnTe on Si(001) and Si(112): Direct MBE growth for large-area HgCdTe infrared focal-plane array applications, J. Electrochem. Soc. **141**, 2888–2893 (1994)

32.213 W.-S. Wang, I. Bhat: Growth of high quality CdTe and ZnTe on Si substrates using organometallic vapor phase epitaxy, J. Electron. Mater. **24**, 451–455 (1995)

32.214 P.S. Wijewarnasuriya, M. Zandian, D.D. Edwall, W.V. McLevige, C.A. Chen, J.G. Pasko, G. Hildebrandt, A.C. Chen, J.M. Arias, A.I. D'Souza, S. Rujirawat, S. Sivananthan: MBE p-on-n Hg$_{1-x}$Cd$_x$Te heterostructure detectors on silicon substrates, J. Electron. Mater. **27**, 546–549 (1998)

32.215 P. Boieriu, G. Brill, Y. Chen, S. Velicu, N.K. Dhar: Hg$_{1-x}$Cd$_x$Te(112) nucleation on silicon composite substrates, SPIE Proc. **4454**, 60–70 (2001)

32.216 W. Kern, D.A. Puotinen: Cleaning solutions based on hydrogen peroxide for use in silicon semiconductor technology, RCA Review **31**, 187 (1970)

32.217 A. Ishikawa, Y. Shiraki: Low temperature surface cleaning of silicon and its application to silicon MBE, J. Electrochem. Soc. **133**, 666–671 (1986)

32.218 P.J. Taylor, W.A. Jesser, M. Martinka, K.M. Singley, J.H. Dinan, R.T. Lareau, M.C. Wood, W.W. Clark III: Reduced carbon contaminant, low-temperature silicon substrate preparation for *defect-free* homoepitaxy, J. Vac. Sci. Technol. A **17**, 1153–1159 (1999)

32.219 N.K. Dhar, P.R. Boyd, M. Martinka, J.H. Dinan, L.A. Almeida, N. Goldsman: Heteroepitaxy on 3"

32.220 Y. Xin, S. Rujirawat, N.D. Browning, R. Sporken, S. Sivananthan, S.J. Pennycook, N.K. Dhar: The effect of As passivation on the molecular beam epitaxial growth of high-quality single-domain CdTe(111)B on Si(111) substrates, Appl. Phys. Lett. **75**, 349–351 (1999)

32.221 M. Jaime-Vasquez, M. Martinka, R.N. Jacobs, M. Groenert: In situ spectroscopic study of the As and Te on the Si(112) surface for high-quality epitaxial layers, J. Electron. Mater. **35**, 1455–1460 (2006)

32.222 P. Sen, I.P. Batra, S. Sivananthan, C.H. Grein, N.K. Dhar, S. Ciraci: Electronic structure of Te- and As-covered Si(211), Phys. Rev. B **68**, 045314 (2003)

32.223 Y.P. Chen, G. Brill, N.K. Dhar: MBE growth of CdSeTe/Si composite substrate for long-wavelength IR HgCdTe applications, J. Cryst. Growth **252**, 270–274 (2003)

32.224 M. Boukerche, P.S. Wijewarnasuriya, S. Sivananthan, I.K. Sou, Y.J. Kim, K.K. Mahavadi, J.P. Faurie: The doping of mercury cadmium telluride grown by molecular-beam epitaxy, J. Vac. Sci. Technol. A **6**, 2830–2833 (1988)

32.225 P.S. Wijewarnasuriya, M.D. Lange, S. Sivananthan, J.P. Faurie: Carrier recombination in indium-doped HgCdTe(211)B epitaxial layers grown by molecular beam epitaxy, J. Appl. Phys. **75**, 1005–1009 (1994)

32.226 M. Boukerche, P.S. Wijewarnasuriya, J. Reno, I.K. Sou, J.P. Faurie: Electrical properties of molecular beam epitaxy produced mercury cadmium telluride layers doped during growth, J. Vac. Sci. Technol. A **4**, 2072–2076 (1986)

32.227 M.L. Wroge, D.J. Peterman, B.J. Feldman, B.J. Morris, D.J. Leopold, J.G. Broerman: Impurity doping of mercury telluride-cadmium telluride superlattices during growth by molecular-beam epitaxy, J. Vac. Sci. Technol. A **7**, 435–439 (1989)

32.228 C.A. Hoffman, J.R. Meyer, F.J. Bartoli, Y. Lansari, J.W. Cook Jr., J.F. Schetzina: Electron mobilities and quantum Hall effect in modulation-doped mercury telluride/cadmium telluride superlattices, Phys. Rev. B **44**, 8376–8379 (1991)

32.229 K.A. Harris, T.H. Myers, R.W. Yanka, L.M. Mohnkern, N. Otsuka: A high quantum efficiency in situ doped midwavelength infrared p-on-n homojunction superlattice detector grown by photoassisted molecular-beam epitaxy, J. Vac. Sci. Technol. B **9**, 1752–1758 (1991)

32.230 A. Sher, M.A. Berding, M. Van Schilfgaarde, A.B. Chen: HgCdTe status review with emphasis on correlations, native defects and diffusion, Semicond. Sci. Technol. **6**, C59–C70 (1991)

32.231 P. Boieriu, C.H. Grein, S. Velicu, J. Garland, C. Fulk, S. Sivananthan, A. Stoltz, L. Bubulac, J.H. Dinan: Effects of hydrogen on majority carrier transport and minority carrier lifetimes in long wavelength infrared HgCdTe on Si, Appl. Phys. Lett. **88**, 062106 (2006)

32.232 M.A. Berding, A. Sher, A.B. Chen: Mercury cadmium telluride, defect structure overview, Mater. Res. Soc. Symp. Proc. **216**, 3–10 (1991)

32.233 M.A. Berding, A. Sher, M. Van Schilfgaarde: Behavior of p-type dopants in (Hg,Cd)Te, J. Electron. Mater. **26**, 625–628 (1997)

32.234 R. Balcerak: Infrared material requirements for the next generation of systems, Semicond. Sci. Technol. **6**, C1–C5 (1991)

32.235 P.A. Bakhitin, S.A. Dvoretskii, V.S. Varavin, A.P. Korabkin, N.N. Mikhailor, I.V. Sabanina, Y.G. Sidorov: Effect of low-temperature annealing on electrical properties of n-HgCdTe, Semiconductors **38**, 1172–1175 (2004)

32.236 P.S. Wijewarnasuriya, M.D. Lange, S. Sivananthan, J.P. Faurie: Minority carrier lifetime in Indium-doped HgCdTe(211)B epitaxial layers grown by molecular beam epitaxy, J. Electron. Mater. **24**, 545–549 (1995)

32.237 D.L. Polla, R.L. Aggarwal, D.A. Nelson, J.F. Shanley, M.B. Reine: Mercury vacancy related lifetime in mercury cadmium telluride ($Hg_{0.68}Cd_{0.32}Te$) by optical modulation spectroscopy, Appl. Phys. Lett. **43**, 941–943 (1983)

32.238 M.E. d'Souza, M. Boukerche, J.P. Faurie: Minority-carrier lifetime in p-type (111)B mercury cadmium telluride grown by molecular-beam epitaxy, J. Appl. Phys. **68**, 5195–5199 (1990)

32.239 R. Fastow, Y. Nemirovsky: The excess carrier lifetime in vacancy-doped and impurity-doped HgCdTe, J. Vac. Sci. Technol. A **8**, 1245–1250 (1990)

32.240 M. Boukerche, S. Sivananthan, P.S. Wijewarnasuriya, I.K. Sou, J.P. Faurie: Electrical properties of intrinsic p-type shallow levels in HgCdTe grown by molecular-beam epitaxy in the (111)B orientation, J. Vac. Sci. Technol. A **7**, 311–313 (1989)

32.241 M. Zandian, A.C. Chen, D.D. Edwall, J.G. Pasko, J.M. Arias: p-type arsenic doping of $Hg_{1-x}Cd_xTe$ by molecular beam epitaxy, Appl. Phys. Lett. **71**, 2815–2817 (1997)

32.242 M. Zandian, E. Goo: TEM investigation of defects in arsenic doped layers grown in-situ by MBE, J. Electron. Mater. **30**, 623–626 (2001)

32.243 M.A. Berding, A. Sher, M. van Schilfgaarde, A.C. Chen: Model for As activation, J. Electron. Mater. **27**, 605 (1998)

32.244 M.A. Berding, A. Sher: Amphoteric behavior of arsenic in HgCdTe, Appl. Phys. Lett. **74**, 685–687 (1999)

32.245 M.A. Berding, A. Sher: Arsenic incorporation during MBE growth of HgCdTe, J. Electron. Mater. **28**, 799–803 (1999)

32.246 H.R. Vydynath: Amphoteric behavior of group V mass action constants for lattice site transfers, Semiconductors **5**, S231 (1990)

32.247 S. Sivananthan, P.S. Wijewarnasuriya, F. Aqariden, H.R. Vydynath, M. Zandian, D.D. Edawll, J.M. Arias: Mode of arsenic incorporation in HgCdTe grown by MBE, J. Electron. Mater. **26**, 621–624 (1997)

32.248 P. Boieriu, C.H. Grein, H.S. Jung, J.W. Garland, V. Nathan: Arsenic activation in molecular beam epitaxy grown, in situ doped HgCdTe(211), Appl. Phys. Lett. **86**, 212106 (2005)

32.249 P.S. Wijewarnasuriya, S. Sivananthan: Arsenic incorporation in HgCdTe grown by molecular beam epitaxy, Appl. Phys. Lett. **72**, 1694–1696 (1998)

32.250 A.C. Chen, M. Zandian, D.D. Edwall, J.M. Arias, P.S. Wijewarnasuriya, S. Sivananthan, M.A. Berding, A. Sher: MBE Growth and characterization of in situ arsenic doped HgCdTe, J. Electron. Mater. **27**, 595–599 (1998)

32.251 J. Wu, F.F. Xu, Y. Wu, L. Chen, Y.Z. Wang, M.F. Yu, Y.M. Qiao, L. He: As-doping HgCdTe by MBE, SPIE Proc. **5640**, 637–646 (2005)

32.252 L.O. Bubulac, W.E. Tennant, D.S. Lo, D.D. Edwall, J.C. Robinson, J.S. Chen, G. Bostrup: Ion implanted junction formation in mercury cadmium telluride ($Hg_{1-x}Cd_xTe$), J. Vac. Sci. Technol. A **5**, 3166–3170 (1987)

32.253 L.O. Bubulac, D.D. Edwall, C.R. Viswanathan: Dynamics of arsenic diffusion in metalorganic chemical vapor deposited mercury cadmium telluride on gallium arsenide/silicon substrates, AIP Conf. Proc. **235**, 1695–1704 (1991)

32.254 L.O. Bubulac, D.D. Edwall, S.J.C. Irvine, E.R. Gertner, S.H. Shin: P-type doping of double layer mercury cadmium telluride for junction formation, J. Electron. Mater. **24**, 617–624 (1995)

32.255 P.S. Wijewarnasuriya, S.S. Yoo, J.P. Faurie, S. Sivananthan: P-doping with arsenic in (211)B HgCdTe grown by MBE, J. Electron Mater. **25**, 1300–1305 (1996)

32.256 M.C. Chen, L. Colombo, J.A. Dodge, J.H. Tregilgas: The minority-carrier lifetime in doped and undoped p-type $Hg_{0.78}Cd_{0.22}Te$ liquid-phase epitaxy films, J. Electron. Mater. **24**, 539–544 (1995)

32.257 S.H. Shin, J.M. Arias, M. Zandian, J.G. Pasko, L.O. Bubulac, R.E. DeWames: Annealing effect on the p-type carrier concentration in low-temperature processed arsenic-doped HgCdTe, J. Electron. Mater. **22**, 1039–1047 (1993)

32.258 C.A. Merilainen, C.E. Jones: Deep centers in gold-doped mercury cadmium telluride, J. Vac. Sci. Technol. A **1**, 1637–1640 (1983)

32.259 M. Brown, A.F.W. Willoughby: Diffusion of gold and mercury self-diffusion in n-type Bridgman-grown mercury cadmium telluride ($Hg_{1-x}Cd_xTe$) ($x = 0.2$), J. Vac. Sci. Technol. A **1**, 1641–1645 (1983)

32.260 L.O. Bubulac, W.E. Tennant, R.A. Riedel, J. Bajaj, D.D. Edwall: Some aspects of lithium behavior in ion-implanted mercury cadmium telluride, J. Vac. Sci. Technol. A **1**, 1646–1650 (1983)

32.261 P.S. Wijewarnasuriya, I.K. Sou, Y.J. Kim, K.K. Mahavadi, S. Sivananthan, M. Boukerche, J.P. Faurie: Electrical properties of lithium-doped mercury cadmium telluride ($Hg_{1-x}Cd_xTe$)(100) by molecular beam, Appl. Phys. Lett. **51**, 2025–2027 (1987)

32.262 D.J. Peterman, M.L. Wroge, B.J. Morris, D.J. Leopold, J.G. Broerman: p-on-n heterojunctions of mercury cadmium telluride by molecular-beam epitaxy, controlled silver doping and compositional grading, J. Appl. Phys. **63**, 1951–1954 (1988)

32.263 A. Uedono, K. Ozaki, H. Ebe, T. Moriya, S. Tanigawa, K. Yamamoto, Y. Miyamoto: A study of native defects in Ag-doped HgCdTe by positron annihilation, Jpn. J. Appl. Phys. **36**, 6661–6667 (1997)

32.264 N. Tanaka, K. Ozaki, H. Nishino, H. Ebe, Y. Miyamoto: Electrical properties of HgCdTe epilayers doped with silver using an $AgNO_3$ solution, J. Electron. Mater. **27**, 579–582 (1998)

32.265 M. Chu, S. Terterian, P.C. Wang, S. Mesropian, H.K. Gurgenian, D.-S. Pan: Au-doped HgCdTe for infrared detectors and focal plane arrays, SPIE Proc. **4454**, 116–122 (2001)

32.266 Y. Selamet, A. Ciani, C.H. Grein, S. Sivananthan: Extrinsic p-type doping and analysis of HgCdTe grown by molecular beam epitaxy, SPIE Proc. **4795**, 8–16 (2002)

32.267 Y. Selamet, R. Singh, J. Zhao, Y.D. Zhou, S. Sivananthan, N.K. Dhar: Gold diffusion in mercury cadmium telluride grown by molecular beam epitaxy, SPIE Proc. **5209**, 67–74 (2003)

32.268 A.I. D'Souza, M.G. Stapelbroek, E.R. Bryan, J.D. Beck, M.A. Kinch, J.E. Robinson: Au- and Cu-doped HgCdTe HDVIP detectors, SPIE Proc. **5406**, 205–213 (2004)

32.269 A.J. Ciani, S. Ogut, I.P. Batra, S. Sivananthan: Diffusion of gold and native defects in mercury cadmium telluride, J. Electron. Mater. **34**, 868–872 (2005)

32.270 C.H. Grein, J.W. Garland, S. Sivanantuan, P.S. Wijewarnasuriya, M. Fuchs: Arsenic Incorporation in MBE Grown $Hg_{1-x}Cd_xTe$, J. Electron. Mater. **28**, 789–792 (1999)

32.271 H.F. Schaake: Kinetics of activation of group V impurities in $Hg_{1-x}Cd_xTe$ alloys, J. Appl. Phys. **88**, 1765–1770 (2000)

32.272 D. Chandra, H.F. Schaake, M.A. Kinch, F. Aqariden, C.F. Wan, D.F. Weirauch, H.D. Shih: Activation of arsenic as an acceptor in $Hg_{1-x}Cd_xTe$ under equilibrium conditions, J. Electron. Mater. **31**, 715–719 (2002)

32.273 T.S. Lee, J.W. Garland, C.H. Grein, M. Sumstine, A. Jandeska, Y. Selamet, L.S. Sirem: Correlation of arsenic incorporation and its electrical activation in MBE HgCdTe, J. Electron. Mater. **29**, 869 (2000)

32.274 H.R. Vydyanath, L.S. Lichtman, S. Sivananthan, P.S. Wijewarnasuriya, J.P. Faurie: Annealing ex-

32.275 L.O. Bubulac, D.D. Edwall, C.R. Viswanathan: Dynamics of arsenic diffusion in metalorganic chemical vapor deposition HgCdTe on GaAs/Si substrates, J. Vac. Sci. Technol. B **9**, 1695–1704 (1991)

32.276 P. Boieriu, Y. Chen, V. Nathan: Low-temperature activation of As in $Hg_{1-x}Cd_xTe$(211) grown on Si by molecular beam epitaxy, J. Electron. Mater. **31**, 694–698 (2002)

32.277 J.W. Han, S. Hwang, Y. Lansari, R.L. Harper, Z. Yang, N.C. Giles, J.W. Cook, J.F. Schetzina, S. Sen: p-type modulation-doped HgCdTe, Appl. Phys. Lett. **54**, 63–65 (1989)

32.278 R.L. Harper, S. Hwang, N.C. Giles, J.F. Schetzina, D.L. Dreifus, T.H. Myers: Arsenic-doped CdTe epilayers grown by photoassisted molecular beam epitaxy, Appl. Phys. Lett. **54**, 170 (1989)

32.279 J.M. Arias, S.H. Shin, D.E. Cooper, M. Zandian, J.G. Pasko, E.R. Gertner, R.E. DeWames, J. Singh: p-type arsenic doping of cadmium telluride and mercury telluride/cadmium telluride superlattices grown by photoassisted and conventional molecular-beam epitaxy, J. Vac. Sci. Technol. A **8**, 1025–1033 (1990)

32.280 J. Arias, M. Zandian, J.G. Pasko, S.H. Shin, L.O. Bubulac, R.E. DeWames, W.E. Tennant: Molecular-beam epitaxy growth and in situ arsenic doping of p-on-n HgCdTe heterojunctions, J. Appl. Phys. **69**, 2143–2148 (1991)

32.281 O.K. Wu, G.S. Kamath, W.A. Radford, P.R. Bratt, E.A. Patten: Chemical doping of HgCdTe by molecular-beam epitaxy, J. Vac. Sci. Technol. A **8**, 1034–1038 (1990)

32.282 O.K. Wu, D.N. Jamba, G.S. Kamath: Growth and properties of In- and As-doped HgCdTe by MBE, J. Cryst. Growth **127**, 365–370 (1993)

32.283 S. Sivananthan, P.S. Wijewarnasuriya, J.P. Faurie: Recent progress in the doping of MBE HgCdTe, SPIE Proc. **2554**, 55–68 (1995)

32.284 P.S. Wijewarnasuriya, F. Aqariden, C.H. Grein, J.P. Faurie, S. Sivananthan: p-type doping with arsenic in (211)B HgCdTe grown by MBE, J. Cryst. Growth **175**, 647–652 (1997)

32.285 P.S. Wijewarnasuriya, S. Sivananthan: Arsenic incorporation in HgCdTe grown by molecular beam epitaxy, Appl. Phys. Lett. **72**, 1694–1696 (1998)

32.286 F. Aqariden, P.S. Wijewarnasuriya, S. Sivananthan: Arsenic incorporation in HgCdTe grown by molecular beam epitaxy, J. Vac. Sci. Technol. B **16**, 1309–1311 (1998)

32.287 Y. Chang, G. Badano, E. Jiang, J.W. Garland, J. Zhao, C.H. Grein, S. Sivananthan: Composition and thickness distribution of HgCdTe molecular beam epitaxy wafers by infrared microscope mapping, J. Cryst. Growth **277**, 78–84 (2005)

32.288 A.R. Beattie, P.T. Landsberg: Auger effect in semiconductors, Proc. R. Soc. Lond. Ser. A **249**, 16–29 (1959)

32.289 J.S. Blakemore: *Semiconductor Statistics* (Pergamon, New York 1962), Chap. 6

32.290 V.C. Lopes, A.J. Syllaios, M.C. Chen: Minority carrier lifetime in mercury cadmium telluride, Semicond. Sci. Technol. **8**, 824–842 (1993), and ref. cit.

32.291 W. Shockley, W.T. Read: Statistics of the recombinations of holes and electrons, Phys. Rev. **87**, 835–842 (1952)

32.292 R.N. Hall: Electron-hole recombination in germanium, Phys. Rev. **87**, 387 (1952)

32.293 C.H. Swartz, S. Chandril, R.P. Tompkins, N.C. Giles, T.H. Myers, D.D. Edwall, E.C. Piquette, C.S. Kim, I. Vurgaftman, J.R. Meyer: Accurate measurement of composition, carrier concentration, and photoconductive lifetime in $Hg_{1-x}Cd_xTe$ grown by molecular beam epitaxy, J. Electron. Mater. **35**, 1360–1368 (2006)

32.294 Y. Chang, J.W. Garland, S. Sivananthan: Infrared optical characterization of the narrow gap semiconductor HgCdTe. In: *Advanced Materials in Electronics*, ed. by Q. Guo (Research Signpost, Trivandrum 2004) pp. 249–264

32.295 Y. Chang, G. Badano, J. Zhao, Y.D. Zhou, R. Ashokan, C.H. Grein, V. Nathan: Near-bandgap infrared absorption properties of HgCdTe, J. Electron. Mater. **33**, 709–713 (2004)

32.296 F. Urbach: The long-wavelength edge of photographic sensitivity and of the electronic absorption of solids, Phys. Rev. **92**, 1324 (1953)

32.297 C.H. Grein, S. John: Temperature dependence of the Urbach optical absorption edge: A theory of multiple phonon absorption and emission sidebands, Phys. Rev. B **39**, 1140–1151 (1989)

32.298 M. Carmody, D. Lee, M. Zandian, J. Phillips, J. Arias: Threading and misfit dislocation motion in molecular-beam epitaxy-grown HgCdTe epilayers, J. Electron. Mater. **32**, 710–716 (2003)

32.299 T. Aoki, D.J. Smith, Y. Y, J. Zhao, G. Badano, C.H. Grein, S. Sivananthan: Mercury cadmium telluride/tellurium intergrowths in HgCdTe epilayers grown by molecular beam epitaxy, Appl. Phys. Lett. **82**, 2275–2277 (2003)

32.300 I.V. Sabinina, A.K. Gutakovsky, Y.G. Sidorov, A.V. Latyshev: Nature of V-shaped defects in HgCdTe epilayers grown by molecular beam epitaxy, J. Cryst. Growth **274**, 339–346 (2005)

32.301 T. Aoki, Y. Chang, G. Badano, J. Zhao, C.H. Grein, S. Sivananthan, D.J. Smith: Electron microscopy of surface-crater defects on HgCdTe/CdZnTe(211)B epilayers grown by molecular beam epitaxy, J. Electron. Mater. **32**, 703–709 (2003)

32.302 M. Zandian, J.M. Arias, J. Bajaj, J.G. Pasko, L.O. Bubulac, R.E. DeWames: Origin of void defects in $Hg_{1-x}Cd_xTe$ grown by molecular beam epitaxy, J. Electron. Mater. **24**, 1207–1210 (1995)

32.303 J.B. Varesi, A.A. Buell, J.M. Peterson, R.E. Bornfreund, M.F. Vilela, W.A. Radford, S.M. Johnson: Performance of molecular-beam epitaxy-grown midwave infrared HgCdTe detectors on four-inch Si substrates and the impact of defects, J. Electron. Mater. **32**, 661–666 (2003)

32.304 J.N. Schulman, T.C. McGill: The CdTe/HgTe superlattice: Proposal for a new infrared material, Appl. Phys. Lett. **34**, 663–665 (1979)

32.305 D. Chandra, F. Aquariden, J. Frazier, S. Gutzler, T. Orent, H.D. Shih: Isolation and control of voids and void-hillocks during molecular beam epitaxial growth of HgCdTe, J. Electron. Mater. **29**, 887–892 (2000)

32.306 J.P. Faurie, A. Million, J. Piaguet: CdTe–HgTe multilayers grown by molecular beam epitaxy, Appl. Phys. Lett. **41**, 713–715 (1982)

32.307 K.A. Harris, R.W. Yanka, L.M. Mohnkern, A.R. Reisinger, T.H. Myers, Z. Yang, Z. Yu, S. Hwang, J.F. Schetzina: Properties of (211)B HgTe–CdTe superlattices grown by photon assisted molecular-beam epitaxy, J. Vac. Sci. Technol. B **10**, 1574–1581 (1992)

32.308 J.P. Faurie: Growth and properties of HgTe-CdTe and other Hg-based superlattices, IEEE J. Quantum Electron. **QE-22**, 1656–1665 (1986)

32.309 J.P. Faurie, J. Reno, M. Boukerche: II–VI semiconductor compounds: New superlattice systems for the future?, J. Cryst. Growth **72**, 111–116 (1985)

32.310 D.K. Arch, J.L. Staudenmann, J.P. Faurie: Layer intermixing in HgTe-CdTe superlattices, Appl. Phys. Lett. **48**, 1588–1590 (1986)

32.311 K. Zanio: The effect of interdiffusion on the shape of HgTe/CdTe superlattices, J. Vac. Sci. Technol. A **4**, 2106–2109 (1986)

32.312 D.J. Leopold, J.G. Broerman, D.J. Peterman, M.L. Wroge: Effect of annealing on the optical properties of HgTe–CdTe superlattices, Appl. Phys. Lett. **52**, 969–971 (1988)

32.313 Y. Kim, A. Ourmazd, M. Bode, R.D. Feldman: Nonlinear diffusion in multilayered semiconductor systems, Phys. Rev. Lett. **63**, 636–639 (1989)

32.314 S. Holander-Gleixner, H.G. Robinson, C.R. Helms: Simulation of HgTe/CdTe interdiffusion using fundamental point diffusion mechanisms, J. Electron. Mater. **27**, 672–679 (1998)

32.315 Y. Selamet, Y.D. Zhou, J. Zhao, Y. Chang, C.R. Becker, R. Ashokan, C.H. Grein, S. Sivananthan: HgTe/HgCdTe superlattices grown on CdTe/Si by molecular beam epitaxy for infrared detection, J. Electron. Mater. **33**, 503–508 (2004)

32.316 J.P. Faurie, S. Sivananthan, J. Reno: Present status of molecular beam epitaxial growth and properties of HgTe-CdTe superlattices, J. Vac. Sci. Technol. A **4**, 2096–2100 (1986)

32.317 T.C. McGill, G.Y. Wu, S.R. Hetzler: Superlattices: Progress and prospects, J. Vac. Sci. Technol. A **4**, 2091–2095 (1986)

32.318 C.A. Hoffman, J.R. Meyer, F.J. Bartoli, Y. Lansari, J.W. Cook Jr., J.F. Schetzina: Electron mobilities and quantum Hall effect in modulation-doped HgTe-CdTe superlattices, Phys. Rev. B **44**, 8376–8379 (1991)

32.319 J.M. Arias, S.H. Shin, D.E. Cooper, M. Zandian, J.G. Pasko, E.R. Gertner, R.E. DeWames: p-type arsenic doping of CdTe and HgTe/CdTe superlattices grown by photoassisted and conventional molecular-beam epitaxy, J. Vac. Sci. Technol. A **8**, 1025–1033 (1990)

32.320 C.R. Becker, L. He, M.M. Regnet, M.M. Kraus, Y.S. Wu, G. Landwehr, X.F. Zhang, H. Zhang: The growth and structure of short period (001) $Hg_{1-x}Cd_xTe$-HgTe superlattices, J. Appl. Phys. **74**, 2486–2493 (1993)

32.321 A. Rogalski, M. Kimata, V.F. Kocherov, J. Piotrowski, F.F. Sizov, I.I. Taubkin, N. Tubouchi, N.B. Zaletaev: *Infrared Photon Detectors* (SPIE Optical Engineering Press, Bellingham 1995)

32.322 P.R. Bratt, T.N. Casselman: Potential barriers in HgCdTe heterojunctions, J. Vac. Sci. Technol. A **3**, 238–245 (1985)

32.323 M.A. Kinch, F. Aqariden, D. Chandra, P.K. Liao, H.F. Schaake, H.D. Shih: Minority carrier lifetime in p-HgCdTe, J. Electron. Mater. **34**, 880–884 (2005)

32.324 S. Velicu, R. Ashokan, C.H. Grein, S. Sivananthan, P. Boieriu, D. Rafol: High-temperature HgCdTe/CdTe/Si infrared photon detectors by MBE, SPIE Proc. **4454**, 180–187 (2001)

32.325 J.F. Piotrowski, A. Rogalski: *High-Operating-Temperature Infrared Photodetectors* (SPIE Optical Engineering, Bellingham 2007)

33. Metalorganic Vapor-Phase Epitaxy of Diluted Nitrides and Arsenide Quantum Dots

Udo W. Pohl

Metalorganic vapor-phase epitaxy offers the ability for controlled layer deposition down to the monolayer range. Versatile application in a wide range of materials and its upscaling ability has established this growth technique in industrial mass production, particularly in the field of semiconductor devices. A topic of current research is the extension of the well-developed GaAs-based technology to the near-infrared spectral range for optoelectronic applications. The complementary approaches of either employing dilute nitrides quantum wells or quantum dots have recently achieved significant advances in the field of laser diodes. This chapter introduces the basics of metalorganic vapor-phase epitaxy and illustrates current issues in the growth of InGaAsN/GaAs quantum wells and InAs/GaAs quantum dots. Section 33.1 gives a brief introduction to the growth technique, exemplified by the classical GaAs epitaxy. Sections 33.2 and 33.3 address two current topics of GaAs-related MOVPE, which are intensely studied for, e.g., datacom laser applications: Epitaxy of dilute nitrides and InGaAs quantum dots.

33.1	**Principle of MOVPE** 1133
	33.1.1 MOVPE Precursors 1133
	33.1.2 Growth Process 1135
33.2	**Diluted Nitride InGaAsN Quantum Wells** .. 1137
	33.2.1 Nitrogen Precursors 1138
	33.2.2 Structural and Electronic Properties of InGaAsN 1139
	33.2.3 Dilute Nitride Quantum Well Lasers. 1141
33.3	**InAs/GaAs Quantum Dots** 1142
	33.3.1 The Stranski–Krastanow 2-D–3-D Transition 1142
	33.3.2 MOVPE of InAs Quantum Dots ... 1144
	33.3.3 Quantum Dot Lasers 1147
33.4	**Concluding Remarks** 1148
References	 1148

33.1 Principle of MOVPE

Metalorganic vapor-phase epitaxy (MOVPE), also termed metalorganic chemical vapor deposition (MOCVD; sometimes O and M in the acronyms are exchanged), is the most frequently applied CVD technique for semiconductor device fabrication. Industrial large-scale reactors presently have the capacity for simultaneous deposition on 50 2inch wafers, and the majority of advanced semiconductor devices are produced using this technique. Applications of MOVPE are not restricted to semiconductors, but also include oxides, metals, and organic materials. The technique emerged in the 1960s [33.1–4], when epitaxy was dominated by liquid-phase epitaxy and chloride vapor-phase epitaxy, and molecular-beam epitaxy (MBE) did not exist in its present form. Complex sample structures with abrupt interfaces and excellent uniformity may today be fabricated using either MOVPE or MBE, though application of MOVPE is advantageous in realizing graded layers or in As-P alloys and nitride materials.

33.1.1 MOVPE Precursors

A common feature of chemical vapor-phase techniques is the transport of the constituent elements in the gas phase to the vapor–solid interface in the form of volatile molecules. In MOVPE these species consist of metalorganic compounds, and the transport is made by a carrier gas such as hydrogen at typically 100 mbar total pressure. The gaseous species dissociate thermally at the growing surface of the heated substrate, thereby releasing the elements for layer growth. The dissociation at the surface is generally assisted by chemical reactions.

The net reaction for the MOVPE of GaAs using the standard source compounds trimethylgallium and arsine reads

$$\mathrm{Ga(CH_3)_3 + AsH_3 \rightarrow GaAs + 3CH_4 \uparrow} \ . \quad (33.1)$$

The reaction is actually much more complicated and comprises many successive steps and species in the chemistry of deposition [33.6] such as, e.g., some steps of precursor decomposition

$$\mathrm{Ga(CH_3)_3 \rightarrow Ga(CH_3)_2 + CH_3}$$
$$\rightarrow \mathrm{GaCH_3 + 2CH_3 \rightarrow Ga + 3CH_3} \ .$$

The source compounds employed for MOVPE must meet some basic requirements. Their stability is low to allow decomposition in the process, but still sufficient for long-term storage. Furthermore their volatility should be high, and a liquid state is favorable to provide a steady-state source flow. Most source molecules have the form MR_n, where M denotes the element used for MOVPE, and R are alkyls such as methyl $\mathrm{CH_3}$. By choosing a suitable organic ligand, the bond strength to a given element M can be selected to comply with the requirements of MOVPE for the solid to be grown. The metal–carbon bond strength depends on the electronegativity of the metal M and the size and configuration of the ligand R [33.7]. As a rule of thumb the bond strength decreases as the number of carbons bonded to the central carbon in the alkyl is increased. This trend is also reflected in the dissociation energy of the first carbon–hydrogen bond, given in Table 33.1 [33.8].

The organic radicals R most frequently used for MOVPE precursors are depicted in Fig. 33.1.

Table 33.1 Dissociation energy of the carbon–hydrogen bond for radicals R used in MOVPE source molecules

R	E (kJ/mol)	R	E (kJ/mol)
Methyl	435	iso-Propyl	398
Ethyl	410	tert-Butyl	381
n-Propyl	410	Allyl	368

Besides metalorganic sources also hydrides such as arsine are employed as precursors. Their use is interesting since they release hydrogen radicals under decomposition that can assist removal of carbon-containing radicals from the surface. A major obstacle is their high toxicity and their very high vapor pressure, requiring extensive safety precautions. To reduce the hazardous potential, hydrides are increasingly replaced by metalorganic alternatives, e.g., arsine by tertiarybutylarsine, where one of the three hydrogen radicals is replaced by a tertiarybutyl radical. Thereby the vapor pressure is strongly reduced, yielding usually liquids at ambient conditions. In addition the toxicity decreases significantly.

Partial pressures for some standard precursors used in the MOVPE of As-related III–V semiconductors are given in Table 33.2. The values are expressed in terms of the parameters a and b to account for the exponential temperature dependence of the vapor pressure according to

$$\log(P_{\mathrm{eq\,MO}}) = a - b/T \ , \quad (33.2)$$

$P_{\mathrm{eq\,MO}}$ and T given in Torr and K, respectively. Hydrides $\mathrm{AsH_3}$ and $\mathrm{PH_3}$ are stored at 20 °C as liquids under pressures of 11 250 and 26 250 Torr, respectively, and introduced as gases to the MOVPE setup.

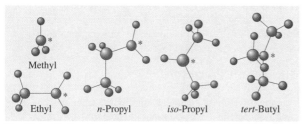

Fig. 33.1 Alkyl radicals used as organic ligands in MOVPE source molecules. *Brown* and *gray spheres* represent carbon and hydrogen atoms, respectively, and the location of a bond to an element M used for epitaxy is indicated by an *asterisk*

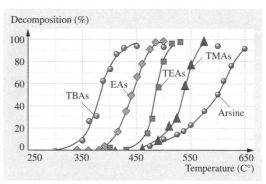

Fig. 33.2 Decomposition of As precursors; the labels TBAs, EAs, TEAs, and TMAs denote *tert*-butyl-As, ethyl-As, triethyl-As, and trimethyl-As, respectively [33.5]

Table 33.2 Equilibrium vapor pressure data of some metalorganic compounds used for III–V MOVPE. (Vapor pressure data taken from data sheets of several precursor suppliers.)

Element	Precursor	Vapor pressure		$P_{eq\,MO}$ (Torr) at 20 °C
		a	b (K)	
Al	Trimethylaluminum	8.224	2135	8.7
Ga	Trimethylgallium	8.07	1703	182
	Triethylgallium	8.083	2162	5.1
In	Trimethylindium	10.520	3014	1.7
P	Tertiarybutylphosphine	7.586	1539	142 (10 °C)
As	Tertiarybutylarsine	7.243	1509	81 (10 °C)
Sb	Trimethylantimony	7.708	1697	83
	Triethylantimony	7.904	2183	2.9
N	Dimethylhydrazine	8.646	1921	123

Note: 1 Torr = 1.333 mbar

Precursor molecules may decompose by a number of pyrolytic mechanisms, the most simple being free-radical homolysis, i. e., simple bond cleavage. Since the M–H bond is generally stronger than the M–C bond, metalorganic alternatives of the stable hydrides decompose at lower temperatures – a further incentive for their use. Results of pyrolysis studies for various As precursors, performed in an isothermal flow tube, are given in Fig. 33.2. The bond strength rule of thumb noted above is well reflected in these curves.

33.1.2 Growth Process

Most metalorganic sources are liquids, which are stored in bubblers. For transport to the reactor a carrier gas with a flow Q_{MO} is introduced by a dip tube ending near the bottom. At a fixed temperature the metalorganic liquid forms an equilibrium vapor pressure $P_{eq\,MO}$ given by (33.2), and the bubbles saturate with precursor molecules. At the outlet port of the bubbler a pressure controller is installed, which acts like a pressure relief valve and allows to define a fixed pressure P_{bub} ($> P_{eq\,MO}$) in the bubbler, thereby decoupling the bubbler pressure from the total pressure P_{tot} in the reactor. The partial pressure of a metalorganic source in the reactor P_{MO} results from the mentioned parameters by

$$P_{MO} = \frac{Q_{MO}}{Q_{tot}} \times \frac{P_{tot}}{P_{bub}} \times P_{eq\,MO} \,. \qquad (33.3)$$

The two fractions in (33.3) are employed to control the partial pressure of the source in the reactor. For sources used as dopants or compounds with very high vapor pressures an additional dilution by mixing with a controlled flux of carrier gas is applied. The gaseous hydrides are directly controlled by their flux Q_{Hyd}, and (33.3) simplifies to

$$P_{Hyd} = \frac{Q_{Hyd}}{Q_{tot}} \times P_{tot} \,. \qquad (33.4)$$

The total flux in the reactor Q_{tot} results from the sum of all component fluxes and the flux of the carrier gas, which is additionally introduced into the reactor by a separate mass flow controller. This flux is generally much higher than that of all sources, and the sum of all source partial pressures P_{MO} and P_{Hyd} is consequently much smaller than the total pressure in the reactor P_{tot}. The reactor pressure P_{tot} is controlled as an independent parameter by a control valve attached to an exhaust pump behind the reactor.

The complete treatment of the MOVPE growth process involves numerous gas-phase and surface reactions, in addition to hydrodynamic aspects. Such complex studies require a numerical approach, and solutions were developed for specific processes such as MOVPE of GaAs from trimethylgallium and arsine [33.6, 9]. We will draw a more general picture of the growth process and outline some relations of growth parameters.

Growth represents a nonequilibrium process. The driving force is given by a drop in the chemical potential μ from the input phase to the solid. For the discussion of the MOVPE process a description by consecutive steps as depicted in Fig. 33.3 is convenient. The reactants in the carrier gas represent the source. Near the solid surface a vertical diffusive transport component originates from reactions of source molecules and incorporation into the growing layer. All processes from adsorption at the surface to the incorporation are sum-

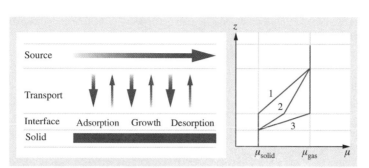

Fig. 33.3 Scheme of the chemical potential μ near the surface of the growing solid during MOVPE. *Path 1* signifies growth controlled by mass transport, *paths 3* and *2* denote growth being limited by interface reactions and the general case, respectively

marized by interface reactions. Finally excess reaction products desorb from the interface by diffusion.

The slowest process of the successive steps limits the growth rate. Without considering mechanisms of growth in detail, processes limited by either transport or kinetics can be well distinguished. Figure 33.4 shows on a logarithmic scale the dependence of the GaAs growth rate on the reciprocal substrate temperature. At low temperature, experiment and simulation show an exponential relation, indicating that thermally activated processes limit the growth rate. Precursor decomposition and interface growth reactions lead to a pronounced temperature dependence, the slope $\propto -\Delta E/(k_\mathrm{B} T)$ yields an activation energy ΔE near

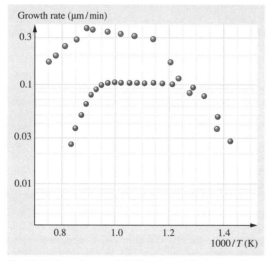

Fig. 33.4 Growth rate of (001)-oriented GaAs as a function of reciprocal temperature. Trimethylgallium and arsine are used as precursors. *Brown* and *gray circles* represent measured data [33.10] and model predictions [33.6], respectively

19 kcal/mole for the given process. This regime is referred to as *kinetically limited growth*. The gas phase supplies precursors to the surface at a rate well exceeding the rate of growth reactions. As the temperature is increased, the growth rate becomes nearly independent of temperature. In this range precursor decomposition and surface reactions are much faster than mass transport from the source to the interface of the growing solid. Since diffusion in the gas phase depends only weakly on temperature, this process is called *transport-limited growth*. Mass transport in this regime depends on the geometry of the reactor, because flow field and temperature profile above the substrate affect cracking and arrival of precursors at the interface; this accounts for the difference in the maximum growth rates in Fig. 33.4. In the high temperature range growth rates decrease due to enhanced desorption and parasitic deposition at the reactor walls, inducing a depletion of the gas phase.

MOVPE is usually performed in the mid-temperature range of transport-limited growth, where variations of the substrate temperature have only a minor effect on growth rate, composition of alloys, and doping. For III–V semiconductors the range is typically 500–800 °C.

Mass transport of the reactants from the gas source to the interface of the growing solid are essentially controlled by diffusion: The mass flow $\boldsymbol{j}_i$ of component i is given by the direct flow component $\boldsymbol{v}$ normal to the interface, the diffusion along the partial pressure gradient $\partial P_i/\partial \boldsymbol{r}$, and the thermodiffusion, according to [33.11]

$$\boldsymbol{j}_i = \frac{P_i \boldsymbol{v}}{k_\mathrm{B} T} - \frac{D_i}{k_\mathrm{B} T}\left(\frac{\partial}{\partial \boldsymbol{r}} P_i + \frac{\alpha_i}{T} P_i \frac{\partial}{\partial \boldsymbol{r}} T\right), \quad (33.5)$$

where D_i and α_i are the diffusion constant and the thermodiffusion factor, respectively. The direct flow component normal to the interface should be negligible in a laminar gas flow. Also thermodiffusion is generally assumed to make no sizeable contribution. Equation (33.5) is hence reduced to the diffusion term.

A simplified one-dimensional model assumes that the partial pressures P_i drop over a so-called diffusion boundary layer of thickness d from their values in the source to values $P_i^{\text{interface}}$ at the interface to the solid [33.7]. Equation (33.5) then reduces to

$$j_i = \frac{D_i}{k_B T d}\left(P_i - P_i^{\text{interface}}\right). \tag{33.6}$$

The factor $D_i/(k_B T d)$ may be considered as an effective coefficient of mass transport for component i.

Due to the supersaturation set to induce growth, the partial pressures of the components at the inlet of the reactor P_i are much higher than the near-equilibrium values at the interface to the solid $P_i^{\text{interface}}$. For III–V compounds such as GaAs this means $P_{\text{III}} P_V \gg P_{\text{III}}^{\text{interface}} P_V^{\text{interface}}$. Furthermore, the group V precursors are far more volatile than the group III species (except for Sb sources); III–V semiconductors are hence usually grown with a large excess of group V species, i.e., $P_V/P_{\text{III}} \gg 1$. These conditions and the requirement of stoichiometric growth lead to the relations of the partial pressures at the interface and the reactor inlet $P_{\text{III}}^{\text{interface}} \ll P_{\text{III}}$, and $P_V^{\text{interface}} \approx P_V$. This means that the growth rate is limited by the flow of group III species, and all group III species arriving at the interface are incorporated into the solid. Equation (33.6) then further reduces to

$$j_{\text{III}} = \frac{D_{\text{III}} P_{\text{III}}}{k_B T d}. \tag{33.7}$$

The transport properties expressed by (33.7) may be related to other growth parameters by applying the *boundary-layer model*. Though being oversimplified, this model provides a reasonable description of basic relations. The model assumes a horizontal flow reactor and considers, that the velocity component of the gas flow parallel to the substrate must be zero at the interface due to friction. In a distance δ above the substrate the velocity flow arrives at a constant value v. The range of δ was interpreted in terms of a stagnant layer for the mass transport to the interface δ may be written [33.7]

$$\delta \cong 5\sqrt{\frac{D}{v}}. \tag{33.8}$$

Substituting d in (33.7) by δ and bearing in mind that the diffusion constant D in the gas phase is inversely proportional to the total pressure P_{tot}, yields for the growth rate

$$r = \text{const}\, P_{\text{III}} \sqrt{\frac{v}{P_{\text{tot}}}}. \tag{33.9}$$

The growth rate shows the proportional dependence on the partial pressure of the group III species already expressed by (33.7). According to (33.3), r is hence proportional to the flow of carrier gas through the bubbler Q_{MO} of the group III source. Equation (33.9) predicts a growth rate that is independent of the total reactor pressure, because both P_{III} (cf., (33.3)) and v (inversely proportional to P_{tot}) are functions of P_{tot}. Moreover, the model yields a square-root decrease of the growth rate as the total flow in the reactor Q_{tot} is increased. Deviations from these reasonable predictions are observed at low pressures and low flow velocities, where the boundary-layer thickness is in the range of the reactor height.

The outline of metalorganic vapor-phase epitaxy given above is intended to provide an insight into the basics of the growth technique. Albeit being established in many fields of materials fabrication, new areas of application are steadily developed. Even the classical GaAs-related materials are widely studied to further extend applications, and two current topics are considered in the following sections.

33.2 Diluted Nitride InGaAsN Quantum Wells

The quaternary dilute nitride alloy InGaAsN (or GINA, GaInNAs), which can be grown lattice-matched to GaAs substrates, has recently gained considerable attention as a promising material for laser diodes in the datacom wavelength range. The constituent binaries GaAs and GaN have large differences in electronegativities and lattice constants, leading to an extraordinary large bowing parameter in GaAsN alloys and a strong bandgap decrease for even small N compositions. The same applies for the In-related binaries, cf. Fig. 33.5; in this connection it must be noted that the InN bandgap energy is still controversial, but serious indications exist for a gap near or below 1 eV [33.12–14]. Large band offsets between InGaAsN and GaAs provide the opportunity for good carrier confinement in quantum structures. The attractive quaternary compound is, however, metastable, and introduction of nitrogen is difficult due to a large miscibility gap. Furthermore, intrinsic

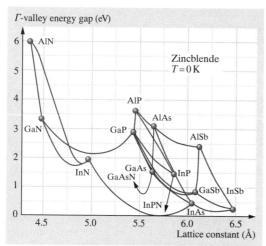

Fig. 33.5 Direct energy gap as a function of lattice constant for zincblende III–V compounds and their alloys; *arrows* indicate limits for predictions with any accuracy [33.15]. For InN more recently a value of 0.78 eV was given [33.13]

regions of high strain and strong localization result from the large differences in lattice parameters and the large energy range induced by statistical composition fluctuations. However, significant advances have been achieved by applying epitaxy at rather low temperatures. This sections addresses the issue of suitable precursors for MOVPE of InGaAsN, the materials properties of epitaxial layers on GaAs substrate, and eventually the current state of laser device applications.

33.2.1 Nitrogen Precursors

The small covalent radius of nitrogen leads to very small solubilities in the conventional III–V compound semiconductors GaAs and InAs, predicted to be only 10^{14} cm^{-3} and 10^{17} cm^{-3}, respectively, at typical growth temperatures [33.17]. To increase the N concentration well above the equilibrium solubility limit, the temperature is lowered to achieve metastable nonequilibrium growth in the range between the concentrations of the spinodal and the stable binodal limit. The latter was determined to be below 2% at 800 K for GaN in GaAs [33.18].

The standard nitrogen source for GaN-related group III nitride semiconductors is ammonia (NH$_3$). Since dissociation of ammonia in MOVPE requires very high growth temperatures, alternative precursors

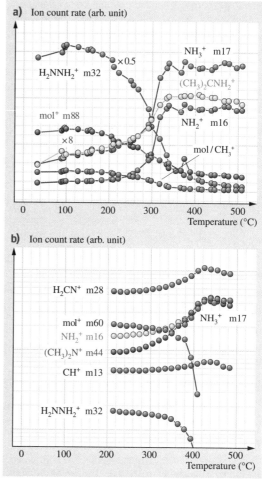

Fig. 33.6a,b Thermal decomposition of the nitrogen sources *tert*-butylhydrazine ((**a**), mass 88) and dimethylhydrazine ((**b**), mass 60), measured by mass spectrometry in an isothermal reaction tube [33.16]

are required for epitaxy of dilute nitride alloys. Hydrazine (H$_2$NNH$_2$) is much less stable than ammonia, has a convenient vapor pressure, and provides reactive nitrogen by N–N bond cleavage at temperatures as low as 400 °C. Though low-temperature growth of GaN was proved [33.19], this source is not used due to its toxicity and very high reactivity. Instead, derivatives with one [33.20] or two methyl [33.21, 22] or with tertiary-butyl [33.23, 24] ligands were introduced. Comparative pyrolysis studies of dimethylhydrazine (CH$_3$)$_2$NNH$_2$ and tertiarybutylhydrazine (C$_4$H$_9$)(H)NNH$_2$ demonstrated that both sources decompose below 400 °C by N–N bond cleavage, enabling low-temperature nitride growth [33.16]. As shown in Fig. 33.6, decomposition of both precursors produce reactive NH_2 radicals (mass 16) required for the growth process.

Dimethylhydrazine, more precisely (1,1)-dimethylhydrazine with both methyl radicals attached to one nitrogen (also referred to as unsymmetric UDMHy), has developed as the standard source for InGaAsN epitaxy. Pressure data are included in Table 33.2. Tertiarybutylarsine is generally employed as As source due to its favorable pyrolysis properties at low temperature, shown in Fig. 33.2. For the same reason triethylgallium is often used as Ga source instead of trimethylgallium employed for standard GaAs MOVPE, while the generally applied trimethylindium precursor for In supply is sufficiently unstable to be also used for InGaAsN growth.

Lattice-match of InGaAsN to GaAs requires an In/N ratio of about 3. Such a ratio cannot be maintained for a quaternary-alloy QW of 1.3 μm devices due to the low miscibility for N alloying. Instead *strained* QWs with ≈ 30% In and up to 4% N are used. Even then, introduction of N is difficult, since In alloying strongly counteracts N incorporation into the solid. The dependence of the nitrogen content on the In concentration is shown in Fig. 33.7 for two growth temperatures, illustrating the benefit of low-temperature growth. For MOVPE of QWs deposition temperatures as low as 525 °C are used, while thicker layers are grown at slightly increased temperatures (e.g., 550 °C) to achieve a higher growth rate.

Composition of the quaternary alloy also depends sensitively on other growth parameters. A high tertiarybutylarsine flow was found to favor N incorporation up to a maximum value near 4% [33.26], and higher In concentration was obtained for lower III/V ratios [33.25]. A particular issue of MOVPE is the incorporation of carbon and hydrogen. p-Type character of as-grown material was assigned to C incorporation, and a type conversion upon annealing was attributed to an N–H complex [33.27]. High C levels were ascribed to the strong N–C bond, and growth using various precursors indicated triethylgallium as a possible source [33.28]. Conditions to supply sufficient other radicals to the surface to prevent bonding of N radicals to C-containing groups are suggested to lower C contamination.

33.2.2 Structural and Electronic Properties of InGaAsN

The lattice parameter of GaAsN calculated from total energy minimum yields excellent agreement with Vegard's law despite the large lattice mismatch of more than 20% [33.18]. Due to the good miscibility with In such linear dependence on the composition also applies for the quaternary InGaAsN. Deviations were observed for high N concentrations exceeding 2.9% and were assigned to interstitial, i.e., nonsubstitutional incorporation [33.29]. In molecular-beam epitaxy Sb was used as a surfactant, significantly improving structural and optical properties of highly strained material [33.30–32].

In and also N were found to be randomly distributed on a large scale. While this applies for In also on a microscopic scale, a slightly enhanced number of nitrogen pairs oriented along the [001] growth direction was found in scanning tunneling micrographs of GaAsN [33.33] and InGaAsN layers [33.34, 35]. Calculations based on the pseudopotential method [33.36] and the Keating valence force field model [33.37] show that such N pairs reduce the strain compared with two isolated N atoms. Moreover, adding of further N atoms to a [001]-oriented chain is found to be energetically favorable. The ordering of nitrogen in the quaternary alloy leads to local nanometer-sized strain fields [33.37, 38]. Figure 33.8 shows a transmission electron microscopy (TEM) image taken from an as-grown QW under strain-sensitive imaging conditions. The apparent columnar structure along the [001] growth direction is also found in thick layers. It is not visible if chemically sensitive imaging conditions are used. The

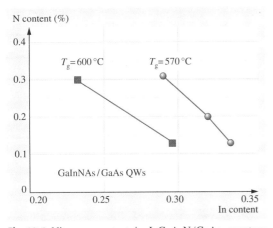

Fig. 33.7 Nitrogen content in InGaAsN/GaAs quantum wells as a function of In content for two deposition temperatures T_g [33.25]

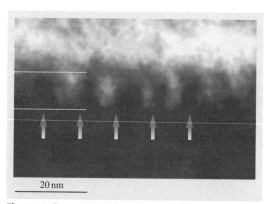

Fig. 33.8 Cross-sectional transmission electron micrograph of an as-grown InGaAsN/GaAs quantum well with 30% In and 2.7% N. *Dashed lines* depict QW boundaries; *arrows* indicate column-like strain fields [33.37]

density of the columns was shown to increase with the N content.

As-grown InGaAsN material with high N content generally shows poor luminescence efficiency, which strongly improves upon annealing. The strain fields visible in TEM images dissolve by annealing. Studies of the nearest-neighbor bonds using x-ray absorption fine structure show that the short-range ordering of the nitrogen chains largely disappears, accompanied by an increase of In–N bonds at the expense of Ga–N bonds [33.39–41]. While N chains build a stable configuration at the growth surface, the highly In-coordinated state of N was calculated to have a lower total energy in the bulk [33.37,42,43]. Strain in the layers does not significantly affect the nearest-neighbor bonding [33.40], but the optimum annealing temperature with respect to photoluminescence (PL) efficiency was found to decrease with strain [33.44]. The origin of the improved luminescence efficiency after anneal was assigned to the removal of nonsubstitutional nitrogen, which builds a complex weakly bonded to Ga and acts as a trap for photoexcited charge carriers [33.29].

Thermal annealing of as-grown InGaAsN induces a blue-shift of the bandgap due to the changed bonding [33.40, 45]. This finding agrees with the described reduction of N pairing and predictions of total energy calculations, which yield a reduced bandgap in case of clustering and also a reduced momentum matrix element (i. e., PL intensity) [33.36]. It was pointed out that the band structure of diluted nitride alloys shows an alloying mechanism different from that of conventional semiconductors [33.46]. Addition of nitrogen induces localized cluster states below the conduction band edge, which are gradually overtaken by states composed of both localized and delocalized states. These perturbed host states form the alloy conduction band edge and lead to the strong, composition-dependent bowing parameter. The effect of a few percent of nitrogen on the valence band is small and neglected in the generally accepted parameterized band anticrossing (BAC) model [33.47, 48]. The splitting into a lower-lying E_- and an E_+ band can be considered as the interaction between the spatially localized N level and the delocalized conduction band states of the host semiconductor, and the respective dependence on the nitrogen fraction x reads

$$E_\pm = \frac{1}{2}\left((E_C + E_N) \pm \sqrt{(E_C - E_N)^2 + 4V^2 x}\right), \tag{33.10}$$

with E_C being the conduction band edge of the unperturbed host semiconductor, E_N the position of the N impurity level in this host, and V the interaction potential between the two bands. Respective model parameters for diluted nitride alloys are given in Table 33.3.

The offsets of valence and conduction bands are basic parameters for fabricating heterostructure devices. Data of InGaAsN/GaAs QW samples were particularly obtained from photoreflectance spectra by evaluating the splitting between the heavy-hole and light-hole ground state [33.49], and from photovoltage spectra by directly measuring the transition energy between confined and free states [33.50]. A type I band alignment is found for electron and heavy hole, with values for $Q_C = \Delta E_C/(\Delta E_C + \Delta E_V)$ of 55–70%. The light-hole valence band offset of a tensile-strained GaAsN QW

Table 33.3 Band anticrossing model parameters for dilute nitride semiconductors; the energy of the isoelectronic nitrogen impurity E_N is given with respect to the valence band maximum [33.15]

Alloy	E_N (eV)	V (eV)
GaAsN	1.65	2.7
InAsN	1.44	2.0
$In_xGa_{1-x}AsN$	$1.65(1-x) + 1.44x - 0.38x(1-x)$	$2.7(1-x) + 2.0x - 3.5x(1-x)$

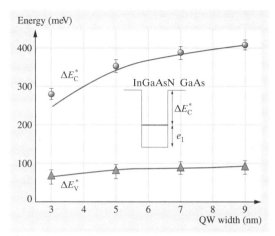

Fig. 33.9 Energy differences ΔE^* between confined states and band edges of $In_{0.3}Ga_{0.7}As_{0.98}N_{0.02}$ quantum wells with various widths, measured using photovoltage spectra; the *continuous lines* represent simulated values [33.50]

with 2% N was reported to be type II, and the counteracting strain introduced by 15% In was expected not to convert the type of alignment [33.49]. Introduction of Sb leads essentially to an increase of the hole QW depth [33.51]. Figure 33.9 shows measured energy differences between confined states and band edges for QWs with 2% N and 30% In. These values are of practical relevance and related to the band offsets by $\Delta E_C = \Delta E_C^* + e_1$, and $\Delta E_V = \Delta E_V^* + h_1$, with e_1 and h_1 being the energies of the first quantized states of electrons and holes, respectively.

33.2.3 Dilute Nitride Quantum Well Lasers

The ability to grow InGaAsN quantum wells lattice matched on GaAs substrates and the formation of a type I band alignment to confine both electrons and holes makes this dilute nitride alloy promising for device applications. Most work focused on fabrication of injection lasers emitting in the 1.3 and 1.55 μm wavelength windows of common datacom fibres, stimulated by the great commercial demand and the compatibility with the well-developed GaAs/AlAs technology. In fact, InP-based device structures used in optical networks suffer from poor thermal conductivity of the substrate material and low refractive index variation of alloys. InGaAsN also gained interest as a 1 eV subcell added to commercially available $InGaP_2$/GaAs/Ge multijunction solar cells. Inserted between GaAs and Ge, conversion efficiencies exceeding 40% are expected [33.52]. In the following some aspects of the work related to edge-emitting lasers are discussed.

InGaAsN QWs for long-wavelength lasers were introduced in the mid 1990s using both gas-source MBE (GSMBE, also denoted metalorganic MBE, MOMBE) and MOVPE [33.53]. The first edge-emitting ridge waveguide lasers used single QWs and achieved emission wavelengths near 1.2 μm at 300 K with high threshold current densities exceeding $1\,kA/cm^2$. Lasers grown using MOVPE have focused since then on the 1.3 μm range, while considerable progress in the 1.55 μm range was achieved with MBE by additionally introducing Sb as a surfactant during growth of the active quantum wells in lasers [33.30]. Most lasers have a single QW with a thickness of 6–8 nm. During the first years of research on diluted nitride lasers a general trend of increasing threshold current densities was found when the emission wavelength was increased to longer wavelengths, irrespective of the growth technique employed, MOVPE or MBE [33.28, 54–56]. Wavelength tuning beyond 1.2 μm requires a sizeable incorporation of nitrogen into InGaAs. This can be achieved by a decreased growth temperature, an increased growth rate, and an increased N/(N+As) ratio [33.54]. The latter is more easily achieved using a low As/(Ga+In) ratio rather than a high flow of the N source, another advantage of employing tertiarybutylarsine over AsH_3 as arsenic source [33.57].

Generally such wavelength tuning was accompanied by material deterioration. A direct correlation of the threshold current density to the concentration of carbon in the QWs was observed [33.28]. For lasers grown using MOVPE the finding could be related to C from the gallium source. The high levels were claimed to be a consequence of the strong C–N bond. A similar trend found in MBE-grown lasers indicated a respective issue also for this growth technique.

A high PL efficiency of the active QW is crucial to obtain a low lasing threshold. As noted in Sect. 33.2.2, the PL, which drops for high N incorporation, largely recovers by annealing. A major improvement of InGaAsN QW lasers was therefore obtained by adding an annealing step to the fabrication procedure [33.58]. This step can also be accomplished in situ during growth of the upper cladding layer at increased temperature [33.57]. Further improvements were achieved by engineering the confinement and strain in the quantum well. Cladding of the compressively strained QW by a strain-compensating tensile

GaAsP barrier layer led to a threshold current density as low as 211 A/cm^2 for lasing at 1295 nm [33.57]. Using tensile-strained GaAsN instead of GaAsP proved to be even more advantageous [33.59, 60]. The conduction band discontinuity from the quaternary QW to the GaAs waveguide is segmented into two steps, thereby reducing the quantum confinement of the electrons and hence inducing a red-shift of the emission. In addition, stronger hole confinement is achieved due to a larger valence band step from the QW to GaAsN with respect to GaAs, resulting in improved high-temperature characteristics of the device. The characteristic temperature T_0 was shown to increase by 15 K for lasers emitting near 1315 nm with thresholds of 210–270 A/cm^2 [33.59]. Using this approach, lasing at the longest wavelength reported to date of 1410 nm was achieved, with a threshold of 1.4 kA/cm^2 [33.60].

Shift of the lasing wavelength to 1.55 μm will certainly be difficult to achieve with InGaAsN QWs. The results obtained using MOVPE are largely comparable to those achieved using MBE of lasers with quaternary QWs [33.61–63]. The most striking difference is a substantially lower temperature during growth of the active QW, ranging between 350 °C and 330 °C in MBE compared with about 530 °C in MOVPE. The longest wavelength reached to date, by applying a carefully adjusted low-temperature MBE, is 1510 nm, with a still very low threshold of 780 A/cm^2 [33.56]. Substantial progress beyond 1.5 μm was recently achieved with quinternary quantum wells in MBE [33.64]. Using a single $In_{0.38}Ga_{0.62}As_{0.943}N_{0.03}Sb_{0.027}$ QW with cladding by strain-compensating $GaAs_{0.96}N_{0.04}$ barriers, lasing was observed at 1.55 μm with a low threshold of 579 A/cm^2.

33.3 InAs/GaAs Quantum Dots

Semiconductor quantum dots (QDs) are nanometer-sized objects in which charge carriers are confined in all three spatial dimensions. Their size is in the range of the de Broglie matter wavelength of the confined particles, and the quantum size effect leads to a delta-function-like electronic density of states. The properties of such zero-dimensional structures resemble those of atoms. The unique electronic properties of QDs, the discovery of self-organization processes which control their coherent (i.e., defect-free) growth, and the ease of incorporating QDs into a semiconductor device led to rapid development in this field. This section focuses on InAs QDs in GaAs matrix material and provides an insight into the basics of their formation, structural, and electronic properties. Finally, results on recent applications in edge-emitting lasers with gain media comprising QDs are reported.

33.3.1 The Stranski–Krastanow 2-D–3-D Transition

Efforts to realize quantum dots in semiconductor heterostructures were initially devoted mainly to lithographic patterning of quantum wells, e.g., [33.65]. Such structures generally suffered from residual damage introduced by nanopatterning that affected the electronic and optical properties. In the 1990s the concept of exploiting the fundamental growth mode named after Stranski and Krastanow was introduced to form large arrays of defect-free quantum dots in a self-organized (also called self-assembled or self-ordered) way. The original paper describes the formation of islands on a flat substrate surface in heteroepitaxy of lattice-matched ionic crystals with different charges [33.66]. Stranski–Krastanow growth can also be induced if a two-dimensional layer is deposited on a crystalline substrate which has a sufficiently large *mismatch* of the lateral lattice constant (typically $\Delta a/a > 2\%$). The layer adopts the lateral atomic spacing of the substrate and hence accumulates strain with increasing thickness. Above a critical layer thickness the strain is relaxed by introduction of misfit dislocations. Below this critical thickness, a considerable part of the strain can be relaxed *elastically*, i.e., without introduction of dislocations, by forming facetted surface structures. A precondition for this 2-D-3-D transition is a total energy gain of the heteroepitaxial system, expressed by the actual surface area A and the change of the areal energy density $\Delta \gamma$. This quantity is given by

$$\Delta \gamma = \gamma_{\text{surface}} + \gamma_{\text{interface}} - \gamma_{\text{substrate}}, \quad (33.11)$$

the three summands being the surface energy of the epitaxial layer, the energy of the interface between layer and substrate, and the surface energy of the substrate, respectively. If $\Delta \gamma < 0$, the layer wets the substrate surface and grows two-dimensionally in the Frank–van-der-Merve layer-by-layer growth mode. If $\Delta \gamma > 0$, the layer tries to leave the substrate uncovered and grows three-dimensionally in the Volmer–Weber island growth mode. In the intermediate case of Stranski–Krastanow growth, $\Delta\gamma < 0$ applies for the first layer(s), which grow two-dimensionally, while $\Delta\gamma > 0$ for subsequently grown layers, e.g., due to accumulated strain. The driving force of the 2-D–3-D transition is therefore minimization of the total strain energy [33.67]. The total energy gain per volume of a single QD can be expressed by [33.68]

$$\frac{E_{\text{total}}}{V} = \varepsilon_{\text{QD}}^{\text{elast}} - \varepsilon_{\text{layer}}^{\text{elast}} + \frac{A\gamma_{\text{facet}} - L^2\gamma_{\text{layer}}(d_0)}{V}$$
$$+ \left(\frac{1}{\rho} - L^2\right)\frac{\gamma_{\text{layer}}(d) - \gamma_{\text{layer}}(d_0)}{V},$$
(33.12)

$\varepsilon_{\text{QD}}^{\text{elast}}$ and $\varepsilon_{\text{layer}}^{\text{elast}}$ being the elastic energy densities of the QD and the uniformly strained layer. The third term describes the change in surface energy due to the QD, with γ_{facet} being the surface energy of the island facets, A their area and L the base length of the QD, which is assumed to have a pyramidal shape. The fourth term accounts for that part of the layer which converts to the QD; ρ is the area density of QDs, $\gamma_{\text{layer}}(d_0)$ and $\gamma_{\text{layer}}(d)$ are the formation energies of the layer as a function of its thickness d. The sum of these contributions is the total energy density, which has an energy minimum for a particular dot size as shown in Fig. 33.10. This minimum causes preferential formation of dots around this size and a stability of the dot ensemble against Ostwald ripening in thermodynamic equilibrium. The ripening still observed in experiments indicates the presence of kinetic barriers under the usually applied growth conditions. It should be noted that, even under equilibrium conditions, the entire material of the two-dimensional layer does not entirely reorganize to QDs; rather, a thin 2-D layer, the wetting layer, remains.

Self-organized growth of InAs QDs on GaAs was first realized using MBE [33.70]. Since then InAs and $\text{In}_{1-x}\text{Ga}_x\text{As}$ [33.71] dots in GaAs matrix have become a model system for self-organized QD growth. The InAs lattice constant exceeds that of GaAs by $\approx 7\%$. Details of QD formation sensitively depend on growth parameters, particularly on temperature, growth rate, and arsenic partial pressure. The 2-D–3-D transition of InAs layers occurs on (001)-oriented GaAs between 1.5 and 1.8 monolayer thickness for typical deposition temperatures of 450–520 °C. Nucleation of 3-D dots starts on top of a 2-D InAs layer which exceeds some critical thickness d_c of coverage. The density ρ of the emerging dots was measured from atomic force microscopy (AFM) images of InAs layers with varied thickness, deposited on (001) GaAs at 530 °C using MBE, cf. Fig. 33.11.

The dot density ρ depicted in Fig. 33.11 follows a dependence on the InAs coverage similar to that of

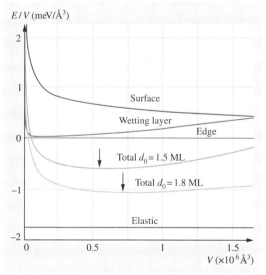

Fig. 33.10 Total energy gain by QD formation from $d_0 = 1.8$ monolayers (ML) and 1.5 monolayers InAs films deposited on GaAs. A QD area density of $\rho = 10^{10}\,\text{cm}^{-2}$ is assumed; notations on the curves denote contributions from the surface, the wetting layer, the edges of the assumed pyramid shape of the QDs, and the elastic relaxation energy [33.68]. *Arrows* mark the minima of the curves

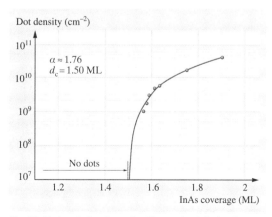

Fig. 33.11 Areal density of self-organized InAs dots on GaAs as a function of deposition thickness d [33.69]

a first-order phase transition [33.69]

$$\rho = \rho_0 (d - d_c)^\alpha \,. \tag{33.13}$$

The least-squares fit represented by the solid line yields $\rho_0 = 2 \times 10^{11}\,\mathrm{cm}^{-2}$, $\alpha = 1.76$, and $d_c = 1.50\,\mathrm{ML}$.

33.3.2 MOVPE of InAs Quantum Dots

Most studies on the MOVPE of quantum dots in the InGaAs system comprise ternary $In_xGa_{1-x}As$ dots with In content x of 30–70% [33.73–75]. The reduced strain in the layer with respect to InAs retards the onset of dot nucleation, facilitating deposition control, and the tendency to form large, dislocated clusters is smaller [33.74]. The growth procedure is often varied in many respects, complicating a comparison of results: The dot material may be deposited on a thin InGaAs buffer layer with lower In content to obtain a high dot density, or overgrown by a respective thin InGaAs layer to induce a red-shift of the emission wavelength [33.76]. In addition, seeding or stacking of dot layers is applied to modify optical properties or increase the dot density in the matrix [33.72, 77], and also atomic layer epitaxy [33.78] or use of a surfactant [33.79] has been reported. Structural and electronic properties of InGaAs QDs were found to differ significantly from those of InAs dots. For InGaAs depositions a depletion of the effective In content in the wetting layer was found [33.80], and an inhomogeneous In distribution in the InGaAs dot was observed [33.81]. To treat a more concise situation, we focus on the MOVPE of binary InAs dots. The precursors used for QD growth are standard sources for III–V epitaxy, namely trimethylindium, trimethylgallium, and either arsine or tertiarybutylarsine.

The general procedure of self-organized QD fabrication starts with the preparation of the surface. The low deposition temperature of typically about $500\,^\circ\mathrm{C}$, which is required for Stranski–Krastanow growth of InAs dots, is much lower than the temperature needed to grow GaAs matrix material with good quality at reasonable growth rate. Therefore first a GaAs buffer layer is grown at a suitable high temperature (e.g., $650\,^\circ\mathrm{C}$), and then the surface is stabilized with arsenic partial pressure while setting up the lowered temperature and eventually adjusting the V/III ratio. Then InAs is deposited, and subsequently a growth interruption is applied. Finally a GaAs cap layer is grown on top. Each of these steps comprises some crucial issues to obtain defect-free structures.

The deposition of InAs must occur with some minimum rate to prevent the formation of large incoherent clusters, in contrast to MBE, where processes can largely be retarded by using very low temperatures. The QD properties depend sensitively on the InAs thickness and the As partial pressure during InAs deposition and the growth interruption. Figure 33.12 shows PL spectra of InAs/GaAs QD samples prepared at $485\,^\circ\mathrm{C}$ and 8 s growth interruption with different depositions thicknesses and V/III ratios [33.72]. The spectra were nonresonantly excited into the matrix and show the near-band-edge emission of GaAs above 1.4 eV, emission from the 2-D wetting layer (WL), and the luminescence of the QD ensemble. For a given InAs deposition thickness d_0, Fig. 33.12a shows the emission from a 2-D wetting layer clad by GaAs (dark line

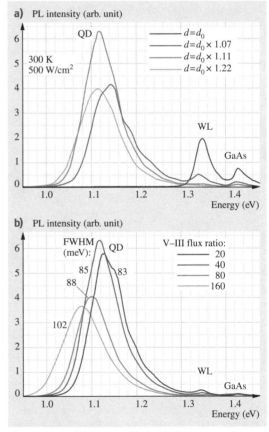

Fig. 33.12a,b PL spectra of InAs/GaAs quantum dot samples with different layer thicknesses (**a**) and different V/III ratios (**b**) applied during InAs deposition [33.72]

$d = d_0$); from the emission energy a thickness of 1.5 ML was evaluated. QDs are formed when this critical value is exceeded, and an optimum at $1.11\,d_0$ was found for the given growth conditions. The decrease of PL intensity for thicker InAs depositions indicates the onset of plastic strain relaxation. Keeping the optimum thickness d_0, the V/III ratio was varied in the series given in Fig. 33.12b by gradually changing the arsine partial pressure from 0.09 to 0.72 mbar. The PL shows a redshift, accompanied by a broadening of the full width at half maximum (FWHM) of the emission. Maximum efficiency and minimum FWHM are obtained for lowest V/III ratios. It must be noted that the effective V/III ratio at the growing surface is much lower than the quoted gas-phase ratios due to a quite incomplete decomposition of the stable arsine at 485 °C, cf. Fig. 33.2. A high V/III ratio apparently degrades the QDs. This applies also for the use of tertiarybutylarsine instead of arsine. Due to the high cracking efficiency, growth temperature and V/III ratio then become largely independent parameters [33.82], and a V/III ratio near unity is usually applied [33.79, 83].

The growth interruption conditions also strongly influence QD properties. In the sample series characterized in Fig. 33.13 the arsine flow was varied during a constant growth interruption (GRI) duration of 14 s [33.84]. Maximum PL efficiency is obtained when arsine is switched off during the first 12 s. In contrast, if the surface is stabilized under an As partial pressure during the entire growth interruption, the PL intensity drops by two orders of magnitude. AFM images of such samples show a high density of large, dislocated clusters. The only slight drop in PL intensity for the intermediate case of 3 s without arsine supply (factor 2) shows that the first stage after InAs deposition is most critical. The presence of arsine during the growth interruption of QDs prepared using MOVPE may obviously affect QD properties. Such behavior is not reported for InAs dots grown using MBE. An enhanced In mobility in MOVPE was suggested to account for this finding [33.84].

A cap layer is deposited after the growth interruption, allowing for integration of QDs into a device structure. The temperature is kept unchanged at least during growth of the first few nanometers due to the instable state of uncovered QDs. GaAs growth at such low temperatures favors the formation of defects, which affect the optical properties of QDs. Since buried InAs QDs are stable below 600 °C [33.85], an overgrowth procedure was established which maintains a high luminescence efficiency [33.86]. Here, the dots are buried by a thin (2–3 nm) GaAs cap at QD growth temperature, and further GaAs overgrowth is accomplished during a temperature increase to 600 °C. Besides an essential improvement of optical properties, a flat growth front is re-established at the otherwise corrugated surface above QDs. This is a precondition for continued QD growth in a stack, made to achieve an increased optical confinement factor and to overcome gain saturation in lasers with QD gain media. Furthermore, the flattening reduces interface roughness for subsequently deposited upper waveguide cladding, and thereby optical cavity losses in edge-emitting lasers.

Detailed insight into the dynamics of formation and evolution of InAs QDs was gained by studying ensembles with a multimodal size distribution [33.87, 88]. The particular feature of this kind of QDs is decomposition of the ensemble luminescence into individual lines as shown in Fig. 33.14. The QDs have the shape of a truncated InAs pyramid with a flat top facet, and the individual lines were shown to originate from subensembles which differ in height by an integer number of InAs monolayers. The clear detection of size quantization allows to trace redistribution of material among the QDs during the growth interruption.

InAs QDs with a multimodal size distribution form after a comparatively fast deposition of a rough, two-dimensional layer with a thickness close to the critical value for the 2-D–3-D transition [33.89]. If no growth interruption is applied, a quantum *well* is formed as shown in the inset of Fig. 33.14. The long exponential low-energy tail of the PL indicates significant thickness

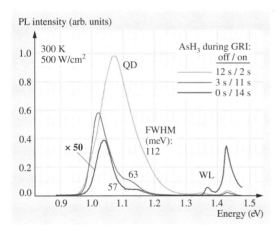

Fig. 33.13 PL spectra of InAs/GaAs QD samples with different arsine fluxes supplied during growth interruption after InAs deposition [33.84]

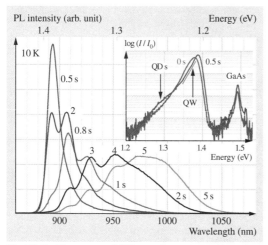

Fig. 33.14 PL spectra of QD samples grown with various growth interruptions (indicated in seconds on the curves); *numbers* denote QD heights in units of monolayers. *Inset*: spectra of samples grown without and with 0.5 s growth interruption [33.91]

fluctuations. Introduction of a short interruption leads to a blue-shift of the QW PL-maximum, i.e., a thinning of the QW, and the appearance of locally thicker regions (QDs). For longer interruptions (≥ 0.8 s) the emission of a multimodal QD ensemble with individual subensemble peaks evolves from shallow localizations in the density-of-states (DOS) tail of the rough QW. The material hence partially concentrates at some QD precursors. It must be noted that the thickness and composition of the wetting layer were found to remain constant during the entire duration of the growth interruption [33.87, 88]. The PL of the QD ensemble experiences a red-shift for longer interruptions due to an increase of the average QD volume, a behavior generally reported for self-organized QDs, cf. Fig. 33.15a. The persistent shift also occurring for very long interruptions indicates the action of kinetic barriers which prevent the achievement of thermal equilibrium on the studied time scale. The drop of PL intensity at the longest growth interruption indicates the onset of plastic relaxation, as confirmed by TEM images. Interestingly, the subensemble peaks shift to the *blue* during QD evolution. This and the PL intensity decrease of a given subensemble peak with prolonged interruption time indicate that the ripening of the QDs if fed by dissolution of smaller QDs. The development of the integral subensemble PL intensities was considered a reasonable measure for the number of QDs with a specific corresponding height, showing directly the formation of subensembles with larger dots by dissolution of smaller QDs [33.90], cf. Fig. 33.15b. The subensemble with 5 ML QD height, e.g., grows during the initial 5 s growth interruption at the expense of those with 2–4 ML height. Then it starts to dissolve, feeding the 6 ML-high subensemble, which after 10 s interruption also starts to dissolve.

The optical study noted above is confirmed by structural data of the InAs QDs. Using strain and chemically sensitive {200} reflections, InAs dots appear as bright central areas in TEM images with a low aspect ratio and flat top and bottom interfaces, cf. Fig. 33.16. Side facets are steep, but less well resolved due to strain perturbations. Analysis of such images show that both the base and the top layer of the flat dots apparently consist of plain, continuous InAs layers [33.87, 88]. This yields QDs with a shape of a truncated pyramid, which

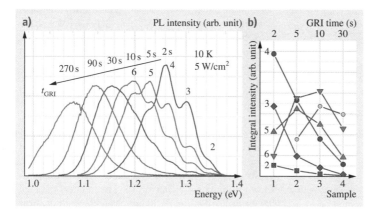

Fig. 33.15 (a) PL spectra of QD samples grown with long growth interruptions t_{GRI}. (b) Integral PL intensity of subensemble peaks as a function of growth interruption time. *Numbers* refer to QD heights as indicated in (a)

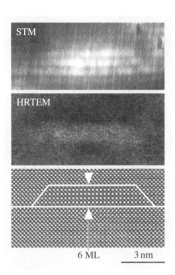

Fig. 33.16 Cross-sectional scanning tunneling micrograph (*top*, [33.92]) and high-resolution transmission electron micrograph [33.87] of InAs/GaAs QDs with 6 ML height; *bottom*: Fourier-filtered image of the transmission electron micrograph

differ in height by integer numbers of InAs monolayers. The same conclusions were drawn from recent scanning tunneling micrographs [33.92].

33.3.3 Quantum Dot Lasers

Work on InGaAs quantum dot lasers is a competing development to InGaAsN quantum well lasers to use GaAs-based laser technology for the long-wavelength datacom range. Zero-dimensional charge-carrier localization in the active region of a semiconductor laser was predicted two decades ago to lead to superior device performance, e.g., with respect to decreased threshold current and high-temperature stability [33.93, 94]. Furthermore, the reduction of lateral charge-carrier diffusion suppresses beam filamentation and strongly enhances robustness against facet degradation during high-power operation. The first realization of a QD injection laser using a single layer of MBE-grown self-organized $In_{0.5}Ga_{0.5}As$ QDs proved the fundamental validity of theoretical predictions [33.95]. A low lasing threshold of $120 A/cm^2$ and a characteristic temperature T_0 as high as $350 K$ were achieved at liquid-nitrogen temperature. At room temperature, lasing occurred at a high threshold of $950 A/cm^2$ from an excited state due to ground-state gain saturation. Generally the lasing threshold is reached if the modal gain g_{mod} just equilibrates the internal losses α_{int} and those of the cavity mirrors α_{mirr},

$$g_{mod} = \Gamma g_{mat} = \alpha_{int} + \alpha_{mirr} \,. \qquad (33.14)$$

The material gain of QD gain media g_{mat} exceeds that of QWs by far [33.94, 96]. The optical confinement factor Γ of a QD layer is, however, much smaller than that of a QW, leading to a largely comparable modal gain of QD and QW lasers. Since the number of states in a QD layer is quite small compared with a QW, low lasing threshold and also a low gain saturation level are easily reached in single active layers. To increase both confinement factor and gain saturation level, actual QD lasers comprise a *stack* of QD layers. The spacers between the QD layers should preferably be thin to position the QDs near the optical field maximum. Further improvements of QD lasers aim to confine charge carriers efficiently in the QDs at energies well below that of the wetting layer.

The first MOVPE-grown QD lasers, reported in 1997, already used a threefold InAs QD stack [33.97] and a tenfold $In_{0.5}Ga_{0.5}As$ QD stack [33.98]. Both approaches achieved ground-state lasing at room temperature, and a T_0 of $385 K$ was demonstrated for the tenfold stack laser up to $50 °C$. The stacks were grown at the same low temperature for QD layers and GaAs spacers. Substantial progress in laser performance has been made since then, particularly by introducing temperature ramping to improve spacer quality and smoothen the interfaces [33.86]. For threefold stacked dot layers emitting at $1.16 \mu m$, threshold and transparency current densities of 110 and $18 A/cm^2$, respectively, with internal quantum efficiency exceeding 90% were achieved [33.99]. Also the internal loss below $1.5 cm^{-1}$ substantiates the benefit of temperature ramping. All these devices were grown using arsine as group V precursor. Good data were likewise reported for lasers grown using tertiarybutylarsine. Devices emitting at $1.1 \mu m$ showed a transparency current below $30 A/cm^2$, 91% internal quantum efficiency, and internal loss of $2.2 cm^{-1}$ [33.82].

Much work has been devoted to extending the emission wavelength towards $1.3 \mu m$. An efficient means is overgrowth of the In(Ga)As QDs by an InGaAs QW with lower In content to form a dot-in-a-well (DWELL) structure [33.76, 100]. Since this QW reduces the strain on the buried QD layer with respect to a GaAs cap, it is also referred to as a strain-reducing layer (SRL). Note that the *overall* strain in the QD stack is increased by the QW. Furthermore, the *local* strain at the large, In-rich QDs required for long-wavelength emission is rather high. This easily leads to formation of large In-rich incoherent clusters, which degrade device characteristics. Using the DWELL approach and carefully adjusting the growth parameters for each individual QD layer in the stack to avoid defect formation, lasing at 1250 nm with a very low threshold current density of 66 A/cm^2 and 94% internal quantum efficiency was recently obtained [33.101]. In a complementary approach a DWELL laser was subjected to a postgrowth annealing procedure at moderate 560 °C, yielding 1280 nm lasing at high threshold [33.102]. As an alternative, strain-compensating GaP layers were introduced into the stacking sequence. Lasing at 1249 nm with a reasonable threshold of 550 A/cm^2 was reported for partial strain compensation using 4 ML-thick GaP layers [33.103]. The presently (2006) most advanced MOVPE-grown lasers used 1.5 nm-thick lattice-matched InGaP buffers below the stacked InGaAs QD layers, yielding 1280 nm lasing at 200 A/cm^2 threshold with $T_0 = 210$ K [33.104]. The target of 1.3 μm lasing has been accomplished to date with DWELL devices grown using MBE, a technique much more widely employed in the field. By applying temperature ramping comparable to that described above for the MOVPE, operation at 1307 nm with an extraordinary low threshold of 33 A/cm^2 (17 A/cm^2 with facet coating) was achieved [33.105].

33.4 Concluding Remarks

Metalorganic vapor-phase epitaxy has become established as a versatile technique for advanced fabrication of heterostructures with thickness control down to the monolayer range. Besides applications in industrial mass production, fast progress is being made in the development of new fields. This chapter intended to deliver an insight into the technique and selected problems. Just two topics on current issues in semiconductor MOVPE were highlighted: epitaxy of GaAs-related dilute nitrides and quantum dots. Both aim at extending the well-developed GaAs-based technology to the long-wavelength spectral range and have recently achieved significant advances in the field of lasers. The performance of these demanding devices is comparable to the best fabricated devices using the complementary molecular-beam epitaxy described in Chap. 32. Both growth techniques presently provide the basis for novel low-dimensional structures in research and development. Note added in proof: Recently ground-state lasing at 1.35 μm and a threshold down to 328 A/cm^2 were accomplished by employing antimony-mediated MOVPE with tenfold stacked InAs/GaAs QDs [33.106].

References

33.1 W. Miederer, G. Ziegler, R. Dötzer: Verfahren zum tiegelfreien Herstellen von Galliumarsenidstäben aus Galliumalkylen und Arsenverbindungen bei niedrigen Temperaturen, German Patent 1176102, filed 25.9.1962; and: Method of crucible-free production of gallium arsenide rods from alkyl galliums and arsenic compounds at low temperatures, US Patent 3226270, filed 24.9.1963

33.2 H.M. Manasevit, W.I. Simpson: The use of metalorganics in the preparation of semiconductor materials on insulating substrates: I. Epitaxial III–V gallium compounds, J. Electrochem. Soc. **12**, 66C (1968)

33.3 R.W. Thomas: Growth of single crystal GaP from organometallic sources, J. Electrochem. Soc. **116**, 1449–1450 (1969)

33.4 H.M. Manasevit: The use of metalorganics in the preparation of semiconductor materials: Growth on insulating substrates, J. Cryst. Growth **13/14**, 306–314 (1972)

33.5 G.B. Stringfellow: Organometallic vapor phase epitaxy reaction kinetics. In: *Handbook of Crystal Growth*, ed. by D.R.T. Hurle (Elsevier, Amsterdam 1994) pp. 491–540

33.6 T.J. Mountziaris, K.F. Jensen: Gas-phase and surface reaction mechanisms in MOCVD of GaAs with trimethyl-gallium and arsine, J. Electrochem. Soc. **138**, 2426–2439 (1991)

33.7 G.B. Stringfellow: *Organometallic Vapor-Phase Epitaxy*, 2nd edn. (Academic, New York 1999)

33.8 R.T. Morrison, R.N. Boyd: *Organic Chemistry*, 5th edn. (Allyn & Bacon, New York 1987)

33.9 K.F. Jensen: Transport phenomena in vapor phase epitaxy reactors. In: *Handbook of Crystal Growth*, ed. by D.R.T. Hurle (Elsevier, Amsterdam 1994) pp. 541–599

33.10 D.H. Reep, S.K. Ghandhi: Deposition of GaAs epitaxial layers by organometallic CVD, J. Electrochem. Soc. **130**, 675–680 (1983)

33.11 R.B. Bird, W.E. Stewart, E.N. Lightfoot: *Transport Phenomena* (Wiley, New York 1962)

33.12 A.G. Bhuiyan, A. Hashimoto, A. Yamamoto: Indium nitride (InN): A review on growth, characterization, and properties, J. Appl. Phys. **94**, 2779–2808 (2003)

33.13 I. Vurgaftman, J.R. Meyer: Band parameters for nitrogen-containing semiconductors, J. Appl. Phys. **94**, 3675–3696 (2003)

33.14 M. Drago, T. Schmidtling, U.W. Pohl, S. Peters, W. Richter: InN metalorganic vapour phase epitaxy and ellipsometric characterization, Phys. Status Solidi (c) **0**(7), 2842–2845 (2003)

33.15 I. Vurgaftman, J.R. Meyer, L.R. Ram-Mohan: Band parameters for III-V compound semiconductors and their alloys, J. Appl. Phys. **89**, 5815–5875 (2001)

33.16 U.W. Pohl, C. Möller, K. Knorr, W. Richter, J. Gottfriedsen, H. Schumann, K. Rademann, A. Fielicke: Tertiarybutylhydrazine: A new precursor for the MOVPE of group III-nitrides, Mater. Sci. Eng. B **59**, 20–23 (1999)

33.17 I. Ho, G.B. Stringfellow: Solubility of nitrogen in binary III-V systems, J. Cryst. Growth **178**, 1–7 (1997)

33.18 J. Neugebauer, C.G. Van de Walle: Electronic structure and phase stability of GaAs$_{1-x}$N$_x$ alloys, Phys. Rev. B **51**, 10568–10571 (1995)

33.19 D.K. Gaskill, N. Bottka, M.C. Lin: Growth of GaN films using trimethylgallium and hydrazine, Appl. Phys. Lett. **48**, 1449–1451 (1986)

33.20 H. Tsuchiya, A. Takeuchi, M. Kurihara, F. Hasegawa: Metalorganic molecular beam epitaxy of cubic GaN on (100) GaAs substrates using triethylgallium and monomethylhydrazine, J. Cryst. Growth **152**, 21–27 (1995)

33.21 S. Miyoshi, K. Onabe, N. Ohkouchi, H. Yaguchi, R. Ito, S. Fukatsu, Y. Shiraki: MOVPE growth of cubic GaN on GaAs using dimethylhydrazine, J. Cryst. Growth **124**, 439–442 (1992)

33.22 H. Sato, H. Takanashi, A. Watanabe, H. Ota: Preparation of GaN films on sapphire by metalorganic chemical vapor deposition using dimethylhydrazine as nitrogen source, Appl. Phys. Lett. **68**, 3617–3619 (1996)

33.23 U.W. Pohl, K. Knorr, C. Möller, U. Gernert, W. Richter, J. Bläsing, J. Christen, J. Gottfriedsen, H. Schumann: Low-temperature metalorganic vapor phase epitaxy (MOVPE) of GaN using tertiarybutylhydrazine, Jpn. J. Appl. Phys. **38**, L105–L107 (1999)

33.24 U.W. Pohl, K. Knorr, J. Bläsing: Metalorganic vapor phase epitaxy of GaN on LiGaO$_2$ substrates using tertiarybutylhydrazine, Phys. Status Solidi (a) **184**, 117–120 (2001)

33.25 Z. Pan, T. Miyamoto, D. Schlenker, S. Sato, F. Koyama, K. Iga: Low temperature growth of GaInNAs/GaAs quantum wells by metalorganic chemical vapor deposition using tertiarybutylarsine, J. Appl. Phys. **84**, 6409–6411 (1998)

33.26 J. Derluyn, I. Moerman, M.R. Leys, G. Patriarche, G. Sęk, R. Kudrawiec, W. Rudno-Rudziński, K. Ryczko, J. Misiewicz: Control of nitrogen incorporation in Ga(In)NAs grown by metalorganic vapor phase epitaxy, J. Appl. Phys. **94**, 2752–2754 (2003)

33.27 S. Kurtz, J.F. Geisz, D.J. Friedman, W.K. Metzger, R.R. King, N.H. Karam: Annealing-induced-type conversion of GaInNAs, J. Appl. Phys. **95**, 2505–2508 (2004)

33.28 K. Volz, T. Torunski, B. Kunert, O. Rubel, S. Nau, S. Reinhard, W. Stolz: Specific structural and compositional properties of (GaIn)(NAs) and their influence on optoelectronic device performance, J. Cryst. Growth **272**, 739–747 (2004)

33.29 S.G. Spruytte, C.W. Coldren, J.S. Harris, W. Wampler, P. Krispin, K. Ploog, M.C. Larson: Incorporation of nitrogen in nitride-arsenides: Origin of improved luminescence efficiency after anneal, J. Appl. Phys. **89**, 4401–4406 (2001)

33.30 X. Yang, M.J. Jurkovic, J.B. Héroux, W.I. Wang: Molecular beam epitaxial growth of InGaAsN:Sb/GaAs quantum wells for long-wavelength semiconductor lasers, Appl. Phys. Lett. **75**, 178–180 (1999)

33.31 L.H. Li, V. Sallet, G. Patriarche, L. Largeau, S. Bouchoule, L. Travers, J.C. Harmand: Investigations on GaInNAsSb quinary alloy for 1.5 μm laser emission on GaAs, Appl. Phys. Lett. **83**, 1298–1300 (2003)

33.32 H.B. Yuen, S.R. Bank, H. Bae, M.A. Wistey, J.S. Harris: The role of antimony on properties of widely varying GaInNAsSb compositions, J. Appl. Phys. **99**, 093504 (2006)

33.33 H.A. McKay, R.M. Feenstra, T. Schmidtling, U.W. Pohl: Arrangement of nitrogen atoms in GaAsN alloys determined by scanning tunneling microscopy, Appl. Phys. Lett. **78**, 82–84 (2001)

33.34 H.A. McKay, R.M. Feenstra, T. Schmidtling, U.W. Pohl, J.F. Geisz: Distribution of nitrogen atoms in dilute GaAsN and InGaAsN alloys studied by scanning tunneling microscopy, J. Vac. Sci. Technol. B **19**, 1644–1649 (2001)

33.35 R. Duca, G. Ceballos, C. Nacci, D. Furlanetto, P. Finetti, S. Modesti, A. Cristofolio, G. Bais, M. Peccin, S. Rubini, F. Martelli, A. Franciosi: In-N and N-N correlation in In$_x$Ga$_{1-x}$As$_{1-y}$N$_y$/GaAs quasi-lattice-matched quantum wells: A cross-sectional scanning tunneling microscopy study, Phys. Rev. B **72**, 075311 (2005)

33.36 L. Bellaiche, A. Zunger: Effects of atomic short-range order on the electronic and optical properties of GaAsN, GaInN, and GaInAs, Phys. Rev. B **57**, 4425–4431 (1998)

33.37 O. Rubel, K. Volz, T. Torunski, S.D. Baranovskii, F. Grosse, W. Stolz: Columnar [001]-oriented nitrogen order in Ga(NAs) and (GaIn)(NAs) alloys, Appl. Phys. Lett. **85**, 5908–5910 (2004)

33.38 K. Volz, T. Torunski, W. Stolz: Detection of nanometer-sized strain fields in (GaIn)(NAs) alloys by specific dark field transmission electron microscopic imaging, J. Appl. Phys. **97**, 014306 (2005)

33.39 G. Ciatto, F. D'Acapito, L. Grenouillet, H. Mariette, D. De Salvador, G. Bisognin, R. Carboni,

33.40 L. Floreano, R. Gotter, S. Mobilio, F. Boscherini: Quantitative determination of short-range ordering in $In_xGa_{1-x}As_{1-y}N_y$, Phys. Rev. B **68**, 161201 (2003)

33.40 V. Lordi, H.B. Yuen, S.R. Bank, M.A. Wistey, J.S. Harris, S. Friedrich: Nearest-neighbor distributions in $Ga_{1-x}In_xN_yAs_{1-y}$ and $Ga_{1-x}In_xN_yAs_{1-y-z}Sb_z$ thin films upon annealing, Phys. Rev. B **71**, 125309 (2005)

33.41 K. Uno, M. Yamada, I. Tanaka, O. Ohtsuki, T. Takizawa: Thermal annealing effects and local atomic configurations in GaInNAs thin films, J. Cryst. Growth **278**, 214–218 (2005)

33.42 E.-M. Pavelescu, J. Wagner, H.-P. Komsa, T.T. Rantala, M. Dumitrescu, M. Pessa: Nitrogen incorporation into GaInNAs lattice-matched to GaAs: The effects of growth temperature and thermal annealing, J. Appl. Phys. **98**, 083524 (2005)

33.43 A.M. Teweldeberhan, S. Fahy: Effect of indium-nitrogen bonding on the localized vibrational mode in $In_yGa_{1-y}N_xAs_{1-x}$, Phys. Rev. B **73**, 245215 (2006)

33.44 H.B. Yuen, S.R. Bank, H. Bae, M.A. Wistey, J.S. Harris: Effects of strain on the optimal annealing temperature of GaInNAsSb quantum wells, Appl. Phys. Lett. **88**, 221913 (2006)

33.45 T. Kitatani, M. Kondow, M. Kudo: Transition of infrared absorption peaks in thermally annealed GaInNAs, Jpn. J. Appl. Phys. **40**, L750–L752 (2001)

33.46 P.R.C. Kent, A. Zunger: Evolution of III–V nitride alloy electronic structure: The localized to delocalized transition, Phys. Rev. Lett. **86**, 2613–2616 (2001)

33.47 J.D. Perkins, A. Mascarenhas, Y. Zhang, J.F. Geisz, D.J. Friedman, J.M. Olson, S.R. Kurtz: Nitrogen-activated transitions, level repulsion, and band gap reduction in $GaAs_{1-x}N_x$ with $x < 0.03$, Phys. Rev. Lett. **82**, 3312–3315 (1999)

33.48 W. Shan, W. Walukiewicz, J.W. Ager III, E.E. Haller, J.F. Geisz, D.J. Friedman, J.M. Olson, S.R. Kurtz: Effect of nitrogen on the band structure of GaInNAs alloys, J. Appl. Phys. **86**, 2349–2351 (1999)

33.49 J.B. Héroux, X. Yang, W.I. Wang: Photoreflectance spectroscopy of strained (In)GaAsN/GaAs multiple quantum wells, J. Appl. Phys. **92**, 4361–4366 (2002)

33.50 M. Galluppi, L. Geelhaar, H. Riechert, M. Hetterich, A. Grau, S. Birner, W. Stolz: Bound-to-bound and bound-to-free transitions in surface photovoltage spectra: determination of the band offsets for $In_xGa_{1-x}As$ and $In_xGa_{1-x}As_{1-y}N_y$ quantum wells, Phys. Rev. B **72**, 155324 (2005)

33.51 R. Kudrawiec, M. Motyka, M. Gladysiewicz, J. Misiewicz, H.B. Yuen, S.R. Bank, H. Bae, M.A. Wistey, J.S. Harris: Band gap discontinuity in $Ga_{0.9}In_{0.1}N_{0.027}As_{0.973-x}Sb_x$/GaAs single quantum wells with $0 \leq x < 0.06$ studied by contactless electroreflectance spectroscopy, Appl. Phys. Lett. **88**, 221113 (2006)

33.52 A.J. Ptak, D.J. Friedman, S. Kurtz, R.C. Reedy: Low-acceptor-concentration GaInNAs grown by molecular-beam epitaxy for high-current p–i–n solar cell applications, J. Appl. Phys. **98**, 094501 (2005)

33.53 M. Kondow, T. Kitatani, S. Nakatsuka, M.C. Larson, K. Nakahara, Y. Yazawa, M. Okai, K. Uomi: GaInNAs: A novel material for long-wavelength semiconductor lasers, IEEE J. Sel. Top. Quantum Electron. **3**, 719–730 (1997)

33.54 S. Sato, S. Sato: Metalorganic chemical vapor deposition of GaInNAs lattice matched to GaAs for long-wavelength laser diodes, J. Cryst. Growth **192**, 381–385 (1998)

33.55 F. Höhnsdorf, J. Koch, S. Leu, W. Stolz, B. Borchert, M. Druminski: Reduced threshold current densities of (GaIn)(NAs)/GaAs single quantum wells for emission wavelengths in the range 1.28–1.38 µm, Electron. Lett. **35**, 571–572 (1999)

33.56 G. Jaschke, R. Averbeck, L. Geelhaar, H. Riechert: Low threshold InGaAsN/GaAs lasers beyond 1500 nm, J. Cryst. Growth **278**, 224–228 (2005)

33.57 N. Tansu, N.J. Kirsch, L.J. Mawst: Low-threshold-current-density 1300-nm dilute-nitride quantum well lasers, Appl. Phys. Lett. **81**, 2523–2525 (2002)

33.58 W. Li, J. Turpeinen, P. Melanen, P. Savolainen, P. Uusimaa, M. Pessa: Growth of strain-compensated GaInNAs/GaAsP quantum wells for 1.3 µm lasers, J. Cryst. Growth **230**, 533–536 (2001)

33.59 N. Tansu, J.-Y. Yeh, L.J. Mawst: Low-threshold 1317-nm InGaAsN quantum-well lasers with GaAsN barriers, Appl. Phys. Lett. **83**, 2512–2514 (2003)

33.60 J.-Y. Yeh, N. Tansu, L.J. Mawst: Long wavelength MOCVD grown InGaAsN-GaAsN quantum well lasers emitting at 1.378–1.41 µm, Electron. Lett. **40**, 739–740 (2004)

33.61 Y.Q. Wei, M. Sadeghi, S.M. Wang, P. Modh, A. Larsson: High performance 1.28 µm GaInNAs double quantum well lasers, Electron. Lett. **41**, 1328–1329 (2005)

33.62 B. Damilano, J. Barjon, J.-Y. Duboz, J. Massies, A. Hierro, J.-M. Ulloa, E. Calleja: Growth and in situ annealing conditions for long-wavelength (GaIn)(NAs)/GaAs lasers, Appl. Phys. Lett. **86**, 071105 (2005)

33.63 M. Hopkinson, C.Y. Jin, H.Y. Liu, P. Navaretti, R. Airey: 1.34 µm GaInNAs quantum well lasers with low room-temperature threshold current density, Electron. Lett. **42**, 923–924 (2006)

33.64 S.R. Bank, H.P. Bae, H.B. Yuen, M.A. Wistey, L.L. Goddard, J.S. Harris: Room-temperature continuous-wave 1.55 µm GaInNAsSb laser on GaAs, Electron. Lett. **42**, 156–157 (2006)

33.65 A. Forchel, H. Leier, B.E. Maile, R. Germann: Fabrication and optical spectroscopy of ultra small III-V compound semiconductor structure. In: *Advances in Solid State Physics*, Vol. 28, ed. by U. Rössler (Vieweg, Braunschweig 1988) pp. 99–119

33.66 I. N. Stranski, L. Krastanow: Zur Theorie der orientierten Ausscheidung von Ionenkristallen aufeinander. In: *Sitzungsberichte der Akademie der Wissenschaften in Wien*, Math.-naturwiss. Klasse, Abt. IIB, Vol. 146 (1937) pp. 797–810, in German

33.67 V.A. Shchukin, N.N. Ledentsov, P.S. Kop'ev, D. Bimberg: Spontaneous ordering of arrays of coherent strained islands, Phys. Rev. Lett. **75**, 2968–2971 (1995)

33.68 L.G. Wang, P. Kratzer, M. Scheffler, N. Moll: Formation and stability of self-assembled coherent islands in highly mismatched heteroepitaxy, Phys. Rev. Lett. **82**, 4042–4045 (1999)

33.69 D. Leonard, K. Pond, P. M. Petroff: Critical layer thickness for self-assembled InAs islands on GaAs, Phys. Rev. B **50**, 11687–11692 (1994)

33.70 L. Goldstein, F. Glas, J.Y. Marzin, M.N. Charasse, G. Le Roux: Growth by molecular beam epitaxy and characterization of InAs/GaAs strained-layer superlattices, Appl. Phys. Lett. **47**, 1099–1101 (1985)

33.71 D. Leonard, M. Krishnamurthy, C.M. Reaves, S.P. Denbaars, P.M. Petroff: Direct formation of quantum-sized dots from uniform coherent islands of InGaAs on GaAs surfaces, Appl. Phys. Lett. **63**, 3203–3205 (1993)

33.72 F. Heinrichsdorff, A. Krost, N. Kirstaedter, M.-H. Mao, M. Grundmann, D. Bimberg, A.O. Kosogov, P. Werner: InAs/GaAs quantum dots grown by metalorganic chemical vapor deposition, Jpn. J. Appl. Phys. **36**, 4129–4133 (1997)

33.73 J. Oshinowo, M. Nishioka, S. Ishida, Y. Arakawa: Area density control of quantum-size InGaAs/Ga(Al)As dots by metalorganic chemical vapor deposition, Jpn. J. Appl. Phys. **33**, L1634–L1637 (1994)

33.74 F. Heinrichsdorff, A. Krost, M. Grundmann, D. Bimberg, F. Bertram, J. Christen, A. Kosogov, P. Werner: Self-organization phenomena of InGaAs/GaAs quantum dots grown by metalorganic chemical vapour deposition, J. Cryst. Growth **170**, 568–573 (1997)

33.75 A.A. El-Emawy, S. Birudavolu, P.S. Wong, Y.-B. Jiang, H. Xu, S. Huang, D.L. Huffaker: Formation trends in quantum dot growth using metalorganic chemical vapor deposition, J. Appl. Phys. **93**, 3529–3534 (2003)

33.76 I.N. Kaiander, F. Hopfer, T. Kettler, U.W. Pohl, D. Bimberg: Alternative precursor growth of quantum dot-based VCSELs and edge emitters for near infrared wavelengths, J. Cryst. Growth **272**, 154–160 (2004)

33.77 N.N. Ledentsov, J. Böhrer, D. Bimberg, I.V. Kochnev, M.V. Maximov, P.S. Kop'ev, I.Z. Alferov, A.O. Kosogov, S.S. Ruvimov, P. Werner, U. Gösele: Formation of coherent superdots using metal-organic chemical vapor deposition, Appl. Phys. Lett. **69**, 1095–1097 (1996)

33.78 K. Mukai, N. Ohtsuka, M. Sugawara: Controlled quantum confinement potentials in self-formed InGaAs quantum dots grown by atomic layer epitaxy technique, Jpn. J. Appl. Phys. **35**, L262–L265 (1996)

33.79 K. Pötschke, L. Müller-Kirsch, R. Heitz, R.L. Sellin, U.W. Pohl, D. Bimberg, N. Zakharov, P. Werner: Ripening of self-organized InAs quantum dots, Physica E **21**, 606–610 (2004)

33.80 A. Krost, F. Heinrichsdorff, D. Bimberg, A. Darhuber, G. Bauer: High-resolution x-ray diffraction of self-organized InGaAs/GaAs quantum dot structures, Appl. Phys. Lett. **68**, 785–787 (1996)

33.81 A. Lenz, R. Timm, H. Eisele, C. Hennig, S.K. Becker, R.L. Sellin, U.W. Pohl, D. Bimberg, M. Dähne: Reversed truncated cone composition distribution of $In_{0.8}Ga_{0.2}As$ quantum dots overgrown by an $In_{0.1}Ga_{0.9}As$ layer in a GaAs matrix, Appl. Phys. Lett. **81**, 5150–5152 (2002)

33.82 R.L. Sellin, I. Kaiander, D. Ouyang, T. Kettler, U.W. Pohl, D. Bimberg, N.D. Zakharov, P. Werner: Alternative-precursor MOCVD of self-organized InGaAs/GaAs quantum dots and quantum dot lasers, Appl. Phys. Lett. **82**, 841–843 (2003)

33.83 L. Höglund, E. Petrini, C. Asplund, H. Malm, J.Y. Andersson, P.O. Holtz: Optimising uniformity of InAs/(InGaAs)/GaAs quantum dots by metal organic vapor phase epitaxy, Appl. Surf. Sci. **252**, 5525–5529 (2006)

33.84 F. Heinrichsdorff, A. Krost, D. Bimberg, A.O. Kosogov, P. Werner: Self-organized defect free InAs/GaAs and InAs/InGaAs/GaAs quantum dots with high lateral density grown by MOCVD, Appl. Surf. Sci. **123/124**, 725–728 (1998)

33.85 F. Heinrichsdorff, M. Grundmann, O. Stier, A. Krost, D. Bimberg: Influence of In/Ga intermixing on the optical properties of InGaAs/GaAs quantum dots, J. Cryst. Growth **195**, 540–545 (1998)

33.86 R. Sellin, F. Heinrichsdorff, C. Ribbat, M. Grundmann, U.W. Pohl, D. Bimberg: Surface flattening during MOCVD of thin GaAs layers covering InGaAs quantum dots, J. Cryst. Growth **221**, 581–585 (2000)

33.87 U.W. Pohl, K. Pötschke, A. Schliwa, F. Guffarth, D. Bimberg, N.D. Zakharov, P. Werner, M.B. Lifshits, V.A. Shchukin, D.E. Jesson: Evolution of a multimodal distribution of self-organized InAs/GaAs quantum dots, Phys. Rev. B **72**, 245332 (2005)

33.88 U.W. Pohl: InAs/GaAs quantum qots with multimodal size distribution. In: *Self-assembled Quantum Dots*, ed. by Z.M. Wang (Springer, New York 2008) pp. 43–66

33.89 U.W. Pohl, K. Pötschke, M.B. Lifshits, V.A. Shchukin, D.E. Jesson, D. Bimberg: Self-organized formation of shell-like InAs/GaAs quantum dot ensembles, Appl. Surf. Sci. **252**, 5555–5558 (2006)

33.90 U.W. Pohl, K. Pötschke, A. Schliwa, M.B. Lifshits, V.A. Shchukin, D.E. Jesson, D. Bimberg: Formation and evolution of multimodal size-distributions of InAs/GaAs quantum dots, Physica E **32**, 9–13 (2006)

33.91 U.W. Pohl, A. Schliwa, R. Seguin, S. Rodt, K. Pötschke, D. Bimberg: Size-tunable exchange interaction in InAs/GaAs quantum dots. In: *Advances in Solid State Physics*, Vol. 46, ed. by R. Haug (Springer, Berlin 2008) pp. 41–54

33.92 R. Timm, H. Eisele, A. Lenz, T.-Y. Kim, F. Streicher, K. Pötschke, U.W. Pohl, D. Bimberg, M. Dähne: Structure of InAs/GaAs quantum dots grown with Sb surfactant, Physica E **32**, 25–28 (2006)

33.93 Y. Arakawa, H. Sakaki: Multidimensional quantum well laser and temperature dependence of its threshold current, Appl. Phys. Lett. **40**, 939–941 (1982)

33.94 M. Asada, Y. Miyamoto, Y. Suematsu: Gain and the threshold of three-dimensional quantum-box lasers, IEEE J. Quantum Electron. **22**, 1915–1921 (1986)

33.95 N. Kirstaedter, N.N. Ledentsov, M. Grundmann, D. Bimberg, V.M. Ustinov, S.S. Ruvimov, M.V. Maximov, P.S. Kop'ev, I.Z. Alferov, U. Richter, P. Werner, U. Gösele, J. Heydenreich: Low threshold, large T_0 injection laser emission from (InGa)As quantum dots, Electron. Lett. **30**, 1416–1417 (1994)

33.96 N. Kirstaedter, O.G. Schmidt, N.N. Ledentsov, D. Bimberg, V.M. Ustinov, A.Y. Egorov, A.E. Zhukov, M.V. Maximov, P.S. Kop'ev, I.Z. Alferov: Gain and differential gain of single layer InAs/GaAs quantum dot injection lasers, Appl. Phys. Lett. **69**, 1226–1228 (1996)

33.97 F. Heinrichsdorff, M.-H. Mao, N. Kirstaedter, A. Krost, D. Bimberg, A.O. Kosogov, P. Werner: Room-temperature continuous-wave lasing from stacked InAs/GaAs quantum dots grown by metalorganic chemical vapor deposition, Appl. Phys. Lett. **71**, 22–24 (1997)

33.98 M.V. Maximov, I.V. Kochnev, Y.M. Shernyakov, S.V. Zaitsev, N.Y. Gordeev, A.F. Tsatsul'nikov, A.V. Sakharov, I.L. Krestnikov, P.S. Kop'ev, I.Z. Alferov, N.N. Ledentsov, D. Bimberg, A.O. Kosogov, P. Werner, U. Gösele: InGaAs/GaAs quantum dot lasers with ultrahigh characteristic temperature ($T_0 = 385$ K) grown by metal organic chemical vapour deposition, Jpn. J. Appl. Phys. **36**, 4221–4223 (1997)

33.99 R.L. Sellin, C. Ribbat, M. Grundmann, N.N. Ledentsov, D. Bimberg: Close-to-ideal device characteristics of high-power InGaAs/GaAs quantum dot lasers, Appl. Phys. Lett. **78**, 1207–1209 (2001)

33.100 L.F. Lester, A. Stintz, H. Li, T.C. Newell, E.A. Pease, B.A. Fuchs, K.J. Malloy: Optical characteristics of 1.24 μm InAs quantum-dot laser diodes, IEEE Photon. Technol. Lett. **11**, 931–933 (1999)

33.101 A. Strittmatter, T.D. Germann, T. Kettler, K. Posilovic, U.W. Pohl, D. Bimberg: Alternative precursor metal-organic chemical vapor deposition of InGaAs/GaAs quantum dot laser diodes with ultralow threshold at 1.25 μm, Appl. Phys. Lett. **88**, 262104 (2006)

33.102 J. Tatebayashi, N. Hatori, M. Ishida, H. Ebe, M. Sugawara, Y. Arakawa, H. Sudo, A. Kuramata: 1.28 μm lasing from stacked InAs/GaAs quantum dots with low-temperature-grown AlGaAs cladding layer by metalorganic chemical vapor deposition, Appl. Phys. Lett. **86**, 053107 (2005)

33.103 N. Nuntawong, Y.C. Xin, S. Birudavolu, P.S. Wong, S. Huang, C.P. Hains, D.L. Huffaker: Quantum dot lasers based on a stacked and strain-compensated active region grown by metal-organic chemical vapor deposition, Appl. Phys. Lett. **86**, 193115 (2005)

33.104 S.M. Kim, Y. Wang, M. Keever, J.S. Harris: High-frequency modulation characteristics of 1.3 μm InGaAs quantum dot lasers, IEEE Photon. Technol. Lett. **16**, 377–379 (2004)

33.105 I.R. Sellers, H.Y. Liu, K.M. Groom, D.T. Childs, D. Robbins, T.J. Badcock, M. Hopkinson, D.J. Mowbray, M.S. Skolnick: 1.3 μm InAs/GaAs multilayer quantum-dot laser with extremely low room-temperature threshold current density, Electron. Lett. **40**, 1412–1413 (2004)

33.106 D. Giumard, M. Ishida, N. Hatori, Y. Nakata, H. Sudo, T. Yamamoto, M. Sugawara, Y. Arakawa: CW lasing at 1.35 μm from ten InAs-Sb:GaAs quantum-dot layers grown by metal-organic chemical vapor deposition, IEEE Photon. Technol. Lett. **20**, 827–829 (2008)

34. Formation of SiGe Heterostructures and Their Properties

Yasuhiro Shiraki, Akira Sakai

The Si/Ge system provides a lot of varieties of materials growth due to the lattice mismatch between Si and Ge. From the point of view of device applications, both pseudomorphic growth and strain-relaxed growth are important. Not only the layer growth but also dot formation is now attracting much attention from both the scientific community and for device applications. Comprehensive studies on the growth mechanisms have resulted in the development of novel formation techniques of SiGe heterostructures and enable us to implement strain effects in Si devices. It is obvious that the device applications largely depend on the material growth, particularly control of surface reaction and formation of dislocations and surface roughness that strongly affect device performances. Here we review the fabrication technology of SiGe heterostructures aiming at growth of high-quality materials. The relaxation of strain of SiGe buffer layers grown on Si substrates is discussed in detail, since many devices are formed on the strain-relaxed buffer layers that are sometimes called *virtual substrates*. Carbon incorporation and dot formation that are now studied to extend the possibilities of SiGe are discussed in this chapter too.

34.1	Background	1153
34.2	Band Structures of Si/Ge Heterostructures	1154
34.3	Growth Technologies	1156
	34.3.1 Molecular-Beam Epitaxy	1156
	34.3.2 Chemical Vapor Deposition	1157
34.4	Surface Segregation	1157
34.5	Critical Thickness	1161
34.6	Mechanism of Strain Relaxation	1163
34.7	Formation of Relaxed SiGe Layers	1165
	34.7.1 Graded Buffer	1165
	34.7.2 Low-Temperature Method	1166
	34.7.3 Chemical–Mechanical Polishing Method	1167
	34.7.4 Ion Implantation Method	1168
	34.7.5 Ge Condensation Method	1169
	34.7.6 Dislocation Engineering for Buffer Layers	1170
	34.7.7 Formation of SiGeC Alloys	1172
34.8	Formation of Quantum Wells, Superlattices, and Quantum Wires	1173
34.9	Dot Formation	1177
34.10	Concluding Remarks and Future Prospects	1184
References		1184

34.1 Background

SiGe heterostructures have great potential to improve state-of-the-art Si devices, particularly very large-scale integrated circuits (VLSIs), and add such new functions as optics. Moreover, they provide a new scientific field of materials growth and characterization related to the lattice mismatch between Si and Ge. Band modification due to the strain coming from the lattice mismatch brings about the increase of the mobility of both electrons and holes. Since heterostructure bipolar transistors (HBTs) became commercially available, many people have become engaged in the research and development of field-effective transistor (FET)-type devices which have a much wider range of applications. SiGe-based optical devices including optical interconnection and optoelectronic integrated circuits (OEICs) will provide new possibilities to enhance the performance and functions of Si VLSIs. Heterostructures such as quantum wells and dots make it possible to realize light-emitting devices even with indirect-bandgap materials. These fascinating applications obviously depend on the material growth, particularly the control of surface reaction and formation of dislocations and surface roughness, which strongly affect device performance.

In this chapter, we review the fabrication technology of SiGe heterostructures developed so far and discuss the physics and chemistry behind the process. The relaxation of strain of SiGe buffer layers grown on Si substrates is very important in this field, since many devices are formed on the strain-relaxed buffer layers which are sometimes called *virtual substrates*. The formation and properties of these layers are discussed here in detail. To extend the possibilities of SiGe, carbon incorporation is applied, which is also very interesting from the point of view of material growth. The large lattice mismatch in this material system causes the formation of misfit dislocations as well as dot formation. The latter is now one of the hottest topics in semiconductor physics and technology and will be discussed in this chapter too.

34.2 Band Structures of Si/Ge Heterostructures

Since there exists about 4.2% lattice mismatch between Si and Ge, the strain induced by the lattice mismatch modifies the band structures of SiGe layers [34.1–3]. The bandgap energy of SiGe layers grown on Si substrates is changed by the strain, as shown in Fig. 34.1 as a function of Ge composition. The bandgap is decreased from the bulk value under lateral compressive strain, and the lowest point of the conduction band of strained SiGe is the delta valley, as in Si crystals, over the whole composition range.

Figure 34.2 shows the change in the conduction and valence bands due to strain. The degenerate sixfold conduction bands are separated into two groups, that is, twofold degenerated bands and fourfold degenerated bands. The degenerated heavy-hole (HH) and light-hole (LH) bands are also separated, and the energy difference of the spin–orbit splitting band is changed. This change effectively modifies the transport properties of electrons and holes, providing opportunities to improve transistor performance. In particular, the mobility enhancement of both electrons and holes is highly attractive from the point of view of device applications such as complementary metal–oxide–semiconductor (MOS) FETs, i.e., CMOS circuits.

The band alignment at Si/Ge heterointerfaces is also significantly modified and type I and type II alignments are realized by changing the strain distribution, as illustrated schematically in Fig. 34.3, i.e., when $Si_{1-x}Ge_x$ ($x < 0.4$) layers are grown on Si substrates and are laterally compressed, the band alignment becomes type I and electrons and holes are confined to the same SiGe region. On the other hand, when Si layers are grown on unstrained SiGe layers and are under lateral tensile strain, type II alignment is realized and electrons and holes are confined sepa-

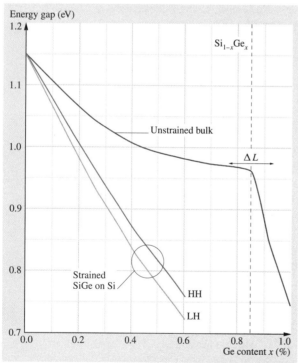

Fig. 34.1 Ge content dependence of the energy bandgap of strained SiGe grown on Si substrates. HH and LH represent bandgaps for heavy and light hole bands. That of unstrained SiGe is also shown as a reference

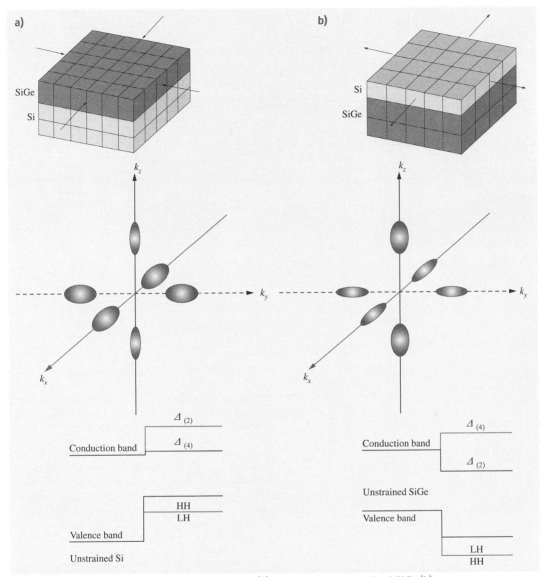

Fig. 34.2a,b Band modification of tensilely strained Si (**a**) and compressively strained SiGe (**b**)

rately. When Ge is laterally compressed by growth on SiGe layers, type II alignment also appears, as shown in the figure. Both band alignments are very useful from the point of view of device applications. Especially, since the band discontinuity at the conduction band of the former is very small, type II alignment is important when band engineering for electrons is intended.

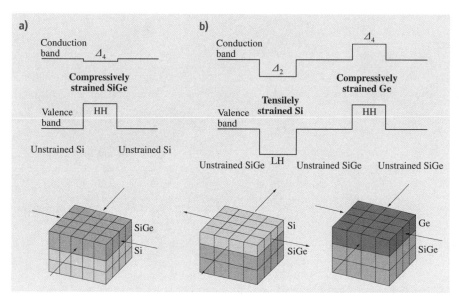

Fig. 34.3a,b Band alignment of Si/Ge heterostructures under various strains: (**a**) compressively strained SiGe on Si substrate (type I), (**b**) tensilely strained Si and compressively strained Ge on unstrained SiGe (type II)

34.3 Growth Technologies

34.3.1 Molecular-Beam Epitaxy

Molecular-beam epitaxy (MBE) is broadly used for depositing semiconductor and metallic materials to form thin films and multilayers. For the growth of Si films by MBE, a Si molecular beam is irradiated onto a clean surface of a Si substrate in a stainless-steel chamber in which the base pressure is reduced to the ultrahigh-vacuum (UHV) range, typically of the order 10^{-10} Torr.

An electron gun evaporator is usually used as a solid source of Si in order to achieve the vapor pressure required for practical growth. The evaporation of Ge is performed using a conventional Knudsen cell (K-cell), which is surrounded by liquid-nitrogen-cooled shrouds to condense unwanted evaporants and improve the vacuum in the sample region. The K-cell is also used for evaporating dopant materials: Ga and B for p-type doping, and Sb for n-type. Solid-source MBE is so simple and safe that it is also widely employed for fundamental research into thin-film growth mechanisms. Due to the reduced-pressure environment during growth, in situ and real-time diagnostic tools, such as reflection high-energy electron diffraction, scanning electron microscopy, and scanning tunneling microscopy, can be readily incorporated into the vacuum system to moni-

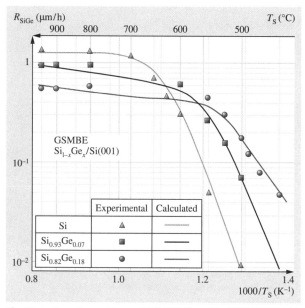

Fig. 34.4 Growth rate R_{SiGe} of GSMBE $Si_{1-x}Ge_x(001)$ layers with $x = 0, 0.07$, and 0.18 as a function of temperature T_s. The *solid lines* are calculated. For details, see [34.4]

tor the dynamics occurring on the surface of a growing film.

On the other hand, growth is often achieved using gas sources of Si and Ge, such as SiH_4, Si_2H_6, and GeH_4, in the same type of UHV chambers as are used for solid-source MBE. An advantage of gas-source MBE is its ability for selective epitaxial growth on SiO_2-mask-patterned Si substrates [34.5, 6]. The growth kinetics of epitaxy on clean Si surfaces has been extensively studied for SiH_4 [34.7, 8], Si_2H_6 [34.4, 9–13], GeH_4 [34.11, 13–15], Ge_2H_6 [34.4, 14], and organosilanes [34.16–18]. Typical trends of SiGe growth on Si in gas-source MBE can be seen in Fig. 34.4 [34.4], which shows the growth rate of SiGe layers on Si(001) substrates as a function of substrate temperature when using Si_2H_6 and GeH_4. In the low-temperature regime, surface-reaction-limited growth takes place and the growth rate increases with increasing Ge concentration. On the contrary, the growth rate saturates at high temperature, where impingement-flux-limited growth occurs, and decreases with increasing Ge concentration. These phenomena are also observed when using monohydride sources of SiH_4 and GeH_4, and can be well modeled in terms of several kinetic parameters such as the sticking probability of gas species at adsorption sites, hydrogen desorption, and Ge segregation [34.4, 19, 20].

34.3.2 Chemical Vapor Deposition

Chemical vapor deposition (CVD) is most frequently used for the growth of semiconductor, metal, and insulator films for device production. In general, either a hot- or cold-wall reactor is used, with pumping and sophisticated gas control systems which enable exact supply of reactive gases onto heated substrates to produce thin, epitaxial SiGe and SiGeC alloy films. CVD can ordinarily be categorized according to operating pressure into: UHV, low pressure (LP), reduced pressure (RP), and atmospheric pressure (AP) CVD.

UHV-CVD was firstly achieved by *Meyerson* et al. for growth of SiGe films on Si substrates using SiH_4 and GeH_4 [34.21]. They developed a hot-wall-type system consisting of a quartz reactor tube surrounded by an electrical heater with pumping systems to realize the UHV condition. Most commercially applicable processing is LP-CVD and RP-CVD, typically operating at pressures ranging from 1 Torr to a few tens of Torr. Gaseous sources of SiH_4, SiH_2Cl_2, SiH_3CH_3, GeH_4, CH_4, and $SiCH_6$ are usually used for growth of Si, SiGe, and SiGeC films, and PH_3 and B_2H_6 for doping [34.22–25].

As mentioned earlier for gas-source MBE, during CVD growth of SiGe layers on Si, dramatic acceleration of the growth rate was also observed with the introduction of GeH_4 into the Si source gas, as compared with that of Si layer growth alone [34.21–23, 25–27]. This is mainly due to the lowering of the hydrogen desorption energy caused by the presence of Ge on the Si surface. Selective epitaxial growth of Si and SiGe on Si surfaces has also been widely studied in CVD. For growth at pressure of more than 10 Torr, besides SiH_2Cl_2 and GeH_4 as gaseous sources, HCl was effectively used to control deposition selectivity with respect to SiO_2-masked regions [34.28, 29], whereas it was achieved without HCl in the case of very low pressure, less than 1 Torr [34.30, 31]. Practical applications of selective epitaxial growth technology have mainly been the formation of SiGe channels for high-performance MOSFETs, elevated source/drain regions in shallow-junction electrode MOSFETs, and base regions of high-speed SiGe heterojunction bipolar transistors.

34.4 Surface Segregation

It is well known that the real heterointerface is not atomically flat and abrupt. There are several effects that deteriorate interface abruptness, of which surface segregation is considered to be the main one. Surface segregation is a reaction between impinging atoms and surface atoms of substrates, which then exchange positions to reduce the total energy of the system. This phenomenon was first recognized to be important when MBE layers were doped with some kinds of impurities [34.32, 33]. It should be pointed out, however, that surface segregation takes place not only in cases of doping, but also during heterointerface formation. Ge and In are well known to segregate when forming Si-on-(Si)Ge and GaAs-on-In(Ga)As heterostructures, respectively.

Figure 34.5 shows a typical example of the surface segregation phenomenon [34.34]. This figure shows secondary-ion mass spectroscopy (SIMS) profiles for the case when Si atoms are deposited on Si substrate covered with submonolayer Sb atoms. It is seen that Sb atoms do not remain at the original position and that

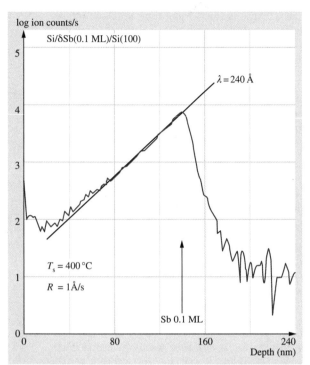

Fig. 34.5 SIMS profile of Sb atoms with 0.1 ML deposited on Si substrates measured after overgrowth of Si epitaxial layer. Due to surface segregation, the profile shows a tail with segregation length of $\lambda = 240$ Å towards the surface

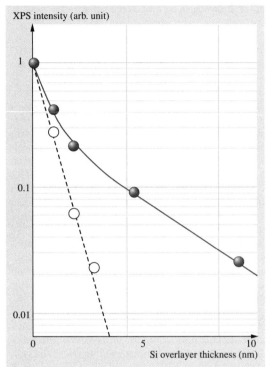

Fig. 34.6 XPS intensity of Ge atoms as a function of the thickness of overgrown Si layer. *Filled circles* are experimental results and the *dotted line* represents the calculated dependence without surface segregation. *Open circles* are for the case of surfactant-mediated growth

they show an exponential distribution towards the surface with a characteristic parameter of the segregation length.

Another example is seen in Fig. 34.6, which shows how surface segregation occurs when Si overlayer is grown on the Ge layers [34.35]. If Ge atoms sat on their original sites and site exchange between Ge and impinging Si did not occur, the x-ray photoemission (XP) intensity of Ge atoms should decrease exponentially with increasing Si overlayer thickness, as shown by the dashed line. It is, however, seen that the x-ray photoemission spectroscopy (XPS) intensity does not follow the exponential decay but is much stronger even after growth of Si layers with thickness of 10 nm, which is much larger than the escape depth of photoexcited electrons.

Surface segregation of Ge is also clearly seen in photoluminescence (PL) of SiGe/Si quantum wells, the details of which will be described later. Due to the quantum confinement effect, the PL peak positions shift to higher energies with decreasing well width. The solid symbols in Fig. 34.7 show no-phonon (NP) and transverse optic (TO) phonon replica peak positions as a function of well width [34.34]. The dashed lines in the figure show the well width dependence of the peaks calculated based on the square well, that is, when surface segregation does not take place. It is seen that the peaks shift to higher energies from the expected positions. This is because the quantum levels are lifted by the deformation of the well shape due to Ge surface segregation.

It is confirmed [34.36] that only the topmost Ge atoms are mainly incorporated in the segregation and that we may neglect underlying atoms to a first approximation. The surface segregation phenomenon is therefore described in terms of a two-state exchange model where only exchange of atoms in surface and subsurface states is taken into account, as shown in

Fig. 34.8 [34.32, 37]. The exchange is described by the following rate equation

$$\frac{dn_1}{dt} = -pn_1 + qn_2, \quad (34.1)$$

$$\frac{dn_2}{dt} = -qn_2 + pn_1, \quad (34.2)$$

$$n_1 + n_2 = n_0,$$

where $p = p_0 \exp(-E_a/k_B T)$, $q = q_0 \exp[-(E_a + E_b)/k_B T]$, E_a is a potential for atoms to jump into the surface state from the underlying site and is less than the bulk thermal diffusion potential, E_b is the energy gain for the surface segregation and a measure of the segregation strength, and p_0 and q_0 are pre-exponential factors corresponding to the attempt frequency of atomic jumps and that may be considered to be on the order of the lattice vibration of 10^{12}–10^{13} s^{-1}. Under the equilibrium condition, the equation can be numerically solved as

$$n_2(t) = n_0 \left[1 - \frac{1}{p+q}\left(q + e^{(-p+q)t}\right)\right]. \quad (34.3)$$

This solution shows the exponential distribution of atoms segregating towards the surface, well representing the results of SIMS experiments. Figure 34.9 shows a logarithmic plot of segregated atoms as a function of inverse temperature, in which the line labeled *equilibrium segregation* corresponds to the above solution. It

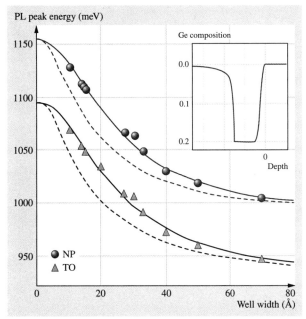

Fig. 34.7 Well width dependence of NP and TO peaks of PL from SiGe/Si quantum wells. *Solid symbols* are experimental results and the *dotted lines* represent the calculated well width dependence without surface segregation. The *solid lines* represent the dependence obtained by taking into account surface segregation in the two-state exchange model. The *inset* shows a schematic of the well shape distorted by surface segregation

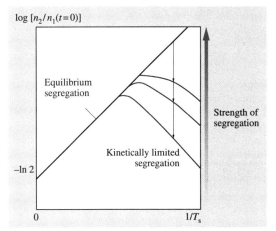

Fig. 34.9 Surface segregation (n_2/n_1) as a function of inverse temperature. Equilibrium segregation occurs under thermal equilibrium conditions and kinetically limited segregation occurs in the case of real crystal growth, particularly at low temperatures

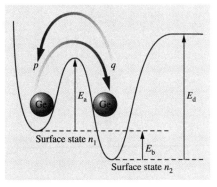

Fig. 34.8 Energy diagram of a two-state exchange model. E_a is the potential for atoms to jump from the subsurface position, and E_b is the energy gain for surface segregation. E_d is the desorption energy

is noted that segregation becomes smaller as the temperature increases. At first sight, this contradicts the observation that segregation is enhanced with increasing temperature. However, it should be pointed out that the experiments are not conducted under conditions of thermal equilibrium but under kinetically limited con-

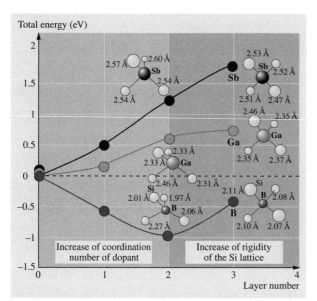

Fig. 34.10 Total energy of the system of Si with impurity atoms as a function of impurity position

ditions, that is, surface atoms are embedded before they reach the equilibrium as the crystal growth proceeds continuously. This situation is called kinetically limited segregation, as shown in Fig. 34.9, and segregation is largely suppressed at lower temperatures.

The solid line in Fig. 34.7 shows the well width dependence of quantum well (QW) edge emissions calculated by taking into account the distortion of the well shape based on the two-state exchange model. It is seen that the energy position is sufficiently represented by the calculation, indicating that the model describes the surface segregation well, even though it is phenomenological.

More microscopically, the local strain energy calculation is very useful to understand the segregation [34.38]. Figure 34.10 shows the total energy of the atomic arrangement in Si as a function of foreign atom site. In the case of Sb in Si, the energy is seen to increase as the site becomes deeper in the Si substrate. This indicates that Sb tends to move to the surface region to reduce the total energy, resulting in surface segregation. Interestingly, the energy minimum of B atoms is the third layer, and therefore B always tends to sit in this layer as crystal growth proceeds.

The trend for surface segregation can be predicted by considering the bond energy with Si atoms. The bond energy of Sb–Si, Ga–Si, and Ge–Si are smaller than that of Si itself. These atoms are, therefore, rejected from the Si matrix to form strong chemical bonds and show surface segregation.

To avoid segregation and obtain abrupt heterointerfaces, low-temperature growth has been thought to be essential. However, low-temperature growth deteriorates crystal quality. Surfactant-mediated growth (SMG) is now well known to avoid interface smearing and obtain abrupt interfaces. In this method, a small amount of foreign atoms that show strong segregation, sometimes called segregants, are deposited before formation of heterointerfaces. For this purpose, As, Sb, Ga, Bi, and H are well known to suppress Ge segregation for the formation of Si/SiGe/Si structures. The principle of this method is as follows: before the Si overlayer of the Si/SiGe/Si heterostructure is grown, the SiGe layer is covered with strong segregants such as Sb so that Ge is no longer the topmost surface atom. When Si atoms are deposited on the Sb-covered Ge surface, the position of Sb becomes the subsurface site and Sb exchanges position with impinging Si atoms sitting at surface sites. However, site exchange between Si and Ge atoms does not follow. This is because Ge atoms do not occupy surface sites or subsurface site, that is, they are in bulk sites once Si is deposited. Between atoms in bulk sites, there is a high potential barrier that they cannot climb at the growth temperature, which is much lower than thermal diffusion temperatures, and therefore they do not exchange their positions.

In Fig. 34.6, the open circles show the XPS intensity of Ge atoms when 0.75 ML of Sb atoms are introduced as suppressors of segregation at the heterointerface before Si layers are overgrown [34.35]. It is seen that the XPS intensity follows the exponential decay and that surface segregation is effectively suppressed. Suppression of Ge segregation is also seen in SIMS profiles of Si/Ge superlattices and, once Sb atoms are deposited on Ge layers, the long tail towards the surface is seen to disappear [34.35]. The amount of Sb atoms required to suppress surface segregation is then an important question, and it was found that the effect becomes pronounced at around 0.5 ML and is optimized at 0.75 ML [34.34].

Gas-source MBE (GSMBE) is another important example of SMG, or pseudo-SMG where atomic hydrogen decomposed from hydrogenated gases such as Si_2H_6 and GeH_4 acts as a suppressor of Ge segregation. The well width dependence of the PL peak positions of QWs grown by GSMBE is found to coincide well with the square-well potential calculation, indicating that GSMBE provides heterostructures without significant interfacial smearing [34.39]. It is also reported that Ge segregation is almost absent in UHV-CVD growth, in which hydrogen and/or hydrogenated compounds are speculated to act as suppressors [34.40]. It is, however, noted that hydrogen desorbs at high temperatures such as those used in GSMBE. Therefore, hydrogen is thought to have a relatively long residence time on Si surfaces before desorption and can stay to act as a surfactant during a period of monolayer growth.

34.5 Critical Thickness

In epitaxial growth of overlayers that have a lattice parameter slightly different from that of the substrate, pseudomorphic (or coherent or commensurate) overlayers are formed when the thickness is not large. Since, in this case, the in-plane lattice constant of the pseudomorphic film is equivalent to that of the substrate, the film is elastically deformed according to Poisson's ratio to accommodate the lattice mismatch between the film and the substrate. The elastic energy due to the strain in the film increases linearly with increasing film thickness. However, there is a film thickness beyond which the introduction of misfit dislocations into the film becomes energetically favorable even though the energy increase due to the self-energy of the dislocations is taken into account. This minimum value of the film thickness is called the *critical thickness*.

Early theoretical works elucidating the critical thickness in lattice-mismatched heteroepitaxial growth were done by *Frank* [34.41–43] and *van der Merwe* [34.44, 45]. They gave a fundamental approach for predicting the critical thickness on the basis of the elastic energy and dislocation energy in strained epitaxial systems. The model proposed by *van der Merwe* was modified to be applied more practically to the diamond lattice in low-lattice-mismatch systems such as SiGe/Si [34.46–48]. Minimization of total energy given as a sum of the elastic energy stored in a homogeneously strained film and the energy associated with the misfit dislocations leads to the following equation for the critical thickness h_c

$$h_c = \frac{b}{8\pi(1+\nu)f}\left[1+\ln\left(\frac{2h_c}{r}\right)\right]. \quad (34.4)$$

In (34.4), b is the magnitude of the active component of the misfit dislocation Burgers vector, ν is Poisson's ratio, and r is the inner cutoff radius of the misfit dislocation. Using the spacing between dislocations d and the elastic strain ε, the misfit of the overlayer with the substrate f is defined to be

$$f = |\varepsilon| + \frac{b}{d}. \quad (34.5)$$

In (34.5), when $d \to \infty$, $\varepsilon = f$. A more detailed expression including numerical values associated with a SiGe/Si(001) system is also given in [34.48]. On the other hand, a simple and helpful model was developed by *Matthews* and *Blakeslee* and has often been used for interpreting various types of lattice-mismatched epitaxial systems [34.49–51]. Contrary to the aforementioned energy balance model, they considered the balance of forces exerted on a propagated dislocation with misfit and threading segments in the strained film, as shown in Fig. 34.11. The physical concept of the derivation of critical thickness in the force balance model is essentially the same as that in the energy balance model. The critical thickness in this case is given as

$$h_c = \frac{b(1-\nu\cos^2\alpha)}{8\pi(1+\nu)f\cos\theta}\left[1+\ln\left(\frac{h_c}{b}\right)\right], \quad (34.6)$$

where α is the angle between the misfit dislocation line and its Burgers vector, and θ is the angle between the misfit dislocation Burgers vector and a line in the interface drawn perpendicular to the dislocation line direction.

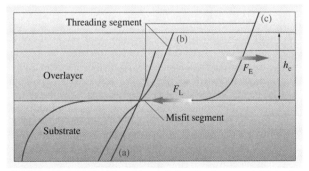

Fig. 34.11 Schematic illustration of the Matthews and Blakeslee model of critical thickness. A pre-existing threading dislocation in a coherent interface (a), critical interface (b), and incoherent interface (c). Critical thickness h_c is determined by the equality of the force exerted in the dislocation line by misfit stress F_E and the line tension in the dislocation line F_L

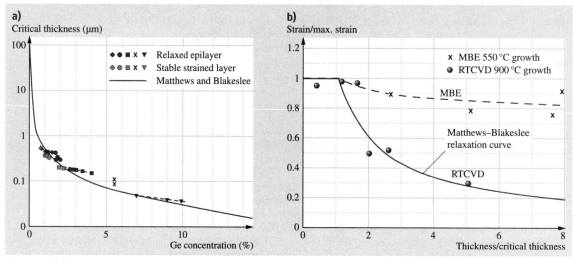

Fig. 34.12 (a) Experimentally determined critical values for thickness and composition obtained from annealed SiGe samples where the thickness and composition were laterally graded. An abrupt transition from a stable strained layer to a relaxed epilayer was detected to give a value of critical thickness at a certain composition. The *solid curve* is based on the Matthews and Blakeslee model, showing critical thickness as strain converted to Ge concentration in SiGe layers [34.56]. (b) Strain in the SiGe film divided by the theoretical maximum strain for that film versus film thickness divided by the critical thickness for that film grown by RTCVD. For comparison, the data of MBE-grown SiGe films [34.52] are also shown [34.57]

The discrepancy between theoretically predicted values and experimentally observed ones for the critical thickness has been reported in early works of molecular-beam epitaxy (MBE) of SiGe films on Si(001) substrates [34.52]. To explain the considerably larger critical thicknesses observed experimentally for SiGe films grown by MBE at 550 °C, *People* and *Bean* [34.53] firstly attempted to derive an expression of the critical thickness by taking into account the extra energy required for the generation of misfit dislocations. They obtained the critical thickness by setting the area density of strain energy associated with a film equal to the energy density of dislocations. Although this model showed quite good agreement with the experimental values of the critical thickness depending on the Ge content in the film, their interpretation is rather ad hoc, and the estimation of dislocation core energy is opposed to the elasticity treatment which is ordinarily employed for conventional semiconductor materials. On the other hand, it was pointed out that finite experimental resolution for detecting dislocation introduction in x-ray diffraction measurements for lattice parameters led to spurious results, and the critical thickness obtained by high-resolution measurements, such as stimulated-emission characteristics and photoluminescence, approaches the equilibrium theory proposed by *Matthews* and *Blakeslee* [34.54]. High-sensitivity detection for misfit dislocations was also performed using electron-beam-induced current imaging in scanning electron microscopy (SEM) and transmission electron microscopy (TEM) [34.55]. The obtained critical thicknesses were revealed to be less than those determined in [34.52], although the SiGe films were grown under similar conditions.

For another dominant reason causing the discrepancy between the theory and experiment, nonequilibrium growth conditions for strain-relaxed SiGe films seemed to be responsible. High-temperature (900 °C) postgrowth annealing for SiGe films grown at 500–600 °C [34.56, 58] and rapid-thermal chemical vapor deposition (RTCVD) at 900 °C [34.57] led to excellent agreement of the experimental critical thickness with the predicted values of the Matthews and Blakeslee model, as shown in Fig. 34.12a,b, respectively. In general, the kinetic barrier for the formation or motion of misfit dislocations gives rise to nonequilibrium behavior of the critical thickness. This kinetic effect on strain-relaxation behavior was the focus of the early stage of development of the Matthews and Blakeslee theory, in which *Matthews* et al. proposed that Peierls stress results in a friction force on the dislocation [34.59]. Contrary to this model, assuming only a velocity-dependent friction force, a model taking the static Peierls barrier into account was developed [34.60]. Other models including precise expression of arrays of misfit dislocations [34.61] and surface-relaxation effects [34.62] have been proposed so far. As mentioned above, prediction of the critical thickness requires the consideration of kinetic effects relevant to practical growth conditions, such as growth temperature, initial dislocation density, and growth rate. This simultaneously means that strain relaxation of SiGe/Si(001) systems is strongly influenced by the kinetics, which will be explained in more detail in the next section.

34.6 Mechanism of Strain Relaxation

SiGe alloy films epitaxially grown on Si substrates have a maximum lattice mismatch of approximately 4.2% (when the film is pure Ge). This lattice mismatch induces strain mainly into the film, and elastic strain energy is accumulated with increasing film thickness. The relief of the elastic energy in the strained film occurs mainly through two mechanisms: elastic deformation accompanying surface evolution of the film, and plastic deformation with the introduction of misfit dislocations. In this section, to explain strain-relaxation mechanisms in SiGe/Si systems, we first mention the former process in which surface perturbation mainly contributes to the reduction of elastic energy and then the latter process including nucleation, propagation, and reaction/multiplication of dislocations.

It is well known that Ge grows on Si in Stranski–Krastanov mode in which layer-by-layer growth is followed by islanding of Ge. In practice, at the initial growth stage of Ge on Si(001) surfaces, dislocation introduction is preceded by the formation of dislocation-free (coherently strained) Ge islands [34.63] and/or *hut clusters* [34.64], i.e., {501}-faceted Ge islands, each of which was revealed by TEM and scanning tunneling microscopy (STM), respectively, and has a slightly different shape with respect to each other. Further TEM analysis revealed the hut cluster to be also a coherently strained Ge island and clarified the interplay between surface morphological variation and defect introduction concomitant with Ge islanding, which is critically dependent on the growth temperature [34.65, 66].

On the other hand, islanding phenomena occurring in SiGe/Si systems can be interpreted as strain relaxation based on Asaro–Tiller–Grinfeld instability in elastically stressed solids [34.67, 68]. The formation of perturbation on the stressed solid surface significantly reduces the stored elastic energy at the peak of perturbation but, at the same time, costs additional surface energy. Thus, the critical wavelength of the surface undulation can be drawn, which determines the stability of perturbation. The critical wavelength λ_c for a semi-infinite uniaxially stressed solid with a sinusoidal surface profile was derived as follows [34.69]

$$\lambda_c = \frac{\pi M \gamma}{\sigma^2}, \tag{34.7}$$

where M is the elastic modulus appropriate to the local surface orientation, γ is the surface tension, and σ is the mean stress in the solid. A more realistic model for the morphological instability of both growing and static epitaxially strained films was further developed, including differences in lattice parameters and elastic constants between the films and the substrate [34.70]. For experimental studies, wafer curvature measurements [34.71], direct observation using scanning transmission electron microscopy (STEM) [34.62], and STM [34.72] have been effectively applied to analyze the strain-related morphological evolution in SiGe/Si(001).

Surface evolution for elastic strain relaxation is generally followed by the introduction of misfit dislocations. A model describing total processes of strain relaxation in SiGe/Si systems was proposed by *Dodson* and *Tsao* [34.73]. On the basis of a phenomenological model of the dislocation dynamics and the plastic flow in bulk diamond-lattice semiconductors [34.74], they established a kinetic model of strain relaxation in which propagation and multiplication of pre-existing dislocations were treated systematically. In their model, a differential equation for the time-dependent degree of strain relaxation $\chi(t)$ is expressed as

$$\frac{d\chi(t)}{dt} = C\mu^2 \left[f_0 - \chi(t) - s(h) \right]^2 \left[\chi(t) + \chi_0 \right]. \tag{34.8}$$

In this equation, f_0 is the misfit, μ is the shear modulus, $s(h)$ is the thickness-dependent homogeneous strain retained by the overlayer, and χ_0 is the dislocation source density. The choice of appropriate C and χ_0 values

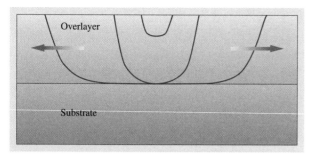

Fig. 34.13 Schematic illustration of nucleation and growth of a dislocation half-loop. Each threading part moves toward the edges of the sample, leaving a misfit segment at the interface, when the film thickness exceeds the critical thickness ◂

as fitting parameters resulted in remarkable agreement with the previous experimental results [34.52, 75]. According to the Dodson and Tsao model, even though the film thickness exceeds the critical thickness, both the effective stress exerted on the overlayer and the density of dislocations available for multiplication are small and this leads to initial slow relaxation of strain. Later on, the strain relaxation proceeds with the increase in the dislocation density exponentially with time, but then becomes sluggish due to the reduction of effective stress accompanied by the strain relaxation. They also explained when predicting the critical thickness that the initial slow relaxation combined with the later sluggish relaxation allows the growth of overlayers with a thickness considerably larger than the equilibrium critical thickness. Quantitative data obtained from in situ TEM analysis were also systematically applied to develop a model of strain relaxation including the effects of dislocation nucleation, propagation, and interaction [34.76].

Although detailed mechanisms for misfit dislocation nucleation in SiGe/Si systems are still controversial, the phenomena generally obey either heterogeneous or homogeneous nucleation mechanism. Earlier works looked for the source of heterogeneous nucleation, such as SiC precipitates formed by incomplete substrate cleaning [34.77], $a/4\langle 114\rangle$ stacking fault regions [34.78, 79], surface half-loops, and faulted dislocation loops associated with metallic contamination [34.80, 81]. However, all of these heterogeneous sites appear dependently of the growth conditions and cannot account for the strain-relaxation behavior occurring in the initial stage because of their relatively low density. An activation barrier for homogeneous nucleation of surface dislocation half-loops was calculated for SiGe/Si(001) systems and was shown to be sensitively influenced by Ge contents, i.e., strain, and the dislocation core energy parameter [34.82]. A con-

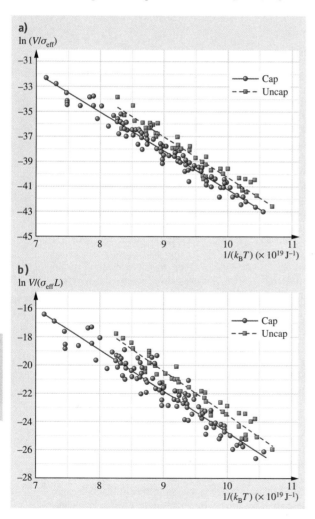

Fig. 34.14a,b Dislocation velocities in capped and uncapped (Si)/Si$_{1-x}$Ge$_x$/Si(001) heterostructures measured by TEM. The measured velocities V are normalized to equivalent velocities in pure Si by assuming a linear interpolation of the activation energy of Si and Ge, i.e., $E_a = (2.2-0.6x)$ eV. The velocities are also normalized to the effective stress σ_{eff} (**a**) and the product $\sigma_{\text{eff}}L$, where L is the length of the propagated dislocation line (**b**) ◂

sensus of qualitative arguments was reached in which homogeneous nucleation of half-loops is physically plausible at moderate and high mismatches, while heterogeneous nucleation relevant to a neighboring defect operates at low mismatch [34.83]. Furthermore, the aforementioned perturbation formed on the film surface plays a crucial role in dislocation nucleation. On the basis of STEM images showing SiGe cusp profiles, the condition that the local stress concentration at the surface cusp overcomes the nucleation barrier of half-loop dislocations was quantitatively discussed and the critical geometry for barrierless nucleation of 60° half-loops was derived [34.82,84]. Similar analysis has been carried out for interfacial prismatic dislocation loops resulting from subnanometer-sized Ge-rich clusters as an efficient nucleation source [34.85].

As shown in Fig. 34.13, once half-loops nucleate on the film surface, they grow until they reach the substrate–overlayer interface. Then, each threading part of the dislocation moves toward the edges of the sample, leaving a misfit dislocation in the plane of the interface. The kinetics of such misfit dislocation propagation has been extensively investigated by two categorized methods: either direct observation of dislocation velocities by TEM or measurements of maximum length of dislocations by Nomarski interference microscopy of defect-etched surfaces [34.86–90]. These studies, irrespective of the difference in the measurements, show reasonable quantitative agreement with each other. In principle, the velocity of dislocation propagation V is of the form of

$$V = V_0 \sigma_{\text{eff}} \exp\left(-\frac{E_a}{k_B T}\right), \quad (34.9)$$

where V_0 is a prefactor containing an attempt frequency, σ_{eff} is the effective stress acting on the threading dislocation arm, and E_a is an activation energy of dislocation glide [34.73]. Systematic measurements of dislocation velocities were achieved using in situ TEM [34.91]. Assuming a linear interpolation of the activation energies of Si and Ge, i.e., $(2.2-0.6x)$ eV, where x denotes the Ge content in $Si_{1-x}Ge_x$, the measured velocity was normalized to an equivalent velocity under a stress of 1 Pa in pure Si. Consequently, fairly good correlation is seen between samples across all ranges of epilayer composition and thickness (Fig. 34.14).

In the final stage of strain relaxation, dislocation multiplication events are dominant. The Hagen–Strunk source [34.92], which was first observed in orthogonal dislocation configurations in Ge/GaAs systems, is commonly cited for dislocation multiplication. However, later on, theoretical and experimental analyses showed that it seems an unlikely source [34.93, 94]. At present, several multiplication sources are proposed, such as so-called modified Frank–Read sources [34.95], cross-slip and pinning [34.96], and cross-slip for branch formation [34.97]; the latter two events are also followed by Frank–Read-type dislocation multiplication mechanisms.

34.7 Formation of Relaxed SiGe Layers

34.7.1 Graded Buffer

There are various types of strain-relaxed SiGe buffer layers developed so far for strained-layer channels in metal–oxide–semiconductor field-effect transistors (MOSFETs) and modulation-doped field-effect transistors (MODFETs). A compositionally graded SiGe buffer layer is most successfully applicable to such devices [34.98–102]. In the formation procedure, the Ge concentration x in the $Si_{1-x}Ge_x$ alloy gradually increases with increasing film thickness. The in-depth composition profile of Ge is generally set to be either linear or step-like. Since this structure is considered to be the sum of low-mismatch interfaces, misfit dislocations are successively introduced during growth, resulting in total strain relaxation of the film. Contrary to SiGe films without a gradient, since each atomic plane tends to have its own equilibrium lattice parameter, the dislocations accommodating the lattice parameter difference between the substrate and the top layer are distributed over the thickness of the graded region. This specific configuration of dislocations results in a much lower density of threading dislocations than that of the constant-composition film. Typically, a threading dislocation density on the order of 10^5-10^7 cm^{-2} is obtained, depending on the grading rate and final Ge concentration in the top layer. One of the reasons for the absence of threading dislocations in the top layer is high dislocation velocities at which the dislocation can bypass pinning points such as a site of dislocation intersection [34.95]. Since all dislocations are not attracted to only a single interface in the compositionally graded film, there is great freedom for a dislocation to move onto another (001) plane reaching the edges of the wafers.

However, severe surface roughness, referred to as a cross-hatch pattern, has been pointed out from the beginning of the research on SiGe graded buffers [34.103–107]. *Fitzgerald* et al. have shown that such cross-hatch patterns are related to the strain field in the epilayer caused by an inhomogeneous distribution of misfit dislocations [34.106]. According to their calculation, the critical thickness h_{gc} for the introduction of dislocations into a graded layer is

$$h_{gc}^2 = \frac{3\mu b}{2\pi Y C_f (1-\nu)} \left(1 - \frac{\nu}{4}\right) \left[1 + \ln\left(\frac{h_{gc}}{b}\right)\right], \quad (34.10)$$

where C_f is the grading rate and Y is the Young's modulus. This formula clearly explains that larger grading rate leads to smaller h_{gc}, meaning that dislocations are introduced much closer to the surface and more strongly affect the evolution of the surface. On the other hand, a different mechanism was proposed, by which surface steps arising from the single and multiple 60° dislocations at the film–substrate interface directly influence the surface morphology of films [34.108]. Note that these two cases employ different growth conditions: a steeper (slower) grading rate of 50% Ge μm^{-1} (10% Ge μm^{-1}) and a lower (higher) growth temperature of 560°C (900°C) for the latter (former). Recently, a simulation study was performed to model the development of cross-hatch patterning on the basis of a combination of dislocation-assisted strain relaxation and surface steps during growth [34.109].

In graded buffers, there is a strong correlation between surface roughening and residual threading dislocation density. In general, the higher the final Ge concentration, the higher the threading dislocation density, despite the same grading rate. This results from a mechanism by which the combination of strain fields of underlying multiple misfit dislocations and the resultant surface roughness blocks the motion of a propagating threading dislocation, leading to dislocation pileups [34.110]. More recently, dislocation glide and blocking kinetics in compositionally graded SiGe buffers grown by UHV-CVD were reported and the dislocation density on the order of 10^4 cm^{-2} for 30% Ge was achieved at relatively high growth temperature of 900°C [34.111].

In principle, graded buffers with low threading dislocation density require slow grading rate, which results in large thickness of the films and requires from time and material. Furthermore, poor thermal conductivity relevant to included Ge severely affects the performance of devices fabricated on graded buffers. Thus several techniques other than growth of graded buffers have also been attempted for strain-relaxed SiGe buffers on Si(001) substrates.

34.7.2 Low-Temperature Method

To reduce the layer thickness and improve the crystal quality of buffer layers, the low-temperature (LT) method was proposed [34.112, 113]. In this method, before growing relatively thin SiGe buffer layers (about 500 nm), a thin Si layer, e.g., 50 nm thick, is grown at a low temperature such as 400°C. Although the thickness of the SiGe buffer is much smaller than that required for the graded buffer method, it was found that the surface roughness is much better and the dislocation density is much smaller than those of the graded buffer method, while the relaxation ratio is almost equal, more than 80% [34.114]. That is, when the Ge composition is 30%, the surface roughness is about 1.5 nm and the dislocation density is around 10^5 cm^{-2}, while they are more than 10 nm and 10^7 cm^{-2} for the graded buffer method. It is known tthe hat surface roughness and re-

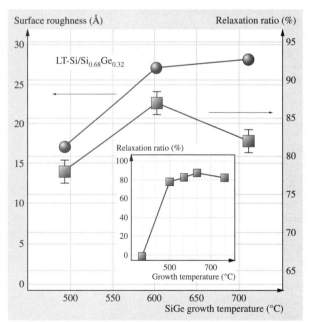

Fig. 34.15 Growth temperature dependence of surface roughness and relaxation ratio of Si$_{0.68}$Ge$_{0.32}$ buffer layers grown on low-temperature-grown Si (LT-Si). The inset shows relaxation ratio in the wider temperature range

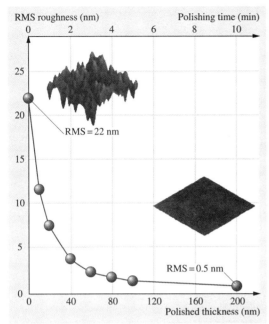

Fig. 34.16 CMP-polished thickness dependence of root-mean-square (RMS) surface roughness of relaxed SiGe buffer layers

laxation ratio do not change much when the growth temperature of LT Si layers is varied between 300 and 400 °C. However, the quality of buffer layers grown on LT Si largely depends on the buffer layer growth temperature. The relaxation ratio increases from 0 to 90% as the temperature increases from 400 to 700 °C and the roughness also increases, from 1 to 3 nm, with increasing temperature as shown in Fig. 34.15.

The quality of buffer layers largely depends on the Ge composition as well. Surface roughness as well as threading dislocation density increase with increasing Ge content up to 40% and then decrease.

From positron annihilation experiments, it is revealed that the Si layer grown at 400 °C contains a lot of defects, particularly vacancy clusters [34.115]. So, it is reasonably considered that these defects act as dislocation sources and confine dislocations in the vicinity of LT Si layers.

When high-Ge-content buffer layers are grown, the two-step growth of LT layers is found to provide strained overgrown layers with higher quality than the one-step growth method. Although the surface roughness of the buffer is hardly distinct between the two- and one-step methods, short-period roughness is found to be superimposed in the case of the one-step method. This roughness may degrade the transport properties of the structures grown on it, and higher mobility was obtained in the case of the two-step method.

34.7.3 Chemical–Mechanical Polishing Method

Since obtaining high-quality relaxed SiGe buffer layers is a critical issue and the quality of the layers obtained by the methods described above is still not satisfactory for production, there have been a lot of attempts to develop better techniques. Chemical–mechanical polishing (CMP) of SiGe buffer layers with large roughness is one of the promising techniques [34.116–120]. CMP consists of mechanical polishing by small particles and chemical etching and is well established for preparation of Si wafers. The surface flatness of SiGe buffers after proper CMP polishing is almost equal to that of Si wafers, as shown in Fig. 34.16 where surface roughness is plotted as a function of polishing time [34.116]. It is demonstrated that the mo-

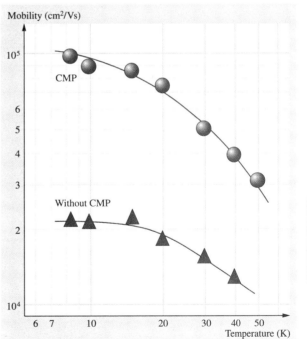

Fig. 34.17 Temperature dependence of electron mobility of SiGe/Si modulation-doped structures with and without CMP polishing

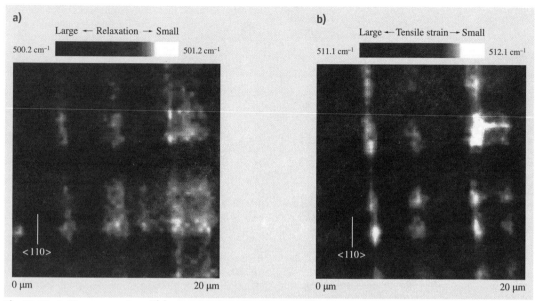

Fig. 34.18a,b Raman mapping of (**a**) relaxed SiGe buffer layer and (**b**) strained Si layer grown on SiGe buffers

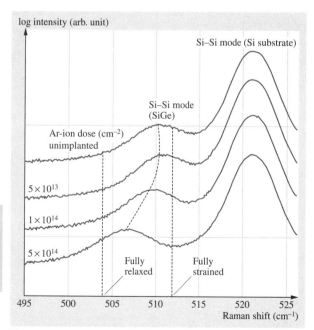

Fig. 34.19 Raman spectra of SiGe buffer layers grown on unimplanted and implanted Si substrates. Si–Si mode of SiGe layer is seen to shift to smaller wavenumber with increasing Ar-ion dose

bility of MOD structures formed on the flat surface is enhanced due to the reduction of interface scattering [34.117]. Figure 34.17 shows the temperature dependence of the mobility of the MOD structures formed on CMP polished and unpolished SiGe buffer layers. It is seen that the CMP sample provides an electron mobility four times higher than that of the unpolished sample. The mobility is shown to increase to about $600\,000\,\text{cm}^2/(\text{V s})$ at very low temperatures. As for hole mobility, it has also been demonstrated recently that p-type Ge channel grown on CMP-polished SiGe buffers provides mobility eight times higher than that on unpolished ones [34.120].

However, it is noted that, even though CMP provides very flat surfaces, there still exists strain fluctuation coming from the underlying misfit dislocations in the SiGe buffer [34.121]. Figure 34.18 shows Raman mapping of SiGe buffer as well as strained Si grown on the buffer, from which it is clear that strain fluctuations similar to the cross-hatch pattern exist. It is also known that this strain fluctuation affects the growth rate of the overlayer and causes surface roughening [34.121].

34.7.4 Ion Implantation Method

Another interesting approach is ion bombardment of Si substrates. It was demonstrated [34.122–126] that proton or helium ion implantation into Si substrates through epitaxially grown SiGe layers and annealing provided good strain relaxation of SiGe layers, and misfit dislocations were generated at the interface thanks to the defects introduced by the ion bombardment. To control defect formation well in the surface region of Si substrates, ion implantation with heavy ions such as Ar and Ge before epitaxial growth of SiGe layers has been shown to be very effective to relax the SiGe buffer layers [34.127–130]. Figure 34.19 shows Raman spectra of implanted and unimplanted samples, showing clearly that the implanted sample shows good strain relaxation, even though the SiGe buffer is much thinner than that produced by the conventional graded method. Many defects, mainly consisting of vacancy clusters, act as nucleation centers as well as dislocation absorbers similar to the role of the LT buffer layers. A thickness of only 100 nm is good enough to obtain fully relaxed SiGe buffer layers, which is attractive from the point of view of production. However, the surface roughness is not very low, being almost the same as that of the LT method.

Surface flatness comparable to that of Si wafers was found to be obtained when thin SiGe layers were pseudomorphically deposited at low temperatures on ion-implanted Si substrates and postannealing was carried out at relatively high temperatures [34.131]. As seen in Fig. 34.20, the surface of the SiGe buffer formed by this method is very smooth, almost the same as for Si wafers, and the relaxation ratio is more than 80%. It is also noted that the strain distribution is much more uniform and no cross-hatch-like pattern is observed. Since the cross-hatch pattern is not seen in this sample, uniform distribution of misfit dislocation without bunching or strain-relief mechanisms due to point defects, different from the conventional misfit dislocation formation, may occur in this sample.

This the ion implantation method is very useful for growth methods such as CVD and GSMBE where low-temperature growth cannot be performed to decompose source gases.

34.7.5 Ge Condensation Method

The ultimate application of strained Si may be as silicon-on-insulator (SOI). SOI has a lot of advantages over the bulk devices, such as suppression of short-channel effects and so on. To realize high-quality strained Si on relaxed SiGe-on-insulator (SGOI), a very unique and attractive method called the Ge condensation method was proposed [34.132–135]. The principle of this method is shown in Fig. 34.21. When SiGe layers are oxidized, only Si atoms are consumed to form SiO_2 while Ge atoms are rejected from the oxide film. Therefore, when one deposits SiGe layers on SOI substrates and oxidizes the SiGe layers, the Ge content of the unoxidized SiGe layers is increased as the oxidation proceeds. It is also found that strain relaxation of SiGe grown on SOI simultaneously takes place during oxidation. As a result, relaxed SiGe layers with higher Ge content are formed on the buried oxide, that

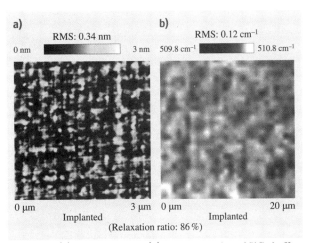

Fig. 34.20 (a) AFM image and (b) Raman mapping of SiGe buffers grown on ion-implanted Si substrate

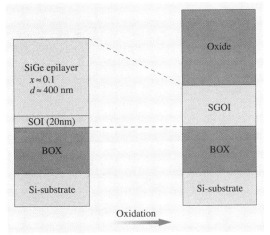

Fig. 34.21 Schematic illustration of Ge condensation method applied on SiGe epitaxial layer grown on SOI substrates. BOX is the buried oxide

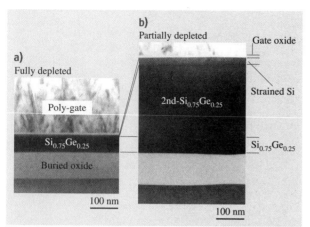

Fig. 34.22a,b TEM images of strained Si grown on (**a**) thin (fully depleted) and (**b**) thick (partially depleted) relaxed SiGe buffer layers formed by Ge condensation method

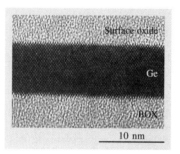

Fig. 34.23 TEM picture of Ge-on-insulator (GOI) structure fabricated by Ge condensation method

cation parallel to the interface, a screw component and components nonparallel to the interface simultaneously induce nonisotropic strain relaxation [34.138], as well as crystallographic tilting and rotation of SiGe. Such crystallographic inhomogeneity severely influences the lattice deformation of the channel Si and eventually leads to nonuniformity of the energy band structure. In this subsection, recent experimental results on epitaxial growth in which dislocation generation and propagation are precisely controlled during strain relaxation of SiGe are presented.

The two-step strain-relaxation procedure [34.139] was demonstrated to realize fully strain-relaxed thin buffer layers with low threading dislocation densities. This procedure consists of, first, annealing of a metastable pseudomorphic SiGe layer with a Si cap layer and, second, subsequent growth of SiGe on that layer. The thin cap layer effectively suppresses surface roughening during the annealing due to reduction of the surface stress of the film [34.140, 141]. More than 90% relaxation was achieved after growth of only a 100 nm thick second $Si_{0.7}Ge_{0.3}$ layer. Figure 34.24 shows a representative cross-sectional TEM image of a sample having a $Si_{0.7}Ge_{0.3}$(200 nm)/Si-cap(5 nm)/$Si_{0.7}Ge_{0.3}$(50 nm)/Si(001) structure. Note that threading dislocations are almost absent from the observed area and almost all misfit dislocations are confined at the first SiGe/Si substrate interface. These dislocations tend to be dispersed at the interface and pileup of the dislocations, which is often observed in compositionally graded layers [34.108], is hardly observed. The observed periodic surface undulation comes

is, SGOI structures are formed. Strained Si layers can, therefore, be grown on this SGOI. Figure 34.22 shows TEM images of SGOI with strained Si layers; both fully (thin) and partially depleted SGOI (thick) structures are seen to be formed [34.136]. In the extreme case, pure Ge layers can be formed by this Ge condensation method and Ge-on-insulator (GOI) is realized, as shown in Fig. 34.23 where the Ge content of the layer on the insulator becomes almost 100% after complete oxidation of SiGe layers [34.137].

34.7.6 Dislocation Engineering for Buffer Layers

In general, glide dislocations, i.e., 60° dislocations, are predominantly introduced into the SiGe/Si(001) interface as a result of the operation of the $\langle 110 \rangle/\{111\}$ slip system. Although misfit strain is mainly relaxed by an edge component of the Burgers vector of the 60° dislo-

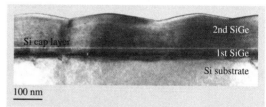

Fig. 34.24 Cross-sectional TEM image of a sample grown by the two-step strain-relaxation procedure. The sample has a $Si_{0.7}Ge_{0.3}$(200 nm)/Si-cap(5 nm)/$Si_{0.7}Ge_{0.3}$(50 nm)/ Si(001) structure. Threading dislocations are almost absent from the observed area and almost all misfit dislocations are dispersively confined at the first SiGe–Si substrate interface. The observed periodic surface undulation comes from aligned SiGe islands formed at the earlier stage on the misfit dislocation network buried at the first SiGe–substrate interface as a template

from aligned SiGe islands formed at the earlier stage on the misfit dislocation network buried at the first SiGe–substrate interface as a template. Similar preferential nucleation over dislocation was reported by *Xie* et al. [34.142]. As mentioned in Sect. 34.6, since a cusp in the surface undulation acts as a preferential nucleation site for misfit dislocations [34.143], dislocation half-loops are likely introduced at every cusp on the surface to relax the strain during the growth of the second layer. Therefore, strain relaxation is dominated by the introduction of new dislocations from the surface, and a regular strain field created by the periodic undulation greatly enhances the propagation of the introduced dislocations so that the threading segments have the opportunity to travel long distances.

Due to the intrinsic structure of a 60° dislocation, a SiGe film strain-relaxed by 60° dislocations often exhibits mosaicity [34.144] and cross-hatch patterns [34.106, 108]. In order to prevent such degradation, the introduction of pure-edge dislocations is crucial. A novel approach based on strain relaxation predominantly by a pure-edge dislocation network buried at the SiGe/Si(001) interface was recently demonstrated [34.145]. Employing pure-Ge thin-film growth prior to SiGe formation effectively restrains the introduction of 60° dislocations; instead, a high density of pure-edge dislocations can be generated [34.146]. Figure 34.25a shows a plan-view TEM image of the Ge layer, which was grown at 200 °C and then subjected to annealing at 700 °C. A dislocation network, consisting of pure-edge dislocations aligned along two orthogonal ⟨110⟩ directions, can be clearly observed at the Ge–Si(001) interface. For forming SiGe layers, solid-phase

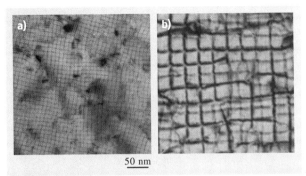

Fig. 34.25 (a) Plan-view TEM image of a Ge layer on Si(001), which was grown at 200 °C and then subjected to annealing at 700 °C, showing a dislocation network consisting of pure-edge dislocations aligned along two orthogonal ⟨110⟩ directions. (b) Plan-view TEM image of an a-Si (17 nm)/Ge(30 nm)/Si(001) sample annealed at 1100 °C (the scale is the same as that in (a)). The morphology of the pure-edge dislocation network is retained even after the high-temperature annealing, but the dislocation spacing increases

intermixing of amorphous Si (a-Si) deposited on the Ge layer was performed. Figure 34.25b shows a plan-view TEM image of an a-Si (17 nm)/Ge(30 nm)/Si(001) sample annealed at 1100 °C. Note that the morphology of the pure-edge dislocation network is explicitly retained even after the high-temperature annealing but the dislocation spacing is found to increase. Figure 34.26a shows an x-ray diffraction two-dimensional reciprocal-space map around a SiGe(115) diffraction peak of the sample annealed at 1100 °C. From the peak position,

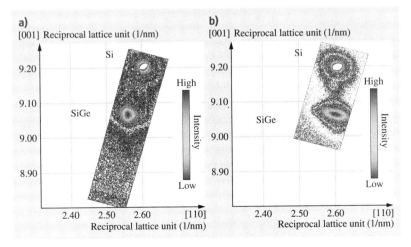

Fig. 34.26a,b XRD two-dimensional reciprocal maps around Si(115) and SiGe(115) diffraction peaks of (a) the sample annealed at 1100 °C, which has a pure-edge dislocation network at the SiGe–Si(001) interface, and (b) the sample with 60° dislocations grown by the two-step strain-relaxation procedure

it is found that an in-plane lattice constant corresponding to that of a SiGe crystal with a Ge fraction of 0.48 is obtained, enabling practical use as a SiGe buffer. A remarkable feature is found in the peak shape being sharp and highly symmetric, in marked contrast to that of the reference sample shown in Fig. 34.26b, which is prepared by the two-step strain relaxation procedure and predominantly has 60° dislocations at the interface. This result clearly demonstrates that the mosaicity, such as lateral finite sizes and microscopic in-plane tilts of the SiGe crystal [34.147], is greatly reduced when introducing pure-edge dislocations.

34.7.7 Formation of SiGeC Alloys

Since the first epitaxial growth of high-quality $Si_{1-y}C_y$ and $Si_{1-x-y}Ge_xC_y$ films on Si(001) was reported in 1992 [34.148–150], the introduction of C into Si and SiGe films has attracted increasing interest for application of these films to electric and optoelectronic devices composed of group IV semiconductor materials. Due to the different covalent radius of Si, Ge, and C of 0.117, 0.122, and 0.077 nm, respectively, substitutional introduction of C atoms into SiGe lattice sites gives rise to a substantial decrease of the lattice parameter and a reduction of strain in SiGe layers on Si substrates [34.151–154]. The addition of C into $Si_{1-x}Ge_x$ films is, therefore, expected to yield several advantageous effects in terms of device performance. In Si/SiGe/Si HBTs, when C is incorporated into a SiGe base layer, the strain caused by the SiGe/Si lattice mismatch can be alleviated and thus the generation of misfit dislocations can be prevented, realizing improved crystalline quality of the base layer and resultant reliability of the devices. Furthermore, C incorporation can significantly suppress boron outdiffusion from a p-type base layer due to the undersaturation of interstitial Si atoms [34.155–158]. On the other hand, it has been reported that the bandgap and band offsets with respect to the conduction band and valence band sensitively vary with C contents and strain in $Si_{1-x-y}Ge_xC_y$ films [34.159–169]. On the basis of the control over the band alignment in $Si_{1-x-y}Ge_xC_y$ heterostructures, $Si_{1-x-y}Ge_xC_y$ epitaxial films were applied to channel layers in MOSFET [34.170, 171], high-electron-mobility transistor (HEMT), and optical devices [34.172–174]. However, substitutional incorporation of C exceeding a few percent is very difficult to achieve. One dominant reason seems to be the thermal equilibrium solubilities of C into Si and Ge, of the order of 10^{17} and 10^8 atoms/cm^3, respectively [34.175, 176]. The crystalline quality of the films is degraded with increasing C fraction and exhibits nonplanar morphology, SiC polytype precipitates, and extended defects such as stacking faults and dislocations. Therefore, film growth techniques in which the growth mode is governed not by thermodynamics but kinetics are now widely employed, such as MBE [34.150, 151, 177, 178], UHV-CVD [34.179, 180], and RT-CVD [34.181, 182]. However, this still remains an essential issue in the growth of high-quality epitaxial $Si_{1-x-y}Ge_xC_y$ films with a substantial C content higher than 3% and this limits the potential of these films to be applied widely to various kinds of devices [34.178, 183]. Enhanced solubility of C was theoretically predicted [34.184], and it was experimentally demonstrated that $Si_{1-y}C_y$ layers with $y \approx 0.2$ can be grown pseudomorphically on Si(001) due to the formation of low-energy ordered structures [34.185]. In the case of $Si_{1-x-y}Ge_xC_y$ films, high repulsive interaction between Ge and C in the Si lattice plays a dominant role in determining the composition profiles of the film; both theoretical and experimental works on this matter have been performed [34.152, 186, 187]. Furthermore, attractive interaction between Si and C affects the final film morphology. Figure 34.27a shows an STM image of a surface of a $Si_{0.478}Ge_{0.478}C_{0.044}$ film with a thickness of four monolayers (ML) [34.188]. The surface exhibits the onset of three-dimensional (3-D) islanding, suggesting the local increase of Ge fraction around the island where the film locally exceeds the critical thickness for islanding. It should be noted that these islands are different from those

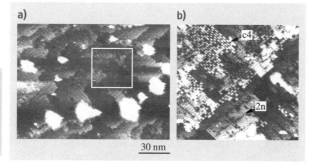

Fig. 34.27a,b STM images of a $Si_{0.478}Ge_{0.478}C_{0.044}$ film of 4 ML thickness grown on Si(001). (a) Islands with a height of about 2 nm on average are frequently observed on top of protruding regions of the terraces. (b) Magnified 30×30 nm^2 image of the *boxed area* in (a). Examples of the $c(4 \times 4)$ and $(2 \times n)$ reconstructions are indicated by the labels "c4" and "2n," respectively

due to C-induced Ge islanding [34.189, 190], since they were not formed at the beginning of the growth. As shown in Fig. 34.27b, a close-up of the fairly flat terraces reveals the $c(4 \times 4)$ reconstructed structure caused by significant C condensation on the growing surface [34.191–193] and the $(2 \times n)$ reconstruction consisting of buckled dimers and missing dimer rows. These results clearly show the formation of C-rich and Ge-rich regions in the film, which is presumably driven by the phase separation between Si-C and Si-Ge during growth when the film contains high concentration of C.

Control over the initial growth stage was performed to improve the film morphology [34.188, 193]. The addition of a thin (1–2 ML) SiGe interlayer between the $Si_{1-x-y}Ge_xC_y$ film and the Si substrate drastically improves the film structure, leading to a planar morphology even with large C fractions present in the film. Figure 34.28 shows an STM image of a sample which has a structure of $Si_{0.473}Ge_{0.473}C_{0.054}$(6 ML)/$Si_{0.5}Ge_{0.5}$(1 ML)/Si(001). Note that a planar surface morphology consisting of a step and terrace structure is formed even though the film contains a 4.6% average C fraction. Three-dimensional islanding partially appears at the step edges with increasing $Si_{1-x-y}Ge_xC_y$ film thickness [34.194, 195] but no $c(4 \times 4)$ reconstructions were observable in any growth stage. This clearly demonstrates that the $Si_{1-x}Ge_x$ interlayer explicitly plays a role in suppressing C condensation and Si-C/Si-Ge phase separation during the film evolution.

More recently, sequential alternate deposition of 1 ML thick $Si_{0.793}Ge_{0.207}$ and 0.048 ML thick C layer on Si(001) was attempted in order to suppress local phase separation, 3-D island growth, and defect formation [34.196]. A comparison of the surface atomic morphologies between the 5 ML thick $Si_{0.769}Ge_{0.183}C_{0.048}$ layer formed by the codeposition of Si, Ge, and C and that by the alternate deposition is shown in Fig. 34.29. In Fig. 34.29a, the 3-D islands, with a height of approximately 2 nm, and a rough surface consisting of small terraces are observed. Many defects, observed as dark spots on the surface, which are sink sites of C atoms, seem to prevent conformal step-flow growth due to the positive (normal) Ehrlich–Schwoebel barrier [34.197–

Fig. 34.28 STM image of a sample with a structure of $Si_{0.473}Ge_{0.473}C_{0.054}$ (6 ML)/$Si_{0.5}Ge_{0.5}$ (1 ML)/Si(001). A planar surface morphology is realized by the formation of the SiGe interlayer, in spite of the large C fraction

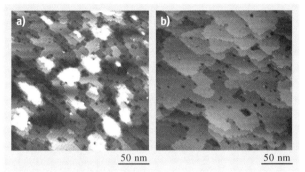

Fig. 34.29a,b STM images of the 5 ML thick $Si_{0.769}Ge_{0.183}C_{0.048}$ layer formed by (a) the codeposition of Si, Ge, and C; (b) the alternating deposition of 1 ML thick $Si_{0.793}Ge_{0.207}$ and 0.048 ML thick C layers. In the codeposition, 3-D islands with a height of approximately 2 nm, a rough surface consisting of small terraces, and defects as dark spots are observed on the surface, while no 3-D islands with reduced defects are seen in the case of the alternating deposition

199]. On the other hand, in Fig. 34.29b, no 3-D islands are observed and aligned steps are still formed. The density of defects is relatively low compared with the codeposition case. In the alternate deposition case, the migration of C on the growing surface is effectively restrained because the Ge atoms, which give rise to the repulsive interaction force to the deposited C atoms, are uniformly distributed on the $Si_{1-x}Ge_x$ surface. This effect leads to suppression of defect formation at the initial stage of $Si_{1-x-y}Ge_xC_y$ growth and consequently increases the critical thickness of layer-by-layer growth of $Si_{1-x-y}Ge_xC_y$.

34.8 Formation of Quantum Wells, Superlattices, and Quantum Wires

The formation of double heterostructures results in quantum well (QW) formation. Type I QWs are easily realized by growing strained SiGe layers on Si substrates and sandwiching it with unstrained Si lay-

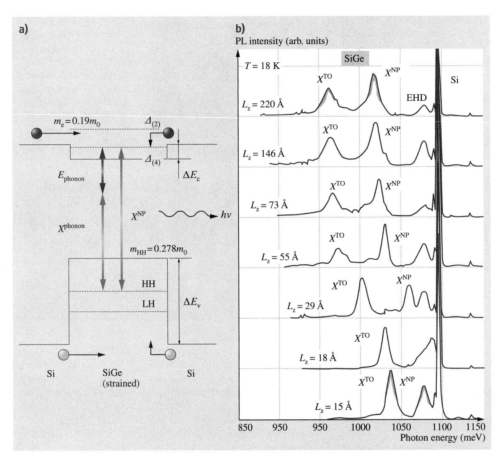

Fig. 34.30a,b Schematics of PL from strained SiGe/Si quantum well (**a**) and well width dependence of PL spectra of the quantum well (**b**). NP represents luminescence peaks without phonon assistance, and phonon and TO represent phonon replica peaks

ers. Peak shift due to the quantum confinement effect characteristic of QWs is clearly seen in the PL spectra of QWs grown by gas-source MBE, as shown in Fig. 34.30 [34.200]. There are two prominent PL peaks, no-phonon (NP) and the TO-phonon replica, which are characteristic of indirect-bandgap transition in this system, and both peaks shift to higher energies with decreasing well width. The theoretical calculation based on square QWs well represents the well width dependence of the quantum confinement energy.

Although the luminescence peaks become broader with increasing temperature, their integrated intensity is stable up to 100 K. Above 100 K, however, the intensity decreases rapidly with an activation energy of around 100 meV [34.34]. This activation energy corresponds to the energy difference between the valence band of Si barriers and the ground quantum level in the SiGe QW. That is, carrier confinement of QWs is quite efficient at low temperatures but holes begin to escape from the well above the temperature corresponding to the confinement energy.

Since the band alignment changes depending on the strain condition, type II QWs are also formed. When tensile-strained Si layer is sandwiched by unstrained SiGe barrier layers, holes are trapped in the Si well layers while electrons are located in the SiGe barrier layers and therefore the transition is indirect in real space as well as in momentum space. However, luminescence, due to both NP and TO peaks, is clearly observed and the energy shift coming from the quantum confinement effect is confirmed [34.201].

Although the band alignment is type I when strained SiGe layers with Ge composition less than 30% is set between unstrained Si layers, the band discontinuity at the conduction band is too small to confine electrons well in the well region. To overcome this problem, there

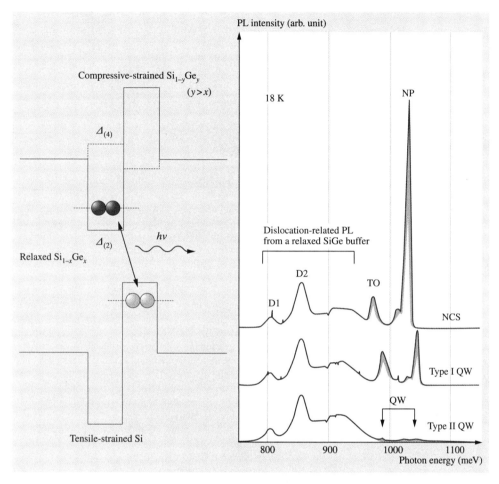

Fig. 34.31 Band structure of Si/SiGe neighboring confinement structure (NCS) and PL spectra of NCS, type I and type II quantum wells

are several ideas to improve the carrier confinement, one of which is the neighboring confinement structure (NCS), in which a pair of compressively strained and tensile-strained layers is sandwiched by strain-relaxed SiGe barrier layers, as shown in Fig. 34.31 [34.202]. Although electrons and holes are separately confined in the well region, the PL intensity, particularly the NP peak, is significantly enhanced in this structure and is stronger than that in conventional type I QWs. This comes from the significant wavefunction overlap between electrons in Si and holes in Ge layers, but it is noted here that the thickness of the pair layers should be thin to keep sufficient wavefunction overlap for the transition.

The coupling of QWs results in the formation of superlattices, which is seen in the energy shift of PL peaks, with the peaks shifting to lower energies with decreasing distance between the QWs [34.203]. The peak energies of coupled QWs agree well with the calculated results based on ideal square-shaped wells, in accordance with the theoretical prediction [34.1] that the band alignment is type I when Ge content is lower than 0.3. This also indicates that wells without deformation due to surface segregation are formed by GSMBE.

The evolution of superlattices is seen when the number of coupled wells is increased [34.204]. Since the increase in the number of coupled wells lowers the ground-state energy and finally forms a miniband, the PL peak energy, corresponding to the miniband edge, decreases with increasing well number. The peak energy is found to follow a simple Kronig–Penny-type calculation for the superlattices [34.204].

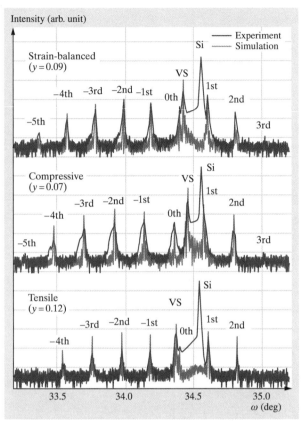

Fig. 34.32 XRD intensity profiles of SiGe/Si multilayers (DBR structures) grown under strain-balanced, compressive, and tensile conditions

example of such structures for optical applications is the distributed Bragg reflector (DBR) mirror, which consists of multilayers of materials with different refractive index, such as GaAs/AlAs and Si/Ge. Since the period of DBR structures should be comparable to the wavelength of light in semiconductors, the thickness of each layer is as large as ten times that in a superlattice. Therefore, the growth of such structures with a strained system is very difficult due to the limitation of the critical thickness. To overcome this problem, strain-balanced structures are very useful [34.205], and DBR mirrors consisting of Si and SiGe layers have been successfully fabricated [34.206–209]. In this method, a pair of layers with lattice constants a_1 and a_2 are grown on a substrate with lattice constant a_0, where $a_1 < a_0 < a_2$. This condition is satisfied when a pair of Si layers with tensile strain and SiGe layers with compressive strain are grown on relaxed SiGe with Ge content less than that of the pair. The thickness and Ge composition are selected so that the strain energy of the structure is minimized.

Figure 34.32 shows x-ray diffraction measurements where the Ge content of the virtual substrate is changed. It is seen that the peak of the sample with $y = 0.09$ coincides well with the position of the strain-balanced condition, showing that strain-balanced structures are formed by carefully choosing the layer thickness and Ge contents. Figure 34.33 shows an SEM image of a DBR structure grown under the strain-balanced condition, and it is seen that high-quality structures without detectable defects are obtained.

Figure 34.34 shows the reflection spectrum of SiGe DBR mirrors fabricated by this method. In this structure, Si and SiGe layers with thickness of 94 and 90 nm, respectively, were grown on fully relaxed SiGe buffer layers to compensate the strain. Reflectivity of about 90%, which is quite large for the strained system, is be obtained.

Selective epitaxial growth, which takes place in growth techniques such as gas-source MBE, leads to the formation of wire and dot structures. This occurs, for instance, between Si and SiO_2 substrates, and epitaxy occurs only on Si. When SiGe/Si QWs are formed on V-groove-patterned Si(100) substrates with (111) facets [34.210, 211], crescent-shaped SiGe features are grown in the bottom of the V-groove. This feature gives rise to PL with large blue-shift in the spectrum compared with the reference QW sample. It is also seen that the cross-sectional emission in the case of the wire

Multilayer structures, in which each layer is much thicker than in superlattices, are also important from the point of view of optical device applications. The typical

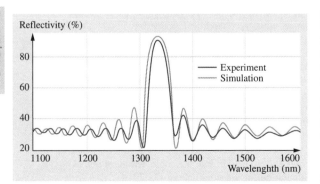

Fig. 34.34 Reflectivity spectrum of SiGe/Si DBR mirror formed by the strain-balanced method ◄

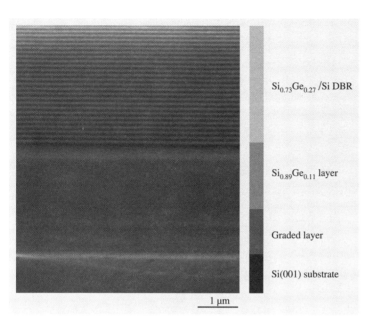

Fig. 34.33 Cross-sectional SEM image of $Si_{0.73}Ge_{0.27}/Si$ DBR mirror grown on $Si_{0.89}Ge_{0.11}$ buffer layers formed by the graded-buffer technique

structure shows polarization characteristics while no polarization is seen in the emission from the QWs. This result suggests that the wire structure in the V-grooved sample is likely to be a quantum wire. However, the energy shift in quantum wires is too large and does not agree with a simple estimation based on their size. The main cause of the energy shift may be the change in the strain distribution. The wire surrounded by Si crystal is considered to be under hydrostatic-like pressure that causes bandgap broadening comparable to the observed energy shift. Since spatial variation of the Ge composition in the wire is also likely to occur, more detailed study is required to clarify the nature of quantum wires in the SiGe/Si system.

34.9 Dot Formation

There are a large number of studies on Ge and SiGe dot formation, which strongly depends on the growth methods and conditions. Since there are several good reviews [34.212–215] concerning Ge dot formation, some general features of the formation, which are observed in the case of gas-source MBE (GSMBE), are described here [34.216].

Figure 34.35 shows the thickness dependence of the PL spectrum from Si/pure-Ge/Si quantum-well structures grown by GSMBE [34.217]. Up to 3.7 ML, the PL spectrum shows a conventional quantum confinement effect of quantum wells and the peaks shift to lower energies with increasing Ge layer thickness. Above 3.7 ML, however, the peaks originating from QWs stop the energy shift and a new broad peak is seen to appear. The appearance of this broad peak corresponds well to the formation of Ge islands observed by TEM measurements. That is, above the critical thickness, Ge begins to form islands on Ge wetting layers to release the strain energy as described in the previous section. It is remarkable that these Ge dots give rise to significant luminescence, and therefore their application as quantum dots is eagerly awaited.

The critical thickness in the case above is 3.7 ML and depends on the growth temperature, as shown in Fig. 34.36. It is seen from this figure that the critical thickness increases with decreasing growth temperature.

It is well known that Ge dot formation takes place in a bimodal fashion, that is, small pyramidal-shape and large dome-shape dots are formed simultaneously, as shown in Fig. 34.37, and that their relative numbers

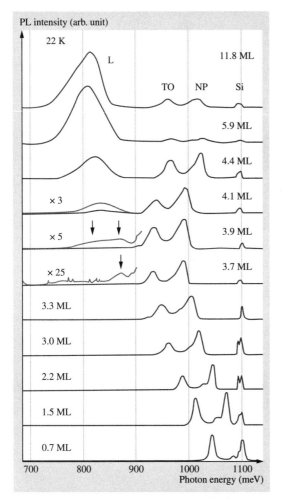

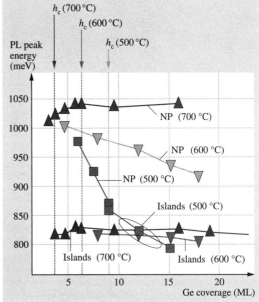

Fig. 34.35 Ge layer thickness dependence of PL spectrum of Si/pure-Ge/Si structures. Above 3.7 ML, a new broad PL peak (L) corresponding to Ge dot formation appears and grows with increasing Ge amount, while the peaks corresponding to quantum wells are seen to stop the energy shift once the dot is formed ◂

Fig. 34.36 Ge coverage dependences of NP and island PL peaks as a function of growth temperature. The critical thickness h_c of Ge dot formation is seen to increase with decreasing growth temperature

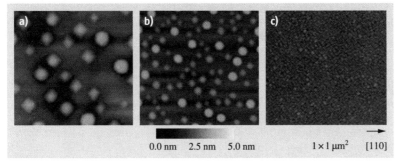

Fig. 34.37a–c AFM images of Ge dots grown at (**a**) 700 °C, (**b**) 600 °C, and (**c**) 500 °C. Scale is $1 \times 1\,\mu m^2$

change depending on the growth conditions. When the growth temperature is decreased, the number of dots becomes larger and their size becomes smaller; in particular, the density of pyramids becomes larger rather

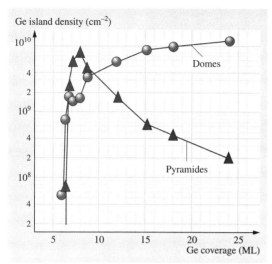

Fig. 34.38 Ge coverage dependence of island density. Morphological change is seen from pyramid to dome shapes when coverage exceeds about 8 ML

than that of domes. At lower temperatures, that is, 500 °C, domes and pyramids disappear and quite a large number of small elongated dots known as *hut clusters* appear, as seen in Fig. 34.37c. Energetically, domes are stable, whereas pyramids are metastable and therefore appear at rather higher temperatures. When the growth temperature is low, atom migration is suppressed and unstable hut clusters are formed.

Figure 34.38 shows the Ge coverage dependence of dot formation. It is interesting that a morphological change from pyramidal to dome shape is observed with increasing coverage. This is probably due to accumulating strain. Since the domes have a much larger degree of strain relaxation than the pyramids, this shape change may occur to reduce the total energy, and it dominates at higher coverage.

The relative number of pyramids to domes is also dependent on the growth rate. The density of pyramids is drastically decreased by decreasing the growth rate, and it is also found that inserting growth interruption or annealing decreases the number of pyramids. This reflects the fact that the dome is energetically more stable than the pyramid, and that the shape change takes place more easily under near-equilibrium conditions.

The formation of SiGe alloy dots is quite different from the case of pure Ge dots at first sight. As seen in Fig. 34.39, a large number of pyramids are formed compared with pure-Ge islands, and domes are hardly observed. The number of pyramids increases with increasing SiGe coverage, in contrast to the case of pure Ge, and the shape change does not take place. This feature is very similar to that of low-temperature growth of pure-Ge dots, in which hut clusters are formed and the shape change does not occur. This difference results from the growth mechanism. It is noted that, when the dots are formed at 600 °C by gas-source MBE with Si_2H_6 and GeH_4, the growth mode is in the reaction-limited regime, where growth is strictly limited by the chemical reaction of impinging gases. In the reaction-limited regime, the surface migration of atoms is suppressed and the shape change from pyramid to dome hardly occurs since the energy barrier is too high compared to the surface migration energy. On the other hand, the growth of pure Ge proceeds in the supply-limited regime at this temperature, and the suppression of surface migration is quite small, which allows the formation of stable domes. If the growth of SiGe dots is performed in the supply-limited regime, that is, at higher growth temperatures, the situation changes greatly and very similar behavior to that of pure Ge dots is seen, as shown in Fig. 34.40. In this figure,

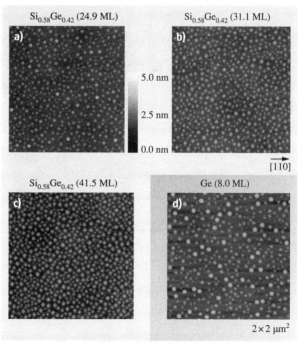

Fig. 34.39a–d Coverage dependence (**a–c**) of AFM images of $Si_{0.58}Ge_{0.42}$ dots and (**d**) Ge dots with 8 ML. Scale is $2 \times 2\,\mu m^2$

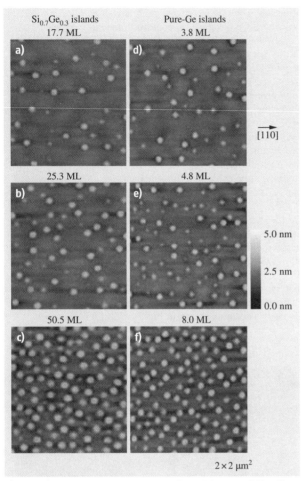

Fig. 34.40a–f Coverage dependence of AFM images of $Si_{0.7}Ge_{0.3}$ dots grown at high temperatures (**a–c**) and pure Ge dots grown at low temperatures (**d–f**). Scale is $2 \times 2\,\mu m^2$

growth and the shape change are very similar to those on Si. The Ge coverage dependence of the dot density is just shifted to lower coverage, which can be understood if we consider the strain energy of the two-dimensional (2-D) underlying strained layers. That is, the strain energy of the Ge thickness difference between SiGe and Si substrates for dot formation is almost equal to that of the underlying strained SiGe layer. This implies that the same dot formation mechanism is in operation and that the strain of the underlying layers contributes to dot formation.

Stacking of dot layers is attracting great attention from the point of view of crystal growth as well as device applications. As mentioned above, the strain originating from the underlying layer affects dot formation greatly. If the thickness of the separation layers is properly selected, it is well known that dots are aligned vertically, as shown in Fig. 34.41. This is because the strain coming from the underlying dots provides energetically favorable sites for dot formation. However, it should be noted that the kinetics is very complicated and that dot alignment does not always occurs, that is, the dot formation strongly depends on the growth conditions, the dot distribution, and the distance to the underlying dot layer. As a typical example of dot alignment, the distance dependence is shown for high-temperature growth in Fig. 34.42 [34.218]. As seen in this figure, the distribution of dot size and position depends on the Si interlayer thickness. It is natural that the distribution is almost similar to that of single-layer formation when the Si spacer is thick, since the strain effect of the underlying layer does not reach the upper surface. In the sample with a thinner Si spacer layer, a very large size distribution and a drastic increase of dome size appear. However, in the case of intermediate thickness such as 39 nm, it is quite interesting that very uniform distribution and ordering of dots are realized. This tendency can also be understood in terms of the change of the strain distribution due to the underlying dots with spacer thickness.

To enhance the quantum effect of Ge dots, small island formation is favorable. For this purpose, the low-temperature growth where small hut-cluster dots are formed is suitable. However, crystal quality is sacrificed at lower growth temperatures. To overcome this problem, two-step growth or stacking of dots is proposed and highly luminescent dots with relatively small size are successfully formed, as seen in Fig. 34.43 [34.219]. Here, the first layer of Ge dots is formed at a low temperature of 500 °C to obtain small dots and the second layer is grown at a higher temperature of 600 °C

bimodal growth of islands and the shape change are clearly seen for SiGe islands. This similarity suggests intermixing of Si and Ge even for pure-Ge dot formation. It is now known that the intermixing effect of Si and Ge, particularly at high temperatures, is important for dot formation and reduces the formation barrier for dots, resulting in a smaller critical thickness than that for low-temperature growth.

The formation of Ge dots on a strained SiGe layer is also important from the device application point of view. Although the critical thickness for dot formation is different from that on Si substrates, the bimodal

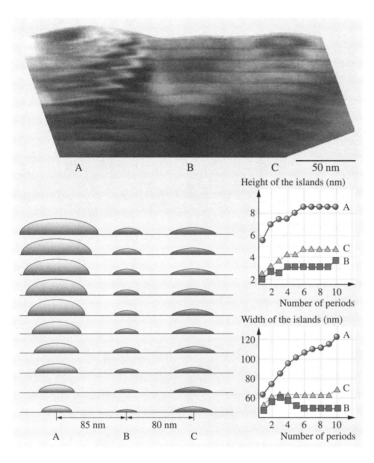

Fig. 34.41 TEM image of stacked Ge dots with 10 nm Si spacer and their schematic illustration. The *right-hand side inset* shows the height and width evolution of the dots as a function of the period number of the Ge layer

to obtain high-quality dots. Thanks to the strain field from the underlying small dots, the size and shape are almost repeated in the second layer, even though the growth is performed at higher temperatures where dome formation is favorable. It is confirmed that these dots have high luminescence efficiency, comparable to that of domes grown at higher temperatures. By using this method, relatively small dots with small size distribution and high luminescence efficiency can be obtained.

However, the size distribution is still large for device applications and the position is not perfectly controlled. Combination of selective epitaxial growth (SEG) and electron-beam (EB) lithography is a promising approach for sufficient control of the dots [34.220, 221]. As mentioned above, GSMBE has the advantage of providing selectivity between Si and SiO_2 surfaces and Ge dots can be grown only on Si surfaces. Therefore, when windows are opened in SiO_2 films on Si substrates, Ge dots are formed only in the windows. If the window size is smaller than the Ge migration length, only one Ge dot grows in a window and the dot size decreases with decreasing window size, as shown in Fig. 34.44 [34.220]. It is noted that the Ge dot is formed on the Ge wetting layer, that is, by the SK growth mode, even in SEG. These controlled Ge dots give rise to luminescence, and two well-resolved peaks are observed in contrast to the disordered Ge dots.

In order to enhance the quantum effects, several attempts to reduce Ge dot size and increase the dot density are now under investigation. Predeposition of elements such as C [34.189] and B [34.222] has been shown to be very effective to reduce the size. Incorporation of C is also found to be effective when dots are formed by using GSMBE [34.223]. Figure 34.45 shows dot formation when $(CH_3)_3SiH$ (TMS) is incorporated into the GeH_4 gas. There are some interesting features, different

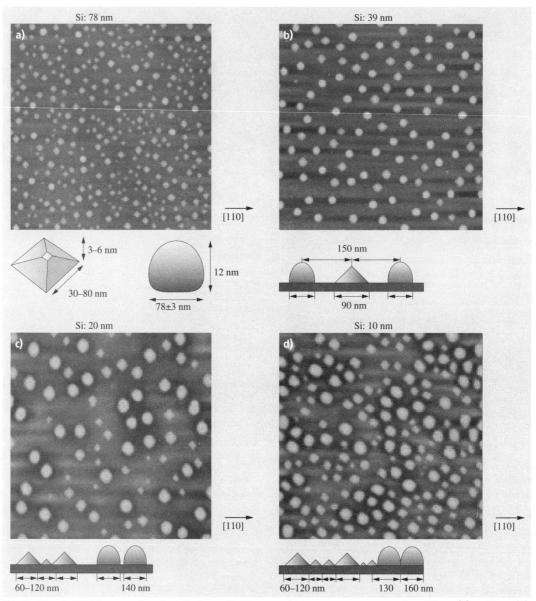

Fig. 34.42a–d AFM images of stacked Ge dots with Si spacer thickness as a parameter, from 78 nm (**a**) to 10 nm (**d**). Scale is $2 \times 2\,\mu m^2$. Subfigures show schematics of formed Ge dots

from pure-Ge dot formation with GeH_4: (1) the critical thickness for dot formation increases to 7.5 ML from about 4 ML, (2) the dots strongly reduce in size and increase in number by as much as three times that of pure-Ge dots and (3) monomodal formation of dome-like dots occurs instead of the bimodal formation of pure-Ge dots. Dots with C provide luminescence as well, but the peak shift as a function of the deposition is different from that of Ge dots grown at the same growth temperature and very similar to the behavior of Ge dots

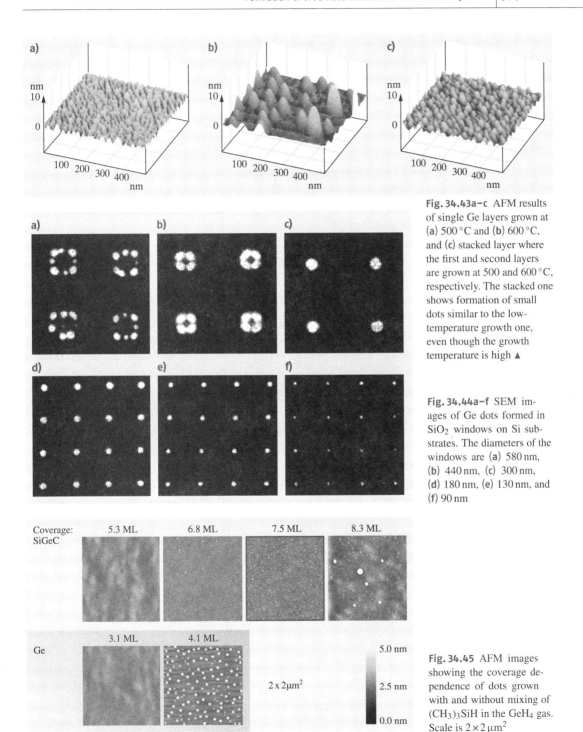

Fig. 34.43a–c AFM results of single Ge layers grown at (**a**) 500 °C and (**b**) 600 °C, and (**c**) stacked layer where the first and second layers are grown at 500 and 600 °C, respectively. The stacked one shows formation of small dots similar to the low-temperature growth one, even though the growth temperature is high ▲

Fig. 34.44a–f SEM images of Ge dots formed in SiO_2 windows on Si substrates. The diameters of the windows are (**a**) 580 nm, (**b**) 440 nm, (**c**) 300 nm, (**d**) 180 nm, (**e**) 130 nm, and (**f**) 90 nm

Fig. 34.45 AFM images showing the coverage dependence of dots grown with and without mixing of $(CH_3)_3SiH$ in the GeH_4 gas. Scale is $2 \times 2\,\mu m^2$

grown at lower temperature. This indicates that migration of atoms is impeded by the presence of C atoms, and therefore the low-temperature growth mode occurs even at higher temperatures. This corresponds well to the small dot formation resulting from the reduction of the migration length of atoms on the surface.

34.10 Concluding Remarks and Future Prospects

The Si/Ge system provides a lot of varieties of materials growth, from pseudomorphic growth and strain-relaxed growth to dot formation, and the strain coming from the lattice mismatch between Si and Ge plays an important role in the growth. Comprehensive studies on the growth mechanisms have resulted in the development of novel formation techniques for SiGe heterostructures and enable us to implement strain effects into Si devices.

It is now widely accepted that the performance of Si VLSIs is greatly improved with the aid of SiGe heterostructures and a lot of work on material growth and device design is being conducted. It is therefore forecast that SiGe heterostructures will be implemented in some important VLSI devices soon.

Recently, much attention has also been paid to the photonic properties of Si-related materials, particularly SiGe heterostructures. This is because the performance of VLSIs is improved by adding optical functions such as optical interconnection and parallel signal processing. In order to realize optical functions in Si VLSIs, various kinds of optical elements based on Si, including light sources, should be developed. Among these, photonic crystals are now attracting much attention, since not only can light emission be well controlled but also light waveguides with high flexibility can be realized. If one can apply the photonic crystal to Si devices, waveguides for optical interconnection may be realized on Si VLSIs. Moreover, by combining VLSIs and sophisticated Si, new devices with optical functions and multiprocessing of signals, which is highly desired for such devices as image processors, can be realized based on Si. In this case, SiGe heterostructures will play the main role, and therefore, more intensive studies on new formation techniques for SiGe heterostructures are required.

References

34.1 C.G. Van de Walle, R.M. Martin: Theoretical calculations of heterojunction discontinuities in the Si/Ge system, Phys. Rev. B **34**, 5621–5634 (1986)

34.2 C.G. Van de Walle, R.M. Martin: Theoretical study of band offsets at semiconductor interfaces, Phys. Rev. B **35**, 8154–8165 (1987)

34.3 C.G. Van de Walle: Band lineups and deformation potentials in the model-solid theory, Phys. Rev. B **39**, 1871–1883 (1989)

34.4 H. Kim, N. Taylor, T.R. Bramblett, J.E. Greene: Kinetics of $Si_{1-x}Ge_x$(001) growth on Si(001) 2×1 by gas-source molecular-beam epitaxy from Si_2H_6 and GeH_4, J. Appl. Phys. **84**, 6372–6381 (1998)

34.5 H. Hirayama, T. Tatsumi, A. Ogura, N. Aizaki: Gas source silicon molecular beam epitaxy using silane, Appl. Phys. Lett. **51**, 2213–2215 (1987)

34.6 H. Hirayama, T. Tatsumi, N. Aizaki: Selective growth condition in disilane gas source silicon molecular beam epitaxy, Appl. Phys. Lett. **52**, 2242–2243 (1988)

34.7 M. Suemitsu, F. Hirose, Y. Takakuwa, N. Miyamoto: Growth kinetics in silane gas-source molecular beam epitaxy, J. Cryst. Growth **105**, 203–208 (1990)

34.8 Y. Tsukidate, M. Suemitsu: Infrared study of SiH_4-adsorbed Si(100) surface: Observation and mode assignment of new peaks, Jpn. J. Appl. Phys. **40**, 5206–5210 (2001)

34.9 H. Hirayama, T. Tatsumi, N. Aizaki: Gas source silicon molecular beam epitaxy using disilane, Appl. Phys. Lett. **52**, 1484–1486 (1988)

34.10 D. Lubben, R. Tsu, T.R. Brambelett, J.E. Greene: Mechanisms and kinetics of Si atomic-layer epitaxy on Si(001) 2×1 from Si_2H_6, J. Vac. Sci. Technol. A **9**, 3003–3011 (1991)

34.11 S.H. Li, S.W. Chung, J.K. Rhee, P.K. Bhattacharya: Gas-source molecular-beam epitaxy using Si_2H_6 and GeH_4 and x-ray characterization of $Si_{1-x}Ge_x$ ($0 < x < 0.33$) alloys, J. Appl. Phys. **71**, 4916–4919 (1992)

34.12 T.R. Bramblett, Q. Lu, T. Karasawa, M.-A. Hasan, S.K. Jo, J.E. Greene: Si(001) 2×1 gas-source molecular-beam epitaxy from Si_2H_6: Growth kinetics and boron doping, J. Appl. Phys. **76**, 1884–1888 (1994)

34.13 R. Chelly, T. Angot, P. Louis, D. Bolmont, J.J. Koulmann: In situ monitoring of growth rate param-

34.13 eters in hot-wire assisted gas source-molecular beam epitaxy using a quartz microbalance, Appl. Surf. Sci. **115**, 299–306 (1997)

34.14 A.M. Lam, Y.-J. Zheng, J.R. Engstrom: Gas-source reactivity in mixed-crystal systems: The reaction of GeH_4 and Ge_2H_6 on Si surfaces, Surf. Sci. **393**, 205–221 (1997)

34.15 T. Murata, M. Suemitsu: GeH_4 adsorption on Si(001) at RT: Transfer of H atoms to Si sites and atomic exchange between Si and Ge, Appl. Surf. Sci. **224**, 179–182 (2004)

34.16 M. Foster, B. Darlington, J. Scharff, A. Campion: Surface chemistry of alkylsilanes on Si(100) 2×1, Surf. Sci. **375**, 35–44 (1997)

34.17 J. Xu, W.J. Choyke, J.T. Yates Jr.: Role of the $-SiH_3$ functional group in silane adsorption and dissociation on Si(100), J. Phys. Chem. B **101**, 6879–6882 (1997)

34.18 K. Senthil, H. Nakazawa, M. Suemitsu: Adsorption and desorption kinetics of organosilanes at Si(001) surfaces, Jpn. J. Appl. Phys. **42**, 6804–6808 (2003)

34.19 S. Gu, R. Wang, R. Zhang, Y. Zheng: Simulation model to very low pressure chemical vapor deposition of SiGe alloy, J. Vac. Sci. Technol. A **14**, 3256–3260 (1996)

34.20 D.J. Robbins, J.L. Glasper, A.G. Cullis, W.L. Leong: A model for heterogeneous growth of $Si_{1-x}Ge_x$ films from hydrides, J. Appl. Phys. **69**, 3729–3732 (1991)

34.21 B.S. Meyerson, K.J. Uram, F.K. LeGoues: Cooperative growth phenomena in silicon/germanium low-temperature epitaxy, Appl. Phys. Lett. **53**, 2555–2557 (1988)

34.22 J. Murota, S. Ono: Low-temperature epitaxial growth of $Si/Si_{1-x}Ge_x/Si$ heterostructure by chemical vapor deposition, Jpn. J. Appl. Phys. **33**, 2290–2299 (1994)

34.23 J. Murota, T. Matsuura, M. Sakuraba: Atomically controlled processing for group IV semiconductors, Surf. Interface Anal. **34**, 423–431 (2002)

34.24 V. Loup, J.M. Hartmann, G. Rolland, P. Holliger, F. Laugier, C. Vannuffel, M.N. Séméria: Reduced pressure chemical vapor deposition of $Si_{1-x-y}Ge_xC_y$ and $Si_{1-y}C_y/Si$ heterostructures, J. Vac. Sci. Technol. B **20**, 1048–1054 (2002)

34.25 J.M. Hartmann, V. Loup, G. Rolland, M.N. Séméria: Effects of temperature and HCl flow on the SiGe growth kinetics in reduced pressure-chemical vapor deposition, J. Vac. Sci. Technol. B **21**, 2524–2529 (2003)

34.26 P.M. Garone, J.C. Sturm, P.V. Schwartz, S.A. Schwartz, B.J. Wilkens: Silicon vapor phase epitaxial growth catalysis by the presence of germane, Appl. Phys. Lett. **56**, 1275–1277 (1990)

34.27 J.L. Hoyt, C.A. King, D.B. Noble, C.M. Gronet, J.F. Gibbons, M.P. Scott, S.S. Laderman, S.J. Roser, K. Kauka, J. Turner, T.I. Kamins: Limited reaction processing: Growth of $Si_{1-x}Ge_x/Si$ for heterojunction bipolar transistor applications, Thin Solid Films **184**, 93–106 (1990)

34.28 T.O. Sedgwick, D.A. Grutzmacher, A. Zaslavsky, V.P. Kesan: Selective SiGe and heavily As doped Si deposited at low temperature by atmospheric pressure chemical vapor deposition, J. Vac. Sci. Technol. B **11**, 1124–1128 (1993)

34.29 S. Bodnar, E. de Berranger, P. Bouillon, M. Mouis, T. Skotnocki, J.L. Regolini: Selective Si and SiGe epitaxial heterostructures grown using an industrial low-pressure chemical vapor deposition module, J. Vac. Sci. Technol. B **15**, 712–718 (1997)

34.30 W.-C. Wang, J.P. Denton, G.W. Neudeck, I.-M. Lee, C.G. Takoudis, M.T.K. Koh, E.P. Kvam: Selective epitaxial growth of $Si_{1-x}Ge_x/Si$ strained-layers in a tubular hot-wall low pressure chemical vapor deposition system, J. Vac. Sci. Technol. B **15**, 138–141 (1997)

34.31 L. Vescan, K. Grimm, C. Dieker: Facet investigation in selective epitaxial growth of Si and SiGe on (001) Si for optoelectronic devices, J. Vac. Sci. Technol. B **16**, 1549–1554 (1998)

34.32 J.J. Harris, D.E. Ashenford, C.T. Foxon, P.J. Dobson, B.A. Joyce: Kinetic limitations to surface segregation during growth of III–V compounds: Sn in GaAs, Appl. Phys. A **33**, 87–92 (1984)

34.33 R.A. Metzger, F.G. Allen: Evaporative antimony doping of silicon during molecular beam epitaxial growth, J. Appl. Phys. **55**, 931–940 (1984)

34.34 S. Fukatsu: Growth of group-IV semiconductor heterostructures with controlled interfaces and observation of band-edge luminescence from strained SiGe/Si quantum wells. Ph.D. Thesis (University of Tokyo, Tokyo 1992)

34.35 K. Fujita, S. Fukatsu, H. Yaguchi, T. Igarashi, Y. Shiraki, R. Ito: Realization of abrupt interfaces in Si/Ge superlattices by suppressing Ge surface segregation with submonolayer of Sb, Jpn. J. Appl. Phys. **29**, L1981–L1983 (1990)

34.36 K. Fujita, S. Fukatsu, H. Yaguchi, Y. Shiraki, R. Ito: Involvement of the topmost Ge layer in the Ge surface segregation during Si/Ge heterostructure formation, Appl. Phys. Lett. **59**, 2240–2241 (1991)

34.37 S. Fukatsu, K. Fujita, H. Yaguchi, Y. Shiraki, R. Ito: Self-limitation in the surface segregation of Ge atoms during Si molecular beam epitaxial growth, Appl. Phys. Lett. **59**, 2103–2105 (1991)

34.38 J. Ushio, K. Nakagawa, M. Miyao, T. Maruizumi: Surface segregation behavior of Ge in comparison with B, Ga, and Sb: Calculations using a first-principles method, J. Cryst. Growth **201/202**, 81–84 (1999)

34.39 S. Fukatsu, N. Usami, Y. Kato, H. Sunamura, Y. Shiraki, H. Oku, T. Ohnishi, Y. Ohmori, K. Okumura: Gas-source molecular beam epitaxy and luminescence characterization of strained $Si_{1-x}Ge_x/Si$ quantum wells, J. Cryst. Growth **136**, 315–321 (1994)

34.40 M. Copel, C. Reuter, E. Kaxiras, R.M. Tromp: Surfactants in epitaxial growth, Phys. Rev. Lett. **63**, 632–635 (1989)

34.41 F.C. Frank, J.H. van der Merwe: One-dimensional dislocations. I. Static theory, Proc. R. Soc. Lond. Ser. A **198**, 205–216 (1949)

34.42 F.C. Frank, J.H. van der Merwe: One-dimensional dislocation. II. Misfitting monolayers and oriented overgrowth, Proc. R. Soc. Lond. Ser. A **198**, 216–225 (1949)

34.43 F.C. Frank, J.H. van der Merwe: One-dimensional dislocation. III. Influence of the second harmonic term in the potential representation on the properties of the model, Proc. R. Soc. Lond. Ser. A **200**, 125–134 (1949)

34.44 J.H. van der Merwe: Crystal interface. Part I. Semi-infinite crystals, J. Appl. Phys. **34**, 117–122 (1963)

34.45 J.H. van der Merwe: Crystal interface. Part II. Finite overgrowths, J. Appl. Phys. **34**, 123–127 (1963)

34.46 E. Kasper, H.-J. Herzog: Elastic and misfit dislocation density in $Si_{0.92}Ge_{0.08}$ films on silicon substrates, Thin Solid Films **44**, 357–370 (1977)

34.47 E. Kasper: Growth and properties of Si/SiGe superlattices, Surf. Sci. **174**, 630–639 (1986)

34.48 E. Kasper, H.-J. Herzog, H. Daembkes, G. Abstreiter: Equally strained Si/SiGe superlattices on Si substrates, Mater. Res. Soc. Proc. **56**, 347–357 (1986)

34.49 J.W. Matthews, A.E. Blakeslee: Defects in epitaxial multilayers. I. Misfit dislocations, J. Cryst. Growth **27**, 118–125 (1974)

34.50 J.W. Matthews, A.E. Blakeslee: Defects in epitaxial multilayers. II. Dislocation pile-ups, threading dislocations, slip lines and cracks, J. Cryst. Growth **29**, 273–280 (1975)

34.51 J.W. Matthews, A.E. Blakeslee: Defects in epitaxial multilayers. III. Preparation of almost perfect multilayers, J. Cryst. Growth **32**, 265–273 (1976)

34.52 J.C. Bean, L.C. Feldman, A.T. Fiory, S. Nakahara, I.K. Robinson: Ge_xSi_{1-x}/Si strained-layer superlattice grown by molecular beam epitaxy, J. Vac. Sci. Technol. A **2**, 436–440 (1984)

34.53 R. People, J.C. Bean: Calculation of critical layer thickness versus lattice mismatch for Ge_xSi_{1-x}/Si strained-layer heterostructures, Appl. Phys. Lett. **47**, 322–324 (1985)

34.54 I.J. Fritz: Role of experimental resolution in measurements of critical layer thickness for strained-layer epitaxy, Appl. Phys. Lett. **51**, 1080–1082 (1987)

34.55 Y. Kohama, Y. Fukuda, M. Seki: Determination of the critical layer thickness of $Si_{1-x}Ge_x$/Si heterostructures by direct observation of misfit dislocations, Appl. Phys. Lett. **52**, 380–382 (1988)

34.56 D.C. Houghton, C.J. Gibbings, C.G. Tuppen, M.H. Lyons, M.A.G. Halliwell: Equilibrium critical thickness for $Si_{1-x}Ge_x$ strained layers on (100) Si, Appl. Phys. Lett. **56**, 460–462 (1990)

34.57 M.L. Green, B.E. Weir, D. Brasen, Y.F. Hsieh, G. Higashi, A. Feygenson, L.C. Feldman, R.L. Headrick: Mechanically and thermally stable Si-Ge films and heterojunction bipolar transistors grown by rapid thermal chemical vapor deposition at 900 °C, J. Appl. Phys. **69**, 745–751 (1991)

34.58 D.C. Houghton, D.D. Perovic, J.-M. Baribeay, G.C. Weatherly: Misfit strain relaxation in Ge_xSi_{1-x}/Si heterostructures: The structural stability of buried strained layers and strained-layer superlattices, J. Appl. Phys. **67**, 1850–1862 (1990)

34.59 J.W. Matthews, S. Mader, T.B. Light: Accommodation of misfit across the interface between crystals of semiconducting elements or compounds, J. Appl. Phys. **41**, 3800–3804 (1970)

34.60 B.A. Fox, W.A. Jesser: The effect of frictional stress on the calculation of critical thickness in epitaxy, J. Appl. Phys. **68**, 2801–2808 (1990)

34.61 T.J. Gosling, S.C. Jain, J.R. Willis, A. Atkinson, R. Bullough: Stable configurations in strained epitaxial layers, Philos. Mag. A **66**, 119–132 (1992)

34.62 A. Fischer, H. Kuhne, M. Eichler, F. Hollander, H. Richter: Strain and surface phenomena in SiGe structures, Phys. Rev. B **54**, 8761–8768 (1996)

34.63 D.J. Eaglesham, M. Cerullo: Dislocation-free Stranski–Krastanov growth of Ge on Si(100), Phys. Rev. Lett. **16**, 1943–1946 (1990)

34.64 Y.-W. Mo, D.E. Savage, B.S. Swartzentruber, M.G. Lagally: Kinetic pathway in Stranski–Krastanov growth of Ge on Si(001), Phys. Rev. Lett. **65**, 1020–1023 (1990)

34.65 A. Sakai, T. Tatsumi: Defect-mediated island formation in Stranski–Krastanov growth of Ge on Si(001), Phys. Rev. Lett. **71**, 4007–4010 (1993)

34.66 A. Sakai, T. Tatsumi: Defect and island formation in Stranski–Krastanov growth of Ge on Si(001), Mater. Res. Soc. Symp. Proc. **317**, 343–348 (1994)

34.67 R.J. Asaro, W.A. Tiller: Interface morphology development during stress-corrosion cracking. Part I. Via surface diffusion, Metall. Mater. Trans. **3**, 1789–1796 (1972)

34.68 M.A. Grinfeld: Instability of the separation boundary between a non-hydrostatically stressed elastic body and a melt, Sov. Phys. Dokl. **31**, 831–834 (1986)

34.69 D.J. Srolovitz: On the stability of surfaces of stressed solids, Acta Metall. **37**, 621–625 (1989)

34.70 B.J. Spencer, P.W. Voorhees, S.H. Davis: Morphological instability in epitaxially strained dislocation-free solid films, Phys. Rev. Lett. **67**, 3696–3699 (1991)

34.71 J.A. Floro, G.A. Lucadamo, E. Chason, L.B. Freund, M. Sinclair, R.D. Twesten, R.Q. Hwang: SiGe island shape transition induced by elastic repulsion, Phys. Rev. Lett. **80**, 4717–4720 (1998)

34.72 G. Medeiros-Ribeiro, A.M. Bratkovski, T.I. Kamins, D.A.A. Ohlberg, R.S. Williams: Shape transition of germanium nanocrystals on a silicon (001) surface from pyramids to domes, Science **279**, 353–355 (1998)

34.73 B.W. Dodson, J.Y. Tsao: Relaxation of strained-layer semiconductor structures via plastic flow, Appl. Phys. Lett. **51**, 1325–1327 (1987)

34.74 H. Alexander, P. Haasen: Dislocations and plastic flow in the diamond structure, Solid State Phys. **22**, 22–158 (1968)

34.75 E. Kasper, H.J. Herzog, H. Kibbel: One-dimensional SiGe superlattice grown by UHV epitaxy, Appl. Phys. **8**, 199–205 (1975)

34.76 R. Hull, J.C. Bean, C. Büscher: A phenomenological description of strain relaxation in Ge_xSi_{1-x}/Si(100) heterostructures, J. Appl. Phys. **66**, 5837–5843 (1989)

34.77 D.D. Perovic, G.C. Whetherly, J.M. Baribeau, D.C. Houghton: Heterogeneous nucleation sources in molecular beam epitaxy-grown $Ge_{1-x}Si_x$/Si strained layer superlattices, Thin Solid Films **183**, 141–156 (1989)

34.78 D.J. Eaglesham, E.P. Kvam, D.M. Maher, C.J. Humphrey, J.C. Bean: Dislocation nucleation near the critical thickness in GeSi/Si strained layers, Philos. Mag. A **59**, 1059–1073 (1989)

34.79 C.J. Humphreys, D.M. Maher, D.J. Eagleshum, E.P. Kvam, I.G. Salisbury: The origin of dislocations in multilayers, J. Phys. III **1**, 1119–1130 (1991)

34.80 V. Higgs, P. Kightley, P. Goodhew, P. Augustus: Metal-induced dislocation nucleation for metastable SiGe/Si, Appl. Phys. Lett. **59**, 829–831 (1991)

34.81 M.D. de Coteau, P.R. Wilshaw, R. Falster: Gettering of copper in silicon – precipitation at extended surface-defects, Inst. Phys. Conf. Ser. **117**, 231–234 (1991)

34.82 R. Hull, J.C. Bean: Nucleation of misfit dislocations in strained-layer epitaxy in the Ge_xSi_{1-x}/Si system, J. Vac. Sci. Technol. A **7**, 2580–2585 (1989)

34.83 U. Jain, S.C. Jain, A.H. Harker, R. Bullough: Nucleation of dislocation loops in strained epitaxial layers, J. Appl. Phys. **77**, 103–109 (1995)

34.84 D.E. Jesson, S.J. Pennycook, J.-M. Bribeau, D.C. Houghton: Surface stress, morphological development, and dislocation nucleation during $Si_{1-x}Ge_x$ epitaxy, Scanning Microsc. **8**, 849–857 (1994)

34.85 D.D. Perovic, D.C. Houghton: The introduction of dislocations in low misfit epitaxial systems, Microsc. Semicond. Mater. 1995, Inst. Phys. Conf. Ser. **146**, 117–134 (1995)

34.86 E.A. Stach, R. Hull, R.M. Tromp, M.C. Reuter, M. Copel, F.K. LeGoues, J.C. Bean: Effect of the surface upon misfit dislocation velocities during the growth and annealing of SiGe/Si (001) heterostructures, J. Appl. Phys. **83**, 1931–1937 (1998)

34.87 D.C. Houghton, C.J. Gibbings, C.G. Tuppen, M.H. Lyons, M.A.G. Halliwell: The structural stability of uncapped versus buried $Si_{1-x}Ge_x$ strained layers through high temperature processing, Thin Solid Films **183**, 171–182 (1989)

34.88 D.C. Houghton: Strain relaxation kinetics in $Si_{1-x}Ge_x$/Si heterostructures, J. Appl. Phys. **70**, 2136–2151 (1991)

34.89 H. Hull, J.C. Bean, D. Bahnck, L.J. Peticolas, K.T. Short, F.C. Unterwald: Interpretation of dislocation propagation velocities in strained Ge_xSi_{1-x}/Si(100) heterostructures by the diffusive kink pair model, J. Appl. Phys. **70**, 2052–2065 (1991)

34.90 Y. Yamashita, K. Maeda, K. Fujita, N. Usami, K. Suzuki, S. Fukatsu, Y. Mera, Y. Shiraki: Dislocation glide motion in heteroepitaxial thin-films of $Si_{1-x}Ge_x$/Si(100), Philos. Mag. Lett. **67**, 165–171 (1993)

34.91 R. Hull, J.C. Bean: New insights into the microscopic motion of dislocations in covalently bonded semiconductors by in-situ transmission electron microscope observations of misfit dislocations in thin strained epitaxial layers, Phys. Status Solidi (a) **138**, 533–546 (1993)

34.92 W. Hagen, H. Strunk: New type of source generating misfit dislocations, Appl. Phys. **17**, 85–87 (1978)

34.93 R. Hull, J.C. Bean, D.J. Eaglesham, J.N. Bonara, C. Büscher: Strain relaxation phenomena in Ge_xSi_{1-x}/Si strained structures, Thin Solid Films **183**, 117–132 (1989)

34.94 A. Lefebvre, C. Herbeaux, J. Di Persio: Interactions of misfit dislocations in $In_xGa_{1-x}As$/GaAs interfaces, Philos. Mag. A **63**, 471–485 (1991)

34.95 F.K. LeGoues, B.S. Meyerson, J.F. Morar, P.D. Kirchner: Mechanism and conditions for anomalous strain relaxation in grade thin films and superlattices, J. Appl. Phys. **71**, 4230–4243 (1992)

34.96 J. Washburn, E.P. Kvam: Possible dislocation multiplication source in (001) semiconductor epitaxy, Appl. Phys. Lett. **57**, 1637–1639 (1990)

34.97 C.G. Tuppen, C.J. Gibbings, M. Hockly, S.G. Roberts: Misfit dislocation multiplication processes in $Si_{1-x}Ge_x$ alloys for $x < 0.15$, Appl. Phys. Lett. **56**, 54–56 (1990)

34.98 Y.J. Mii, Y.-H. Xie, E.A. Fitzgerald, F.B.E. Weir, L.C. Feldman: Extremely high electron mobility in Si/Ge_xSi_{1-x} structures grown by molecular beam epitaxy, Appl. Phys. Lett. **59**, 1611–1613 (1991)

34.99 J.J. Welser, J.L. Hoyt, S. Takagi, J.F. Gibbons: Strain dependence of the performance enhancement in strained-Si n-MOSFETs, Tech. Dig. Int. Electron Device Meet. (1994) pp. 373–376

34.100 K. Ismail, M. Arafa, K.L. Saenger, J.O. Chu, B.S. Meyerson: Extremely high electron mobility in Si/SiGe modulation-doped heterostructures, Appl. Phys. Lett. **66**, 1077–1079 (1995)

34.101 A.C. Churchill, D.J. Robbins, D.J. Wallis, N. Griffin, D.J. Paul, A.J. Pidduck: High-mobility two-dimensional electron gases in Si/SiGe heterostructures on relaxed SiGe layers grown at high temperature, Semicond. Sci. Technol. **12**, 943–946 (1997)

34.102 S. Bozzo, J.-L. Lazzari, C. Coudreau, A. Ronda, F. Arnaud d'Avitaya, J. Derrien, S. Mesters, B. Holländer, P. Gergaud, O. Thomas: Chemical vapor deposition of silicon-germanium heterostructures, J. Cryst. Growth **216**, 171–184 (2000)

34.103 K.H. Chang, R. Gibala, D.J. Srolovitz, P.K. Bhattacharya, J.F. Mansfield: Crosshatched surface morphology in strained III–V semiconductor films, J. Appl. Phys. **67**, 4093–4098 (1990)

34.104 A.J. Pidduck, D.J. Robbins, A.G. Cullis, W.Y. Leong, A.M. Pitt: Evolution of surface morphology and strain during SiGe epitaxy, Thin Solid Films **222**, 78–84 (1992)

34.105 J.W.P. Hsu, E.A. Fitzgerald, Y.-H. Xie, P.J. Silverman, M.J. Cardillo: Surface morphology of related Ge_xSi_{1-x} films, Appl. Phys. Lett. **61**, 1293–1295 (1992)

34.106 E.A. Fitzgerald, Y.-H. Xie, D. Monroe, P.J. Silverman, J.M. Kuo, A.R. Kortan, F.A. Thiel, B.E. Weir: Relaxed Ge_xSi_{1-x} structures for III–V integration with Si and high mobility two-dimensional electron gasses in Si, J. Vac. Sci. Technol. B **10**, 1807–1819 (1992)

34.107 S.Y. Shiryaev, F. Jensen, J.W. Petersen: On the nature of cross-hatch patterns on compositionally graded $Si_{1-x}Ge_x$ alloy layers, Appl. Phys. Lett. **64**, 3305–3307 (1994)

34.108 M.A. Lutz, R.M. Feenstra, F.K. LeGoues, P.M. Mooney, J.O. Chu: Influence of misfit dislocations on the surface morphology of $Si_{1-x}Ge_x$ films, Appl. Phys. Lett. **66**, 724–726 (1995)

34.109 A.M. Andrews, J.S. Speck, A.E. Romanov, M. Bobeth, W. Pompe: Modeling cross-hatch surface morphology in growing mismatched layers, J. Appl. Phys. **91**, 1933–1943 (2002)

34.110 E.A. Fitzgerald, S.B. Samavedam: Line, point and surface defect morphology of graded, relaxed GeSi alloys of Si substrates, Thin Solid Films **294**, 3–10 (1997)

34.111 C.W. Leitz, M.T. Currie, A.Y. Kim, J. Lai, E. Robbins, E.A. Fitzgerald, M.T. Bulsara: Dislocation glide and blocking kinetics in compositionally graded SiGe/Si, J. Appl. Phys. **90**, 2730–2736 (2001)

34.112 H. Chen, L.W. Guo, Q. Cui, Q. Hu, Q. Huang, J.M. Zhou: Low-temperature buffer layer for growth of a low-dislocation-density SiGe layer on Si by molecular-beam epitaxy, J. Appl. Phys. **79**, 1167–1169 (1996)

34.113 J.H. Li, C.S. Peng, Y. Wu, D.Y. Dai, J.M. Zhou, Z.H. Mai: Relaxed $Si_{0.7}Ge_{0.3}$ layers grown on low-temperature Si buffers with low threading dislocation density, Appl. Phys. Lett. **71**, 3132–3134 (1997)

34.114 T. Ueno, T. Irisawa, Y. Shiraki: p-type Ge channel modulation doped heterostructures with very high room-temperature mobilities, Physica E **7**, 790–794 (2000)

34.115 T. Ueno, T. Irisawa, Y. Shiraki, A. Uedono, S. Tanigawa, R. Suzuki, T. Ohdaira, T. Mikado: Characterization of low temperature grown Si layer for SiGe pseudo-substrates by positron annihilation spectroscopy, J. Cryst. Growth **227–228**, 761–765 (2001)

34.116 K. Sawano, K. Kawaguchi, T. Ueno, S. Koh, K. Nakagawa, Y. Shiraki: Surface smoothing of SiGe strain-relaxed buffer layers by chemical mechanical polishing, Mater. Sci. Eng. B **89**, 406–409 (2002)

34.117 K. Sawano, Y. Hirose, S. Koh, K. Nakagawa, T. Hattori, Y. Shiraki: Mobility enhancement in strained Si modulation-doped structures by chemical mechanical polishing, Appl. Phys. Lett. **82**, 412–414 (2003)

34.118 K. Sawano, K. Arimoto, Y. Hirose, S. Koh, N. Usami, K. Nakagawa, T. Hattori, Y. Shiraki: Planarization of SiGe virtual substrates by CMP and its application to strained Si modulation-doped structures, J. Cryst. Growth **251**, 693 (2003)

34.119 K. Sawano, K. Kawaguchi, S. Koh, Y. Hirose, T. Hattori, K. Nakagawa, Y. Shiraki: Surface planarization of strain-relaxed SiGe buffer layers by CMP and post cleaning, J. Electrochem. Soc. **150**, G376–G379 (2003)

34.120 K. Sawano, Y. Abe, H. Satoh, K. Nakagawa, Y. Shiraki: Mobility enhancement in strained Ge heterostructures by planarization of SiGe buffer layers grown on Si substrates, Jpn. J. Appl. Phys. **44**, L1320–L1322 (2005)

34.121 K. Sawano, N. Usami, K. Arimoto, S. Koh, K. Nakagawa, Y. Shiraki: Observation of strain field fluctuation in SiGe relaxed buffer layers and its influence on overgrown structures, Mater. Sci. Semicon. Process. **8**, 177–180 (2005)

34.122 R. Hull, J.C. Bean, J.M. Bonar, G.S. Higashi, K.T. Short, H. Temkin, A.E. White: Enhanced strain relaxation in $Si/Ge_xSi_{1-x}/Si$ heterostructures via point-defect concentrations introduced by ion implantation, Appl. Phys. Lett. **56**, 2445–2447 (1990)

34.123 B. Holländer, S. Mantl, R. Liedtke, S. Mesters, H.-J. Herzog, H. Kibbel, T. Hackbarth: Enhanced strain relaxation of epitaxial SiGe layers on Si(100) after H^+ ion implantation, Nucl. Instrum. Methods B **148**, 200–205 (1999)

34.124 H. Trinkaus, B. Holländer, S. Rongen, S. Mantl, H.-J. Herzog, J. Kuchenbecker, T. Hackbarth: Strain relaxation mechanism for hydrogen-implanted $Si_{1-x}Ge_x/Si(100)$ heterostructures, Appl. Phys. Lett. **76**, 3552–3554 (2000)

34.125 M. Luysberg, D. Kirch, H. Trinkaus, B. Holländer, S. Lenk, S. Mantl, H.-J. Herzog, T. Hackbarth, P.F.P. Fichtner: Effect of helium ion implantation and annealing on the relaxation behavior of pseudomorphic $Si_{1-x}Ge_x$ buffer layers on Si (100) substrates, J. Appl. Phys. **92**, 4290–4295 (2002)

34.126 L.-F. Zou, Z.G. Wang, D.Z. Sun, T.W. Fan, X.F. Liu, J.W. Zhang: Characterization of strain relaxation in As ion implanted $Si_{1-x}Ge_x$ epilayers grown by gas

34.126 source molecular beam epitaxy, Appl. Phys. Lett. **72**, 845–847 (1998)

34.127 K. Sawano, Y. Hirose, S. Koh, K. Nakagawa, T. Hattori, Y. Shiraki: Relaxation enhancement of SiGe thin layers by ion implantation into Si substrates, J. Cryst. Growth **251**, 685–688 (2003)

34.128 K. Sawano, Y. Hirose, Y. Ozawa, S. Koh, J. Yamanaka, K. Nakagawa, T. Hattori, Y. Shiraki: Enhancement of strain relaxation of SiGe thin layers by pre-ion-implantation into Si substrates, Jpn. J. Appl. Phys. **42**, L735–L737 (2003)

34.129 K. Sawano, Y. Hirose, S. Koh, K. Nakagawa, T. Hattori, Y. Shiraki: Formation of thin SiGe virtual substrates by ion implantation into Si substrates, Appl. Surf. Sci. **224**, 99–103 (2004)

34.130 K. Sawano, S. Koh, Y. Shiraki, Y. Ozawa, T. Hattori, J. Yamanaka, K. Suzuki, K. Arimoto, K. Nakagawa, N. Usami: Fabrication of high-quality strain-relaxed thin SiGe layers on ion-implanted Si substrates, Appl. Phys. Lett. **85**, 2514–2516 (2004)

34.131 K. Sawano, Y. Ozawa, A. Fukumoto, N. Usami, J. Yamanaka, K. Suzuki, K. Arimoto, K. Nakagawa, Y. Shiraki: Strain-field evaluation of strain-relaxed thin SiGe layers fabricated by ion implantation method, Jpn. J. Appl. Phys. **44**, L1316–L1319 (2005)

34.132 T. Tezuka, N. Sugiyama, S. Takagi: Fabrication of strained Si on an ultrathin SiGe-on-insulator virtual substrate with a high-Ge fraction, Appl. Phys. Lett. **79**, 1798–1800 (2001)

34.133 T. Mizuno, N. Sugiyama, T. Tezuka, S. Takagi: Relaxed SiGe-on-insulator substrates without thick SiGe buffer layers, Appl. Phys. Lett. **80**, 601–603 (2002)

34.134 T. Tezuka, N. Sugiyama, S. Takagi, T. Kawakubo: Dislocation-free formation of relaxed SiGe-on-insulator layers, Appl. Phys. Lett. **80**, 3560–3562 (2002)

34.135 N. Sugii, S. Yamaguchi, K. Washio: SiGe-on-insulator substrate fabricated by melt solidification for a strained-silicon complementary metal-oxide-semiconductor, J. Vac. Sci. Technol. B **20**, 1891–1896 (2002)

34.136 T. Mizuno, N. Sugiyama, T. Tezuka, T. Numata, S. Takagi: High performance CMOS operation of strained-SOI MOSFETs using thin film SiGe-on-insulator substrate, Dig. Tech. Pap. Symp. VLSI Technology 2002, pp. 106–107

34.137 S. Nakaharai, T. Tetsuka, N. Sugiyama, Y. Moriyama, S. Takagi: Characterization of 7-nm-thick strained Ge-on-insulator layer fabricated by Ge-condensation technique, Appl. Phys. Lett. **83**, 3516–3518 (2003)

34.138 T. Egawa, A. Sakai, T. Yamamoto, N. Taoka, O. Nakatsuka, S. Zaima, Y. Yasuda: Strain-relaxation mechanisms of SiGe layers formed by two-step growth on Si(001) substrates, Appl. Surf. Sci. **224**, 104–107 (2004)

34.139 A. Sakai, K. Sugimoto, T. Yamamoto, M. Okada, H. Ikeda, Y. Yasuda, S. Zaima: Reduction of threading dislocation density in SiGe layers on Si(001) using a two-step strain-relaxation procedure, Appl. Phys. Lett. **79**, 3398–3400 (2001)

34.140 A. Sakai, T. Tatsumi, K. Aoyama: Growth of strain-relaxed Ge films on Si(001) surfaces, Appl. Phys. Lett. **71**, 3510–3512 (1997)

34.141 N. Ikarashi, T. Tatsumi: Suppression of surface roughening on strained Si/SiGe layers by lowering surface stress, Jpn. J. Appl. Phys. **36**, L377–L379 (1997)

34.142 Y.H. Xie, S.B. Samavedam, M. Bulsara, T.A. Langdo, E.A. Fitzgerald: Relaxed template for fabricating regularly distributed quantum dot arrays, Appl. Phys. Lett. **71**, 3567–3568 (1997)

34.143 D.E. Jesson, S.J. Pennycook, J.-M. Baribeau, D.C. Houghton: Direct imaging of surface cusp evolution during strained-layer epitaxy and implications for strain relaxation, Phys. Rev. Lett. **71**, 1744–1747 (1993)

34.144 P.M. Mooney, F.K. LeGoues, J.O. Chu, S.F. Nelson: Strain relaxation and mosaic structure in relaxed SiGe layers, Appl. Phys. Lett. **62**, 3464–3466 (1993)

34.145 N. Taoka, A. Sakai, T. Egawa, O. Nakatsuka, S. Zaima, Y. Yasuda: Growth and characterization of strain-relaxed SiGe buffer layers on Si(001) substrates with pure-edge misfit dislocations, Mater. Sci. Semicond. Process. **8**, 131–135 (2005)

34.146 T. Yamamoto, A. Sakai, T. Egawa, N. Taoka, O. Nakatsuka, S. Zaima, Y. Yasuda: Dislocation structures and strain-relaxation in SiGe buffer layers on Si (001) substrates with an ultra-thin Ge interlayer, Appl. Surf. Sci. **224**, 108–112 (2004)

34.147 P.F. Fewster: *X-Ray Scattering from Semiconductors* (Imperial College Press, World Scientific, Singapore 2000)

34.148 S.S. Iyer, K. Eberl, M.S. Goorsky, F.K. LeGoues, J.C. Tsang, F. Cardone: Synthesis of $Si_{1-y}C_y$ alloys by molecular beam epitaxy, Appl. Phys. Lett. **60**, 356–358 (1992)

34.149 K. Eberl, S.S. Iyer, J.C. Tsang, M.S. Goorsky, F.K. LeGoues: The growth and characterization of $Si_{1-y}C_y$ alloys on Si(001) substrate, J. Vac. Sci. Technol. B **10**, 934–936 (1992)

34.150 K. Eberl, S.S. Iyer, S. Zollner, J.C. Tsang, F.K. LeGoues: Growth and strain compensation effects in the ternary $Si_{1-x-y}Ge_xC_y$ alloy system, Appl. Phys. Lett. **60**, 3033–3035 (1992)

34.151 B. Cordero, V. Gomez, A.E. Platero-Prats, M. Reves, J. Echeverria, E. Cremades, F. Barragan, S. Alvarez: Covalent radii revisited, J. Chem. Dalton Trans., 2832–2838 (2008)

34.152 H.-J. Osten, E. Bugiel, P. Zaumseil: Growth of an inverse tetragonal distorted SiGe layer on Si(001) by adding small amounts of carbon, Appl. Phys. Lett. **64**, 3440–3442 (1994)

34.153 P.C. Kelires: Monte Carlo studies of ternary semiconductor alloys: Application to the $Si_{1-x-y}Ge_xC_y$ system, Phys. Rev. Lett. **75**, 1114–1117 (1995)

34.154 P.C. Kelires: Short-range order, bulk moduli, and physical trends in c-$Si_{1-x}C_x$ alloys, Phys. Rev. B **55**, 8784–8787 (1997)

34.155 M. Berti, D. De Salvador, A.V. Drigo, F. Romanato, J. Stangl, S. Zerlauth, F. Schäffler, G. Bauer: Lattice parameter in $Si_{1-y}C_y$ epilayers: Deviation from Vegard's rule, Appl. Phys. Lett. **72**, 1602–1604 (1998)

34.156 H.-J. Osten, B. Heinemann, D. Knoll, G. Lippert, H. Rücker: Effects of carbon on boron diffusion in SiGe: Principles and impact on bipolar devices, J. Vac. Sci. Technol. B **16**, 1750–1753 (1998)

34.157 H. Rücker, B. Heinemann, K.D. Bolze, D. Knoll, D. Krüger, R. Kurps, H.-J. Osten, P. Schley, B. Tillack, P. Zaumseil: Dopant diffusion in C-doped Si and SiGe: Physical model and experimental verification, Tech. Dig. Int. Electron Device Meet. (1999) pp. 345–348

34.158 A. Biswas, P.K. Basu: Estimated effect of germanium and carbon on the early voltage of a $Si_{1-x-y}Ge_xC_y$ heterojunction bipolar transistor, Semicond. Sci. Technol. **16**, 947–953 (2001)

34.159 K. Oda, E. Ohue, I. Suzumura, R. Hayami, A. Kodama, H. Simamoto, K. Washio: High performance self-aligned SiGeC HBT with selectively grown $Si_{1-x-y}Ge_xC_y$ base by UHV/CVD, IEEE. Trans. Electron. Dev. **50**, 2213–2220 (2003)

34.160 P. Boucaud, C. Francis, F.H. Julien, J.-M. Lourtioz, D. Bouchier, S. Bodnar, B. Lambert, J.L. Regolini: Band-edge and deep level photoluminescence of pseudomorphic $Si_{1-x-y}Ge_xC_y$ alloys, Appl. Phys. Lett. **64**, 875–877 (1994)

34.161 K. Brunner, W. Winter, K. Eberl: Spatially indirect radiative recombination of carriers localized in $Si_{1-x-y}Ge_xC_y/Si_{1-y}C_y$ double quantum well structure on Si substrates, Appl. Phys. Lett. **69**, 1279–1281 (1996)

34.162 B.A. Orner, J. Olowolafe, K. Roe, J. Kolodzey, T. Laursen, J.W. Mayer, J. Spear: Band gap of Ge rich $Si_{1-x-y}Ge_xC_y$ alloys, Appl. Phys. Lett. **69**, 2557–2559 (1996)

34.163 B.A. Orner, J. Kolodzey: $Si_{1-x-y}Ge_xC_y$ alloy band structures by linear combination of atomic orbitals, J. Appl. Phys. **81**, 6773–6780 (1997)

34.164 B.L. Stein, E.T. Yu, E.T. Croke, A.T. Hunter, T. Laursen, A.E. Bair, J.W. Mayer, C.C. Ahn: Band offsets in $Si/Si_{1-x-y}Ge_xC_y$ heterojunctions measures by admittance spectroscopy, Appl. Phys. Lett. **70**, 3413–3415 (1997)

34.165 O.G. Schmidt, K. Eberl: Photoluminescence of tensile strained, exactly strain compensated, and compressively strained $Si_{1-x-y}Ge_xC_y$ layers in Si, Phys. Rev. Lett. **80**, 3396–3399 (1998)

34.166 D.V. Singh, K. Rim, T.O. Mitchell, J.L. Hoyt, J.F. Gibbons: Admittance spectroscopy analysis of the conduction band offsets in $Si/Si_{1-x-y}Ge_xC_y$ and $Si/Si_{1-y}C_y$ heterostructures, J. Appl. Phys. **85**, 985–993 (1999)

34.167 C.L. Chang, L.P. Rokhinson, J.C. Sturm: Direct optical measurement of the valence band offset of $p^+Si_{1-x-y}Ge_xC_y/p^-Si(100)$ by heterojunction internal photoemission, Appl. Phys. Lett. **73**, 3568–3570 (1998)

34.168 H.-J. Osten: Band-gap changes and band offsets for ternary $Si_{1-x-y}Ge_xC_y$ alloys on Si(001), J. Appl. Phys. **84**, 2716–2721 (1998)

34.169 K. Brunner, O.G. Schmidt, W. Winter, K. Eberl, M. Glück, U. König: SiGeC: Band gaps, band offsets, optical properties, and potential applications, J. Vac. Phys. Technol. B **16**, 1701–1706 (1998)

34.170 S. Galdin, P. Dollfus, V.A. Fortuna, P. Hesto, H.J. Osten: Band offset predictions for strained group IV alloys: $Si_{1-x-y}Ge_xC_y$ on Si(001) and $Si_{1-x}Ge_x$ on $Si_{1-z}Ge_z$(001), Semicond. Sci. Technol. **15**, 565–572 (2000)

34.171 S. John, S.K. Ray, E. Quinones, S.K. Oswal, K. Banerjee: Heterostructure P-channel metal–oxide–semiconductor transistor utilizing a $Si_{1-x-y}Ge_xC_y$ channel, Appl. Phys. Lett. **74**, 847–849 (1999)

34.172 E. Cassan, P. Dollfus, S. Galdin: Effect of doping profile on the potential performance of buried channel SiGeC/Si heterostructure MOS devices, Physica E **13**, 957–960 (2002)

34.173 M. Glück, U. König, W. Winter, K. Brunner, K. Eberl: Modulation-doped $Si_{1-x-y}Ge_xC_y$ p-type heteroFETs, Physica E **2**, 768–771 (1998)

34.174 R.A. Soref: Silicon-based group IV heterostructures for optoelectronic applications, J. Vac. Sci. Technol. A **14**, 913–918 (1996)

34.175 A.S. Amour, L.D. Lanzerotti, C.L. Chang, J.C. Sturm: Optical and electrical properties of $Si_{1-x-y}Ge_xC_y$ thin films and devices, Thin Solid Films **294**, 112–117 (1997)

34.176 O. Madelung (Ed.): *Semiconductors-Basic Data*, 2nd edn. (Springer, Berlin 1996) p. 22

34.177 M. Okinaka, Y. Hamana, T. Tokuda, J. Ohta, M. Nunoshita: MBE growth mode and C incorporation of GeC epilayers on Si(001) substrates using an arc plasma gun as a novel C source, J. Cryst. Growth **249**, 78–86 (2003)

34.178 M.W. Dashiell, L.V. Kulik, D.A. Hits, J. Kolodzey, G. Watson: Carbon incorporation in $Si_{1-y}C_y$ alloys grown by molecular beam epitaxy using a single silicon-graphite source, Appl. Phys. Lett. **72**, 833–835 (1998)

34.179 J.P. Liu, H.-J. Osten: Substitutional carbon incorporation during $Si_{1-x-y}Ge_xC_y$ growth on Si(100) by molecular-beam epitaxy: Dependence on germanium and carbon, Appl. Phys. Lett. **76**, 3546–3548 (2000)

34.180 Y. Kanzawa, K. Nozawa, T. Saitoh, M. Kubo: Dependence of substitutional C incorporation on Ge content for $Si_{1-x-y}Ge_xC_y$ crystals grown by ultrahigh vacuum chemical vapor deposition, Appl. Phys. Lett. **77**, 3962–3964 (2000)

34.181 V. LeThanh, C. Calmes, Y. Zheng, D. Bouchier, V. Fortuna, J.-C. Dupuy: In situ RHEED monitoring of carbon incorporation during SiGeC/Si(001) growth in a UHV-CVD system, Mater. Sci. Eng. B **89**, 246–251 (2002)

34.182 J. Mi, P. Warren, P. Letourneau, M. Judelewicz, M. Gailhanou, M. Dutoit, C. Dubois, J.C. Dupuy: High quality $Si_{1-x-y}Ge_xC_y$ epitaxial layers grown on (100) Si by rapid thermal chemical vapor deposition using methylsilane, Appl. Phys. Lett. **67**, 259–261 (1995)

34.183 W.K. Choi, J.H. Chen, L.K. Bera, W. Feng, K.L. Pey, J. Mi, C.Y. Yang, A. Ramam, S.J. Chua, J.S. Pan, A.T.S. Wee, R. Liu: Structural characterization of rapid thermal oxidized $Si_{1-x-y}Ge_xC_y$ alloy films grown by rapid thermal chemical vapor deposition, J. Appl. Phys. **87**, 192–197 (2000)

34.184 V. Loup, J.M. Hartmann, G. Rolland, P. Holliger, F. Laugier, M.N. Séméria: Growth temperature dependence of substitutional carbon incorporation in SiGeC/Si heterostructures, J. Vac. Sci. Technol. B **21**, 246–253 (2003)

34.185 J. Tersof: Enhanced solubility of impurities and enhanced diffusion near crystal surfaces, Phys. Rev. Lett. **74**, 5080–5083 (1995)

34.186 H. Rücker, M. Methfessel, E. Bugiel, H.-J. Osten: Strain-stabilized highly concentrated pseudomorphic $Si_{1-x}C_x$ layers in Si, Phys. Rev. Lett. **72**, 3578–3581 (1994)

34.187 P.C. Kelires: Theoretical investigation of the equilibrium surface structure of $Si_{1-x-y}Ge_xC_y$ alloys, Surf. Sci. **418**, L62–L67 (1998)

34.188 H. Jacobson, J. Xiang, N. Herbots, S. Whaley, P. Ye, S. Hearne: Heteroepitaxial properties of $Si_{1-x-y}Ge_xC_y$ on Si(001) grown by combined ion- and molecular-beam deposition, J. Appl. Phys. **81**, 3081–3091 (1997)

34.189 A. Sakai, Y. Torige, M. Okada, H. Ikeda, Y. Yasuda, S. Zaima: Atomistic evolution of $Si_{1-x-y}Ge_xC_y$ thin films on Si(001) surfaces, Appl. Phys. Lett. **79**, 3242–3244 (2001)

34.190 O.G. Schmidt, C. Lange, K. Eberl, O. Kienzle, F. Ernst: Formation of carbon-induced germanium dots, Appl. Phys. Lett. **71**, 2340–2342 (1997)

34.191 O. Leifeld, A. Beyer, E. Müller, K. Kern, D. Grützmacher: Formation and ordering effects of C-induced Ge dots grown on Si(001) by molecular beam epitaxy, Mater. Sci. Eng. B **74**, 222–228 (1999)

34.192 R.I.G. Uhrberg, J.E. Northrup, D.K. Biegelsen, R.D. Bringans, L.-E. Swartz: Atomic structure of the metastable $c(4 \times 4)$ reconstruction of Si(100), Phys. Rev. B **46**, 10251–10256 (1992)

34.193 H. Nörenberg, G.A.D. Briggs: The Si(001) $c(4 \times 4)$ surface reconstruction: a comprehensive experimental study, Surf. Sci. **430**, 154–164 (1999)

34.194 O. Leifeld, D. Grützmacher, B. Müller, K. Kern, E. Kaxiras, P.C. Kelires: Dimer pairing on the C-alloyed Si(001) surface, Phys. Rev. Lett. **82**, 972–975 (1999)

34.195 S. Ariyoshi, S. Takeuchi, O. Nakatsuka, A. Sakai, S. Zaima, Y. Yasuda: Influence of $Si_{1-x}Ge_x$ interlayer on the initial growth of SiGeC on Si(100), Appl. Surf. Sci. **224**, 117–121 (2004)

34.196 S. Zaima, A. Sakai, Y. Yasuda: Control in the initial growth stage of heteroepitaxial $Si_{1-x-y}Ge_xC_y$ on Si(001) substrates, Appl. Surf. Sci. **212–213**, 184–192 (2003)

34.197 S. Takeuchi, O. Nakatsuka, Y. Wakazono, A. Sakai, S. Zaima, Y. Yasuda: Initial growth behaviors of SiGeC in SiGe and C alternate deposition, Mater. Sci. Semicond. Process. **6**, 5–9 (2005)

34.198 G. Ehrlich, F. Hudda: Atomic view of surface self-diffusion: Tungsten on tungsten, J. Chem. Phys. **44**, 1039–1049 (1966)

34.199 R.L. Schwoebel, E.J. Shipsey: Step motion on crystal surfaces, J. Appl. Phys. **37**, 3682–3686 (1966)

34.200 J. Mysliveček, C. Schelling, G. Springholz, F. Schäffler, B. Voigtländer, P. Šmilauer: On the origin of the kinetic growth instability of homoepitaxy on Si(001), Mater. Sci. Eng. B **89**, 410–414 (2002)

34.201 S. Fukatsu, H. Yoshida, N. Usami, A. Fujiwara, T. Takahashi, Y. Shiraki, R. Ito: Quantum size effect of excitonic band-edge luminescence in strained $Si_{1-x}Ge_x$/Si single quantum well structures grown by gas-source Si molecular beam epitaxy, Jpn. J. Appl. Phys. **31**, L1319–L1321 (1992)

34.202 D.K. Nayak, N. Usami, S. Fukatsu, Y. Shiraki: Band-edge photoluminescence of SiGe/strained-Si/SiGe type-II quantum wells on Si(100), Appl. Phys. Lett. **63**, 3509–3511 (1993)

34.203 N. Usami, Y. Shiraki, S. Fukatsu: Spectroscopic study of Si-based quantum wells with neighbouring confinement structure, Semicond. Sci. Technol. **12**, 1596–1602 (1997)

34.204 S. Fukatsu, Y. Shiraki: Interwell coupling in strained $Si_{1-x}Ge_x$/Si quantum wells, Ext. Abs. 1993 Int. Conf. Solid State Devices Mater. (Makuhari, 1993) p. 895

34.205 S. Fukatsu: Luminescence investigation on strained $Si_{1-x}Ge_x$/Si modulated quantum wells, Solid-State Electron. **37**, 817–823 (1994)

34.206 K. Kawaguchi, Y. Shiraki, N. Usami, J. Zhang, N.J. Woods, G. Breton, G. Parry: Fabrication of strain-balanced $Si/Si_{1-x}Ge_x$ multiple quantum wells on $Si_{1-y}Ge_y$ virtual substrates and their optical properties, Appl. Phys. Lett. **79**, 344–346 (2001)

34.207 K. Kawaguchi, S. Koh, Y. Shiraki, J. Zhang: Fabrication of strain-balanced $Si_{0.73}Ge_{0.27}$/Si distributed Bragg reflectors on Si substrates, Appl. Phys. Lett. **79**, 476–478 (2001)

34.208 K. Kawaguchi, S. Koh, Y. Shiraki, J. Zhang: Fabrication of strain-balanced $Si_{0.73}Ge_{0.27}$/Si-distributed Bragg reflectors on Si substrates for optical device applications, Physica E **13**, 1051–1054 (2002)

34.209 K. Kawaguchi, K. Konishi, S. Koh, Y. Shiraki, Y. Kaneko, J. Zhang: Optical properties of strain-balanced $Si_{0.73}Ge_{0.27}$ planar microcavities on Si substrates, Jpn. J. Appl. Phys. **41**, 2664–2667 (2002)

34.210 K. Kawaguchi, M. Morooka, K. Konishi, S. Koh, Y. Shiraki: Optical properties of strain-balanced SiGe planar microcavities with Ge dots on Si substrates, Appl. Phys. Lett. **81**, 817–819 (2000)

34.211 N. Usami, T. Mine, S. Fukatsu, Y. Shiraki: Realization of crescent-shaped SiGe quantum wire structures on a V-groove patterned Si substrate by gas-source Si molecular beam epitaxy, Appl. Phys. Lett. **63**, 2789–2791 (1993)

34.212 N. Usami, T. Mine, S. Fukatsu, Y. Shiraki: Optical anisotropy in wire-geometry SiGe layers grown by gas-source selective epitaxial growth technique, Appl. Phys. Lett. **64**, 1126–1128 (1994)

34.213 B. Teichert: Self-organization of nanostructures in semiconductor heteroepitaxy, Phys. Rep. **365**, 335–432 (2002)

34.214 Z. Zhang, M.G. Lagally: *Morphological Organization in Epitaxial Growth and Removal* (World Scientific, Singapore 1998)

34.215 K. Brunner: Si/Ge nanostructures, Rep. Prog. Phys. **65**, 27–72 (2002)

34.216 J.-M. Baribeau, N.L. Rowell, D.J. Lockwood: Self-assembled $Si_{1-x}Ge_x$ dots and islands. In: *Self-organized Nanoscale Materials*, ed. by M. Adachi, D.J. Lockwood (Springer, New York 2006)

34.217 M. Miura: Studies of formation process and optical properties of self-assembled Ge/Si nanostructures. Ph.D. Thesis (University of Tokyo, Tokyo 2001)

34.218 M. Miura, J.M. Hartmann, J. Zhang, B. Joyce, Y. Shiraki: Formation process and ordering of self-assembled Ge islands, Thin Solid Films **369**, 104–107 (2000)

34.219 H. Takamiya, M. Miura, J. Mitsui, S. Koh, T. Hattori, Y. Shiraki: Size reduction of the Ge islands by utilizing the strain fields from the lower-temperature-grown hut-clusters buried in the Si matrix, Mater. Sci. Eng. B **89**, 58–61 (2002)

34.220 E.S. Kim, N. Usami, Y. Shiraki: Control of Ge dots in dimension and position by selective epitaxial growth and their optical properties, Appl. Phys. Lett. **72**, 1617–1619 (1998)

34.221 E.S. Kim, N. Usami, Y. Shiraki: Selective epitaxial growth of dot structures on patterned Si substrates by gas source molecular beam epitaxy, Semicond. Sci. Technol. **14**, 257–265 (1999)

34.222 H. Takamiya, M. Miura, N. Usami, Y. Shiraki: Drastic modification of the growth mode of Ge quantum dots on Si by using boron adlayer, Thin Solid Films **369**, 84–87 (2000)

34.223 S. Koh, K. Konishi, Y. Shiraki: Small and high-density GeSiC dots stacked on buried Ge hut-clusters in Si, Physica E **21**, 440–444 (2004)

35. Plasma Energetics in Pulsed Laser and Pulsed Electron Deposition

Mikhail D. Strikovski, Jeonggoo Kim, Solomon H. Kolagani

35.1	Energetic Condensation in Thin Film Deposition 1193
35.2	PLD and PED Techniques 1194
35.3	Transformations of Atomic Energy in PLD and PED 1195
	35.3.1 Plasma Formation of Vaporized Material 1196
	35.3.2 Plasma Formation in PED 1198
	35.3.3 Expansion of Plasma and Particle Acceleration 1199
	35.3.4 Deceleration of Plasma in Background Gas 1202
35.4	Optimization of Plasma Flux for Film Growth 1204
	35.4.1 Ion Current of Plasma Propagating in Ambient Gas 1205
	35.4.2 Optimization of Growth of GaN Films – A Materials Example 1206
35.5	Conclusions 1208
	References 1209

Surface bombardment by energetic particles strongly affects thin-film growth and allows surface processing under non-thermal-equilibrium conditions. Deposition techniques enabling energy control can effectively manipulate the microstructure of the film and tune the resulting mechanical, electrical, and optical properties. At the high power densities used for depositing stoichiometric films in the case of pulsed ablation techniques such as pulsed laser deposition (PLD) and pulsed electron deposition (PED), the initial energetics of the material flux are typically on the order of 100 eV, much higher than the optimal values (≤ 10 eV) required for high-quality film growth. To overcome this problem and to facilitate particle energy transformation from the original as-ablated value to the optimal value for film growth, one needs to carefully select the ablation conditions, conditions for material flux propagation through a process gas, and location of the growth surface (substrate) within this flux. In this chapter, we discuss the evolution of the material particles energetics during the flux generation and propagation in PLD and PED, and identify critical control parameters that enable optimum thin-film growth. As an example, growth optimization of epitaxial GaN films is provided.

PED is complementary to PLD and exhibits an important ability to ablate materials that are transparent to laser wavelengths typically used in PLD. Some examples include wide-bandgap materials such as SiO_2, Al_2O_3, and MgO. Both PLD and PED can be integrated within a single deposition module. PLD–PED systems enable in situ deposition of a wide range of materials required for exploring the next generation of complex structures that incorporate metals, complex dielectrics, ferroelectrics, semiconductors, and glasses.

35.1 Energetic Condensation in Thin Film Deposition

Thin-film deposition under energetic particle bombardment is well known. Originally suggested in 1938 for improving film density, the energetic process was further developed as *ion plating* in 1963 [35.1]. A variety of techniques to produce energetic conditions are currently in practice [35.2]. These include plasma-assisted film deposition techniques [35.3], magnetron sputtering [35.4], and ion-beam-assisted deposition [35.5, 6]. A common factor accompanying all energetic condensation processes is that the film growth surface is under constant bombardment by an energetic particle flux as it receives the desired material for film growth.

Under the bombardment by energetic particles, surfaces and subsurface layers of crystals are in an unusual (nonequilibrium) state controlled by the particle flux parameters rather than by deposition conditions such as substrate temperature. In this state, the equilibrium dynamics are no longer applicable to the dynamics of the crystal surface modification processes, which include implantation of atoms, radiation-enhanced diffusion, point-defect interaction, sputtering, ion mixing, adsorption, adatom movement, and film growth. The modified layers retain a *memory* of the conditions of their origin and exhibit unusual properties. This can be beneficial for some applications of the modified layers, and undesirable for others. Control of the bombardment conditions is critical for optimizing a specific process. In general, the particle energy of interest is in the range of $10-100\,\text{eV}$, which is well above the common processing temperature of $< 0.1\,\text{eV}$ ($\approx 1200\,\text{K}$). In some sense, the particle bombardment substitutes for the higher temperature that is problematic to achieve under equilibrium conditions.

Pulsed laser deposition (PLD) and pulsed electron deposition (PED) are relatively new energetic condensation techniques with unique plasma flux parameters. In PLD and PED, a pulsed laser or a pulsed electron beam rapidly vaporizes a thin section of a target material, providing a stream of energetic plasma flux under highly nonequilibrium conditions. This process is known as ablation. The film growth conditions are intricately linked to and controlled by the conditions of the flux generation and its propagation in an ambient process gas (or in vacuum). This chapter will consider and compare these two techniques, emphasizing the critical role of plasma energetics in thin-film formation. The objective is to follow the energy balance through the entire process sequence from flux generation at the target surface to flux arrival at the growth surface (substrate), and to draw conclusions on the possibilities of controlling and optimizing the deposition conditions in PLD and PED.

We limit our consideration to the most commonly used laser pulses of nanosecond duration. Also we leave beyond the scope of this chapter the subject of macrodefects (droplets) in deposited films.

35.2 PLD and PED Techniques

The idea of utilizing the exceptionally high power density of a laser beam for material processing has been tested contemporarily with the development of lasers since the 1960s. The effects of heating a material to temperatures well above an *evaporation temperature* and transforming the vapor into a state of plasma have been studied extensively. In their pioneering work, *Smith* and *Turner* [35.7] explored plasma condensation to form a film layer on a substrate placed in a vacuum. Since then, a variety of PLD processes have been considered and summarized in several excellent reviews [35.6, 8–10]. In the conventional PLD process, shown schematically in Fig. 35.1, a high-power pulsed laser beam is used as an external energy source to rapidly vaporize the target material. When a pulse of submicrosecond duration is used (accomplished in Q-switched solid-state or excimer lasers, for example), the general process is not limited to evaporation but is accompanied by strong absorption of the laser beam in the vapor and its transformation into plasma. The interaction of the laser beam with the target and the generated plasma is a very complex physical process, and depends on the properties of both the target and the laser beam.

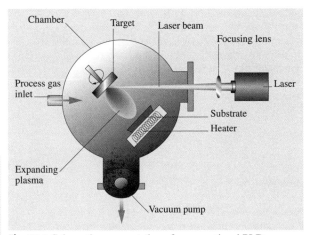

Fig. 35.1 Schematic representation of a conventional PLD process

In PED, the general scheme and ablation processes are very similar to in PLD, although a significant difference in the nature of the pulse energy source exists. In the PED process, the energy for ablation is delivered to the target by pulsed electrons rather than photons (as in PLD). The electron beam interacts differently with the target and the evaporant it generates. However, similarly dense plasma is formed at the target surface by both the pulsed electron and the pulsed laser beam. This dense plasma is the primary precursor source material for film deposition. The dense, high-temperature, strongly ionized plasma layer at the target surface propagating in a direction perpendicular to the target surface is known as the *plasma plume* in the published literature.

35.3 Transformations of Atomic Energy in PLD and PED

In energetic deposition processes, atoms from the target surface pass through several energy transformation phases before arriving at a substrate surface. Starting from their evaporation, the atoms are involved in a series of energy transfer processes controlled by different mechanisms. These atoms experience significant changes in their temperature, ionization state, as well as kinetic energy. To evaluate the possibilities of controlling the energetics in the film deposition process, it is critical to follow the dynamics of the energetics of the atoms. For every phase, there are key parameters enabling control of the transient processes. Ultimately, this enables control over the energy of the atoms arriving at the substrate, which is important for the film growth. PLD and PED processes, viewed as a sequence of several phases, are shown schematically in Fig. 35.2.

The pulse width of the laser or electron beam defines the duration of phase 1 in Fig. 35.2. During this period, most of the beam energy is transformed into the internal energy (enthalpy) of the atoms in the plasma. This includes energies of evaporation and ionization, the thermal energy of material atoms, as well as the energy of electrons in the plasma. Typically, the thickness of the plasma layer near the target surface is below 0.1 mm during this phase.

Under the action of high internal pressure, the plasma layer ejects itself in a direction perpendicular to the target surface (phase 2, Fig. 35.2). The plasma pressure is orders of magnitude larger than any practical gas pressure in the process chamber. Thus, the plasma ejection is not affected by the presence of process gases. The plasma experiences continuous acceleration by the pressure gradient supported through recombination of atoms and electrons. During this phase, the original energy spectrum of the plasma flux is formed. A typical plasma size in this phase is on the order of several times the spot size on the target. A three-dimensional expansion begins within this distance. In a typical PLD or PED process, this expansion is ≈ 10 mm in length. If deposition is carried out in a vacuum (10^{-6} Torr or below), the plasma expands and the original energy spectrum is *frozen*, and can be analyzed by an appropriate time-of-flight technique.

If deposition takes place in a background gas (phase 3, Fig. 35.2), the flux atoms collide with the gas atoms. As a result, the initial energetic spectrum of the atoms in the plasma evolves on a characteristic length scale related to the length scale of the mean free path at a given pressure. If the number of ablated atoms is sufficient to involve a comparable number of gas atoms in the mutual hydrodynamic movement, this *snowplow* effect drives most of the gas atoms towards the substrate, leaving behind a gas-rarefied region (shown pale brown in Fig. 35.2). Simultaneously, as the entire ensemble decelerates during this phase, the kinetic energy of both the target and gas atoms decreases.

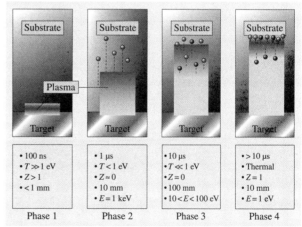

Fig. 35.2 Schematic representation of PLD and PED processes. Conventions: *brown circles*: fast atoms of target material, *black circles*: gas atoms, *pale brown area*: depleted gas volume behind the plasma front

In the next phase (phase 4, Fig. 35.2), the directional velocity of plasma flux becomes comparable to the thermal velocity of its particles (i.e., the flux transforms from supersonic to subsonic states). Plasma expansion nearly stops (seen as a range for its visible radiation *plume*), and becomes nearly isotropic. In most cases, an optimal substrate location for film growth can be found in the near vicinity of this range.

35.3.1 Plasma Formation of Vaporized Material

The dominant process in phase 1 is the evaporation and ionization of a small amount of material. First, at the typical beam power density used in ablation ($Q \approx 10^9$ W/cm^2), the target surface is rapidly heated, attaining temperatures well above the common *evaporation* temperature of all the elements in the target. The evaporation can be so intense that the vapor density at the surface can approach the limiting value of the atomic concentration in a solid ($n_L \approx 5 \times 10^{22}$ cm^{-3}). Second, in the field of the intense laser beam, the vapor effectively ionizes (transforms into plasma) and absorbs the majority of the incoming energy from the laser, preventing further heating of the target. Several key parameters that control the process are, pulse duration τ, power density Q [W/cm^2], and energy density Q_τ [J/cm^2]. The laser wavelength λ in PLD and the energy of each electron [keV] in PED define the absorption depth of the beam with respect to the target material.

The pulse energy is distributed in the target over a depth $(D + D_T)$, and the rate of surface temperature rise can be expressed as

$$\frac{dT}{dt} \sim \frac{Q}{D + D_T}, \quad (35.1)$$

where D is the absorption length (photon or electron range) and D_T is the thermal diffusion length. D_T can be expressed as $D_T = 2(at)^{1/2}$, where a is the thermal diffusivity and t is the time. D_T is also the characteristic depth of exponential decay of the temperature distribution, $T(x,t) \propto \exp(-x^2/4at)$ for one-dimensional (x) heat diffusion from a surface source [35.12].

For $\tau = 10$ ns, a pulse width typical in PLD, D_T is about 0.2 μm for most common dielectrics. This value is less than or comparable to the photon absorption depth D, and D_T can be neglected in (35.1). Thus the heating rate is controlled by the target optical properties at a given wavelength, and increases *proportionally* with the beam power.

This is not the case in PED. The penetration depth (or absorption range) of an electron in a solid depends on its energy. In PED, both the electron beam power Q and the electron range depend on the electron source voltage V. Since the number of electrons in the pulse is proportional to the total beam current density I, and since the average energy of each electron ($\sim$ eV) is directly related to the pulsed electron beam source (PEBS) voltage V, the power density carried by the beam is $Q \approx IV$, where $I = I(V)$. The $I(V)$ relation characterizes a PEBS design. For 10–100 kV electrons, the range D changes with voltage as $D \propto V^2$, as shown in Fig. 35.3. If $D > D_T$, the $D \propto V^2$ dependence controls the denominator in (35.1), and thus strongly affects the heating rate in PED. In some sense, changing the voltage in the PEBS is equivalent to changing the laser wavelength (and absorption depth) in PLD. The interplay of $Q(V)$ and $D(V)$ controls the surface heating rate in PED. The expected heating rate for PED has been examined elsewhere [35.13]. Experiments show that the beam power is not able to increase as fast as the beam penetration length at $>$ 15 kV, which leads to a maximum in the heating rate at $\approx$ 15 kV (Fig. 35.4). Therefore, it is not always beneficial to use the source at higher voltage (and accordingly with higher energy electrons). A more efficient way to improve the heating rate in PED is via a larger current *density* at the target surface by minimizing the beam cross-section. In order to increase the current density, the target has to

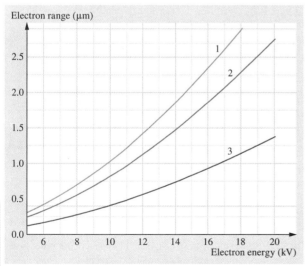

Fig. 35.3 Calculated penetration depth (range) of energetic electrons in Si (1), Al$_2$O$_3$ (2), and YBa$_2$Cu$_3$O$_x$ (3) (after [35.11])

be as close to the tube output as is practically possible since the electron beam diverges with distance from the output channel tube.

In both PLD and PED, once vapor with a sufficient density is formed at the target surface, the beam ionizes it. PLD differs from PED in the target surface heating dynamics and the mechanism of ionization. In PLD, free electrons gain energy in the oscillating electric field of a laser and ionize atoms by impact. This avalanche-like process is similar to the *optical breakdown* of gases by intense laser beams. The breakdown physically means that, once transformed into a strongly ionized plasma state, the plasma becomes nontransparent to the laser beam. The plasma shields the target material from the beam, and absorbs the majority of the pulse energy. This plasma shielding strongly affects the energy balance in PLD and PED, the primary topic of interest in this chapter. Well above the ablation threshold, the fraction of the beam energy used for target heating is relatively small. The majority of the beam energy is preserved within the plasma [35.14]. The net energy balance of the entire ablation process can be roughly understood as

$$Qt = E_1 N, \qquad (35.2)$$

where E_1 is the *total* average energy per evaporated particle (ion or atom) and N is the total number of atoms or ions per unit area. The energy E_1 eventually transforms into kinetic energy $E \leq E_1$ of atoms ejected from the target. The balance indicates that higher particle energy E_1 is expected in an ablation process in which a smaller number (N) of ablated particles enables a strong shielding effect.

The total number N of atoms absorbing the beam energy can be estimated as $N \approx nL$, where n is the atomic concentration in the shielding plasma and L is the thickness of the plasma layer. Some understanding of $n(\lambda)$ scaling laws for different laser wavelength λ (frequency $\omega = c/\lambda$) can be gained by considering the conditions for optical breakdown of gases. Under a crude approximation, the gas concentration most favorable for developing the electron avalanche corresponds to the condition where the frequency Ω of electron collisions with atoms is on the same order as the beam frequency ω. The critical concentration $n = n_c$ can be found from the condition $\Omega(n_c) \approx \omega$. The physical reason for the optimum is that, for a given ω of the laser, the rate of electron energy gain in the electric field of the beam has a maximum at this value of collision frequency [35.15]. The gas ionization develops via inverse bremsstrahlung absorption. The

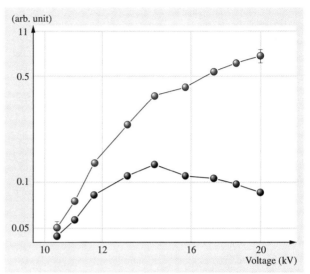

Fig. 35.4 Estimated surface heating rate (*black circles*) and measured beam power (*brown circles*) as functions of the electron energy

collision frequency relates directly to the gas (vapor) concentration as $\Omega = \alpha n$. As an example, for normal conditions of 760 Torr ($n = 2.47 \times 10^{19}$ cm^{-3}) of Ar gas ($\alpha = 1.6 \times 10^{-7}$ cm^{3}s^{-1}), one finds $\Omega = 4 \times 10^{12}$ s^{-1}, and $\lambda = 75\,\mu$m radiation would be optimal to create plasma.

As a wide range of atom concentration in the vapor is available in PLD at the prebreakdown stage, the above condition can be satisfied at least for λ down to $\approx 1\,\mu$m.

The condition gives an estimate for the initial atomic concentration to develop the shielding plasma. It is inversely proportional to the laser wavelength, $n_c \propto \lambda^{-1}$. The relations above describe proper conditions to initiate plasma at microwave frequencies, where the pressure of ≈ 2 Torr is optimal for 2.8 GHz ($\lambda = 1.0$ mm). It is also suitable for CO_2 laser with $\lambda = 10.6\,\mu$m. The calculated value of 15 atm ($n_c = 3.8 \times 10^{20}$ cm^{-3}) as the optimal pressure for Xe gas plasma formation is close to the experimental value of 20 atm [35.15]. The breakdown threshold power density Q^* of the CO_2 pulsed laser is as low as $\approx 5 \times 10^7$ W/cm^2. For radiation of an Nd:yttrium aluminum garnet (YAG) laser ($\lambda = 1.06\,\mu$m), it can be estimated that the optimal vapor concentration is $n_c = 3.8 \times 10^{21}$ cm^{-3}. For a shorter-wavelength laser, the plasma starts to develop in a higher atomic concentration in the vapor. For the PLD process, this suggests that more material has to be evaporated before shielding occurs.

For much shorter wavelengths such as those of ultraviolet (UV) excimer lasers, the vapor breakdown condition and mechanism are more complex. Theoretical estimation from the above relations implies that the estimated optimal atomic concentration in a vapor of $n_c = 1.6 \times 10^{22}\,\text{cm}^{-3}$ becomes unrealistically high at $\lambda = 248\,\text{nm}$ (KrF laser), for example. The vapor breakdown threshold intensity of the beam, according to the classical theory, should also increase sharply as $Q^* \propto \lambda^{-2}$ [35.16]. The scaling from $10.6\,\mu\text{m}$ leads to $Q^* \approx 9 \times 10^{10}\,\text{W/cm}^2$. These estimations, however, have not been supported by experimental results. Experimental values as low as $\approx 1 \times 10^8\,\text{W/cm}^2$ for Al and Cu targets has been found for $248\,\text{nm}$ [35.17]. The suggested explanation of the reduced breakdown threshold for UV lasers is that a multiphoton photoelectric effect takes over from the atomic ionization process as the high energy of UV quantum becomes comparable to the atomic excitation and ionization potentials [35.15, 18]. The threshold can actually *decrease* with wavelength at $\lambda < 1\,\mu\text{m}$. The low thresholds indicate that a very dense vapor is not needed to form UV laser plasma. Thus, the atomic concentration on the order of $n \approx 1 \times 10^{21}\,\text{cm}^{-3}$ in the vapor may be a reasonable estimate for typical breakdown conditions for wavelengths in the range of $\lambda = 1.0$–$0.25\,\mu\text{m}$ in PLD. However, the concentration does change with λ. Data for the laser wavelength dependence of the ablated material mass [35.19] indicate that the plasma can still be created in vapor with *higher* density for shorter-wavelength lasers. Following the considerations mentioned above, in this case the energy balance for the PLD process is shifted towards a smaller number of ablated atoms and a larger energy of ejected atoms/ions. This trend is indeed seen in the energy spectrum of ejected ions [35.20, 21].

Experiments suggest that, at $3 \times 10^8\,\text{W/cm}^2$ ($\lambda = 248\,\text{nm}$), breakdown starts at the beginning of a $30\,\text{ns}$ pulse and takes just $\sim 5\,\text{ns}$ to develop a plasma absorbing $\approx 85\%$ of the beam intensity [35.17]. The majority of the beam energy is loaded into the vapor. The plasma thickness L can be estimated as $v_T t \approx 10^{-3}\,\text{cm}$ where $v_T \approx 10^5\,\text{cm/s}$ is the thermal velocity of vapor expanding from the target surface. The plasma contains $nL \approx 1 \times 10^{18}\,\text{/cm}^2$ atoms, which would originally occupy a $\approx 2 \times 10^{-5}\,\text{cm}$ deep layer in a solid target. Note that, for typical cohesive energy of $\approx 4\,\text{eV/atom}$ in solids, the evaporation of this number of atoms requires $\approx 0.8\,\text{J/cm}^2$. This is typically a small fraction of the beam energy in intensive ablation conditions ($> 3\,\text{J/cm}^2$). Most of the pulse energy is loaded into the total energy (enthalpy) of plasma, including the thermal energy of ions and electrons as well as the ionization energy.

With increasing laser intensity Q, both N and E increase, as confirmed by numerous data on the Q dependence of evaporated materials, deposited film thickness, and ion/neutral atom kinetic energy. Although most of these data were obtained in conditions of one-dimensional plasma expansion (plasma layer thickness smaller than the focal spot size of the laser beam), the dependencies generally follow the scaling of the model [35.22] developed for the regime with partial target shielding by the flowing plasma. Such a self-regulating regime (*optical plasmatron*) is established once the laser pulse duration is much longer than the time of the plasma expansion to the characteristic scale of the ablation spot diameter, as has been observed in $\geq 1\,\mu\text{s}$ long-pulse (transversely excited atmospheric pressure (TEA) CO_2 laser, [35.23, 24]) or small-spot nanosecond-pulse ablation [35.12, p. 243].

Based on mass and energy conservation laws and based on the assumption of a three-body plasma recombination process, the *steady-state* ablation model suggests that the ablation rate can be expressed as

$$\frac{dN}{dt} \propto \lambda^{-4/9} Q^{5/9}. \qquad (35.3)$$

Accordingly, to satisfy the $E_1\,dN/dt = Q = \text{const.}$ condition, the energy per atom can be expressed by

$$E_1 \propto \lambda^{4/9} Q^{4/9}. \qquad (35.4)$$

These relations again point to the tendency of a shorter-wavelength laser beam to evaporate more material from a target, providing each atom with a lower energy. As a crude approximation, one can conclude that the average kinetic energy E of particles produced by laser ablation scales as $E \propto \lambda$ for 1.06–$10.6\,\mu\text{m}$ wavelength lasers and as $E \propto \lambda^{1/2}$ for 0.2–$1.0\,\mu\text{m}$ wavelength lasers. Choice of the laser is one of the possibilities to control the energy spectrum of the plasma flux in PLD.

35.3.2 Plasma Formation in PED

In PED, the electron beam energy of 5–$15\,\text{keV}$ is well above the threshold potential for atomic excitations and ionizations. Fast electrons collide directly with the deep-level electrons of target material atoms, experience broad-angle scattering, and eventually stop at a certain depth from the target surface (electron range). This stopping at a certain depth requires a certain number of collisions to happen. This number is practically independent of the specific state of the atoms, with atoms in the solid lattice scattering electrons as effectively as atoms in a gas phase (vapor). An important consequence of this fact is that, for the electron beam, there is no transition to the vapor plasma shielding effect that is characteristic for the laser ablation. Throughout the entire pulse length, the electron beam continues to interact with the same number of atoms as contained within the beam absorption range D in the target. As $D \propto 1/\rho$ (ρ [g/cm^3] is the target material's density), the total mass ablated by the intense beam is expected to be independent of materials ablated, $m = \rho D = $ const. [35.11]. As mentioned above, the electron range is strongly dependent on the electron energy ε [kV]. The absorption range can be expressed as

$$D \text{ [µm]} \approx 5.37\varepsilon \left(1 - \frac{0.9815}{1 + 0.0031\varepsilon}\right) \rho^{-1}.$$

Figure 35.3 shows the range of 0.5–2.0 µm for a typical PED electron beam energy of 10–15 keV. Even for dense materials (such as YBCO with a density of ≈ 6 g/cm^3), the range of ≈ 1.0 µm is greater than the typical depth of ≈ 0.2 µm for laser ablation (using a 248 nm KrF laser at 5 J/cm^2) [35.12]. If all the material within the penetration depth is PED-ablated, it follows, from (35.2), that the average kinetic energy of particles in the plasma flux can be smaller in PED than in PLD. For lower-density materials, a higher PED ablation rate and accordingly smaller energy per ablated atom could be expected. This is actually the case. However, the particle energy spectrum in PED is broader than in PLD due to the presence of a small amount of highly energetic ions, up to several keV. Some specific mechanism of ion acceleration could be responsible for the fast ion generation in PED.

35.3.3 Expansion of Plasma and Particle Acceleration

A dense layer of highly ionized high-pressure plasma is formed at the target surface by the focused laser pulse. The plasma expands and accelerates primarily in a direction normal to the target surface (Fig. 35.2, phase 2). During this process, the energy E_1 (the total energy per atom, i.e., the enthalpy per atom) is transformed into an asymptotic kinetic energy E of atoms [35.25], which is the energy of interest in the present discussion, as this flux of energetic particles arrives at the substrate for film formation (when PLD is carried out in vacuum). Gradient of plasma pressure is the driving force for the acceleration of particles. Note that, even when PLD is carried out in a typical pressure background gas, due to the large difference between the plasma pressure at the ablation spot and the gas pressure, the initial energy spectrum of E is established at a distance of a few millimeters (≈ 3 times the spot diameter) from the target and is not affected by the presence of the process gas.

For a vacuum process (or at a low gas pressure), the ablation particle energy spectrum established soon after leaving the spot region remains mostly unchanged during the following geometrical expansion. The ion condition is *frozen* due to the low thermal energy of the plasma and the lack of efficient collisions. Laser-produced plasma retains a significant portion of the atoms in the ionized state. Film growth in vacuum is accompanied by the impact of particles of the largest possible energy. This feature also makes the laser flux an efficient source of ions for analytical laser plasma spectroscopy or for further acceleration of ions in special applications.

Specific conditions of the laser-produced plasma facilitate kinetic energy of particles well above the plasma temperature. At the end of the laser pulse, the total internal energy per plasma atom, $I + (1 + Z)T$, is stored in the thermal energy of the ion, Z electrons, and the total potential energy I of the atomic ionization. During the plasma expansion, electrons contribute to the acceleration pressure. The plasma temperature is also supported by release of the potential energy by three-body recombination. Thus the resulting kinetic energy of the flux particles can greatly exceed the original thermal energy of the plasma atom $\sim T$ [eV].

The model for the steady plasma flow regime (pulse length > 1000 ns) [35.22] predicts for the asymptotic particle energy

$$E = 5(1 + Z)T. \tag{35.5}$$

The functional relation $Z(I, T)$ can be used for atoms of a specific target material. As the beam energy stored per a plasma atom is ≈ 3 times greater than the target atom cohesion energy ≈ 4 eV (see before), a plasma temperature of > 10 eV can be estimated for a typical ablation with 3 J/cm^2. The potential energy of the first ionization I_1 is usually within the range 7–15 eV, and the $Z = 1$ ionization is already significant at temperature $T \approx I_1/5 = 1.4$–3 eV [35.16]. At $T \approx 2$–3 eV, the generation of $Z = 2$ ions is already possible. $Z = 2$ ions have been routinely observed under PLD ablation conditions. Equation (35.5) estimates $E = 30$–45 eV for $Z = 2$ ions according to the model [35.22] for a steady flow ablation regime.

The steady regime is not usually realized in a practical excimer-laser-based PLD configuration with spot sizes of ≈ 1 mm, where the ≈ 30 ns laser pulse is too short in comparison with the time to establish the steady regime. However, this regime has been realized in small-spot nanosecond-pulse ablation [35.12, p. 243], and ablation with a long TEA CO_2 laser ($\lambda = 10.6\,\mu$m) pulse of $\approx 2\,\mu$s duration [35.23,24]. Its specific features are an increased average ablation rate, and narrower energy spectrum $dN/dE = f(E)$ relative to the unsteady pulsed regime.

The narrow ion energy spectrum $dN/dE = f(E)$ observed [35.23, 24] can be explained by the fact that the majority of the registered ions have been generated under nearly similar, quasistationary evaporation–absorption–acceleration conditions during most of the pulse duration. The average particles energy is laser wavelength dependent. For a $\lambda = 10.6\,\mu$m CO_2 laser with $Q = 10^9$ W/cm^2, a narrow spectrum with a maximum at ≈ 400 eV was measured. This energy is about a factor of ≈ 2.5 larger than that observed for the $\lambda = 1.06\,\mu$m laser, and ≈ 5 times larger than that for $0.248\,\mu$m radiation, in accordance with the prediction of (35.4). Another feature of the steady regime is that the average energy of neutrals is nearly as high as that of ions [35.23, 24]. This reflects the fact that both the neutrals and the ions experience the same acceleration conditions.

Thus, at the same beam energy density of a 11 ns duration excimer laser pulse, the ablation rate (μm/pulse) increases by several times if the spot size is reduced from 83 to 24 μm [35.12]. This can be explained by a transition to a steady regime with a plasma more transparent to the laser beam than in the case of one-dimensional expansion (large spot). In accordance with the energy balance, one can expect a lower average kinetic energy of ablated particles to result.

If the laser beam power density is high enough, such an important feature of PLD as congruent evaporation of complex materials is preserved in the quasistationary ablation regime. Indeed, the basic requirement for stoichiometric material removal is that the volume of ablated material is much greater than the volume of remaining heated material. That can be expressed as $v\tau \gg a/v$, where the ablation rate v is the velocity of the ablation surface propagation into a target, and a/v is the thickness of the heated material in front of it. As mentioned above, for a given power density, v is even higher in stationary than in pulsed ablation, and the condition is satisfied. Experimentally, for example, stoichiometric ablation in the quasistationary regime has been demonstrated by fabricating high-quality YBCO films using 2 μs long pulses from a TEA CO_2 laser [35.26].

In an *unsteady* pulsed regime, the energy spectrum broadens in both directions compared with the energy spectrum for the steady flow regime. The leading ions on the front of the plasma layer are in favorable conditions to experience more effective acceleration, and attain greater energy. On the other hand, at the end of the laser pulse, the plasma pressure build up is terminated, and the ions at the surface created in this time are not able to accelerate. Thus, the leading ions with higher energy are present on the front of the expanding plasma stream. The effect is well known in classical gas dynamics in unsteady (abruptly started) gas flow in vacuum [35.16]. The energy of the leading particles E_{max} is $\sim 2/(\gamma - 1)$ times larger than the energy from the steady flow. The additional kinetic energy is acquired at the expense of heat energy in the neighboring particles. For plasma at the ablation conditions, the exponent γ can be approximated to 1.3–1.4. The maximal energy of the accelerated particles is $E_{max} = 10T(1+Z)/(\gamma - 1) = 100$–$200$ eV at $Z = 2$ and $T = 2$–3 eV.

The velocity or energy of neutral components of the flux is generally somewhat lower, but still close to that of the ions. A possible explanation is that a significant number of the neutrals is in effect the former ions fraction, which has recombined. This fact makes it possible to apply the ion time-of-flight technique to probe the energetics of the laser-producing plasma fluxes. Data of the time-of-flight mass spectrometry and time-resolved emission spectroscopy applicable to trace the neutrals dynamics [35.27, 28] show a neutral's velocity of ≈ 0.6–0.75 for ions (energy difference about a factor of 2–3). Still, the ion probe measurements are representative of the general flux energetics and represent a good estimate of the upper energy limit of the particles.

Films grow in vacuum under the impact of the particles of largest possible energy. The ion probe (Faraday cap) is a convenient and frequently used technique to characterize the dynamics of the plasma arriving at a point a distance L from the target. In this technique, a slightly negatively biased probe registers the current of the ions in the flux. For ablation in vacuum (or at low gas pressure), due to a large mean free path, the ions travel the distance ballistically, and the delay time t for ions arriving at the probe directly relates to their speed $v = L/t$ and kinetic energy $mV^2/2$. Thus, the ionic energy spectrum can be obtained as $dN/dE = f(E)$. The spectrum for $f(E)$ can be calculated from the ion probe current density $j = eZ(dN/dt)$ using the relation $dN/dE = (dN/dt)(dE/dt)^{-1}$, where $E = M(L/t)^2/2$

is the kinetic energy of an ion with mass M arriving at time t.

For ablation in an ambient background gas pressure, collisions with the gas alter the velocity of flux species and the probe signal. Since the relation $v = L/t$ is not valid any more due to the velocity change with distance, the probe signal cannot be used to extract the arriving ion's energy spectrum. The relation overestimates the ion's kinetic energy at distance L, as the original velocity is greater than L/t and as the velocity at arrival is smaller than L/t. A proper particle deceleration model can be used to describe the special evolution of their average energy.

A typical energy spectrum of Si ions is given in Fig. 35.5 for the plasma flux created by KrF laser pulse (wavelength 248 nm, 20 ns pulse, 2.5 J/cm² energy density) in vacuum. The insert shows the original ion current signal obtained at a distance of 11 cm. The median energy is ≈ 60 eV, large enough not only to significantly change the conditions of the surface movement of the arriving atoms but also to introduce property changes of the previously deposited layers of the growing film [35.29]. Characteristically, the spectrum exhibits an extended lip towards energies up to ≈ 200 eV. As discussed above, the fastest ions are those that are on the front of the expanding plasma cloud, where the largest pressure gradient exists for an extended time. Two additional factors can also contribute to the increased ion signal amplitude at the higher end of the spectrum and extending from the median energy value. First, multiple-ionized ions create a larger probe signal by a factor of Z (ion charge) than single-charged ones. Second, the energy of leading ions can be additionally increased by the effect of electrostatic acceleration of ions in a double layer of space charge existing at the front of the plasma [35.30].

Figure 35.6 compares the energy spectra of CeO_2 plasma flux ions obtained by PLD and PED at a distance of 14.5 cm at a similar energy density. The relatively small oxygen gas pressure of 1 mTorr does not alter the spectra significantly, at this distance, in comparison with the vacuum conditions. The spectra are normalized to probe signal maximum. A somewhat surprising result of this comparison is that the energy of the fastest ions in the PED-generated plasma flux is larger than in PLD. Possibly, some additional mechanism of ion acceleration exists in PED, the origin of which is not fully understood at the present time. One possible explanation is that the negative charging of the plasma cloud leads to a stronger expulsion of the electrons and to a stronger effect of electrostatic acceleration

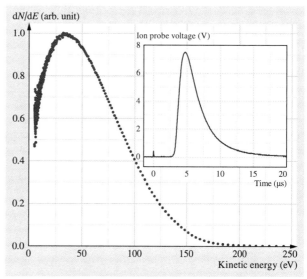

Fig. 35.5 Typical energy spectrum of Si ions generated by a KrF (248 nm) pulsed laser in vacuum. The *inset* shows the original ion probe signal at a distance of 11 cm from the Si target

of ions on the plasma front. At the same time, the PED spectrum is broader, and relatively more atoms with low energy are in ionized state. The average energy of the PED flux particles can be smaller than the

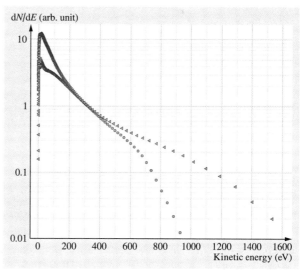

Fig. 35.6 Energy spectra of ions produced by pulsed laser (*closed circles*) and pulsed electron beam (*open triangles*) in oxygen pressure of 1 mTorr

energy in PLD. More studies are needed to fully understand the dynamics of PED-generated plasmas and their properties.

35.3.4 Deceleration of Plasma in Background Gas

In PED and PLD, the presence of background gases during ablation can significantly change both the average energy of the particles arriving at a substrate and the general attributes of the energy spectrum. When the mass of the evaporated atoms and of the incorporated gas atoms become comparable, hydrodynamic models can be used to describe deceleration of this ensemble and its thermalization. These are the phases 3 and 4 shown in Fig. 35.2. Introduction of a reactive or inert buffer gas into the deposition chamber is an effective tool to control the energy of species during film growth. Unlike vacuum PLD, plasma flux parameters change drastically with distance L from the target, which is critical for film growth. At a characteristic distance L_0, the flux loses its unidirectional velocity, as it undergoes scattering, thermalization, and finally deceleration. It has been experimentally discovered that optimum conditions for film growth exist in the vicinity of this distance [35.31–33].

In the presence of background gases, the ion probe signal is not useful to calculate the arriving particles' energy as their velocity changes with distance. If calculated as the ratio L/t, the velocity (and the energy) would be overestimated. A simple model has been suggested to describe the dynamics of plasma deceleration, the average particle energy at a distance L_0, and the scaling of L_0 as a function of ablation and gas parameters [35.34]. In this model, the ensemble of initially ablated N_0 atoms, with an average velocity v_0, is considered as a piston, incorporating and driving the gas atoms as it propagates. The essential processes here are energy transfer from the initial kinetic energy of the N_0 atoms to the thermal energy of the atom and the gas cloud (i.e., deceleration and thermalization). From the conservation of momentum and energy in an adiabatic process, the following dependencies are obtained for the velocity v of the directed movement of the ensemble and the characteristic thermal velocity $v_T = (k_B T/M)^{1/2}$ of the particles.

$$\frac{v}{v_0} = (1+x^3)^{-1},$$

$$\frac{v_T}{v_0} = x^{3/2}[3(1+x^3)(1+\mu x^3)]^{-1/2},$$

where $\mu = M/m$ is the atomic mass ratio between the target material and the gas, and $x = L/R$ is the normalized distance. R has a clear physical meaning: after the ensemble covers the distance R, the total mass of the gas atoms becomes equal to the mass of ejected N_0 atoms.

Comparison of the velocities $v(x)$ and $v_T(x)$ (Fig. 35.7) shows that there is a characteristic distance L_0 where the thermal expansion rate of the ensemble exceeds its forward movement rate. It is reasonable to view the distance L_0 as the *range* of the flux defined according to the condition

$$v(L_0) = v_T(L_0).$$

By the definition of v_T, half of the ensemble atoms are not moving towards the substrate at a distance L_0. The optimal position of the substrate for growth of high-quality films is usually found experimentally to be at around the end of the luminous flux area (*plume*). It is reasonable to associate this position with the characteristic range L_0 of the plasma flux. Let us consider three issues regarding this location important for film growth: the average particle energy $E(L_0) = Mv^2(L_0)/2$ at the distance L_0, the dependence of the distance L_0 on the gas pressure and intensity of ablation N_0, and the deposition rate $h(L_0)$ [nm/pulse] at this location.

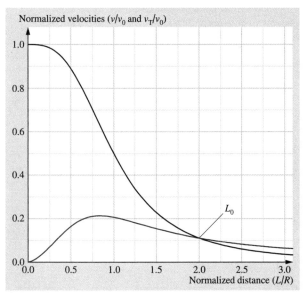

Fig. 35.7 Evolution of the directional v and thermal v_T velocities of plasma flux particles as a function of normalized distance L/R from the target. Velocities are normalized to the initial velocity v_0 of particles. L_0 indicates the location of the plasma range, where $v(L_0) = v_T(L_0)$.

Note that the flux range L_0 does not depend on the initial velocity v_0. However, an increase in the initial energy $E_0 = (Mv_0^2)/2$ of ejected atoms makes the process faster and increases the kinetic and thermal energies of particles at distance L_0. As can be seen from Fig. 35.7, the ensemble velocity in the vicinity of the flux range drops to $\approx 10-20\%$ of the initial velocity, i.e., $E(L_0) \approx (0.01-0.04)E_0$ at this point. Thus, for a flux with an average initial energy of particles of $\approx 200\,\text{eV}$, the expected energy of particles arriving at the substrate surface at the distance L_0 is 2–8 eV. Particles with this energy are acceptable for film growth as they are able to nonthermally activate the film surface without damaging it.

The distance L_0 itself is controlled by the ratio of the ablated material amount to the background gas concentration n_g (gas pressure $P \propto n_g$)

$$L_0 \propto \left(\frac{N_0}{n_g}\right)^{1/3}.$$

To maintain the substrate in the optimal position, the ejected material amount N_0 should be changed proportionally with the changing pressure. The parameter N_0 characterizes a specific PLD system and can be easily measured. To do so, it is sufficient to measure the thickness (per pulse) h of the film deposited with the system (at some target-to-substrate distance d) in a vacuum or at a pressure low enough to assure that all the ejected atoms reach the substrate. By definition, N_0 is directly related to h as $N_0 = hn_L d^2$, where n_L is the concentration of deposited atoms ($n_L \approx 5 \times 10^{22}\,\text{cm}^{-3}$ for most solids). Numerically, $N_0 = 5 \times 10^{16}$ for a system delivering $h = 0.1\,\text{nm/pulse}$ at $d = 10\,\text{cm}$. For typical PLD and PED of dielectric materials, the deposition rates at 5 cm distance are within the range 0.01–0.1 nm/pulse. The dependence $L_0(n_g)$ of the range on gas pressures is shown in Fig. 35.8. The process pressure for PED (in the range 1–10 mTorr) is about two orders of magnitude lower than that for PLD in the case of several dielectrics. This leads to larger target-to-substrate distances that are expected to be optimal in PED relative to PLD.

The deposition rate at the optimum distance $h(L_0)$ turns out to be only weakly dependent on the ablation intensity N_0

$$h(L_0) \propto n_g^{2/3} N_0^{1/3}. \tag{35.6}$$

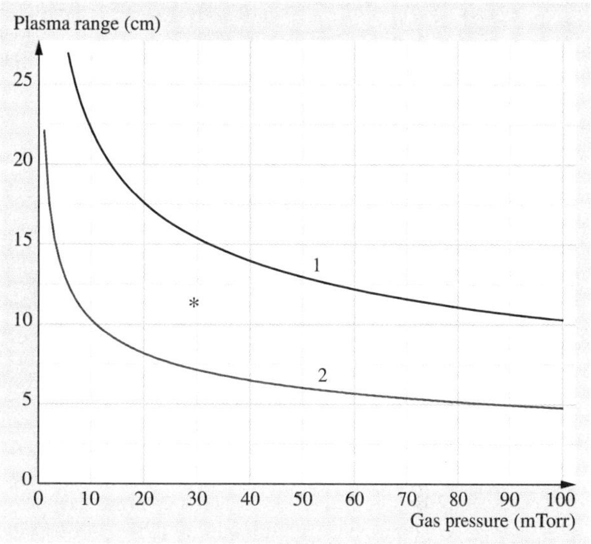

Fig. 35.8 Plasma propagation range L_0 as a function of gas pressure for ablation intensities N_0 of 0.1 nm/pulse (1) and 0.01 nm/pulse (2) at 10 cm distance. The *asterisk* represents an experimental optimal condition for PED growth of YBCO films

This result is remarkable, since it explains why different PLD systems (with different N_0 values) exhibit quite similar optimal deposition rates. Indeed, the reported deposition rates of YBCO films were close to 0.1 nm/pulse in a number of publications [35.35–39]. Thus, that deposition rate could be considered typical for PLD in a background gas pressure of 100–200 mTorr.

On the other hand, as follows from (35.6), the deposition rate at an optimal distance scales rather strongly with the gas pressure. The optimal deposition rates in PED (at 5 mTorr) are expected to be about ten times lower than in PLD (at 150 mTorr). Note here that all these differences follow from the difference in operational pressures. The dynamics of the deceleration of plasmas produced with pulsed lasers and pulsed electron beams is the same. Lower-pressure operation generally leads to a longer flux range, larger optimal film deposition distance, and larger deposition area. The overall *productivity* of the film fabrication (grams of material per second) is controlled by the average power of the beam (pulsed laser or pulsed electron).

35.4 Optimization of Plasma Flux for Film Growth

As the models predict, and experimental data confirm, the particle energy generated by lasers (not considering exotic high-power, subnanosecond pulse lasers) and intensive electron beams can cover a wide energy spectrum range of 3–3000 eV. Depending on the particle energy, their interaction with crystal surface can have very different results. Due to the significant dispersion of the spectrum, it is difficult to predict the cumulative integrated effect of the energetic particle flux on a crystal (substrate) surface. However, to fit the spectrum and the desired result for a particular application, an optimal type of laser (λ, τ), and beam intensity Q can be chosen accordingly to the discussions presented above.

High-energy (> 1000 eV) particles cause shallow implantation on a crystal surface, resulting in generally improved coating adhesion through ion mixing. Nonequilibrium generation of lattice vacancies promotes radiation-enhanced diffusion into the subsurface layer. Crystal surface compensation or doping by the laser-ablated materials can be realized [35.40].

When the incoming flux has a large fraction of energetic ions ≥ 200 eV, this can strongly affect the average deposition rate of metallic films where self-sputtering becomes significant. The effect is especially strong for large-atomic number metals such as Cu, Ag or Au [35.41]. For these materials, the self-sputtering coefficient of unity is reached at the ion energy as low as 100 eV [35.42, 43]. When the high-energy fraction in the deposition flux is large enough, no film can be formed. In some instances, much thinner film (or no film at all) is obtained on the axis of the deposition flux (where most of the higher-energy particles are concentrated) than in off-axis directions. Attempt to reduce the self-sputtering effect by adding a background gas leads to a maximum in the deposition rate (at ≈ 100 mTorr) dependence on pressure [35.44]. The effect can be understood as due to the interplay between two kinds of rate reductions. One is due to scattering or stopping of the flux material at higher pressures while the other is due to partial self-sputtering of films by the energetic portion of plasma flux ions at lower pressures. It is interesting that the maximum deposition rate was obtained in He background gas, which was able to decelerate fast ions without causing their effective large-angle scattering. The film stress also showed strong gas pressure dependence.

The concurrent bombardment of growing film promotes the growth of dense films. The increase in film density is a major factor in modifying film properties such as hardness, electrical resistivity, optical properties, and corrosion resistance. At the same time, it can introduce high compressive stresses. Especially strong is the effect of stress level on the electrical properties of conducting oxide films where their charge carrier concentration and mobility critically depend on the film lattice parameters.

Enhancement of film growth and film properties by the energetic particle interaction with film surfaces is a well-known phenomenon. Low-energy (≈ 5 eV) bombardment promotes the surface mobility of adatoms that is used for epitaxial growth [35.45]. Properties of films grown from wide-energy-spectrum plasma fluxes are the integrated result of all the effects above. Especially important process parameters are the average energy of arriving atoms and the ratio of high- and low-energy particles. According to the well-known Thornton model, the energy per deposited atom should be about 20 eV to complete the disruption of the columnar morphology of the growing film for the maximum density [35.46, 47]. A variety of techniques have been developed to provide additional nonthermal activation at the film growth surface. In the low-pressure process of ion-beam-assisted deposition (IBAD), an auxiliary source of energetic ions is used to concurrently bombard the surface of the growing film. The effectiveness of IBAD combined with PLD has also been demonstrated elsewhere [35.48]. In plasma-based techniques such as direct-current (DC) or radiofrequency (RF) sputtering deposition, substrate biasing is used to extract and accelerate the energetic ions directly from the deposition plasma [35.29].

The above considerations show several means that can be used to control the energetics of the PLD- and PED-generated plasma flux for thin-film deposition. Accepting particles with energy of ≈ 10 eV as the most favorable for nonthermally activated film growth, one can see that most of the plasmas produced by laser or electron beam pulses are excessively energetic, especially at the higher-energetic end. The reason for this higher-energy tail is that, especially in the common nonsteady pulsed ablation regime, much larger energy/atom than the target's atom cohesive energy of $\approx 5-10$ eV is loaded into the target. This is necessary to facilitate reproducible stoichiometric ablation, far from its threshold. The pulsed nature of the ablation leads to a broad energy spectrum that includes a significant portion of the extra-energetic particles generated at the front of generated plasma flux.

Film deposition with both the PLD and PED techniques includes two processes that are very different in their requirements, and yet both have to be facilitated. First is the ablation of the target material, which is necessarily a highly nonequilibrium process. Second is the film condensation process, which is desirable to be close to equilibrium (i. e., providing the surface with nonthermal excitation without causing surface damage). Thus, the general objective in optimizing the film deposition process is to transform the original energetic ablation plasma flux into a flux most suitable for the film growth.

As discussed above, some control over the initial particles' energy can be accomplished at the level of the energy source. Minimization of the particles energy leads to the choice of a shorter-wavelength laser (248 nm KrF excimer, for example) and energy density not too far above the ablation threshold. Another possibility is to thermalize the flux with ambient gases. Proper choices of gas pressure and target-to-substrate distance can also drastically improve the resultant film quality.

35.4.1 Ion Current of Plasma Propagating in Ambient Gas

The ion probe signal reflects the dynamics of the ionized component arriving at a substrate surface, regardless of whether it is in a vacuum or an ambient gas. Typically, the maximum intensity of the ion current is created by the relatively small amount of faster-moving ions on the front of the plasma flux. Therefore, the maximum probe signal represents the dynamics of this flux component rather than that of the bulk plasma. The probe signal from the relatively slow but numerous ions with a lower energy is small, especially for ablation in a background gas. Still, those particles contribute to most of the film formation. The average energy of the arriving particles cannot be obtained directly from the probe signal when measured in a background gas pressure. However, the dependence of the probe signal on the gas pressure provides useful information on the degree of interaction between the particles of the plasma flux and the background gas atoms.

A typical evolution of the probe signal is given in Fig. 35.9 for GaN film formation with a KrF laser at a distance of 17.5 cm from the target. The signal amplitude decreases with pressure, as expected. However, the energy of the faster ions (taken as the flight time of the ions contributing to the signal at half-maximum) is nearly independent of the pressure. This can be understood by considering that the number of the fastest

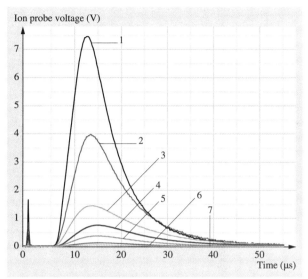

Fig. 35.9 Evolution of ion probe signal of ions propagating in a background gas with different pressure of 6 (1), 10 (2), 15 (3), 17 (4), 20 (5), 25 (6), and 30 mTorr (7) of N_2 (GaN target, KrF 248 nm laser, 17.5 cm target-to-probe distance)

ions is small, and they propagate individually, unable to create a hydrodynamic ensemble with gas atoms. The fraction of ions passing a distance L without a collision is $F \propto \exp(-\alpha PL)$, where the constant α is directly related to an effective cross-section of ion collisions with gas atoms. Evidently, the *surviving* ions arrive at the probe with their original energy. However their number exponentially decreases with pressure. The $F(P)$ dependence can be used to estimate the cross-section. This portion of ballistically moving ions is also seen as *fast components* in the emission studies of the plasma dynamics [35.49, 50]. A single collision of a fast ion significantly reduces the energy, and most likely changes its direction away from the aperture of the ion probe.

From the viewpoint of film growth, this implies that the fast ions cannot be completely removed from the plasma flux arriving at the substrate. Their number can only be exponentially decreased by increasing the gas pressure and/or distance from the plasma source. At some degree of filtering out the fast ions by the ambient gas, their fraction in the film growth flux can become acceptably small.

Propagation of the bulk plasma becomes hydrodynamic as it incorporates a gas mass comparable to the plasma mass. Dynamics of the deceleration in this slow flux component can be seen in time-resolved emission measurements [35.49]. Due to the much lower ion velocity (ion current), the probe signal generated by the bulk plasma is smaller than the first peak. However, at some conditions, the decelerated ions can be seen as a second peak in the probe signal (plume splitting) [35.51, 52]. The amount of ion energy lost per collision depends on the mass ratio of the ion and the gas atom. For heavier gases, a single collision can significantly drive down the ion energy and it appears in the slower, second peak. The bulk plasma propagation distance increases with time as $\sim t^{2/5}$ in a point-blast model [35.16, 53] or $\sim t^{1/4}$ in the simple model [35.34].

PLD and PED share the common feature of energetic particles in the as-generated plasma flux. Thus, both techniques require some *dumping* of the energy in a process gas of suitable pressure to adjust it for optimal film growth conditions. The main difference between the methods is in the dynamic range of the gas pressure, and the ability to ablate different materials. In contrast to PLD, with its very broad range of allowed gas pressures, PED operates in a rather narrow pressure interval of 5–10 mTorr. Accordingly, it requires more attention to optimizing target-to-substrate distance.

The main advantageous feature of PED is its ability to ablate materials based on wide-bandgap (SiO_2, Al_2O_3, MgO, etc.) dielectrics, which is difficult for PLD due to their transparency at (excimer) laser wavelengths. Due to strong absorption of electrons and relatively low thermal conductivity, PED ablation of these dielectrics results in films of higher quality and with fewer droplets. Furthermore, PED easily creates plasmas of polymers.

Thus, PLD and PED can be considered as complementary techniques, sharing a similar arrangement, and able to be explored in a unified deposition chamber. A PLD–PED deposition system enable in situ deposition of a wider range of materials, which is important the for exploration of complex structures, incorporating metals, complex dielectrics, ferroelectrics, semiconductors, glasses, etc.

35.4.2 Optimization of Growth of GaN Films – A Materials Example

Wide-bandgap GaN thin-film growth has been intensively researched due to this material's excellent blue-light emitting characteristics. Solid-state lighting-based applications have fueled a large amount of research in this area. A variety of deposition techniques have been used, including chemical vapor deposition (CVD) [35.54], molecular-beam epitaxy

Table 35.1 PLD process parameters for AlN and GaN films on c-axis Al_2O_3 substrate

Parameter	AlN	GaN
Temperature	930 °C	950 °C
Target distance	20 cm	20 cm
Deposition rate	0.1 Å per pulse	0.08 Å per pulse
Energy density	2.5 J/cm^2	3.1 J/m^2
Gas pressure	0.5 mTorr NH_3	1–50 mTorr N_2

(MBE) [35.55], vapor-phase epitaxy (VPE) [35.56], pulsed laser deposition (PLD) [35.57], and sputtering (SP) [35.58]. PLD-produced GaN films have been reported with extremely high crystalline quality, but only a few groups have been successful in demonstrating photoluminescence in these films, usually at low temperature (≈ 12 K) only [35.57], or in films with an associated broadband *yellow* emissions indicative of impurity states in these films [35.59].

While PLD has the advantage of producing a very energetic plasma [35.60], the kinetic energy of particles in the plasma flux can be too high in the as-generated plasmas and could potentially deteriorate the electronic properties of the deposited films. To demonstrate the importance of controlling the plasma energetics in tailoring the electronic properties, a material example is presented in this section. The process optimization for the fabrication of GaN films, performed at Neocera, is presented. The plasma energy was optimized by a systematic variation of gas pressure at fixed ablation intensity and target-to-substrate distance. Room-temperature photoluminescence accompanied by excellent crystalline quality indicate that careful optimization of plasma energetics during film growth is critical for realizing optimum electronic and structural properties during PLD film growth.

GaN films by PLD were fabricated on single-crystalline c-axis-oriented Al_2O_3 substrates with an AlN buffer layer. A background of NH_3 and N_2 was used during depositions. For process optimization, the plasma energy was tailored to an optimal level by gas-phase collisions with the ambient gas. Excellent photoluminescence was observed at room temperature for GaN films fabricated under optimized conditions. The general PLD process conditions are presented in Table 35.1.

During optimization of film growth, increasing the nitrogen gas pressure during film growth systematically reduced the plasma energy. The best GaN films were obtained at nitrogen gas pressure of 30 mTorr. Both the AlN buffer layers and GaN layers were grown epi-

Table 35.2 Impact of gas pressure on GaN film parameters

Process pressure (mTorr)	ω-Scan width (deg)	Room-temperature photoluminescence
1	0.47	no
15	0.45	no
30	0.38	yes
50	1.78	no

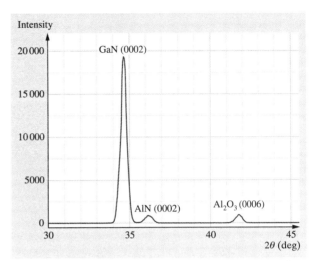

Fig. 35.10 θ–2θ scan of GaN grown under the conditions of Table 35.1 on (0001)Al$_2$O$_3$ substrate with AlN buffer layer

taxially on the c-axis-oriented single-crystalline Al$_2$O$_3$ substrates. For 450 nm thick GaN films with AlN buffer layers, the film crystal structures were measured by four-circle x-ray diffractometer; the θ–2θ scan is illustrated in Fig. 35.10.

c-Axis-oriented GaN(0002) and AlN(0002) peaks were clearly visible at 2θ of 34.65 and 36.18°, respectively. Comparing with the Al$_2$O$_3$ single-crystal peak ($2\theta = 41.75°$), the GaN and AlN peaks were strongly intensified, suggesting high quality of the GaN and AlN layers. In the optimized films deposited at 30 mTorr N$_2$, full-width at half-maximum (FWHM) of 0.38 and 0.48° was obtained from ω-scans of GaN(0002) and AlN(0002) respectively, indicating highly c-axis-oriented layers. In addition, Φ-scans were performed for GaN(10$\bar{1}$1), AlN(10$\bar{1}$1), and Al$_2$O$_3$(11$\bar{2}$3) peaks, as illustrated in Fig. 35.11. The repeated peak-to-peak 60° distance indicates the hexagonal structure of the film layers, and the clear matching peak positions of GaN, AlN, and Al$_2$O$_3$ confirm that GaN and AlN layers are epitaxial with regards to the (0001)Al$_2$O$_3$ substrate.

Photoluminescence (PL) was measured at room temperature and is illustrated in Fig. 35.12. A nitrogen laser with wavelength of 340 nm was used as the excitation source, and room-temperature PL intensity was plotted with respect to wavelength. For GaN films deposited at pressures lower than 30 mTorr excellent crystallinity was exhibited, but no PL was observed at room temperature. Room-temperature PL was only observed when the chamber pressure for GaN deposition was 30 mTorr. Blue luminescence was detected at 370 nm with significant peak intensity. The yellow emission at 545 nm was very weak, which was mainly attributed to structural imperfections such as grain boundaries or dislocations [35.59, 61]. The peak-to-peak ratio of blue emission to yellow emission is superior to other results from GaN films processed by CVD [35.61], MBE [35.55], VPE [35.62], SP [35.63], and in several nonoptimal PLD films [35.57, 59].

Table 35.2 summarizes the influence of ambient gas pressures on the GaN film quality as measured by FWHM of ω-scans and room-temperature photoluminescence (RTPL). RTPL appears to be extremely sensitive to the energy of plasma particles, being observed only from films deposited at pressure

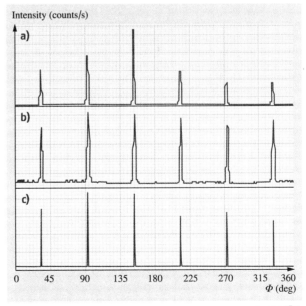

Fig. 35.11a–c Φ-scans for GaN/AlN/Al$_2$O$_3$ structure: GaN(10$\bar{1}$1) peak (**a**), AlN(10$\bar{1}$1) peak (**b**), Al$_2$O$_3$(11$\bar{2}$3) peak (**c**)

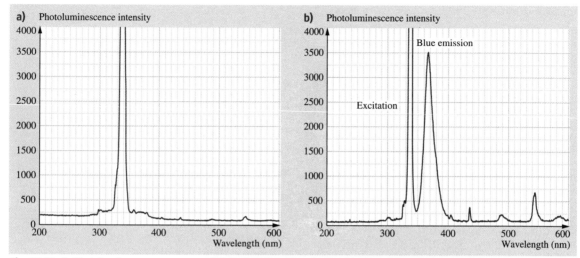

Fig. 35.12a,b Photoluminescence spectrum of PLD GaN films deposited at 14 mTorr (**a**), and 30 mTorr (**b**)

≈ 30 mTorr. There was no RTPL in GaN deposited at lower pressures. This can be related to the excessive energy of arriving particles. However, remarkably, room-temperature PL was not observed in film deposited at 50 mTorr. Thus, some degree of nonthermal activation (present in the 30 mTorr process) is indeed necessary to obtain optimal film properties, such as photoluminescence at room temperature. ω-Scan width is another parameter exhibiting the pressure dependence, with the best performance in the 30 mTorr process. As seen from the Table 35.2, ω-scan has the minimal FWHM at this pressure, implying that GaN crystallinity was significantly degraded at higher pressure of 50 mTorr.

Comparison of the optimal conditions with the expected range (Fig. 35.8, asterisk) of the plasma flux (Table 35.1) in 30 mTorr nitrogen shows that the substrate position of 20 cm corresponds to a distance from the target greater than the expected ≈ 12 cm. Thus, the optimal average energy of the arriving particles can be estimated as lower than expected from the model [35.34]. The initial mean energy of the Ga ions is ≈ 50 eV (from Fig. 35.9 data at low pressure). The mean energy at the 12 cm is $(0.02-0.04) \times 50$ eV $\approx 1-2$ eV, according to the model. The estimated mean energy at the optimal 20 cm distance is lower but roughly agrees with that predicted by the model.

35.5 Conclusions

Thin-film deposition by condensation of energetic plasma combines two events. First, material from a target is ablated in a highly nonequilibrium process, resulting in the generation of particles with kinetic energies orders of magnitude above those observed in equilibrium evaporation. This rapid (pulsed) evaporation of material is critically needed to achieve a congruent, stoichiometric flux from a multicomponent target. The second event is the film formation on a substrate that requires critical control of the plasma energetics leading to an optimal material flux arriving at the substrate. The optimal films can be deposited at energetic conditions that are not too far from equilibrium. Lower deposition rates and particle energies below the point-defect generation threshold seem to aid the formation of optimal films. In a vacuum process, these two contradicting requirements (high-energy ablation plasma near the target surface and a lower-energy materials flux at the substrate surface) can only be satisfied partially by choosing a shorter-wavelength laser (in the case of PLD) and lower beam intensity near the ablation threshold. A smaller material flux and particle energy can be achieved in this way. An appropriate process gas in a deposition chamber can serve as a more effective tool for converting the high-density high-energy ablation plasma flux into the required low-density low-energy condensation flux. By controlling the deposition conditions, low-temperature film growth activated by particles with optimal energy can be achieved. Careful optimization of the entire deposition system design and process parameters facilitates growth of high-quality single- and multicomponent films by both PLD and PED. The main advantageous feature of PED is its ability to ablate materials based on wide-bandgap (SiO_2, Al_2O_3, MgO, etc.) dielectrics, which are difficult for PLD due to their transparency at (excimer) laser wavelengths. Thus, PLD and PED can be considered as complementary techniques, sharing a similar arrangement, and able to be explored in a unified deposition chamber. PLD–PED deposition systems enable in situ deposition of a wider range of materials, which is important for the exploration of complex structures that incorporate metals, complex dielectrics, ferroelectrics, semiconductors, and glasses.

References

35.1 D.M. Mattox, J.E. McDonald: Interface formation during thin film deposition, J. Appl. Phys. **34**, 2493–2496 (1963)

35.2 J.S. Colligon: Energetic condensation: processes, properties, and products, J. Vac. Sci. Technol. A **13**, 1649–1657 (1995)

35.3 H. Fehler: Vacuum **45**, 997–1000 (1994)

35.4 W.D. Sproul: Mater. Sci. Eng. A **163**, 187–190 (1993)

35.5 J.M.E. Harper, J.J. Kuomo, H.R. Kaufman: J. Vac. Sci. Technol. **21**, 737 (1982)

35.6 D.B. Chrisey, G. Hubler (Eds.): *Pulsed Laser Deposition of Thin Films* (Wiley, New York 1994)

35.7 H.M. Smith, A.F. Turner: Vacuum deposited thin films using a ruby laser, Appl. Opt. **4**, 147–148 (1965)

35.8 K.L. Saenger: Pulsed laser deposition, Part 1, Process. Adv. Mater. **2**, 1–24 (1993)

35.9 K.L. Saenger: Pulsed laser deposition, Part 2, Process. Adv. Mater. **3**, 63–82 (1993)

35.10 J. Schou: Laser beam–solid interactions: fundamental aspects. In: *Materials Surface Processing by Direct Energy Techniques*, ed. by Y. Pauleau (Elsevier, Oxford 2006) pp. 35–66

35.11 G. Müller, M. Konijnenberg, G. Krafft, C. Schultheiss: Thin film deposition by means of pulsed electron beam ablation. In: *Science and Technology of Thin Films*, ed. by F.C. Matacotta, G. Ottaviani (World Scientific, Singapore 1995) pp. 89–119

35.12 D. Bäuerle: *Laser Processing and Chemistry* (Springer, Berlin, Heidelberg 2000)

35.13 M. Strikovski, K.S. Harshavardhan: Parameters that control pulsed electron beam ablation of materials and film deposition processes, Appl. Phys. Lett. **82**, 853–855 (2003)

35.14 A. Bogaerts, Z. Chen: Effect of laser parameters on laser ablation and laser-induced plasma formation, Spectrochim. Acta A **60**, 1280–1307 (2005)

35.15 Y.P. Raizer: *Gas Discharge Physics* (Springer, Berlin, Heidelberg 1997)

35.16 Y.B. Zel'dovich, Y.P. Raizer: *Physics of Shock Waves and High-Temperature Hydrodynamic Phenomena* (Academic, New York 1966)

35.17 H. Schittenhelm, G. Callies, P. Berger, H. Hügel: Investigation of extinction coefficient during excimer laser ablation and their interpretation in terms of Rayleigh scattering, J. Phys. D Appl. Phys. **29**, 1564–1575 (1996)

35.18 V.I. Mazhukin, V.V. Nossov, M.G. Nickiforov: Optical breakdown in aluminum vapor induced by ultraviolet laser radiation, J. Appl. Phys. **93**, 56–66 (2003)

35.19 C. Geertsen, P. Mauchien: Optical spectrometry coupled with laser ablation for analytical applications on solids. In: *Application of Beams in Materials Technology*, ed. by P. Misaelides (Kluwer, Dordrecht 1995) pp. 237–258

35.20 S. Metev: Process characteristics and film properties in pulsed laser deposition. In: *Pulsed Laser Deposition of Thin Films*, ed. by D.B. Chrisey, G. Hubler (Wiley, New York 1994) pp. 255–264

35.21 L. Torrisi, S. Gammino, L. Ando, V. Nassisi, D. Doria, A. Pedone: Comparison of nanosecond laser ablation at 1064 and 308 nm wavelength, Appl. Surf. Sci. **210**, 262–273 (2003)

35.22 H. Puell, H.J. Neusser, W. Kaiser: Heating of laser plasma generated at plane solid targets, Z. Naturforsch. A **25**, 1807–1815 (1970)

35.23 S.V. Gaponov, M.D. Strikovski: Formation of plasma during vaporisation of materials by the radiation of a CO_2 TEA laser, Sov. Phys. Tech. Phys. **27**(9), 1127–1130 (1982)

35.24 N. Arnold, J. Gruber, J. Heitz: Spherical expansion of the vapor into ambient gas: an analytical model, Appl. Phys. A **69**, s87–s93 (1999), (suppl.)

35.25 J. Stevefelt, C.B. Collins: Modelling of a laser plasma source of amorphous diamond, J. Phys. D Appl. Phys. **24**, 2149–2153 (1991)

35.26 M. Strikovski: unpublished (1987)

35.27 R. Teghil, L. D'Alessio, A. Santagata, M. Zaccagnino, D. Ferro, D.J. Sordelet: Picosecond and femtosecond pulsed laser ablation and deposition of quasicrystals, Appl. Surf. Sci. **210**, 307–317 (2003)

35.28 L. D'Alessio, A. Galasso, A. Santagata, R. Teghil, A.R. Villani, P. Villani, M. Zaccagnino: Plume dynamics in TiC laser ablation, Appl. Surf. Sci. **208/209**, 113–118 (2003)

35.29 D.M. Mattox: *Handbook of Physical Vapor Deposition (PVD) Processing* (Noyes, Westwood 1998)

35.30 N.M. Bulgakova, A.V. Bulgakov, O.F. Bobrenok: Double layer effects in laser-ablation plasma plumes, Phys. Rev. E **62**, 5624–5635 (2000)

35.31 H.S. Kim, H.S. Kwok: Correlation between target substrate distance and oxygen pressure in pulsed laser deposition of $YBa_2Cu_3O_7$, Appl. Phys. Lett. **61**, 2234–2236 (1992)

35.32 P.E. Dyer, A. Issa, P.H. Key: An investigation of laser ablation and deposition of Y-Ba-Cu-O in an oxygen environment, Appl. Surf. Sci. **46**, 89–95 (1990)

35.33 H.S. Kwok, H.S. Kim, D.H. Kim, W.P. Chen, X.W. Sun, R.F. Xiao: Correlation between plasma dynamics and thin film properties in pulsed laser deposition, Appl. Surf. Sci. **109/110**, 595–600 (1997)

35.34 M. Strikovski, J. Miller: Pulsed laser deposition of oxides: Why the optimum rate is about 1 Å per pulse, Appl. Phys. Lett. **73**, 1733–1735 (1998)

35.35 C.C. Chang, X.D. Wu, R. Ramesh, X.X. Xi, T.S. Ravi, T. Venkatesan, D.M. Hwang, R.E. Muenchausen, S. Foltyn, N.S. Nogar: Origin of surface roughness for c-axis oriented Y-Ba-Cu-O superconducting films, Appl. Phys. Lett. **57**, 1814–1816 (1990)

35.36 A. Gupta, B.W. Hussey: Laser deposition of $YBa_2Cu_3O_{7-x}$ films using a pulsed oxygen source, Appl. Phys. Lett. **58**, 1211–1213 (1991)

35.37 S.J. Pennycook, M.F. Chisholm, D.E. Jesson, R. Feenstra, S. Zhu, X.Y. Zheng, D.J. Lowndes: Growth and relaxation mechanisms of $YBa_2Cu_3O_{7-x}$ films, Physica C **202**, 1–11 (1992)

35.38 A.T. Findikoglu, C. Doughty, S.M. Anlage, Q. Li, X.X. Xi, T. Venkatesan: DC electric field effect on the microwave properties of $YBa_2Cu_3O_7/SrTiO_3$ layered structures, J. Appl. Phys. **76**, 2937–2944 (1994)

35.39 W. Zhang, I.W. Boyd, M. Elliott, W. Herrenden-Harkerand: Transport properties and giant magneto resistance behavior in La-Nd-Sr-Mn-O films, Appl. Phys. Lett. **69**, 1154–1156 (1996)

35.40 Y.A. Bityurin, S.V. Gaponov, E.B. Klyuenkov, M.D. Strikovsky: GaAs compensation by intense fluxes of low energy particles, Solid State Commun. **45**, 997–1000 (1983)

35.41 S. Fähler, K. Sturm, H.U. Krebs: Resputtering during the growth of pulsed-laser-deposited metallic films in vacuum and in ambient gas, Appl. Phys. Lett. **75**, 3766–3768 (1999)

35.42 A. Anders: Observation of self-sputtering in energetic condensation of metal ions, Appl. Phys. Lett. **85**, 6137–6139 (2004)

35.43 Oak Ridge National Laboratory: *Atomic Data for Fusion*, Vol. 3 (1985), http://www-cfadc.phy.ornl.gov/redbooks/three/a/3a18.html

35.44 T. Scharf, J. Faupel, K. Sturm, H.-U. Krebs: Pulsed laser deposition of metals in various inert gas atmospheres, Appl. Phys. A **79**, 1587–1589 (2004)

35.45 T. Ohmi, T. Shibata: Advanced scientific semiconductor processing based on high-precision controlled low-energy ion bombardment, Thin Solid Films **241**, 159–162 (1993)

35.46 J.A. Thornton: The influence of bias sputter parameters on thick cupper coatings deposited using a hollow cathode, Thin Solid Films **40**, 335–340 (1977)

35.47 D.R. Brighton, G.K. Hubler: Binary collision cascade prediction of critical ion-to-atom arrival ratio in the production of thin films with reduced intrinsic stress, Nucl. Instrum. Methods Phys. Res. B **28**, 527–530 (1987)

35.48 K.S. Harshavardhan, H.M. Christen, S.D. Silliman, V.V. Talanov, S.M. Anlage, M. Rajeswari, J. Claasen: Low-loss $YBa_2Cu_3O_7$ films on flexible, polycrystalline-yttria-stabilized zirconia tapes for cryoelectronic applications, Appl. Phys. Lett. **78**, 1888–1890 (2001)

35.49 S.S. Harilal, C.V. Bindhu, M.S. Tillack, F. Najmabadi, A.C. Gaeris: Internal structure and expansion dynamics of laser ablation plumes into ambient gases, J. Appl. Phys. **93**, 2380–2388 (2003)

35.50 S.S. Harilal, B. O'Shay, Y. Tao, M.S. Tillack: Ambient gas effects on the dynamics of laser-produced tin plume expansion, J. Appl. Phys. **99**, 083303–1–083303–10 (2006)

35.51 R.F. Wood, J.N. Leboeuf, K.R. Chen, D.B. Geohegan, A.A. Puretzky: Dynamics of plume propagation, splitting, and nano-particle formation during pulsed-laser ablation, Appl. Surf. Sci. **127–129**, 151–158 (1998)

35.52 S. Amoruso, B. Toftman, J. Schou: Broadening and attenuation of UV laser ablation plumes in background gases, Appl. Surf. Sci. **248**, 323–328 (2005)

35.53 N. Arnold, J. Gruber, J. Heitz: Spherical expansion of the vapor into ambient gas: An analytical model, Proc. COLA'99, 5th Int. Conf. Laser Ablation, Göttingen (Springer, Berlin, Heidelberg 1999)

35.54 J. Han, M.H. Crawford, R.J. Shul, J.J. Figiel, M. Banas, L. Zhang, Y.K. Song, H. Zhou, A.V. Nurmikko: AlGaN/GaN quantum well ultraviolet light emitting diodes, Appl. Phys. Lett. **73**, 1688–1690 (1998)

35.55 D. Doppalapudi, E. Iliopoulos, S.N. Basu, T.D. Moustakas: Epitaxial growth of gallium nitride thin films on a-Plane sapphire by molecular beam epitaxy, J. Appl. Phys. **85**, 3582–3589 (1999)

35.56 T. Nishida, H. Saito, N. Kobayashi: Efficient and high-power AlGaN-based ultraviolet light-emitting diode grown on bulk GaN, Appl. Phys. Lett. **79**, 711–712 (2001)

35.57 M. Cazzanelli, D. Cole, J.F. Donegan, J.G. Lunney, P.G. Middleton, K.P. O'Donnell, C. Vinegoni, L. Pavesi: Photoluminescence of localized excitons in pulsed-laser-deposited GaN, Appl. Phys. Lett. **73**, 3390–3392 (1998)

35.58 M.P. Chowdhury, R.K. Roy, S.R. Bhattacharyya, A.K. Pal: Stress in polycrystalline GaN films prepared by R.F. Sputtering, Eur. Phys. J. B **48**, 47–53 (2005)

35.59 T. Venkatesan, K.S. Harshavardhan, M. Strikovski, J. Kim: Recent advances in the deposition of multi-component oxide films by pulsed energy deposition. In: *Thin Films and Heterostructures for Oxide Electronics*, ed. by S.B. Ogale (Springer, New York 2005) pp. 385–413

35.60 S. Ito, H. Fusioka, J. Ohta, H. Takahshi, M. Oshima: Effect of AlN. Buffer Layers on GaN/MnO Structure, Phys. Status Solidi (c) **0**, 192–195 (2002)

35.61 A.N. Red'kin, V.I. Tatsii, Z.I. Makovei, A.N. Gruzintsev, E.E. Yakimov: Chemical vapor deposition of GaN from gallium and ammonium chloride, Inorg. Mater. **40**, 1049–1053 (2004)

35.62 P.R. Tavernier, P.M. Verghese, D.R. Clarke: Photoluminescence from laser assisted debonded epitaxial GaN and ZnO films, Appl. Phys. Lett. **74**, 2678–2680 (1999)

35.63 T. Miyazaki, K. Takada, S. Adachi, K. Ohtsuka: Properties of radio-frequency-sputter-deposited GaN films in a nitrogen/hydrogen mixed gas, J. Appl. Phys. **97**, 093516–093518 (2005)